Diseases of Woody Ornamentals *and* Trees in Nurseries

Edited by

Ronald K. Jones
North Carolina State University, Raleigh

and

D. Michael Benson
North Carolina State University, Raleigh

APS PRESS

The American Phytopathological Society
St. Paul, Minnesota

Reference in this publication to a trademark, proprietary product, or company name by personnel of the U.S. Department of Agriculture or anyone else is intended for explicit description only and does not imply approval or recommendation to the exclusion of others that may be suitable.

Library of Congress Control Number: 2001089055
International Standard Book Number: 0-89054-264-3

Printed in the United States of America on acid-free paper

The American Phytopathological Society
3340 Pilot Knob Road
St. Paul, Minnesota 55121-2097, USA

Preface

Over 20 years ago, participants in the Ornamentals Workshop at Crossnore, North Carolina, worked together under the editorship of Ronald K. Jones (North Carolina State University, Raleigh) and Robert C. Lambe (Virginia Tech, Blacksburg) to write an extension publication entitled *Diseases of Woody Ornamental Plants and Their Control in Nurseries.* The Ornamentals Workshop has continued on a biennial basis and has evolved into the national working group for ornamental diseases. Under the present editorship, the group planned this volume, *Diseases of Woody Ornamentals and Trees in Nurseries.*

The content has expanded to include ornamental crops grown in nurseries throughout the United States. Diseases on over 65 crops grown in nurseries, including shrubs, ground covers, subtropical ornamentals, and shade trees, are described, along with methods for their control. A feature of many chapters on individual crops is information on cultivar resistance to diseases. This information has been scattered in a wide variety of sources, many of which are not readily available to nurserymen and those interested in disease control until now. Uniquely, for each disease, a summary gives the nurseryman a quick reference to timely control measures. These measures, part of an integrated disease management approach, are explained in detail in the text. Since not all diseases occur on a given crop across the United States, the USDA Plant Hardiness Zone Map (facing page 68) is used as a reference for geographical production zones in which the nurserymen might expect diseases to occur. In addition, the contributors have developed a list of the occurrence of diseases in each state, presented in the appendix. Color plates illustrate many of the diseases discussed. A glossary of plant pathological terms is provided.

In addition to the latest information on diseases of specific ornamental crops, the book also introduces the reader to the general nature of plant diseases through chapters on abiotic causes of diseases (nonliving disease agents) and biotic pathogens. The chapter on abiotic diseases includes an extensive list of herbicides that may cause injury in ornamental crops if misused, or if ornamentals are planted in former agricultural soils containing these materials.

Chapters on management strategies give nurserymen and other readers an in-depth guide to integrated disease management, including cultural control, sanitation, resistance, fungicides, and bactericides, and chapters on control of various pathogen groups are also included. In addition, information is included on the role of recycled water in disease development and ways to manage pathogens in recycled irrigation water. The latest horticultural practices are discussed in relation to plant diseases. Tissue culture techniques to eliminate pathogens from propagation stock are described, as are the role of plant diagnostic clinics in assisting the nurseryman in disease diagnosis.

This book would not have been possible without the cooperation and dedication of a number of people. The editors would like to thank all the contributors who took time to help plan the initial outline and then volunteer to write their chapters. Without their contributions, this book would not have been possible. The editors thank former APS Press editor-in-chief Kurt Leonard for his encouragement of the project. The editors are grateful to APS Press senior editor Robert G. Linderman for his time to edit the manuscript and to offer valuable suggestions for improvement. Finally, we express our most deeply felt thanks to Lori Force, Department of Plant Pathology, North Carolina State University, for typing the manuscript throughout the many revisions, and we sincerely appreciate Kina Jordan for typing the final version.

Ronald K. Jones
D. Michael Benson

Raleigh, North Carolina
March 2001

Contributors

Cynthia L. Ash, American Phytopathological Society, St. Paul, Minnesota
Thomas J. Banko, Virginia Polytechnic Institute and State University, Hampton Roads Agricultural Research and Extension Center, Virginia Beach
Edward Barnard, Florida Department of Agriculture and Consumer Services, Division of Forestry, Gainesville
Larry W. Barnes, Agricultural Extension Service, Texas A&M University, College Station
Luther W. Baxter, Jr., Department of Plant Pathology, Clemson University, Clemson, South Carolina
Julie Beale, Plant Disease Diagnostic Lab, University of Kentucky, Lexington
D. Michael Benson, Department of Plant Pathology, North Carolina State University, Raleigh
Ted E. Bilderback, Department of Horticultural Science, North Carolina State University, Raleigh
James H. Blake, Agricultural Service Laboratory, Clemson University, Clemson, South Carolina
Walter Bliss, Agdia, Inc., Elkhart, Indiana
Lawrence G. Brown, U.S. Department of Agriculture, Animal and Plant Health Inspection Service, Center for Plant Health Science and Technology, Raleigh, North Carolina
Diane E. Brown-Rytlewski, Department of Plant Pathology, University of Wisconsin, Madison
Jillanne R. Burns, Department of Plant Pathology, North Carolina State University, Raleigh
Thomas J. Burr, Department of Plant Pathology, Cornell University, New York State Agricultural Experiment Station, Geneva
Ralph S. Byther, Research and Extension Center, Washington State University, Puyallup
A. R. Chase, Chase Research Gardens, Inc., Mt. Aukum, California
Gary A. Chastagner, Research and Extension Center, Washington State University, Puyallup
David L. Clement, Home Garden Center, University of Maryland, Ellicott City
Phillip Colbaugh, Texas A&M Agricultural Experiment Station, Dallas
Tom Creswell, Department of Plant Pathology, North Carolina State University, Raleigh
Margery L. Daughtrey, Cornell University, Long Island Horticultural Research Laboratory, Riverhead, New York
Robert A. Dunn, Department of Entomology and Nematology, University of Florida, Gainesville
Ethel Dutky, Department of Entomology, University of Maryland, College Park
Brian C. Eshenaur, Cornell Cooperative Extension Service, Rochester, New York
Loren J. Giesler, Department of Plant Pathology, University of Nebraska, Lincoln
Mark Gleason, Department of Plant Pathology, Iowa State University, Ames
Ann Brooks Gould, Department of Plant Pathology, Rutgers University, New Brunswick, New Jersey
Austin K. Hagan, Department of Plant Pathology, Auburn University, Auburn, Alabama
Larry Hanning, Missouri Department of Agriculture, Division of Plant Industries, Jefferson City
John Hartman, Department of Plant Pathology, University of Kentucky, Lexington
Mary Francis Heimann, University of Wisconsin, Madison
Charles Hodges, Department of Plant Pathology, North Carolina State University, Raleigh
Elizabeth Hudgins, Department of Entomology and Plant Pathology, Oklahoma State University, Stillwater
George W. Hudler, Department of Plant Pathology, Cornell University, Ithaca, New York
Karel A. Jacobs, The Morton Arboretum, Lisle, Illinois
Steven N. Jeffers, Department of Plant Pathology, Clemson University, Clemson, South Carolina
David Johnson, Missouri Department of Agriculture, Division of Plant Industries, Jefferson City
Ronald K. Jones, Department of Plant Pathology, North Carolina State University, Raleigh
Laura R. Kabrick, Department of Plant Pathology, University of Missouri, Columbia
T. Michael Likins, Virginia Department of Agriculture, Richmond
Robert G. Linderman, U.S. Department of Agriculture, Agricultural Research Service, Horticultural Crops Research Unit, Corvallis, Oregon
James D. MacDonald, Department of Plant Pathology, University of California, Davis
Walter F. Mahaffee, U.S. Department of Agriculture, Agricultural Research Service, Horticultural Crops Research Unit, Corvallis, Oregon
Deborah D. Miller, Genesee Plant Diagnostics Inc., Burton, Michigan
John Miller, Florida Department of Agriculture and Consumer Services, Division of Plant Industry, Gainesville

R. Walker Miller, Department of Plant Pathology, Clemson University, Clemson, South Carolina
Margaret T. Mmbaga, Tennessee State University, Nursery Crop Research Station, McMinnville
Larry W. Moore, Department of Botany and Plant Pathology, Oregon State University, Corvallis
Gary W. Moorman, Department of Plant Pathology, Pennsylvania State University, University Park
Jacqueline Mullen, Department of Plant Pathology, Auburn University, Auburn, Alabama
Robert P. Mulrooney, Department of Plant and Soil Sciences, University of Delaware, Newark
John W. Olive, Auburn University, Ornamental Horticulture Substation, Mobile, Alabama
Judith O'Mara, Department of Plant Pathology, Kansas State University, Manhattan
Kala C. Parker, Department of Plant Pathology, North Carolina State University, Raleigh
Paul C. Pecknold, Department of Botany and Plant Pathology, Purdue University, West Lafayette, Indiana
John D. Peplinski, Pennsylvania State University, University Park
George Philley, Research and Extension Center, Texas A&M University, Overton
Charles C. Powell, Jr., Plant Health Advisory Services, Worthington, Ohio
J. E. Preece, Plant Soil and General Agriculture, Southern Illinois University, Carbondale
Jay W. Pscheidt, Department of Botany and Plant Pathology, Oregon State University, Corvallis
Karen K. Rane, Department of Botany and Plant Pathology, Purdue University, West Lafayette, Indiana
Balakrishna Rao, Davey Tree Co., Kent, Ohio
Diane M. Reaver, Department of Plant Pathology, Physiology and Weed Science, Virginia Polytechnic Institute and State University, Blacksburg
Gail E. Ruhl, Department of Botany and Plant Pathology, Purdue University, West Lafayette, Indiana
Heather J. Scheck, Office of the Agricultural Commissioner, Santa Barbara, California
Timothy S. Schubert, Florida Department of Agriculture and Consumer Services, Division of Plant Industry, Gainesville
Anni Self, Tennessee Department of Agriculture, Nashville
James L. Sherald, Center for Urban Ecology, National Park Service, Washington, D.C.
Gary W. Simone, Department of Plant Pathology, University of Florida, Gainesville
Wayne A. Sinclair, Department of Plant Pathology, Cornell University, Ithaca, New York
Karen Snover, Plant Disease Diagnostic Lab, Department of Plant Pathology, Cornell University, Ithaca, New York
Thomas Stebbins, University of Tennessee, Nashville
Marcia A. Stefani, Virginia Polytechnic Institute and State University, Hampton Roads Agricultural Research Extension Center, Virginia Beach
R. Jay Stipes, Department of Plant Pathology, Virginia Polytechnic Institute and State University, Blacksburg
Nancy J. Taylor, Department of Plant Pathology, Ohio State University, Columbus
Julia W. Thompson, Missouri Department of Agriculture, Division Plant Industries, Jefferson City
Timothy E. Tidwell, California Department of Food and Agriculture, Plant Diagnostics Center, Sacramento
Ned A. Tisserat, Department of Plant Pathology, Kansas State University, Manhattan
R. N. Trigiano, Department of Entomology and Plant Pathology, University of Tennessee, Knoxville
Stephen R. Vann, University of Arkansas, Maumelle, Arkansas
Sharon L. von Broembsen, Department of Entomology and Plant Pathology, Oklahoma State University, Stillwater
Jerry T. Walker, Department of Plant Pathology, University of Georgia, Griffin
Sarah Walker, Florida Department of Agriculture and Consumer Services, Gainesville
John E. Watkins, Department of Plant Pathology, University of Nebraska, Lincoln
Robert L. Wick, Department of Microbiology, University of Massachusetts, Amherst
Margaret R. Williamson, Plant Problem Clinic, Clemson University, Clemson, South Carolina
Jean L. Williams-Woodward, Extension Plant Pathology, University of Georgia, Athens
Alan S. Windham, University of Tennessee, Nashville
Mark Windham, Department of Entomology and Plant Pathology, University of Tennessee, Knoxville

Contents

Disease Management

Diseases of Woody Ornamentals *and* Trees in Nurseries

CHAPTER
1

Ronald K. Jones ▪ North Carolina State University, Raleigh

Introduction

The nursery industry is increasing in the volume of production, in the number of plant species produced, and in the variety of production techniques used in the United States. The retail trade in woody ornamentals demands high-quality products. However, diseases can kill plants or detract from their appearance and reduce their growth rate, thus lowering their quality. Diseases are a constant threat to nursery production of woody ornamentals. The introduction of plant species and the adoption of new production techniques has created conditions for the spread of new diseases and new outbreaks of old diseases. As plants are distributed for planting, pathogens may be introduced into areas of the country where they were not previously present and may impose a serious limitation on the growth and survival of the plants in the landscape. Disease development can be greatly influenced by the plant species involved and production practices followed.

Damage due to disease, reduced quality and growth rate, and plant death reduce the operating efficiency of the nursery, and application of pesticides increases the cost of production. A disease management program must be practiced constantly and must mesh with the production program.

This book will explain the nature of plant diseases; the causes of plant diseases, both living (biotic) and nonliving (abiotic); conditions that favor disease development; and symptoms of diseases; and it will describe all of the components of an integrated disease management program.

The information presented in this book will help in the diagnosis of diseases, evaluation of their potential severity, and development of an integrated management strategy to reduce or eliminate losses from diseases. Management strategies include sanitation, cultural practices, regulatory restrictions, planting resistant species or cultivars, and (as a last resort) chemical treatment. Diseases cannot be managed in a nursery without a complete program. Chemical, nonchemical, and biological management strategies will be presented. Specific recommendations for chemical treatment to manage any disease must be obtained locally.

Diseases of woody ornamental plants do not occur uniformly in all states. The occurrence of diseases in different states is presented in the appendix. Many states do not have current indexes of diseases that have occurred, particularly in numerous species of woody ornamentals. The information in the appendix is based on the authors' memory and some available clinic records.

This publication reports research involving antimicrobial substances. It does not contain recommendations for their use, nor does it imply that the uses discussed here have been registered. All uses of pesticides must be registered by appropriate state and federal agencies before they can be recommended.

CHAPTER 2

Ronald K. Jones • North Carolina State University, Raleigh

Sarah Walker • Florida Department of Agriculture and Consumer Services, Gainesville

Steven N. Jeffers • Clemson University, Clemson, South Carolina

Plant Disease Development

A plant disease is a progressive condition in which the normal function, structure, or appearance of the plant is disrupted or changed, usually in an undesirable manner, reducing the value of the plant. This abnormal condition is the result of an interaction between the plant (the host) and a living (biotic) or a nonliving (abiotic) causal agent. A disease can be identified by characteristic signs and symptoms, often referred to as the disease syndrome or signature. A sign is the visible presence of a causal organism, such as spore-producing structures of a fungus. Symptoms are changes in the plant, such as leaf spots or galls, in response to a disease-causing agent. This book will deal more with diseases caused by living agents but will also cover some common abiotic problems.

Biotic causal agents, known as pathogens, include fungi, bacteria, nematodes, viruses, phytoplasmas, and seed-producing plants such as dodder and mistletoe. Pathogens are infectious, which means that they can spread or be spread from plant to plant.

Abiotic diseases or disorders are noninfectious conditions, brought about by nonliving causal agents, which can be grouped in three broad categories: environmental, chemical, and mechanical. Environmental factors include conditions such as extremes of temperature, deficiency or excess of moisture, hail, and lightning. Chemical injury, often termed phytotoxicity, may be due to unfavorable pH of the soil or growth medium, nutritional problems, fertilizer injury, air pollution, or pesticide burn. Mechanical injury includes damage due to animals, insects, machinery, plant ties, and other agents. Abiotic disorders can be one-time events, such as mechanical injury, or ongoing problems, such as nutrient deficiency. Sometimes an abiotic causal agent leaves a sign, such as a visible chemical residue.

Abiotic disorders may also cause wounds through which biotic pathogens can enter plants. Fertilizer injury to the root system, due to a high level of soluble salts, may impose stress on the plant and increase its susceptibility to root rot pathogens.

Symptoms of abiotic and biotic plant diseases may be the same or very similar, and at times symptoms may be caused by more than one agent. Proper identification of each and all causes of one or more symptoms can be a difficult challenge, requiring time and several types of tests or assays, and incurring costs. For instance, is chlorosis caused by a lack of nutrients in the medium (an abiotic disorder) or the result of root rot (a biotic disease) that prevents the absorption of nutrients in the medium? It may take several tests to identify the true cause or the relative involvement of each. This can be a diagnostic nightmare for a grower. The correct cause must be identified, however, before the correct management strategy can be selected.

The Disease Pyramid

The development of an infectious disease in a crop is the result of an interaction of four factors: (1) a susceptible host plant, (2) a pathogen capable of infecting the host, (3) environmental conditions that favor infection and subsequent reproduction and spread of the pathogen, and (4) enough time for the host to react to the pathogen and for the pathogen to multiply. The longer conditions remain favorable for disease development, the more severe and widespread the disease will be, and the greater the economic loss. The relationship of these four factors is often represented as the *disease pyramid* (Fig. 2.1).

To fully understand the concepts of plant disease development, it is necessary to study the four factors in the disease pyramid and how they interact. Understanding these principles and their relationships will help growers to forecast disease development and know when to implement an appropriate management strategy. To some extent, these factors can also be manipulated by growers in an integrated disease management program.

The Host

A susceptible plant is called a host because it provides food for a pathogen. Large numbers of host plants are produced in a limited space in nurseries. Some species of plants are more susceptible to disease than others. Likewise, some cultivars of a species are more susceptible to a disease than other cultivars. Bradford pear is highly resistant to fire blight, for example, while many other cultivars of flowering pear are highly susceptible. A plant can be more susceptible to a pathogen at one stage of growth than at other times: new, succulent leaves can be more susceptible to some pathogens than fully expanded, mature leaves. Some parts of a plant (such as the petals) may be susceptible to a pathogen while other parts (such as the leaves) are not. How plants are grown can also influence their susceptibility to disease. If growing conditions in the nursery are optimum for a specific plant, it may be more resistant to a pathogen than it would be under less favorable conditions.

The Pathogen

A pathogen is a living organism that causes a disease. It must be present for the disease to develop. While a host is always present in a nursery, a pathogen may not be. The quantity of the pathogen and its virulence (ability to cause disease) directly affect the

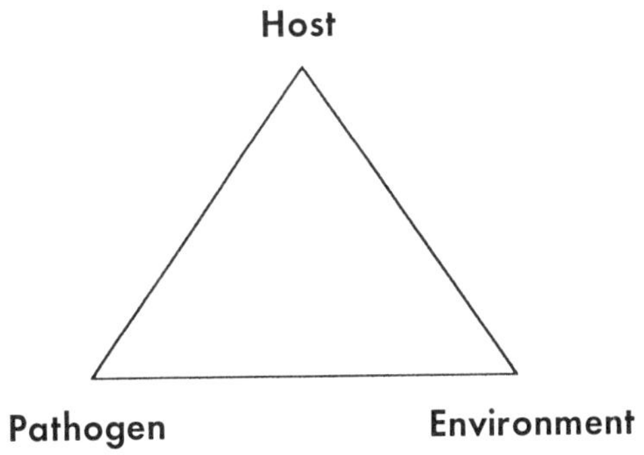

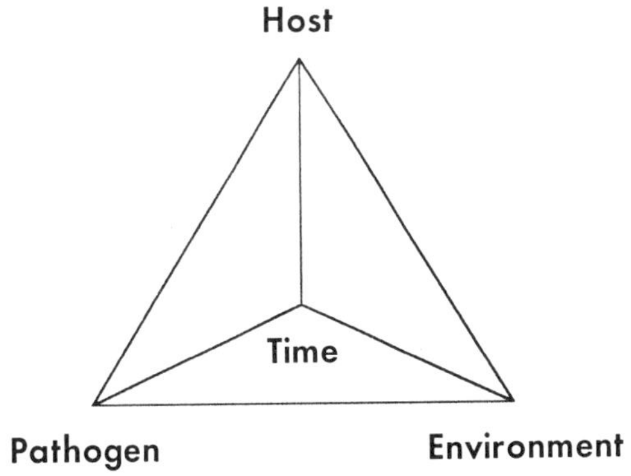

Fig. 2.1 Disease development as the result of the interaction of a susceptible host and a pathogen under prevailing environmental conditions is often represented as the *disease triangle* (left). The addition of time as a factor in the interaction forms a *disease pyramid* (right).

occurrence and severity of disease. Some pathogens can cause disease in many different species of plants, while others are very specific and cause disease in only one host species.

Pathogens need a means of entering a host, reproducing, and spreading to other plants. Some pathogens produce spores that are windborne. Others are spread in splashing rain or irrigation water. Still others are spread on or in plant parts during propagation or other nursery operations. The most efficient means of spreading a pathogen is by asexual propagation and rapid transport of diseased plants from one grower to another. This is normal in the nursery business.

The Environment

The environment includes all the factors and conditions that affect the growth and development of living organisms, both host plants and pathogens. The most obvious factors are moisture, temperature, light, and nutrition. The total environment includes nursery employees, mechanical equipment, insects, and other factors. All of these environmental factors influence both the host and the pathogen and affect their interaction. If environmental factors are more favorable for the growth and development of the pathogen, then disease develops in the host. If environmental factors favor the growth and development of the host, then healthy, salable plants are produced in the nursery. In some cases, however, environmental conditions unfavorable to both the host and the pathogen can result in severe disease, if the host has no resistance to the pathogen.

Moisture in the form of rain or irrigation and humidity is critical in the development of most infectious diseases. There is an optimum amount of moisture for good growth of the host plant. Increasing the volume of moisture and the frequency of wetting beyond the optimum tends to favor the pathogen. More diseases develop under wet conditions than under dry conditions. Water or high humidity is often required for entry of a plant by a pathogen, and many pathogens spread from plant to plant during wet weather. Irrigation can also influence disease development. For example, the amount of overhead irrigation and the time of day when it is applied affect how long leaves and stems remain wet. The ratio of air space to water in a potting medium influences the development of many root diseases.

Temperature also affects the growth and development of both the host plant and the pathogen. Just as different plants grow best at different temperatures, many pathogens are also affected by temperature. Some pathogens are most active in cool spring weather, while others are most active at high temperatures in the summer. Many diseases develop in the spring, because temperatures are moderate, rainfall is usually frequent, and plants are producing new growth. Most pathogens cause disease when several environmental conditions occur simultaneously, and such diseases can thus be very restricted in occurrence. Other pathogens are active over a wider range of environmental conditions, and the diseases they cause can occur during several seasons.

Infection by root pathogens is influenced primarily by soil moisture and temperature. Once root infection occurs, the pathogen can spread within an infected root under a much wider range of moisture conditions, but is it still influenced by soil temperature. Root rot may kill roots for several months or years before aboveground symptoms become obvious. Aboveground symptoms of root rot diseases often show up during or just after a growth flush or a period of hot, dry weather. This can confuse growers, when the book says that root rot develops in cool, wet weather.

Environmental conditions in nurseries can be more favorable for disease development than a natural setting. Large, dense blocks of genetically uniform plants in nurseries are watered and fertilized for maximum growth. Water, which is necessary for the development of many diseases, is seldom a limiting factor in nurseries.

Time

Time influences disease development in several ways in woody plants, which often remain in the nursery for one or more years. The longer a plant remains in the nursery, the greater the chance that all of the conditions needed for disease development will occur simultaneously. A pathogen needs time to infect a host and develop within the host to a stage at which it can spread to other leaves or other plants. The longer conditions remain favorable for the pathogen, the more severe and widespread the disease will be.

Plant Disease Development Calendar

Many diseases develop at a certain time of year, and the pathogens causing them are dormant or inactive the rest of the year. The development of any disease is predictable to some degree. The seasonal plant disease development calendar (Fig. 2.2) is intended to help predict the occurrence of certain biotic diseases. Only

PLANT DISEASE DEVELOPMENT CALENDAR

Plants	Development or Disease	Pathogen Scientific Name	Plant Parts Affected	Severity	Distribution	Jan.	Feb.	March	April	May	June	July	Aug.	Sept.	Oct.	Nov.	Dec.
Trees																	
Crab apple	Flowering																
	Cedar apple rust	*Gymnosporangium* sp.	leaves	++	G												
	Fire blight	*Erwinia amylovora*	shoots	+++	G												
	Powdery mildew	*Podosphaera leucotricha*	shoots	++	G												
	Scab	*Venturia inaequalis*	fruit, leaves	+++	G												
Dogwood	Flowering																
	Anrhracnose	*Discula destructiva*	leaves, twigs	++++	M												
	Leaf spot	*Septoria floridae*	leaves	+	G												
	Scorch	Physiological	leaves	+	G												
	Spot anthracnose	*Elsinoe corni*	leaves, flowers	++	G												
Oak	Flowering																
	Anthracnose	*Gnomonia veneta*	leaves	+	G												
	Leaf blister	*Taphrina caerulescens*	leaves	+	G												
	Rust	*Cronartium* sp.	leaves	+	G												
Pine	Flowering																
	Eastern gall rust	*Cronartium quercuum*	branches, trunk	+++	G												
	Fusiform rust	*Cronartium fusiforme*	branches, trunk	+++	G												
	Needle cast	*Hypoderma lethale*	needles	+	G												
	Needle rust	*Coleosporium* sp.	needles	+	G												
Redbud	Flowering																
Redcedar	Cedar apple rust	*Gymnosporangium* sp.	branches	+	G												
Redmaple	Flowering																
	Anthracnose	*Gloeosporium apocryptum*	leaves	+	G												
Riverbirch	Anthracnose	*Discula betulina*	leaves	++	G												
Saucer magnolia	Flowering																
Sycamore	Anthracnose	*Gnomonia platani*	shoots	+	G												
	Scorch	*Xylella fastideosa*	leaves, twigs	++++	G												

Woody Plants

Azalea	Flowering				
	Leaf gall	*Exobasidium vaccinii*	leaves	+	G
	Petal blight	*Ovulinia azaleae*	petals	++	G
Camellia japonica	Flowering				
	Flower blight	*Sclerotinia camelliae*	flower	++	G
	Leaf gall	*Exobasidium camelliae*	shoot	+	G
Camellia sasanqua	Flowering				
	Leaf gall	*Exobasidium camelliae*	shoot	+	G
Crape myrtle	Flowering				
	Powdery mildew	*Erysiphe lagerstroemiae*	leaves, flowers	+	G
Forsythia	Flowering				
Photinia	Leaf spot	*Entomosporium mespili*	leaves	++++	G
Pyracantha	Fire blight	*Erwinia amylovora*	shoots	+	M, P
Rhododendron	Flowering				
	Leaf gall	*Exobasidium vaccinii*	shoots	+	M
	Shoot blight	*Botryosphaeria dothidea*	young shoots	+++	G
Rose	Flowering				
	Black spot	*Diplocarpon rosae*	leaves	++++	G
	Botrytis	*Botrytis cinerea*	flowers, stems	+++	G
	Downy mildew	*Peronospora sparsa*	leaves, stems	++++	G
	Powdery mildew	*Sphaerotheca pannosa*	leaves	+++	G

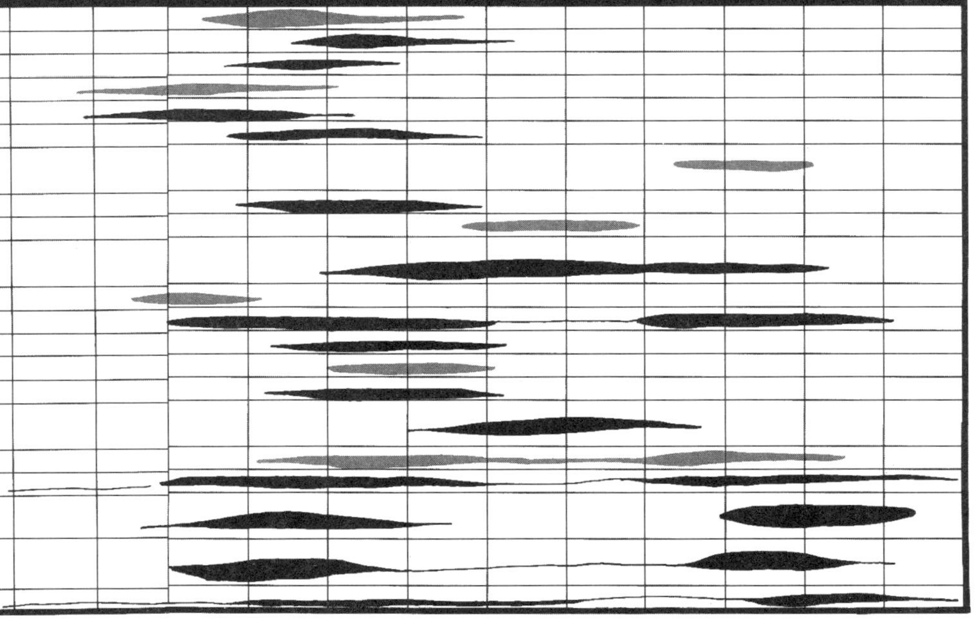

Key to Severity Range:
+ Very slight or no damage
++++ Very severe: kills or severely weakens plant

Key to Distribution:
G General
C Coastal
CP Coastal Plain
P Piedmont

Note: This calendar is intended to help anticipate the occurrence of certain diseases. The irregularly shaped horizontal black lines indicate when each disease is likely to occur and when the pathogen may be active. The wider the line, the greater the probability of the disease occurring during that particular time period. The flowering periods, shown in red, are included as a point of reference to help adjust the disease dates to different locations in the state and to years with unusually early or late seasons. The dates given are for the middle of North Carolina; spring diseases will occur earlier in southeastern counties and later in western counties.

The calendar does not list all diseases that can occur in North Carolina, only those that usually occur at a particular time of year or at a particular stage of plant growth. Some of the diseases listed do not occur every year, and some do not occur across the entire state. The diseases may vary in severity from year to year. Some of the diseases that affect woody tissue may cause symptoms that remain after the calendar indicates that the disease is no longer active; examples are fusiform rust on pine and fire blight on pear.

Prepared by
Ronald K. Jones, Department Extension Leader, Extension Plant Pathology

Fig. 2.2 Plant disease development calendar for central North Carolina, in USDA hardiness zone 8. Areas originally printed in red are reproduced here in gray. (Reprinted, by permission, from North Carolina Cooperative Extension Service, AG-135) diseases that occur at a particular time of year or at a particular stage of growth of the host plant are listed. The lines on the calendar indicate when pathogens are likely to be active and diseases are likely to develop. The wider the line, the greater the probability of disease. The flowering periods of several plants are shown as a seasonal point of reference, to help adjust the dates for different locations or for years with unusually early or late seasons. For instance, plant growth and disease development begin later in the spring in Virginia than in Florida and even later in New York.

The calendar is based on conditions in central North Carolina, in USDA hardiness zone 8, with a frost-free date of April 27. Some of the diseases listed do not occur every year, some do not occur across the entire United States, and the diseases vary in severity from year to year. Some diseases that affect woody plants cause symptoms, such as dead branches, that remain after the calendar indicates the pathogen is no longer active; other examples are fusiform rust galls on pine and fire blight of pear. Persistent symptomatic tissue is often a source of the pathogen for the next disease cycle.

Each grower could create a customized calendar of this type, to help predict disease development at a specific location and allow implementation of a timely disease management strategy. The descriptions of diseases in this book present information on periods when pathogens are active, but only as a generalization. Periods of pathogen activity should be determined especially for each nursery.

Summary

The interactions of all of the factors in plant disease development are very complex, but they need to be understood to fully comprehend disease development and to sort out the various causes and their relative importance, so that effective management strategies can be implemented. For most diseases only a few factors play a significant role. The better the grower understands them, the easier it is to determine which contributing factors the grower can control in developing a management strategy.

CHAPTER 3

Timothy S. Schubert ▪ Florida Department of Agriculture, Gainesville

Jerry T. Walker ▪ University of Georgia, Griffin

Abiotic Diseases of Woody Ornamentals

Abiotic diseases of plants are those caused by nonliving agents. As such, they are nontransmissible, and in some circles they are referred to as environmentally induced or physiological disorders. Abiotic diseases are probably much more common than most horticulturists might think. Informal personal consultation with many plant pathologists responsible for plant disease diagnosis in clinics around the United States reveals that about half of the specimens submitted to their clinics show indications of primary abiotic disease. Consider that abiotic agents may predispose plants to attack by biotic pathogens, and the importance of abiotic diseases escalates even more.

Diagnosis of abiotic diseases is a formidable challenge for several reasons. First, there is no living pathogen, and signs of the causal agent often are absent. Certain injurious pollutants or phytotoxic pesticides leave a telltale residue, but in most cases of abiotic disease no detectable signs of the causal agent are present, leaving the diagnostician with only symptoms and circumstantial evidence as clues.

Second, the syndromes induced by abiotic diseases can be very similar to those of biotic or other abiotic diseases, to the point of being indistinguishable at times. This requires that the diagnostician be familiar with a wide range of syndromes and the many agents that can cause them, and then resort to a long process of elimination to decide on one or more likely causal agents. Diagnosis by process of elimination often lacks the level of certainty and satisfaction of diagnosis by isolation or detection of the pathogen causing a biotic disease.

Third, diagnosis of abiotic disease carries with it a strict timetable for gathering pertinent details of disease development. Once the window of opportunity closes, there is no certain recreation of the environmental conditions prevailing at the time the damage occurred. In cases of chronic injury from an abiotic agent, the timetable is not nearly so critical, but an abiotic disease occurring irregularly can pose an especially perplexing diagnostic challenge. Enormous amounts of information are needed to support or refute the candidate causal agents. Collecting this vital information can be frustrating and very time consuming, and in the end most of the information collected may be of no direct help in discovering what induced the disease.

The visual aid of the disease pyramid (Fig. 2.1) can be applicable to abiotic diseases, with some modifications for the nature of the nonliving causal agent. The susceptible or predisposed plant, sufficient time, and a conducive environment are components of abiotic disease, and in certain instances it helps to conceptualize the causal agent as a noncontagious fourth component of the pyramid. This is especially helpful in cases of pesticide phytotoxicity or air pollution damage. Another helpful modification of the concept is to regard the environment as the causal agent, as in cases of damage from adverse weather. This reduces the graphic to only three components, forming a disease triangle rather than a pyramid (Fig. 2.1). The graphic emphasizes that a host, a causal agent, and an environment conducive to disease must exist concurrently and interact for a sufficient time in order for a disease (in this chapter, an *abiotic* disease) to occur. Furthermore, disease management clues are inspired by conceptualizing a disease event in this manner, just as with biotic diseases.

Rationale for Presentation

There are a number of good references on abiotic diseases in general and on specific diseases and their diagnosis. The difficulty generally lies in compiling a concise and affordable library of the better references for consultation in the diagnostic clinic. Rather than reiterate various abiotic syndromes and diagnostic tips in this chapter, we will provide references to various works that should be considered for inclusion in a clinic library or reference collection. This collection can be customized according to the types and species of plants normally encountered in your area, the environmental conditions that prevail in your region, and the budget available for the library.

Plant pathology resources available on the Internet can provide additional assistance. APSnet (www.APSnet.org) provides links to other phytopathology Web sites around the world, which could be consulted for diagnostic assistance.

Main Categories of Abiotic Diseases

Abiotic causal agents are often excesses or deficiencies of substances or conditions that are necessary for plant life. The major categories of abiotic agents of plant disease are the following, with allowances for some overlap and interaction between categories:

weather-related events: extremes of temperature, precipitation, moisture, and light; lightning; hail; wind; ultraviolet (UV) radiation

extremes in the soil environment: soluble salts, soil-applied fertilizers, soil pH, water, and soil gas exchange; problems resulting from poor-quality irrigation water
phytotoxicity (injury induced by pesticides, especially herbicides)
airborne pollutants: sulfur dioxide, ozone, hydrogen fluoride, chlorine, peroxyacetyl nitrates, oxides of nitrogen, ethylene, particulates
mechanical injury: e.g., root circling, girdling roots, rough handling, constricting ties; weather-related events, such as hail, wind, and lightning, can cause mechanical injury
nutrient deficiencies and toxicities

Weather-Related Events and Environmental Factors

Extremes of Temperature

Plants have temperature preferences and tolerances, as do all living organisms. The USDA Plant Hardiness Zone Maps along with detailed horticultural descriptions of particular plants are useful guides for determining the likelihood of cold damage (Putnam and Lang, 1991). A new guide addressing heat tolerance in plants is helpful for predicting a plant's adaptability, especially to southern climes (Cathey, 1998). At the extremes of a plant's range of temperature tolerances, reduced growth and unthriftiness are the major manifestations of plant stress. Extended periods near the extremes of the temperature range can actually toughen the plant and lessen the negative response slightly in some circumstances. Once the range is exceeded (temperature too high or too low for a sufficient period of time), more dramatic symptoms appear.

Most woody ornamentals will display split bark when subjected to sudden freezing temperatures prior to hardening off for the winter, and deep stem cracking occurs ("frost ribs") when plants are subjected to extreme cold after unseasonably warm weather. Unseasonably cold weather in spring after budbreak, however, is likely to simply kill back the new flush's tender growth. Late spring frosts often cause mystery diebacks of this sort, which can be spotty, reflecting the topography or overstory in the microenvironment. Fertilization with nitrogen or pruning late in the growing season can promote off-season growth flushes, which may suffer cold injury with the onset of freezing weather. To harden off or acclimate for winter, such tissues need about six weeks or more of exposure to gradually reducing low temperatures after the flush has concluded. Another manifestation of cold injury in evergreen woody ornamentals is leaf scorch, caused by the continued drying effects of winter winds coupled with the inability of the roots to absorb water from frozen soil.

At the other end of the spectrum, extreme high temperatures will induce premature senescence, chlorosis, and wilting of foliage. High-temperature injury to whole plants is less likely to occur outdoors unless some environmental modifications concentrate heat in unusual ways, such as under a large expanse of poorly ventilated shade cloth, on dark-colored soils or black plastic at the base of tender transplants, or in proximity to south- or west-facing reflective or heat-collecting wall surfaces. In thin-barked species, branches that have been recently exposed by pruning or loss of shade from a defoliating storm are subject to sunburn, especially when the exposed areas are oriented perpendicular to intense direct sunlight.

Extremes of Moisture

Too little water supplied to a plant in the short term will predictably cause wilt, and eventually in the long term cause early senescence, leaf scorch, defoliation, root death and sloughing, progressive branch tip dieback, and then death of the plant. Many cases of suspected root rot likely arise as a result of temporary water deficiency that initiates root death. Chronic sublethal water deficiency may simply cause stunting. Note that belowground symptoms are part of the moisture extremes syndrome (and other abiotic diseases), so it is important to *examine the whole plant* whenever possible. This means removing a plant from its container or excavating part of the root system, especially on the symptomatic side of a plant, to assess the condition of the roots and to obtain samples for lab work (see Chapter 99, "Plant Problem Diagnosis and Plant Diagnostic Clinics," for complete coverage of this topic).

When potentially hydrophobic components, such as ground bark or peat, are used in a potting mix, care must be taken to avoid extreme drying of the mix and root ball. Rewetting such a mix after it dries may require the use of drip applicators, a surfactant to reduce surface tension, or both. Water applied by overhead sprinklers can simply channel around the root ball and drain out the bottom of the container without providing enough water for sustained plant growth. Components can be incorporated into the mix at a formulation to minimize the hydrophobicity phenomenon. Dry areas in media after rain or irrigation are quite visible.

Excess moisture in the root zone suffocates roots by inhibition of oxygen uptake needed for respiration. This initiates root death and, if prolonged, results in senescence of the whole plant, ending in death. The early stages of the syndrome induced by flooding or excess water can resemble wilt from a water deficiency, since water uptake requires respiration. Especially vigorous plants may attempt to form adventitious roots above the soil level when they are overwatered.

In addition to the suffocation effect, excess water promotes the growth of anaerobic microbes, and by-products of their metabolism in the rhizosphere can be phytotoxic to roots. Bacterial growth especially is enhanced in waterlogged soil. The problem is worse at elevated temperatures, because bacterial growth is favored and oxygen requirements for root respiration increase with temperature (Lucas et al., 1995). Once root cells have lost their selective permeability from the combined effects of suffocation or exposure to toxic metabolites, these toxic substances and perhaps toxic metals move by osmosis into plant tissues and are distributed in the sap stream, causing further damage (Agrios, 1997).

Foliage of some woody ornamentals responds to excess moisture by producing small, blister-like outgrowths, usually on the undersurface of the leaves. This symptom, called edema, is most likely to occur under cloudy, humid conditions with abundant soil moisture. Such conditions create low transpirational demand on plants that are oversupplied with water. The outgrowths are composed of undifferentiated callus-like cells, which eventually burst through the epidermis. Fresh edema lesions are white or light green. As these tender cells dehydrate and die, the lesions turn corky and tan brown. Woody ornamental genera especially prone to edema are *Pittosporum, Rhaphiolepis, Camellia, Ligustrum, Hibiscus, Hedera,* and *Fatshedera.*

Some woody ornamentals, such as alder, bald cypress, willows, river birch, wax myrtle, black gum, sweet gum, larch, black spruce, and red maple, can tolerate and even thrive in saturated soil. In most cases, however, established plants tolerant of saturated soil adapt and grow best if soil moisture increases from normal to wet gradually, not abruptly.

Lists of plants tolerant of dry and wet soil conditions have been presented by Chaplin (1994), Hightshoe (1988), and Putnam and Lang (1991), and a list of common woody ornamentals that use low, moderate, and high amounts of irrigation has been published by the Southern Nurserymen's Association (Yeager et al., 1997).

The consequences of excess water aboveground also merit attention. As a general rule, prolonged wetting of foliage is undesirable, because it creates conditions necessary for invasion by fungal and bacterial pathogens and some foliar nematodes. Therefore, if overhead watering is unavoidable, the application should be timed so as to minimize the length of the period in which the foliage will stay wet. This may mean irrigating during the dew period, when foliage is wet anyway, or during a warmer, breezier time of day to encourage rapid drying. Better still, drip irrigation or aimed microemitters can eliminate altogether the problem of leaf wetness due to irrigation.

Extremes of Light

Inadequate light causes a condition known as etiolation, in which plants are pale and spindly, with long internodes. At the opposite extreme, exposure to excessively intense solar radiation can result in pale yellow or red-pigmented, thickened foliage. This is due to destruction of chlorophyll, leaving the yellow or red pigments visible. More extreme overexposure to light can cause sunburned necrotic areas on tissues most directly exposed to the sun. Sunburn is probably more accurately classified as heat damage that arises from direct exposure to intense sunlight. Transpiration fails to cool tissues adequately under such conditions. Sunburn of tender foliage and fruits is worse under water stress.

Woody ornamentals can be broadly classified according to their shade tolerance. Shade tolerance ratings are offered in many publications (Sinnes, 1982; Hightshoe, 1988; Chaplin, 1994). Light intensity decreases with increasing latitude in the United States, so much that widely temperature-adapted plants that can withstand full sun in the middle latitudes may require some protection from full sun in the southern latitudes.

Few woody ornamentals grown for their flowers can be expected to bloom freely in moderate to dense shade, so low light intensity should be considered when investigating the cause of a lack of reproductive growth. Some exceptions are *Rhododendron, Camellia,* and *Hydrangea* spp., which require no more than moderate to light shade to flower satisfactorily.

Not only can the intensity of light influence plant growth, but the duration of the photoperiod also plays an important role. Stray light has been known to disrupt flowering of trees and shrubs with strict photoperiods (Hightshoe, 1988). Disrupted photoperiods may also cause plants to fail to harden off in the fall in preparation for winter cold, and they may suppress or delay normal flowering and foliation in the spring. Such influences can subsequently predispose plants to biotic diseases.

Abrupt changes in light levels can be damaging. Foliage and even exposed branches can be sunburned when plants are moved from low to high light, or when overstory or a tree's own canopy is suddenly removed. Conversely, early senescence and leaf drop frequently occur when containerized plants are relocated from a setting with high light to one with low light.

Lightning

In lightning-prone areas of the country, one has the opportunity to learn about both the manifest and the inconspicuous destructiveness of lightning as it affects plants. Lightning-struck trees are generally tall, mature trees in exposed locations. Diagnosis of lightning damage is sometimes very straightforward, because a lightning discharge often leaves an obvious longitudinal scar of missing bark down the side of its primary target. Bark is blasted away along the path of the electrical discharge, as the electrical energy is dissipated into heat energy, which turns water in sap into steam. Lightning damage underground is difficult to visualize and assess, but damage to root systems can be extensive.

In other instances, no outwardly visible scarring occurs, and a sudden or gradual decline in plant vigor may ensue after the discharge (DuCharme, 1974). In some cases, trees afflicted without scarring were in the discharge zone in the soil but were not the primary target of the lightning bolt. This zone is usually more or less circular and may be 15 m or more in diameter (Lucas et al., 1985). Even short-statured woody ornamentals and herbaceous vegetation in the vicinity of primary lightning targets may die suddenly after a strike, without obvious involvement of pathogens, presumably secondary victims of lightning. Proximity to the primary target and the likelihood of root grafts or intermingled root systems are factors to consider in such a diagnosis. Smaller woody plants injured by lightning may display a streak of necrosis up the side nearest the discharge, terminating along the midrib of the leaf. Pith in the damaged plant may show signs of scorching or other disruption (Lucas et al., 1985).

Authorities generally agree that certain trees, such as oaks, pines, elms, tulip poplars and poplars, are lightning-prone, because they are commonly in exposed locations in open fields or on hilltops, or because they are the dominant trees in a woodland stand. Some authorities have noted that trees most likely to be damaged are also rough-barked (Gram and Weber, 1953). Smooth-barked trees, such as beech, birch, sycamore, and horse-chestnut, seem to suffer less lightning injury, perhaps because the trunk wets quickly in a rainstorm, providing an alternative path for the electrical discharge outside the tree trunk (Tattar, 1978; Smith, 1970; Sinclair et al., 1987). Lightning protection in the form of copper cables strung through the tree canopy and attached to grounding rods in the earth is available for especially valuable trees at high risk (Tattar, 1978).

Hail Damage and Sheet Ice Damage

Hail damage leaves a signature scarring pattern on its victims, which can yield an accurate diagnosis even in the absence of complete weather records. Immediately after a storm with small hailstones, shredded foliage or holes and tears ripped through leaf blades can be useful indicators of hail damage. Larger hailstones cause small, circular to elliptical scars at the impact point on the more horizontally oriented upper branches of trees and shrubs. If hail is accompanied by high winds, the observant diagnostician may even be able to reconstruct the wind direction by the orientation of the scars. Scars on woody plant parts persist for years. Twig dieback can result from hail damage, especially if the damage occurs during the dormant season. Furthermore, canker fungi can take advantage of hail wounds as infection courts. Prevention of hail injury is impractical, but protective fungicide applications or pruning after hail injury can help prevent canker.

Sheet ice damage is due to the extra weight of frozen precipitation or irrigation water on the plant canopy. Flexible stems may exhibit a series of parallel cracks perpendicular to the axis of the stem. Once the weight of accumulated ice becomes great enough, whole limbs may split or break out of the canopy entirely. Snow in evergreens (or in trees in leaf at the time of an unseasonable snowfall) can impose a similar load.

Wind Damage

Wind damage is most commonly observed in tender new shoot growth during windy spring weather. The symptoms usually suggest either rapid drying, with marginal leaf scorch, or bruising and battering from mechanical injury. To confirm that wind is to blame, one would expect to see fairly uniform damage on all exposed shoot growth of the same age, even in different species, while shoots in wind-protected areas would be symptomless. Short-statured plants and seedlings in exposed soil can be abraded by sand and debris driven by the wind, causing small scratches or contusions on the leaf surface, visible under a dissecting microscope. Directional damage is normally evident from windblown particles.

Wind damage on the grander scale, caused by the 40- or 50-mile-per-hour winds of thunderstorms or by high winds associated with tornadoes and hurricanes, is usually easy to diagnose, because inanimate structures suffer mechanical damage to a similar degree.

Ultraviolet Light Damage

With the predicted thinning and deterioration of the earth's upper atmospheric ozone layer over the last couple of decades, the risk damage to plants due to ultraviolet (UV) light, particularly UV-B (radiation with wavelengths of 280–315 nm) becomes a concern. Some of this concern is evidently premature, since levels of UV-B reaching the earth's surface in temperate and tropical climates in the middle latitudes have not yet shown a general trend toward the predicted higher levels. This is generally attributed to the interception of UV radiation by particulate pollution, especially smoke from burning to clear tropical rainforests.

If elevated UV-B does begin to impact woody ornamental plants, the immediate effects will most likely be expressed as nothing more than a general thickening and more intense pigmentation of the leaf tissue (Lumsden, 1997; Greenberg et al., 1997). UV-B is energetic enough to damage DNA in plant cells, but the repair mechanisms for DNA in plant cells are very efficient. Therefore, mutations from exposure to elevated UV-B are not likely to result in obvious symptoms in exposed plants.

Excess Soluble Salts in the Soil

Excess soluble salts in the soil will first cause wilting of plants and then the development of scorched margins on the leaves. Tip dieback and root degeneration will soon follow if the salt concentration in the root zone is not lowered by leaching with fresh water or by repotting the plant in fresh potting medium. Chronic sublethal problems with excess soil soluble salts will result in poor root establishment and stunted top growth. Certain plants (Table 3.1) are especially sensitive to high levels of soluble salts in the soil, and extra care should be taken when fertilizing them. A more extensive list of salt-sensitive and salt-tolerant plants has been presented by Shurtleff and Averre (1997).

Excess soil salts create a hypertonic bathing solution that osmotically draws water out of the plant's root system, causing symptoms of water stress. The condition most often arises from overfertilization, poor placement of fertilizer, or naturally high salt levels in irrigation water. Deicing salts used on roads and sidewalks also creates conditions in which high levels of soluble salts can accumulate in the vicinity of treated areas or in the path of runoff water from a treated surface. Foliage of plants some distance away from the treated surface can be damaged by salt spray from the atomization of salty liquid by passing vehicles.

Levels of soluble salts in the soil are easily monitored with conductivity meters designed for this purpose. These meters measure the electrical conductivity of a soil solution extracted in a prescribed manner recommended by the manufacturer. Readings performed using different equipment and methodology may not be directly comparable. Several good references are available for a thorough discussion of the measurement of soil soluble salts (Lang, 1996; Shurtleff and Averre, 1997).

Table 3.1 Some woody ornamentals sensitive to high levels of soluble salts

Woody plant species	Common name
Abies balsamea	Balsam fir
Acer negundo	Box elder
Acer pseudoplatanus	Sycamore maple
Acer rubrum	Red maple
Acer saccharum	Sugar maple
Alnus spp. (except *A. glutinosa*)	Alder
Amelanchier laevis	Serviceberry
Buxus sempervirens	Boxwood
Carpinus caroliniana	American hornbeam
Carya ovata	Shagbark hickory
Cornus stolonifera	Red-stemmed dogwood
Cycas spp.	Cycads
Euonymus alatus	Spindle-tree
Fagus grandifolia	American beech
Juglans nigra	Black walnut
Malus spp.	Apple
Morus alba	White mulberry
Ostrya virginiana	Eastern hop-hornbeam
Oxydendrum arboreum	Sourwood
Picea glauca	White spruce
Picea pungens	Blue spruce
Pinus resinosa	Red pine
Pinus strobus	Eastern white pine
Pinus sylvestris	Scots pine
Populus nigra var. *italica*	Lombardy poplar
Rhododendron spp.	Azalea
Rosa spp. (except *R. rugosa*)	Rose
Spiraea spp.	Bridal wreath
Taxus spp.	Yew
Tilia spp.	Basswood
Tsuga canadensis	Eastern hemlock
Ulmus americana	American elm
Vaccinium spp.	Blueberry
Viburnum spp.	Viburnum
Zamia spp.	Zamia

Table 3.2 Plants tolerant of salt spray or high levels of soluble salts in the soil

Woody plant species	Common name
Acer pseudoplatanus	Sycamore maple
Alnus glutinosa	Black alder
Avicennia germinans	Black mangrove
Baccharis spp.	Groundsel-bush
Coccoloba uvifera	Sea-grape
Conocarpus erectus	Buttonwood
× *Cupressocyparis leylandii*	Leyland cypress
Elaeagnus angustifolia	Russian olive
Elaeagnus pungens	Silverthorn
Forestiera segregata	Florida privet
Juniperus silicicola	Southern red cedar
Laguncularia racemosa	White mangrove
Leptospermum laevigatum	Coast tea-tree
Myrica cerifera	Wax myrtle
Pinus clausa	Sand pine
Pinus eliottii	Slash pine
Pinus thunbergii	Japanese black pine
Pittosporum tobira	Pittosporum
Populus alba	White poplar
Populus tremula	European aspen
Prunus maritima	Beach plum
Prunus spinosa	Blackthorn
Quercus geminata	Sand live oak
Quercus myrtifolia	Myrtle oak
Quercus virginiana	Live oak
Rhaphiolepis umbellata	Indian hawthorn
Rhizophora mangle	Red mangrove
Sabal palmetto	Sabal palm
Serenoa repens	Saw palmetto
Taxus baccata	English yew

Soil salt can be leached down to acceptable levels for plant growth after damage has occurred, a factor which can confound the diagnosis. Leaf scorch, tip dieback, and root damage in the absence of any clear biotic pathogens give reason to suspect injury from excess soluble salts, even if the levels determined by testing (normally expressed in parts per million, millimhos, or Siemens) are acceptable. Reviewing the history of fertilizer applications (including the amount and form of fertilizer, concentration of nutrients in the fertilizer, timing, placement, and irrigation frequency) could clinch the diagnosis.

Certain plants can tolerate an environment with high levels of soluble salts, such as the seaside, where salts may be present at high levels in the soil and also in spray that contacts foliage. Table 3.2 lists some representative woody ornamentals for sites where high salts are unavoidable; these plants are less likely to exhibit damage from excess soluble salts in the soil and in salt spray.

Phytotoxicity

For purposes of this discussion, we will restrict the concept of phytotoxicity to damage that is induced by pesticides or similar agricultural chemicals, whether intentionally applied or drifted onto nontarget plants.

Pesticide manufacturers are reasonably certain that the products they formulate are not damaging to the crops they are designed to treat and for which registration has been obtained from the U.S. Environmental Protection Agency. In the ornamental industry in particular, however, it is practically impossible for a product to be evaluated under wide-ranging environmental conditions on every conceivable plant species and cultivar that may be grown commercially. Therefore, it behooves the grower to test each new pesticide under consideration on a small portion of the crop, to determine its potential for phytotoxicity. This is especially important for ornamental plants, since the plant value is primarily aesthetic, based on its appearance, and there is little room for profit-reducing blemishes in the finished crop.

The following general guidelines can serve to minimize problems with pesticide phytotoxicity in woody ornamentals:

1. Oil-based pesticides that require an emulsifier in the formulation to put the product into suspension in water tend to cause more contact phytotoxicity (damage to tissues at sites where pesticide droplets land on them) than water-based pesticides.
2. Applications during hot, dry weather are more likely to cause contact phytotoxicity. It is preferable to make applications in the early morning, in the cooler part of the day, and on days when the temperature is not likely to rise into the unseasonable range.
3. Forcefully applied sprays are more likely to cause damage than gently applied sprays. Mist blowers can be problematic in this regard.
4. Pesticide tank mixes can create problems, even with pesticides that are normally safe when used alone. Use a jar compatibility test before mixing an entire tank. If the shaken, combined pesticides do not remain commingled in suspension for 20–30 minutes, or if a precipitate or a thick gel forms, do not tank mix them.
5. Do not use the same spray equipment for both herbicide applications and insecticide or fungicide applications. Use dedicated equipment to avoid cross-contamination. If dedicated equipment is not possible, very thoroughly clean the equipment according to the procedures recommended by the herbicide manufacturer.
6. Employ nozzles and pressure to create a droplet size appropriate for the application. Larger droplets (150–200 µm in diameter) minimize the risk of drift, whereas smaller droplets provide better coverage.
7. Avoid applications when the wind speed is greater than 10 mph. Drift to nontarget plants is likely under windy conditions.
8. Use caution when adding spreader-stickers, fertilizer, or adjuvants to pesticides. Such amended mixtures should be evaluated for phytotoxicity on a small sample of the target crop before full-scale use. The jar compatibility test (see above, item 4) can be useful in evaluating spreader-stickers and adjuvants.

A diagnosis of pesticide phytotoxicity requires a review of crop history, usually combined with a careful on-site and clinical examination of the damaged crop. In most cases of contact phytotoxicity, some residue of the offending agent can be seen under a dissecting microscope in a pattern matching the damage, if the sample is collected promptly after the damage appears.

Herbicide damage constitutes a special class of pesticide phytotoxicity in ornamentals. The potential for unintentional damage to a crop is obviously much greater with pesticides that are designed to kill plants or inhibit seed germination. Herbicides that damage woody ornamentals can be classified according to their intended uses:

Table 3.3 Herbicide injury in woody ornamentals[a]

Herbicide family	Common name (trade name)	Mechanism of action	Symptoms	Uses and notes
Arsenicals	cacodylic acid (Pytar, Quickpick) MSMA (Bueno, Ansar) DMSA (several products) CAMA (Super Crab-E-Rad-Calar)	Interference with ATP production, inhibition of certain enzymes, disruption of mitosis	Chlorosis, growth cessation, desiccation, death; symptoms appear within 5–7 days	Postemergence control of grass and some broadleaf weeds in noncropland, cotton, and turf Not leached, not absorbed by roots, short residual
Aryl-oxyphenoxys	diclofop (Hoelon) fenoxaprop (Whip, Acclaim) fluazifop-P (Fusilade DX and T/O) quizalofop (Assure)	Inhibition of fatty acid synthesis in grasses Translocated in the plant	Slow death over 1–3 weeks; foliage red, chlorotic, blackened at nodes	Mostly postemergence foliar application for grass control Soil persistence for 1–4 weeks, longer at high rates and in sandy soils Fusilade causes leaf spots and tip blight in some azalea cultivars
Benzoic acids	dicamba (Banvel, Banvel II, Clarity)	Interference with DNA and RNA synthesis and protein synthesis	Downward curving and twisting of stems within 1 or 2 days; leaves may curve up or down and may pucker In grasses, tightly rolled leaves and brittle stems	Postemergence mostly broadleaf control in grains, pasture, turf, and noncropland Readily leached, especially in sand; microbial breakdown in 1–6 weeks
Benzonitriles	dichlobenil (Casuron, Norosac, Dyclomec)	Interference with cell plate formation in cell division Translocated in the plant	Inhibition of root and shoot growth; marginal chlorosis and necrosis	Preemergence control of annual grasses, nutsedge, and broadleaves in fruits, woody ornamentals, noncropland, and aquatics Persistence for 1–12 months; volatile, especially in hot weather; not for use on sandy soils; low water solubility
	bromoxynil (Buctril, Brominal)	Inhibition of electron transport in photosynthesis and respiration Not translocated in the plant	General discoloration, mottling, shriveling, disintegration within 4–7 days	Postemergence control of broadleaf seedlings in grains and turf No soil activity
Benzothiadiazoles	bentazon (Basagran)	Inhibition of light-dependent phases of photosynthesis	Leaf chlorosis followed by leaf spotting and death in 2–7 days	Postemergence broadleaf and yellow nutsedge control in agronomic crops, woody ornamentals, and turf Persistence less than 1 month; no significant soil activity
Bipyridiliums	paraquat (Gramoxone, Starfire) diquat (Reward, Torpedo, Aquacide)	Interruption of photosynthetic electron transport, causing formation of free radicals and membrane disruption Contact action only; absorbed through leaves, green stems, thin bark	Wilt, desiccation within a few hours, leaf spotting at point of contact	Postemergence total weed control in fruit, vegetable, agronomic, noncropland, aquatics, and ornamentals No residual; binds to soil
Carbamates	asulam (Asulox)	Inhibition of mitosis by disruption of microtubule arrangement	Chlorosis, growth cessation, death within 2–3 weeks	Postemergence control of certain broadleaf and grass weeds in turf, ornamentals, and noncropland Short residual
	desmedipham (Betanex) phenmedipham (Spin-Aid)	Blockage of electron transport in photosystem II	Chlorosis, tipburn in 4–8 days	Postemergence foliar application for control of certain broadleaf weeds in row and vegetable crops Short residual; water solubility < 1 ppm

Cyclohexanediones	sethoxidim (Poast, Vantage) clethodim (Select, Envoy)	Inhibition of fatty acid synthesis	Reddening, chlorosis, blackening and breaking of nodes	Systemic postemergence control of grasses in vegetables, fruits, agronomic crops, and ornamentals Short residual
Dinitroanilines	oryzalin (Surflan) trifluralin (Treflan, Preen) prodiamine (Factor, Endurance, Barricade) benefin (Balan) pendimethalin (Prowl, Stomp, Pendulum, OH2 [with oxyfluorfen], many others)	Inhibition of cell division by interference with tubulin formation; these compounds are known as "mitotic poisons" Absorbed by roots and shoots, but no foliar activity	Growth cessation, pruned or stubby roots, lack of lateral roots, thickened stems	Preemergence control of annual grasses and certain broadleaf weeds in agronomic, vegetable, and ornamental crops, turf, and noncropland Differential placement is important; strongly adsorbed by soil, but some varying volatility risks; low leaching potential; soil persistence for 4–6 months
Diphenyl ethers	oxyfluorfen (Goal, Rout [with oryzalin], OH2 [with pendimethalin])	Inhibition of photosynthesis or respiration (or both), causing accumulation of free radicals in tissues and membrane disruption	In susceptible plants, wilting, browning, death within 24 hours In tolerant plants, spotting, puckering, darkening of leaves	Pre- and postemergence broadleaf control in ornamentals, vegetables, and tree crops Some volatility problems; water solubility = 0.1 ppm; residual for 30–60 days
	acifluorfen (Blazer, Tackle)			Pre- and postemergence broadleaf control in agronomic crops and strawberries
	lactofen (Cobra)			
Imidazolinones	imazamethabenz (Assert) imazapyr (Arsenal) imazaquin (Scepter, Image) imazethapyr (Pursuit) imazameth (Cadre, Plateau)	Inhibition of synthesis of certain amino acids Transported to meristems	Stunting, shortening of internodes, reddening of grasses, chlorosis developing from leaf tips backward; growth cessation within 7–10 days; chlorosis and necrosis of growing points within 2–4 weeks	Pre- and postemergence control of grasses, broadleaves and nutsedge in agronomic crops (Assert, Scepter, Pursuit), turf (Image, Plateau), woody ornamentals (Image), peanuts (Cadre), and noncropland and forest land (Arsenal) Both foliar and root activity; not very mobile in soil, but more mobile at soil pH > 6.5; relatively long persistence
Oxydiazoles	oxydiazon (Ronstar)	Inhibition of photosynthesis	Leaf spotting	Preemergence control of small-seeded broadleaves and annual grasses in ornamentals and turf Persistence for 3–6-months; moderate volatility; low water solubility
Phenoxys	2,4-D (many products) 2,4-DB 2,4-DP MCPP	Interference with nucleic acid and protein synthesis, causing uncontrolled growth Translocated in the plant	Epinasty, twisting, abnormal leaf development, tip chlorosis and dieback, death within 1–3 weeks Symptoms are difficult to distinguish from those caused by picolinic and benzoic acid herbicides	Postemergence broadleaf control in rights-of-way, turf, pasture, agronomic crops, and noncropland Drift problems occur, especially in ester formulation, which volatilizes readily; soil persistence for 1–4 weeks
Phosphoric acid compounds	fosamine (Krenite)	Probably, interference with cell division in new growth Slowly absorbed by leaves, with little translocation	Formation of small, spindly leaves at the next foliation; symptoms are very slow to appear	Herbaceous and woody brush control; site preparation for conifer planting Good coverage is essential for control; not much leaching, because of quick degradation in soil

(continued on next page)

Table 3.3 (continued) Herbicide injury in woody ornamentals[a]

Herbicide family	Common name (trade name)	Mechanism of action	Symptoms	Uses and notes
Phosphoric acid compounds (continued)	glufosinate (Finale, Rely)	Inhibition of the enzyme that converts ammonia into amino acids Absorbed by leaves, with little translocation	Chlorosis in 2 days; desiccation in 2–5 days; no root kill in perennial weeds and woody plants	Postemergence nonselective weed control in ornamentals, fruit and nut crops, and noncropland Not mobile in soils; quickly degraded Avoid all contact with thin bark or green tissues of desirable plants
	glyphosate (Roundup, Roundup-Pro, Roundup Ultra, Accord, Rodeo, Glyphos)	Inhibition of amino acid and protein synthesis Absorbed by leaves and green stems and translocated to roots	Slow wilt, chlorosis, and necrosis of new growth, gradually killing the whole plant, over 10–14 days Regrowth of woody and some perennial plants is distorted, stunted, and wrinkled, often with chlorotic witches'-broom	Postemergence nonselective weed control in many crops and noncropland No soil residue Avoid all contact with thin bark and green tissues of desirable plants
	sulfosate (Touchdown)	Inhibition of amino acid and protein synthesis Absorbed by leaves and green stems and translocated to roots	Slow wilt, chlorosis and necrosis of new growth, gradually killing the whole plant over 10–14 days Regrowth of woody and some perennial plants is distorted, stunted, and wrinkled, often with chlorotic witches'-broom	Postemergence, nonselective control of annual and perennial weeds in noncropland and agronomic crops Very similar to glyphosate
Pyridazinones	fluridone (Sonar)	Inhibition of carotenoid biosynthesis Absorbed by roots and translocated throughout the plant	Gradual starvation of the plant from chlorophyll destruction; meristems and interveinal areas bleached white, sometimes red pink	Control of submersed and some emersed aquatic weeds Effective at low rate; low toxicity to fish; slow action prevents depletion of oxygen in water
	norflurazon (Zorial, Solicam)			Grass and broadleaf control in peanuts, cotton, and soybeans (Zorial) and in fruits, nuts, and ornamentals (Solicam) Soil persistence for 12–18 months
Substituted amides	acetochlor (Surpass, Harness) alachlor (Lasso, others) dimethenamid (Frontier) metolachlor (Dual, Pennant) pronamide (Kerb) propachlor (Ramrod) propanil (Stam)	Probably, interference with nucleic acid and protein synthesis Absorbed by roots and shoots of seedlings	Inhibition of root and shoot growth In broadleaves, leaves are small and wrinkled In grasses, leaves are rolled and fail to unfurl	Broad-spectrum seedling weed control, mostly preemergence (except propanil) in agronomic, vegetable, fruit, turf, and ornamental crops Low leaching potential; persistence varies from a few days to up to a year, but usually 2–4 months
	napropamide (Devrinol)	Inhibition of photosynthesis	Interveinal chlorosis	
Substituted ureas		Inhibition of electron transport phases of photosynthesis	Chlorosis and death of seedlings soon after emergence; veinal chlorosis may occur; woody plants may defoliate repeatedly before death Symptoms are sometimes difficult to distinguish from those of triazines and uracils	

	diuron (Karmex)			Preemergence weed control in fruits, noncropland, aquatics Persistence for 4–12 months
	fluometuron (Cotoran, Meturon, Flo-Met)			Pre- and early postemergence in cotton for broadleaf and grass control Water solubility = 105 ppm; persistence for 1.5–6 months
	linuron (Lorox, Linex, Linuron)			Agronomic and vegetable crops Water solubility = 75 ppm; persistence for 2–4 months
	tebuthiuron (Spike)			Pre- and postemergence broad-spectrum weed control in pastures and noncropland Persistence for 1–5 years; moderate leaching potential
	siduron (Tupersan)	Inhibition of root growth	Stubby roots	Turf Not leached; residual for 4–6 weeks
Sulfonyl ureas		Inhibition of acetolactate synthase, preventing synthesis of the amino acids leucine, isoleucine, and valine; inhibition of cell division and growth Translocated to meristems	In broadleaves, severe stunting, shortening of internodes, chlorosis and necrosis from growing tips downward In grasses, red purple discoloration of leaves, death in 1–4 weeks	Pre- and postemergence broad-spectrum control with individual selectivities and soil persistence times Active at extremely low levels; generally, soil persistence increases at soil pH near 7 and up, at lower soil temperatures, and with higher levels of soil organic matter Values given for soil persistence are the longest periods for most sensitive plants or the half-life of the product
	bensulfuron (Londax)			Rice; persistence for 4 months
	chlorimuron (Classic)			Soybeans, peanuts; not for use on alkaline soils; soil persistence for 3–30 months
	chlorsulfuron (Glean, Telar)			Wheat, barley, oats; soil persistence for 4–18 months
	ethametsulfuron (Muster)			Canola
	halosulfuron (Permit, Manage)			Corn (Permit); turf and woody ornamentals (Manage); soil persistence for 1–36 months; do not use treated turf clippings for mulch around ornamentals
	metsulfuron (Escort, Ally)			Noncropland (Escort); Bermuda grass pastures (Ally); soil persistence for 1–34 months
	nicosulfuron (Accent)			Corn; soil persistence for 2 weeks to 18 months
	primisulfuron (Beacon)			Corn; soil persistence for 2 weeks to 18 months
	prosulfuron (Peak)			Cereal grains; soil persistence for 1–34 months
	rimsulfuron (Matrix)			Potatoes; soil persistence for 9–18 months
	sulfometuron (Oust)			Noncropland and forest land; half-life is 4 weeks in summer
	triasulfuron (Amber)			Wheat, barley; soil persistence for 2–6 months
	tribenuron (Express)			Wheat, barley; soil persistence for 2 months
	triflusulfuron (Debut, Upbeet)			Sugar beets; soil persistence for 2 months

(continued on next page)

Table 3.3 (continued) Herbicide injury in woody ornamentals[a]

Herbicide family	Common name (trade name)	Mechanism of action	Symptoms	Uses and notes
Thiocarbamates		Inhibition of gibberellin and lipid synthesis (mode of action not completely understood)	Small, puckered, wrinkled leaves; shortened internodes In broadleaves, stunting In grasses, onion-leaf and twisted, looped leaves Symptoms are much like those of substituted amides	Highly volatile, must be incorporated into soil; soil persistence 3–6 weeks; subject to leaching
	butylate (Sutan+, Genap+)			Corn
	cycloate (Ro-Neet)			Sugar beets and table beets
	EPTC (Eptam)			Legumes, ornamentals, and potatoes
	EPTC + protectant (Eradicane)			Corn
	metam sodium (Vapam)			Preplant fumigation
	molinate (Ordram)			Rice
	pebulate (Tillam)			Sugar beets, tobacco, tomato
	thiobencarb (Bolero, Abolish)			Lettuce, rice
	triallate (Far-Go)			Barley, wheat, peas
	vernolate (Vernam)			Peanut, soybeans
Triazines		Inhibition of electron transport in photosystem II Absorbed primarily by roots (absorbed by leaves if used with adjuvants) and translocated upward	Death of seedlings after exposure to light; interveinal chlorosis, with veins remaining green, then marginal scorch and death Symptoms are similar to those of substituted ureas and uracils	
	atrazine (AAtrex, many others)			Pre- and early postemergence control of annual grasses and broadleaf weeds in agronomic crops, turf, pastures, forest land, and noncropland Persistence for 6–12 months; water solubility = 33 ppm
	simazine (Princep, many others)			Preemergence control of annual grasses and broadleaf weeds in fruits, ornamentals, and corn Persistence for 6–12 months; water solubility = 5 ppm
	cyanazine (Bladex)			Preemergence use on agronomic crops and fallow land Persistence for 2–4 months; water solubility = 171 ppm
Triazines (asymmetrical)		Inhibition of photosynthesis, similar to the action of symmetrical triazines	Rapid chlorosis and contact necrosis; chlorosis of growing points resulting from root uptake	
	metribuzin (Lexone, Sencor)			Pre- and postemergence use on agronomic and vegetable crops Persistence for 2–4 months; water solubility = 1,200 ppm

	ametryn (Evik)			Pre- and postemergence use in corn and tropical fruits Water solubility = 190 ppm
	prometon (Pramitol)			Pre- and postemergence use on noncropland Persistence for 2–6 months
Uracils		Inhibition of light-dependent phases of photosynthesis Absorbed primarily by roots and translocated in the plant	Veinal, interveinal, marginal, or general chlorosis Symptoms are like those of triazines and substituted ureas	Pre- and postemergence Easily leached and very persistent (up to 24 months); not for use on sandy soils with low levels of organic matter; slow-acting
	bromacil (Hyvar)			Control of annual and perennial weeds and some woody species in citrus and noncropland Water solubility = 815 ppm
	terbacil (Sinbar)			Control of annual and perennial weeds in alfalfa and fruit and nut crops
Miscellaneous and unclassified compounds	amitrole, a triazole compound (Amitrole)	Inhibition of carotenoid synthesis, exposing chlorophyll to photooxidation; inhibition of cell division and elongation Absorbed by leaves, trunks, stems, and roots and transported to meristems	Bleaching (whitening) of leaves; leaf margins and petioles may turn red	Postemergence use on noncropland Persistence for 2 weeks
	bensulide, a sulfonamide compound (Betasan, Bensumac, Pre-San)	Inhibition of cell division and root formation Absorbed by roots and seeds, not translocated	Stunting, lack of root development, poor germination	Preemergence control of grasses and some broadleaf weeds in turf and ornamentals Low water solubility; soil persistence for 6–8 months
	bensulide (Prefar)			Preemergence use on vegetables
	clomazone, an isoxazolidinone compound (Command)	Inhibition of phytol tail formation of certain hormones and pigments Absorbed by roots and translocated to leaves	White to yellow bleaching and chlorosis of foliage	Preemergence control of annual grasses and some broadleaves in some vegetables, cotton, and soybeans Very volatile; mobile in soil; moderate soil persistence, especially at low soil pH; not for use within 1,000 feet of desirable plants
	copper sulfate (many products) Cu-triethanolamine (K-Tea) Cu-ethylenediamine (Komeen) Cu-ethanolamine (Cutrine, Aquatrine)	Interference with photosynthesis by blocking enzymes that need metal cofactors	Growth cessation, stunting, necrosis	Control of algae and selected aquatic weeds Activity increases at higher temperatures; readily absorbed by algae and some aquatic plants; phytotoxic to some aquatic plants at concentrations less than 1.0 ppm; directly toxic to fish and indirectly toxic when plant decay causes depletion of oxygen in the water
	dazomet (Basamid, Mylone)	Conversion to methyl isothiocyanate in the soil, which is absorbed by plant and seed tissues and causes general enzyme inhibition	Stunting, growth cessation, death	Seedbed fumigation for general soil sterilization, including broad-spectrum preemergence weed control Persistence for 1 week or more
	difenzoquat, a pyrazolium compound (Avenge)	Inhibition of nucleic acid formation, photosynthesis, and ATP production Absorbed by foliage and translocated to meristems	General plant decline	Postemergence control of wild oats in grains Not leached; soil persistence for less than 1 month

(continued on next page)

Table 3.3 (continued) Herbicide injury in woody ornamentals[a]

Herbicide family	Common name (trade name)	Mechanism of action	Symptoms	Uses and notes
Miscellaneous and unclassified compounds (continued)	dithiopyr, a pyridine compound (Dimension, Stakeout)	Inhibition of cell division Absorbed mostly by roots and accumulated in meristems	Stunting, clubbed roots Symptoms are similar to those of dinitroanilines	Preemergence and some postemergence control of annual grasses and some broadleaves in rice, ornamentals, and turf Soil persistence for 3 months
	endothall, a carboxylic acid compound (Endothall, Accelerate, Des-I-Cate, Aquathol, Hydrothol)	Contact action (mode of action not clearly understood)	Quick desiccation and death	Control of aquatic weeds and general weed control in turf and sugar beets; crop desiccant in cotton, clover, potatoes, and alfalfa Readily leached; soil and water persistence for 3–4 weeks
	ethofumesate, a benzofuran compound (Nortran SC, Prograss)	Inhibition of photosynthesis and respiration Absorbed by young shoots and roots	Death of seedlings	Selective pre- and postemergence control of annual grasses and some broadleaves in sugar beets and turf Control lasts 10 weeks; soil persistence for up to 12 months; adsorbed by soil organic matter; not leached
	fenarimol, a pyrimidine compound (Rubigan)	Inhibition of gibberellin synthesis	Severe stunting and growth cessation of annual bluegrass only	Systemic fungicide for preemergence control of annual bluegrass in turf
	flumetsulam, a pyrimidine compound (Broadstrike)	Inhibition of acetolactate synthase, in the same manner as sulfonyl ureas and imidazolinones Absorbed mostly by roots and accumulated in meristems	Stunting, interveinal chlorosis, red purple discoloration of veins; chlorosis and necrosis of growing points	Premixed only for preplant or preemergence control of annual grasses and broadleaves in soybeans and corn More active at high soil pH; soil persistence for up to 18 months; low leaching potential
	flumiclorac, a dicarboximide compound (Resource)	Contact activity, inhibition of photosynthesis, similar to the action of diphenyl ethers	Rapid desiccation and death within 24 hours	Postemergence broadleaf control in corn and soybeans Tightly bound to soil, not leached; very short soil persistence
	isoxaben, an amide compound (Gallery, Snapshot)	Inhibition of cell wall synthesis Quickly absorbed by roots (little foliar uptake) and translocated to leaves	Stunting, reduced root growth Following foliar application, stunting, stem and petiole cracking, puckering of foliage, and tip necrosis may occur	Preplant and postemergence broadleaf control in turf, ornamentals, Christmas trees, and noncropland Slight leaching potential; control lasts 5–6 months
	metaborate, an inorganic salt (Polybor, others)	Activity appears to affect calcium levels in plant tissues Absorbed by leaves and roots	Marginal leaf scorch, total foliar desiccation	Total vegetation control Best on young plants; used in combination with other herbicides; can persist for several years; very easily leached
	naptalam, a phthalic acid compound (Analap)	Antiauxin activity in meristems, where it accumulates Absorbed by all plant parts, with little translocation	Stem and leaf epinasty; roots may grow upward Symptoms are similar to those of phenoxy or benzoic acid compounds	Selective preemergence control of young broadleaves in cucurbits and ornamental nursery stock Soil persistence for 6–8 weeks; leachable in coarse soils
	oxydiazon, an oxydiazole compound (Ronstar)	Contact action, interference with photosynthesis, damage to membranes	Desiccation and death of tissues within 1–2 days Symptoms are similar to those of diphenyl ethers	Selective preemergence control of annual grasses and broadleaves in turf and ornamentals Not leached; half-life is 3–6 months

pyrazon, a pyridazine compound (Pyramin)	Inhibition of photosynthesis, similar to the action of ureas and triazines Absorbed by roots and translocated to leaves, with some absorption by leaves	Interveinal or veinal chlorosis, marginal scorch, then death	Selective pre- and early postemergence control of broadleaves in beets and sugar beets
pyridate, a pyridazine compound (Tough, Lentagram)	Inhibition of photosynthesis, similar to the action of triazines and ureas Rapidly absorbed by leaves, not translocated	Leaf desiccation, chlorosis, and death within 5–10 days	Postemergence broadleaf control in peanuts, corn, and cole crops Tightly bound in soil; not absorbed by roots; minimal soil residual
pyrithiobac, a benzoate compound (Staple)	Inhibition of acetolactate synthase, similar to the action of sulfonyl ureas and imidazolinones	Slowly developing chlorosis and stunting, death within 2–4 weeks	Postemergence control of young broadleaves in cotton Slowly degraded in soil and active at very low rates, thus potentially causing problems in crop rotation
quinclorac, a quinoline compound (Facet)	Auxin disruption and interference with cell wall synthesis and ethylene and cyanide production (action not well understood) Absorbed by all tissues and translocated to meristems	Twisting, stem swelling, growth inhibition, chlorosis, then death	Selective preplant and pre- and postemergence control of grasses and broadleaves in rice Variable leaching potential; moderate soil persistence
sodium chlorate, an inorganic salt (several products)	Strong oxidizing activity, causing rapid desiccation Readily absorbed by foliage and roots	Browning and desiccation of tissues, stunting	Nonselective control of emerging seedlings in noncropland; desiccant in cotton, soybeans, and corn Usually combined with other herbicides; leachable and persistent in soil
sulfcarbamide, an organic salt (Enquick, Wilthin)	Rapid breakdown, in the plant, into urea and sulfuric acid, which destroy cell walls and membranes Readily absorbed by all plant tissues, causing immediate damage	Wilt and desiccation within hours of treatment	Nonselective contact herbicide and preharvest desiccant for vegetable, agronomic, and fruit crops, and grass grown for seed; also used for thinning peach blooms Good coverage is essential
sulfentrazone, an analide compound (Authority)	Activity causing a buildup of light-absorbing chlorophyll intermediates (free radicals), which destroy membrane integrity, in the same manner as diphenyl ethers Absorbed by roots and foliage	Chlorosis, bronzing, scorching in a very short time after exposure to light	Selective preplant and pre- or postemergence control of annual grasses, broadleaves, and yellow nutsedge in tobacco and soybeans Fairly long soil persistence, longer in sandy soils; moderately mobile in soil

[a] The authors gratefully acknowledge the helpful suggestions of Tim Murphy, University of Georgia, for much of the information presented in this table.

1. Nursery and landscape herbicides
2. Food crop herbicides (used on agronomic, fruit, and vegetable crops)
3. Turf herbicides for annual grass control
4. Turf herbicides for broadleaf and dicot weed control
5. Herbicides for total vegetation control
6. Aquatic weed herbicides
7. Forestry herbicides

Some herbicides are used in more than one category. Nursery and landscape herbicides (category 1) have been shown to be generally quite safe, with respect to phytotoxicity, *when used as directed* on the label. Problems arise if a product is misapplied (used at too high a rate, placed in direct contact with foliage or on thin bark, applied at the wrong growth stage, etc.) or applied to sensitive plant species. It is not enough to be familiar with the symptoms of herbicide damage from products labeled for use on woody ornamentals. Herbicides in any of these categories can easily find their way into the woody ornamental arena in three ways:

1. Drift, volatilization, or leaching, of any type of herbicide, carrying it from the target area to nontarget plants. For example, total vegetation control herbicides can move downslope to affect nontarget plants.
2. Carryover (of any type of herbicide) to nontarget plants at the planting site, in the soil, on containers, in application equipment, in contaminated agricultural products (such as fertilizer or other pesticides), or in water contaminated by one crop and applied to another. This category includes the application of turf herbicides that also impact woody ornamentals in the same landscape and the use of total vegetation control herbicides in root zones shared with desirable plants.
3. Overapplication of herbicides labeled for ornamental plant use, due to overlapping application patterns, errors in calculating application rates or in mixing, poor application technique, especially sloppy herbicide placement, and poor timing of application in relation to the growth stage of the treated plant.

Another helpful classification method for the diagnostician is based on the mode of action of herbicides that can impact woody ornamental production and the symptoms resulting from that mode of action. Several good references are available to help the diagnostician decide what type of herbicidal mode of action is likely to cause the symptoms in question (Derr and Appleton, 1988; Chemical Publishing Company, 1981; MacDonald et al., 1997). Symptoms of herbicide injury, listed by herbicide families that are fairly commonly used in agriculture and could conceivably impact woody ornamentals, are described in Table 3.3.

Air Pollution Injury

Plants are excellent indicators of air pollutants, to the extent that certain species have been utilized to detect the presence of ozone, peroxyacetyl nitrate, sulfur dioxide, fluoride, heavy metals, and ethylene (Manning and Feder, 1980). As growers contemplate our future environments, it may be in their interest to include a few pollutant-sensitive species as part of their operations, just to determine if air pollutants are present in their area at dosages sufficient to cause chronic or acute symptoms. More information is available on the effects of air pollutants on crops and forest trees (Laurence and Weinstein, 1981; Pye, 1988; Schmeiden and Wild, 1995) than on woody ornamentals, but this does not imply that these plants are not affected by air pollutants. A database on effects of air pollution on vegetation, covering over 2,000 vascular plant species, revealed that *Picea abies* (Norway spruce) and *Phaseolus vulgaris* (bean) were the most studied vascular plants (Bennett and Buchen, 1995).

Both primary and secondary air pollutants affect vegetation. Primary gaseous pollutants are those emitted directly into the air by artificial or natural sources, such as sulfur dioxide. Secondary pollutants are those resulting from gaseous reactions in the atmosphere under various of conditions, such as ozone (O_3). Both types can be local or transported over large areas or regions.

The most phytotoxic and most common widespread air pollutant is ozone, which forms from a reaction between oxides of nitrogen (NO_x) and hydrocarbons in the presence of sunlight. It has been known to cause plant injury since the mid-1940s and 1950s, when it was recognized as a component of the Los Angeles smog. Ozone enters leaves through stomates, typically causing white or red to purplish black stippling on the leaf surfaces. Older leaves generally show more injury than young leaves. In tobacco, the injury was referred to as "weather flecking." Ash, aspen, black cherry, birch, crabapple, eastern white pine, ponderosa pine, snowberry, linden, bridal wreath, grape, privet, sycamore, lilac, silver maple, and catalpa have been shown to develop these symptoms from ozone exposure. Euonymus, Pfitzer juniper, and yew appear to be resistant, since they did not develop symptoms when exposed to concentrations as high as 100 pphm (parts per hundred million) (Jacobson and Hill, 1970). Table 3.4 lists woody plants known to be sensitive to ozone. Recent episodes of high ozone concentrations in certain regions of United States may offer opportunities to identify other species which are injured or not injured by ozone.

Open-top chamber studies comparing yields in filtered and unfiltered air led the U.S. Environmental Protection Agency (1996) to conclude that visible injury due to O_3 exposure reduces the market value of certain crops and ornamentals and that injury occurs at O_3 concentrations that presently occur in the United States (0.04 to 0.10 ppm). The response of certain trees, e.g., poplars and black cherry, indicates they are as sensitive to O_3 as are annual plants; exposure to 80 ppb (0.8 ppm) for 7 hours per day for eight to 12 weeks reduced biomass of these two species (Davis and Skelly, 1992). Exposure to ozone reduces carbohydrate production, thus diminishing the resources needed for plant growth processes. The physiological processes involved in ozone injury have been reviewed by Sandermann (1996).

Air pollutants such as ozone can have direct effects, causing visible foliar injury, and potential indirect effects, such as predisposing plants to biotic or other abiotic stresses (Chappelka and Freer-Smith, 1995; Treshow and Anderson, 1989).

Sulfur dioxide (SO_2) alone is not the problem it was formerly, except as a component of acidic deposition. The bleaching of interveinal areas on abaxial and adaxial leaf surfaces is a common symptom on broadleaf plants. Brown, necrotic tissue along leaf margins is not unusual. Necrotic areas may fall out, giving the leaves a tattered leaf appearance. In extremely severe injury, defoliation may occur. Injury may not always occur, because plants can convert much SO_2 to the nontoxic sulfate (SO_4) form. The interaction of SO_2 with other pollutants remains a concern. Anhydrous ammonia, iron deficiency, residual herbicides, and drought can mimic SO_2 injury.

Ash, birch, aspen, pine, apple, crabapple, and Norway spruce

Table 3.4 Woody plants sensitive to ozone

Woody plant species	Common name	Reference
Acer rubrum	Red maple	Davis and Skelly, 1992a,b
Ailanthus altissima	Tree-of-heaven	Davis and Coppolino, 1976
Betula papyrifera	White birch	Jensen and Masters, 1975
Betula pendula	Silver birch	Skelly et al., 1996
Buddleia davidii hybrids	Butterfly-bush	Findley et al., 1997
Royal Red		
Pink Delight		
Nanho Blue		
Black Knight		
Liriodendron tulipifera	Tuliptree	
Morus nigra	Black mulberry	Skelly et al., 1996
Parthenocissus quinquefolia	Virginia creeper	
Physocarpus opulifolius	Dwarf ninebark	
Populus tremula	European aspen	Matyssek et al., 1993
Prunus serotina	Black cherry	Davis and Skelly, 1992a,b
Quercus rubra	Red oak	
Rhododendron spp.	Azalea	Davis and Coppolino, 1976
Glacier		
Hinodegiri		
Korean		
Pink Gumpo		Sanders and Reinert, 1982
Mme. Pericat		
Red Wing		
Red Luann		
Hershey Red		
Rhus typhina	Staghorn sumac	Davis and Coppolino, 1976
Rosa multiflora	Multiflora rose	
Sambucus canadensis	American elder	
Cornus sericea	Redosier dogwood	
Symphoricarpos orbiculatus	Indian currant, coralberry	
Viburnum lantana	Wayfaring tree	Skelly et al., 1996

are sensitive to atmospheric SO_2, as are many other taxa. Among the conifers, eastern white pine is considered one of the most sensitive to SO_2, which causes needle tips to turn red. The chlorotic dwarf syndrome of this species is caused by a combination of SO_2 and O_3. Not all clones of white pine are sensitive, but for those that are, symptoms can be severe and tree death can result.

Leaf tissue analyses for sulfur content can be a useful indicator of accumulation of sulfur from SO_2, which enters leaves via stomates or is absorbed through the cuticle. Certain precautions must be exercised in interpreting results, because sulfur is an essential element for plants. The background sulfur level for the species at the given location must be known.

Atmospheric fluoride is not a common problem in the United States today, although it occurs in both the gaseous and the particle phases (Krupa, 1997), usually from point sources. Symptoms on plants sensitive to fluorides usually include a reddening of leaf tissue. Red suture on peach fruits is known to be caused by fluoride.

Plants known to be injured by hydrogen fluoride include Chinese apricot, Douglas-fir, blueberry, eastern white pine, gladiolus, ponderosa pine, tulip, and western larch. American holly, white birch, London plane, flowering dogwood, magnolia, mulberry, apple, and Norway maple were extremely resistant to fluoride damage in New Jersey, whereas several oak species were severely injured (Rhoads and Brennan, 1975).

Peroxyacetyl nitrate (PAN), occurs at lower concentrations than ozone, yet produces severe injury symptoms on the undersides of leaves. It is produced through photochemical processes similar to those that produce ozone. Injury to vegetables is characterized by a glazing or silvery glistening of the leaf tissue. Trees reportedly are relatively resistant. Because of its explosive nature, not many laboratories are involved in research with PAN.

Other gaseous air pollutants that may cause plant injury include volatile organic compounds (VOCs), ammonia, and chlorine. Often these chemicals result in acute injury to vegetation because of excessive concentrations at a local point source or around urban industrial centers. The effect of air pollutant mixtures on plants has been reviewed by Reinert (1984).

Considerable information has accumulated during the last few decades on the effects of both wet and dry atmospheric deposition on crops and forest trees. Studies by various authors, dealing with the combination of ozone and acid rain, have demonstrated various growth responses under field or chamber conditions. Simulated acidic rains at pH 2 resulted in acute leaf injury to *Syringa vulgaris, Elaeagnus angustifolia, Ligustrum japonicum,* and *Forsythia intermedia* in Georgia (Walker, 1989) accompanied by an increase in elemental sulfur concentrations in leaf tissue.

Particulates are receiving more attention as they impact plant foliage, disrupting photosynthesis, respiration, and transpiration, and causing visible injury to vegetation (Farmer, 1995). Local weather patterns and emission concentrations will determine the degree of injury to foliage.

What action can be taken by producers to counteract the detrimental effects of pollutants or diminish their losses from pollutant injury to plants? Treatment of plants with antioxidants, such as ethylene diurea (EDU), for ozone protection has been somewhat effective in an experimental setting (Lee et al., 1992; Long and Davis, 1991), but it is difficult to demonstrate a dose response (Heagle, 1989). The future development of such materials will depend on economics, the degree of injury that growers are willing to accept before taking steps to minimize injury, and whether or not air pollution becomes more serious.

The answer to the question posed in the preceding paragraph depends on the development of control strategies that are socially, economically, and environmentally feasible. One strategy that is always considered for disease control is plant resistance. A research program for determining resistance to O_3 in crop plants is being conducted by the U.S. Department of Agriculture (Tyson, 1997). Perennial ornamental species provide a breadth of germ plasm that must not be overlooked in the search for resistance to air pollutants.

REFERENCES

Agrios, G. N. 1997. Plant Pathology. 4th ed. Academic Press, San Diego, Calif.

Bennett, J. P., and Buchen, M. J. 1995. Bioleff: Three databases on air pollution effects on vegetation. Environ. Pollut. 88:261–265.

Cathey, H. M. 1998. Heat Zone Gardening. Time-Life Books, Alexandria, Va.

Chaplin, L. T. 1994. The Southern Gardener's Book of Lists. Taylor Publishing, Dallas, Tex.

Chappelka, A. H., and Freer-Smith, P. H. 1995. Predisposition of trees by air pollutants to low temperatures and moisture stress. Environ. Pollut. 87:105–117.

Chemical Publishing Company. 1981. Diagnosis of Herbicide Damage to Crops. Chemical Publishing, New York.

Davis, D. D., and Coppolino, J. B. 1976. Ozone susceptibility of selected woody shrubs and vines. Plant Dis. Rep. 60:876–878.

Davis, D. D., and Skelly, J. M. 1992a. Foliar sensitivity of eight eastern hardwood tree species to ozone. Water Air Soil Pollut. 62:269–277.

Davis, D. D., and Skelly, J. M. 1992b. Growth response of four species of eastern hardwood tree seedlings exposed to ozone, acidic precipitation, and sulfur dioxide. J. Air Waste Manage. Assoc. 42:309–311.

Derr, J. F., and Appleton, B. L. 1988. Herbicide Injury to Trees and Shrubs. Blue Crab Press, Virginia Beach, Va.

Dirr, M. A. 1998. Manual of Woody Landscape Plants: Their Identification, Ornamental Characteristics, Culture, Propagation and Uses. 5th ed. Stipes Publishing, Champaign, Ill.

DuCharme, E. P. 1974. Lightning – A predator of citrus trees in Florida. Proc. Annu. Tall Timbers Fire Ecol. Conf. 13:483–496.

Farmer, A. M. 1995. The effects of dust on vegetation – A review. Environ. Pollut. 79:63–75.

Findley, D. A., Keever, G. J., Chappelka, A. H., Gilliam, C. H., and Eakes, D. J. 1997. Screening *Buddleia* cultivars for acute ozone sensitivity. J. Environ. Hortic. 15:142–145.

Gram, E., and Weber, A. 1953. Plant Diseases. Philosophical Library, New York.

Greenberg, B. M., Wilson, M. I., Huang, X.-D., Duxbury, C. L., Gerhardt, K. E., and Gensemer, R. W. 1997. The effects of ultraviolet-B radiation on higher plants. Pages 2–35 in: Plants for Environmental Studies. W. Wang, J. W. Gorsuch, and J. S. Hughes, eds. CRC Press, Boca Raton, Fla.

Heagle, A. S. 1989. Ozone and crop yield. Annu. Rev. Phytopathol. 27: 397–423.

Hightshoe, G. L. 1988. Native Trees, Shrubs, and Vines for Urban and Rural America. Van Nostrand Reinhold, New York.

Jacobson, J. S., and Hill, A. C., eds. 1970. Recognition of Air Pollution Injury to Vegetation: A Pictorial Atlas. Air Pollution Control Association, Pittsburgh, Pa.

Jensen, K. F., and Masters, R. G. 1975. Growth of six woody species fumigated with ozone. Plant Dis. Rep. 59:760–762.

Krupa, S. V. 1997. Air Pollution, People, and Plants. American Phytopathological Society, St. Paul, Minn.

Lang, H. J. 1996. Growing media testing and interpretation. Pages 123–139 in: Water, Media, and Nutrition for Greenhouse Crops. D. W. Reed, ed. Ball Publishing, Batavia, Ill.

Laurence, J. A., and Weinstein, L. H. 1981. Effects of air pollutants on productivity. Annu. Rev. Phytopathol. 19:257–271.

Lee, E. H., Kramer, G. F., Rowland, R. A., and Agrawal, M. 1992. Antioxidants and growth regulators counter the effects of O_3 and SO_2 in crop plants. Agric. Ecosyst. Environ. 38:99–106.

Long, R. P., and Davis, D. D. 1991. Black cherry growth response to ambient ozone and EDU. Environ. Prot. 70:241–254.

Lucas, G. B., Campbell, C. L., and Lucas, L. T. 1985. Introduction to Plant Diseases. AVI Publishing, Westport, Conn.

Lumsden, P. J. 1997. Plants and UV-B. Cambridge University Press, Cambridge.

MacDonald, G., Murphy, T. R., Brown, S. M., and Vencill, W. K. 1997. Know your herbicide. Univ. Ga. Coop. Ext. Serv. Bull. 682.

Manning, W. J., and Feder, W. A. 1980. Biomonitoring Air Pollutants with Plants. Applied Science Publishers, Essex, England.

Matyssek, R., Keller, T., and Koike, T. 1993. Branch growth and leaf gas exchange of *Populus tremula* exposed to low ozone concentrations throughout two growing seasons. Environ. Pollut. 79:1–7.

Putnam, C., and Lang, S., eds. 1991. Ortho's Plant Selector. Ortho Books, San Ramon, Calif.

Pye, J. M. 1988. Impact of ozone on the growth and yield of trees: A review. J. Environ. Qual. 17:347–360.

Reinert, R. A. 1984. Plant response to air pollutant mixtures. Annu. Rev. Phytopathol. 22:421–442.

Rhoads, A. F., and Brennan, E. 1975. Fluoride damage to woody vegetation in New Jersey in 1974. Plant Dis. Rep. 59:427–429.

Sandermann, H. 1996. Ozone and plant health. Annu. Rev. Phytopathol. 34:347–366.

Sanders, J. S., and Reinert, R. A. 1982. Screening azalea cultivars for sensitivity to nitrogen dioxide, and ozone alone and in mixtures. J. Am. Soc. Hortic. Sci. 107:87–90.

Schmieden, U., and Wild, A. 1995. The contribution of ozone to forest decline. Physiol. Plant. 94:371–378.

Shurtleff, M. C., and Averre, C. W., III. 1997. The Plant Disease Clinic and Field Diagnosis of Abiotic Disease. American Phytopathological Society, St. Paul, Minn.

Sinclair, W. A., Lyon, H. H., and Johnson, W. T. 1987. Diseases of Trees and Shrubs. Comstock Publishing Associates, Cornell University Press, Ithaca, N.Y.

Sinnes, A. C. 1982. Shade Gardening. K. Burke, ed. Ortho Books, San Francisco, Calif.

Skelly, J. M., Innes, J. L., Savage, J. E., and Snyder, K. R. 1996. Foliar ozone injury on native vegetation in Switzerland. (Abstr.) Phytopathology 86:124.

Smith, W. H. 1970. Tree Pathology. Academic Press, New York.

Tattar, T. A. 1978. Diseases of Shade Trees. Academic Press, New York.

Treshow, M., and Anderson, F. K. 1989. Plant Stress from Air Pollutants. John Wiley & Sons, Chichester, U.K.

Tyson, R. A. E. 1997. Ozone resistance sought in crop varieties to combat pollutant damage. Environ. Sci. Technol. 31:508A.

U.S. Environmental Protection Agency. 1996. Air Quality Criteria for Ozone and Related Photochemical Oxidants. Vol. 2. EPA/600/P-93/004bF.

Walker, J. T. 1989. Influence of simulated acidic rain on growth and elemental concentrations in leaf tissue of woody ornamentals. Univ. Ga. Agric. Exp. Stn. Res. Rep. 572.

Yeager, T., Gilliam, C., Bilderback, T., Fare, D., Niemiera, A., and Tilt, K. 1997. Best Management Practices: Guide for Producing Container-Grown Plants. Southern Nurserymen's Association, Marietta, Ga.

CHAPTER 4

Karel A. Jacobs ▪ Morton Arboretum, Lisle, Illinois

Fungi

Nature and Classification of Fungi

A group of diverse microorganisms have traditionally been classified as fungi because of similar morphology (a filamentous, spore-forming body, or thallus) and biology (a dual life cycle, in which both asexual and sexual reproduction can occur) and a number of characteristics distinguishing them from plants. For example, fungi lack chlorophyll and are heterotrophic, so they cannot perform photosynthesis, as green plants do; fungi have adsorptive nutrition; and, in some cases, they have a unique biochemistry, including chitinous cell walls and the steroid ergosterol. Fungi generally have fine, threadlike hyphae, intertwined to form mycelium, which may be visible on the surface of infected plants. Spores are often produced by specialized hyphae that arise from mycelium.

Advances in nucleic acid technologies, such as DNA sequencing, have allowed the evolutionary relationships of fungi to be studied more accurately than in the past. As a result of this work, some microorganisms formerly placed in the kingdom Fungi are now placed elsewhere. For example, the class Oomycetes, comprised of the downy mildew fungi and water molds, such as *Pythium* and *Phytophthora* spp., is now grouped with brown algae rather than fungi. The current concept of true fungi divides them into four classes:

- class Basidiomycetes: mushrooms, conks, and similar organisms that produce sexual spores on top of clublike structures
- class Ascomycetes: microorganisms that produce small, pimple-like fruiting structures and sexual spores inside sacs
- class Zygomycetes: microorganisms that form nonsegmented (coenocytic) mycelia and dark, thick-walled sexual spores
- class Chytridiomycetes: microorganisms that form coenocytic mycelium and motile spores with a "tail," or flagellum

Fungi that produce only asexual spores (conidia) are, with few exceptions, related to fungi in the class Ascomycetes, but they are usually called imperfect fungi (referring to the lack of a perfect, or sexual, state). Plant pathogens are found in all of the classes of true fungi and in the Oomycetes. Table 4.1 presents a classification of some common fungal pathogens of woody ornamentals and the diseases they cause.

Fungi as Plant Pathogens

Fungi exist primarily as decomposers of organic matter. However, a small percentage are capable of causing diseases in plants and animals. Of the 69,000 species of fungi that have been identified, approximately 8,000 are plant pathogens. More diseases of plants are caused by fungi than by bacteria, nematodes, viruses, and phytoplasmas combined. A good understanding of fungal biology, including nutritional preference and survival and dispersal mechanisms, can help in disease control and management.

The capacity of a fungus to cause disease is influenced by its nutritional preference, i.e., whether it uses dead tissue or living tissue as a food source. Fungi that live strictly on dead plant material are considered saprobes (or saprophytes) and are generally not a threat to living plants. Exceptions are the wood decay fungi that can degrade xylem in living trees. Fungi that use living hosts to obtain nutrients or to complete their life cycles are parasites, and if they cause disease, they are deemed pathogens. The terms *parasite* and *pathogen* overlap and are sometimes interchanged.

Fungi may be divided further into facultative and obligate parasites. Facultative parasites are predominantly saprobic, but they have the capacity to live on living tissue. Anthracnose fungi are examples of facultative parasites. Fungi that are obligate parasites live exclusively on living host tissue. Rust and powdery mildew are caused by fungi that are obligate parasites. Most fungal diseases of plants are caused by facultative parasites, as obligate parasites have a vested interest in keeping their hosts alive.

Facultative parasites form survival structures that remain dormant outside of a living host. Survival structures are also called overwintering or oversummering structures, depending on the climate, and they generally consist of thick-walled, darkened cells or groups of cells. Some examples are the microsclerotia of *Verticillium dahliae,* chlamydospores of *Phytophthora cinnamomi,* rhizomorphs of *Armillaria mellea,* cleistothecia of powdery mildew fungi, and pycnidia of *Sphaeropsis sapinea* and similar canker-causing fungi. Fungi may also persist inside living plants or in dead plant tissues, as mycelium or fruiting bodies. Survival structures can remain dormant or quiescent for months to years, until environmental conditions and the presence of a host support their renewed growth and development.

Reproduction and Dispersal of Fungi

Like all living organisms, fungi need to reproduce. They do so mostly by forming microscopic spores sexually (in a process involving meiotic division and nuclear fusion) or asexually (in a process involving mitotic division). A tremendous diversity of spore morphology and dispersal mechanisms has evolved in fungi,

Table 4.1 Classification of some common fungal pathogens of woody ornamentals and the diseases they cause

Kingdom and class	Examples[a]	Diseases
Kingdom Fungi		
Basidiomycetes	*Armillaria*	Root rot, butt rot
	Ganoderma	Wood decay
	Gymnosporangium	Rust of junipers and plants in the family Rosaceae
	Rhizoctonia	Damping-off, web blight
	Sclerotium	Southern blight
Ascomycetes and most imperfect fungi		
Apothecial fruiting body	***Botrytis***	Gray mold
	Diplocarpon	Black spot of rose
	Entomosporium	Leaf spot
	Lophodermium	Needle cast
	Marssonina	Leaf spot
	Monilinia	Brown rot
Locular or stromatic fruiting body	***Alternaria***	Leaf spot, leaf blight
	Apiosporina	Black knot
	Botryosphaeria	Canker
	Septoria	Leaf spot
	Sphaeropsis	Tip blight, canker
	Venturia	Apple scab
Perithecial fruiting body	***Cylindrocladium***	Root rot
	Cytospora	Canker
	Discula	Anthracnose
	Fusarium	Damping-off, wilt
	Nectria	Canker
	Ophiostoma	Dutch elm disease
	Verticillium	Wilt
Cleistothecial fruiting body	***Aspergillus***	Stem, leaf, and fruit rots
	Sphaerotheca	Powdery mildew
No fruiting body	*Taphrina*	Leaf curl and blister
Kingdom Chromista		
Oomycetes	*Phytophthora*	Root rot, collar rot
	Plasmopara	Downy mildew
	Pythium	Damping-off, root rot

[a] Only a few common fungal pathogens are listed. The names of asexual (imperfect) fungi are in boldface type.

and the latter are an important target of successful disease control.

Some spores (e.g., conidia of many canker-causing fungi and anthracnose fungi) are produced in a slimy matrix, which hastens dispersal in water or by insects.

Other spores (e.g., conidia of the gray mold and powdery mildew fungi) remain dry and rely on air currents for passive dispersal.

Some spores are forcibly discharged from the fruiting body to facilitate their transport. For example, sexual spores (ascospores) of the apple scab fungus, *Venturia inaequalis,* are forcibly ejected from leaf litter in the spring and carried by air currents to emerging leaves of host plants.

Insects such as bees and bark beetles and small animals can serve as vectors and transport fungal spores between hosts. A well-known example of an insect vector is the elm bark beetle, which carries spores of *Ophiostoma ulmi,* the cause of Dutch elm disease, from infected elms to healthy elms as the beetles feed on young twig crotches.

In total, fungal pathogens can be dispersed in myriad ways: in water (such as splashing rain and irrigation), by wind, on cultivation tools, by insects and other vectors (including humans), and in the form of survival structures carried in plant debris and soil.

Fungi infect all parts of woody plants: roots, crowns, trunks, branches, twigs, leaves, flowers, and fruits. Frequently, more than one part of a plant is colonized by the same fungus, and more than one fungal species can colonize the same plant. This makes for a confusing diagnosis, especially because symptoms caused by fungi can sometimes be confused with symptoms caused by other disease agents or abiotic factors. Nonetheless, fungi usually produce diagnostic signs, such as mycelia, conidia, sclerotia, and fruiting bodies, and may develop in predictable patterns. In the end, effective control of fungal diseases is based on good diagnosis of symptoms and accurate identification of the pathogen, which in turn reveals knowledge of the pathogen's biology.

REFERENCES

Agrios, G. N. 1997. Plant Pathology. 4th ed. Academic Press, San Diego, Calif.

Alexopoulos, C. J., Mims, C. W., and Blackwell, M. 1996. Introductory Mycology. 4th ed. John Wiley & Sons, New York.

CHAPTER 5

A. R. Chase ▪ Chase Research Gardens, Mt. Aukum, California
D. Michael Benson ▪ North Carolina State University, Raleigh

Bacteria

Bacteria have been known as important plant pathogens since the late 1800s, when the cause of fire blight of pears and apples was discovered to be a bacterium, later identified as *Erwinia amylovora*. It remains one of the most common diseases of woody ornamentals today. By the 1920s, many more bacterial diseases had been described, followed by increased interest in the 1960s, as improved methods for culturing and differentiation were developed.

Nature of Bacteria

Plant-pathogenic bacteria are single-celled, rod-shaped microorganisms. They lack chlorophyll, and therefore, like fungi, they require an external source of nutrients. They are 1–3 μm long and can easily be seen with a standard light microscope following staining. Bacterial pathogens enter plants primarily through wounds or natural openings, such as stomates and hydathodes. They can survive in plant tissue for various periods of time in an inactive form, eliciting a disease response only when environmental conditions are favorable.

Another group of bacterial pathogens is the fastidious organisms, such as *Xylella fastidiosa*. They are obligate parasites, growing only in xylem cells of their hosts. These bacteria have a rod-shaped, rippled cell wall and are about 0.3 μm wide and 3 μm long. They induce a scorch disease in oleander, oaks, and others crops. Xylem-limited bacteria are transmitted primarily by insect vectors and vegetative propagation. Fortunately for nurserymen, scorch diseases are usually seen only in landscape plants.

Identification of Bacteria

Many methods are available for identification of plant-pathogenic bacteria. Some methods in the 1960s involved physiological tests (for example, oxygen utilization and enzyme production); determining optimum temperatures for growth, salt tolerance, and host range; and elicitation of a hypersensitive reaction in indicator plants. These methods were time-consuming and rarely led to a diagnosis of a previously described pathogen for some bacterial groups. Identification to the level genus was common, but little methodology had been developed for identification of certain groups to the level of species, subspecies, or pathovar.

The difficulty in identification of bacterial pathogens continued until the 1980s, when antibody techniques and carbon source utilization were incorporated into a system to rapidly and reliably read small differences in response based on the transmission of specific wavelengths of light. The use of computers made possible the rapid comparison of unidentified bacteria with large libraries, and thus a series of methods for bacterial identification was practical. The most common limit of these systems remains the reliability of the identification of the strain used to generate the original library for comparison.

Nevertheless, over the past decade many bacteria have been reclassified in new genera and species, and reclassification has usually been based on a large number of *in vitro* tests, including DNA homology, carbon source utilization, monoclonal antibodies, and fatty acid analyses. Current bacterial nomenclature can be found on the home page if the International Society for Plant Pathology (http://www.isppweb.org/).

Identification of bacteria by means of reactions elicited in different hosts appears to be out of favor, or at least of minimal interest at this time, probably because of the high cost of this method, in the space and time it requires. Despite these drawbacks, host reaction remains an important criterion for identification of new or undescribed bacterial diseases.

Symptoms of Bacterial Diseases

Plants with bacterial diseases exhibit a range of symptoms. Depending on the particular pathogen and host involved, infected plants may have soft rots, galls, leaf spots, cankers, leaf and stem blight, wilts, or shoot and tip dieback. For example, *Agrobacterium tumefaciens* causes crown gall of roses and other plants in the family Rosaceae. Galls develop on roots and lower stems and, over time, can get quite large and woody, while the foliage exhibits symptoms associated with water stress, such as poor shoot growth and chlorosis. Soft rot symptoms, which include water-soaked, mushy tissue and the collapse of stems of cuttings, may develop in many different crops during propagation, when temperature and moisture conditions in the propagation house are favorable for infection.

Bacterial pathogens often cause plants to express symptoms rapidly, because bacteria can multiply quickly, forming large populations in host tissues within one or two days after infection. *Pseudomonas syringae* is a good example of a bacterial pathogen that induces rapid symptom expression, causing leaf spot or tip dieback in most hosts. Bacterial soft rots are also examples of rapid disease development.

The fire blight bacterium, *Erwinia amylovora,* causes flower blight, twig blight, and shoot dieback during the initial phases of infection. As the pathogen persists in the host, such as flowering crabapple or flowering pear, extensive shoot dieback, shepherd's-crook (downward curving of infected shoots), and perennial cankers can develop.

Determining the difference between a bacterial leaf spot and a fungal leaf spot is not an easy matter. Symptoms are commonly very similar, and culturing often yields more than one potential pathogen, including both bacterial and fungal pathogens. Many xanthomonads (leaf rot bacteria) are difficult to culture, compared to most pseudomonads (tip dieback, canker, and wilt bacteria), and all are easily confused with saprophytic bacteria, which grow in profusion on culture media. Long-term and intimate experience with bacterial pathogens is often the only way to be completely equipped to identify familiar as well as new bacterial diseases.

Epidemiology

Bacterial pathogens that attack woody ornamentals survive on or within their hosts. The only exception is the crown gall bacterium, *A. tumefaciens,* which survives well in field soil where roses and other susceptible crops are grown. Bacterial pathogens require natural openings in host tissue or wounds for infection and spread. Free moisture on plant surfaces is critical during the infection process. Therefore, cultural practices associated with crop production can have an important impact on the epidemiology and development of bacterial diseases in susceptible crops.

Bacteria are transferred readily from plant to plant by handling or cutting instruments, and they are commonly carried in splashing water. They also are found as contaminants of seeds and other propagative materials. The use of infected cuttings and the transfer of bacteria by contaminated cutting instruments are the two most common means of disease transmission in woody ornamentals. Once a bacterial pathogen is established in a landscape or production area, it may be nearly impossible to eradicate and can spread easily in water from overhead irrigation or rainfall.

Conclusions

As new methods come into play, our understanding of bacterial relationships and thus taxonomy continues to improve. We can expect better descriptions of many "new" diseases during the next 10–20 years and extensive revision of names and relationships of diseases we have previously come to know. The attention received by bacterial pathogens will no doubt be altered as other, more economically important pathogens are described and controlled, leaving the lowly bacterium a more attractive target for research.

REFERENCES

Klement, Z., Rudolph, K., and Sands, D. C. 1990. Methods in Phytobacteriology. H. Stillman Publishers, Boca Raton, Fla.

Schaad, N. W., Jones, J. B., and Chun, W., eds. 2001. Laboratory Guide for Identification of Plant Pathogenic Bacteria. 3rd ed. American Phytopathological Society, St. Paul, Minn.

Sinclair, W. A., Lyon, H. H., and Johnson, W. T. 1987. Diseases of Trees and Shrubs. Comstock Publishing Associates, Cornell University Press, Ithaca, N.Y.

CHAPTER 6

Robert A. Dunn ▪ University of Florida, Gainesville

Plant-Parasitic Nematodes

Plant nematodes attack most nursery and agricultural crops grown in the United States. Damage from nematodes is most severe in sandy soils, but they are a problem in soil of all types. Nematodes cause farmers and nurserymen millions of dollars in crop loss annually; they also cause problems in the urban world, by damaging turfgrasses, ornamentals, and home gardens. We are often unaware of losses caused by nematodes, since much of the damage is so subtle that it goes unnoticed or is attributed to other causes.

Some scientists estimate that there are over 1 million kinds of nematodes, making them second only to the insects in numbers. However, few people are aware of nematodes or have seen any, for several reasons: most nematodes are very small, even microscopic, and colorless; most live hidden in soil, in water, or in the plants and animals which they parasitize; and few nematodes have obvious direct effects on humans or their activities.

Of the known nematodes, about 50% are small animals living in marine environments and 25% live in soil or fresh water and feed on bacteria, fungi, other decomposer organisms, small invertebrates, or organic matter. About 15% are parasites of animals, ranging from small insects and other invertebrates to domestic and wild animals and humans. Some of the parasites of animals are the largest nematodes known: a nematode parasite of grasshoppers can be several inches long, and one from whales can reach lengths of over 20 feet! Only about 10% of known nematodes are parasites of plants.

Morphology and Anatomy

Plant-parasitic nematodes are tiny worms, mostly 0.25–3.0 mm (1/100–1/8 inch) long, and cylindrical, tapering toward the head and tail. Females of a few species lose their worm shape as they mature, becoming pear-, lemon-, or kidney-shaped in host tissue. Plant-parasitic nematodes possess all of the major organ systems of higher animals except respiratory and circulatory systems. The body of a nematode is covered by a transparent cuticle, which bears surface marks that are helpful for identifying nematode species.

Life Cycle and Reproduction

The life cycle of a plant-parasitic nematode has six stages: egg, four juvenile stages, and adult. Males and females occur in most species, but reproduction without males is common, and some species are hermaphroditic ("females" produce both sperm and eggs). Egg production completes the cycle. In most species, an individual female produces between 50 and 500 eggs, depending on the nematode species and the environment, but some can produce more than 1,000. The length of the life cycle varies considerably, depending on the nematode species, the host plant, and the temperature of the habitat. During the summer, when soil temperatures are in the 80°F range, many plant nematodes complete their life cycles in about four weeks.

Nematode Feeding and Host-Parasite Relationships

Plant-parasitic nematodes feed on living plant tissues, using an oral stylet or spear (somewhat like a hypodermic needle) to puncture host cells (Fig. 6.1). Most (probably all) plant nematodes inject enzymes into a host cell before feeding, to partially digest the cell contents before they are sucked into the gut. Most of the injury that nematodes cause to plants is related in some way to the feeding process.

Some nematodes (ectoparasites) feed on plant tissues from outside the plant, and some (endoparasites) feed inside the host. In some species the adult female moves freely through the soil or plant tissues; these species are said to be *migratory.* In other species the adult female becomes swollen and permanently immobile in one place in or on a root; these species are termed *sedentary.* Migratory endoparasitic and ectoparasitic nematodes generally deposit their eggs singly as they are produced, wherever the female happens to be in the soil or plant. Sedentary nematodes, such as root-knot nematodes (*Meloidogyne* spp.), cyst nematodes (*Heterodera* spp.), reniform nematodes (*Rotylenchulus* spp.), and the citrus nematode (*Tylenchulus semipenetrans*) produce large numbers of eggs, which remain in the body or accumulate in masses attached to the body of the female.

The feeding habits of nematodes and their relationships with their hosts affect sampling methods and the success of management practices. Ectoparasitic nematodes which never enter roots may be recovered only from soil samples. Endoparasitic nematodes are often most easily detected in samples of the tissues in which they feed and live (burrowing and lesion nematodes), but some occur more commonly as migratory stages in the soil (root-knot and reniform nematodes).

Endoparasitic nematodes inside root tissues may be protected

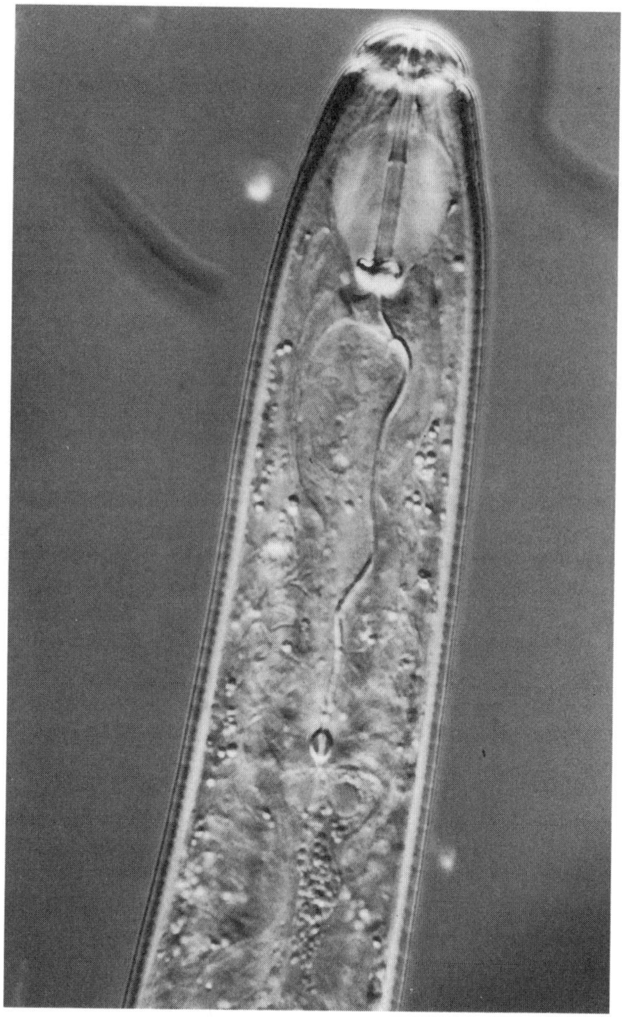

Fig. 6.1 Head region of a plant-parasitic nematode with a stylet, the spear-shaped tube used to penetrate and feed from host cells. (Courtesy Plant Pathology Photo Library, North Carolina State University)

from pesticides which do not penetrate roots. Root tissues may also shield them from many microorganisms which attack nematodes in the soil. Ectoparasites are more exposed to pesticides and natural antagonistic agents in the soil.

Foliar nematodes (*Aphelenchoides* spp.) are migratory species which feed on or inside the leaves and buds of azalea, ferns, hosta, chrysanthemums, and many other ornamentals. They cause distortion or death of buds, leaf distortion, or yellow to dark brown lesions between major veins of leaves. Other nematodes which attack plants aboveground may cause leaf or seed galls, and still others cause deterioration of bulbs.

Diagnosing Nematode Problems

Determining whether nematodes are involved in a plant growth problem is difficult, because few nematodes cause distinctive diagnostic symptoms. A sound diagnosis should be based on as many of the following as possible: aboveground and belowground symptoms, field history, and laboratory assay of soil and plant samples.

Aboveground Symptoms

Aboveground symptoms are rarely, if ever, sufficient evidence for diagnosis of nematodes affecting plant roots. However, aboveground symptoms are important because they are almost always the first symptoms to be noticed. Since most plant nematodes affect root functions, most symptoms associated with them are the result of inadequate water supply or mineral nutrition to the tops: chlorosis (yellowing) or other abnormal coloration of foliage, stunting of top growth, failure to respond normally to fertilizers, small or sparse foliage, a tendency to wilt more readily than healthy plants, and slower recovery from wilting. Woody plants in advanced stages of decline caused by nematodes may exhibit dieback of progressively larger branches. Melting out, or gradual decline, is typical of nematode-injured turf and pasture. Plantings stunted by nematodes are less capable of competing with weeds and often have worse weed problems than crops without nematode damage.

The distribution of nematodes at any site is very irregular, so the shape, size, and distribution of areas with the most severe damage from nematodes will be erratic within the field. Nematodes move very small distances on their own (a few feet per year). In the undisturbed soil of nurseries, visible symptoms of nematode injury normally appear in round, oval, or irregularly shaped areas, which gradually increase in size year by year. In cultivated land, nematode-infested soil often has elongated spots in the direction of cultivation, because nematodes are moved by machinery. Erosion, land leveling, and any other force which moves masses of soil or plant parts can also spread a nematode infestation much more rapidly than occurs by natural movement of nematodes. Nematode damage is often seen first and most pronounced in areas under stress, such as areas with heavy traffic, areas with excessive drainage because of slope or soil, and dry areas outside regular irrigation patterns.

Belowground Symptoms

Belowground symptoms may be more useful than top symptoms for diagnosing nematode problems. Galls caused on roots by root-knot nematodes, abbreviated roots or stunted root growth, necrotic lesions in the root cortex, and root rotting may all be symptoms of nematode problems.

An experienced observer can often see cyst nematodes (*Heterodera, Globodera,* and *Cactodera* spp.) on the roots of their hosts without magnification. The young adult females are visible as tiny white beads, about the size of a period on this page. After a female cyst nematode dies, her white body wall is tanned to a tough brown capsule containing several hundred eggs.

Field History

Field history can provide valuable clues in identifying problems caused by nematodes and other pests. A nematode which has been present in the field in recent years is probably there yet and is likely to cause injury to a susceptible crop if environmental conditions are favorable. Production records which show a gradual decline in yield over a period of years despite no change in cultural practices may indicate the progressive development of a nematode problem. Nematode infestation in a new field usually begins in a small area. It gradually intensifies in the original spot and is spread through the field by cultivation, harvest, erosion, and other means by which infested soil or plant parts are spread.

Therefore, the total effect of a recently introduced nematode is a gradual production decline for the field, as the percentage of the field that is involved and the severity of the damage in any given area in the field increase over the years.

Laboratory Assay

Laboratory analysis of soil and plant tissue samples is often necessary to complete a diagnosis. In the lab, nematodes are extracted from soil and plant tissues, identified, and counted. The results can be compared with research and field observations to determine whether or not the crop is likely to be injured by the estimated population under current conditions. In some cases, steps to reduce the population of a nematode species or limit its effects are recommended only if the population density exceeds some predetermined level, which represents the threshold of economic loss in that crop. Such thresholds are determined by the long-term experience of nematologists working with a specific pest and crop in growers' operations and in controlled experiments.

Summary

Plant-parasitic nematodes occur in most nursery soils in the United States. Nematodes are wormlike animals, most of which are microscopic. Some species of plant-parasitic nematodes are ectoparasitic (feeding on plant tissues from outside the host), and some are endoparasitic (feeding from within host tissues). Nematodes feed on plant tissues by means of a stylet, and the enzymes produced during the feeding process injure host cells.

Symptoms of nematode disease include poor plant growth, chlorosis of foliage, poorly developed root systems, lesions on roots, and root galls. Evidence of nematode disease in field-produced nursery stock may first appear in limited areas where plants grow poorly.

Diagnosis of nematode disease is based on aboveground and belowground symptoms, field history, and laboratory assays to identify and count the nematodes present in plant or soil samples.

CHAPTER
7

Lawrence G. Brown ▪ U.S. Department of Agriculture, Raleigh, North Carolina
Walter Bliss ▪ Agdia, Elkhart, Indiana

Viruses

Nature of Viruses

A virus is composed of genetic material (DNA or RNA) that is protected by a coat of protein. Viruses are able to multiply only within living cells and cannot live and multiply on their own. They can enter plant cells only through wounds or by being placed in them by another organism, called a vector. Virus vectors include insects, mites, nematodes, fungi, and parasitic plants, such as dodder. Some viruses do not have vectors, and they enter cells of plants that have been subjected to mechanical injury, such as plants with damaged roots, broken leaf or root hairs, or leaves wounded by rubbing together. Seed and pollen can move viruses over long distances. However, the primary means by which viruses move into new areas is the mass production of virus-infected plants. In the nursery, viruses can be efficiently transmitted during grafting and other forms of asexual propagation.

Viruses in woody ornamentals are subtle, and they are a challenge to manage, because they are spread so readily by vegetative propagation. When an entire crop is infected, the losses may go unnoticed because there is no opportunity to compare the "hidden" loss to "clean" or uninfected plants. Furthermore, in some plants, the presence of a virus provides the unique trait that makes them interesting to the horticulturist. The yellow venation of the leaves of honeysuckle (*Lonicera japonica*) is the result of infection by honeysuckle vein mosaic virus, and plants that are free of the virus are an uninteresting green. The diagnosis of viruses in woody plants is a relatively young discipline in plant pathology.

Symptoms of Virus Infection

Symptoms provide an important, but incomplete, guide for diagnosis, because similar symptoms can be produced by different viruses. Symptoms of virus infection are extremely variable. The same virus can produce a range of symptoms, depending on the environment and the host genotype. In latent or masked infections, symptoms are absent. When several viruses infect one plant, the effects may be additive, synergistic, or antagonistic.

A good description of symptoms is critical to a correct diagnosis, however. Discoloration affects leaves, flowers, fruits, and roots. Chlorosis, bleaching, yellowing, reddening, browning and bronzing may be evenly distributed. *Chlorosis* is a weakening of the green color. Mineral deficiencies cause a similar appearance. *Bleaching* is a disappearance of all color. *Yellowing* appears as chlorosis with a dominance of yellow pigments. *Reddening* is the result of abnormal anthocyanin production; it can be caused by mineral deficiencies as well as viral infection. *Browning* and *blackening* are the result of the pathogen-incited production of dark, melanin-like substances. *Bronzing* is the result of the necrosis and collapse of epidermal cells covering apparently healthy mesophyll. Injury due to mites, thrips, and air pollution can mimic bronzing symptoms. Usually a hand lens is sufficient to detect the presence of mites and thrips.

Mosaic, mottle, ring spots, and streaking are irregular discolorations. *Mosaic* consists of pale green or yellow, chlorotic areas that are sharply bordered by small veins, which give the discolored area an angular shape (Plate 21). *Mottle* is discolored areas of various shapes, often with diffuse borders. *Ring spots* are circular areas of chlorosis with green centers. *Streaking* is manifested as elongated, sharply defined chlorotic patches.

Certain leaf parts are uniformly discolored by some virus infections. *Vein yellowing,* due to lack of chlorophyll, is accented by the reds and yellows of carotenes and xanthophylls. *Vein clearing* causes veins to be translucent rather than yellow. *Vein banding* consists of discolored areas accompanying the veins. *Vein necrosis* is the death of vascular tissues, resulting in browning.

Malformations resulting from virus infection can distort, dwarf, and thicken leaves and cause them to curl upward or downward. *Enations* are tissue malformations or overgrowths, which may be caused by virus infection.

Infected flowers may have color deviations, intensifications, or pigment changes. *Color breaks* consist of flecks, streaks, or sectors of abnormally colored tissue. They may reduce plant value, but sometimes these imperfections are prized. For example, infection by camellia yellow mottle virus gives rise to a unique yellow variegation of leaves.

Virus-infected fruits and roots may be discolored and may have necrotic lesions. Flowers are similarly affected, with distortions, dwarfing, and formation of abnormal parts. Fruits may be dwarfed or deformed and may have tumorous swellings or aborted seed. Stems may also be distorted and may have shortened internodes. Roots may decay and die back and may form tumors or a proliferation of side roots in the absence of other pathogens.

Viral symptoms that can be confused with those of other pathogens are wilting, defoliation, premature leaf drop, deviation in flower number, premature or delayed flowering, gummosis, bark scaling, and graft incompatibility. In the ornamental industry, many plants with chlorotic (yellow) or variegated foliage are selected and propagated. These unique patterns are usually due to

Table 7.1 Characteristics of persistent, semipersistent, and nonpersistent transmission of viruses by aphids[a]

	Persistent transmission	Semipersistent transmission	Nonpersistent transmission
Acquisition period	Several hours to days	Several hours to days	A few seconds
Latent period	Hours to days	None	None
Inoculation	Several hours to days	Several hours to days	A few seconds
Retention	Lifetime of the aphid	1–3 days	A few minutes to several hours
Vector ecology	Colonizing	Colonizing	Often noncolonizing
Location in plants	Usually phloem-associated	Generally distributed	Generally distributed

[a] Courtesy of Susan Halbert, Florida Department of Agriculture and Consumer Services, Division of Plant Industry, Gainesville.

genetic abnormalities, often called sports, and may or may not be stable in propagation.

Transmission of Viruses

Many viruses in woody ornamentals are transmitted by vegetative propagation from virus-infected stock plants. Viruses may also be transmitted from diseased to healthy plants by vectors, such as aphids and nematodes. Characteristics of virus infection resulting from transmission by aphid vectors are summarized in Table 7.1. Some viruses can be transmitted in seed and pollen. In woody ornamentals most viruses are systemic in the plant. For many viruses a vector has yet to be discovered. A few viruses reach such high concentrations in the plant that they can be transmitted by mechanical injury and plant-to-plant contact. Virus particles on the surface of or within pollen grains and seeds can be carried over a long distance. However, the primary method of virus movement into new areas is by the introduction of virus-infected plants.

Viruses cannot enter new hosts by themselves. They enter only through wounds or with the help of other organisms (vectors). A vector acquires a virus by feeding on an infected plant and then introduces the virus into a new plant. Insects and mites are responsible for transmitting more than 26 of the 81 plant viruses reported in woody ornamentals, while nematodes transmit 14 of them, and there are 14 aphid-transmitted and two whitefly-transmitted viruses that are important to woody ornamentals (Table 7.2). All the whitefly-transmitted viruses of ornamentals are spread by persistent transmission. Cacao swollen shoot badnavirus is semipersistently transmitted by mealybugs. Cherry mottle leaf (?) trichovirus is transmitted by mites, and okra mosaic tymovirus by a beetle. Many reports of viruses in woody plants are from older plants in the landscape and seldom occur or cause damage in nursery production.

Table 7.3 lists some woody ornamentals that are not susceptible to certain viruses.

Virus Detection

Nurserymen can detect viruses in woody ornamentals by visual inspection. The characteristic symptoms described above are often indicative of virus infection. However, visual inspection alone is insufficient in propagation operations where virus-free stocks are maintained. Two techniques commonly used to detect and identify viruses are more accurate than visual inspection: bioassay, in which a suspected virus is transmitted from a plant to another plant that shows typical symptoms when infected, and enzyme-linked immunosorbent assay (ELISA), a serological test.

Molecular analyses, based on nucleic acid analysis, are increasingly being used. These include hybridization of viral nucleic acids, to mark their presence in the plant, and polymerase chain reaction (PCR), a technique that allows almost infinite amplification of a segment of DNA associated with the virus under study.

Bioassay

Bioassay is a technique in which indicator plants are inoculated with material being tested. A virus will cause characteristic symptoms in indicator plants if present. Some inoculation methods are (1) rubbing leaf disks or Carborundum-coated leaves with buffered sap, (2) grafting, and (3) transmitting viruses by means of pollen or vectors, including arthropods, nematodes, and dodder. Knowledge of the viruses commonly found in the plant of interest and the symptoms caused in indicator plants is required.

ELISA

ELISA is the second commonly used technique. An animal, often a rabbit, is inoculated with a purified virus to produce antibodies against the virus in its blood serum. The antibodies are extracted and used in ELISA to detect the virus in the plants of interest. This is accomplished in commercially produced plates containing an array of test wells. First, antibodies are coated on the test well surfaces. Then a leaf extract is incubated in the test well, and viruses that the antibody recognizes will be captured. Next, the extract is washed out, and more antibodies with an enzyme attached to them are added. These enzyme-linked antibodies (enzyme conjugates) bind with the captured virus. Any unattached antibody enzyme conjugate is then washed from the plate, and a substrate solution that produces color in the presence of the enzyme is added. If color is produced, a positive result is indicated. The intensity of the color depends on the amount of virus present and the ability of the antibodies to bind. This technique is good for processing large numbers of samples and for confirming the presence of both introduced viruses and viruses that are difficult to diagnose by other means.

Inclusion Bodies

Inclusion bodies may be observed in plant cells when a virus is present. They are crystal-like structures with specific patterns, depending on which virus or viruses are infecting the plant. Inclusion bodies may aggregate in plant cells in numbers large enough that they can be seen with a regular light microscope. This very practical method can be the first method used on plants with virus-like symptoms in a clinic, because it gives direct evidence that a virus is present. The type of inclusion body formed and the loca-

Table 7.2 Viruses confirmed in woody ornamental plants[a]

Ornamental species	Common name	Virus	Transmission[b]	Symptoms
Abelia grandiflora	Glossy abelia	Abelia latent virus	M	None
Abutilon spp.	Abutilon (several species)	Abutilon mosaic virus	Whitefly (P)	Yellow mosaic
Albizia julibrissin	Mimosa	Mimosa striped chlorosis virus[c]	M, S	Striped chlorosis
Astilbe spp.	Astilbe	Arabis mosaic virus	M, aphid (N), Nem, S	Mosaic, ringspots, some necrosis
Berberis spp.	Berberis	Cucumber mosaic virus	M, aphid (N), Nem	Foliar mosaic and reddish blotches
Betula spp.	Betula	Cherry leaf roll virus	M, Nem, Pol, S	Chlorotic ringspot, yellow vein netting
		Prunus necrotic ringspot virus	M	Chlorotic ringspot
Buddleia davidii	Butterfly-bush	Alfalfa mosaic virus	Aphid (N), M, Nem	Mosaic
		Cucumber mosaic virus	Aphid (N), M, Nem	Yellow and chlorotic arc and ring patterns
Buxus sempervirens	Boxwood	Arabis mosaic virus	Aphid (N), M, Nem, S	Mosaic, ringspots, some necrosis
Camellia japonica	Camellia	Camellia yellow mottle virus	Grafting only	Chlorotic leaf fleck, flower break
Camellia sasanqua	Sasanqua camellia	Camellia yellow mottle virus	Grafting only	Chlorotic leaf fleck, flower break
Catalpa bignonioides	Catalpa	Broad bean wilt virus[d]	M	Chlorotic leaf spotting
		Scrophularia mottle virus[e]	Beetle (N), M	Mottling
Chamaecyparis lawsoniana	Port Orford cedar	Arabis mosaic virus	Aphid (N), M, Nem, S	Mosaic, ringspots, some necrosis
Corchorus olitorius	Nalta jute	Okra mosaic virus[f]	Beetle (N), M	Vein chlorosis, veinbanding
Corchorus spp.	Corchorus	Cacao swollen shoot virus[f]	M; mealybugs, family Pseudococcidae (SP)	No information
Cornus florida	Dogwood	Broad bean wilt virus	M	Mosaic
		Cherry leaf roll virus[g]	Unknown	Yellow vein netting
		Cucumber mosaic virus	Aphid (N), M, Nem	Mosaic
		Dogwood mosaic virus[g]	M	Mosaic
		Tobacco ringspot virus[g]	Unknown	Mosaic
		Tomato ringspot virus[g]	Unknown	Mosaic
Crataegus spp.	Hawthorn	Apple chlorotic leaf spot virus	M, Nem	Chlorotic ringspots
		Apple stem pitting virus	M	Latent infection
Cycas revoluta	Cycas	Cycas necrotic stunt virus[h]	M, Nem, S	Mild mosaic
Cydonia oblonga	Quince	Apple chlorotic leaf spot virus	M, Nem	Chlorotic rings and spots
Daphne mezereum	February daphne	Arabis mosaic virus	Aphid (N), M, Nem, S	Mosaic, ringspots, some necrosis
Daphne odora	Winter daphne	Carnation mottle virus	C, M,	Mottle
		Daphne S virus[i]	Aphids (N)	Down curling of leaf edges
		Daphne X virus[j]	M	None
		Daphne Y virus[j]	M	Leaf chlorosis, distortion, reduced flower quality
Dodonaea viscosa	Hop bush	Dodonaea yellows-associated virus	Grafting only	Witches'-broom
Euonymus europaeus	Spindletree	Arabis mosaic virus	Aphid (N), M, Nem, S	Mosaic, ringspots, some necrosis
		Strawberry latent ringspot virus	M, Nem, S	Yellow mottle
Euonymus japonicus	Euonymus	Euonymus fasciation virus[k]	Grafting only	Fasciation
	Euonymus	Euonymus virus[l]	Grafting only	Vein yellowing, chlorotic spots
Ficus carica	Fig	Fig virus[l]	Aphids, M	Leaf chlorosis, distortion
		Fig S virus[h]	Grafting only	None reported
Ficus spp.	Ficus	Citrus enation–woody gall virus	Aphids (P)	Woody galls
Forsythia × intermedia	Forsythia	Arabis mosaic virus	Aphid (N), M, Nem, S	Mosaic, ringspots, some necrosis
Fraxinus americana	Ash	Tobacco mosaic virus	C, M, S[m]	Not described
		Tobacco ringspot virus	M	Chlorotic veinbanding, oak leaf patterns
Fraxinus excelsior	European ash	Arabis mosaic virus	Aphid (N), M, Nem, S	Mosaic, ringspots, some necrosis
Hedera spp.	Ivy	Arabis mosaic virus	Aphid (N), M, Nem, S	Chlorotic, yellow banding, ringspots
		Ivy vein clearing virus[d]	M	Seasonally variable

(continued on next page)

Table 7.2 (continued) Viruses confirmed in woody ornamental plants[a]

Ornamental species	Common name	Virus	Transmission[b]	Symptoms
Hibiscus rosa-sinensis	Hibiscus	Abutilon mosaic virus	M, whitefly (P)	Variable
		Eggplant mottled dwarf virus[n]	M	Vein yellowing, distortion
		Hibiscus chlorotic ringspot virus	M	Mottle, rings, veinbanding
		Hibiscus latent ringspot virus[n]	M, Nem	Leaf chlorosis (sometimes)
		Pittosporum vein yellowing virus[o]	C, M, Pol	Vein yellowing
		"Black Splash" viruslike agent[p]	Unknown	Black lesions on bark, premature leaf drop
Hibiscus syriacus	Rose-of-Sharon	Abutilon mosaic virus	M, whitefly (P)	Variable
Hibiscus spp.	Hibiscus	Abutilon mosaic virus	M, whitefly (P)	Variable
		Cotton leaf crumple virus[f]	Whitefly (P)	Mosaic, distortion
		Hibiscus yellow mosaic virus[h]	M	Yellow mottling
Hydrangea spp.	Hydrangea	Elm mottle virus	M, S(?)	Leaf chlorosis
		Hydrangea mosaic virus	M, S(?)	Distortion
		Hydrangea ringspot virus	C, M	Distortion
		Tomato ringspot virus	M, Nem, Pol	Leaf chlorosis
Jasminum officinale	Jasmine	Arabis mosaic virus	Aphid (N), M, Nem, S	Mosaic, ringspots, some necrosis
Juglans regia	English walnut	Cherry leaf roll virus	M, Nem	Leaf pattern, black line
Kalmia latifolia	Mountain laurel	Rhododendron necrotic ringspot virus	Grafting only	Concentric rings
Laburnum anagyroides	Golden-chain	Laburnum yellow vein virus[d]	Grafting only	Chlorotic veinbanding
Ligustrum vulgare	Privet	Arabis mosaic virus	Aphid (N), M, Nem, S	Mosaic, ringspots, some necrosis
Lonicera japonica	Honeysuckle	Tobacco leaf curl virus	Whitefly (P)	Vein yellowing
Lonicera spp.	Lonicera	Eggplant mottled dwarf virus[n]	M	Yellow veinbanding
		Honeysuckle latent virus[d]	Aphids (N), M	None
Maclura pomifera	Osage orange	Maclura mosaic virus[l]	Aphids (N), M	Mosaic
Malus angustifolia, M. pumila, M. × platycarpa	Apple	See Cooper (1993)		
Malus sylvestris	Crabapple	See Cooper (1993)		
Mimosa sensitiva	Sensitive plant	Mimosa mosaic virus[q]	Insects (SP), M	Mosaic
Morus alba	Mulberry	Citrus enation–woody gall virus	Aphids (P)	Woody galls
		Mulberry latent virus[h]	M	None
		Mulberry ringspot virus[h]	M, Nem, S	Mosaic, ringspot
Nandina domestica	Heavenly bamboo	Mosaic virus	Aphids (N), S?	Distortion
		Nandina mosaic virus[g]	M	Mosaic
		Nandina stem pitting virus[g]	Grafting only	Mosaic, leaf cupping, distortion, stem pitting
Olea europaea	Olive	Cherry leaf roll virus	M, Nem	None
		Olive latent 1 virus[o]	M	None
		Olive latent 2 virus[o]	M	None
		Olive latent ringspot virus[o]	M	None
Pentas spp.	Pentas	Tomato ringspot virus	M	Mild mosaic
Populus balsamifera	Balsam poplar	Poplar vein yellowing virus[r]	M	Veinclearing, vein yellowing
Populus spp.	Poplar	Poplar decline virus	M	Chlorosis, necrosis, death
		Poplar mosaic virus	M, Pol	Leaf spot, decline
Prunus avium	Cherry, wild	Arabis mosaic virus	Aphid (N), M, Nem, S	Mosaic, ringspots, some necrosis
		Cherry leaf roll virus	M, Nem	Leaf roll and death
		Cherry mottle leaf virus	M, mite (*Eriophyes*)	Irregular mottle, leaf distortion
		Cherry rasp leaf virus	M, Nem, S	Enations, stunting
		Epirus cherry virus[o]	M, S	Enations on underside of leaf, stunting
		Myrobalan latent ringspot virus[k]	M, Nem	Enations on underside of leaf
		Petunia asteroid mosaic virus[d,r]	M, S	Veinal necrosis, distortion, stunting of shoots
Prunus cerasifera	Plum, bronze cultivar	Cherry leaf roll virus	M, Nem	Leaf roll and death
		Cherry mottle leaf virus	M, mites	Irregular mottle, leaf distortion
		Myrobalan latent ringspot virus[k]	M, Nem	None

(continued on next page)

Table 7.2 (continued) Viruses confirmed in woody ornamental plants[a]

Ornamental species	Common name	Virus	Transmission[b]	Symptoms
Prunus serrulata	Flowering oriental cherry	Cherry mottle leaf virus	M, mites	Irregular mottle, leaf distortion
		Plum American line pattern virus[g,r]	M	Lines, chlorotic rings
Pyrus communis	Pear	Apple chlorotic leaf spot virus	M, Nem	Ring patterns, mosaic
		Apple stem pitting virus	M	Vein yellowing
Quercus velutina	Black oak	Oak ringspot virus[g]	Grafting only	Chlorotic ringspots, veinbanding, oak leaf patterns on leaves
Rhododendron spp.	Rhododendron	Rhododendron necrotic ringspot (?) virus	Grafting only	Concentric rings
Robinia pseudoacacia	Black locust	Robinia true mosaic virus	Aphids (N), M	Mosaic, mottling
		Tomato black ring virus	M, Nem, Pol, S	Necrotic and chlorotic rings, spots, flecks, distortion
Rosa setigera	Prairie rose	Tobacco streak virus	M, Pol, S, thrips[m]	Vein yellowing
Rosa spp.	Rose	Apple mosaic virus	M, Pol	Necrotic ringspots
		Arabis mosaic virus	Aphid (N), M, Nem, S	Mosaic, ringspots, some necrosis
		Citrus enation–woody gall virus	Aphids (P)	Woody galls
		Prunus necrotic ringspot virus	M, Pol, S	Chlorotic lines and rings, decline
		Rose mosaic virus	M	Necrotic rings
		Rose virus[s]	M	Flower break
Sambucus spp.	Elderberry	See Cooper (1993)		
Syringa oblata var. *affinis*	Syringa	Lilac mottle virus	Aphids (N), M	Mottle
Syringa vulgaris	Lilac	Arabis mosaic virus	Aphid (N), M, Nem, S	Mosaic, ringspots, some necrosis
		Elm mottle virus	M, S	White mosaic
		Lilac chlorotic leafspot virus[t]	M	Chlorotic lesions
		Lilac ring mottle virus	C, M, S[m]	Mottle, usually in the spring
		Lilac ringspot virus[h]	Grafting only	None reported
Ulmus americana	American elm	Cherry leaf roll virus	M, Nem	Mosaic, rings, dieback
Ulmus spp.	Elm	Citrus enation–woody gall virus	Aphids (P)	Woody galls
		Elm mosaic virus	M, S	White mosaic
		Elm mottle virus	M, S	Mottling, oak-leaf pattern
Viburnum opulus	European cranberry-bush	Alfalfa mosaic virus	Aphids (N), M, Pol, S	Variable
Vitis spp.	Grape	Grapevine fanleaf virus	M, Nem, S	Yellow mosaic, ring and line patterns
		Sowbane mosaic virus	Aphids (N), M, Pol, S	Latent
		Tomato black ring virus	M, Nem, Pol, S	Chlorotic and necrotic rings, spots, flecks, vein yellowing, distortion
Washingtonia robusta	Mexican fan palm	Palm mosaic virus[g]	Not M	Mosaic, ringspot, stunting
Wisteria floribunda	Japanese wisteria	Wisteria vein mosaic virus[d]	Aphids (N), M	Mosaic, yellowing, veinclearing, distortion
Wisteria sinensis	Wisteria	Wisteria vein mosaic virus[d]	Aphids (N), M	Mosaic, yellowing, veinclearing, distortion

[a] Data from Alfieri et al. (1993), Brunt et al. (1996–), and Cooper (1993).
[b] C = transmission by contact; M = mechanical transmission in the laboratory; N = nonpersistent transmission by insects; Nem = transmission by nematodes; P = persistent transmission by insects; Pol = transmission by pollen; S = transmission by seed; SP = semipersistent transmission by insects.
[c] Reported on this host in Arkansas.
[d] Reported on this host in Europe.
[e] Reported on this host in Germany.
[f] Reported on this host in Africa.
[g] Reported on this host in the United States.
[h] Reported on this host in Japan.
[i] Reported on this host in New Zealand.
[j] Reported on this host in Australia and New Zealand.
[k] Reported on this host in France.
[l] Reported on this host in Yugoslavia.
[m] Seedborne in indicator plants only.
[n] Reported on this host in the Mediterranean region.
[o] Reported on this host in Greece.
[p] A viruslike agent is considered the cause when a transmissible disease condition exists but no recognizable pathogen can be detected. This is a convenient and temporary term to describe a disease of unknown etiology that requires further study.
[q] Reported on this host in Brazil.
[r] Reported on this host in Canada.
[s] Reported on this host in the United Kingdom.
[t] Reported on this host in southeast England.

Table 7.3 Plants not susceptible to certain viruses

Ornamental species	Common name	Family	Viruses not infecting the plant
Bauhinia purpurea	Purple bauhinia	Leguminosae (Fabaceae)	Bean common mosaic potyvirus Tephrosia symptomless (?)[a] carmovirus
Berberis darwinii	Barberry	Berberidaceae	Carnation mottle carmovirus
Berberis thunbergii	Barberry	Berberidaceae	Carnation mottle carmovirus
Catalpa bignonioides	Catawba	Bignoniaceae	Belladonna mottle tymovirus Cardamine latent (?) carlavirus Dulcamara mottle tymovirus Erysimum latent tymovirus Turnip yellow mosaic tymovirus
Clematis occidentalis	Clematis	Ranunculaceae	Ranunculus mottle potyvirus
Lonicera periclymenum	Lonicera	Caprifoliaceae	Carnation mottle carmovirus
Lonicera spp.	Lonicera	Caprifoliaceae	Tomato yellow leaf curl bigeminivrius
Malus domestica	Apple	Rosaceae	Plum pox potyvirus
Malus sylvestris	Crabapple	Rosaceae	Heracleum latent trichovirus Lilac chlorotic leafspot capillovirus Plum pox potyvirus
Nerium oleander	Oleander	Apocynaceae	Tomato yellow leaf curl bigeminivrius
Prunus avium	Sweet cherry	Rosaceae	Plum pox potyvirus
Prunus serrulata	Oriental cherry	Rosaceae	Grapevine fanleaf nepovirus Plum pox potyvirus
Pyrus communis	Pear	Rosaceae	Plum pox potyvirus
Saxifraga cordifolia	Siberian-tea	Saxifragaceae	Aster chlorotic stunt (?) carlavirus

[a] Question marks indicate viruses that are only tentatively associated with a virus group.

tion of inclusion bodies in the plant can indicate which virus group is present. The major hurdle in using this technique is learning to recognize the many types of inclusions.

ACKNOWLEDGMENTS

We thank and the following personnel of the Florida Department of Agriculture and Consumer Services, Division of Plant Industry: Dr. Nancy Coile for critically reading the tables, Dan Hudson for organizing the tables, Dr. Susan Halbert for the information in Table 7.1 and for critically reading the manuscript, and Dr. Wayne Dixon and Dr. Tim S. Schubert for critically reading the manuscript.

REFERENCES

Agrios, G. N. 1997. Plant Pathology. 4th ed. Academic Press, San Diego, Calif.

Alfieri, S. A., Jr., Langdon, K. R., Kimbrough, J. W., El-Gholl, N. E., and Wehlburg, C. 1994. Diseases and disorders of plants in Florida. Fla. Dep. Agric. Consumer Serv. Bull. 14.

Brunt, A. A., Crabtree, K., Dallwitz, M. J., Gibbs, A. J., Watson, L., and Zurcher, E. J., eds. 1996–. Plant Viruses Online: Descriptions and Lists from the VIDE Database. Version 16 January 1997. Australian National University, Canberra. http://biology.anu.edu.au/Groups/MES/vide/

Christie, R. G., and Edwardson, J. R. 1986. Light microscopic techniques for detection of plant virus inclusions. Plant Dis. 70:273–279.

Cooper, J. L. 1993. Virus Diseases of Trees and Shrubs. Chapman and Hall, New York.

El-Gholl, N. E., Schubert, T. S., and Coile, N. C. 1997. Diseases and disorders of plants in Florida. Fla. Dep. Agric. Consumer Serv., Div. Plant Ind. Bull. 14, Suppl. 1.

Green, S. K. 1991. Guidelines for diagnostic work in plant virology. 2nd ed. Asian Vegetable Res. Dev. Cent. Tech. Bull. 15.

Horst, R. K., and Klopmeyer, M. J. 1993. Viral diseases. Pages 286–297 in: Geraniums 4. J. W. White, ed. 4th ed. Ball Publishing, Geneva, Ill.

Nameth, S. T., and Adkins, S. T. 1993. Viral diseases. Pages 267–275 in: Geraniums 4. J. W. White, ed. 4th ed. Ball Publishing, Geneva, Ill.

Ravelonandro, M., Scorza, R., Bachelier, J. C., Labonne, G., Levy, L., Damsteegt, V., Callahan, A. M., and Dunez, J. 1997. Resistance of transgenic *Prunus domestica* to plum pox infection. Plant Dis. 81:1231–1235.

CHAPTER

8

Wayne A. Sinclair ▪ Cornell University, Ithaca, New York

Phytoplasmas

Nature and Occurrence of Phytoplasmas

Phytoplasmas, formerly known as mycoplasmalike organisms, are pleomorphic prokaryotes that lack cell walls. They are among the smallest organisms known. Phytoplasmas parasitize angiospermous plants and certain insects that act as vectors (Kirkpatrick, 1992; Bové and Garnier, 1998; Lee et al., 1998a; McCoy et al., 1989). They colonize plants systemically in phloem sieve tubes, which are the only plant cells commonly invaded. They are, up to now, unculturable and can be visualized in detail only by electron microscopy (Fig. 8.1), but they can be detected by various microscopic, immunological, and recombinant DNA techniques. Evidence of their pathogenicity in plants is multifaceted, but direct proof of pathogenicity awaits tests with pure cultures.

Phytoplasmas are a discrete group within the class Mollicutes, order Mycoplasmatales. They apparently evolved from an ancestor of the contemporary genus *Acholeplasma* (Gundersen et al., 1994; Seemüller et al., 1994, 1998). The genetic code of a phytoplasma is carried on a single chromosome ranging from 600 to 1,185 kilobases in length, which is smaller than the chromosomes of walled bacteria but comparable in length to those of culturable mollicutes.

Phytoplasmas have not been formally classified and named, because pure cultures are necessary for the physiological and metabolic tests required for taxonomic description. However, systematists have approved a scheme that permits provisional classification and naming of uncultured microorganisms that are known (by their nucleic acid sequences) to be distinct from previously described species. Unculturable microbes may thus be given *Candidatus* names (Murray and Stackebrandt, 1995). A provisional genus, "*Candidatus* Phytoplasma," initially with approximately 15 species, will accommodate the phytoplasmas. Among the groups (putative species) of phytoplasmas that will be named are those typified by pathogens that cause the following diseases (asterisks denote phytoplasma groups that include parasites of diverse woody plants): *apple proliferation, ash yellows, *aster yellows, *elm yellows, lethal yellowing of palms, and *X-disease of *Prunus* (IOM, 1995).

Prominent phytoplasmal diseases of woody plants in North America include ash yellows, blueberry stunt, bunch diseases of pecan and walnut, elm yellows, lethal yellowing of palms, lilac witches'-broom, pear decline, and X-disease of *Prunus* (Hiruki, 1988; McCoy et al., 1989). The word *yellows* appears in names of various phytoplasmal diseases, because the symptoms and causal agents are similar to those of the much-studied aster yellows. Phytoplasmal diseases are rare, or at least rarely diagnosed, in nurseries. Possible reasons for this rarity are that plant populations change regularly, recently infected plants may be asymptomatic or have subtle symptoms, scattered symptomatic plants may be rogued without diagnosis, nursery habitats may be unsuitable for phytoplasma vectors, and vectors may be suppressed by routine nursery pest management.

Transmission

Phytoplasmas in nature are transmitted primarily by insect vectors. Over 120 species of hemipteran and homopteran insects, mostly leafhoppers, but also planthoppers, psyllids, and a stink

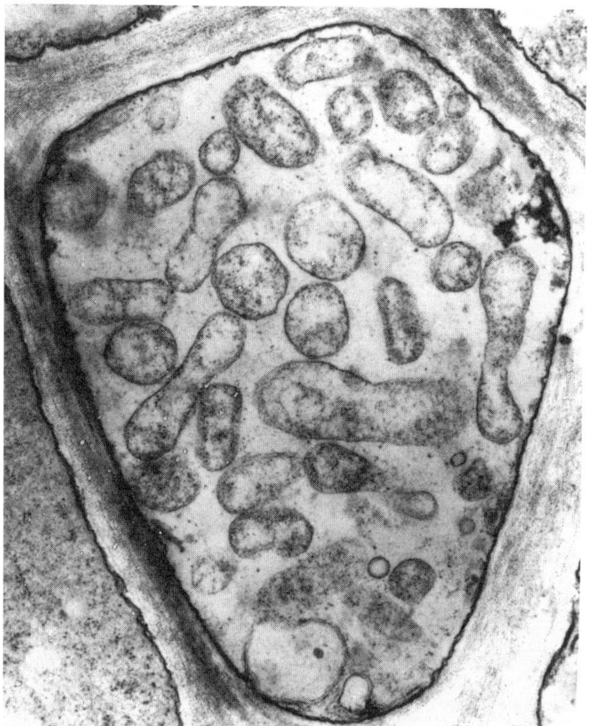

Fig. 8.1 Electron micrograph of an elm sieve tube (cross section) containing the membrane-bound profiles of phytoplasmas. Bar = 1 µm. (Reprinted, by permission, from Sinclair et al., 1987)

bug, are reported vectors of phytoplasmas (Nielsen, 1979; Nault and Rodriguez, 1985). Some vector species transmit phytoplasmas associated with several diseases, and some phytoplasmas have several vectors. The vectors of many phytoplasmas are still unknown.

Both nymphs and adults of a given vector species acquire phytoplasmas, but adults are responsible for most transmission, because they fly from plant to plant, and because phytoplasmas cannot be transmitted until approximately three weeks after a vector acquires them. This interval is required for multiplication of the phytoplasma and its circulation through the insect's body into a salivary gland, from which it may be injected into a plant as the insect feeds. Multiplication in the vector ensures that the phytoplasma can be transmitted as long as the insect lives. Usually phytoplasmas do not infect the eggs of vectors and so are not transmitted directly to offspring (vertical transmission). However, a report of transovarial transmission of an aster yellows phytoplasma (Alma et al., 1997) has reopened the question of vertical transmission.

Phytoplasmas are transmitted between experimental hosts by grafting or by various dodders (*Cuscuta* spp.), as well as by vector insects. Inadvertent transmission in nurseries may occur during propagation by grafting. Phytoplasmas do not infect plant embryos and thus are not seedborne.

Recognition of Phytoplasmal Diseases

Phytoplasmal infections of trees and shrubs are usually incurable. They cause symptoms ranging from slow growth and partial to complete sterility to rapid decline ending in death. Other symptoms include foliar yellowing or reddening and dwarfing, precocious or dwarfed flowers, phyllody, virescence, diminished or lost apical dominance, abnormally upright and spindly twigs, witches'-brooms, phloem hyperplasia, and phloem necrosis (Plates 1–3) (Hiruki, 1988; McCoy et al., 1989). None of these symptoms alone is diagnostic, because other agents and disorders can cause similar symptoms. However, experienced observers often make reliable field diagnoses. Cryptic infections and those causing only retarded growth can be diagnosed by means of microscopic or molecular tests.

Diagnosis

This brief account mentions only the most commonly used methods for detecting phytoplasmas and identifying them to the level of group (putative species).

Nonspecific Detection

Phytoplasmas in plants can be detected by various microscopic and molecular tests. The most commonly used method involves extraction of DNA from a diseased plant, amplification of a diagnostic fragment of phytoplasmal DNA by polymerase chain reaction (PCR), and identification of the fragment by comparison to size standards during gel electrophoresis (Fig. 8.2). Several primer sequences have been published that mediate the amplification of DNA fragments from ribosomal RNA genes of all phytoplasmas but do not amplify fragments of the same size from other organisms (Smart et al., 1996). This method of detection is preferred because of its objectivity and sensitivity.

Nonspecific microscopic detection is also possible using the DAPI (4′,6-diamidino-2-phenylindole · 2HCl) fluorescence test (Schaper and Converse, 1985) (Fig. 8.3). This test has the advantages of being quick and simple and the disadvantages of being less sensitive than PCR, of requiring an experienced observer, and of being inapplicable in certain plants that produce fluorescence-quenching chemicals.

Identification

The appropriate method for phytoplasmal identification depends on circumstance. If phytoplasmas have been detected by nonspecific DNA amplification, then the amplified fragment can be analyzed by sequencing or by digestion with restriction enzymes and comparison of the sizes of resulting fragments to those from similarly treated DNA of known phytoplasmas (Fig. 8.2) (Lee et al., 1998b). The latter procedure, called restriction fragment length polymorphism (RFLP) analysis, is preferred in most laboratories.

If a particular phytoplasma is suspected to be present in a plant sample (e.g., ash yellows phytoplasma in a lilac) then one of the following methods can be used.

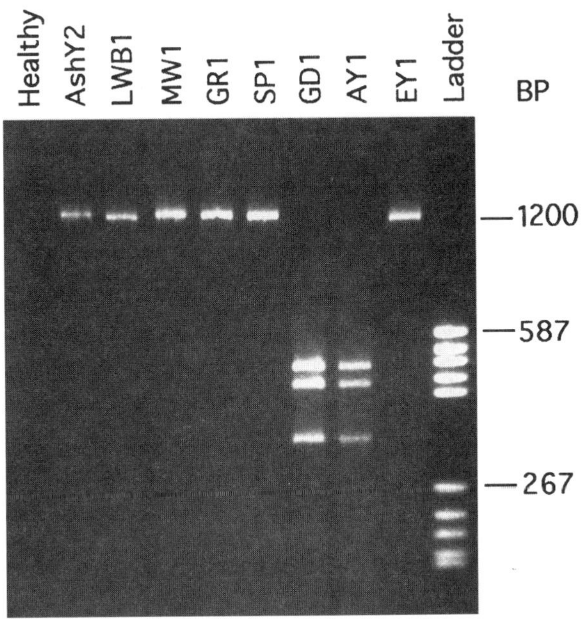

Fig. 8.2 Phytoplasma detection by polymerase chain reaction (PCR) and differentiation by restriction enzyme analysis. A segment, approximately 1,200 base pairs (BP) long, of the 16S ribosomal RNA gene of each of eight phytoplasma strains was amplified by PCR. PCR products were digested with restriction enzyme *Kpn*I and subjected to electrophoresis in an agarose gel, which was then stained with ethidium bromide and photographed under UV illumination. The amplified DNA segment from six of the phytoplasma strains was not cut by this enzyme, so a single band of DNA from each of these strains is visible in the gel. Phytoplasmas belonging to the aster yellows group (GD1 and AY1) have two recognition sites for *Kpn*I in the amplified DNA segment, resulting in three fragments after digestion, which distinguish these phytoplasmas from the others tested. The phytoplasmas are designated as follows: AshY, ash yellows; AY, aster yellows; EY, elm yellows; GD, dogwood stunt; GR, goldenrod yellows; LWB, lilac witches'-broom; MW, milkweed yellows. The lane on the far left contains DNA from a healthy plant (no PCR product obtained). The lane on the far right contains molecular size standards.

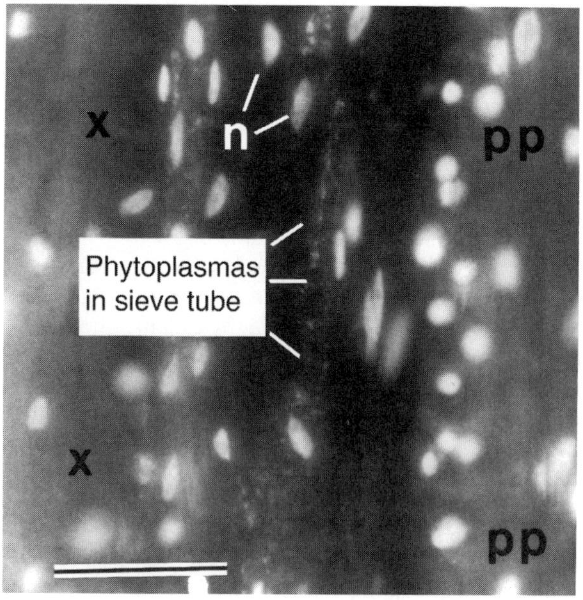

Fig. 8.3 Phytoplasma detection with the DAPI (4′,6-diamidino-2-phenylindole · 2HCl) fluorescence test. A longitudinal section of phytoplasma-infected ash rootlet was treated with dilute DAPI solution (0.4 μg · ml^{-1}) and observed with a fluorescence microscope. The brightly fluorescing objects (n) are plant nuclei. Xylem (x) and phloem fibers (pp) are autofluorescing. DNA of phytoplasmas in phloem sieve tubes appears as fluorescent specks. The sieve tubes, which lack nuclei when mature, are otherwise invisible in this test. Bar = 50 μm. (Reprinted, by permission, from W. A. Sinclair et al., 1994, Journal of Arboriculture)

1. A fragment of phytoplasmal DNA is amplified from diseased plant DNA using phytoplasma-universal PCR primers as noted above. Then the amplified DNA is used as the template for a second PCR using primers specific for phytoplasmas of a particular group, in a technique called nested PCR (Lee et al., 1994; Smart et al., 1996). The nested PCR product is subjected to gel electrophoresis to detect and verify the size of the amplified fragment. This approach has the advantage that the PCR product obtained in the nonspecific step can be subjected to multiple specific tests if necessary.

2. Ribosomal DNA is amplified from diseased plant DNA using primers that only mediate the amplification of DNA from phytoplasmas of a particular group. The PCR product is visualized and its size verified during gel electrophoresis in comparison with known standards. Primer sequences appropriate for detection of various specific phytoplasmas have been published (Davis and Lee, 1993; Lee et al., 1994; Smart et al., 1996).

3. Specific phytoplasmas in plant phloem can be detected by means of immunofluorescence microscopy, provided that appropriate antibodies are available. This requirement has been fulfilled for relatively few phytoplasmas.

REFERENCES

Alma, A., Bosco, D., Danielli, A., Bertaccini, A., Vibio, M., and Arzone, A. 1997. Identification of phytoplasmas in eggs, nymphs and adults of *Scaphoideus titanus* Ball reared on healthy plants. Insect Mol. Biol. 6: 115–121.

Bové, J. M., and Garnier, M. 1998. Walled and wall-less eubacteria from plants: Sieve-tube-restricted plant pathogens. Plant Cell Tissue Organ Cult. 52:7–16.

Davis, R. E., and Lee, I.-M. 1993. Cluster-specific polymerase chain reaction amplification of 16S rDNA sequences for detection and identification of mycoplasmalike organisms. Phytopathology 83:1008–1011.

Gundersen, D. E., Lee, I.-M., Rehner, S. A., Davis, R. E., and Kingsbury, D. T. 1994. Phylogeny of mycoplasmalike organisms (phytoplasmas): A basis for their classification. J. Bacteriol. 176:5244–5254.

Hiruki, C., ed. 1988. Tree Mycoplasma Diseases and Epidemiology. University of Alberta Press, Edmonton.

International Organization for Mycoplasmology (IOM). 1995. International Research Programme on Comparative Mycoplasmology (IRPCM) of the IOM. Report of Consultations, Bordeaux, France, July 1994.

Kirkpatrick, B.C. 1992. Mycoplasma-like organisms – Plant and invertebrate pathogens. Pages 4050–4067 in: The Prokaryotes. 2nd ed. Vol. 4. A. Balows, H. G. Trüper, M. Dworkin, W. Harder, and K. H. Schleifer, eds. Springer-Verlag, New York.

Lee, I.-M., Gundersen, D. E., Hammond, R. W., and Davis, R. E. 1994. Use of mycoplasmalike organism (MLO) group-specific oligonucleotide primers for nested-PCR assays to detect mixed-MLO infections in a single host plant. Phytopathology 84:559–566.

Lee, I.-M., Gundersen-Rindal, D. E., and Bertaccini, A. 1998a. Phytoplasma: Ecology and genomic diversity. Phytopathology 88:1359–1366.

Lee, I.-M., Gundersen-Rindal, D. E., Davis, R. E., and Bartoszyk, I. M. 1998b. Revised classification scheme of phytoplasmas based on RFLP analyses of 16S rDNA and ribosomal protein gene sequences. Int. J. Syst. Bacteriol. 48:1153–1169.

McCoy, R. E., Caudwell, A., Chang, C. J., Chen, T. A., Chiykowski, L. N., Cousin, M. T., Dale, J. L., De Leeuw, G. T. N., Golino, D., Hackett, K. J., Kirkpatrick, B. C., Marwitz, R., Petzold, H., Sinha, R. C., Sugiura, M., Whitcomb, R. F., Yang, I. L., Zhu, B. M., and Seemüller, E. 1989. Plant diseases associated with mycoplasmalike organisms. Pages 545–640 in: The Mycoplasmas. Vol. 5, Spiroplasmas, Acholeplasmas, and Mycoplasmas of Plants and Arthropods. R. F. Whitcomb and J. G. Tully, eds. Academic Press, New York.

Murray, R. G. E., and Stackebrandt, E. 1995. Taxonomic note: Implementation of the provisional status *Candidatus* for incompletely described procaryotes. Int. J. Syst. Bacteriol. 45:186–187.

Nault, L. R., and Rodriguez, J. G., eds. 1985. The Leafhoppers and Planthoppers. John Wiley & Sons, New York.

Nielsen, M. W. 1979. Taxonomic relationships of leafhopper vectors of plant pathogens. Pages 3–27 in: Leafhopper Vectors and Plant Disease Agents. K. Maramorosch and K. L. Harris, eds. Academic Press, New York.

Schaper, U., and Converse, R. H. 1985. Detection of mycoplasmalike organisms in infected blueberry cultivars by the DAPI technique. Plant Dis. 69:193–196.

Seemüller, E., Schneider, B., Mäurer, R., Ahrens, U., Daire, X., Kison, H., Lorenz, K.-H., Firrao, G., Avinent, L., Sears, B. B., and Stackebrandt, E. 1994. Phylogenetic classification of phytopathogenic mollicutes by sequence analysis of 16S ribosomal DNA. Int. J. Syst. Bacteriol. 44:440–446.

Seemüller, E., Marcone, C., Lauer, U., Ragozzino, A., and Göschl, M. 1998. Current status of molecular classification of the phytoplasmas. J. Plant Pathol. 80:3–26.

Sinclair, W. A., Lyon, H. H., and Johnson, W. T. 1987. Diseases of Trees and Shrubs. Comstock Publishing Associates, Cornell University Press, Ithaca, N.Y.

Sinclair, W. A., Griffiths, H. M., and Lee, I.-M. 1994. Mycoplasmalike organisms as causes of slow growth and decline of trees and shrubs. J. Arboric. 20:176–189.

Smart, C. D., Schneider, B., Blomquist, C. L., Guerra, L. J., Harrison, N. A., Ahrens, U., Lorenz, K.-H., Seemüller, E., and Kirkpatrick, B. C. 1996. Phytoplasma-specific PCR primers based on sequences of the 16S-23S rRNA spacer region. Appl. Environ. Microbiol. 62:2988–2993.

CHAPTER

9

Gary W. Moorman ▪ Pennsylvania State University, University Park

Gary A. Chastagner ▪ Washington State University, Puyallup

Botrytis Blight (Gray Mold)

The plant-pathogenic fungus *Botrytis cinerea* (sexual stage, *Botryotinia fuckeliana*) is found virtually everywhere plants are grown. It is fast-growing, utilizing many different sources of nutrients, and attacks many different types of plants. The disease caused by *Botrytis* is commonly called Botrytis blight or gray mold.

Signs and Symptoms

Botrytis at first appears on plants as a white growth but very soon darkens to gray as spores are formed. The smoky gray, "dusty" spores are spread by air currents, rain, or splashing water. In greenhouses, any activity that causes a sudden fluctuation in relative humidity will result in a release of spores. Even automated trickle irrigation systems, when turned on, trigger a release of spores. The spores can remain dormant on plant surfaces for weeks or can germinate and invade immediately.

Fading flowers and damaged succulent plant parts are particularly susceptible to attack by this fungus. In the spring during wet conditions or when sprinkler irrigation is used, flowers can become covered with fungal growth and masses of gray spores. If wet conditions persist, whether outdoors or in misted propagation beds, succulent new growth can be killed. When plants are dug in the autumn and placed in cold storage at high humidity, gray mold can severely damage twigs and leaves. *Botrytis* is not a pathogen of woody tissue. As twigs become woody, they are not susceptible to *Botrytis.* Rose stems can be susceptible, however.

Botrytis forms two types of resting structures on and in infected plant tissue: (1) very dark brown or black, multicellular structures called sclerotia and (2) single-celled, thick, dark-walled chlamydospores. In either form, the fungus can persist in the greenhouse for long periods in the absence of plants.

Botrytis is a relatively weak pathogen, and it must have a food source before it can invade vigorous, healthy plant tissue other than flowers. The fungus uses nutrients leaking from wounded plant parts or the base of cuttings or from dying tissue, such as old flower petals, as a food base. With these nutrients, it becomes more aggressive and invades healthy tissue. A dark to light brown rot forms in diseased tissue. High humidity (relative humidity above 93%) favors the growth of the fungus on plants.

The fungus produces massive numbers of asexual spores (conidia) on the surface of infected tissue. These spores are at peak concentration in the air in midmorning and midafternoon. They may germinate immediately on plant surfaces or may remain dormant for up to three weeks before germinating. Germination can occur over a very wide range of temperatures (38–77°F). Conidia can germinate at the low temperatures typically maintained for holding and shipping plants. Succulent or injured plants held for

Botrytis Blight (Gray Mold)

Geographic occurrence Virtually everywhere plants are grown

Seasonal occurrence Year-round, except in the coldest months; can occur in greenhouse overwintering structures and in cold storage

DISEASE FREQUENCY	DISEASE SEVERITY
5 annual	5 plants killed
4	**4**
3	**3**
2	**2**
1 rare	1 very little damage

CHEMICAL TREATMENT	CULTURAL PRACTICES
3 used every year	**3 very important**
2	2
1 not used	1 not important

SANITATION	RESISTANT CULTIVARS
3 very important	3 many cultivars
2	**2**
1 not important	1 no resistance

sale or shipped at high humidity in the presence of condensed water are very susceptible to gray mold.

Disease Management

Growers should learn which plants are most susceptible to gray mold. In the greenhouse, geraniums are probably the most susceptible crop and can be the source or massive numbers of spores. Therefore, keep geraniums in structures separate from woody plant material that is susceptible to gray mold and away from mist propagation beds. Keep susceptible species of plants in areas where the relative humidity can be kept low.

The most effective way to manage gray mold is to keep the relative humidity low and avoid wetting plant surfaces. Sprinkler irrigation favors infection of leaves, flowers, and twigs by *Botrytis* and should be applied early in the day, so that plant surfaces dry quickly in the sun. In mist beds where cuttings are being rooted, use as little mist as feasible. As soon as possible after rooting begins, reduce or cease misting. Ventilate cold frames and greenhouses to reduce the number of hours of high humidity. This may only require venting early in the day when moisture has condensed on plant surfaces and before sunlight has warmed the air. Venting in the evening before sunset avoids exposing the plants to high humidity throughout the night. Even lowering the humidity slightly can have a significant effect on *Botrytis*. Outdoor and indoor plantings should be planned to provide good air circulation among the plants. This is the most important means of inhibiting *Botrytis* activity.

Sanitation is also important in managing gray mold, because the fungus survives in and sporulates on plant debris. Dead or dying leaves and petals should be removed from plants and from under benches, mist beds, and walkways. Reducing the food base available to the fungus will reduce the production of new spores; it also limits the number of places the fungus can survive under adverse environmental conditions.

Sanitation alone, however, is not sufficient to control *Botrytis*. This fungus can produce 60,000 or more spores on a piece of plant tissue the size of your small fingernail. Even one spore can infect a plant and cause disease. Therefore, other steps must also be taken:

1. Avoid injuring plants in any way.
2. Do not leave large stubs of tissue on stock plants when taking cuttings or pruning twigs.
3. Apply chemical protection. Mixtures of nonsystemic fungicides applied before infection have long residual activity and provide protection equal to that of systemic fungicides. Some *Botrytis* populations have been found to be resistant to systemic fungicides, particularly compounds in the benzimidazole class (such as benomyl and thiophanate-methyl) and the dicarboximide class (such as iprodione and vinclozolin) when used exclusively over a long period of time. *Botrytis* populations have not developed resistance to nonsystemic protectants.

REFERENCES

Coley-Smith, J. R., Verhoeff, K., and Jarvis, W. R. 1980. The Biology of *Botrytis*. Academic Press, New York.

Delp, C. J. 1980. Coping with resistance to plant disease control agents. Plant Dis. 64:652–658.

Gullino, M. L., and Garibaldi, A. 1982. Use of mixture or alternation of fungicides with the aim of reducing the risk of appearance of strains of *Botrytis cinerea* resistant to dicarboximides. EPPO Bull. 12:151–56.

Hausbeck, M. K., and Pennypacker, S. P. 1991. Influence of grower activity and disease incidence on concentrations of airborne conidia of *Botrytis cinerea* among geranium stock plants. Plant Dis. 75:798–803.

Moorman, G. W., and R. J. Lease. 1990. Residual activity of fungicides applied to geraniums in the greenhouse. Phytopathology 80:979.

Salinas, J., Glandorf, D. C. M., Picavet, F. D., and Verhoeff, K. 1989. Effects of temperature, relative humidity, and age of conidia on the incidence of spotting on gerbera flowers by *Botrytis cinerea*. Neth. J. Plant Pathol. 95:51–64.

Sirjusingh, C., and Sutton, J. C. 1996. Effects of wetness duration and temperature on infection of gernaium by *Botrytis cinerea*. Plant Dis. 80: 160–165.

Vali, R., and Moorman, G. W. 1990. Effects of fungicide spray regimes on disease control and development of dicarboximide resistance in *Botrytis cinerea*. Phytopathology 80:124.

Zhang, P. G., Sutton, J. C., and Hopkin, A. A. 1994. Evaluation of microorganisms for biocontrol of *Botrytis cinerea* in container-grown black spruce seedlings. Can. J. For. Res. 24:1312–1316.

CHAPTER

10

Thomas J. Burr ▪ Cornell University, Ithaca, New York

John Miller ▪ Florida Department of Agriculture, Gainesville

Crown Gall

Crown gall, caused by the gram-negative bacterium *Agrobacterium tumefaciens,* is one of the most common bacterial diseases of plants worldwide. *A. tumefaciens* causes crown gall on numerous species of dicotyledonous plants. The disease can be particularly severe in deciduous tree and berry crops and woody ornamentals including cypress, euonymus, forsythia, hibiscus, lilac, flowering *Prunus,* privet, rose, viburnum, and willow. Plants that have been recently reported as hosts of *A. tumefaciens* include weeping fig and chrysanthemum.

Crown galls are most commonly formed on roots and crowns of plants but may also occur on stems, trunks, and branches (Plate 4). They usually form at wound sites. The gall is composed of fleshy, disorganized plant tissue, which is white to cream-colored for the first few months after infection. It darkens in the autumn and dormant seasons, and much of the gall tissue dies. The presence of galls may restrict the vascular flow of nutrients in plants, resulting in reduced root and shoot development and generally unthrifty growth of infected plants.

Crown gall can result in great economic losses in nurseries. Plants infected when they are young may not produce adequate growth, and the appearance of galls alone can make them unsalable. Large numbers of infected plants have been discarded in some nurseries.

Disease Development

A. tumefaciens can survive systemically in plant material, in soil, and in association with plant roots (in the rhizosphere). Therefore inoculum can be carried with plant material, and plants may become infected after they are planted in contaminated soil. In general, infections occur at wound sites. However, it has also been reported that the pathogen may enter plants through lenticels on roots.

Feeding on roots by parasitic-plant nematodes can increase the severity of crown gall. It is not known if the nematodes enhance the spread of the pathogen or simply cause wounds that stimulate infection, as described below.

The mechanism by which *A. tumefaciens* infects plants has been studied extensively, although there are still large gaps in our knowledge of the infection process. It is the only known (naturally occurring) example of the transfer of DNA from a bacterium to a plant and expression of the DNA in the plant. The process begins when a plant is wounded. The wound may result from mechanical injury, such as a pruning cut, or from environmental causes, such as freeze injury. The injured plant cells emit chemicals that are sensed by *A. tumefaciens* in surrounding plant material or soil. Some chemicals from the plant stimulate the bacterium to move toward the plant, and it attaches to wounded plant cells. Then other plant chemicals exuded from the wounded cells trigger the induction of pathogenicity genes in the bacterium. The main pathogenicity genes of *A. tumefaciens* are located on a plasmid in the bacterium, and part of the plasmid is transferred to the plant. The transferred DNA, or T-DNA, carries genes that are responsible for the formation of crown galls. These genes cause infected plant cells to produce abnormal amounts of hormones (auxin and cytokinin), resulting in disorganized growth and the formation of galls.

Different strains of *A. tumefaciens* exist in nature. Some strains are host-specific, infecting certain plant species but not others. As researchers look more closely at different plant hosts, it is apparent that *A. tumefaciens* strains can be quite diverse.

Disease Management

Several strategies have been tried for controlling crown gall. Applications of antibiotics and other bactericides have generally not been used. When tested experimentally, they have not provided adequate control, probably because of their short-lived activity in soil and because the bacterium may reside within plant tissue where it cannot be reached by antibiotics. Soil fumigation has been tested in nurseries, with poor success. Broad-spectrum fumigants kill bacteria in the soil, but not bacteria in plant tissues in the soil. In some cases an increased incidence of crown gall was apparent after fumigation; the increase may have occurred because competing bacteria in soil were killed, allowing the remaining crown gall bacteria to multiply unimpeded.

Biological control of crown gall by a nonpathogenic strain of *Agrobacterium* discovered in an Australian nursery in the mid-1970s is one of the most successful examples of biological control of plant diseases. This strain, known as K84, was found to protect plant wounds from infection by *A. tumefaciens.* Plants are treated by dipping them in a suspension of K84 prior to planting.

K84 produces an antibiotic (agrocin 84) which is lethal to most, but not all, strains of *A. tumefaciens.* Agrocin 84 is probably a major factor in providing biological control, but it is not responsible for the entire mechanism, since some pathogenic strains that are insensitive to agrocin 84 are still controlled. One possibility is that K84 produces other antibiotics that act on the pathogenic strains, and in fact two other antibiotics have been found to be produced by this strain. It is not clear whether other factors, such as competition between pathogenic strains on the plant, may be involved in the control. Research continues in attempts to more completely understand the mechanism by which K84 controls crown gall. K84 is registered with the U.S. Environmental Protection Agency and is sold commercially in the United States (see Chapter 98, "Biological Control of Woody Ornamental Diseases").

CHAPTER

11

Larry W. Barnes ▪ Texas A&M University, College Station

Robert G. Linderman ▪ U.S. Department of Agriculture, Corvallis, Oregon

Diseases Caused by *Cylindrocladium*

Cylindrocladium species are important fungal pathogens infecting a range of herbaceous and woody ornamental plant species. Several species of *Cylindrocladium* can be pathogenic in ornamental hosts. *C. scoparium, C. floridanum, C. theae,* and *C. crotalariae* are the most common of these pathogens in nursery stock. *C. scoparium* has caused serious disease problems in bare-root forest seedling nurseries. Several species of *Cylindrocladium* became a major concern in the forcing azalea industry in the late 1960s, with the rapid expansion of the industry and the introduction of new, highly susceptible cultivars and extensive interstate shipment of symptomless plant material carrying the pathogen in roots. Since then, these pathogens have become established in production areas of many types of ornamentals, and constant vigilance and active management programs are required to limit the diseases they cause.

Cylindrocladium diseases have not become established in landscape situations, although many landscape plant species are susceptible. These diseases occur primarily during nursery propagation, whether from cuttings in mist beds or in bare-root seedling beds. Frequently, newly propagated plants die and are discarded, while many neighboring plants near infection loci become infected but remain symptomless for some time during production and shipping, becoming symptomatic at a later stage. By the time symptoms appear, it is too late to apply control measures, because none are known to eradicate infections and the pathogens. Control measures must be applied before or at the time of propagation in order to prevent infection. This means fumigation of nursery beds, use of pathogen-free propagation and growth media, use of pathogen-free propagation material or treatment of the material to sanitize and protect it, and separation of infected from uninfected plant material to avoid contamination of production areas.

Damage from *Cylindrocladium* diseases can occur at all stages of plant production, but disease development during propagation can result in total crop loss.

Causal Agents

Several species of *Cylindrocladium* can be involved in diseases of woody ornamentals, including *C. scoparium* (the most frequently encountered species), *C. floridanum, C. theae,* and *C. crotalariae.* All these species except *C. scoparium* produce a sexual stage, *Calonectria,* on infected plant tissue. Perithecia of the sexual stage are red to orange and bear asci containing ascospores, which can ooze under high humidity and are forcibly discharged when the relative humidity drops below 100%. Most species produce distinct microsclerotia or clustered pigmented hyphae in infected tissue. These structures are capable of germinating directly by germ tubes or indirectly by producing conidiophores, which generate masses of septate, elongate conidia. The surface of infected tissue may be covered with masses of conidial clusters, which give it a buff-colored appearance to the unaided eye. The conidiophores typically form a hyphal stipe with a terminal cell that may be swollen in an elongate or globose shape. Species of *Cylindrocladium* are distinguished by the size and septation of conidia and the shape of the terminal cell of the stipe.

Symptoms

Cylindrocladium species are capable of infecting all host tissues, including flowers, leaves, stems, crowns, and roots, causing a range of disease symptoms.

Flowers can develop both necrotic spot and petal blight symptoms. Leaf spots are tan to brown necrotic areas surrounded with yellow or red halos on the upper sides of leaves, depending on the plant species and the flower color. On the undersides of infected leaves, leaf veins radiating from the spots are red to pink in red- or pink-flowered cultivars of azalea. Twig dieback can also result from the spread of leaf infections. Leaf infection in leucothoe is characterized by a brownish spot, commonly surrounded by alternating light and dark rings. In many plants, severe leaf infection can result in extensive defoliation due to high levels of ethylene induced by the infection.

The major effects of *Cylindrocladium* infection in woody plants are root and crown rots. Symptoms of root and crown infection include dark brown cortical root rot, stunting, discoloration (chlorosis) of foliage, defoliation, damping-off, wilting, cutting rot, and mortality. The symptoms depend on the host species, the stage of development of the plant, and environmental conditions. Root infections are cortical rots that destroy roots. However, new roots can form as fast as the rot progresses, and thus plants may remain symptomless until moisture stress occurs. Root infection may involve both lateral and tap roots, resulting in dark brown to black, longitudinal, cracklike lesions, which may coalesce to completely destroy the root.

Root infection can spread to the crown, and subsequently the plant may suddenly wilt and die, often within a matter of days. This mortality is the result of infection that began during cutting propagation. Cuttings are most vulnerable to infection during propagation, when high temperatures, moisture from frequent misting, and the close spacing common in propagation favor infection and disease development. Under these conditions, rapid and extensive losses can occur. Probably more important in the long run, however, are cuttings that root and survive but become symptomless carriers of the disease.

Epidemiology

The foliage and root phases of *Cylindrocladium* diseases are often related, although they appear to be distinct because they may occur in geographically separate locations. This is because often cuttings (especially azaleas and roses) are propagated in one location and liner plants grown in another. The disease begins when cuttings have inconspicuous leaf spots or are contaminated with spores that have not initiated infections. Under the warm, moist conditions in which rooting is conducted, the pathogens can initiate infections in contaminated cuttings, causing cutting rot, root rot, and mortality. Spores produced in those infections can be splashed to adjacent cuttings or wash down onto newly formed roots. However, cuttings infected in this manner may remain symptomless during propagation and the growing-on stage when environmental stresses are minimal. Infected liner plants may thus be shipped to other areas for further growing and forcing. Shipping and handling conditions as well as accelerated growth conditions appear to hasten the development and spread of the pathogen in host roots, and plants may suddenly wilt and die. Growers may be confused by the suddenness of plant collapse from infections that began during the cutting propagation phase.

Infections can also be initiated by inoculum that is carried over in infected tissue residue, whether from dead cuttings or plants, or from infected roots, leaves, or flowers in the propagation or production areas. Conidia and ascospores (produced by *C. theae, C. crotalariae,* and *C. floridanum,* which form the perithecial stage, *Calonectria*) are dispersed by splashing water and by wind and are distributed to adjacent plant surfaces. Conidia originate from conidiophores and ascospores from perithecia that arise from microsclerotia embedded in infected tissues. The microsclerotia are dark-pigmented resistant structures that can survive long periods until conditions are right, when they sporulate or germinate directly to infect roots. Warm, moist, humid conditions favor sporulation, plant infection, and disease development.

Cylindrocladium species sporulate readily in lesions on leaves and stems and in infected crown tissue of some hosts. However, sporulation is more abundant in infected leaf litter, which collects at the base of plants as a result of infection-induced defoliation. Splashing water from rainfall, overhead irrigation, and mist propagation not only disseminates conidia but also provides an environment conducive to spore germination and infection. Microsclerotia embedded in infected plant material can sporulate repeatedly over time, so spores can be quite abundant in the production areas. When spores land on plants that will be used for cuttings, the pathogen is introduced into the propagation area, where conditions are generally perfect for infection to occur. The close spacing of stuck cuttings may obscure infected and rotting cuttings, and those cuttings provide inoculum that can infect adjacent cuttings and their roots. Cuttings may also become inoculated by workers who have acquired spores by handling infected cuttings or plants.

The microsclerotia produced by *Cylindrocladium* species help these fungi resist heat, desiccation, and other adverse environmental conditions. Microsclerotia can develop in infected flower, leaf, root, stem, and bark tissue and can infest soil in growing areas, and they can maintain populations for several to many years. Exudates from roots growing near microsclerotia stimulate the microsclerotia to germinate and cause root infections.

An important component of the epidemiology of *Cylindrocladium* diseases is the saprophytic phase of the pathogens. Plant debris from pruning or shearing of plants can become colonized by spores of the pathogens, and after the fungi grow saprophytically throughout the tissue, they form microsclerotia that can carry them over for long periods. The microsclerotia can produce abundant conidiophores in clusters on tissue surfaces, which can be readily disseminated to living plant tissues and cause disease. Thus, the removal of plant debris from pot or bed surfaces is an essential component of the management of *Cylindrocladium* pathogens in the nursery.

Host Range

The host range of *Cylindrocladium* species is wide, including such commercially important woody ornamentals as azalea, camellia, conifer seedlings, gardenia, rose, holly, lilac, kinnikinnick, mahonia, peach, *Pieris,* plum, bottlebrush, redbud, rhododendron, sweet gum, *Vaccinium,* and yellow poplar. Within susceptible plant groups, there is probably some variation in susceptibility, but information on cultivar resistance or susceptibility is largely unavailable. However, some larger propagation nurseries have breeding and selection programs to identify and eliminate highly susceptible cultivars from the nursery trade. In the long run, this approach will be critical to reducing the incidence of *Cylindrocladium* diseases.

Disease Management

Management of *Cylindrocladium* requires aggressive sanitation measures, the use of healthy stock plants for propagation, sanitization of cutting material with disinfectant or fungicide dips or soaks, the use of sound horticultural production practices, and judicious use of effective fungicides. Application of fungicides to sanitize and protect cuttings is probably the most efficient and effective use of chemical pesticides, because of direct, thorough delivery to the infection site and contaminating propagules. Failure to select pathogen- or disease-free cuttings or to sanitize them chemically has resulted in severe losses in propagation. Application of fungicides as sprays or drenches after cuttings are stuck is frequently not completely effective, because of incomplete coverage of cutting surfaces where the pathogen may be located.

Production areas of plants susceptible to *Cylindrocladium* diseases should be regularly scouted for signs and symptoms of infection. Infected plants should be carefully rogued from the production area, with precautions to prevent spreading spores while removing infected tissue. Infected leaf litter should be carefully collected and removed from the production area. Importation of new cultivars into the nursery is risky from the standpoint of introducing new strains of the pathogens. Therefore, new plants brought to the nursery should be quarantined, segregated from production areas, in case they are infected or inoculum is carried in their pots. Cuttings from those plants should be thoroughly sanitized before propagation to avoid introducing the pathogens.

Drench applications of fungicides have proved highly effective in helping to manage *Cylindrocladium* diseases when used in conjunction with a thorough sanitation program. Drenches should be applied at the time of sticking of cuttings and as frequently thereafter as allowed on the label if infection is likely. Waiting too long between applications can open the door for the pathogens to invade the propagation area before the next treatment.

For nursery bed preplant treatments, soil fumigation can be effective, although high rates are usually needed to eradicate the pathogens, because of the durability of the microsclerotia. Some organic amendments have also been shown to reduce populations of *Cylindrocladium* propagules in the soil, but this biological approach has not been widely tested.

REFERENCES

Horst, K. R., and Hoitink, H. A. J. 1968. Occurrence of Cylindrocladium blight on nursery crops and control with fungicide 1991 on azalea. Plant Dis. Rep. 52:615–617.

Linderman, R. G. 1972. Isolation of *Cylindrocladium* from soil or infected azalea stems with azalea leaf traps. Phytopathology 62:736–739.

Linderman, R. G. 1973. Formation of microsclerotia of *Cylindrocladium* spp. in infected azalea leaves, flowers, and roots. Phytopathology 63: 187–191.

Linderman, R. G. 1974. The role of abscised *Cylindrocladium*-infected azalea leaves in the epidemiology of Cylindrocladium wilt of azaleas. Phytopathology 64:481–485.

Linderman, R. G. 1986. Cylindrocladium blight and wilt. Pages 17–20 in: Compendium of Rhododendron and Azalea Diseases. D. L. Coyier and M. K. Roane, eds. American Phytopathological Society, St. Paul, Minn.

Sinclair, W. A., Lyon, H. H., and Johnson, W. T. 1987. Diseases of Trees and Shrubs. Comstock Publishing Associates, Cornell University Press, Ithaca, N.Y.

Sobers, E. K., and Alfieri, S. A., Jr. 1972. Species of *Cylindrocladium* and their hosts in Florida and Georgia. Proc. Fla. Hortic. Soc. 85:366–369.

Timonin, M. I., and Self, R. L. 1955. *Cylindrocladium scoparium* Morgan on azaleas and other ornamentals. Plant Dis. Rep. 37:860–865.

CHAPTER

12

Gary W. Moorman ▪ Pennsylvania State University, University Park

Damping-Off of Seeds and Seedlings and Cutting Rot

Damping-off is the rotting of seeds in the soil and death of newly emerged seedlings caused by fungi. Seeds may decay before germination, or seedlings may rot just below the soil line. Stem cuttings of woody plants infected by fungi rot at the base and usually fail to root. Sometimes cuttings produce roots on the stem, just above the rotted area. Plants that survive damping-off and stem cuttings that are only partially rotted may carry the fungi and later rot and die. The fungi that most often cause damping-off and stem cutting rots belong to the genera *Pythium, Fusarium,* and *Rhizoctonia.* These fungi attack a wide range of plants growing outdoors in seedbeds and seedlings and cuttings in the greenhouse.

Symptoms and Signs

In seedbeds where the seed is broadcast, groups or patches of plants are killed soon after emergence or fail to emerge. In beds where seeds are sown in rows, damping-off results in spaces or gaps in the rows. Similarly in stem cutting beds, groups of cuttings may die. However, if much of the rooting medium is infested with one of these fungi, large numbers of the cuttings throughout the bed may be rotted.

Fusarium forms masses of white or pink spores on the surface of the plant material. The microscopic spores are canoe-shaped and multicelled.

Rhizoctonia does not form spores but can produce weblike threads, which extend from plant to plant or from infected tissue to moist soil. When an infected plant is pulled from the soil, particles of soil attached to the plant by these fungal threads dangle in midair.

Pythium appears as a fluffy white growth on the plant surface under very moist conditions, if the plant tissue is very heavily infected.

Pythium, Rhizoctonia, and *Fusarium* all reside in soil and survive there very well. It is difficult to eliminate these fungi from field soil. Pathogenicity is generally favored by excessive soil moisture, excessive overhead misting, low soil temperatures before germination (below 20°C, or 68°F), high soil temperatures after emergence (above 25°C, or 77°F), and overcrowding of stem cutting beds or seedbeds.

Control in the Greenhouse

Seeds and cuttings should be started in pasteurized soil (heated to 60°C, or 140°F, for 30 minutes) or a pathogen-free soilless mix. New or sanitized flats or pots should be used. Keep soil and containers off floors, and use clean tools and equipment to handle them, to avoid contaminating them with soilborne pathogens. Encourage rapid maturation of seedlings and cuttings by supplying bottom heat, so that the soil in the containers is warm (70–75°F, or 22–24°C). Cold irrigation water (below 10°C, or 50°F) greatly slows plant growth for a few hours following each watering, and this retardation of growth will prolong the maturation period and lengthen the time during which the plants are susceptible to damping-off organisms. If necessary, warm the irrigation water to 68–77°F (20–25°C).

If damping-off is a persistent problem, the entire production program should be examined for conditions that favor the pathogens. Fungicides can be applied to seed to inhibit damping-off until seedlings mature and gain some resistance to the fungi. Growers who decide to treat seed with a fungicide should be certain that it is registered for use on the particular crop being treated. This information is on the fungicide label. It has been demonstrated that certain fungicides in some rooting mixes can inhibit root initiation and elongation of cuttings of some cultivars.

If possible, avoid the use of pesticides on cuttings being rooted. Rely instead on pasteurized rooting and growing media and clean containers. If fungicides are required, it is suggested that only a portion of the crop be treated initially, to observe any injury. If no injury is noted, it can be assumed that the fungicide is safe to use on that particular cultivar and medium. If different cultivars are treated or if a different rooting medium is used, very different results may occur.

Biological control of Pythium and Rhizoctonia damping-off is now feasible with commercially available formulations of the fungus *Trichoderma.* Also, potting soils suppressive to *Pythium* are being produced.

Five Steps to Avoiding Damping-Off

1. Pasteurize the soil or potting medium. If the soil is dark and heavy, it may stay too wet and may be difficult to effectively pasteurize.
2. Use new flats and pots if possible, because used flats and pots can be very difficult to sanitize.
3. Schedule irrigations to provide optimum moisture for germination and seedling growth, but avoid excessive application, because wet growing media favor damping-off.
4. Apply fertilizers carefully, as high levels of soluble salts can slow seedling growth, cause injury to seedlings, and favor damping-off.

5. Avoid growing seedlings under low-intensity light, which can increase the succulence of the seedlings and increase the incidence of damping-off.

REFERENCES

Baker, R., and Martinson, C. A. 1970. Epidemiology of diseases caused by *Rhizoctonia solani.* Pages 172–188 in: *Rhizoctonia solani:* Biology and Pathology. J. R. Parmeter, ed. University of California Press, Berkeley.

Bateman, D. F. 1961. The effect of soil moisture upon development of poinsettia root rots. Phytopathology 51:445–451.

Bateman, D. F., and Dimock, A. W. 1959. The influence of temperature on root rots of poinsettia caused by *Thielaviopsis basicola, Rhizoctonia solani,* and *Pythium ultimum.* Phytopathology 49:641–647.

Beach, W. S. 1949. The effects of excess solutes, temperature and moisture upon damping-off. Pa. Agric. Exp. Stn. Bull. 509.

Chase, A. R., and Poole, R. T. 1984. Investigations into the roles of fertilizer level and irrigation frequency on growth, quality, and severity of Pythium root rot of *Peperomia obtusifolia.* J. Am. Soc. Hortic. Sci. 109:619–622.

Engelhard, A. W., Miller, H. N., and DeNeve, R. T. 1971. Etiology and chemotherapy of Pythium root rot on chrysanthemums. Plant Dis. Rep. 55:851–855.

Farr, D. F., Bills, G. F., Chamuris, G. P., and Rossman, A. Y. 1989. Fungi on Plants and Plant Products in the United States. American Phytopathological Society, St. Paul, Minn.

Gutierrez, W. A., Shew, H. D., and Melton, T. A. 1997. Sources of inoculum and management of *Rhizoctonia solani* damping-off on tobacco transplants under greenhouse conditions. Plant Dis. 81:604–606.

Hoitink, H. A. J., Inbar, Y., and Boehm, M. J. 1991. Status of compost-amended potting mixes naturally suppressive to soilborne diseases of floricultural crops. Plant Dis. 75:869–873.

Hoppe, P. E. 1966. *Pythium* species still viable after 12 years in air-dried muck soil. Phytopathology 56:1411.

Moorman, G. W. 1986. Increased mortality caused by Pythium root rot of poinsettias associated with high fertilization rates. Plant Dis. 70:160–162.

Moulin, F., Lemanceau, P., and Alabouvette, C. 1994. Pathogenicity of *Pythium* species on cucumber in peat-sand, rockwool and hydroponics. Eur. J. Plant Pathol. 100:3–7.

Stanghellini, M. E., and Kronland, W. C. 1986. Yield loss in hydroponically grown lettuce attributed to subclinical infection of feeder rootlets by *Pythium dissotocum.* Plant Dis. 70:1053–1056.

Vanachter, A. 1979. Fumigation against fungi. Pages 163–183 in: Soil Disinfestation. D. Mulder, ed. Elsevier Scientific, New York.

CHAPTER
13

D. Michael Benson ▪ North Carolina State University, Raleigh

Nematode Diseases

Plant-parasitic nematodes cause diseases in a wide range of field-grown ornamentals. Over the past 40 years, extensive host lists have been published, detailing the occurrence of nematode genera and species associated with the root systems of woody ornamentals in nurseries. Some of the most important nematodes affecting woody ornamentals are species of *Aphelenchoides, Criconemella, Meloidogyne, Pratylenchus, Tylenchorhynchus,* and *Xiphinema.* Nematodes in these genera attack many agronomic crops, in addition to ornamentals, as well as native plants worldwide.

Distribution of Nematodes

Nematodes occur in most of the nursery growing regions of the United States. Typically, they are more damaging to plants in areas with sandy soils and hot climates, such as along the Gulf Coast and in the southwest and the coastal plain of the southeastern United States. In heavier-textured soils and cooler climates, nematodes no doubt limit plant vigor and growth, but severe symptoms may not be as apparent.

Nematodes are well adapted as soilborne and foliar pathogens of ornamentals. They reproduce rapidly on host plants under favorable conditions, and they survive long periods of unfavorable conditions in the absence of a host. Populations of these pathogens in the root zone of ornamentals increase dramatically during the growing season but drop precipitously during the winter, only to rebound the following year.

Even though nematodes are widespread in nursery soils, the distribution of nematodes at a given site is usually aggregated or patchy. That is, they tend to be clustered, in clumps from 6 to 15 inches deep in soil, scattered somewhat erratically across the field. Tillage and cultivation practices that move soil down rows and across drainage features can redistribute nematodes in a field. Since nematodes can move only short distances in water films under their own power, long-distance movement of nematodes within a field and between fields is often associated with infested soil carried on equipment, introduced on infected planting stock, or carried passively in runoff water. They can also be moved in windblown soil particles and by animals.

Symptoms

Foliar symptoms of nematode infection may be apparent in nursery crops during the first growing season, but usually plant decline occurs over several seasons, as nematode damage to the root system restricts uptake of water and nutrients. Symptoms induced by root infection include lack of new growth, chlorosis, necrosis, defoliation, miniature leaves, and (in some crops, such as boxwood) bronzing of foliage. Lack of overall foliar growth at budbreak is one of the most common symptoms. Irregular or spotty occurrence of poor growth in a field is a primary indicator of nematode damage in large plantings, reflecting the aggregated distribution of nematode populations.

Root symptoms precede foliage symptoms but are difficult to detect until severe foliar symptoms are present and one begins to wonder what is wrong with the crop. Damage due to nematode feeding may cause the entire root system of an infected plant to be smaller than that of a healthy plant. Probably the most familiar root symptom caused by root-knot nematodes is root galls, which range from a quarter of an inch to almost three-quarters of an inch in diameter or larger on some ornamentals. Other nematodes cause root tips to branch abnormally and limit the growth of branched roots, resulting in "stubby root." Some nematodes feed through the outer root tissues and cause lesions, which become necrotic, along the roots. When feeding damage is severe, plants do not have enough normally functioning roots to provide adequate uptake of water and nutrients from the soil. Thus infected plants may die during a drought when nematode-free plants survive.

Nematodes Attacking Woody Ornamentals in Nurseries

Root-Knot Nematodes

The root-knot nematodes *Meloidogyne arenaria* and *M. incognita* attack over 2,000 species of plants worldwide, including ornamentals in many different genera (Table 13.1). They are the most damaging nematodes in woody ornamentals. They are widely distributed in the southeastern United States, the Gulf Coast states, Arizona, and California, and this distribution probably accounts for their importance in ornamentals. Root-knot nematodes produce distinct root galls along infected roots where they feed. The growth of infected plants may be severely restricted, compared to nematode-free plants at the nursery site.

Disease development Root-knot nematodes are infective only as second-stage juveniles. At this stage of development, they penetrate between cells at the root tip, migrate a short distance behind the root tip, and establish themselves near vascular tissues. They then develop into the nonfeeding third- and fourth-stage juveniles. Female fourth-stage juveniles molt into female adults, which feed within root tissues and grow larger, developing a pear-shaped body. Only the head region and stylet can move, and so the nematode is limited to feeding on three to six root cells. These cells become enlarged as the nematode "saliva" induces the cell nuclei to divide, and adjoining cell walls may dissolve, forming three to six *giant cells* at a feeding site. As giant cells enlarge, they tend to crush xylem tissue, so that the vascular tissue loses the ability to conduct water. When several nematodes are present in a root segment, galls develop along the root, giving it a knotty appearance.

A female root-knot nematode, once fully developed, lays about 300 to 1,000 eggs, which are deposited in a mass attached to the body. The body of the nematode, as it enlarges, may protrude through the root surface, and its eggs can hatch in the soil. Second-stage juveniles from these eggs find and infect nearby root tips. If an egg mass is confined within root tissue, the second-stage juveniles hatched from it establish new feeding sites near the mother and develop into females. At 27°C eggs can be produced within 25 days of the hatching of second-stage juveniles.

Male root-knot nematodes leave the root and live in the soil and cause little damage to root tissues. They are also rarely involved in reproduction of these species.

Lesion Nematode

The lesion nematode *Pratylenchus vulnus* is not as widespread and devastating as root-knot nematodes, but it can have a tremendous effect on plant growth in susceptible taxa (Table 13.1). Symptoms may not include the typical chlorosis and necrosis associated with some nematode species, but plant growth is often dramatically poor, compared to that of nematode-free plants in the field (Plate 70). Roots develop lesions that turn necrotic in areas where the nematodes are active. Secondary fungi and bacteria may invade infected tissues and cause additional necrosis of the roots.

Disease development Lesion nematodes are endoparasitic and migratory in host roots. Second-stage juveniles form in eggs laid or released in soil around infected roots. They pass through two more stages before maturing as adults. The nematode can infect roots at all stages of development. It invades roots by persistent thrusting of the head and stylet against root tissues. Once inside, it remains in the cortical cell region and aligns its body more or less parallel to the vascular tissues. Lateral roots may form above the site of feeding activity. As the nematode feeds on cortical cells, the cell wall and cytoplasm turn brown in reaction to the nematode saliva. Unlike root-knot nematodes, the lesion nematode continues to move or migrate between cortical cells, creating tunnels between cells as it moves. Thus, necrotic areas develop in the root along the path of the nematode. The lesion nematode can complete its life cycle in the root; therefore, infected roots may contain the nematode in all stages of development. Adults and juveniles can leave roots and infect other nearby roots. The lesion nematode overwinters in infected roots or in soil in the form of eggs, juveniles, or adults.

Ring Nematode

The ring nematode *Criconemella xenoplax* damages many cultivars of *Ilex crenata,* some cultivars of *I. cornuta,* and *Aucuba japonica* (Table 13.1). Poor plant growth and other symptoms of nematode infection described above are typical of ring nematode infection. Necrotic areas may form on the root surface where ring nematodes have been feeding. The root systems of infected plants may not be as extensive as those of nematode-free plants. Ring nematodes takes their name from the concentric ring pattern in the cuticle (skin) on the surface of the body.

Disease development Ring nematodes are ectoparasitic, feeding at the root surface by means of a long, needle-like stylet, which is thrust into root cells as deep as the vascular tissues in small-diameter roots. They do not enter roots but instead move in water films around soil particles and the root surface. After a suitable feeding period, the adult female lays eggs in the soil. The eggs pass through a juvenile stage of development, and the nematodes emerge as second-stage juveniles, which possess a stylet and can begin feeding on plant roots. Juveniles progress through two more stages, in which they continue to feed on roots, before developing into adults. Each of the juvenile stages and the adult stage cause disease as they feed on root cells, primarily as a result of saliva injected into the cells to break down cell contents for ingestion. Populations of ring nematodes can build up dramatically during the growing season, as more and more root tissue is colonized by each new generation of juveniles and adults. With the onset of cold weather, soil populations of these nematodes decline. However, since host roots are perennial, the nematode population spirals ever higher in each new growing season. Infected plants develop more severe foliar symptoms as the population increases over time.

Stunt Nematode

Populations of the stunt nematode *Tylenchorhynchus claytoni* increase in the root zone of azalea, aucuba, rose, and some cultivars of Chinese and Japanese hollies, in which they cause a range of symptoms, from chlorosis to severe lack of growth compared to nematode-free plants (Table 13.1). This nematode was first recognized in woody ornamentals as a problem in azaleas (Jenkins, 1956; Sher, 1958; Barker et al., 1965). Other woody ornamentals are no doubt also susceptible. *T. claytoni* is distributed worldwide. It is known in eastern areas of the United States and also occurs on azalea in California. The stunt nematode is named for the stunting symptom induced when infected plants fail to grow.

Disease development Stunt nematodes are ectoparasitic, like ring nematodes. All stages of juveniles as well as adults can feed on roots, with reproduction taking place in as little as 33 days after eggs hatch. Populations of stunt nematodes build up in the root zone during the growing season but decline during winter, only to resurge the following growing season. Feeding sites are mostly limited to epidermal cells. Feeding does not necessarily lead to necrotic lesions. Plants decline over a period of several growing seasons.

Dagger Nematode

The dagger nematode *Xiphinema diversicaudatum* was found to cause damage in field-grown roses in the 1950s (Schindler, 1957). It has an exceptionally long and powerful stylet, for which it is named. Infected plants show typical foliar symptoms, including chlorosis and low vigor. This nematode induces root galls on the ends of infected roots, which may be confused with those caused by root-knot nematodes. *Xiphinema* spp. are known as

Table 13.1 Reactions of some ornamentals to four common plant-parasitic nematodes[a]

	Reaction[b] to:			
Host	**Root-knot nematode**	**Stunt nematode**	**Lesion nematode**	**Ring nematode**
Aucuba japonica	HS	S	. . .[c]	S
Azalea	T	S	. . .	T
Buxus microphylla (Japanese boxwood)	S	T	S	T
Buxus sempervirens (American boxwood)	. . .	T	HS	. . .
Camellia japonica	T	T	. . .	. . .
Camellia sasanqua	T	T	. . .	. . .
Gardenia jasminoides	S	T	T	T
Gardenia radicans	HS	T	T	T
Ilex cornuta (Chinese holly)				
Burfordi	T	T	. . .	. . .
Rotunda	S	S	. . .	S
Ilex crenata (Japanese holly)				
Compacta	HS	T	T	S
Convexa	HS	T	. . .	S
Helleri	HS	S	. . .	S
Rotundifolia	HS	S	. . .	S
Ilex vomitoria nana (Yaupon holly)	T	T	. . .	T
Juniperus spp.				
Blue Rug	T	T	HS	T
Shore	T	T	. . .	T
Spiny Greek	T	T	S	T
Ligustrum lucidum	T	T	. . .	T
Nandina domestica	T	T	T	T
Photinia × *fraseri*	T	T	T	T
Rose	S	S	S	T

[a] Adapted from Jones et al. 1982.
[b] HS = highly susceptible (severe stunting, branch dieback, and death). S = susceptible (some stunting). T = tolerant (satisfactory plant growth).
[c] Not tested.

vectors of many different plant viruses, and thus plant growth may be further affected if virus infection occurs.

Disease development Dagger nematodes are ectoparasitic feeders with a long stylet. Populations of dagger nematodes on rose can increase 100-fold within three growing seasons. Juveniles and adults can feed on roots. Individuals may feed for several days at the same location on a root. Plants decline over several growing seasons, so roses in field production for one or two years may not exhibit noticeable symptoms prior to sale in the wholesale market.

Foliar Nematodes

Foliar nematodes, *Aphelenchoides* spp., have only recently been recognized as serious pathogens of some ornamentals. They are most common in herbaceous hosts, such as begonias, hostas, and chrysanthemums, and woody plants, such as *Buddleia,* but they also have been observed on evergreen azaleas. Foliar nematodes, as the name implies, feed on leaves and buds, causing damage to foliage. This characteristic distinguishes them from other nematodes parasitic on woody ornamentals, which mostly infect and feed on root tissue.

Leaf symptoms of foliar nematodes begin as yellowish blotches restricted by the main veins of the leaf. The blotches turn brownish green and finally dark brown. Eventually, new infection sites develop across the leaf blade and coalesce, resulting in leaf necrosis. Dead, shriveled leaves hanging down on stems may extend to the top of the plant. Foliar nematodes feeding in buds may kill the growing point and prevent flowering or cause emerging leaves to be malformed.

Disease development Foliar nematodes can complete their entire life cycle in 10–14 days within the foliage of their host. The female nematode lays eggs in the intercellular spaces of infected leaves. The second-stage juvenile emerges from the egg case and develops through two more juvenile stages before becoming an adult, all within leaf cells. Foliar nematodes may leave an infected area of a leaf and migrate across the leaf surface in a film of water, reenter the leaf through stomata, and establish new infection sites in the intercellular spaces of the leaf tissue. They can overwinter as adults in leaves and buds and possibly in infested leaf debris on the soil surface. Adults infect healthy plants by swimming up stems from the soil surface in a film of water formed during rain, irrigation, or humid periods. The lowest leaves are thus infected first, but eventually the entire plant, from bottom to top, may develop symptoms.

Control of foliar nematodes is difficult once plants become infected, as only systemic nematicides translocated to the leaves by the plant are effective. Management practices to limit the introduction of infected planting and propagating stock are most important, along with cultural practices to avoid irrigation that allows foliage to remain wet overnight.

Other Nematodes Associated with Ornamentals

Plant-parasitic nematodes in several other genera have been associated with ornamentals. Some are severe pathogens of particular crops, such as *Bursaphelenchus xylophilus,* the pine wilt nematode, which is vectored by a bark beetle from one pine tree to another. This nematode feeds on the twigs and stems of pine trees in the landscape and native stands but it is usually not a problem in pine tree nurseries. *Rotylenchus buxophilus* is a pathogen of boxwood, but its distribution is limited in the United States. Nematodes in several other genera, including *Helicotylenchus, Paratrichodorus,* and *Trichodorus,* build up tremendous populations in the root zones of several ornamentals, including hollies and boxwoods, without disturbing the normal growth of the plants.

REFERENCES

Aycock, R., Barker, K. R., and Benson, D. M. 1976. Susceptibility of Japanese holly to *Criconemoides xenoplax, Tylenchorhynchus claytoni,* and certain other plant-parasitic nematodes. J. Nematol. 8:26–31.

Barker, K. R., and Benson, D. M. 1977. Japanese hollies: Intolerant host of *Meloidogyne arenaria* in microplots. J. Nematol. 9:330–334.

Barker, K. R., Worf, G. L., and Epstein, A. H. 1965. Nematodes associated with the decline of azaleas in Wisconsin. Plant Dis. Rep. 49:47–49.

Barker, K. R., Benson, D. M., and Jones, R. K. 1979. Interactions of Burfordi, Rotunda, and dwarf yaupon hollies and aucuba with selected plant-parasitic nematodes. Plant Dis. Rep. 63:113–116.

Benson, D. M., and Barker, K. R. 1982. Susceptibility of Japanese boxwood, dwarf gardenia, compacta (Japanese) holly, Spiny Greek and blue rug junipers, and nandina to four nematode species. Plant Dis. 66:1176-1179.

Benson, D. M., Barker, K. R., and Aycock, R. 1976. Effects of density of *Helicotylenchus dihystera* and *Pratylenchus vulnus* on American boxwood growing in microplots. J. Nematol. 8:322–326.

Benson, D. M., Jones, R. K., and Barker, K. R. 1982. Disease loss assessment for azalea, rhododendron, and Japanese holly in North Carolina nurseries. Plant Dis. 66:125–128.

Heald, C. M., and Jenkins, W. R. 1964 Aspects of the host-parasite relationship of nematodes associated with woody ornamentals. Phytopathology 54:718–722.

Jenkins, W. F. 1956. Decline of azaleas, a possible new nematode disease. Md. Florist 35:1–2.

Jones, R. K., Benson, D. M., and Barker, K. R. 1982. Nematodes and their control on woody ornamentals in the nursery. N.C. State Coop. Ext. Serv. Plant Pathol. Inf. Note 5.

Nemec, S., and Struble, F. B. 1968. Response of certain woody ornamental plants to *Meloidogyne incognita.* Phytopathology 58:1700–1703.

Osborne, W. W., and Jenkins, W. R. 1962. Pathogenicity of *Pratylenchus vulnus* on boxwood, *Buxus sempervirens* var. *arborescens.* Plant Dis. Rep. 46:712–714.

Schindler, A. F. 1957. Parasitism and pathogenicity of *Xiphinema diversicaudatum,* and ectoparasitic nematode. Nematologica 2:25–31.

Sher, S. A. 1958. The effect of nematodes on azaleas. Plant Dis. Rep. 42: 84–85.

Springer, J. K. 1964. Nematodes associated with plants in cultivated woody plant nurseries and uncultivated woodland areas in New Jersey. N.J. Dep. Agric. Circ. 429.

Southey, J. F. 1966. New host-plant records for plant-parasitic nematodes. Plant Pathol. 15:46.

Southey, J. F. 1968. New or unusual host-plant records for plant-parasitic nematodes. Plant Pathol. 17:95.

Southey, J. F. 1971. New or unusual host-plant records for plant-parasitic nematodes, 1968–1970. Plant Pathol. 20:96.

Southey, J. F. 1974. New or unusual host-plant records for plant-parasitic nematodes, 1971–1974. Plant Pathol. 23:45–46.

Stessel, G. J. 1961. Nematodes of woody plants in Rhode Island. R.I. Agric. Exp. Stn. Bull. 360.

CHAPTER

14

D. Michael Benson ▪ North Carolina State University, Raleigh

Sharon von Broembsen ▪ Oklahoma State University, Stillwater

Phytophthora Root Rot and Dieback

Phytophthora spp. are soilborne plant pathogens that cause disease in a wide variety of crops and ornamentals around the world. Common species causing root rot and dieback in ornamentals are *P. cactorum, P. cinnamomi, P. citricola, P. citrophthora, P. cryptogea, P. drechsleri, P. lateralis, P. megasperma, P. parasitica,* and *P. syringae* (which causes dieback only). Root rot is the most common disease caused by *Phytophthora* spp. in nursery crops. An aerial form of disease can develop when inoculum from the soil or contaminated irrigation water is splashed on susceptible stem and leaf tissue. In this phase, shoot tips and stems become infected, and dieback occurs.

Occurrence and Distribution of *Phytophthora* spp.

Phytophthora spp. cause disease in field, fruit, vegetable, and ornamental crops and in forest trees around the world. *P. cinnamomi,* a common cause of root rot in ornamental nurseries, has been reported on over 900 hosts worldwide. This introduced pathogen has become established in soils of the southeastern and Pacific coast regions of the United States. Other species of *Phytophthora* also occur widely in nature, both in the United States and worldwide.

Temperature and moisture limit the survival and distribution of *Phytophthora* spp. Generally, species that form resistant spores, such as chlamydospores and oospores, can persist under extreme soil conditions. Many *Phytophthora* spp. cannot survive under extreme soil conditions in nature but can survive on root systems of plants growing in containers that are protected from extreme temperatures. Even species which do not form chlamydospores or oospores are able to survive at low levels in infected roots and in soil and gravel when protected by overwintering houses or straw in winter or by shade cloth in summer.

Phytophthora Root Rot

Since the *Phytophthora* spp. that cause root rot are soilborne pathogens capable of long-term survival in soils, ornamental crops are threatened throughout the production cycle. These pathogens are associated with diseased plants grown under wet conditions in nursery beds or in containers of media that drain slowly. *Phytophthora* spp. produce motile zoospores in films of water around roots. Zoospores then infect roots and establish the pathogen in the root system. As the root system is attacked by the pathogen, root rot symptoms develop. *Phytophthora* spp. attack many important nursery crops, including andromeda (*Pieris*), arborvitae, aucuba, azalea, *Camellia, Chamaecyparis, Cunninghamia, Daphne,* deodar cedar, dogwood, forsythia, Fraser fir, hemlock, Japanese holly, juniper, pine, *Pittosporum, Podocarpus, Rhododendron, Stewartia, Taxus,* and others.

Symptoms

The severity of Phytophthora root rot and hence the expression of symptoms associated with the disease vary with the host. Hosts that limit parasitism and colonization of the root system by *Phytophthora* spp. may exhibit only mild symptoms of root rot, such as slight stunting of the foliage and necrosis (death) of only a small proportion of feeder roots. In this case, the disease may go unnoticed by the grower. When root rot is severe, aboveground symptoms become more noticeable. Symptoms on foliage include chlorosis (yellowing), small-diameter stems, poor shoot growth, wilting (in some crops), and death. Temporary disappearance of interveinal chlorosis may occur if foliar applications of iron are provided, even when adequate iron is available in the soil, since infected roots cannot adequately supply iron to the foliage. Plants may be stunted overall, and individual leaves may be dwarfed. Wilting often occurs just prior to plant death and results in leaves that droop on the stems, even when water is provided to the plant. Cutting through the bark on the main stem of a wilted plant at the soil line often will expose a reddish brown discoloration of the wood, a symptom of advanced root rot.

Belowground symptoms are also evident in plants with severe root rot. Removal of potting mix from roots with water may allow better observation of the discoloration resulting from root rot. Roots of diseased plants are reddish brown to dark brown, depending on the host. Feeder roots may be completely lacking, and coarse roots and the lower stem may be discolored. In plants with fleshy roots, such as aucuba and camellia, diseased roots may appear brown and water-soaked. Roots of healthy plants appear white to light tan, depending on the host.

Detection of *Phytophthora*

Root samples from plants suspected to have Phytophthora root rot must be tested to detect the pathogen, because other factors, such as excess fertilizer, might damage and discolor roots. Detection of *Phytophthora* spp. in root samples is based on culture of the fungus or enzyme-linked immunosorbent assay (ELISA). In cornmeal agar culture, for example, *P. cinnamomi* from diseased roots has characteristic hyphal swellings and chlamydospores, which appear in two or three days. The formation of sporangia can be induced by cultural manipulation, and these structures and other characteristics can be used to identify the fungus microscopically. Cultural methods of detecting *Phytophthora* spp. require the use of selective media and take several days to complete, and the tests must be performed by a trained mycologist with specialized equipment, since each species has unique characters.

ELISA kits are commercially available and can be used by nursery personnel to detect *Phytophthora* spp. in diseased root or crown tissue in as little as 15 minutes. However, they do not detect *Phytophthora* before root rot symptoms are apparent; therefore this method is used to confirm suspected root rot disease. ELISA kits also detect dead fungus and therefore cannot be used to evaluate whether control measures have been effective or not. Current ELISA kits do not distinguish different species of *Phytophthora.*

Disease Development

Phytophthora spp. overwinter as mycelium in infected plant tissue and as oospores and chlamydospores in soil, infected roots, and infested crop debris. Germination of chlamydospores and oospores begins in the spring when the soil temperature reaches 50°F and continues through the growing season (some species, such as *P. lateralis,* germinate only in cooler periods, with temperatures of 59–76°F, or 15–25°C) when soil moisture is near saturation (–0.025 to –0.100 mbar of matric potential). Both spore types germinate to form sporangia in water-filled pores around soil particles. Sporangia can form in as little as 4 to 8 hours under optimum conditions (in nearly saturated soil at 68°F). They persist in soil or infected host tissue for variable periods of a few days to several weeks, depending on soil conditions and the species of *Phytophthora.* Sporangia release zoospores when the soil becomes saturated (0 bar of matric potential), such as after heavy rainfall or excessive irrigation. Each sporangium is capable of releasing 20–50 motile zoospores, which swim a short distance (usually less than an inch) in static water films around soil particles but may be carried longer distances by moving water, e.g., surface drainage water, irrigation water, and percolating water in the soil profile. Zoospores may be released in 20–60 minutes when soils become saturated.

Zoospores are the infective propagule of *Phytophthora* spp. and can actually sense carbohydrates (i.e., food) diffusing away from the root tips of a host. They encyst in the region just behind the root tip and form infection hyphae, which penetrate between cells at the root tip to begin the process of infecting the root system.

Once infection occurs, the rate of spread and development of root rot varies among hosts, depending on susceptibility and the size of the root system. Since the pathogen penetrates feeder roots first and then grows into larger roots, the size of the root system and the total number of infection sites determine how fast the disease develops. For instance, rooted cuttings of azaleas died in 30 days after infection, whereas one-year-old plants in 1-gallon containers died five to seven months after infection. Landscape-sized plants may decline over a period of several years before they die from Phytophthora root rot.

Hosts vary in their susceptibility to *Phytophthora* spp. For instance, rhododendrons are more susceptible to root rot caused by *P. cinnamomi* than azaleas, and Fraser fir is more susceptible to the same pathogen than white pine. Certain cultivars and hybrids of these ornamentals are more resistant than others (see the chapters on individual crops for information on resistant cultivars).

Survival and Dispersal of Pathogens

Chlamydospores, oospores, and mycelium in infected tissue are the predominant means of survival of *Phytophthora* spp. Chlamydospores and oospores form in and on infected host roots and later are dispersed to soil as the infected root decomposes. *Phytophthora* spp. can spread in the nursery when soil containing spores is moved about. Spores of *Phytophthora* can be introduced into susceptible nursery stock during any phase of nursery production, including propagation. Since these pathogens overwinter in the soil, infection can occur whenever soil is introduced during production. Spores can be splashed from the soil surface to propagation beds or container-grown stock.

Once plants become infected, they serve as a source of inoculum (spores) that can be splashed to nearby plants or move in water films through drainage holes in containers. During heavy rains or excessive irrigation, zoospores may be carried passively in runoff water over considerable distances. In field-grown nursery stock, it is not uncommon for spores to be carried down rows, but the disease usually develops first in poorly drained areas of the field. A *Phytophthora* species may be spread over an entire nursery operation if runoff water containing spores is recycled from retention basins.

Disease Management

Integrated pest management (IPM) is needed for control of Phytophthora root rot in container- and field-grown nursery stock. The IPM scheme needs to be based on the factors involved in disease development and the survival and dispersal of the pathogen. No one management tool will control the disease. Selecting resistant cultivars whenever possible (see the chapters on individual crops) is an important step in IPM.

Given the widespread distribution of *Phytophthora* spp. in the United States, most nurserymen who grow susceptible crops should assume that the pathogen is lurking in the soil on the premises. Therefore, initial design of the operation is critical in managing the disease in the years ahead. The design of water management and drainage systems is critical. Engineering consultants may be helpful in this regard.

Field Production

Fields need to be laid out with well-designed drainage features, such as raised planting beds and adequate ditching to handle the heaviest of rains. Some growers use tile drains buried below beds to accomplish this. However, even the best-designed field will not be able to drain quickly enough after a heavy rain to prevent saturation of the soil for a period of time. Sporangia form in 4–8 hours and release zoospores shortly thereafter. Therefore, field production should possibly be limited to certain regions of the country where *P. cinnamomi* does not survive in soil during the winter or regions with cool summers, where infection and disease do not develop rapidly enough to threaten the crop.

Container Production

Pads for containers should be crowned for drainage to ditches that are adequate during heavy rains. Ground cloths or gravel beds are needed to move water away quickly from individual containers and to prevent the splashing of *Phytophthora* spores onto plants. Containers used in potting operations should be new or cleaned and fumigated or steamed prior to reuse. Avoid storing new containers directly on the ground. A well-drained soilless potting mix is absolutely essential for growing ornamentals susceptible to *Phytophthora*. Generally, soil should be avoided because it retains water too long, may contain *Phytophthora* spores, and makes containers too heavy to handle conveniently.

Tree barks are widely used as the major component of nursery potting media for container-grown stock. Hardwood bark is readily available in the midwestern and northeastern United States, while pine bark is most abundant in the southeast. Fir bark may predominant in the mixes available in the western United States. Hardwood bark must be composted prior to use, but an additional benefit is derived in the composting process: certain inhibitory factors that suppress root rot fungi, including *Phytophthora* spp., are formed during composting of hardwood bark. New products containing composted hardwood bark fortified with biological control agents are coming on the market and promise to provide an even greater range of root rot prevention.

Tree barks are advantageous as potting components for root rot suppression, because media containing them drain well and hence provide fewer periods when conditions are favorable for the formation, release, and dispersal of zoospores. Media containing bark combined with sand in (a ratio of three to five units of bark per unit of sand) and appropriate lime, fertilizer, and micronutrients have desirable physical and chemical properties for good plant growth.

Irrigation of container-grown plants can have a dramatic effect on the incidence and severity of Phytophthora root rot. Several irrigation designs to avoid overwatering have been introduced for irrigating individual containers, including drip and spray sticks. Overhead impact sprinklers may be the most economical to install, but they allow the grower much less control over the amount of water applied to plants. The volume of water delivered by impact sprinklers can also fluctuate widely across a nursery block of containers. Arranging plants by container size helps to prevent overwatering of small plants, which do not need as much water as larger plants. Separating crops that are resistant from those that are susceptible to Phytophthora root rot can give growers flexibility in irrigation frequency. See Chapter 88, "Horticultural Practices to Reduce Disease Development," for more information on managing irrigation.

Nurseries with a history of Phytophthora root rot should develop an IPM plan that includes the use of fungicides to prevent the spread of *Phytophthora* spp. among susceptible crops in the nursery. Several systemic fungicides are available for management of Phytophthora root rot. Since these prevent infection, they must be applied before the crop is exposed to inoculum of *Phytophthora.* Normally, this would not be before liners were potted from propagation for placement in grow-out locations. First applications each year would be needed when the container mix temperature reaches 50°F and then continued as directed on the label.

To minimize the use of fungicides, it may be practical and economical to collect runoff water from a block of plants where *Phytophthora* infection is suspected and test the water for the pathogen. Positive results would be a signal that *Phytophthora* is active and a fungicide is needed. Only plants in blocks where the disease is developing would need to be treated.

Propagation Area

In the propagation area, strict sanitation practices should be followed routinely to avoid contamination and infestation during propagation. Debris from previous crops should be removed. Prior to the filling of benches and the sticking of cuttings, all walls, benches, and concrete floors should be scrubbed down with a disinfectant, to remove any residual inoculum of *Phytophthora* spp. A soilless propagation mix should be used, and benches should not be in contact with a soil surface where inoculum of *Phytophthora* spp. might persist. Preparation of cuttings should be done on a surface that is cleaned with disinfectant several times each day. Cuttings should be taken high on the stock plants, where soil has not splashed, to avoid introducing the pathogen on the cuttings. Workers taking cuttings should not be preparing or sticking cuttings without changing field shoes or completely cleaning their shoes.

Should diseased cuttings develop in a propagation bed, care must be taken to place them in a plastic bag as they are removed from the bed, to avoid distributing spores to other areas of the bench. A fungicide drench or spray should then be used to prevent the subsequent spread of *Phytophthora* spores to cuttings adjoining the diseased area.

Potting Area

As in the propagation area, strict sanitation practices need to be established for the potting area in a nursery. The money spent on concrete to store mix ingredients with enough room left to manipulate loaders will more than come back to the operation in the form of increased production as a result of lower disease losses. Be sure to lay out the concrete pad so that no water runs onto it from other areas of the nursery. If possible, locate the potting area where it can be reached by suppliers delivering materials such as tree bark without driving through the production area. Avoid deliveries on rainy days, so that vehicles with mud-caked tires are not driven on the concrete pad. Devise a layout so that trailers taking potted plants to growing locations do not travel across the same concrete used for storage piles. Keep the loader on the slab with the storage pile.

Phytophthora Dieback

In addition to root rot, many species of *Phytophthora,* including *P. cactorum, P. citricola, P. citrophthora, P. cryptogea, P. heveae, P. ilicis, P. syringae,* and *P. parasitica,* cause an aerial blight of plant foliage when spores splash from the soil or container surface to stems and leaves or are present in contaminated overhead irrigation water. The resulting dieback of tissue may be limited to a few leaves and stems, or it can progress rapidly, killing the plant, even though the root system is still healthy. Phytophthora dieback occurs mainly when overhead irrigation is excessive; after a thunderstorm followed by cloudy, humid weather; or when contaminated irrigation water is applied by overhead irrigation.

Symptoms

The first symptoms of dieback are small, water-soaked areas on expanding leaves and tender stems. In susceptible hosts, such as hybrid rhododendron, water-soaked areas on expanding leaves turn necrotic within 48 hours, and the entire leaf blade is destroyed. Lesions may initially form at the leaf margin and then progress toward the center of the leaf. The pathogen colonizes the leaf as the lesions expand and then often moves into and through the midrib tissue to reach the stem. Stem lesions can remain discrete if the weather turns unfavorable for disease development, or they can progress rapidly through young stem tissues, killing the foliage. Small plants with only three or four succulent stems may be killed within a week of initial infection. In larger plants with woody stems, dieback may be limited to the current season's growth, depending on the *Phytophthora* species involved.

Disease Development

Initial inoculum for Phytophthora dieback infection comes from sporangia or zoospores splashed onto foliage during rainstorms or irrigation. Thus, the disease is restricted to warm but not hot portions of the growing season and to climates with frequent thunderstorms (*P. syringae* is the exception, preferring cool, winter conditions). Inoculum has been known to splash 60 cm (24 inches) above the container pad to infect rhododendron foliage. Once spores have splashed to leaf tissue, a film of water on leaves overnight is sufficient for infection. Lesions develop rapidly on succulent leaves and stems. Excess nitrogen fertilizer enhances host susceptibility.

Secondary inoculum, consisting of sporangia and oospores, develops in lesions on leaves that remain wet overnight. Subsequent irrigation or rainstorms can then disperse spores to adjacent foliage or plants, where new infections develop, initiating an epidemic. Thus, once inoculum has been splashed from the container pad or nursery bed, Phytophthora dieback becomes independent of soil conditions.

P. syringae causes a dieback of field-grown rhododendrons in the Pacific Northwest. Phytophthora dieback caused by *P. syringae* occurs only during the winter, when temperatures are cool and rain is frequent. Both leaves and stems are infected by sporangia and zoospores splashed during rain showers. Typically, infected areas on stems have a black, shiny appearance. During warmer parts of the year, *P. syringae* does not infect rhododendrons.

Survival and Dispersal of Pathogens

Depending on the *Phytophthora* species, chlamydospores, oospores, and mycelium in crop debris can serve as overwintering inoculum for survival of the pathogen. Contaminated irrigation water applied through sprinklers can also be a source of inoculum at any time of the season. Wet conditions that result in standing water on container pads, in nursery beds, or in containers contribute to the dispersal of the pathogen. Spore germination and dispersal can occur in just a few hours during rainy weather. Normal irrigation during dry periods can usually be managed to prevent films of water on foliage during the evening and thus limit the dispersal of *Phytophthora* spp. causing dieback and the incidence of the disease.

Disease Management

Many of the management practices described for the prevention of Phytophthora root rot apply equally well to Phytophthora dieback. A ground cover, such as gravel over ground cloth on the container area, is essential to prevent ponding of water during irrigation and thunderstorms and thus dispersal of inoculum to foliage. It is especially important that irrigation water free of *Phytophthora* spp. be used on crops that are prone to dieback. Avoidance of excessive fertility levels in plants is important, to limit lesion expansion in succulent tissues. Fungicides applied to aerial parts can prevent spore infection and production and thus limit epidemics. Repeated applications may be necessary as long as succulent tissue is present, but fungicides with different modes of action should be rotated in blocks of two to four sprays, to prevent a buildup of resistant strains.

REFERENCES

Blaker, N. S., and MacDonald, J. D. 1981. Predisposing effects of soil moisture extremes on the susceptibility of rhododendron to Phytophthora root and crown rot. Phytopathology 71:831–834.

Benson, D. M. 1980. Chemical control of rhododendron dieback caused by *Phytophthora heveae.* Plant Dis. 64:684–686.

Benson, D. M., and Cochran, F. D. 1980. Resistance of evergreen hybrid azaleas to root rot caused by *Phytophthora cinnamomi.* Plant Dis. 64: 214–215.

Benson, D. M., and Hoitink, H. A. J. 1986. Phytophthora dieback. Pages 12–15 in: Compendium of Rhododendron and Azalea Diseases. D. L. Coyier and M. K. Roane, eds. American Phytopathological Society, St. Paul, Minn.

Benson, D. M., and Jones, R. K. 1980. Etiology of rhododendron dieback caused by four species of *Phytophthora.* Plant Dis. 64:687–691.

Benson, D. M., Jones, R. K., and Daughtry, B. I. 1978. Ground covers that restrict disease spread. Amer. Nurseryman 148(4):19, 143–144.

Duniway, J. M. 1975. Limiting influence of low water potential on the formation of sporangia by *Phytophthora drechsleri* in soil. Phytopathology 65:1089–1093.

Erwin, D. C., and Ribeiro, O. K. Phytophthora Diseases Worldwide. American Phytopathological Society, St. Paul, Minn.

Gerlach, W. W. P., Hoitink, H. A. J, and Schmitthenner, A. F. 1976. *Phytophthora citrophthora* on *Pieris japonica:* Infection, sporulation, and dissemination. Phytopathology 66:302–308.

Hoitink, H. A. J. 1980. Composted bark, a lightweight growth medium with fungicidal properties. Plant Dis. 64:142–147.

Hoitink, H. A. J., and Schmitthenner, A. F. 1974. Resistance of *Rhododendron* species and hybrids to Phytophthora root rot. Plant Dis. Rep. 58:650–653.

Hoitink, H. A. J, Daft, G., and Gerlach, W. W. P. 1975. Phytophthora shoot blight and stem dieback of azalea and pieris and its control. Plant Dis. Rep. 59:235–237.

Hoitink, H. A. J., Benson, D. M., and Schmitthenner, A. F. 1986. Phytophthora root rot. Pages 4–8 in: Compendium of Rhododendron and Azalea Diseases. D. L. Coyier and M. K. Roane, eds. American Phytopathological Society, St. Paul, Minn.

Hoitink, H. A. J., Watson, M. E., and Faber, W. R. 1986. Effect of nitrogen concentration in juvenile foliage of rhododendron on Phytophthora dieback severity. Plant Dis. 70:292–294.

Kirby, H. W., and Grand, L. F. 1975. Susceptibility of *Pinus strobus* and *Lupinus* spp. to *Phytophthora cinnamomi.* Phytopathology 65:693–695.

Kliejunas, J. T., and Ko, W. K. 1976. Dispersal of *Phytophthora cinnamomi* on the island of Hawaii. Phytopathology 66:457–460.

Kuske, C. R., and Benson, D. M. 1983. Survival and splash dispersal of *Phytophthora parasitica,* causing dieback of rhododendron. Phytopathology 73:1188–1191.

Kuske, C. R., and Benson, D. M. 1983. Overwintering and survival of *Phytophthora parasitica,* causing dieback of rhododendron. Phytopathology 73:1192–1196.

Kuske, C. R., and Benson, D. M. 1983. A gravel container base for control of Phytophthora dieback in rhododendron nurseries. Plant Dis. 67:1112–1113.

Linderman, R. G. 1986. *Phytophthora syringae* blight. Pages 15–17 in: Compendium of Rhododendron and Azalea Diseases. D. L. Coyier and M. K. Roane, eds. American Phytopathological Society, St. Paul, Minn.

McIntosh, D. L. 1977. *Phytophthora cactorum* propagule density levels in orchard soil. Plant Dis. Rep. 61:528–532.

Mircetich, S. M., and Zentmyer, G. A. 1966. Production of oospores and chlamydospores of *Phytophthora cinnamomi* in roots and soil. Phytopathology 56:1076–1078.

Rao, B., Schmitthenner, A. F., and Hoitink, H. A. J. 1978. A simple axenic mycelial disk salt soaking method for evaluating effects of composted bark extracts on sporangia and zoospores of *Phytophthora cinnamomi.* Proc. Am. Phytopathol. Soc. 4:174.

CHAPTER
15

D. Michael Benson ▪ North Carolina State University, Raleigh

Powdery Mildew

Powdery mildew is probably the most easily recognized disease, because the white, powdery-looking colonies on plant foliage are easy to see. The powdery mildew fungi are obligate parasites, which means that they are strictly dependent on a living host for nutrients, growing only in constant association with the host. Unlike most other fungal pathogens, 99% of the mycelium of powdery mildew fungi is restricted to the surface of the plant foliage.

Since powdery mildew fungi are obligate parasites, diseases caused by these pathogens generally do not kill their hosts, although severe infections of emerging shoots may lead to shoot death. Even so, the extensive white mycelium growing on leaf, stem, and flower surfaces makes plants unsightly and thus unsalable. Powdery mildew diseases occur throughout nursery production areas in the United States as well as worldwide but may be most severe in nurseries employing drip irrigation in arid climates.

Pathogens

The most common powdery mildew fungi infecting woody ornamentals and trees in nurseries and in the landscape are, in the sexual stage, members of the genera *Erysiphe, Microsphaera, Phyllactinia, Podosphaera, Sphaerotheca,* and *Uncinula* and, in the asexual stage, members of the genera *Acrosporium, Oidiopsis, Oidium,* and *Ovulariopsis.* These fungi form cleistothecia containing asci and ascospores in the sexual stage and conidia in the asexual stage. When a fungus is found in its sexual stage on a host, the genus name associated with the sexual stage should be used to refer to the pathogen. If only conidia (asexual spores) are found on a host, the pathogen can be identified by conidiophore and conidial morphology, and the genus name of the asexual stage should be used.

Symptoms

Infected plants may be affected with chlorosis, crinkled and distorted leaves, aborted flowers and fruit, and premature senescence. In most hosts, young succulent tissues and water sprouts are most susceptible to infection.

Powdery mildew does not develop on large dogwoods until the last two or three leaves are emerging in the late spring. Infection of the young terminal leaves, however, results in dwarfed, crinkled, and distorted leaves, which are easily observed, giving the plant an unsightly appearance. Dogwood seedlings in nurseries can be severely stunted or killed by powdery mildew infection, however.

In some hosts, such as *Leucothoe* and *Photinia,* leaf tissue in the areas under mildew colonies turns red as the green chlorophyll is lost.

In many cases, the first sign of powdery mildew infection is the white mycelium with hyphae and conidiophores bearing conidia (the asexual stage of the fungus), which develops on the surface of host tissues and often times takes on a powdery appearance. The mycelium may first develop as distinct colonies or spots on the leaf surface or may ramify completely across a leaf. In time the mycelium may turn tan or brown in the central part of a mildew colony, leaving a white-fringed margin. In late summer or early fall, sexual reproduction may occur, resulting in the formation of structures known as cleistothecia (described in the following section), which can be seen as tiny black specks in the colony. In some tree species infected with powdery mildew, very little white mycelium is visible on the underside of the leaf, but colonies are covered with large numbers of cleistothecia.

Epidemiology

In regions with a year-round mild climate, such as southern California, southern Arizona, southern Texas, and southern Florida, powdery mildew fungi may remain active throughout the year. In regions with a dormant season, these fungi overwinter as dormant mycelium in buds or on shoot tips, or they survive in the form of cleistothecia. Infections initiated by cleistothecia result from the rupture of the cleistothecium and the liberation of ascospores, which are dispersed by wind and cause infections on new growth. Dormant mycelium that renews growth in the spring is also a source of inoculum, producing new mildew colonies as plant tissues develop.

Soon after the white mycelium develops on host tissues, the fungus begins to sporulate, producing conidia on conidiophores that arise from the mycelium. Conidia are released from the conidiophore at maturity when a rapid decrease in relative humidity occurs. Rapid increases in relative humidity or temperature, however, do not affect conidial release in the rose powdery mildew pathogen (Coyier, 1985). The conidia are windborne. When they land on a susceptible plant, they germinate, producing hyphae that grow across the plant surface. The conidia of the powdery mildew fungi are unique in that they do not require a film of water on the leaf to germinate. Frequent periods of leaf wetness actually inhibit formation and dispersal of conidia.

Powdery mildew fungi penetrate host epidermis and produce a specialized hyphal structure called a haustorium, which forms between the epidermal cell wall and the membrane of a host cell, to absorb nutrients from the cytoplasm of the cell. The haustorium is the only portion of the pathogen that penetrates host tissues. As a mildew colony develops, conidia are produced throughout the growing season, so that secondary cycles of disease can occur if environmental conditions remain favorable.

After a mildew colony is well established on host tissue, it may undergo sexual reproduction in the late summer or early fall, producing ascospores in an ascus within a protective structure called a cleistothecium. Young cleistothecia are clear to light yellow; as they mature, the yellow gives way to an amber color, and finally they become jet black at maturity. Cleistothecia are large enough that they give a mildew colony a pepper-specked appearance. A cleistothecium bears many short-stalked hyphae known as appendages, and the cleistothecium of each genus of powdery mildew fungus can be uniquely identified by the shape of its appendages and the number asci, or sacs containing ascospores.

Cleistothecia also function as a means of overwintering for powdery mildew fungi. They are dispersed by rain and wind at maturity in the fall and latch on to shoots. The appendages then serve to hold them to shoots of a host until next year. Cleistothecia of the grape powdery mildew fungus that fail to lodge on a shoot and fall to the ground are consumed by soil microbes and thus fail to overwinter. Cleistothecia that remain in mildew colonies on fallen leaves are usually immature and do not overwinter (Cortesi et al., 1995).

Table 15.1 Host ranges of some powdery mildew fungi [a]

Pathogen	Hosts [b]
Erysiphe spp.	**Crapemyrtle,** gardenia, *Lonicera,* oak, *Rhododendron,* tuliptree
Microsphaera spp.	Ash, beech, birch, **dogwood,** elm, *Euonymus,* holly, *Kalmia, Leucothoe, Lonicera, Ligustrum,* **lilac,** magnolia, maple, **oak,** *Rhododendron, Spiraea,* **sycamore,** *Viburnum*
Phyllactinia spp.	Ash, *Berberis,* birch, elm, **crabapple, crapemyrtle,** hawthorn, holly, magnolia, maple, **oak,** *Prunus,* **rose,** tuliptree
Podosphaera spp.	*Cotoneaster,* **crabapple,** hawthorn, *Photinia, Prunus, Pyracantha, Spiraea, Vaccinium*
Sphaerotheca spp.	**Crabapple,** maple, *Photinia, Prunus,* **rose**
Uncinula spp.	**Elm,** basswood, maple, poplar, willow

[a] Data from Farr et al. (1989).

[b] Names of hosts on which powdery mildew is common are in boldface type.

Host Range

As a group, the various powdery mildew fungi attack a wide range of woody ornamentals and trees as well as other crops. However, most species of these pathogens are host-specific. Therefore, even if mildew is developing in one crop in the nursery, it does not necessarily lead to an outbreak in other hosts that are susceptible to other genera or species of mildew fungi. Table 15.1 presents a partial listing of susceptible ornamentals and the pathogens reported to cause powdery mildew in them.

Disease Management

Powdery mildew diseases are normally easy to manage in the nursery. Most of the fungus is exposed on the leaf surface, so that fungicides are very effective. Numerous fungicides are labeled and effective for powdery mildew. Some are eradicants, which can eliminate mildew colonies after they first start to form, while others prevent initial germination and infection by conidia or ascospores. However, weekly to biweekly fungicide applications are needed to prevent powdery mildew. Fungicide resistance can develop when the same protective fungicide is used repeatedly (see Chapter 93, "Fungicides for Ornamental Crops in the Nursery").

Cultural practices can also be used effectively in the nursery to manage powdery mildew diseases. Since conidia of powdery mildew fungi are inhibited by free water on the leaf surface, overhead irrigation will suppress disease development. Powdery mildew may be particularly severe in plants grown under drip irrigation. Shade cloth tends to maintain higher relative humidity around plant foliage and thus increases the severity of powdery mildew. Good circulation of air, as in shade structures with open sides, should reduce the severity of disease. Fertilization rates should be adjusted to avoid producing succulent water sprouts, which are extremely susceptible to powdery mildew infection.

Sanitation is an important management tactic in removing sources of inoculum from the crop. Since mycelium of powdery mildew fungi can overwinter in dormant buds and shoots, pruning to remove heavily infected and damaged shoots can eliminate inoculum that could induce new infections in the spring. However, removal of fallen leaves is probably not as important as once thought, because cleistothecia that remain attached to fallen leaves may not produce any ascospore inoculum in the spring.

Even though these diseases are fairly easily controlled in the nursery, the nurseryman can provide the best product service by producing cultivars resistant to powdery mildew. Resistant cultivars of a wide range of ornamentals have been identified, including flowering crabapple, dogwood, crapemyrtle, and others. Lists of resistant cultivars are presented in the chapters covering diseases of individual crops.

REFERENCES

Cortesi, P., Pearson, R. C., Seem, R. C., and Gadoury, D. M. 1995. Distribution and retention of cleistothecia of *Uncinula necator* on the bark of grapevines. Plant Dis. 79:15–19.

Coyier, D. L. 1985. Powdery mildews. Pages 103–140 in: Diseases of Floral Crops. Vol. 1. D. L. Strider, ed. Praeger Publishers, New York.

Farr, D. F., Bills, G. F., Chamuris, G. P., and Rossman, A. Y. 1989. Fungi on Plants and Plant Products in the United States. American Phytopathological Society, St. Paul, Minn.

CHAPTER 16

J. W. Pscheidt ▪ Oregon State University, Corvallis
Larry W. Moore ▪ Oregon State University, Corvallis
Heather J. Scheck ▪ Office of the Agricultural Commissioner, Santa Barbara County, Santa Barbara, California

Diseases Caused by *Pseudomonas syringae*

The bacterium *Pseudomonas syringae* is responsible for a number of economically important diseases throughout the United States. It can infect a wide variety of ornamental plants as well as fruits and vegetables. Frequent and severe losses occur in apricot, blueberry, cherry laurel, flowering cherry, flowering dogwood, flowering pear, *Forsythia,* lilac, maple, magnolia, mountain ash, and *Viburnum.*

Disease severity and host range can be highly variable within and between growing seasons. For example, in the early 1980s several nurseries in the Pacific Northwest reported linden trees with severe symptoms. However, other nurseries in the same area reported very serious disease in Japanese lilac and two cultivars of red maple but not in linden trees, still other nurseries reported the disease in *Laburnum* (golden chain tree) and Bradford pear, while another nursery found damage was mostly light and primarily affected the leaves of a few plum cultivars and Norway, red, and silver maples. In the early 1990s, some nurseries in the Pacific Northwest were unable to grow lilacs because of disease due to *P. syringae.*

Several important factors about diseases caused by *P. syringae* must be considered before control measures can be implemented. This chapter will focus on symptoms, plant susceptibility, and sources and spread of the bacterium. Once these are known, cultural, biological, and chemical control methods can be effectively integrated.

Symptoms

Various symptoms are associated with infection of woody plants by *P. syringae* pv. *syringae:*

- flower blast, in which flowers or flower buds turn brown to black
- death of dormant buds, a common symptom in flowering cherries
- necrotic leaf spots
- discolored or blackened leaf veins and petioles resulting from systemic invasion and infection
- spots and blisters on fruit
- shoot tip dieback, with dead, blackened twig tissue extending some distance down from the tip, a common symptom in maples and other seedlings
- stem cankers (depressed areas in the bark, which become darkened with age) and, in fruiting and flowering stone fruits, gummosis (exudation of a gummy substance from cankers); cankers may enlarge and girdle the stem, subsequently killing the branch or the entire plant; when the outer tissues of a canker are cut away, the tissue underneath is discolored reddish brown, and discoloration may also occur as vertical streaks in the vascular tissue

The kinds of symptoms and symptom development depend on the species of plant infected, the plant part infected, the strain of *P. syringae,* and the environment. More than one symptom can occur simultaneously in a single plant.

Shoot tip dieback was the most common symptom observed in 40 woody deciduous plants collected in a survey of Pacific Northwest nurseries (Canfield et al., 1986). *P. syringae* was isolated from all plants with the tip dieback symptom. The plants most commonly and most severely affected were aspen, blueberry, dogwood, hazelnut, lilac, linden, magnolia, maple, and oriental pear. This survey was not exhaustive but shows the widespread nature of the disease.

Plant Susceptibility

Most researchers consider *P. syringae* a weak pathogen, that is, an opportunist that capitalizes on hosts weakened by some predisposing condition. A number of factors reportedly make plants more susceptible to infection; the foremost is freeze damage. Freezing wounds the plant, allowing the bacterium to get into and destroy plant cells. Numerous workers have reported that symptom development in the field was related to low temperatures.

Ironically, many strains of *P. syringae* catalyze the formation of ice crystals (ice nucleation) on and in plant tissues (Lindow, 1983); organisms having this capability are generally described as ice nucleation–active (INA). Their presence on a plant raises the freezing temperature of sensitive plant tissues above that at which they would normally freeze. For example, in the absence of biological ice nucleators many plants can be supercooled to below –10°C, while the presence of INA bacteria can cause freezing between –2 and –10°C. Most frost-sensitive plants have no significant mechanism of frost tolerance and must be protected from ice formation to avoid frost injury.

Other predisposing factors include wounding, plant dormancy, soil factors, and infection by other pathogens.

Wounding Wounding of any kind, mechanical or environmental (such as frost injury), seems to play a major role in initiating disease development. Wounds have been shown to predispose trees to blossom blight and bacterial canker. Pruning wounds not only allow the bacterium to enter but also aid infection by fungi such as *Cytospora* and *Nectria.*

Plant dormancy Dormancy may also predispose susceptible trees to damage from *P. syringae.* Semidormant fruit trees are reportedly more susceptible to the disease than active ones.

Soil factors Factors such as improper soil pH (either too high or too low) and deficiencies or excesses of mineral nutrients may also predispose plants to *P. syringae* infection.

Dual infections Disease severity is greater when a plant is attacked by more than one pathogen. Diagnosis of symptoms due to *P. syringae* on maple in Oregon nurseries has been complicated by the presence of Verticillium wilt. In maple, symptoms of these diseases can be similar, and dual infections have been observed. The relative contribution of each pathogen to the total impact on the tree is unknown.

Sources of Inoculum

There are several potential sources of inoculum of *P. syringae.* The relative contribution of each source to overall disease development remains unknown.

Buds Buds are considered a major overwintering site of *P. syringae.* The bacterium has been detected inside apparently healthy apple and pear buds during both the growing season and the dormant season.

Cankers Cankers from a previous year's infection have long been thought to be the primary source of inoculum.

Epiphytes *P. syringae* exists on the surfaces of many plants and therefore is in a position to cause infection if the right environmental conditions develop. Monitoring surface populations associated with nursery trees in Oregon showed that the population increased rapidly and peaked during the first two to three weeks after budbreak, declined during the summer, increased a small amount in October, and often was undetectable December through February.

An antibiotic-resistant strain of *P. syringae* has been reisolated from symptomless maple trees up to 10 months after application, showing that a particular strain of *P. syringae* can survive successfully over the summer and winter. Epiphytic *P. syringae* may be important for the survival and spread of the pathogen during the growing season, but it may not be important relative to survival and overwintering of primary inoculum.

Latent infections Establishment of *P. syringae* inside symptomless tissues during the summer could represent an important source of primary inoculum.

Weeds and grasses Some weeds are hosts of *P. syringae.* Workers in Michigan, Oregon, Poland, and South Africa have reported that *P. syringae* was isolated from weeds or grasses during the growing season.

Systemic invasion In South Africa, *P. syringae* introduced into leaves and leaf petioles of plum and cherry trees during the growing season invaded leaves and shoots and caused disease (Hattingh et al., 1989). *P. syringae* has also been isolated from interior tissues of several different fruit trees, some with and some without symptoms of disease.

Soil It is generally accepted that *P. syringae* survives poorly in soil, but neither the soil phase nor the potential survival of the bacterium in the rhizosphere of roots has been studied in much depth.

Spread of *Pseudomonas syringae*

P. syringae can be moved from place to place by wind, rain, and insects and in infested budwood and nursery stock. Mechanical equipment and pruning tools are a frequently overlooked means of dispersal or of generating aerosols containing the bacterium. The number of INA bacteria captured on petri plates of agar in citrus groves greatly increased when nearby alfalfa fields were being harvested. *P. syringae* has also been recovered from the air above and next to bean fields.

Pathogenicity Factors

The ability of *P. syringae* to cause disease is increased by two common genetic traits. Most isolates produce a powerful plant toxin, syringomycin, which destroys plant tissues as the bacterium multiplies inside wounds. Many isolates also produce a protein that acts as an ice nucleus, increasing frost wounds, which are easily colonized and expanded by the bacterium.

Disease Management

A variety of methods have been tested for management of *P. syringae* in commercial plantings, including cultural management, disease resistance, biological control with microbial antagonists, and chemical control. Efforts have been targeted primarily either to control disease or to reduce the risk of frost damage from INA *P. syringae.*

Cultural Management

Eliminating or reducing the predisposing factors that favor disease development can help in the overall management of diseases caused by *P. syringae.* Frost prevention, minimal fertility, maintaining correct soil or media conditions (such as pH), and delaying the time of pruning can all help.

Frost prevention Growers should avoid planting stock in fields with frost pockets. Keeping plants under plastic shelters when young tissues are most susceptible provides protection from frost and rain and thus reduces disease severity. In a field test, lilacs under plastic shelters had as little disease as plants that received the best chemical treatment (Scheck and Pscheidt, 1998).

Fertility High nitrogen fertility that produces a lot of early- or late-season growth should be avoided. Generally, plants that receive late-season application of nitrogen are much more sensitive to winter injury and thus more susceptible to attack by *Pseudomonas.*

Soil conditions Liming the soil at the planting site reportedly promoted peach tree growth and vigor. Altering soil pH affected the susceptibility of peach to *P. syringae* (Weaver and Wehunt, 1975). The nature of the soil at the planting site can affect tree susceptibility in an indirect way. Clay soils in South Carolina are typically free of nematodes, and losses from peach tree short life in these soils have been minimal. Soil amendments with iron, calcium, and magnesium reportedly affect tree susceptibility to bacterial canker and short life disease.

Pruning Pruning in the fall and early winter has been reported to predispose trees to severe damage from *P. syringae* infections and the short life syndrome (Chandler and Daniell, 1976).

Cankers from *P. syringae* infections were longer when twigs were pruned from trees in December than when they were pruned in January or February. Where trees are threatened by *Cytospora* sp. and *P. syringae,* pruning in early spring, when they are more resistant to *Cytospora,* may be of benefit.

Disease Resistance

Genetic resistance to bacterial blight has not been a large part of woody ornamental breeding programs. Most cultivars of lilac (*Syringa vulgaris*) are very susceptible. However, less disease has been observed in the cultivars Edith Cavell, Glory, and Pink Elizabeth when planted in a garden. Other *Syringa* species may have some resistance to bacterial blight. In trials in western Washington, *S. josikaea, S. komarowii, S. microphylla, S. pekinensis,* and *S. reflexa* had less disease than *S. vulgaris.* Most maples (*Acer* spp.) are susceptible to bacterial blight, except for sugar maple (*A. saccharum*). Japanese maple (*A. palmatum*) cultivars Sango Kaku and Oshi Beni are highly susceptible.

Successful control of *P. syringae* has been obtained using resistant plant germ plasm in the tree fruit industry. The cherry rootstock F-12-1 is reportedly resistant to *P. syringae.* Unfortunately, the cultivars of sweet cherry are not resistant. However, by budding the scion high on the F-12-1 rootstock scaffold, growers have been able to avoid total death of orchard trees, since cankers usually stop at the graft union. Diseased limbs are removed, leaving other limbs to produce a crop.

In Germany, some cultivars of sour cherry have been observed to be resistant to bacterial canker caused by *P. syringae.* There has been an active program in England to select rootstocks resistant to *P. syringae* pv. *morsprunorum.* Breeding woody trees for resistance is a slow process, both because of the time involved for tree growth and the ability of *P. syringae* to adapt genetically and infect new germ plasm. As a result, plants resistant to this bacterium may be a long time coming.

Biological Management with Bacterial Antagonists

Biological control has been directed almost entirely at frost control using bacterial antagonists to prevent the buildup of INA populations of *P. syringae* (Lindow, 1983). Many experimental controls have been successful, and a product called BlightBan, containing *P. fluorescens* strain A506, is now commercially available. It is currently used for frost and fire blight control, but some laboratory studies have found activity against diseases caused by *P. syringae.*

Chemical Treatment

Fixed copper compounds (such as Bordeaux mixture and copper hydroxide), streptomycin (an antibiotic), and coordination products (such as Bravo CM and ManKocide) are registered and have been used to control *P. syringae* with various degrees of success. Adding spreader-stickers to some of these bactericides has resulted in longer-lasting control under the cool, wet conditions of the Pacific Northwest.

Current chemical recommendations for control of bacterial blight include a fall application of copper sulfate or copper hydroxide. This type of program is designed to kill the overwintering bacterium and protect wounds and leaf scars from infection. Most of the copper applied in the fall will be washed away by the time the bacterium starts to multiply and spread in the spring. An additional application of copper at delayed dormant, just before budbreak, provides the best level of disease control. Copper must be applied to young tissues with care after budbreak, because of the risk of burning foliage.

Copper bactericides come in many different formulations. Pesticide labels list the active ingredient as metallic copper, but this must be dissolved in water to be toxic. Most of the copper sprayed on plants will not be able to kill bacteria, because it does not easily dissolve. Using a higher than recommended rate of copper or spraying more often will not necessarily fix this problem or improve control of copper-resistant bacteria.

In laboratory and field experiments with lilacs, all copper formulations tested gave better disease control of copper-sensitive bacteria than streptomycin sulfate. For copper hydroxide, wettable powder formulations were better than liquid and dry flowable formulations.

Some growers in Oregon have reported poor control of *P. syringae* following either copper or streptomycin sprays. Poor control may be due to copper- or streptomycin-resistant strains, poor chemical coverage, inadequate timing, or reinfection from outside sources.

Bactericide resistance Resistance to copper and streptomycin is widespread in commercial ornamental nurseries in the Pacific Northwest (Scheck et al., 1996). Of bacteria collected in 1982–83, 25% were copper-resistant, 7% were streptomycin-resistant, and 68% were sensitive to both chemicals. Of bacteria collected in 1992–93, 24% were copper-resistant, 6% were streptomycin-resistant, 24% were resistant to both, and only 46% were sensitive to both. The amount of copper that bacteria can resist more than doubled between 1983 and 1993.

Some copper formulations have proved to be better than others against copper-resistant bacteria when tested on lilac in tissue culture. Basic copper sulfate plus hydrated lime (Bordeaux 4-4-100), tribasic copper sulfate, copper hydroxide tank-mixed with mancozeb, and copper hydroxide plus ferric chloride gave the best blight control.

Improved control of *P. syringae* pv. *tomato,* the cause of bacterial speck of tomato, has been reported when a mixture of copper and either maneb or mancozeb (fungicides) was applied (Conover and Gerhold, 1981). This mixture produces a copper carbamate, which is more effective than copper alone. Commercial products containing these substances have been developed, including ManKocide (a mixture of copper hydroxide and mancozeb) and Bravo CM (a mixture of chlorothalonil, copper oxychloride, and maneb). Many fungal diseases are also controlled by these treatments.

Summary

P. syringae causes serious diseases affecting a large number of ornamentals grown in nurseries. Common disease symptoms include shoot tip dieback, death of dormant buds, flower and shoot blast, blackened twigs, black or brown leaf veins or spots, and stem cankers. Because the pathogen can survive on healthy plants and weeds, inoculum is always present in the nursery. In spring, when weather conditions are favorable, bacterial populations can increase very rapidly, and disease losses can be severe. Factors that worsen disease include frost during early shoot development; cool, wet weather in spring or fall; pruning wounds; and heavy vegetative growth early or late in the season, promoted by high levels of nitrogen. Management of these diseases involves integration of cultural practices, chemical treatment, and disease resistance.

REFERENCES

Baca, S., and Moore, L. W. 1987. *Pseudomonas syringae* colonization of grass species and cross-infectivity to woody nursery plants. Plant Dis. 71:724–726.

Canfield, M. L., Baca, S., and Moore, L. W. 1986. Isolation of *Pseudomonas syringae* from 40 cultivars of diseased woody plants with tip dieback in Pacific Northwest nurseries. Plant Dis. 70:647–650.

Chandler, W. A., and Daniell, J. W. 1976. Relation of pruning time and inoculation with *Pseudomonas syringae* van Hall to short life of peach trees growing on old beachland. HortScience 11:103–104.

Conover, R. A., and Gerhold, R. R. 1981. Mixtures of copper and maneb and mancozeb for control of bacterial spot of tomato and their compatibility for control of fungus diseases. Proc. Fla. State Hortic. Soc. 94: 154–156.

Endert, E., and Ritchie, D. F. 1984. Overwintering and survival of *Pseudomonas syringae* pv. *syringae* and symptom development in peach trees. Plant Dis. 68:468–470.

Hattingh, M. J., Roos, I. M. M., and Mansvelt, E. L. 1989. Infection and systemic invasion of deciduous fruit trees by *Pseudomonas syringae* in South Africa. Plant Dis. 73:784–789.

Hawkins, J. E. 1976. A cauterization method for the control of cankers caused by *Pseudomonas syringae* in stone fruit trees. Plant Dis. Rep. 60:60–61.

Lindow, S. E. 1983. The role of bacterial ice nucleation in frost injury to plants. Annu. Rev. Phytopathol. 21:363–384.

Scheck, H. J., and Pscheidt, J. W. 1998. Effect of copper and bactericides on copper-resistant and sensitive strains of *Pseudomonas syringae* pv. *syringae*. Plant Dis. 82:397–406.

Scheck, H. J., Pscheidt, J. W., and Moore, L. W. 1996. Copper and streptomycin resistance in strains of *Pseudomonas syringae* from Pacific Northwest nurseries. Plant Dis. 80:1034–1039.

Weaver, D. J., and Wehunt, E. J. 1975. Effect of soil pH on susceptibility of peach to *Pseudomonas syringae*. Phytopathology 65:984–989.

CHAPTER

17

D. Michael Benson ▪ North Carolina State University, Raleigh

Ronald K. Jones ▪ North Carolina State University, Raleigh

Rhizoctonia Web Blight

Rhizoctonia web blight is a foliar disease of many ornamental crops. It develops during warm, humid periods, when sclerotia or hyphae of the pathogen surviving in debris are splashed into the plant canopy. Hyphae can also grow up the stem from the soil surface. The disease is known by several names – aerial blight, foliar blight, Rhizoctonia blight, and web blight. The accepted common name of the disease in azalea is *web blight* (Benson, 1985).

Web blight is most commonly reported in the southern and southeastern United States where periods of high humidity persist for extended periods during the growing season. Although the pathogens that cause web blight are distributed throughout most of the United States as well as worldwide, weather conditions are seldom ideal for disease development outside the South.

Symptoms

Brown hyphae of *Rhizoctonia* growing in the canopy of infected plants are a common sign of web blight during and after periods of warm, humid weather. Leaves may become water-soaked and necrotic, and defoliation may occur. Hyphae of the pathogen grow so quickly during warm, humid periods that defoliated leaves may be webbed to the stems and one another, and hence the name *web blight.* Leaves that are not matted by hyphae in the canopy accumulate on the soil surface. Normally, the interior portion of the plant canopy is affected by web blight when conditions are ideal for disease development. However, in a block of plants spaced too closely, the entire canopy of plants in the interior of the block may be affected.

Rhizoctonia web blight can also cause severe defoliation and death of cuttings under mist.

Pathogens

Rhizoctonia web blight is caused by binucleate *Rhizoctonia* spp. and by the multinucleate *R. solani.* Only binucleate *Rhizoctonia* spp. have been reported in azalea, but *R. solani* caused severe infection in Satsuki azalea in inoculation experiments (Frisina and Benson, 1987). Binucleate species were found to cause web blight in nursery-grown *Cotoneaster, Ilex crenata,* and a *Pittosporum* sp., in addition to azalea (Frisina and Benson, 1987). They have also been reported on *Juniperus chinensis,* a *Ligustrum* sp., *Pittosporum tobira,* and *Rhaphiolepis indica.* In a study of over 300 *Rhizoctonia* isolates from ornamentals in Florida, 129 multinucleate and 180 binucleate isolates were found (Chase, 1991).

Weber and Roberts (1951) identified *R. ramicola,* a binucleate species, as the cause of a web blight disease of *Elaeagnus pungens.* They reported that the pathogen produced no sclerotia in culture or on infected plants and that other hosts included *Erythrina herbacea,* a *Feijoa* sp., *Ilex crenata, Lagerstroemia indica,* and *Pittosporum tobira.*

R. solani has been isolated from foliage of plants with web blight, including *Ilex crenata, Impatiens,* and a *Pittosporum* sp. The potential of this fungus to cause severe web blight is enormous: 64% of *R. solani* isolates from ornamentals were from infected stem or leaf samples (Chase, 1991), and the pathogen has been reported in ornamental plants from over 360 genera (Farr et al., 1989).

Detection of *Rhizoctonia*

Rhizoctonia spp., including *R. solani,* are easy to identify, because of their distinct whitish to brownish hyphae on infected foliage during periods of high humidity. Infected leaves placed in a humidity chamber will usually yield a conspicuous growth of hyphae within 24 hours. Colonies of the fungus in culture are smooth and whitish to brownish, and sclerotia may develop in a few days.

Commercial kits are available for the detection of *Rhizoctonia* by enzyme-linked immunosorbent assay (ELISA) They detect both multinucleate and binucleate *Rhizoctonia* spp., but the kit reaction (color change) is stronger with the multinucleate *R. solani.*

Separation of multinucleate from binucleate isolates is difficult. Nuclei in hyphae must be stained with DAPI (4′,6-diamidino-2-phenylindole) or trypan blue and observed with a microscope. However, separation of multinucleate from binucleate isolates is not necessary, as the management approaches described below work equally well for both groups of pathogens.

Disease Development

Rhizoctonia web blight develops in the canopy of susceptible hosts during warm periods with high humidity following thunderstorms. Overhead irrigation also promotes the development of the disease. Without careful regular scouting, web blight is usually first observed when closely spaced plants are separated for sale or relocation in the nursery. Lesions develop on the petiole or leaf blade of leaves in the interior in the plant canopy. The lesions may be water-soaked and discolored. Under conditions favorable for disease, the entire leaf blade becomes necrotic, and the leaf abscises. Mycelium of the pathogen growing across foliage is often evident as a sign of the disease. The pathogen grows rapidly, producing webs of mycelium that trap abscised leaves, holding them to the stems and giving the typical web blight appearance. In severely infected plants, the center of the canopy may be completely defoliated.

Rhizoctonia spp. causing web blight are limited to infection of leaf tissue, so stem lesions and dieback are not associated with the disease. Infected plants with extensive loss of canopy are unsalable, but the lack of stem lesions allows new leaves to develop during the next growing cycle.

Survival and Dispersal of *Rhizoctonia*

Rhizoctonia spp. survive as hyphae and sclerotia associated with infected crop debris. These fungi are also able to grow on dead organic matter on the surface of soilless potting media, thus extending their survival between crops and over the seasons. Infection is initiated when sclerotia or hyphae associated with debris are splashed into the canopy from the soil, medium, or container surface, or when hyphae from the surface of the potting medium grow up the stem to infect leaves. Once web blight develops in one plant, it can spread to the canopies of adjacent plants. Web blight fungi can be widely dispersed in production plants when cuttings are infected during propagation and when infested crop debris is transported by various means, including runoff irrigation water.

Disease Management

Plants spaced pot-to-pot in the spring should be spaced farther apart just before weather conditions become favorable for web blight development (mid-June to mid-July, in the South). Growers who produce susceptible plants under shade protection may want to space plants farther apart than the normal spacing in open sun, to promote air movement around and through the canopy of plants. Irrigation cycles should be timed to allow the plant canopy to dry completely before sunset, and early morning irrigation that extends moist nighttime conditions should be delayed until the sun has dried the canopy.

In highly susceptible crops, nursery records on the disease may dictate the use of preventative fungicide applications prior to the onset of conditions favorable for web blight. However, tight canopies and a canopy structure that makes sufficient distribution of fungicide spray impractical are the same conditions that favor web blight. Spray equipment must have enough pressure to get good penetration of the canopy. Spacing the plants promotes faster drying and better coverage.

REFERENCES

Benson, D. M. 1985. Common names for plant diseases: Azalea (*Rhododendron* spp.). Plant Dis. 69:651–652.

Benson, D. M. 1991. Control of Rhizoctonia stem rot of poinsettia during propagation with fungicides that prevent colonization of rooting cubes by *Rhizoctonia solani.* Plant Dis. 75:394–398.

Benson, D. M., and Cartwright, D. K. 1996. Ornamental diseases incited by *Rhizoctonia* spp. Pages 303–314 in: *Rhizoctonia* Species: Taxonomy, Molecular Biology, Ecology, Pathology and Disease Control. B. Sneh, S. Jabaji-Hare, S. Neate, and G. Dijst, eds. Kluwer Academic Publishers, Boston.

Benson, D. M., Jones, R. K., and Frisina, T. 1986. Rhizoctonia web blight. Pages 20–21 in: Compendium of Rhododendron and Azalea Diseases. D. L. Coyier and M. K. Roane, eds. American Phytopathological Society, St. Paul, Minn.

Burpee, L. L., Sanders, P. L., Cole, H., Jr., and Sherwood, R. T. 1980. Anastomosis groups among isolates of *Ceratobasidium cornigerum* and related fungi. Mycologia 72:689–701.

Castillo, S., and Peterson, J. L. 1990. Cause and control of crown rot of New Guinea impatiens. Plant Dis. 74:77–79.

Chase, A. R. 1991. Characterization of *Rhizoctonia* species isolated from ornamentals in Florida. Plant Dis. 75:234–238.

Chase, A. R., and Conover, C. A. 1987. Temperature and potting medium effects on growth of Boston fern infected with *Rhizoctonia solani.* HortScience 22:65–67.

Chase, A. R., and Poole, R. T. 1990. Effect of air and growing medium temperatures on Rhizoctonia foot rot of *Epipremnum aureum.* J. Environ. Hortic. 8:139–141.

Farr, D. F., Bills, G. F., Chamuris, G. P., and Rossman, A. Y. 1989. Fungi on Plants and Plant Products in the United States. American Phytopathological Society, St. Paul, Minn.

Frisina, T. A., and Benson, D. M. 1987. Characterization and pathogenicity of binucleate *Rhizoctonia* spp. from azaleas and other woody ornamentals with web blight. Plant Dis. 71:977–981.

Frisina, T. A., and Benson, D. M. 1988. Sensitivity of binucleate *Rhizoctonia* spp. and *R. solani* to selected fungicides in vitro and on azalea under greenhouse conditions. Plant Dis. 72:303–306.

Frisina, T. A. and Benson, D. M. 1989. Occurrence of binucleate *Rhizoctonia* spp. on azalea and spatial analysis of web blight in container-grown nursery stock. Plant Dis. 73:249–254.

Hointink, H. A. J. 1980. Composted bark, a lightweight growth medium with fungicidal properties. Plant Dis. 64:143–147.

Sneh, B., Burpee, L., and Ogoshi, A. 1991. Identification of *Rhizoctonia* Species. American Phytopathological Society, St. Paul, Minn.

Stephens, C. T., and Stebbins, T. C. 1985. Control of damping-off pathogens in soilless container media. Plant Dis. 69:494–496.

Stephens, C. T., Herr, L. F., Schmitthenner, A. F., and Powell, C. C. 1982. Characterization of *Rhizoctonia* isolates associated with damping-off of bedding plants. Plant Dis. 66:700–703.

Stephens, C. T., Herr, L. F., Schmitthenner, A. F., and Powell, C. C. 1983. Sources of *Rhizoctonia solani* and *Pythium* spp. in a bedding plant greenhouse. Plant Dis. 67:272–275.

Weber, G. F., and Roberts, D. A. 1951. Silky threadblight of *Elaeagnus pungens* caused by *Rhizoctonia ramicola* n. sp. Phytopathology 41:615–621.

Wehlburg, C., and Cox, R. S. 1966. Rhizoctonia leaf blight of azalea. Plant Dis. Rep. 50:354–355.

CHAPTER 18

Austin K. Hagan ▪ Auburn University, Auburn, Alabama
Gary W. Simone ▪ University of Florida, Gainesville

Southern Blight

Southern blight was first described as a disease of tomato in Florida in the late 1800s. Subsequently, the causal fungus, *Sclerotium rolfsii,* has been reported as an aggressive pathogen on over 500 species of annual, perennial, and woody plants. It is distributed worldwide in temperate, subtropical, and tropical regions wherever summer weather patterns are hot and wet. In the United States, outbreaks of southern blight in container- and field-grown shrubs and trees are largely confined to the Southeast, California, and Arizona, but sporadic outbreaks have also been noted in bulb crops and woody shrubs in the Pacific Northwest and North Central states.

Southern blight is much more prevalent in plants grown in well-aerated soil or potting medium with high sand content than in those produced in poorly drained clay or silt soils. Juvenile or mature woody trees and shrubs that lack a thickened, corky bark in the area of the collar-root interface are most susceptible to attack by *S. rolfsii.* Once the protective bark begins to form, susceptible shrubs and trees become virtually immune to southern blight unless they are mechanically damaged or injured by herbicide drift.

Among the nursery crops that are hosts of *S. rolfsii* are ajuga, apple, aucuba, azalea, butterfly iris (*Dietes* and *Moraea* spp.), caladium, crabapple, winter daphne, *Ficus,* hydrangea, liriope, loquat, peach, phlox, pittosporum, quince, Russian olive, schefflera, and black walnut. In Alabama, severe outbreaks of southern blight in container-grown butterfly-bush (*Buddleia davidii*), forsythia, hosta, and Prague viburnum (*Viburnum* × *pragense*) have recently been observed. Southern stem rot does not have nearly as much economic impact as Phytophthora root rot on the production of nursery crops, but significant losses can occur in some container- and field-grown woody shrubs and trees.

Symptoms

In nearly all nursery crops, the most noticeable symptom of southern blight is the sudden wilting or flagging of leaves and succulent shoots, which is usually followed by rapid plant death. In early stages of the disease, foliage may recover somewhat at night, but within a few days wilting becomes permanent, and the plant dies. Succulent plants are rapidly killed, but a month or more may pass before small trees such as apple succumb to the disease. Colonization of the stem or root collar by *S. rolfsii* usually occurs an inch or two above or below the surface of the medium or soil. Brown, water-soaked, and sometimes sunken girdling cankers often mark the site of fungal colonization on the stem or root collar.

The superficial, fanlike mycelial mat of *S. rolfsii* is typically found on the surface of the soil or medium, fallen leaves or other crop debris, and the lower stem of the target plant. Tightly spaced container-grown ornamentals may produce high-humidity microclimates, which allow *S. rolfsii* mycelia to ascend plant stems to a height of 8–10 inches above the crown. At times, the mycelial mat may be seen before the stem or root collar has been colonized. When the surface dries, white mycelial strands or threads of *S. rolfsii* may be found intertwined among plant roots in moist soil to a depth of 2–3 inches.

Clusters of round sclerotia, which are the size of a mustard seed, form on the mycelial mat. They are initially white but turn medium to dark brown within a few days. Over time, the white mycelial mat disintegrates when conditions no longer favor pathogen activity, but the diagnostic sclerotia remain.

S. rolfsii biotype *delphinii* occurs on bulb crops. Its sclerotia are large, not uniformly shaped, and purple brown.

Disease Development

Sclerotia are the primary means of survival of *S. rolfsii.* They can persist in soil or potting media for several years. The mycelia of this fungus are viable in the soil and on host tissues for only a few weeks. The sexual stage (teleomorph), *Athelia rolfsii,* apparently has little or no influence on the spread of southern blight.

S. rolfsii is a native soil resident across a sizable portion of the United States, and it is easily introduced into container and field nurseries as sclerotia or mycelia on transplants or rootstock, in soil or partially decomposed plant debris, on tools or tillage equipment, and by flowing water. In container nurseries in the southern United States, sclerotia-contaminated soil mixed with bark media and recycled containers are common sources of *S. rolfsii.* The pathogen can also be introduced into propagation beds on infested or infected cuttings taken close to the ground.

Sclerotial germination is often triggered by volatile organic compounds released by decaying plant material. The fungus often colonizes fallen leaves and other debris prior to attacking nearby host plants.

Southern blight occurs over a wide range of temperatures and moisture conditions, but it is typically considered a hot-, humid-weather disease. Outbreaks of southern blight in nursery crops usually occur from late spring to early fall, when day and night temperatures exceed 86 and 70°F, respectively. Sclerotial germination and pathogen growth are favored by high temperatures during the day. Southern blight may develop at slightly lower temperatures, but the elapsed time between infection and plant death is longer and levels of plant mortality are often lower. Heavy losses to southern blight have often occurred during extended periods of excessive rainfall or overwatering. Explosive outbreaks of the disease in field plantings have been noted when heavy rains have followed a period of hot, dry weather.

Disease Management

Sanitation and cultural practices that exclude the pathogen are critical to avoiding significant losses of nursery crops susceptible to southern blight. If *S. rolfsii* is accidentally introduced, timely fungicide drenches are an effective option for disease control.

To minimize the risk of accidentally introducing this pathogen into container-grown nursery crops, store bulk bark media and containers on a raised asphalt or concrete pad and not on bare ground. Do not locate bulk storage areas on sites that are prone to flooding. If possible, place bagged potting media in a covered shed with a gravel or concrete floor. Never add nonsterile soil or recycled potting medium to a fresh bark medium. Recycled containers must be carefully rinsed of any old potting media and then disinfested. When soil is added to potting medium, it must be sterilized, preferably by steam, before it is mixed with other components. Clear all fallen leaves from container production areas. Collect and discard symptomatic container and field stock off-site. Locate dump piles and compost piles well away from production areas.

Container production beds must be crowned to prevent flooding or puddling of water, then covered with black plastic or weed barrier, and then topped with crushed rock or a similar coarse material. A drainage system should be designed to rapidly channel runoff from irrigation and heavy rains away from production beds. To avoid over- or underirrigating, block plant material by container size and water need. Account for rainfall when scheduling irrigation. Fertilize according to need, and maintain a slightly acid soil pH.

Rooted liners straight from propagation are very vulnerable to southern blight. When liner trays are moved outside and placed on ground beds, widespread disease can develop from soilborne inoculum. Nursery bed areas with a history of *S. rolfsii* should be raked or blown free of residual media and plant debris. Where possible these beds should be fumigated under tarps to eradicate the fungus at the level of the ground cloth and below. An alternative to fumigation is the elevation of liner trays of sensitive plant species off the ground cloth, on pallets, PVC pipe, or similar support materials. This prevents direct root contact with the infested ground bed through continual air pruning of roots from the liner trays.

Particularly sensitive plant species can sometimes benefit from a modification of the microclimate to minimize the incidence and spread of *S. rolfsii.* The prostrate-growing ground cover *Ficus pumila* (grown in zone 9 and higher) is particularly damaged by *S. rolfsii.* Grower modifications in production have led to staking or trellising this species, to raise the canopy above the container surface, thus hastening evaporation from the medium and drastically reducing the relative humidity of the container microenvironment, in which moist air would otherwise be trapped, because of the prostrate growth habit of the plant.

For effective control in susceptible nursery crops, begin fungicide drenches in the spring, just prior to the onset of hot, humid weather, and repeat treatments until temperatures are lower again in late summer or early fall. Refer to the product label for optimum application rates and treatment intervals for control of southern blight. For additional information concerning the control of southern stem rot of nursery crops, contact your county extension agent or state specialist responsible for diseases of nursery crops.

REFERENCES

Aycock, R. 1966. Stem rot and other diseases caused by *Sclerotium rolfsii.* N.C. Agric. Exp. Stn. Tech. Bull. 174.

Hagan, A. K., and Olive, J. 1995. Screening fungicides for the control of southern stem rot on aucuba. 1994. Fungic. Nematic. Tests 50:381.

McRitchie, J. J. 1983. Southern blight of ajuga. Fla. Dep. Agric. Consumer Serv. Plant Pathol. Circ. 252.

Sinclair, W. A., Lyon, H. H., and Johnson, W. T. 1987. Diseases of Trees and Shrubs. Comstock Publishing Associates, Cornell University Press, Ithaca, N.Y.

CHAPTER

19

Cynthia L. Ash ▪ American Phytopathological Society, St. Paul, Minnesota

Verticillium Wilt

Verticillium wilt, caused by the soilborne fungus *Verticillium dahliae,* is found worldwide in cultivated soils. Early literature often cited *V. albo-atrum* as the pathogen, but today it is considered a separate species, and *V. dahliae* is more commonly reported in woody hosts.

Verticillium persists indefinitely in the soil in the form of resistant structures called microsclerotia. Plant exudates from root tips and other intact or injured areas of the root stimulate nearby microsclerotia to germinate. The fungus penetrates the epidermis, colonizes cortical tissues, moves into the xylem, and progresses upward in the plant. In the vascular tissue, chemical and physical responses by the host attempt to limit movement of the pathogen. Blockage of the xylem results in reduced water flow, and symptoms develop above the point of damage.

Plant species which limit the movement of *Verticillium* in vascular tissue exhibit some degree of resistance. However, the conditions under which *Verticillium* becomes a serious pathogen or is limited in its movement are poorly understood in woody plants. The disease can be symptomless in trees for several years before causing serious wilting and dieback.

Host Range

V. dahliae has a very wide host range, including annuals, herbaceous perennials, vegetables, shrubs, and trees. As early as 1914, it was reported in weeds, such as cocklebur (*Xanthium* sp.) and velvetleaf (*Abutilon* sp.).

Sinclair et al. (1987) listed the following woody hosts as susceptible to *Verticillium:* ash, avocado, azalea, barberry, boxwood, brambles, buckeye, camphor tree, carob, carrotwood, catalpa, ceanothus, cherry and other stone fruit trees, cork tree, creosote bush, currant and gooseberry, daphne, elder, elm, erigonum, weeping fig, flannelbush, golden-rain tree, grapevine, guayule, heath, hebe, hibiscus, honeysuckle, hop seed bush, horse-chestnut, India-hawthorn, Japanese pagoda tree, jasmine, Kentucky coffee tree, lilac, black locust, magnolia, maple, nandina, olive, osage orange, osmanthus, peony, pepper tree, persimmon, photinia, pistache, privet, rabbit-brush, redbud, rock-rose, rose, Russian olive, sage brush, salt bush, sassafras, serviceberry, smoke tree, spirea, sumac, tree-of-heaven, tulip tree, tupelo, viburnum, wiegela, winter fat, and yellowwood.

Sinclair et al. (1987) listed the following as resistant or immune, as determined by testing or practical experience: all gymnosperms, all monocots, and, among the dicots, apple and crabapple, mountain ash, beech, birch, boxwood, butternut, ceanothus, chestnut, citrus, dogwood, eucalyptus, firethorn, sweet gum, hackberry, hawthorn, certain species of hebe, hickory, holly, katsura tree, California laurel, linden, honey locust, manzanita, mulberry, oak, oleander, pawpaw, pear, pecan, plane tree and sycamore, poplar, quince and flowering quince, rhododendron, certain species of rock rose, sugarberry, walnut, willow, and Japanese zelkova.

Symptoms

In the nursery, symptoms tend to show up in 2- to 3-inch-caliper trees but may also be present in seedling stock or young grafted liners. Symptoms are variable among species and even within species and are affected by environmental and soil conditions. High summer temperatures have been reported to eliminate visible field symptoms. Seasonal variation in sap nutrient levels may be a factor in susceptibility. Top chlorosis affecting 5% of 11,000 green and white ash seedlings and young trees in a nursery and nearly half of 5,000 ash trees in another nursery with severe scorch and branch dieback was attributed to *V. dahliae* (Worf et al., 1994).

Sudden wilting and scorching or defoliation of a single branch or portion of the tree are often the first visible symptoms. In some species, such as green ash, the foliage is a light green to yellow, often mottled, prior to becoming necrotic, and defoliation occurs without wilting. Ash may also drop leaflets before noticeable yellowing. Leaves may be abnormally red or yellow, smaller in size, and exhibit patchy interveinal or marginal scorching. Branches that are symptomatic early in the growing season may give rise to healthy new growth. Occasionally, some species develop elongate dead areas of bark on diseased branches or trunks.

Infected plants may not exhibit symptoms annually, but the infection reduces plant vitality. Symptoms associated with chronic infections include reduced branch length and caliper growth, fewer leaves, leaves with mottled color patterns, failure of buds to break in the spring, dieback, the presence of secondary insects and diseases, and wilting. Partial or complete tree death can occur following acute or chronic symptoms.

Brown, gray, or green streaking may be present in the xylem of affected branches or in the lower portion of the tree supporting those branches. Infected root tissue will also have streaking. Green ash does not typically develop discolored xylem, making field diagnosis difficult. Vascular discoloration is fairly indicative of Verticillium wilt, but other agents can cause it. Laboratory tests are required to confirm *Verticillium* as the causal agent; the tests require branches with vascular streaking or symptomatic leaves and can take one to four weeks. Check with your local lab for instructions on sample collection before collecting and submitting samples.

Virulence

Isolates of *V. dahliae* vary in the severity of the disease they cause. Isolates belonging to some vegetative compatibility groups tend to be more virulent in some hosts. Recent research indicates that the isolates commonly associated with woody ornamentals are in a group different from those found in potatoes and other herbaceous plants, suggesting that established populations in nurseries are an important source of inoculum of the pathogen in urban areas.

Environmental conditions and host vitality also affect the severity of Verticillium wilt. Plants under water stress or other stresses are more susceptible to infection. In some studies more infection occurred under wet conditions, but more wilt symptoms occurred under drought.

Disease Management

V. dahliae can be a serious problem in nursery fields, and measures should be taken to reduce its introduction into the field or its inoculum level in the field. Increasing disease incidence and severity can be correlated with increasing inoculum levels in the soil. Rotate all fields on a regular basis. Do not plant susceptible plants in fields where *Verticillium* is known to be present in the soil. Eliminate weeds which allow *Verticillium* to persist and increase inoculum levels in the soil. Have a soil test run by a university lab or private company that assays soils to detect the presence of *Verticillium*. This is especially important before buying new land. However, because of mixtures of pathotypes, inoculum density determinations can be difficult to interpret.

Soil type can affect Verticillium wilt development. In general, the disease has been reported to be more severe in sandy loam, loam, and clay soils and in soils high in organic mater. Heavy soils tend to be cooler and more conducive to infection. However, frequent irrigation of sandy soils apparently makes them as favorable for Verticillium wilt as clay soils.

Addition to the soil of materials with a high carbon–nitrogen ratio may reduce microsclerotial germination, on account of the immobilization of nitrogen during microbial decomposition. Conflicting information has been reported on the effects of applications of macro- and micronutrients on disease. The interaction of nutrients with numerous soil factors makes generalizations difficult. It appears that low to moderate levels of nitrogen fertilization have the best results. Adequate levels of potassium are necessary for disease defense, but that defense can be overcome by high levels of inoculum.

Keep good disease records, and avoid stock from wholesalers with a history of Verticillium wilt. Be conscious of spreading *Verticillium* into the landscape in infested soil on machinery and plant materials. Nematodes can also travel with soil, and several species, notably the lesion nematode and root-knot nematode, can increase the damage caused by Verticillium wilt.

Resistant or tolerant species and cultivars are the most promising way of managing this disease. Fungicides are not generally effective or practical in a nursery setting. Soil fumigants, where they are an option, may offer some control.

REFERENCES

Bedwell, J. L., and Childs, T. W. 1938. Verticillium wilt of maple and elm in the Pacific Northwest. Plant Dis. Rep. 22:22–23.

Bom, G. L. 1974. Root infection of woody hosts with *Verticillium albo-atrum*. Ill. Nat. Hist. Surv. Bull., vol. 31, art. 6, pp. 209–248.

Chen, W. 1994. Vegetative compatibility groups of *Verticillium dahliae* from ornamental woody plants. Phytopathology 84:214–219.

Pennypacker, B. W. 1989. The role of mineral nutrition in the control of Verticillium wilt. Pages 33–45 in: Soilborne Plant Pathogens: Management of Diseases with Macro- and Microelements. A. W. Engelhard, ed. American Phytopathological Society, St. Paul, Minn.

Schnathorst, W. C. 1981. Life cycle and epidemiology of *Verticillium*. Pages 81–111 in: Fungal Wilt Diseases of Plants. M. E. Mace, A. A. Bell, and C. H. Beckman, eds. Academic Press, New York.

Schrieber, L. R., and Mayer, J. S. 1992. Seasonal variation in susceptibility and in internal inoculum densities in maple species inoculated with *Verticillium dahliae*. Plant Dis. 76:184–187.

Sinclair, W. A., Smith, L. L., and Larsen, A. O. 1981. Verticillium wilt of maples: Symptoms related to movement of the pathogen in stems. Phytopathology 71:340–345.

Sinclair, W. A., Lyon, H. H., and Johnson, W. T. 1987. Diseases of Trees and Shrubs. Comstock Publishing Associates, Cornell University Press, Ithaca, N.Y.

Townsend, A. M., Schreiber, L. R., Hall, T. J., and Bentz, S. E. 1990. Variation in response of Norway maple cultivars to *Verticillium dahliae*. Plant Dis. 74:44–46.

Worf, G. L., Spear, R. N., and Heimann, M. F. 1994. *Verticillium*-induced scorch and chlorosis in ash. J. Environ. Hortic. 12(3):124–130.

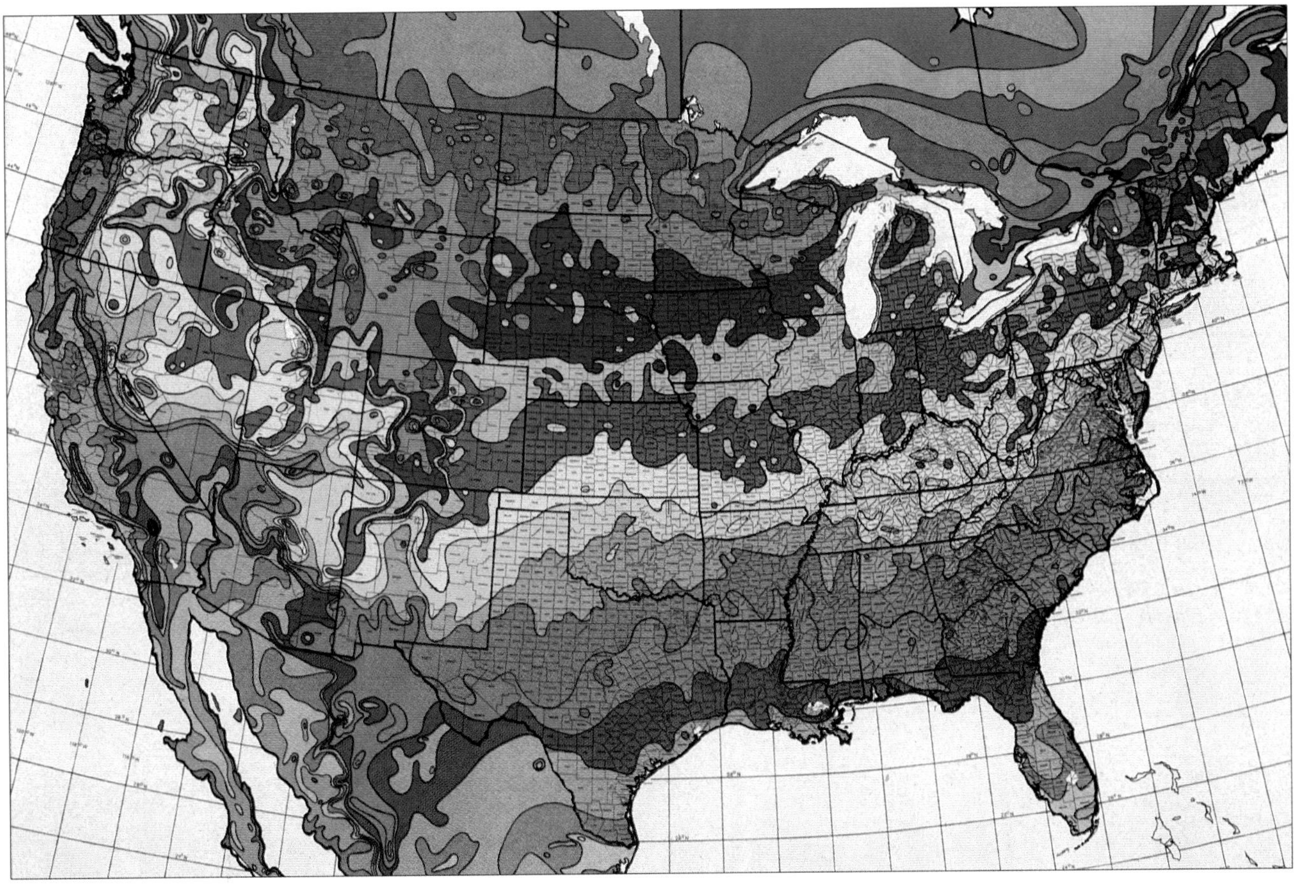

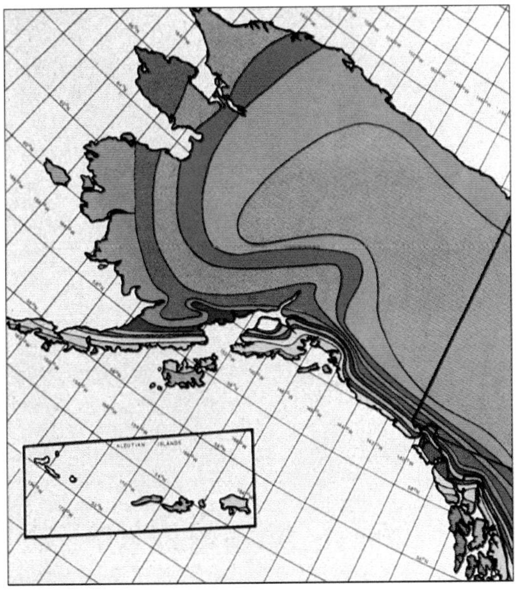

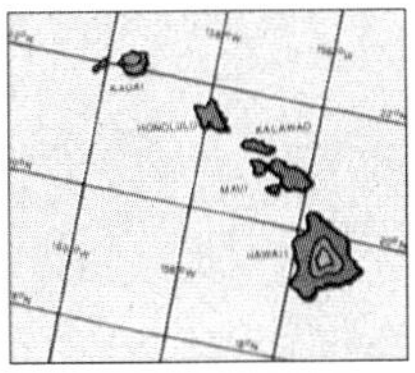

Average Annual Minimum Temperature

Temperature (°C)	ZONE	Temperature (°F)
-45.6 and Below	1	Below -50
-42.8 to -45.5	2a	-45 to -50
-40.0 to -42.7	2b	-40 to -45
-37.3 to -40.0	3a	-35 to -40
-34.5 to -37.2	3b	-30 to -35
-31.7 to -34.4	4a	-25 to -30
-28.9 to -31.6	4b	-20 to -25
-26.2 to -28.8	5a	-15 to -20
-23.4 to -26.1	5b	-10 to -15
-20.6 to -23.3	6a	-5 to -10
-17.8 to -20.5	6b	0 to -5
-15.0 to -17.7	7a	5 to 0
-12.3 to -15.0	7b	10 to 5
-9.5 to -12.2	8a	15 to 10
-6.7 to -9.4	8b	20 to 15
-3.9 to -6.6	9a	25 to 20
-1.2 to -3.8	9b	30 to 25
1.6 to -1.1	10a	35 to 30
4.4 to 1.7	10b	40 to 35
4.5 and Above	11	40 and Above

USDA Plant Hardiness Zone Map

(R. Jordan, U.S. National Arboretum)

1. Severe infection of Madagascar periwinkle (*Catharanthus roseus*) by a phytoplasma of the elm yellows group, resulting in dwarfed shoots with light green to yellow leaves and absence of flowers (right), and an asymptomatic branch of the same plant (left). (Reproduced, by permission, from W. A. Sinclair)

2. Ash yellows symptoms: dwarfed shoots with simple, chlorotic leaves growing from the base of a green ash sapling (*Fraxinus pennsylvanica*). (Reproduced, by permission, from W. A. Sinclair)

3. Lilac witches'-broom, caused by phytoplasmas of the ash yellows group, in a hybrid lilac, *Syringa* × *prestoniae* 'Hiawatha,' with dwarfed shoots growing from axillary buds (top left). (Reproduced, by permission, from W. A. Sinclair)

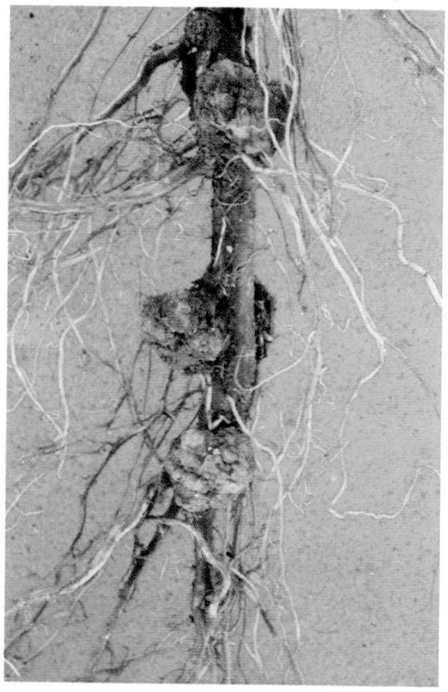

4. Crown gall symptoms on roots, caused by *Agrobacterium tumefaciens*. (A. K. Hagan)

5. Damping-off of seedlings (G. W. Moorman)

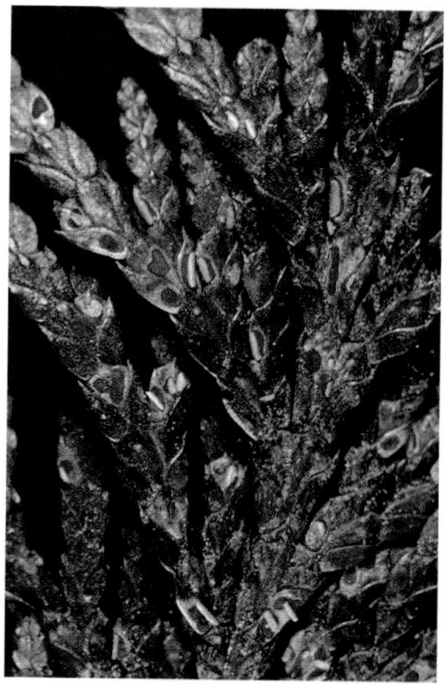

6. Leaf blight of arborvitae, caused by *Didymascella thujina*. (R. S. Byther)

7. Tip blight of arborvitae, caused by *Coryneum berckmansii*. (R. S. Byther)

8. Ash anthracnose, caused by *Gnomoniella fraxini* (asexual stage, *Discula fraxinea*). (C. L. Ash)

9. Verticillium wilt of ash, caused by *Verticillium dahliae*. (C. L. Ash)

10. Ash canker. (C. L. Ash)

11. Phytophthora root rot of azalea (right), caused by *Phytophthora cinnamomi*, and a healthy plant (left). (D. M. Benson)

12. Phytophthora root rot of azalea, caused by *Phytophthora cinnamomi*. (R. K. Jones)

13. Black sclerotia of *Ovulinia azaleae* in a blighted azalea flower. (R. K. Jones)

14. Web blight of azalea, caused by *Rhizoctonia solani*. (R. K. Jones)

15. Leaf gall of azalea, caused by *Exobasidium vaccinii*. (R. K. Jones)

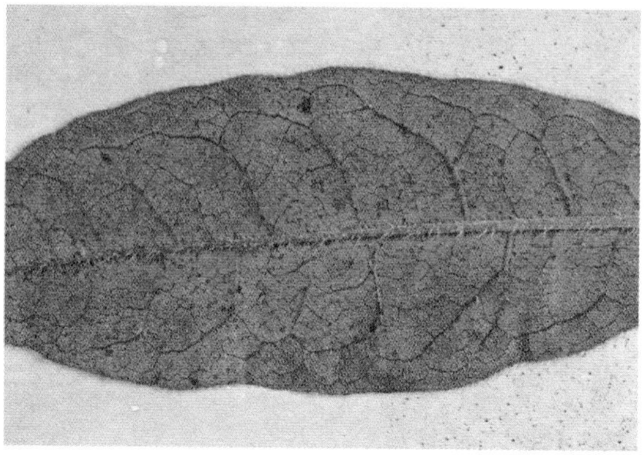

16. Signs of the rust fungus *Pucciniastrum vaccinii* on the underside of a leaf of Exbury azalea. (A. Windham)

17. Anthracnose of river birch, caused by *Discula betulina*. (R. K. Jones)

18. Root-knot nematode (*Meloidogyne incognita*) symptoms in Japanese boxwood. (R. K. Jones)

19. Lesion nematode (*Pratylenchus vulnus*) symptoms in American boxwood. (R. K. Jones)

20. Lesion nematode (*Pratylenchus vulnus*) symptoms in American boxwood (black label), and a healthy plant (red label). (D. M. Benson)

21. Symptoms of cucumber mosaic virus and alfalfa mosaic virus on butterfly-bush (*Buddleia davidii*). (S. W. Scott)

22. Camellia flower blight, caused by *Ciborinia camelliae*. (R. K. Jones)

23. Apothecium of *Ciborinia camelliae*. (R. K. Jones)

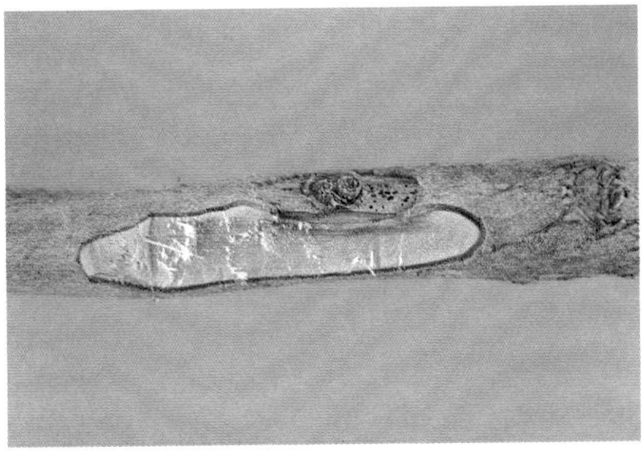

24. Canker of camellia, caused by *Glomerella cingulata*. (R. K. Jones)

25. Camellia yellow mottle symptoms. (R. K. Jones)

26. Signs of the powdery mildew fungus *Erysiphe lagerstroemiae* on crapemyrtle. (A. S. Windham)

27. Cercospora leaf spot of crapemyrtle, caused by *Cercospora lythracearum*. (A. K. Hagan)

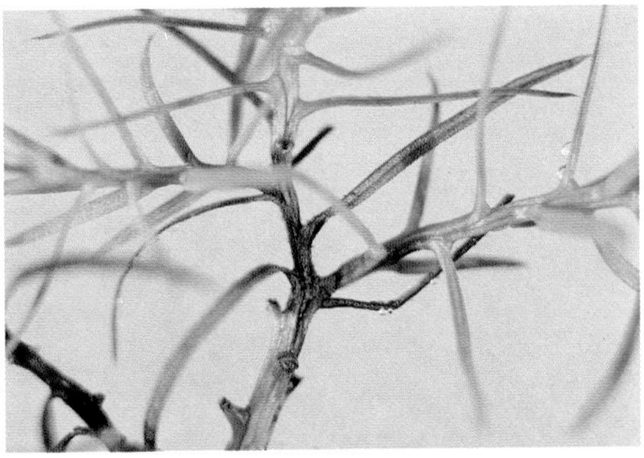

28. Cercospora blight of Japanese cryptomeria, caused by a *Cercospora* sp. (R. L. Wick)

29. Anthracnose of flowering dogwood, caused by *Discula destructiva*. (A. S. Windham)

30. Septoria leaf spot of flowering dogwood, caused by a *Septoria* sp. (R. K. Jones)

31. Powdery mildew fungus, *Oidium* sp., on flowering dogwood. (A. S. Windham)

32. Powdery mildew of Cherokee Sunset flowering dogwood, caused by an *Oidium* sp. (M. Daughtrey)

33. Spot anthracnose of flowering dogwood, caused by *Elsinoe corni*. (R. K. Jones)

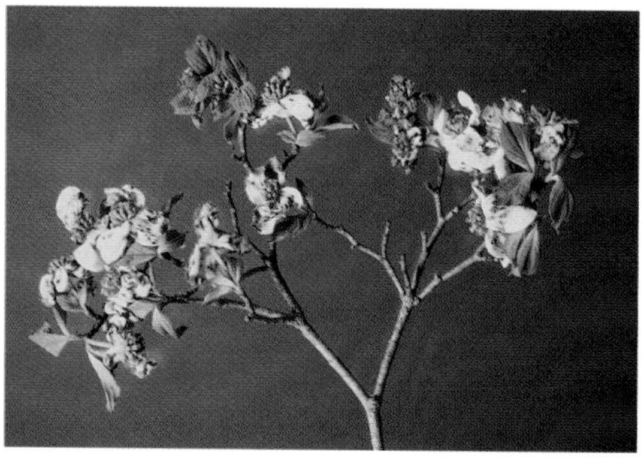

34. Spot anthracnose of flowering dogwood, caused by *Elsinoe corni*. (R. K. Jones)

35. Canker of flowering dogwood (cause unknown). (R. K. Jones)

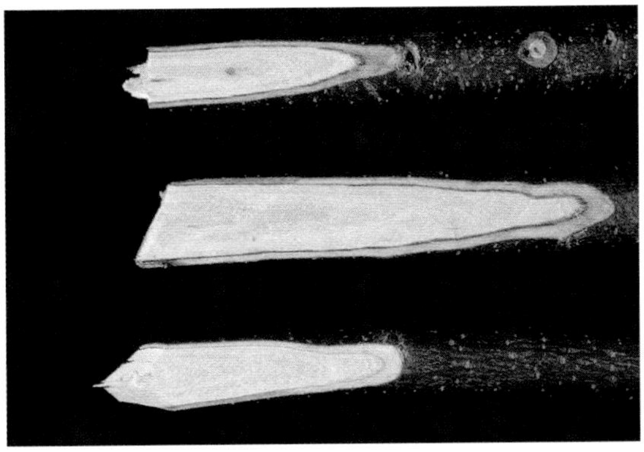

36. Dutch elm disease symptoms, caused by *Ophiostoma ulmi,* in stems of American elm (top and middle), with a healthy stem (bottom). (R. K. Jones)

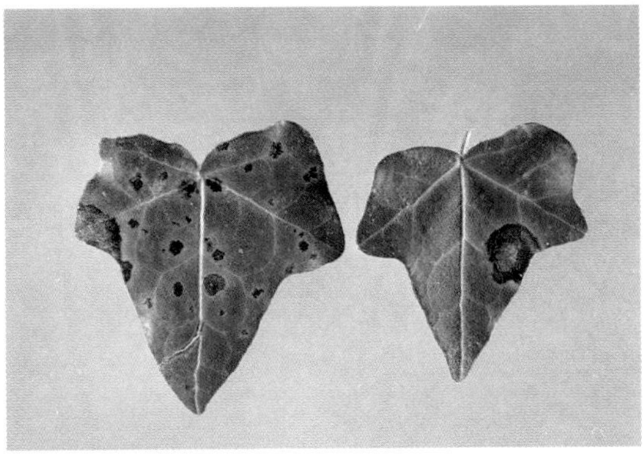

37. Anthracnose of English ivy, caused by *Colletotrichum trichellum.* (R. K. Jones)

38. Bacterial leaf spot of English ivy, caused by *Xanthomonas campestris* pv. *hederae.* (A. S. Windham)

39. Canker and leaf spot of euonymus, caused by *Colletotrichum gloeosporioides.* (R. L. Wick)

40. Powdery mildew fungus (*Oidium* sp.) on euonymus. (J. W. Olive)

41. Crown gall of euonymus, caused by *Agrobacterium tumefaciens.* (M. Daughtrey)

42. Leaf spot of euonymus, caused by a *Cercospora* sp. (R. L. Wick)

43. Leaf spot of fatsia caused by an *Alternaria* sp. (G. W. Simone)

44. Leaf spot of fatsia, caused by a *Xanthomonas* sp. (A. R. Chase)

45. Root rot of fatsia, caused by a *Phytophthora* sp. (G. W. Simone)

46. Fusarium hypocotyl rot of fir seedlings, caused by a *Fusarium* sp. (G. A. Chastagner)

47. Phytophthora root rot of Fraser fir, caused by *Phytophthora cinnamomi*. (R. K. Jones)

48. Phytophthora root rot of Noble fir, caused by *Phytophthora cinnamomi*. (G. A. Chastagner)

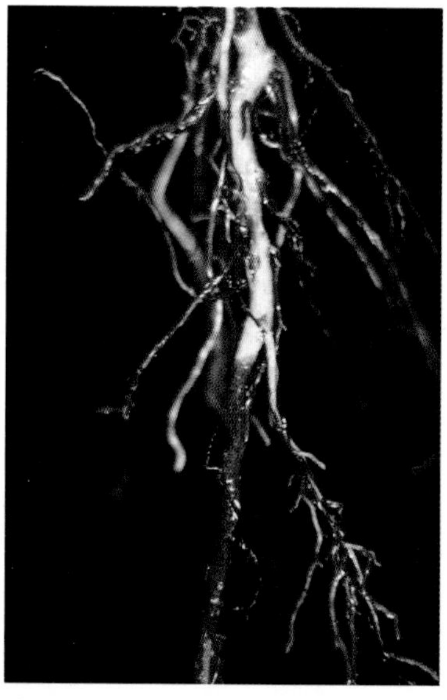

49. Root symptoms of Phytophthora root rot of Noble fir, caused by *Phytophthora cinnamomi*. (G. A. Chastagner)

50. Botrytis blight of Fraser fir, caused by *Botrytis cinerea,* affecting new shoots. (G. A. Chastagner)

51. Botrytis blight of fir, caused by *Botrytis cinerea.* (G. A. Chastagner)

52. Scab of crabapple, caused by *Venturia inaequalis.* (A. S. Windham)

53. Fire blight of crabapple, caused by *Erwinia amylovora.* (A. S. Windham)

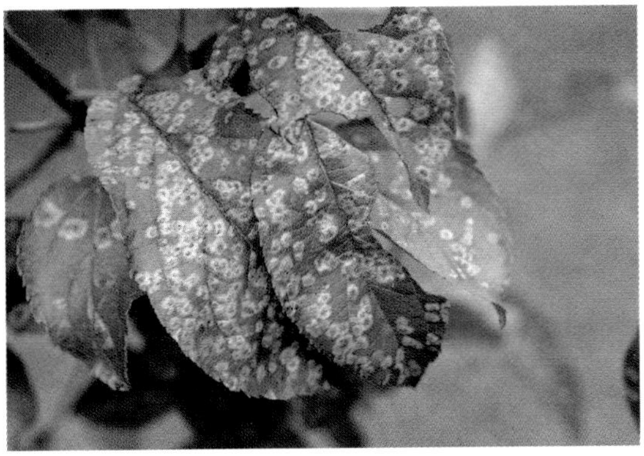

54. Cedar-apple rust of crabapple, caused by *Gymnosporangium juniperi-virginianae.* (A. S. Windham)

55. Leaf spot of Bradford pear, caused by *Entomosporium mespili.* (A. K. Hagan)

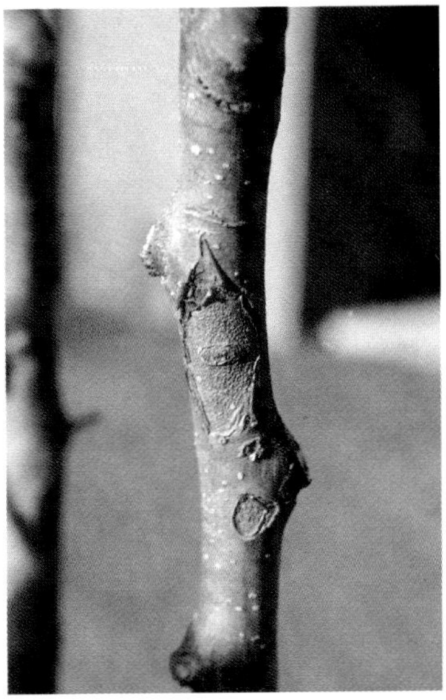

56. Canker of Bradford pear, caused by *Botryosphaeria dothidea.* (A. K. Hagan)

57. Southern blight of forsythia, caused by *Sclerotium rolfsii.* (A. K. Hagan)

58. Phytophthora root rot of gardenia: three plants inoculated with isolates of *Phytophthora parasitica* (right) and a healthy plant (left). (R. L. Wick)

59. Bacterial leaf spot of hibiscus, caused by *Xanthomonas campestris* pv. *malvacearum.* (A. R. Chase)

60. Black root rot on Japanese holly, caused by *Thielaviopsis basicola.* (A. S. Windham)

61. Root symptoms of black root rot of Japanese holly, caused by *Thielaviopsis basicola.* (R. K. Jones)

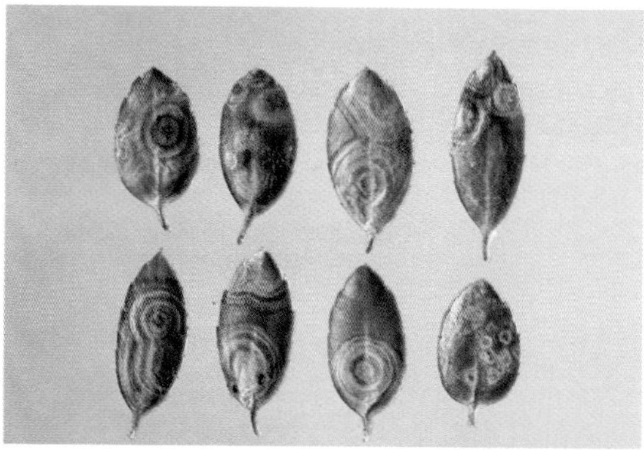

62. Web blight of Japanese holly, cultivar Helleri, caused by *Rhizoctonia solani.* (R. K. Jones)

63. Web blight of Japanese holly, cultivar Helleri, caused by *Rhizoctonia solani.* (R. K. Jones)

64. Root-knot nematode (*Meloidogyne incognita*) symptoms in Japanese holly, cultivar Rotundifolia (foreground), with a healthy plant (red label). (R. K. Jones)

65. Winter injury in Japanese holly. (R. K. Jones)

66. Leaf spot of Indian hawthorn, caused by *Entomosporium mespili*. (R. K. Jones)

67. Phomopsis tip blight of eastern red cedar seedlings, caused by *Phomopsis juniperovora*. (N. A. Tisserat)

68. Kabatina tip blight of juniper, caused by *Kabatina juniperi*. (N. A. Tisserat)

69. Cedar-apple rust fungus, *Gymnosporangium juniperi-virginianae*, on juniper. (N. A. Tisserat)

70. Lesion nematode (*Pratylenchus* sp.) symptoms in Blue Rug juniper (foreground), with a healthy plant (rear). (R. K. Jones)

71. Seiridium canker of Leyland cypress, caused by *Seiridium unicorne*. (A. S. Windham)

72. Leaf spot of ligustrum, caused by a *Cercospora* sp. (G. W. Simone)

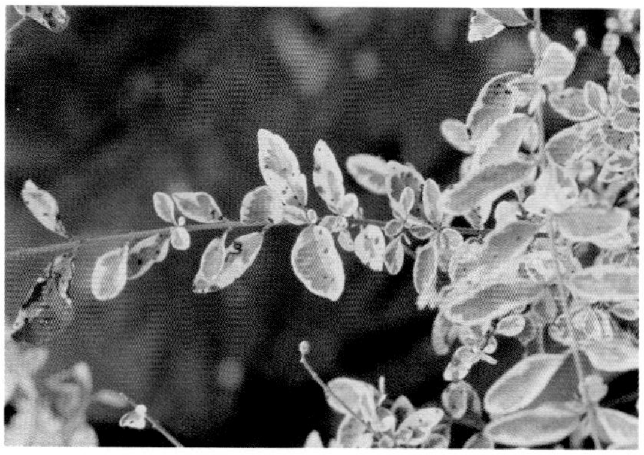

73. Leaf spot of *Ligustrum sinense,* caused by a *Corynespora* sp. (G. W. Simone)

74. Edema of ligustrum. (G. W. Simone)

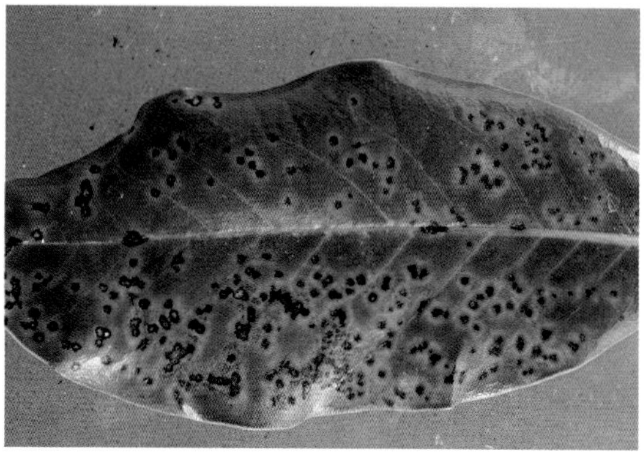

75. Bacterial leaf spot of southern magnolia, caused by a *Pseudomonas* sp. (A. K. Hagan)

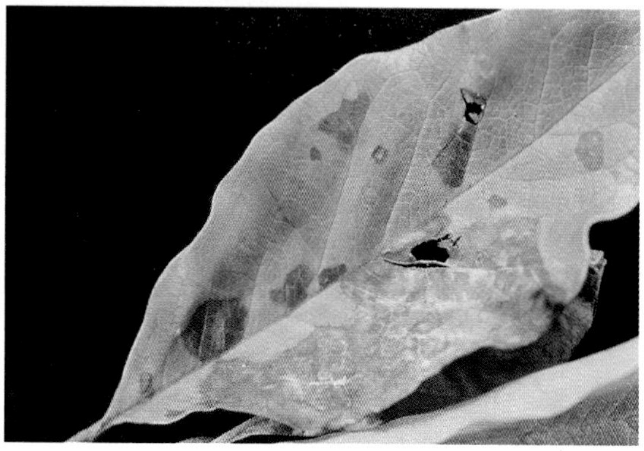

76. Bacterial leaf spot of Japanese magnolia, caused by *Pseudomonas syringae.* (A. K. Hagan)

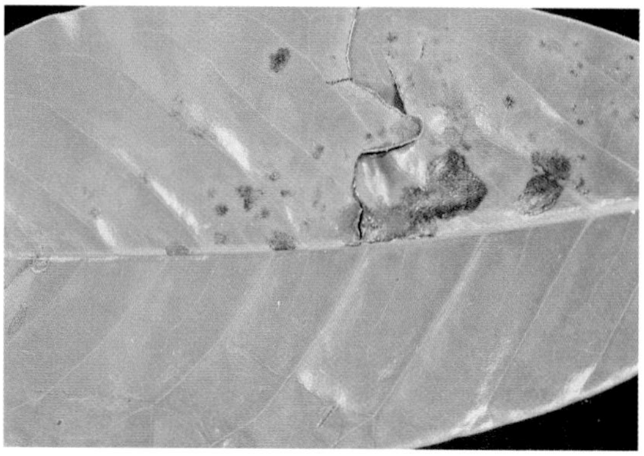

77. Anthracnose of southern magnolia, caused by a *Colletotrichum* sp. (A. K. Hagan)

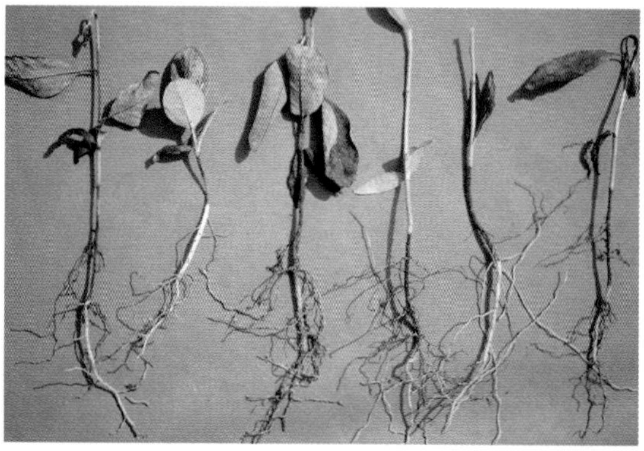

78. Damping-off of sweet bay magnolia seedlings, caused by *Rhizoctonia solani.* (A. K. Hagan)

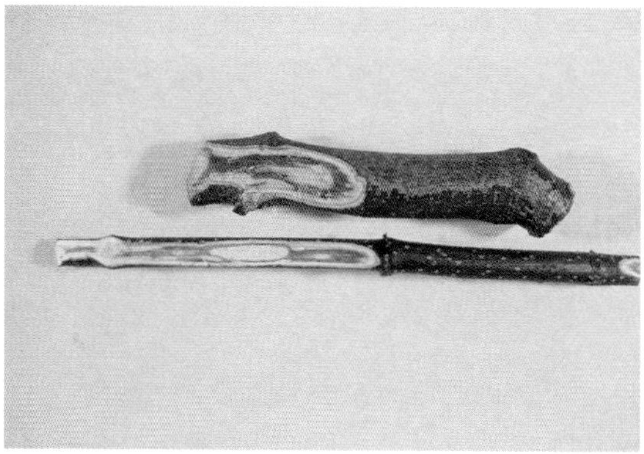

79. Streaking of vascular tissue of maple due to Verticillium wilt, caused by *Verticillium dahliae.* (J. W. Pscheidt)

80. Nectria canker of Crimson King maple, caused by *Nectria cinnabarina.* (J. W. Pscheidt)

81. Canker caused by *Cytospora chrysosperma* on a Norway maple stem. (Reproduced, by permission, from G. W. Hudler)

82. Canker caused by a *Cryptosporiopsis* sp. in red maple at the site of tree cricket injury. (Reproduced, by permission, from G. W. Hudler)

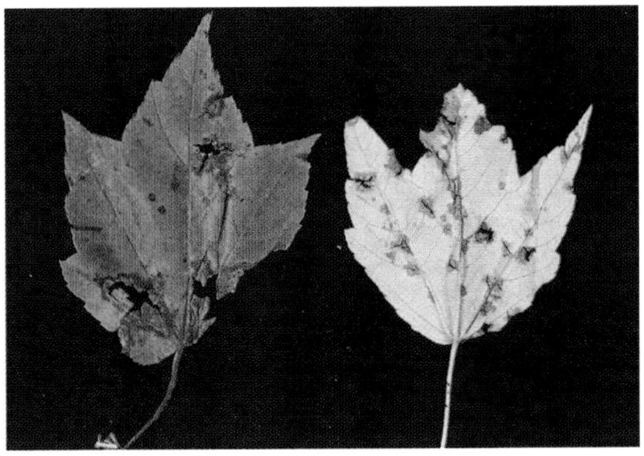

83. Anthracnose of maple, caused by a *Discula* sp. (R. K. Jones)

84. Leaf spot of Japanese maple, caused by *Kabatiella apocrypta*. (M. Daughtrey)

85. Tar spot of maple, caused by a *Rhytisma* sp., with signs of the pathogen. (Reproduced, by permission, from G. W. Hudler)

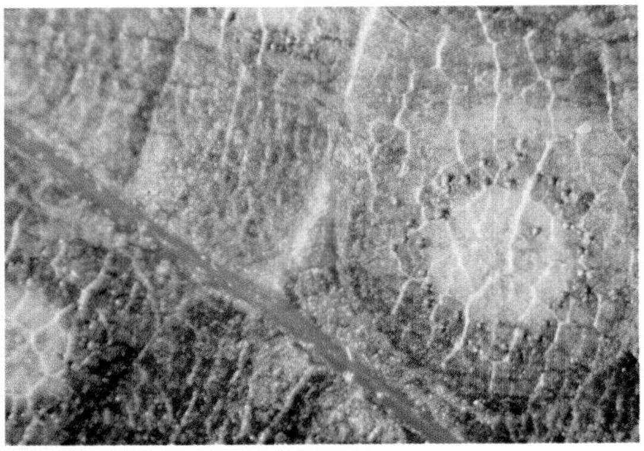

86. Leaf spot of maple, caused by a *Phyllosticta* sp., with signs of the pathogen. (M. Daughtrey)

87. Leaf spot of nandina, caused by a *Colletotrichum* sp. (G. W. Simone)

88. Leaf spot of nandina, caused by *Pseudocercospora nandinae*. (G. W. Simone)

89. Virus symptoms in dwarf nandina. (R. K. Jones)

90. Web blight of nandina, caused by *Rhizoctonia solani*. (G. W. Simone)

91. Bacterial knot of oleander, caused by *Pseudomonas syringae* pv. *savastanoi*. (G. W. Simone)

92. Dieback of oleander, caused by *Botryosphaeria dothidea*. (G. W. Simone)

93. Leaf spot of oleander, caused by a *Septoria* sp. (G. W. Simone)

94. Web blight of oleander, caused by *Rhizoctonia solani.* (G. W. Simone)

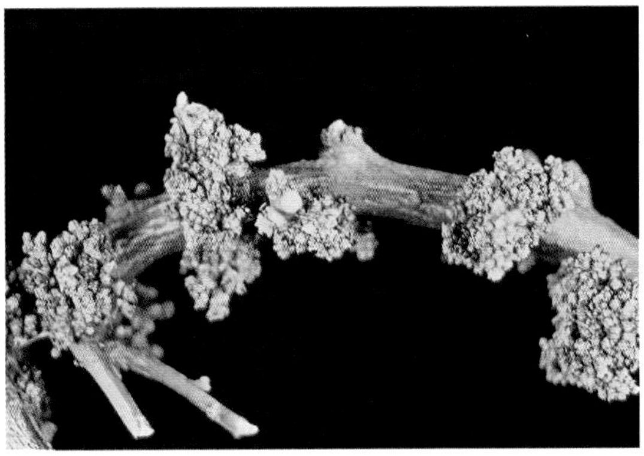

95. Gall on oleander, caused by a *Sphaeropsis* sp. (G. W. Simone)

96. Leaf spot of osmanthus, caused by a *Colletotrichum* sp. (G. W. Simone)

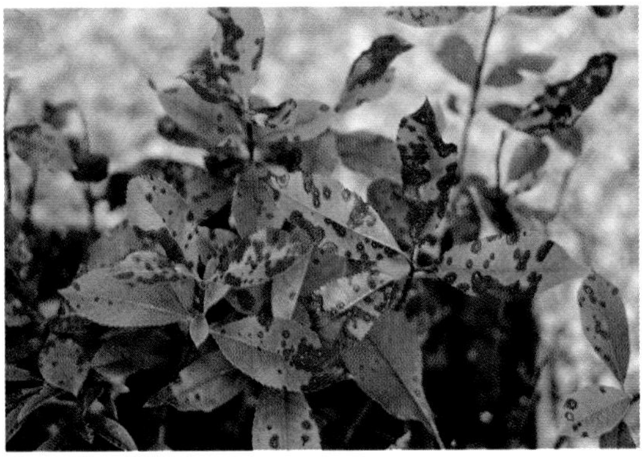

97. Leaf spot of red-tip photinia liners, caused by *Entomosporium mespili.* (A. S. Windham)

98. Fire blight of photinia, caused by *Erwinia amylovora.* (A. K. Hagan)

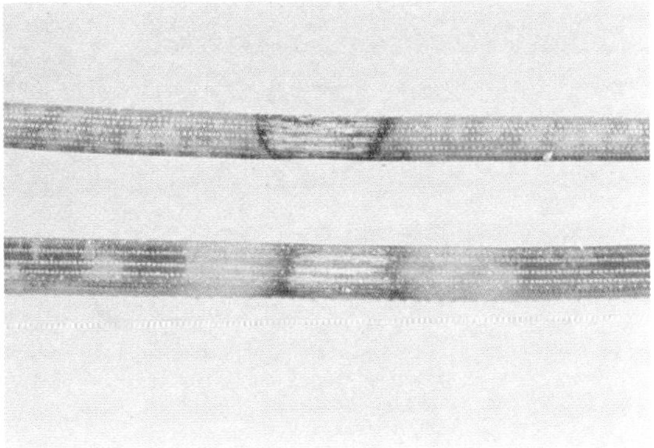

99. Brown spot needle blight of pine, caused by *Mycosphaerella dearnessii.* (Reprinted, by permission, from Hansen and Lewis, 1997)

100. Brown spot needle blight of Scots pine, caused by *Mycosphaerella dearnessii.* (Reprinted, by permission, from Hansen and Lewis, 1997)

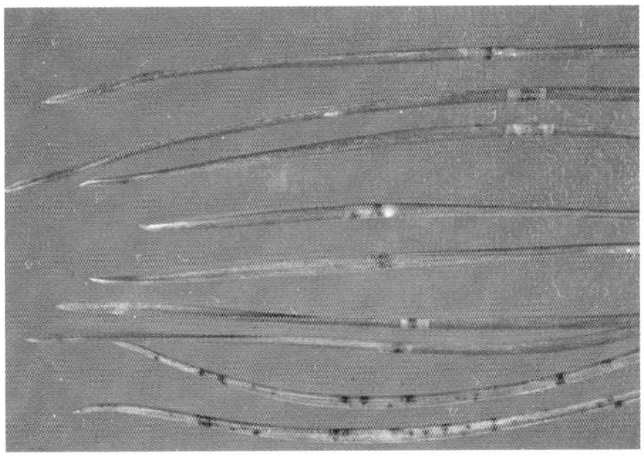

101. Dothistroma needle blight of Austrian pine, caused by *Mycosphaerella pini.* (Reprinted, by permission, from Hansen and Lewis, 1997)

102. Lophodermium needle cast of mugo pine, caused by *Lophodermium pinastri.* (R. K. Jones)

103. Fruiting structures of three *Lophodermium* spp. on pine (left to right): unidentified species, *L. pinastri,* and *L. seditiosum.* (Reprinted, by permission, from Hansen and Lewis, 1997)

104. Hysterothecia of *Lophodermella morbida* on ponderosa pine. (Reprinted, by permission, from Hansen and Lewis, 1997)

105. Rust pustules on pine, caused by a *Coleosporium* sp. (Reprinted, by permission, from Hansen and Lewis, 1997)

106. Needle rust fungus, *Melampsora* sp., on pine. (G. A. Chastagner)

107. Pycnial stage of the eastern gall rust fungus, *Cronartium quercuum,* on jack pine. (Reprinted, by permission, from Hansen and Lewis, 1997)

108. Fusiform rust fungus, *Cronartium quercuum* f. sp. *fusiforme,* on pine. (J. R. Hartman)

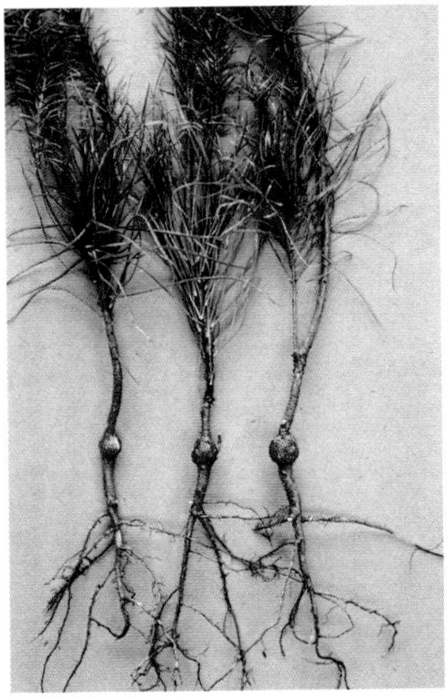

109. Galls of the western gall rust fungus, *Endocronartium harknessii,* on jack pine. (Reprinted, by permission, from Hansen and Lewis, 1997)

110. Aecial stage of the white pine blister rust fungus, *Cronartium ribicola,* on western white pine. (Reprinted, by permission, from Hansen and Lewis, 1997)

111. Pycnidia of *Sphaeropsis sapinea* on a pine needle. (Reprinted, by permission, from Hansen and Lewis, 1997)

112. Tip blight of Austrian pine, caused by *Sphaeropsis sapinea.* (M. Daughtrey)

113. Resinous canker caused by *Sphaeropsis sapinea* on pine. (Reprinted, by permission, from Hansen and Lewis, 1997)

114. Pycnidia of *Sphaeropsis sapinea* on pinecone scales. (Reprinted, by permission, from Hansen and Lewis, 1997)

115. Pitch canker of pine, caused by *Fusarium circinatum.* (Reprinted, by permission, from Hansen and Lewis, 1997)

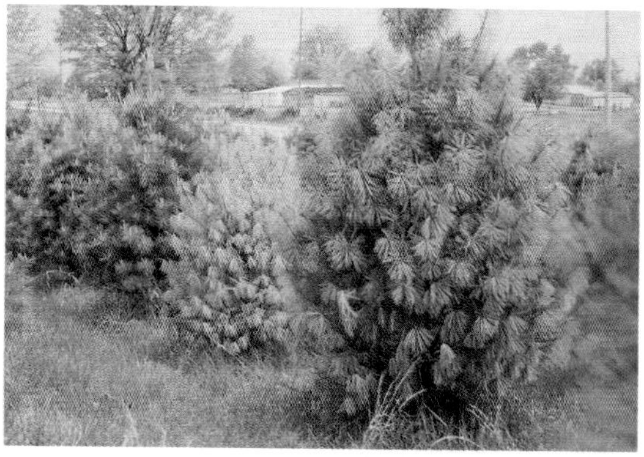

116. Procerum root disease, caused by *Leptographium procerum.* (J. R. Hartman)

117. Stain and resinosis associated with infection of pine by *Leptographium procerum.* (Reprinted, by permission, from Hansen and Lewis, 1997)

118. Oregon fir sawyer (*Monochamus scutellatus oregonensis*), a possible vector of the pinewood nematode. (Reprinted, by permission, from Hansen and Lewis, 1997)

119. Leaf spot of *Pittosporum,* caused by an *Alternaria* sp. (G. W. Simone)

120. Leaf spot of *Pittosporum,* caused by a *Cercospora* sp. (G. W. Simone)

121. Kutilakesa gall of *Pittosporum,* caused by *Kutilakesa pironii* (sexual stage, *Nectriella pironii*). (G. W. Simone)

122. Limb blight of *Pittosporum,* caused by a *Corticium* sp. (G. W. Simone)

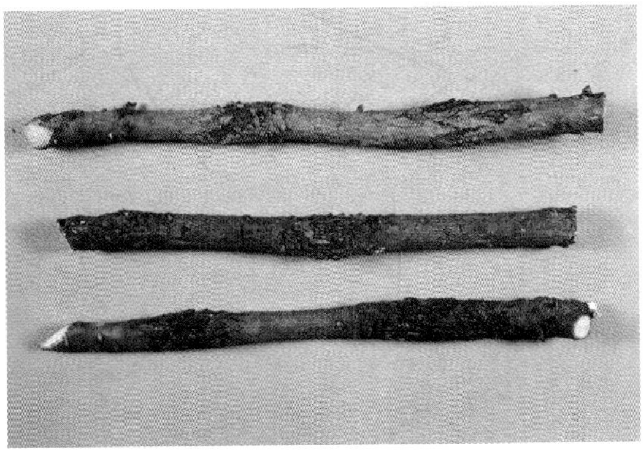

123. Rough bark of *Pittosporum* (cause unknown). (G. W. Simone)

124. Southern blight of *Pittosporum,* caused by *Sclerotium rolfsii.* (G. W. Simone)

125. Southern blight of *Pittosporum,* caused by *Sclerotium rolfsii.* (G. W. Simone)

126. Web blight of *Pittosporum,* caused by *Rhizoctonia solani.* (G. W. Simone)

127. Dieback of *Podocarpus,* caused by a *Botryodiplodia* sp. (G. W. Simone)

128. Dieback of *Podocarpus,* caused by a *Botryodiplodia* sp. (G. W. Simone)

129. Rust fungus, *Melampsora* sp., on poplar leaves. (G. A. Chastagner)

130. Leaf spot of poplar, caused by a *Venturia* sp. (G. A. Chastagner)

131. Shoot and leaf blight of hybrid poplar, caused by a *Venturia* sp. (G. A. Chastagner)

132. Leaf spot of silver poplar, caused by *Marssonina castagnei*. (R. S. Byther)

133. Canker caused by *Botryosphaeria dothidea* on redbud. (A. S. Windham)

134. Canker caused by *Botryosphaeria dothidea* on hybrid redbud. (K. A. Jacobs)

135. Phytophthora root rot of rhododendron, caused by *Phytophthora cinnamomi*. (R. K. Jones)

136. Dieback of hybrid rhododendron, caused by *Phytophthora parasitica*. (R. K. Jones)

137. Dieback of hybrid rhododendron, caused by *Botryosphaeria dothidea*. (M. Daughtrey)

138. Petal blight of rhododendron, caused by *Ovulinia azaleae*. (R. K. Jones)

139. Tissue proliferation in rhododendron. (M. Daughtrey)

140. Black spot of rose, caused by *Marssonina rosae*. (A. S. Windham)

141. Powdery mildew fungus, *Sphaerotheca* sp., on rose. (G. L. Philley)

142. Powdery mildew fungus, *Sphaerotheca* sp., on a rose stem. (R. K. Jones)

143. Foliar symptoms of downy mildew of rose, caused by *Peronospora sparsa.* (G. L. Philley)

144. Angular lesions of downy mildew on a rose leaf, caused by *Peronospora sparsa.* (G. L. Philley)

145. Defoliation typical of downy mildew of rose, caused by *Peronospora sparsa.* (G. L. Philley)

146. Cylindrocladium stem rot of rose, caused by *Cylindrocladium scoparium.* (G. L. Philley)

147. Cankers girdling rose stems at the surface of the medium, caused by *Cylindrocladium scoparium.* (G. L. Philley)

148. Shoot dieback associated with Botrytis blight of rose, caused by *Botrytis cinerea.* (A. K. Hagan)

149. Cane canker associated with Botrytis blight of rose, caused by *Botrytis cinerea*. (R. K. Jones)

150. Root-knot nematode (*Meloidogyne* sp.) symptoms in rose. (G. L. Philley)

151. Rose mosaic, caused by apple mosaic virus and Prunus necrotic ringspot virus. (G. L. Philley)

152. Rose mosaic. (R. K. Jones)

153. Leaf spot of sourwood, caused by a *Cercospora* sp. (T. J. Banko)

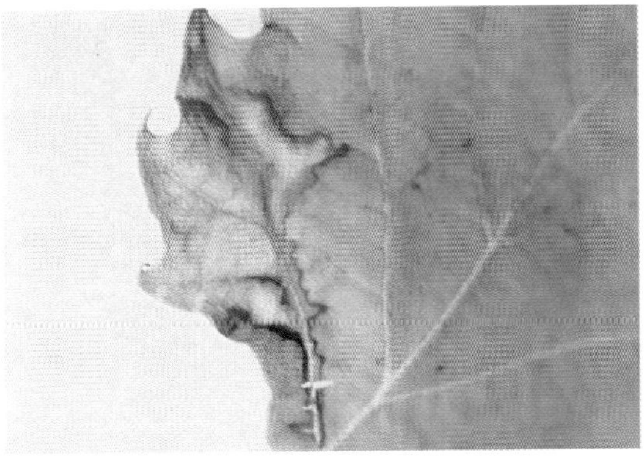

154. Anthracnose of sycamore, caused by *Discula platani*. (J. R. Hartman)

155. Anthracnose canker on a sycamore stem, caused by *Discula platani*. (J. R. Hartman)

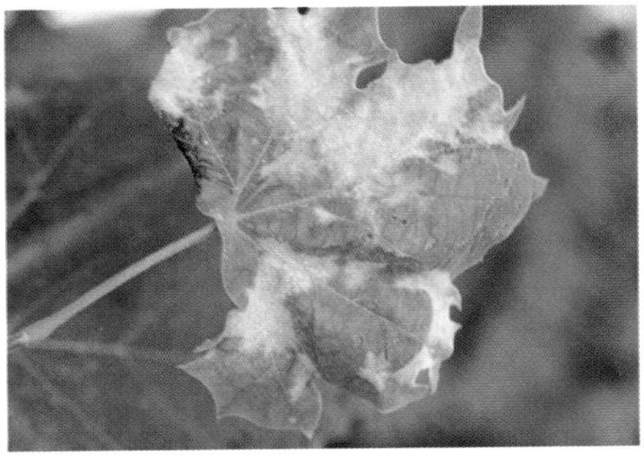

156. Powdery mildew fungus, *Microsphaera* sp., on sycamore. (R. K. Jones)

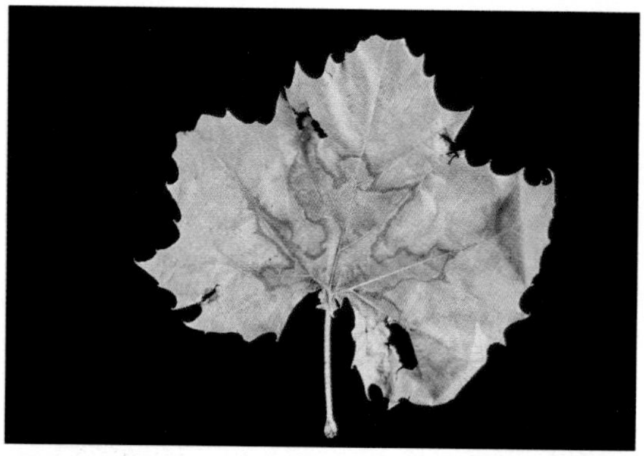

157. Bacterial leaf scorch of sycamore, caused by *Xylella fastidiosa.* (R. K. Jones)

158. Phytophthora root rot of *Taxus,* caused by *Phytophthora cinnamomi.* (R. K. Jones)

159. Phytophthora root rot of *Taxus,* caused by *Phytophthora cinnamomi.* (R. K. Jones)

160. *Taxus* inoculated with *Phytophthora citrophthora* (center) and *P. cinnamomi* (right) and an uninoculated control (left). (R. L. Wick)

CHAPTER 20

Ralph S. Byther ▪ Washington State University, Puyallup

Arborvitae Diseases

Geographic production USDA zones 2–6

Family Cupressaceae

Genus *Thuja*

Species

T. occidentalis	American arborvitae, northern white cedar
T. orientalis	oriental arborvitae
T. plicata	western red cedar, giant arborvitae, giant cedar

Arborvitaes are planted extensively in the Midwest, East, and Pacific Northwest. They are small to midsize evergreens with scale-like foliage, generally developing a pyramidal shape. Arborvitaes provide evergreen foliage during winter, but the foliage is often discolored, lost, or thin in some species and cultivars. Leaf blight and tip blight may be common in the nursery.

Leaf Blight

***Didymascella thujina* (syns. *Fabrella thujina, Keithia thujina*)**

Leaf blight caused by the fungus *Didymascella thujina* is most commonly observed in western red cedar (*Thuja plicata*), but it also occurs in American and oriental arborvitaes. The pathogen is endemic in North America and can be found wherever its hosts are grown. The disease affects arborvitaes in landscape and forest settings as well as in nurseries.

Symptoms

The initial symptoms are small, bleached spots on the upper surface of one-year-old leaves (Plate 6). The infection can spread to encompass the entire leaf, which turns brown. Fruiting bodies (apothecia) of the pathogen appear as slightly swollen, reddish brown areas under the epidermis, eventually turning black. When the apothecia mature, the epidermis will split open in damp weather, revealing the olive-colored spore-producing layer. Dead leaves that are not cast eventually turn gray. Spent apothecia and the surrounding tissue often fall out, leaving a conspicuous dark cavity.

Epidemiology

Ascospores of *D. thujina* are forcibly released from June to October during moist weather when the temperature is above 10°C. They are dispersed by splashing water for short distances and by wind over much longer distances. The long-lived ascospores are coated with a sticky film, enabling them to remain attached to leaf surfaces for extended periods. Different reports have been given regarding the period when most germination and infection take place. Infection can occur during the summer and fall, when spores are released, or during the following spring, when moisture and temperature are favorable. Symptoms, however, are not apparent until the growing season following ascospore release.

Leaf Blight

Geographic occurrence USDA zones 2–6

Seasonal occurrence June to October

DISEASE FREQUENCY	DISEASE SEVERITY
5 annual	5 plants killed
4	4
[3]	**[3]**
2	2
1 rare	1 very little damage

CHEMICAL TREATMENT	CULTURAL PRACTICES
3 used every year	3 very important
[2]	**[2]**
1 not used	1 not important

SANITATION	RESISTANT CULTIVARS
[3 very important]	3 many cultivars
2	2
1 not important	**[1 no resistance]**

Host Range

Leaf blight has been reported in *T. occidentalis, T. orientalis* (*Platycladus orientalis*), and *T. plicata.*

Management

Plant debris from previous infected crops should be removed and destroyed. Disease-free stock plants should be used for propagation. Plant spacing and irrigation schedules should be adjusted to keep the plants as dry as possible. Fungicide applications are recommended to protect plants during periods of anticipated wet conditions.

Tip Blight

Coryneum berckmansii

Tip blight, also known as Berckman's blight in the Pacific Northwest, is an important disease in both nurseries and landscape plantings of certain cultivars of *T. orientalis.* The pathogen, *Coryneum berckmansii,* is unable to invade woody tissue, but repeated defoliation can kill severely infected plants. A closely related fungus, *C. cardinale* (syn. *Seiridium cardinale*), is weakly pathogenic to western red cedar, but it causes an important disease of Italian and Monterey cypress.

Tip Blight

Geographic occurrence USDA zones 2–6

Seasonal occurrence September to March

DISEASE FREQUENCY	DISEASE SEVERITY
5 annual	5 plants killed
4	**[4]**
[3]	3
2	2
1 rare	1 very little damage

CHEMICAL TREATMENT	CULTURAL PRACTICES
3 used every year	3 very important
[2]	**[2]**
1 not used	1 not important

SANITATION	RESISTANT CULTIVARS
3 very important	3 many cultivars
[2]	2
1 not important	**[1]** no resistance

Symptoms

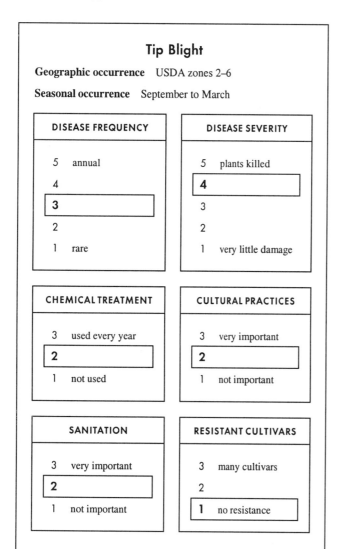

Blighting of young, tender foliage at branch tips is the first indication of the disease (Plate 7). Diseased tissues turn from green to a light gray. As the invading fungus moves down the small branchlets, larger branchlets are girdled, resulting in a reddish brown dieback in late spring when the weather becomes warmer and dryer. Many infected leaves and branchlets eventually fall, leaving a tangle of dead gray stems. The diagnostic fruiting bodies (acervuli) of the pathogen appear as tiny, black pustules on infected foliage and small girdled stems still enclosed by leaf scales. Continuous reinfection leading to severe defoliation eventually kills the plant.

Epidemiology

During moist weather in the early fall, spore pustules on diseased tissue are activated to produce the six-celled spores which are responsible for primary infection. Spores can be spread by splashing water, wind, and insect activity. Further stem colonization and secondary infections can take place during the fall, winter, and early spring when the weather is mild and moist. When the weather turns warmer and dryer, the fungus ceases activity. However, small branches that have been girdled continue to die, giving the false impression of continued fungal activity.

Host Range

Ornamental varieties of *T. orientalis* are the principal hosts of *C. berckmansii. T. orientalis* var. *conspicua* 'Berckmansii' is very susceptible. The golden arborvitae and several globular or pyramidal varieties are also susceptible. *T. orientalis* 'Beverleyensis' 'Compacta,' 'Elegantissima,' and 'Stricta,' have been reported to be naturally infected. *T. orientalis* 'Pyramidalis' and other American and European species are not susceptible. Tip blight has also been observed in Italian cypress (*Cupressus sempervirens*).

Management

Establish plantings with disease-free arborvitae propagated from cuttings. Diseased leaves, twigs, and branchlets on older plants should be removed and destroyed. A fungicide application should be made just prior to the fall rainy season. An additional application in late fall (early November) and again in early spring (February to March) is recommended in the Pacific Northwest.

Nursery Blight and Twig and Tip Dieback

Alternaria
Botrytis
Pestalotiopsis
Phomopsis

Numerous fungi are associated with leaf blights and stem dieback of *Thuja* species. Arborvitae seedlings and young nursery plants can suffer significant damage. Infection of older, established trees is usually not regarded as a serious problem. The fungi discussed in this section are those most likely to be encountered causing arborvitae leaf blight and dieback, but they are not the only pathogens associated with these diseases.

Symptoms

Succulent leaves and stem tissues initially turn a dull gray green. These tissues then often turn reddish brown and eventually gray. The damage is often observed as tip dieback. Cankers may also form, resulting in girdling and further dieback.

Nursery Blight

Geographic occurrence USDA zones 2–6

Seasonal occurrence April to June

DISEASE FREQUENCY	DISEASE SEVERITY
5 annual	5 plants killed
4	4
3	**[3]**
[2]	2
1 rare	1 very little damage

CHEMICAL TREATMENT	CULTURAL PRACTICES
3 used every year	3 very important
[2]	**[2]**
1 not used	1 not important

SANITATION	RESISTANT CULTIVARS
3 very important	3 many cultivars
[2]	2
1 not important	**[1]** no resistance

Sporulation on affected tissues distinguishes the specific fungal pathogen causing the infection. Access to a microscope and some training in mycology aid in making these determinations. *Phomopsis* produces tiny, black fruiting bodies embedded in the surface of dead tissues and is characterized by two types of microscopic, single-celled, colorless spores (ellipsoid and filamentous) produced in these fruiting bodies. Characteristic five-celled spores produced in tiny, dark pustules indicate *Pestalotiopsis.* Abundant production of gray brown, powdery spores on diseased tissues under moist or humid conditions suggests *Botrytis.* Multiple-celled dark spores formed in chains are characteristic of *Alternaria.*

Epidemiology

Infection can take place whenever succulent young tissue and adequate moisture are present throughout the growing season. Free moisture is necessary for spore germination, and high humidity is conducive to spore production. Under unfavorable environmental conditions, the fungal pathogens survive in infected plants and plant debris. *Botrytis* also produces sclerotia (long-term survival structures), which enable it to survive unfavorable conditions.

Host Range

Phomopsis juniperovora, in addition to causing a blight of *Thuja,* has been reported on many other conifers. It is considered to be responsible for a major disease of junipers.

Botrytis has a wide host range, including many conifers and other hosts.

Four different species of *Alternaria* are pathogenic in trees and woody shrubs, but the only conifers reported as hosts are *Thuja* and juniper.

Pestalotiopsis (*Pestalotia*) species are found on many conifer and deciduous hosts and are often associated with plants suffering from environmental stresses or tissues damaged by more aggressive pathogens.

Management

Infected plants and plant debris should be removed and destroyed. Plant spacing and irrigation schedules should be arranged to minimize the duration of wetness of plant surfaces. Adequate plant care should be provided to avoid environmental and cultural stress. When periods of high disease pressure are expected, protective fungicide applications are recommended.

REFERENCE

Dirr, M. A. 1998. Manual of Woody Landscape Plants: Their Identification, Ornamental Characteristics, Culture, Propagation and Uses. 5th ed. Stipes Publishing, Champaign, Ill.

CHAPTER

21

Cynthia L. Ash ▪ American Phytopathological Society, St. Paul, Minnesota

Ash Diseases

Geographic production USDA zones 3–9

Family Oleaceae

Genus *Fraxinus*

Species			
	F. americana	white ash	zones 3–9
	F. excelsior	common ash	zones 4–9
	F. mandschurica		zones 4–9
	F. nigra	black ash	zones 2–9
	F. pennsylvanica	green ash	zones 2–9
	F. quadrangulata	blue ash	zones 4–9
	F. velutina	velvet ash	zones 6–9

Ash species are important landscape and timber trees, adaptable and hardy throughout much of the eastern United States and southern Canada. Major production of ash occurs in the Midwest and Pacific Northwest, but ash is also grown in nurseries scattered throughout the Northeast, the South, and California. Ash species are susceptible to many insect pests, diseases, and abiotic influences, but few threaten their health. Nursery production of ash can be threatened by several important diseases.

Anthracnose

Gnomoniella fraxini
(asexual stage, *Discula fraxinea*)

Anthracnose, caused by the fungus *Gnomoniella fraxini*, can cause extensive defoliation of ash in seedling beds, fields, and landscapes under persistent cool, moist conditions. Leaves and twigs of black, green, velvet, and white ash are susceptible.

Symptoms

Symptoms develop on the very susceptible new leaves and shoots as they emerge and on older, more resistant foliage. Under conditions favorable for disease development, new leaves are infected and shed as they emerge. The damage may be confused with frost injury but is distinguished from it by the presence of pinpoint-sized, purple to brown lesions on fallen leaflets and rachises. Lesions on older foliage are blotchy and black if the leaves are still succulent and tan if they are mature (Plate 8). Lesions on expanding leaves result in leaf distortion. Infection of young twigs results in small, elliptical cankers and twig dieback.

Epidemiology

Infection immediately following budbreak may occur throughout the canopy of the tree but mostly occurs in the lower portion. Later infections are concentrated in the inner and lower canopy, because of the persistence of moisture there, due to reduced penetration of air and light. Fungal fruiting structures form and pro-

Anthracnose

Geographic occurrence Wherever ash is grown; especially prevalent in the upper Midwest

Seasonal occurrence From spring into summer, with most damage in early spring at or immediately following budbreak and through the leaf expansion stage, under average yearly conditions in Minnesota (adjust for your hardiness zone)

DISEASE FREQUENCY

5 annual
4
3
2
1 rare

DISEASE SEVERITY

5 plants killed
4
3
2
1 very little damage

CHEMICAL TREATMENT

3 used every year
2
1 not used

CULTURAL PRACTICES

3 very important
2
1 not important

SANITATION

3 very important
2
1 not important

RESISTANT CULTIVARS

no data

duce spores during wet weather, further spreading the disease. *G. fraxini* overwinters on infected twigs and petioles, both in the tree and on the ground. Wet conditions in the spring induce spore production and are favorable for spore dispersal and infection, and the disease cycle begins again.

Host Range

Anthracnose has been reported in *F. americana, F. nigra, F. pennsylvanica,* and *F. velutina.*

Management

Sanitation, crop rotation, and provisions for minimizing poor air circulation reduce the amount of anthracnose. Protective fungicides are occasionally used in seedling beds but are seldom warranted in the field.

Verticillium Wilt

Verticillium dahliae

Verticillium wilt of ash, caused by the fungus *Verticillium dahliae,* is of greater importance in nursery and landscape trees than previously recognized. The lack of recognition of the disease is due largely to the absence of the distinctive vascular discoloration associated with the disease in other trees. For detailed information, refer to Chapter 19, "Verticillium Wilt."

Symptoms

Acute symptoms include sudden scorching and leaf loss, often on a single branch or portion of the tree (Plate 9). Leaf wilt is rare. Affected branches occasionally refoliate, but more often they die. Chronic symptoms develop more slowly, as diffuse, patchy chlorosis and reduced growth, often leading to necrosis, defoliation, and branch dieback. Discoloration of the vascular tissue is rarely present, making an assumptive field diagnosis difficult and necessitating laboratory confirmation.

Worf et al. (1994) identified *V. dahliae* as the cause of an unusual top chlorosis problem affecting nearly 5% of 11,000 seedlings and young trees of mixed white and green ash cultivars in a Midwest nursery. Yellowing, bronzing, and leaf blotching progressed from the top down in affected trees. In the white ash cultivar Autumn Purple, foliage in the upper portion of the tree turned purple, and the associated bark was generally darker, somewhat mottled, and wrinkled. Tree tops were dead the following spring.

Epidemiology

Verticillium wilt has been found in trees in nursery and landscape plantings but only rarely on undisturbed sites. *V. dahliae* is soilborne, persisting indefinitely on numerous woody and herbaceous plants, which may or may not exhibit symptoms. Infection occurs near the root tips when growing roots come in contact with sclerotia (long-term survival structures) in the soil. The fungus colonizes the cortical tissues and penetrates the endodermis to enter the vascular tissue. Once in the xylem, it inhibits the movement of water in the plant. Symptom expression is erratic and dependent on temperature and other cultural factors that are not clearly defined.

Host Range

Verticillium wilt has been reported in *F. americana* and *F. pennsylvanica.* Data on other species are lacking.

Management

Trees infected in the nursery often develop symptoms within one to three years after planting but can remain symptomless for longer periods. Hot, dry conditions limit disease expression. In the nursery, infected trees should be destroyed and the soil tested for *Verticillium.* Landscape trees may benefit by regular watering and low nitrogen fertilization. Fungicides are not effective. Research is currently under way to determine the persistence of *V. dahliae* in wood chips (Ash et al., 1996). Preliminary results suggest that fresh wood chips from *Verticillium*-infected trees could introduce the fungus to susceptible plants.

Verticillium Wilt

Geographic occurrence Data are lacking, but Verticillium wilt can be expected to occur in ash species in nurseries where *Verticillium dahliae* is present in the soil

Seasonal occurrence Verticillium wilt can occur at any time; green leaf drop may occur in late spring, and scorch and defoliation are most evident in August and September, under average yearly conditions in Minnesota (adjust for your hardiness zone)

DISEASE FREQUENCY[a]	DISEASE SEVERITY[b]
5 annual	**5 plants killed**
4	**4**
3	**3**
2	**2**
1 rare	**1 very little damage**

[a] If *V. dahliae* is present in seedling beds or fields.

[b] Severity is variable.

CHEMICAL TREATMENT[c]	CULTURAL PRACTICES
3 used every year	**3 very important**
2	2
1 not used	1 not important

[c] Soil fumigation.

SANITATION	RESISTANT CULTIVARS
3 very important	
2	**no data**
1 not important	

Ash Rust

Puccinia sparganioides

Ash rust can be prevalent in nurseries close to natural stands of marsh grass (*Spartina* sp.) and cord grass (*Distichlis spicata*). The disease is caused by the fungus *Puccinia sparganioides,* which is distributed throughout most of eastern North America and in Brazil. The fungus requires both the grass host and the tree host to complete its life cycle. While the disease does not appear damaging to its grass host, it can cause serious losses in tree growth, even death, if severe over several years.

Symptoms

Infected leaves, petioles, and green twigs develop yellow spots, which become markedly swollen and distorted, often bending sharply and developing gall-like growths. Clusters of bright orange, cuplike fungal structures appear on the swollen, distorted areas about two weeks after infection. These structures eventually turn brown and wither, leaving cankers on the twigs, and leaves fall to the ground.

Epidemiology

Fruiting bodies of the rust fungus overwinter on grass hosts and release spores in the spring. Wind-blown spores infect new leaves, petioles, and developing twigs of most *Fraxinus* species. *Forestiera* spp. (privet) are also susceptible. In late spring to early summer spores are blown from ash to the alternate host, where they cause infection and overwinter.

Host Range

Ash rust has been reported in *F. americana, F. berlandierana, F. caroliniana, F. nigra, F. pennsylvanica, F. quadrangulata,* and *F. velutina.*

Management

In areas where infection is severe, a fungicide should be applied at two-week intervals starting at budbreak. Some degree of

Ash Rust

Geographic occurrence Nova Scotia to Saskatchewan and south into Montana; Texas; Florida

Seasonal occurrence Spring and early summer, under average yearly conditions in Minnesota (adjust for your hardiness zone)

DISEASE FREQUENCY	DISEASE SEVERITY
5 annual	5 plants killed
4	4
3	3
2	2
1 rare	1 very little damage

CHEMICAL TREATMENT	CULTURAL PRACTICES
3 used every year	3 very important
2	2
1 not used	1 not important

SANITATION	RESISTANT CULTIVARS
3 very important	no data
2	
1 not important	

Ash Yellows

Geographic occurrence Nineteen contiguous states, from Montana and Nebraska east to New England; Quebec and Ontario; scattered localities in the Southwest; rare in shade tree nurseries

Seasonal occurrence Symptoms may be expressed year-round, under average yearly conditions in Minnesota (adjust for your hardiness zone)

DISEASE FREQUENCY[a]	DISEASE SEVERITY
5 annual	5 plants killed
4	4
3	3
2	2
1 rare	1 very little damage

[a] The disease is systemic.

CHEMICAL TREATMENT	CULTURAL PRACTICES
3 used every year	3 very important
2	2
1 not used	1 not important

SANITATION	RESISTANT CULTIVARS
3 very important	no data
2	
1 not important	

control might be achieved by controlling the alternate grass host near nurseries.

Ash Yellows

Phytoplasma

Ash yellows is a systemic disease caused by an organism called a phytoplasma, which is spread by phloem-feeding insects such as leafhoppers and spittlebugs (for detailed information, refer to Chapter 8, "Phytoplasmas"). Sinclair et al. (1996) reported that the disease is rare in shade tree nurseries, presumably because of unsuitable habitats for vectors. However, an outbreak occurred in one midwestern nursery as a result of grafting diseased scion material.

Symptoms

Trees of all ages are susceptible. Growth is markedly reduced radially and terminally. Root growth is also affected, resulting in top dieback, decline, and death. Aboveground symptoms include minimal internodal elongation, chlorotic, undersized foliage, loss of apical dominance, thin crowns, basal frost cracks, and early fall coloration. Of significant diagnostic importance are witches'-brooms that form on the trunk, especially near the base, on a small percentage of infected trees. Leaves in brooms tend to be narrow, chlorotic, and simple (not compound), which helps to distinguish them from epicormic sprouts. These general symptoms may be caused by various agents.

Host Range

Most species of ash and lilac are susceptible, with reactions ranging from mild to severe in different species. Susceptible ash species include *F. americana, F. angustifolia, F. bungeana, F. excelsior, F. latifolia, F. nigra, F. ornus, F. pennsylvanica, F. potamophila, F. profunda, F. quadrangulata,* and *F. velutina.* White ash, *F. americana,* sustains the most damage. Green ash, *F. pennsylvanica,* although susceptible, is more tolerant. The data are insufficient to rank the susceptibility of other species.

Management

Ash yellows is systemic in infected trees. Stress on infected trees may increase symptom development and disease progression. Infected trees in the nursery should be destroyed. Not enough information on the spread of the pathogen by vectors is currently available to suggest control strategies. However, regular observation of trees used for budwood and rootstocks, and laboratory testing of suspect trees is recommended.

Leaf Spots

Mycosphaerella effigurata
Mycosphaerella fraxinicola

Several fungi cause leaf spots on ash. Mycosphaerella leaf spot (formerly called Piggotia leaf spot), caused by the fungus *Mycosphaerella effigurata,* can be severe in seedling beds and in small trees. In early summer, numerous necrotic lesions, less than 3 mm across, coalesce to form larger necrotic areas. Another species, *M. fraxinicola,* causes a disease formerly called Phyllosticta leaf spot, resulting in irregularly shaped necrotic lesions, 5–15 mm in diameter, which also coalesce. Severe infections by these fungi cause foliage to be shed four to six weeks earlier than usual. Little damage is usually done to established trees, and premature defoliation of ash seedlings in the nursery accelerates the development of seasonal dormancy, enabling early lifting for storage.

Removal of plant debris and incorporation of remaining debris may help to reduce the amount of inoculum available at the beginning of the next season, as these fungi overwinter on fallen leaves. Irrigation of seedling beds should be done early in the day and as infrequently as possible.

Trees affected by leaf spots are frequently also affected by ash plant bug, *Tropidosteptes* species. Feeding damage appears as light green to white flecking concentrated irregularly over the leaf. Heavy feeding can cause areas of the leaf to turn brown, especially the tips, but rarely causes defoliation. For additional evidence of ash plant bug, check the backsides of leaves for shiny, black dots of excrement.

Cankers

Cytospora sp.
Fusicoccum sp.
Nectria sp.

Several canker-causing fungi have been reported on ash, including *Fusicoccum, Cytospora,* and *Nectria.* These fungi are weak pathogens of trees following exposure to stress or mechanical damage. Cankers can be identified as discolored, sometimes sunken areas with a distinct border, on branches and stems (Plate 10). Callus tissue often forms at canker margins. The inner bark and cambium are dead and turn dark brown. Pimple-like fruiting bodies develop in and push through the outer bark to release spores during wet weather. Cankers may girdle small branches, resulting in branch dieback.

Minimize stress and mechanical damage during production and transplanting. Remove infected areas during dry weather, using proper pruning techniques. Fungicides are not necessary or effective.

Powdery Mildew

Phyllactinia guttata

Powdery mildew of ash, caused by the fungus *Phyllactinia guttata,* is a minor problem in nursery and landscape plantings, often appearing late in the season and causing no significant damage. White patches of powdery fungal growth occur on the upper and lower surfaces of leaves. Tiny, black, spherical fruiting bodies form in the white patches. Close examination with a hand lens may be needed to distinguish powdery mildew from ash plant bug damage. Severe infections can cause chlorosis, necrosis, and premature defoliation.

Initial infection is due to spores released from fallen leaves. Spores continue to spread the disease for the entire season. Cool to warm, humid weather with little irrigation or rainfall encourages infection.

Sanitation may be helpful in the control of powdery mildew. Fungicides are generally not needed.

Damping-Off

Damping-off can be a serious problem in ash seedlings when seedling bed conditions are less than optimal. Moisture management is critical. For information on symptoms, fungi associated with damping-off, and disease management, refer to Chapter 12, "Damping-Off of Seeds and Seedlings and Cutting Rot."

REFERENCES

Ash, C. L., and French, D. W. 1992. Detecting Verticillium wilt on *Fraxinus pennsylvanica.* (Abstr.) Phytopathology 82:1097.

Ash, C. L., Courneya, D. F., and Larsen, P. O. 1996. Transmission of *Verticillium dahliae* with fresh wood chips from an infected sugar maple to eggplant. (Abstr.) Phytopathology (Suppl.) 86:S98.

Redlin, S. C. 1988. The biology and taxonomy of *Gnomoniella fraxini,* cause of ash anthracnose. Ph.D. dissertation. North Dakota State University, Fargo.

Riffle, J. W., and Peterson, G. W. 1986. Diseases of trees in the Great Plains. U.S. Dep. Agric. For. Serv., Rocky Mountain For. Range Exp. Stn. (Ft. Collins, Colo.), Gen. Tech. Rep. RM-129.

Sinclair, W. A., Lyon, H. H., and Johnson, W. T. 1987. Diseases of Trees and Shrubs. Comstock Publishing Associates, Cornell University Press, Ithaca, N.Y.

Sinclair, W. A., Griffiths, H. M., and Davis, R. E. 1996. Ash yellows and lilac witches'-broom: Phytoplasmal diseases of concern in forestry and horticulture. Plant Dis. 80:468–475.

Solomon, J. D., Leininger, T. D., Wilson, A. D., Anderson, R. L., Thompson, L. C., and McCracken, R. I. 1993. Ash pests: A guide to major insects, diseases, air pollution injury and chemical injury. U.S. Dep. Agric. For. Serv., South. For. Exp. Stn. (New Orleans), Gen. Tech. Rep. SO-96.

Wolf, F. A. 1939. Leaf spot of ash and *Phyllosticta viridis.* Mycologia 31:258–266.

Wolf, F. A., and Davidson, R. W. 1941. Life Cycle of *Piggotia fraxini,* causing leaf disease of ash. Mycologia 33:526–539.

Worf, G. L., Spear, R. N., and Heimann, M. F. 1994. *Verticillium*-induced scorch and chlorosis in ash. J. Environ. Hortic. 12(3):124–130.

CHAPTER
22

D. Michael Benson ▪ North Carolina State University, Raleigh
Gary W. Simone ▪ University of Florida, Gainesville

Aucuba (Japanese Laurel) Diseases

Geographic production USDA zones 7–10

Family Cornaceae

Genus *Aucuba*

Species *A. japonica* Japanese aucuba, Japanese laurel, gold-dust plant

Aucuba japonica was introduced into the United States from Japan in 1783. There are several cultivars, of which the variegated forms are the most popular. Aucuba is susceptible to Phytophthora root rot, Sclerotium blight, and root-knot nematodes. In addition, it is subject to leaf spot diseases due to *Colletotrichum* and many other fungi. Sunscald and cold injury are common on plants grown in full sun without adequate protection. Most aucubas in production are container-grown.

Phytophthora Root Rot

Phytophthora cinnamomi
Phytophthora citricola

Symptoms

Roots of plants infected with *Phytophthora* spp. become necrotic, and the outer root tissue may slough off the stele easily when squeezed between the fingers during examination. Foliar symptoms include off-color leaves and stems, wilting, and necrosis. Slow progression of the disease results in lower leaf yellowing, necrosis, and a gradually thinning canopy from the crown to the meristem. Disease development during periods when the soil is saturated may result in a progressive daily wilting and eventual death of plants, with little leaf discoloration. Plants in field production may be dwarfed and decline gradually, whereas container-grown plants may collapse rapidly.

Epidemiology

P. cinnamomi and *P. citricola* overwinter as chlamydospores, oospores, sporangia, or hyphae associated with debris from previously diseased crops. Infection of root tips by zoospores initiates the disease. Soil moisture near saturation is required for the formation of sporangia followed by brief periods (1–4 hours) of soil saturation for the release and dispersal of zoospores. Extensive root infection results in foliar symptoms due to impaired root function (see Chapter 14, "Phytophthora Root Rot and Dieback").

Host Range

All cultivars are susceptible to Phytophthora root rot. The host range *P. cinnamomi* includes plants in more than 90 genera of plants in the United States, while the range of *P. citricola* is lim-

Phytophthora Root Rot

Geographic occurrence USDA zones 7–10

Seasonal occurrence April to October, with an optimum period in June–July, under average yearly conditions in the southern United States (adjust for your hardiness zone)

DISEASE FREQUENCY	DISEASE SEVERITY
[5 annual]	**[5 plants killed]**
4	4
3	3
2	2
1 rare	1 very little damage

CHEMICAL TREATMENT	CULTURAL PRACTICES
[3 used every year]	**[3 very important]**
2	2
1 not used	1 not important

SANITATION	RESISTANT CULTIVARS
[3 very important]	3 many cultivars
2	2
1 not important	**[1 no resistance]**

ited to plants in a dozen or so genera, including *Abies, Persea, Pieris, Prunus, Quercus,* and *Rhododendron.*

Management

Soilless container media should be maintained on concrete pads prior to potting, to avoid Phytophthora root rot. Because aucuba needs a shaded growing area for best root growth, avoid overwatering that would favor zoospore production and root infection. Containers should be placed on a ground cover or ground cloth to avoid splashing *Phytophthora* spores into them or introducing zoospores through their drainage holes during heavy rain or irrigation. Particular care should be exercised with liners when they are moved out of propagation houses to ground beds. Avoid placement in areas that are poorly drained or subject to surface runoff or rapid elevation of the water table. Fungicides may be appropriate as protectants, if Phytophthora root rot has been a problem in previous aucuba crops.

Field production of aucuba may be difficult due to its requirement of shade and a serious threat of Phytophthora root rot in native soils. Successful production will require very well drained beds.

Sclerotium Blight

Geographic occurrence USDA zones 8–10

Seasonal occurrence June to October, with an optimum period in July–August, under average yearly conditions in the southern United States (adjust for your hardiness zone)

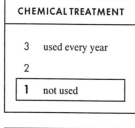

DISEASE FREQUENCY	DISEASE SEVERITY
5 annual	**5 plants killed**
4	4
3	3
2	2
1 rare	1 very little damage

CHEMICAL TREATMENT	CULTURAL PRACTICES
3 used every year	3 very important
2	**2**
1 not used	1 not important

SANITATION	RESISTANT CULTIVARS
3 very important	3 many cultivars
2	2
1 not important	**1 no resistance**

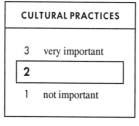

Sclerotium Blight (Southern Blight)

Sclerotium rolfsii

Symptoms

Rapid collapse of foliage is a common symptom of aucuba affected with Sclerotium blight, also known as Southern blight, caused by *Sclerotium rolfsii.* Infected plants wilt rapidly, and their foliage becomes discolored. The pathogen attacks aucuba at the soil line, so stem tissue becomes water-soaked and brown. The pathogen is visible as coarse, white, fanlike mycelia growing across the surface of the medium and upward on lower stem tissue. Small, white sclerotia, the size of mustard seed, form on the hyphae within several days. The sclerotia mature, turning a tan color, in two to three days. Roots will normally be healthy-appearing, with white tips, until stem infection prevents the transport of sugars to them.

Epidemiology

S. rolfsii survives as sclerotia on or near the surface of the soil or container medium. The tan-colored sclerotia are smaller than coated fertilizer products but could be confused with them without close examination. They are usually associated with infected crop debris. A moist microclimate, created by dense foliage close to the soil or container surface or by fallen leaves, is required for the germination of sclerotia. When they germinate, sclerotia produce a dense, coarse, white mycelium, which grows across the soil surface and around the lower stems of plants. As the stem is infected and colonized near the soil line, the foliage begins to wilt and collapse.

Losses from Sclerotium blight are generally low, with the disease developing in only a few random plants. Hot, wet periods of summer favor the development of the disease.

Host Range

All cultivars of aucuba are susceptible to *S. rolfsii.* Other nursery crops, including *Ajuga, Caladium, Crataegus, Ficus, Forsythia, Liriope, Phlox, Pittosporum, Rhaphiolepis, Thuja,* and *Yucca,* are also susceptible, as are many other annual ornamentals and vegetable and fruit crops.

Management

Generally, management of Sclerotium blight is not needed. However, if the disease develops, prompt removal of infected plants and debris in the container area should minimize further losses. Do not recycle media or pots infested with *S. rolfsii* unless the media are heat-sterilized or fumigated and the pots are thoroughly washed and disinfested.

Root-Knot Nematodes

Meloidogyne arenaria
Meloidogyne incognita

Symptoms

Root-knot nematodes can infect aucuba in field production, inducing galls on roots and poor top growth. Infected plants may be sensitive to sunscald and cold injury. Root galls can be observed by examination of root samples.

Epidemiology

Root-knot nematodes survive as eggs and larvae in soil. Larvae enter roots behind the root tip or at sites of lateral root emergence and migrate toward the stelar tissues of the root. As the nematodes develop, the males leave the roots, and the females become nonmotile and encysted, forming specialized feeding cells known as giant cells. The growth of giant cells and multiple infections by female nematodes results in root galls, which may grow quite large and disrupt the uptake and translocation of water.

Root-knot nematodes are common in nursery soils, particularly in coastal plain soils. They are rare in container production with soilless media.

Host Range

All cultivars of aucuba are susceptible to root-knot nematodes. Root-knot nematodes have an extremely wide host range among woody ornamentals and fruit and vegetable crops.

Management

Avoid fields known to be infested with root-knot nematodes. Preplant treatment (soil fumigation) may be required where production fields are limited. Nematicides are generally of limited efficacy in controlling root-knot nematodes. Infected plants should be rogued and destroyed.

Leaf Spot

***Alternaria* sp.**
***Cercospora* sp.**
***Colletotrichum* sp.**
Phomopsis aucubae
Phyllosticta aucubae

Symptoms

Many leaf spot fungi attack aucuba, particularly under suboptimal growing conditions. Symptoms may begin as dark spots, which develop tan or necrotic centers as they enlarge. They may be surrounded with a purple margin. Leaf spots may expand, although they may be limited by leaf veins or margins. Black spore-producing structures, such as acervuli and pycnidia, may be present in the centers of leaf spots, depending on the pathogen involved. Spore-producing structures of some leaf spot pathogens develop in concentric ring patterns.

Epidemiology

Spores of the pathogens are dispersed from leaf spots by splashing water and require a film of water on the aucuba leaf to

Root-Knot Nematode

Geographic occurrence USDA zones 7–10

Seasonal occurrence May to October, with an optimum period in July–August, under average yearly conditions in the southern United States (adjust for your hardiness zone)

DISEASE FREQUENCY	DISEASE SEVERITY
[5] annual	5 plants killed
4	**[4]**
3	3
2	2
1 rare	1 very little damage

CHEMICAL TREATMENT	CULTURAL PRACTICES
3 used every year	3 very important
[2]	**[2]**
1 not used	1 not important

SANITATION	RESISTANT CULTIVARS
3 very important	3 many cultivars
[2]	2
1 not important	**[1] no resistance**

Leaf Spot

Geographic occurrence USDA zones 7–10

Seasonal occurrence May to October, with an optimum period in July–August, under average yearly conditions in the southern United States (adjust for your hardiness zone)

DISEASE FREQUENCY	DISEASE SEVERITY
5 annual	5 plants killed
4	4
[3]	3
2	**[2]**
1 rare	1 very little damage

CHEMICAL TREATMENT	CULTURAL PRACTICES
3 used every year	3 very important
[2]	**[2]**
1 not used	1 not important

SANITATION	RESISTANT CULTIVARS
3 very important	3 many cultivars
2	2
[1] not important	**[1] no resistance**

establish a new infection. The disease often develops in plants damaged by cold injury or sunscald. Leaf spot pathogens generally do not cause problems under optimal growing conditions.

Host Range

No differences in the susceptibility of aucuba cultivars to leaf spot fungi have been reported.

Management

Protective fungicides can be used to control leaf spots, but the pathogens usually require no control effort.

Cold Injury and Sunscald

Symptoms

In aucubas damaged by cold injury or sunscald, irregular black sections develop in leaves, particularly along the margins near the tip. Affected tissue may drop out, leaving an irregularly shaped leaf with gaping holes in the margin, or it may remain intact. The black, crinkled tissue is very unsightly.

Epidemiology

Cold injury and sunscald are abiotic diseases. Aucubas grown in full sun without adequate protection are susceptible to these problems. Leaf spot pathogens are often associated with plants suffering cold injury or sunscald.

Management

Cold injury can be avoided by providing adequate winter protection for plants, such as bunching up pots and mulching, providing overhead lath or shade cloth, and putting plants in overwintering structures.

Sunscald can be avoided by growing plants under lath, screen cloth, or pine trees. However, plants grown under pine trees may be affected by aphids that excrete droppings on aucuba leaves, and the black sooty mold fungus will grow in the droppings. Sooty mold, though not a serious disease, can render plants unsalable.

REFERENCES

Barker, K. R., Benson, D. M., and Jones, R. K. 1979. Interactions of Burfordi, Rotunda, and Dwarf yaupon hollies and aucuba with selected plant-parasitic nematodes. Plant Dis. Rep. 63:113–116.

Spencer, S., and Benson, D. M. 1981. Root rot of *Aucuba japonica* caused by *Phytophthora cinnamomi* and *P. citricola* and suppressed with bark media. Plant Dis. 65:918–921.

CHAPTER 23

D. Michael Benson ▪ North Carolina State University, Raleigh

Jean L. Williams-Woodward ▪ University of Georgia, Athens

Azalea Diseases

Geographic production USDA zones 6–10

Family Ericaceae

Genus *Rhododendron*

Species many

Hybrids many groups

Azaleas are among the most common woody ornamentals grown in nursery production. Both deciduous and evergreen cultivars are widely grown which offer a diversity of plant forms and color. Azaleas may be produced in the field or in containers. Production in the field is popular in some mid-Atlantic states, such as Virginia and Maryland. However, production in containers is more common across much of the southeastern United States. Each method of production may favor important diseases of azalea roots and foliage. For lists of azalea taxa and cultivars of the various hybrid groups, see Dirr (1998) and Galle (1987).

Phytophthora Root Rot

Phytophthora cinnamomi
Phytophthora parasitica

The single most important disease of evergreen azaleas is root rot caused by *Phytophthora cinnamomi* and *P. parasitica* (syn. *P. nicotianae*). Phytophthora root rot occurs in azaleas at all growth stages in both container and field production. Along the Gulf Coast, *P. parasitica* is the dominant species attacking azalea. *P. cinnamomi* predominates in the rest of the South, in the mid-Atlantic states, and on the West Coast. The temperature optima of the two species probably play a major role in their distribution, as the optimum temperature for *P. parasitica* is higher than that of *P. cinnamomi.*

Symptoms

Azaleas suffering from Phytophthora root rot may exhibit a range of symptoms, depending on the cultivar and the growing conditions. Typically, symptoms on Kurume hybrid azaleas include chlorosis, dwarf leaves, smaller than normal stem diameters, and off-color stems during active growth periods (Plates 11 and 12). Root symptoms include reddish brown fine roots and lack of new root growth. Under similar growing conditions, a healthy plant may produce 6 to 8 inches of new growth, while a diseased plant may make only 1/2 to 1 inch of growth. Infected plants seldom wilt or defoliate during the growing season, but severely infected plants may die. Infected plants of cultivars with red flowers frequently produce excessive amounts of red pigments in the leaves in the fall and winter whereas Snow and other white-flowered cultivars produce yellow leaves in the fall. Fall defolia-

Phytophthora Root Rot

Geographic occurrence USDA zones 6–10

Seasonal occurrence April to September, with an optimum period in July–August, under average yearly conditions in zone 8 (adjust for your hardiness zone)

DISEASE FREQUENCY	DISEASE SEVERITY
[5 annual]	5 plants killed
4	**[4]**
3	3
2	2
1 rare	1 very little damage

CHEMICAL TREATMENT	CULTURAL PRACTICES
[3 used every year]	**[3 very important]**
2	2
1 not used	1 not important

SANITATION	RESISTANT CULTIVARS
3 very important	**[3 many cultivars]**
[2]	2
1 not important	1 no resistance

tion, which produces naked stems with a tuft of dwarf leaves around the terminal bud, is much more extensive on infected plants.

Epidemiology

See Chapter 14, "Phytophthora Root Rot and Dieback."

Host Range

The host ranges of *P. cinnamomi* and *P. parasitica* are enormous. More than 1,900 species of plants, in over 85 genera, have been reported as hosts of *P. cinnamomi*. Plants in over 80 genera are susceptible to *P. parasitica*.

Management

In addition to the general suggestions for management of Phytophthora root rot (see Chapter 14), cultivar selection is important for azaleas when problems with the disease occur in the nursery. In general, the Indica hybrids have the highest level of resistance to Phytophthora root rot and the Kurume hybrids the least resistance, but cultivar response within hybrid groups can vary considerably. For instance, the Kurume azalea Morning Glow was rated moderately resistant, while the Kurume azalea Hino Crimson was rated very susceptible (Benson and Cochran, 1980). Several cultivars were identified with good resistance to Phytophthora root rot. These included Formosa, an Indica hybrid; Fakir, a Glenn Dale hybrid; and Corrine Murrah, a Back Acres hybrid.

One azalea species, *R. poukhanense*, is highly resistant to Phytophthora root rot (Benson and Cochran, 1980). It grows vigorously, is very hardy, and blooms well.

Petal Blight

Ovulinia azaleae

Petal blight, caused by the fungus *Ovulinia azaleae*, is a serious disease of azaleas in landscapes and greenhouses. It also occurs in container nurseries but is often overlooked, because the flowers are the only part affected by the disease. Infection in nurseries is believed to be a major source of disease in landscapes. Petal blight occurs primarily in the warmer regions of the Southeast and mid-Atlantic states but has been reported occurring outdoors as far north as Connecticut, New York, and Rhode Island. Most azalea and rhododendron cultivars are susceptible to the disease. Indian and Kurume type azaleas are the most severely affected. If the environmental conditions are favorable, the disease may spread so rapidly as to completely destroy flowers in two to four days.

Symptoms

Water-soaked spots on the petals are first apparent when they are about the size of a pinhead. They are pale or whitish on colored flowers and rust-colored on white flowers. The spots enlarge rapidly into irregular blotches, becoming tan to light brown. Eventually, the entire corolla turns prematurely brown and collapses within two to three days of initial infection. Infected petals are slimy and fall apart readily if rubbed gently between the fingers. This test distinguishes diseased flowers from those injured by low temperature, insects, or other causes. Diseased flowers dry and cling to the plant for some time, presenting an unsightly appearance, whereas healthy flowers of Indian azaleas fall from the plant while still displaying color, normal shape, and turgor. Small, curved (concave), black sclerotia form on infected flowers six to eight weeks after infection (Plate 13).

Epidemiology

O. azaleae produces hard, black structures known as sclerotia in blighted flowers. Blighted flowers may fall to the ground or remain attached to the plant until flower budbreak the following year. Small, tan, cup-shaped reproductive structures (apothecia) develop from sclerotia in the spring. The spores that cause primary infection (ascospores) are propelled from apothecia to lower flower buds or are carried by air currents to adjacent plants. Infection is favored by frequent precipitation, irrigation, or heavy dew and warm weather during flowering. Secondary spores (conidia) are produced in large numbers on infected petals. Successive populations of conidia are produced on infected flowers every three to four days. Conidia are windborne and insect-borne and are responsible for widespread outbreaks of flower blight.

Sclerotia produced in diseased petals drop to the ground and remain undetected in surface litter. Unsold container-grown aza-

Petal Blight

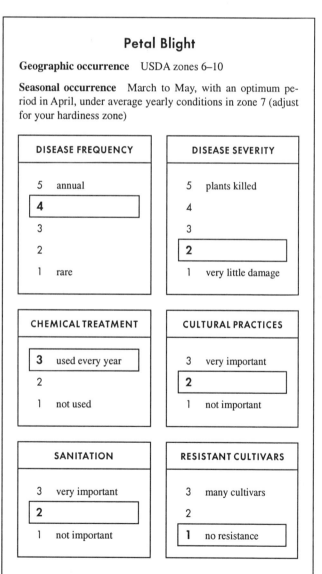

Geographic occurrence USDA zones 6–10

Seasonal occurrence March to May, with an optimum period in April, under average yearly conditions in zone 7 (adjust for your hardiness zone)

DISEASE FREQUENCY	DISEASE SEVERITY
5 annual	5 plants killed
4 [selected]	4
3	3
2	**2** [selected]
1 rare	1 very little damage

CHEMICAL TREATMENT	CULTURAL PRACTICES
3 used every year [selected]	3 very important
2	**2** [selected]
1 not used	1 not important

SANITATION	RESISTANT CULTIVARS
3 very important	3 many cultivars
2 [selected]	2
1 not important	**1** no resistance [selected]

leas held over for forcing the following year serve as a source of primary inoculum if they are carrying sclerotia. Infested container-grown azaleas sold and set in the landscape carry sclerotia and thus provide primary inoculum for future epidemics.

Management

Since the pathogen can destroy individual flowers in two to three days, management in the nursery is important to maintain sales during flowering. In nurseries, fungicide sprays will protect flower buds from infection if they are applied at weekly intervals beginning when the first buds start to show color and continuing until the last buds open. In the landscape, removing and destroying affected flowers and replacing the surface litter under infected plants with uncontaminated material are means of reducing the sources of primary infection in the spring.

Salt Injury Due to Excess Soluble Salt

Azaleas are very sensitive to excess soluble salts in the root zone, which result from overfertilization or improper release of nutrients. A burn on the margins of lower leaves is one symptom of excess salinity. Many cultivars will drop oldest leaves in response to salt injury. A red coloration of the leaves of some cultivars of evergreen azaleas is also a symptom. Some cultivars, such as Hexe, Lenthegruss, and Vervaeneana, are reported to be less salt-sensitive than others. Such injury can occur on one side of the root system in container-grown plants if fertilizer is deposited in one spot on the surface of the medium. See Chapter 87, "Sanitation: Plant Health from Start to Finish."

Rhizoctonia Web Blight

Binucleate *Rhizoctonia* spp.
Rhizoctonia solani

Rhizoctonia web blight of azaleas is often not detected until tight blocks of plants are respaced or selected for sale, because only the innermost leaves may be affected (Plate 14). Since the pathogen does not kill plants, growers may not realize that plants are infected until closer examination. The disease may develop in azaleas grown in full sun or in shady conditions. Only binucleate *Rhizoctonia* spp. have been isolated from azaleas with web blight. However, cross-inoculation studies with *R. solani,* a multinucleate species, have demonstrated that the disease due to *R. solani* is more severe than that due to binucleate *Rhizoctonia* spp.

For more information on symptoms, epidemiology, and management of web blight in azaleas, see Chapter 17, "Rhizoctonia Web Blight."

Leaf and Flower Gall

Exobasidium vaccinii

Leaf and flower gall, caused by the fungus *Exobasidium vaccinii,* is a very common disease of azalea and, to a lesser extent, rhododendron and blueberry. Closely related species of *Exobasidium* cause the same type of gall formation in other plants, such as *Arbutus, Camellia, Ledum,* and *Leucothoe.* The disease occurs in a wide range of azalea cultivars in nurseries and landscapes. Some species and cultivars are more susceptible than others. The Indica azalea group and *R. maximum, R. catawbiense,* and their hybrids are the most susceptible to infection. The disease is usually not a serious threat to azalea production. However, galls may become abundant under very humid, wet conditions, resulting in reduced plant vigor, and plants may be unsalable if some control measures are not implemented.

Symptoms

The disease causes individual leaves to swell, curl, or form fleshy, bladder-like galls (Plate 15). The galls are pale green to white or pink during the early stages of the disease. As the season progresses, they become covered with a whitish mold-like growth and then turn brown and hard. Infected flowers are fleshy, waxy, and swollen. Galls are made up of abnormal leaf and flower tissue. The lower leaves on plants are usually the most seriously damaged portion, but under humid conditions and in shaded locations, galls may occur at the ends of top branches.

Epidemiology

The occurrence and intensity of the disease is dependent upon weather conditions and upon a source of the fungal spores. Spores

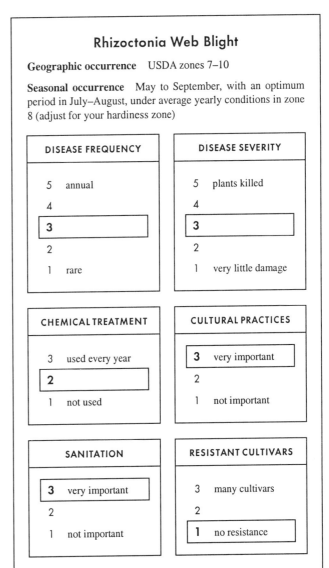

Rhizoctonia Web Blight

Geographic occurrence USDA zones 7–10

Seasonal occurrence May to September, with an optimum period in July–August, under average yearly conditions in zone 8 (adjust for your hardiness zone)

DISEASE FREQUENCY	DISEASE SEVERITY
5 annual	5 plants killed
4	4
[3]	**[3]**
2	2
1 rare	1 very little damage

CHEMICAL TREATMENT	CULTURAL PRACTICES
3 used every year	**[3] very important**
[2]	2
1 not used	1 not important

SANITATION	RESISTANT CULTIVARS
[3] very important	3 many cultivars
2	2
1 not important	**[1] no resistance**

overwinter in bud scales of host plants. Spores germinate and infect expanding leaf and flower buds in the spring. Older leaves are resistant to infection. Infection is dependent on high humidity or moisture at budbreak. Fungal spores are produced in the whitish mold on the surface of the galls. Spores are either blown or carried by splashing water to expanding leaf and flower buds, where they cause secondary infections, or they adhere to newly formed buds, overwinter, and infect buds the following spring.

Management

Where only a few plants are involved, as in a home planting or a small greenhouse, the disease can be kept in check by picking the galls and destroying them before they turn white. Reducing humidity, increasing aeration around plants, and avoiding prolonged leaf wetness can reduce infection. Fungicide control is not always successful. To prevent infection, fungicides need to be applied to the leaf and flower buds prior to budbreak and reapplied as the buds expand. Some evergreen azalea varieties are less susceptible than others.

Rust

Pucciniastrum vaccinii

Rust, caused by the fungus *Pucciniastrum vaccinii* (syn. *P. myrtilli*), can be a serious disease of deciduous azaleas and the native *R. canadense*, *R. nudiflorum*, *R. ponticum*, and *R. viscosum*. The alternate host is hemlock, *Tsuga canadensis*.

Table 23.1 Resistance to rust, caused by *Pucciniastrum vaccinii*, in deciduous azalea cultivars[a]

High	Moderate	Low
Gibraltar	Brazil	Klondyke
Red Letter	Clarice	Peachy Keen
Balzae	Exbury Crimson	Pink William
	Homebush	Primrose
	Oxydol	Rufus
	Peach Sunset	Sunrise

[a] Adapted from Bir et al., 1981.

Leaf and Flower Gall

Geographic occurrence USDA zones 6 and 7

Seasonal occurrence April to June, with an optimum period in May, under average yearly conditions in zone 7 (adjust for your hardiness zone)

DISEASE FREQUENCY
- 5 annual
- **4** (boxed)
- 3
- 2
- 1 rare

DISEASE SEVERITY
- 5 plants killed
- 4
- 3
- **2** (boxed)
- 1 very little damage

CHEMICAL TREATMENT
- 3 used every year
- **2** (boxed)
- 1 not used

CULTURAL PRACTICES
- 3 very important
- **2** (boxed)
- 1 not important

SANITATION
- **3 very important** (boxed)
- 2
- 1 not important

RESISTANT CULTIVARS
- 3 many cultivars
- **2** (boxed)
- 1 no resistance

Azalea Rust

Geographic occurrence USDA zones 6 and 7

Seasonal occurrence July to September, with an optimum period in August, under average yearly conditions in zone 7 (adjust for your hardiness zone)

DISEASE FREQUENCY
- 5 annual
- 4
- **3** (boxed)
- 2
- 1 rare

DISEASE SEVERITY
- 5 plants killed
- 4
- **3** (boxed)
- **2** (boxed)
- 1 very little damage

CHEMICAL TREATMENT
- **3 used every year** (boxed)
- 2
- 1 not used

CULTURAL PRACTICES
- 3 very important
- **2** (boxed)
- 1 not important

SANITATION
- **3 very important** (boxed)
- 2
- 1 not important

RESISTANT CULTIVARS
- 3 many cultivars
- **2** (boxed)
- 1 no resistance

The first symptoms appear as small, circular, chlorotic spots on the upper leaf surface. The fungus produces abundant yellow to orange spots on the lower leaf surface (Plate 16). The disease usually appears in late summer and fall. In highly susceptible deciduous azalea cultivars, the lower leaf surfaces can be completely covered with spore masses, and early fall defoliation can be severe.

Rust can be controlled by growing resistant cultivars (Table 23.1) and by weekly applications of a fungicide beginning with the first appearance of the disease in susceptible cultivars.

Powdery Mildew

Microsphaera penicillata
Microsphaera vaccinii
Sphaerotheca pannosa

Evergreen azaleas are occasionally infected with powdery mildew fungi classified as *Erysiphe polygoni* in previous taxonomy but now placed in the genera *Microsphaera* and *Sphaerotheca.* Certain deciduous azalea cultivars also are infected. Overcast conditions with high humidity but low rainfall are favorable for mildew development.

Powdery Mildew

Geographic occurrence USDA zones 6–9

Seasonal occurrence July to September, with an optimum period in August

DISEASE FREQUENCY	DISEASE SEVERITY
5 annual	5 plants killed
4	4
3	3
2	2
1 rare	**1 very little damage**

CHEMICAL TREATMENT	CULTURAL PRACTICES
3 used every year	3 very important
2	2
1 not used	**1 not important**

SANITATION	RESISTANT CULTIVARS
3 very important	3 many cultivars
2	**2**
1 not important	1 no resistance

Symptoms

Signs of powdery mildew on young leaves first appear as white, powdery spots. With continued favorable environment, the fungus covers the entire lower or upper surface of the leaf with whitish growth. Symptoms of mildew are generally rare but may include development of reddish pigments in leaves of Kurume hybrid azaleas.

Epidemiology

Spores of powdery mildew fungi produced by the white mycelial growth are carried by the wind to new leaf tissue, where they infect the leaf and spread the disease. Powdery mildew fungi from azalea can infect rhododendron and vice versa. Mildew has been found to develop more rapidly under warm conditions (25 and 15°C, respectively, during the day and at night) than under cool conditions (20 and 10°C, respectively), but the damage was more severe at the lower temperatures. Young foliage is more susceptible than the old. Plants grown in greenhouses, in shade, or with high rates of nitrogen fertilizer are more susceptible. Black cleistothecia, the overwintering stage of these fungi, form on deciduous azaleas, but their role in the survival of the fungi has not been determined. Dormant mycelium in infected buds may be important in the overwintering of the pathogens.

Powdery mildew seldom causes significant damage to evergreen azaleas. Some deciduous azalea cultivars appear to have some resistance and offer the best method of avoiding the disease.

Host Range

M. penicillata has a very wide host range among ornamental plants and trees. Although many powdery mildew pathogens are biologically specialized, *M. penicillata* apparently is not. Thus, many other taxa in the nursery and landscape could serve as a source of inoculum for powdery mildew infections in azalea.

Management

See Chapter 15, "Powdery Mildew."

Botrytis Gray Mold

Botrytis cinerea

Botrytis gray mold may develop on young leaves and stems of azaleas that are overwintered in plastic houses with minimal heat if humid conditions or films of moisture develop on the foliage during abnormally rainy periods in the winter. Since azaleas are normally well hardened before they are placed in overwintering structures, and since these structures are usually unheated (except by solar heating), the disease should be observed only in rare situations. The fungal pathogen, *Botrytis cinerea,* does not normally invade healthy plant tissue. It prefers to colonize dead or dying plant tissue, such as fallen leaves and petals, and this material may serve as a source of inoculum during overwintering.

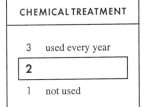

In the spring, when azaleas flower, gray mold may develop on blossoms during cool, wet periods, but petal blight (caused by *Ovulinia azaleae*) is more prevalent in the South. During the summer growing period, conditions for gray mold development occur only rarely, and then only in the northern range of azalea production.

For information on symptoms, epidemiology, and management, see Chapter 9, "Botrytis Blight (Gray Mold)."

Phomopsis Dieback

***Phomopsis* sp.**

Phomopsis dieback develops on stems of azaleas under environmental stress, such as drought. However, this disease rarely occurs in nursery production. In the landscape, large plants of the Southern Indica type are the most susceptible.

The primary symptoms of dieback are the death of leaves and stems on portions of the top and a reddish brown discoloration of the wood in diseased stems. Necrotic leaves may remain attached to stems for one season.

The pathogen is a *Phomopsis* sp., which is chiefly a wound pathogen affecting stem tissue. Stem wounds up to eight days old are susceptible. After stem infection, the disease progresses up and down the stem, killing entire branches. Pruning wounds are probably the most important infection sites.

Preventing moisture stress and stem splitting due to cold injury are important control methods. Pruning diseased stems is effective for individual plants.

Anthracnose

Colletotrichum gloeosporioides

Anthracnose, caused by the fungus *Colletotrichum gloeosporioides* (syn. *C. azaleae*), causes leaf spotting and defoliation of Indica and Kurume hybrid azaleas. This disease is uncommon in evergreen azaleas unless the plants are under stress.

The first symptoms of leaf spotting appear on young leaves. Mature leaves are generally not infected, but young shoots can be severely defoliated. The pathogen forms fruiting bodies (acervuli) containing asexual spores (conidia) in young leaves after they have fallen. Therefore, control should include removal of dead leaves from the surface of the potting medium.

Anthracnose and leaf spots caused by *Cercospora, Pestalotia,* and other fungi are common during fall and winter on senescent leaves of plants weakened by root rot.

Botrytis Gray Mold

Geographic occurrence USDA zones 6–10

Seasonal occurrence December to May, with an optimum period in April, under average yearly conditions in zone 8 (adjust for your hardiness zone)

DISEASE FREQUENCY	DISEASE SEVERITY
5 annual	5 plants killed
4	4
3	3
2	**[2]**
[1] rare	1 very little damage

CHEMICAL TREATMENT	CULTURAL PRACTICES
3 used every year	3 very important
[2]	**[2]**
1 not used	1 not important

SANITATION	RESISTANT CULTIVARS
3 very important	3 many cultivars
[2]	2
1 not important	**[1] no resistance**

Phomopsis Dieback

Geographic occurrence USDA zones 7–10

Seasonal occurrence April to October, with an optimum period in July–August, under average yearly conditions in zone 8 (adjust for your hardiness zone)

DISEASE FREQUENCY	DISEASE SEVERITY
5 annual	5 plants killed
4	4
3	**[3]**
2	2
[1] rare	1 very little damage

CHEMICAL TREATMENT	CULTURAL PRACTICES
3 used every year	3 very important
2	2
[1] not used	**[1] not important**

SANITATION	RESISTANT CULTIVARS
3 very important	3 many cultivars
2	2
[1] not important	**[1] no resistance**

Preventative fungicide applications have been effective in control in nurseries where anthracnose has been a problem.

Cylindrocladium Blight and Root Rot

Cylindrocladium scoparium

Cylindrocladium blight and root rot is primarily a disease of azaleas produced for the florist trade. It is rare in nursery-grown azaleas.

In the greenhouse, the fungus *Cylindrocladium scoparium* attacks leaves, stems, and roots of evergreen azaleas. Cuttings become infected during propagation when they are collected from stock plants infected with Cylindrocladium leaf spot. During rooting under mist, conidia of *Cylindrocladium* from infected cuttings are dispersed to healthy cuttings. Symptomless but infected cuttings spread the pathogen. Infected azalea leaves fall to the surface, and spores (conidia) formed on them serve as inoculum for subsequent infection of new roots.

The disease is most severe under humid conditions and often appears as a leaf spot, but diseased plants may also be affected with root rot or a sudden wilting of the top. It has been reported that the wilt phase of the disease is aggravated in plants subjected to overwatering, overfertilizing, high levels of salts, or other stress factors.

Disease-free cuttings should be placed in rooting medium that is free of the pathogen. Healthy cuttings can be collected from clean stock plants. If the rooting medium is to be used again, it should be sterilized after the crop of rooted cuttings has been removed. Cuttings that are contaminated by conidia without causing infections should be sanitized prior to sticking for propagation under warm, moist conditions.

Phytophthora Dieback

Phytophthora parasitica

Phytophthora dieback of azalea, caused by the fungus *Phytophthora parasitica* (syn. *P. nicotianae*), occurs in nurseries in Florida and elsewhere in the Southeast. The disease does not appear to occur widely or frequently.

Foliar symptoms appear as dark brown to black, irregularly shaped lesions on leaves. From the leaves, the pathogen invades shoots, causing dieback and eventual death of the shoots. Infected

Anthracnose

Geographic occurrence USDA zones 7–10

Seasonal occurrence May to October, under average yearly conditions in zone 9 (adjust for your hardiness zone)

DISEASE FREQUENCY	DISEASE SEVERITY
5 annual	5 plants killed
4	4
3	3
2	2
1 rare	**1 very little damage**

CHEMICAL TREATMENT	CULTURAL PRACTICES
3 used every year	3 very important
2	2
1 not used	**1 not important**

SANITATION	RESISTANT CULTIVARS
3 very important	3 many cultivars
2	2
1 not important	**1 no resistance**

Cylindrocladium Blight and Root Rot

Geographic occurrence USDA zones 6–10

Seasonal occurrence April to September, under average yearly conditions in zone 9 (adjust for your hardiness zone)

DISEASE FREQUENCY	DISEASE SEVERITY
5 annual	5 plants killed
4	4
3	**3**
2	2
1 rare	1 very little damage

CHEMICAL TREATMENT	CULTURAL PRACTICES
3 used every year	3 very important
2	**2**
1 not used	1 not important

SANITATION	RESISTANT CULTIVARS
3 very important	3 many cultivars
2	2
1 not important	**1 no resistance**

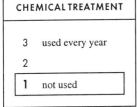

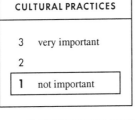

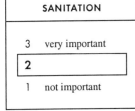

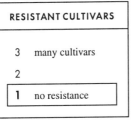

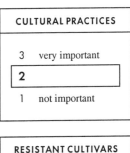

Phytophthora Dieback

Geographic occurrence USDA zones 8–10

Seasonal occurrence July to August

DISEASE FREQUENCY	DISEASE SEVERITY
5 annual	**5 plants killed**
4	**4**
3	**3**
2	2
1 rare	1 very little damage

CHEMICAL TREATMENT	CULTURAL PRACTICES
3 used every year	3 very important
2	2
1 not used	**1 not important**

SANITATION	RESISTANT CULTIVARS
3 very important	**3 many cultivars**
2	2
1 not important	1 no resistance

leaves tend to remain attached to the stems and eventually dry up completely. In small plants, shoot infection may kill the entire plant in a few weeks.

The pathogen infects leaves when it is splashed from the surface of potting mix in containers during thunderstorms. Cloudy weather following a thunderstorm promotes infection and stem dieback.

The cultivar Hershey Red is particularly susceptible to dieback.

For more information, see Chapter 14, "Phytophthora Root Rot and Dieback."

REFERENCES

Benson, D. M. 1985. Phytophthora dieback of azalea and spatial pattern of diseased plants in North Carolina nurseries. (Abstr.) Phytopathology 75:1382.

Benson, D. M., and Cochran, F. D. 1980. Resistance of evergreen hybrid azaleas to root rot caused by *Phytophthora cinnamomi.* Plant Dis. 64: 214–215.

Bir, R. E., Jones, R. K., and Benson, D. M. 1981. Susceptibility of selected deciduous azalea cultivars to azalea rust. (Abstr.) Proc. South. Nurserymen's Assoc. Res. Conf. 26:116–117.

Cox, R. S. 1969. *Cylindrocladium scoparium* on azalea in south Florida. Plant Dis. Rep. 53:139.

Dirr, M. A. 1998. Manual of Woody Landscape Plants: Their Identification, Ornamental Characteristics, Culture, Propagation and Uses. 5th ed. Stipes Publishing, Champaign, Ill.

Farr, D. F., Bills, G. F., Chamuris, G. P., and Rossman, A. Y. 1989. Fungi on Plants and Plant Products in the United States. American Phytopathological Society, St. Paul, Minn.

Frisina, T. A., and Benson, D. M. 1987. Characterization and pathogenicity of binucleate *Rhizoctonia* spp. from azaleas and other woody ornamentals with web blight. Plant Dis. 71:977–981.

Frisina, T. A., and Benson, D. M. 1988. Sensitivity of binucleate *Rhizoctonia* spp. and *R. solani* to selected fungicides in vitro and on azalea under greenhouse conditions. Plant Dis. 72:303–306.

Frisina, T. A., and Benson, D. M. 1989. Occurrence of binucleate *Rhizoctonia* spp. on azalea and spatial analysis of web blight in container-grown nursery stock. Plant Dis. 73:249–254.

Galle, F. 1987. Azaleas. Timber Press, Portland, Ore.

Jones, R. K., Bir, R. E., and Benson, D. M. 1981. Rust of deciduous azaleas. (Abstr.) Proc. South. Nurserymen's Assoc. Res. Conf. 26:122–123.

Linderman, R. G. 1974. The role of abscised *Cylindrocladium*-infected azalea leaves in the epidemiology of Cylindrocladium wilt of azaleas. Phytopathology 64:481–485.

Miller, S. B., and Baxter, L. W., Jr. 1970. Dieback in azaleas caused by *Phomopsis* sp. Phytopathology 60:387–388.

Peterson, J. L. 1986. Leaf and flower gall. Pages 21–22 in: Compendium of Rhododendron and Azalea Diseases. D. L. Coyier and M. K. Roane, eds. American Phytopathological Society, St. Paul, Minn.

Peterson, J. L. 1986. Ovulinia petal blight. Pages 22–24 in: Compendium of Rhododendron and Azalea Diseases. D. L. Coyier and M. K. Roane, eds. American Phytopathological Society, St. Paul, Minn.

Sinclair, W. A., Lyon, H. H., and Johnson, W. T. 1987. Diseases of Trees and Shrubs. Comstock Publishing Associates, Cornell University Press, Ithaca, N.Y.

Stathis, P. D., and Plakidas, A. G. 1958. Anthracnose of azaleas. Phytopathology 48:256–260.

Strider, D. L. 1976. Increased prevalence of powdery mildew of azalea and rhododendron in North Carolina. Plant Dis. Rep. 60:149–151.

Wehlburg, C., and Cox, R. S. 1966. Rhizoctonia leaf blight of azalea. Plant Dis. Rep. 50:354–355.

CHAPTER 24

Kala C. Parker · North Carolina State University, Raleigh

Barberry Diseases

Geographic production USDA zones 3–8

Family Berberidaceae

Genus *Berberis*

Species	*B. candidula*	paleleaf barberry	zones 5–8
	B. koreana	Korean barberry	zones 3–7
	B. × *mentorensis*	mentor barberry	zones 5–8
	B. thunbergii	Japanese barberry	zones 4–8

Barberries are sturdy, adaptable, and versatile ornamental landscape shrubs. They are used most often as a hedge plant; however, some cultivars are suitable in mass plantings, as ground covers, or as specimen plants. Barberries may be deciduous or evergreen, and they vary in growth habit and leaf color, but they are all sturdy plants with few disease problems.

Anthracnose, Phytophthora root rot, Verticillium wilt, and rust have all been reported in barberry but are unlikely to cause problems in nurseries where plants are vigorous. There is little chance of disease when plants are grown in a suitable climate and care is taken to avoid flooding, drought, mechanical injury, overfertilization, and untimely pruning.

Phytophthora Root Rot

Phytophthora cinnamomi
Phytophthora parasitica

Phytophthora root rot can cause gradual dieback of branches and can eventually kill the entire plant. The pathogen, *Phytophthora cinnamomi,* occurs in warm regions of the world. It is not widespread on barberry in nurseries but is more common on *B. thunbergii* (Japanese barberry).

Symptoms

Initial symptoms are similar to those of drought: chlorosis and bronzing of leaves and wilting. They are due to necrosis of small, absorbing roots and infection of progressively larger roots. Individual branches die, and eventually the entire plant can be killed.

Epidemiology

P. cinnamomi can survive in soil for several years, primarily as chlamydospores. Under favorable conditions (soil temperature above 50°F and soil moisture near saturation) chlamydospores germinate to form sporangia, which release zoospores, by which the pathogen is dispersed. Zoospores travel in films of water in poorly drained soils, in splashing rain, and in irrigation water. See Chapter 14, "Phytophthora Root Rot and Dieback."

Management

Cultural controls are very effective in the control of *P. cinnamomi.* Using clean media, pots, and planting stock and avoiding recycled water will ensure that the pathogen is not introduced into the nursery. Maintaining vigorous plants, especially when well-

Phytophthora Root Rot

Geographic occurrence USDA zones 7–9

DISEASE FREQUENCY	DISEASE SEVERITY
5 annual	**[5] plants killed**
4	4
3	3
2	2
[1] rare	1 very little damage

CHEMICAL TREATMENT	CULTURAL PRACTICES
3 used every year	**[3] very important**
[2]	2
1 not used	1 not important

SANITATION	RESISTANT CULTIVARS
[3] very important	3 many cultivars
2	**[2]**
1 not important	1 no resistance

drained potting mix is used, will go far in avoiding root rot of barberry caused by *P. cinnamomi.* Several fungicides are effective against *P. cinnamomi.*

REFERENCES

Agrios, G. N. 1988. Plant Pathology. 3rd ed. Academic Press, San Diego, Calif.

Dirr, M. A. 1998. Manual of Woody Landscape Plants: Their Identification, Ornamental Characteristics, Culture, Propagation and Uses. 5th ed. Stipes Publishing, Champaign, Ill.

Sinclair, W. A., Lyon, H. H., and Johnson, W. T. 1987. Diseases of Trees and Shrubs. Comstock Publishing Associates, Cornell University Press, Ithaca, N.Y.

CHAPTER 25

David L. Clement ▪ University of Maryland, Ellicott City

Birch Diseases

Geographic production USDA zones 2–9, depending on the species and cultivar

Family Betulaceae

Genus *Betula*

Species

B. lenta	sweet birch	zones 3–7
B. nigra	river birch	zones 4–9
B. papyrifera	paper birch	zones 2–6(7)
B. pendula	European white birch	zones 2–6(7)
B. populifolia	gray birch	zones 3–6(7)

Birches are often planted for the texture and color of their bark. River birch has gained significant interest in the South, so there is great demand for it in nurseries. Several diseases are common in birch, including diebacks, cankers, leaf spots, anthracnose, rust, and wood decay.

Upper Branch Dieback and Canker Diseases

Botryosphaeria dothidea
Botryosphaeria obtusa
Nectria cinnabarina
Nectria galligena
***Valsa* spp.**

Several fungal pathogens, which are distributed worldwide, cause canker diseases of twigs and small branches of birches. These pathogens are opportunistic and are frequently associated with trees stressed by drought, defoliation, wind damage, nutrient imbalances, flooding, compacted soils, ice and snow damage, hail, improper pruning cuts, mechanical damage, transplanting stresses, and injuries from animals, such as birds and insects (e.g., wood borers). These diseases generally cause only minor dieback, but they may be indicative of overall poor health.

Symptoms

Young cankers are slightly darker than adjacent healthy bark and appear slightly sunken. As they enlarge, they kill living woody tissue within the branch or trunk. Canker growth may cause the bark along the edges to crack and fall away, exposing dead wood underneath. After a canker enlarges enough to girdle a branch or trunk, the portion above the canker dies. Small twigs are killed more quickly than larger branches. Symptoms may include progressive upper branch dieback, disfigured branch growth, and target-shaped areas with concentric rings of dead bark on trunks.

Epidemiology

Cankers caused by *Nectria* often begin on small branch stubs or wounds. The cankers are perennial and slowly enlarge over

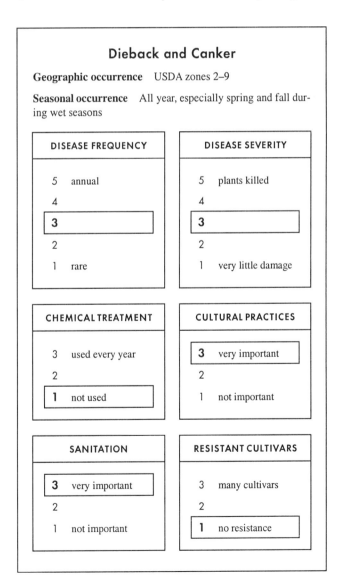

Dieback and Canker

Geographic occurrence USDA zones 2–9

Seasonal occurrence All year, especially spring and fall during wet seasons

DISEASE FREQUENCY	DISEASE SEVERITY
5 annual	5 plants killed
4	4
3 (boxed)	**3** (boxed)
2	2
1 rare	1 very little damage

CHEMICAL TREATMENT	CULTURAL PRACTICES
3 used every year	**3 very important** (boxed)
2	2
1 not used (boxed)	1 not important

SANITATION	RESISTANT CULTIVARS
3 very important (boxed)	3 many cultivars
2	2
1 not important	**1 no resistance** (boxed)

time. As a canker enlarges, the fungus attacks new callus tissue at the edge of the canker, creating a target pattern. In the spring and early summer, the pathogen produces spores (conidia) in a gelatinous mass in sporodochia, which are typically coral-colored. Conidia are spread from cankers by water movement. Reddish brown fruiting bodies (perithecia) are formed in late summer and fall and release ascospores, which are dispersed by water.

Cankers caused by *Botryosphaeria* typically grow rapidly and cause girdling cankers on small branches. Within cankers, the fungus produces fruiting bodies (pycnidia), which exude conidia in sticky spore tendrils. Conidia are spread by moisture or insects to new infection sites. Where pycnidia break the surface, they cause a conspicuous roughening and darkening of the bark. Perithecia are produced within cankers and release ascospores, generally after the second growing season, and predominately in the spring.

Cankers caused by *Valsa* are sunken and discolored and frequently form at the base of twigs or branches. Conidia and ascospores are produced within cankers and are released during wet weather.

Management

There are no chemical controls for these canker diseases, and they cannot be stopped once they become extensive. Pruning affected branches back to healthy wood is the only control measure available. Stressed trees should be fertilized and watered during dry periods, to promote better tree vigor. Proper plant selection and site location will promote better tree vigor and growth.

Leaf Spots, Leaf Blotches, Anthracnose, and Leaf Blister

Cylindrosporium betulae
Discula betulina
Glomerella cingulata
(asexual stage, *Colletotrichum gloeosporioides*)
Marssonina betulae
Septoria betulae
Taphrina carnea
Taphrina flava

Leaf spots, leaf blotches, anthracnose, and leaf blister are widespread and can cause severe problems during wet weather. Generally, these diseases are not severe enough in the landscape to warrant spraying.

Symptoms

Leaf diseases of birch can be caused by several different fungi.

Septoria and *Cylindrosporium* produce small spots without definite borders.

Colletotrichum causes larger brown spots with brown to black borders.

Discula and *Marssonina* cause anthracnose diseases, characterized by broader large spots called leaf blotches, which under optimum disease conditions run together to form larger blotches. Leaf blotches tend to have indefinite margins, and the remaining leaf tissue turns yellow (Plate 17). Other leaf symptoms range from small distinct spots to raised or puckered areas. Severe anthracnose can cause premature defoliation.

Taphrina carnea causes reddish leaf blisters and curling. *T. flava* causes light green to yellow leaf blisters on birch leaves in the spring. Later in the summer these blistered or puckered areas usually turn brown and fall out, leaving holes in the leaves.

Management

Leaf diseases of birch in the landscape are usually not severe enough to warrant chemical spraying for disease control. Symptoms are more severe in cool, wet spring weather. For young trees in the nursery, on which it is more critical to maintain an adequate number of healthy leaves, sprays of copper or sulfur fungicides during budbreak and leaf expansion can lessen disease severity. Practical control also includes raking up fallen leaves.

Leaf Rust

Melampsoridium betulinum

Leaf rust, caused by the fungus *Melampsoridium betulinum*, can be easily identified by the powdery, bright orange yellow pustules that form on the lower surfaces of leaves. The symptoms are generally angular brown spots with yellow borders on the upper surfaces of leaves. Spots may run together to form large lesions, resulting in defoliation. The pathogen also causes a blister rust of larch. In the forest both hosts may be severely affected. Leaf rust

Leaf Spots

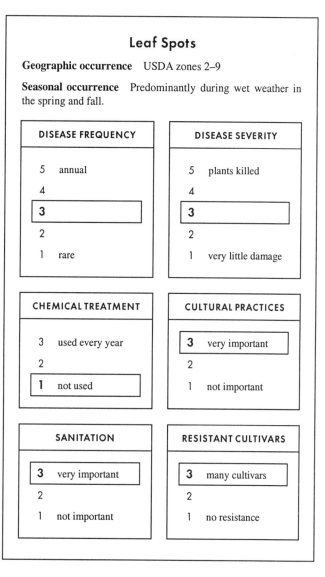

Geographic occurrence USDA zones 2–9

Seasonal occurrence Predominantly during wet weather in the spring and fall.

DISEASE FREQUENCY	DISEASE SEVERITY
5 annual	5 plants killed
4	4
3	**3**
2	2
1 rare	1 very little damage

CHEMICAL TREATMENT	CULTURAL PRACTICES
3 used every year	**3 very important**
2	2
1 not used	1 not important

SANITATION	RESISTANT CULTIVARS
3 very important	**3 many cultivars**
2	2
1 not important	1 no resistance

can persist indefinitely in birches independent of larches. Symptoms in birch are more severe in the spring and fall.

Epidemiology

M. betulinum has four spore stages and forms four types of spore-bearing fruiting bodies: spermagonia and aecia, which are produced on larch needles, and uredinia and telia, which are produced on birch. Urediniospores can infect birch leaves during cool weather. Periods of rain or dew can provide the necessary moisture for spore germination and infection. Normally, birch rust does not increase in severity until early autumn, when cooler conditions favor urediniospore germination and infection. Both urediniospores on bud scales and teliospores on fallen leaves can overwinter and serve as a source of inoculum for new infections in the spring.

Management

Cleanup of fallen leaves and sprays of copper-based fungicides in the spring during budbreak and leaf expansion will help to lessen disease severity.

Wood Decay Fungi

Fomes fomentarius
Ganoderma applanatum
Inonotus obliquus
Phellinus laevigatus
Piptoporus betulinus

Wood decay fungi usually attack weakened or stressed trees and live in the wood where they cause decay. They rarely attack birch in the nursery.

Symptoms

Dead or dying branches become rotten, and shelf- or hoof-shaped structures develop, growing out of the bark.

Management

There are no chemical sprays to control wood decay fungi. Prune back infected portions to healthy wood, avoid wounding, and maintain trees in good vigor.

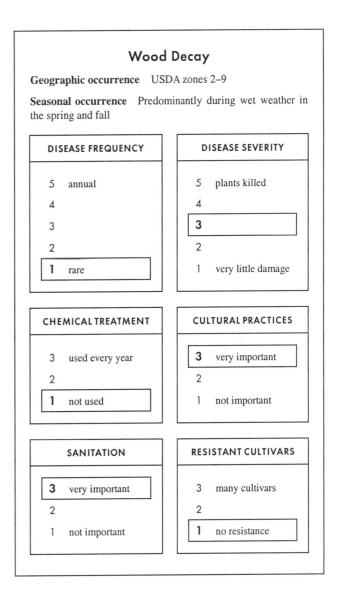

Leaf Rust

Geographic occurrence USDA zones 2–7 in California

Seasonal occurrence Predominantly during wet weather in the spring and fall.

DISEASE FREQUENCY	DISEASE SEVERITY
5 annual	5 plants killed
4	4
3	**3**
2	2
1 rare	1 very little damage

CHEMICAL TREATMENT	CULTURAL PRACTICES
3 used every year	**3 very important**
2	2
1 not used	1 not important

SANITATION	RESISTANT CULTIVARS
3 very important	**3 many cultivars**
2	2
1 not important	1 no resistance

Wood Decay

Geographic occurrence USDA zones 2–9

Seasonal occurrence Predominantly during wet weather in the spring and fall

DISEASE FREQUENCY	DISEASE SEVERITY
5 annual	5 plants killed
4	4
3	**3**
2	2
1 rare	1 very little damage

CHEMICAL TREATMENT	CULTURAL PRACTICES
3 used every year	**3 very important**
2	2
1 not used	1 not important

SANITATION	RESISTANT CULTIVARS
3 very important	3 many cultivars
2	2
1 not important	**1 no resistance**

REFERENCES

Dirr, M. A. 1998. Manual of Woody Landscape Plants: Their Identification, Ornamental Characteristics, Culture, Propagation and Uses. 5th ed. Stipes Publishing, Champaign, Ill.

Rane, K., and Pataky, N. 1997. Managing infectious plant diseases. Chapter 7 in: Plant Health Care for Woody Ornamentals, A Professional's Guide to Preventing and Managing Environmental Stresses and Pests. University of Illinois, Urbana.

Sinclair, W. A., Lyon, H. H., and Johnson, W. T. 1987. Diseases of Trees and Shrubs. Comstock Publishing Associates, Cornell University Press, Ithaca, N.Y.

CHAPTER 26

Larry W. Barnes ▪ Texas A&M University, College Station
Robert L. Wick ▪ University of Massachusetts, Amherst
D. Michael Benson ▪ North Carolina State University, Raleigh

Boxwood Diseases

Geographic production USDA zones 5–9

Family Buxaceae

Genus *Buxus*

Species		
B. microphylla	littleleaf boxwood	zone 6
B. microphylla var. *japonica*	Japanese boxwood	zones 6–9
B. microphylla var. *koreana*	Korean boxwood	zone 6
B. sempervirens	common boxwood	zones 5(6)–8
B. sempervirens var. *suffruticosa*	English box	zone 6

Boxwoods are grown for their color, texture, form, and aroma. They have been planted extensively in formal gardens of the Atlantic Coast states dating back to colonial days. The most important diseases of boxwoods in the nursery and landscape are Phytophthora root rot and infection by nematodes.

Phytophthora Root and Crown Rot

Phytophthora cinnamomi
Phytophthora parasitica

Phytophthora root and crown rot occurs in all boxwood species. The disease is especially prevalent in the southeastern and Gulf Coast states but can occur wherever conditions allow the production of boxwood.

Symptoms

Initial symptoms include dull, off-color foliage that subsequently develops a tan, desiccated appearance. Leaf tissue may become twisted and distorted as it desiccates. Individual sections of the plant canopy may brown as root and crown rot progresses.

Lower stem and crown tissue turns from the normal cream white to colors ranging from dull gray to dark brown. Roots turn dark brown. As they deteriorate, the cortex commonly sloughs from the root cylinder. Plant death follows advanced deterioration of the root and crown.

Epidemiology

Wet, poorly drained soils and irrigation scheduling that promotes prolonged soil saturation are conducive to plant infection and disease development. Although *Phytophthora* spp. are commonly soilborne, recycled irrigation water can disseminate them. Plant infection and disease development are greater at soil temperatures above 85°F (30°C) in infections by *P. parasitica* and above 72°F (22°C) in infections by *P. cinnamomi.* See Chapter 14, "Phytophthora Root Rot and Dieback."

Phytophthora Root and Crown Rot

Geographic occurrence USDA zones 7 and 8, with some occurrence in zone 6

Seasonal occurrence May to August, with an optimum period in July

DISEASE FREQUENCY	DISEASE SEVERITY
[5 annual]	**[5 plants killed]**
4	4
3	3
2	2
1 rare	1 very little damage

CHEMICAL TREATMENT	CULTURAL PRACTICES
3 used every year	**[3 very important]**
[2]	2
1 not used	1 not important

SANITATION	RESISTANT CULTIVARS
[3 very important]	3 many cultivars
2	2
1 not important	**[1 no resistance]**

Host Range

All boxwood species are susceptible to Phytophthora root and crown rot.

Management

Careful attention to sanitation and cultural practices are important. Fungicide drenches may be useful to augment cultural procedures but should not be relied on as an exclusive means of disease management.

Cultural practices Healthy, pathogen-free liners should be planted in a well-drained growing medium. Schedule irrigation based on plant water consumption, to avoid overwatering. Production beds should be crowned and graveled to eliminate standing water.

Sanitation Avoid reuse of pots from previous crops. If pots must be reused, wash off all soil and organic debris from the previous crop, and soak the pots in a sanitizing solution prior to reuse. Periodically scout the production area, and remove symptomatic plants from the growing area.

Resistance No resistance to Phytophthora root and crown rot is known.

Chemical treatment Application of fungicide drenches to well-drained media, with careful irrigation scheduling and other cultural practices, can aid in *Phytophthora* disease management. Drenches should be applied in sufficient liquid volume to ensure that the entire soil profile and the associated root mass is thoroughly wetted by the fungicide.

Root-Knot Nematodes

Meloidogyne arenaria
Meloidogyne hapla
Meloidogyne incognita
Meloidogyne javanica

Root-knot nematodes are important pathogens of boxwood, especially in southern regions of the United States. Several species of *Meloidogyne* induce root knots or galls in many species of plants. Infection by root-knot nematodes can contribute to or aggravate several fungal diseases of boxwood.

Symptoms

Root-knot nematode infection results in stunting, chlorosis, nutritional deficiencies, wilting, reduced vigor, progressive plant decline, and other symptoms typical of plants with compromised root systems (Plate 18). Typically, in *B. microphylla* var. *japonica* infected with root-knot nematodes, particularly *M. arenaria,* inner leaves in the plant canopy are orange to bronze-colored. Overall, symptom expression is influenced by the nematode population, the extent of root infection and root damage, and the impact of environmental stresses on plants with reduced functional root capacity.

Root symptoms include the formation of swollen galls throughout the root system. The galls are usually firm, white, and round to irregularly elongated. Old galls may deteriorate and leave discolored areas on the root.

Epidemiology

Infested soil or infected plants are common sources of initial root-knot nematode introduction into previously nematode-free production areas. The nematodes survive in egg masses laid by females and as larval stages in the soil and infected roots. Splashing or flowing rain or irrigation water can disseminate root-knot nematodes over large distances.

Infested soil particles or infected plant debris can also introduce root knot nematodes into or throughout a production area. A new generation of root-knot nematodes can be produced in three to four weeks when temperatures are between 77 and 86°F (25–30°C).

Host Range

Root-knot nematodes are most commonly found on Japanese boxwood, *B. microphylla* var. *japonica.*

Management

Strict sanitation is the most important consideration in preventing or managing root-knot nematodes.

Sanitation Components of potting media should be periodically tested to ensure that they are nematode-free. Store media on concrete pads and not in direct contact with soil. For in-ground

Root-Knot Nematodes

Geographic occurrence USDA zones 7 and 8

Seasonal occurrence New infections in late spring (May to June), with subsequent infection cycles every three to four weeks through August–September; root galls are perennial

DISEASE FREQUENCY	DISEASE SEVERITY
5 annual	5 plants killed
4	**4**
3	**3**
2	**2**
1 rare	**1** very little damage

CHEMICAL TREATMENT	CULTURAL PRACTICES
3 used every year	**3** very important
2	2
1 not used	1 not important

SANITATION	RESISTANT CULTIVARS
3 very important	3 many cultivars
2	2
1 not important	**1** no resistance

production, soil should be tested and found to be free of root-knot nematodes. Tools and implements should be washed free of adhering soil and particles of potting media when they are moved from site to site across the nursery, to avoid moving root-knot nematodes on them. Avoid reusing pots, containers, and trays.

Resistance No resistance to root-knot nematodes is known.

Chemical treatment Areas of field production should be fumigated prior to planting. In containerized production, containers can be drenched with materials having nematicidal properties, but the efficacy of the control can be highly variable.

Lesion Nematode

Pratylenchus vulnus

The lesion nematode *Pratylenchus vulnus,* also called the meadow nematode, can devastate plantings of boxwood in field production sites and in the landscape. Nematodes typically induce more damage to plants growing in sandy soils, but lesion nematodes are prevalent and damaging in boxwoods in loam soils of the Piedmont region of the eastern United States.

Lesion Nematode

Geographic occurrence USDA zones 6–8

Seasonal occurrence Infection occurs from May to June, with repeated cycles of eggs and infectious larvae produced throughout the growing season

DISEASE FREQUENCY	DISEASE SEVERITY
5 annual	5 plants killed
4	**4**
3	3
2	2
1 rare	1 very little damage

CHEMICAL TREATMENT	CULTURAL PRACTICES
3 used every year	**3 very important**
2	2
1 not used	1 not important

SANITATION	RESISTANT CULTIVARS
3 very important	3 many cultivars
2	**2**
1 not important	1 no resistance

Symptoms

Unlike the root-knot nematode, lesion nematodes live both outside and inside host roots. Inside the roots they create necrosis along the root as they move and feed. This disrupts the normal uptake of water and minerals by the plant, so that decline symptoms develop. Foliar symptoms induced by lesion nematodes are similar to those caused by root-knot nematodes (Plates 19 and 20). Infected boxwood does not grow when even a small population of lesion nematodes is present in the root system. Leaves may turn orange or bronze. Dieback may occur. Affected plants may decline markedly.

Epidemiology

Eggs of the lesion nematode hatch in soil, and the nematode infects roots in its larval stages. Lesion nematodes are migratory and endoparasitic. As soon as soils warm up in the spring, the larvae become active. They enter root tissues by using their stylet to separate epidermal and cortical cells. Once inside they continue to move throughout the root system. Although the number of lesion nematodes in the soil declines during the winter, enough of them survive in infected roots to continue damaging the plant during the next growing season.

Host Range

Among boxwood species, *B. sempervirens* is the most susceptible to the lesion nematode. Other woody ornamental hosts include rose and juniper cultivars Blue Rug and Spiny Greek (see Table 13.1).

Management

Lesion nematode damage to boxwoods in the nursery can be avoided by use of a nematode-free soilless potting mix. Containers should be placed on ground cloth or gravel to prevent splashing of soil, which can disperse nematodes. Most postplant nematicides are ineffective, so strict sanitation is critical.

In field production of boxwoods, avoiding lesion nematodes may be difficult, because they are associated with the root systems of many other nursery crops and native plants in woodlands, as well as boxwoods. Movement of soil between fields or from natural areas should be avoided. Drainage patterns should be modified to prevent runoff that could carry nematodes from one field to another. Rotation of production fields to nonsusceptible hosts may help reduce damage from lesion nematodes. Preplant fumigation is an expensive option but may be required in regions where lesion and root-knot nematodes are prevalent.

English Boxwood Decline

Unknown etiology

English boxwood decline is a disease of *B. sempervirens* var. *suffruticosa* 'English.' It first became evident in 1970 in Virginia, when historic plantings of boxwoods were found to have severe dieback. Boxwood decline is not limited to old, established plantings, but it usually does not occur in container-grown nursery stock.

A considerable effort was made at Virginia Polytechnic Institute and State University to identify the cause of English boxwood decline. *Paecilomyces buxi, Fusarium,* and *Phoma* are readily cultured from roots and crowns of plants exhibiting decline symptoms. *P. buxi* has often been presented as the likely cause of boxwood decline, but conclusive evidence is lacking. Root-lesion nematodes (*Pratylenchus*) and a spiral nematode (*Rotylenchus buxophilus,* the boxwood nematode) may also play a role in the decline. Drought, mineral deficiencies, and pruning practices have all been suggested as possible causes of English boxwood decline. It is likely that the disease has a complex etiology.

Symptoms

Dieback occurs in sectors of individual plants and randomly in plantings. Leaves fall from stems, and stems are desiccated and die back to the main stem or crown. Roots and crowns show various amounts of rot and streaking of the wood. *P. buxi* is easily cultured from these tissues; young cultures look very much like those of *Verticillium.*

Management

An integrated approach to management is important. Prevent drought stress, apply adequate nutrients, and follow appropriate cultural practices. No fungicides have been identified that will control English boxwood decline.

Volutella Leaf and Stem Blight
Volutella buxi

Volutella leaf and stem blight is a common disease of English and American boxwoods, resulting in defoliation and death of stems. It is common in unthrifty boxwood but occasionally occurs in apparently healthy specimens.

Symptoms

The disease is easily recognized by the appearance of cream-colored to light pink, mealy-appearing growth on the lower surfaces of leaves. On blighted stems, domelike spore masses, approximately 1/16 inch in diameter, erupt through the epidermis. Infected leaves fall, and stems die back.

Epidemiology

The pathogen, *Volutella buxi,* can survive in soil and has been reported to cause root and crown rot. There is very little information regarding the epidemiology of this disease.

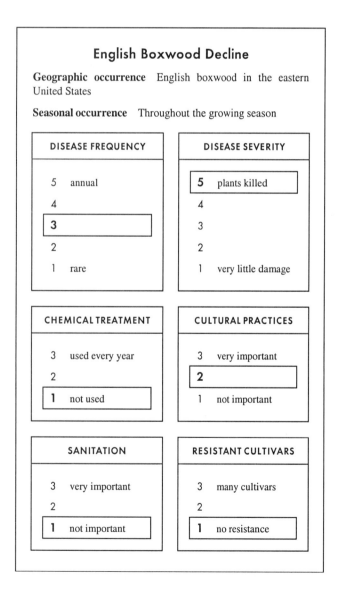

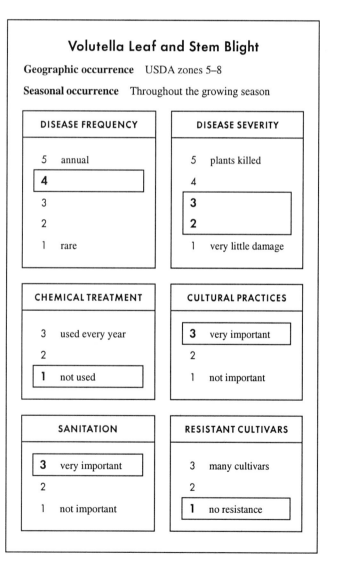

English Boxwood Decline

Geographic occurrence English boxwood in the eastern United States

Seasonal occurrence Throughout the growing season

DISEASE FREQUENCY
- 5 annual
- 4
- **3**
- 2
- 1 rare

DISEASE SEVERITY
- **5 plants killed**
- 4
- 3
- 2
- 1 very little damage

CHEMICAL TREATMENT
- 3 used every year
- 2
- **1 not used**

CULTURAL PRACTICES
- 3 very important
- **2**
- 1 not important

SANITATION
- 3 very important
- 2
- **1 not important**

RESISTANT CULTIVARS
- 3 many cultivars
- 2
- **1 no resistance**

Volutella Leaf and Stem Blight

Geographic occurrence USDA zones 5–8

Seasonal occurrence Throughout the growing season

DISEASE FREQUENCY
- 5 annual
- **4**
- 3
- 2
- 1 rare

DISEASE SEVERITY
- 5 plants killed
- 4
- **3**
- **2**
- 1 very little damage

CHEMICAL TREATMENT
- 3 used every year
- 2
- **1 not used**

CULTURAL PRACTICES
- **3 very important**
- 2
- 1 not important

SANITATION
- **3 very important**
- 2
- 1 not important

RESISTANT CULTIVARS
- 3 many cultivars
- 2
- **1 no resistance**

Management

Prevent winter desiccation, and maintain plant vigor through the growing season. No fungicides are available for the control of this disease.

Macrophoma Leaf Spot

***Macrophoma* sp.**

Macrophoma leaf spot is common in boxwoods of low vigor. The fungal pathogen, a *Macrophoma* sp., occasionally colonizes stem tissue. The disease is not usually observed in vigorous plants.

Symptoms

The disease is easy to recognize by the spore-bearing pycnidia of the pathogen, which appear as conspicuous black dots on leaves, produced fairly uniformly throughout the leaf.

Epidemiology

Like *Phoma, Macrophoma* can survive in soil. Very little is known about the epidemiology of this disease.

Management

Prevent winter desiccation, and maintain plant vigor through the growing season. No fungicides are available for the control of this disease.

REFERENCES

Bell, D. K., and Haasis, F. A. 1967. Etiology and epiphytology of root rot, stem necrosis and foliage blight of boxwood caused by *Phytophthora parasitica* Dastur. N.C. Agric. Exp. Stn., Raleigh, Tech. Bull. 177.

Benson, D. M., and Barker, K. R. 1982. Susceptibility of Japanese boxwood, dwarf gardenia, compacta (Japanese) holly, Spiny Greek and blue rug junipers, and Nandina to four nematode species. Plant Dis. 66:1176–1179.

Benson, D. M., Barker, K. R., and Aycock, R. 1976. Effects of density of *Helicotylenchus dihystera* and *Pratylenchus vulnus* on American boxwood growing in microplots. J. Nematol. 8:322–326.

Haasis, F. A. 1961. *Phytophthora parasitica,* the cause of root rot, canker, and blight of boxwood. Phytopathology 51: 734–736.

CHAPTER 27

Steven N. Jeffers ▪ Clemson University, Clemson, South Carolina

Luther W. Baxter, Jr. ▪ Clemson University, Clemson, South Carolina

Camellia Diseases

Geographic production USDA zones 7–9

Family *Theaceae*

Genus *Camellia*

Species *C. japonica*
C. sasanqua
C. hiemalis
C. oleifera
C. reticulata
several others, including hybrids

Cultivars thousands, too numerous to list

Camellias are evergreen shrubs with dark green leaves. They are particularly popular in the Southeast, Gulf Coast, and Pacific Coast regions, where winters are mild. They can be trained to a single stem (or leader) and, therefore, can grow into relatively tall trees. Consequently, camellias are quite versatile for creating screens, borders, and backdrops, although they grow relatively slowly. Camellias are grown primarily for their large, colorful, and showy flowers, which bloom in the late fall through early spring.

Camellia sinensis, the commercial tea plant, is grown occasionally as an ornamental plant.

Flower Blight

Ciborinia camelliae

Flower blight affects only open flowers. It does not affect leaves, stems, roots, unopened buds, or fruit. The disease is caused by the ascomycete fungus *Ciborinia camelliae,* previously known as *Sclerotinia camelliae.* This pathogen is usually active from January through April, which coincides with the normal flowering period of *Camellia japonica.*

Symptoms

Infected flowers develop light brown spots a few days after infection, depending on the temperature. The lesions are particularly noticeable on white and pink flowers and less noticeable on red flowers because of their darker color. As the disease progresses, the entire flower becomes brown, and gray mycelium of *C. camelliae* forms at the base of the petals near the calyx (Plate 22). Eventually some of the mycelium develops into small, black, hard sclerotia, which are the survival structures of the pathogen. The presence of sclerotia is a diagnostic characteristic of flower blight.

Two other fungal pathogens, *Sclerotinia sclerotiorum* and *Botrytis cinerea,* occasionally attack camellia flowers and cause similar symptoms, but neither of them produces sclerotia on camellia flowers.

Flower blight can be distinguished from frost injury by the brown color of diseased petals; in contrast, petals injured by frost are white to light tan. However, severe freezing injury results in a color similar to that resulting from flower blight. The disease can

Flower Blight

Geographic occurrence Wherever camellias are grown

Seasonal occurrence The blossoming period of *Camellia japonica,* usually January to April

DISEASE FREQUENCY		DISEASE SEVERITY	
5	**annual**	5	plants killed
4		4	
3		3	
2		2	
1	rare	**1**	**very little damage**

CHEMICAL TREATMENT		CULTURAL PRACTICES	
3	used every year	**3**	**very important**
2		2	
1	not used	1	not important

SANITATION		RESISTANT CULTIVARS	
3	**very important**	3	many cultivars
2		2	
1	not important	**1**	**no resistance**

be distinguished from severe freezing injury by the number of open flowers affected: all flowers in an area are usually affected by a severe freeze, whereas with flower blight some flowers almost always escape infection and remain healthy.

Certain cultivars of *C. japonica* (e.g., Betty Sheffield) develop brown spots when contacted by water. These spots can resemble flower blight lesions, but they do not enlarge and no sclerotia develop in them.

Epidemiology

The development of *C. camelliae* in nature has evolved to coincide very closely with that of its primary host, *C. japonica:* spores of the fungus are released during the blossom period, when susceptible plant tissue is available. Most cultivars of *C. japonica,* the most popular camellia species in the landscape and nursery, bloom from January through March or April; however, the blossom period of any one cultivar will vary, depending on the geographic location. Regardless of the time when camellias reach peak bloom, the pathogen is active and sporulating if it is present.

C. camelliae survives as sclerotia in the soil or in soilless mix in nursery containers. Sclerotia germinate to produce one to several apothecia, which are small (5–10 mm in diameter), tan, saucer-shaped fruiting bodies (Plate 23). Apothecia produce ascospores, which are ejected into the air and blown by the wind to surrounding plants. Some of the spores land on susceptible, open camellia flowers, where they can infect and cause flower blight. Millions of spores are released from each apothecium over a period of several days; most are deposited relatively close to the apothecium, but some can be blown up to one mile away. Ascospores are the only type of inoculum that infects camellia flowers; there is no subsequent production of conidia or recurring infection cycles. However, mycelium of the fungus may move from flower to flower, if the flowers are in direct contact. Eventually, new sclerotia are produced at the base of infected flowers when they fall to the ground. Sclerotia can survive in and on the soil or soilless container mix for several years; however, they usually germinate the following year and produce a fresh supply of apothecia and ascospores – which will initiate the flower blight cycle again.

Host Range

C. camelliae is host-specific, attacking only *Camellia* spp. All species of camellia can be affected, but flower blight tends to be most severe in *C. japonica,* because the development of flowers of this species and the release of ascospores by the pathogen are closely synchronized. *C. sasanqua* and other species, although susceptible, often avoid infection, because their flowers are open before ascospores are released.

Management

Cultural practices and sanitation provide the most effective management for flower blight in the nursery. Fungicides are useful only in situations where infestations have been or are expected to be severe. The best management strategy for this disease is to keep the pathogen out of the nursery, so that flower blight never gets established. Ascospores can be blown in from surrounding locations if the disease is present in nearby nurseries or landscapes, but this usually is unlikely.

Cultural practices Whenever possible, do not bring plants into the nursery that have or have had infected flowers. If plants are brought in from places where flower blight occurs or has occurred, insist on bare-rooted plants, which are less likely to harbor sclerotia of the pathogen. Keep weeds and other ground covers out from under or around camellia plants to prevent diseased flowers and sclerotia from being overlooked. Ground covers also may provide a moist environment that enhances sclerotium germination and apothecium production. If flower blight has been present, remove old mulch or the top layer of container mix from pots in early fall, to remove dormant sclerotia that might be present, and then add a thick layer of fresh container mix or mulch to the pots and fresh mulch to the ground around pots to cover any remaining sclerotia and prevent ascospore release. Prune lower branches so that air can circulate freely beneath the plants, which promotes drying; it also permits easy collection of senescent flowers.

Sanitation Collecting and destroying diseased flowers is an effective method of eliminating the pathogen from the nursery and preventing recurrence of the disease the following year. It is very important to be thorough in removing flowers from plants, collecting flowers that have dropped to the ground, and then destroying them. Even a single large petal can support sclerotium production if not removed. On plants too small to sell or where flowers are not needed to promote sales, all flower buds should be removed before they open to avoid infection altogether.

Resistance All cultivars of *C. japonica* and other species appear to be susceptible; no sources of resistance have been identified. However, the search for resistance to flower blight is a high priority among camellia enthusiasts.

Chemical treatment Fungicides are another option for managing flower blight in the nursery, but *only if the disease is or has been severe.* Fungicides can be applied to the ground around plants and to the surface of the container mix in pots to prevent the development of apothecia and subsequent release of ascospores. Proper timing of applications and thorough coverage are essential for this treatment to be effective. Applications should be made in the fall or early winter prior to flowering by thoroughly drenching soil and container mix surfaces around camellia plants. A repeat application may be necessary if the blossom period extends for more than four weeks. Fungicides also can be applied directly to open flowers to protect petals from infection by ascospores. However, because new flowers keep opening during the lengthy blossom period, frequent applications are necessary to maintain adequate protection – which may be neither economical nor feasible.

Dieback (Twig Blight or Canker)

Glomerella cingulata
(asexual stage, *Colletotrichum gloeosporioides*)

Dieback, also called twig blight or canker, can be a very serious disease, killing shoots, stems, larger branches, and even entire plants. It is most severe in warm, humid growing regions, where environmental conditions favor infection. It usually is not a serious problem on the West Coast, where semiarid conditions prevail.

Symptoms

The most prominent symptom is the death of new, succulent shoots in the spring. Initially, symptoms appear as wilting of newly formed leaves, followed by premature leaf drop. Later in the season, entire twigs and shoots turn brown and die, especially in the heat of the summer. Dead shoots and twigs are very prominent and produce a flagging on diseased plants. At the base of dead twigs, cankers form on woody stems and branches that are two years old or older (Plate 24). Infected shoots on one-year-old stems may be too small or die too quickly for cankers to develop. On very susceptible cultivars or species (e.g., *C. oleifera*), the disease may progress so quickly that larger stems (e.g., 1–2 cm in diameter) are invaded and killed before cankers form. In more resistant cultivars and species (e.g., *C. hiemalis* 'Kanjiro'), cankers may form but then heal within a year. Active cankers containing the pathogen can persist on woody stems for several years. Most camellias are propagated vegetatively, and dieback can kill cuttings and newly grafted plants if the pathogen invades wounds created during propagation.

Epidemiology

The asexual stage (anamorph) of the pathogen, *Colletotrichum gloeosporioides,* is the one that typically occurs on camellias. Perithecia, the sexual fruiting bodies of *Glomerella cingulata,* rarely are found on camellia cankers. Diseased plants with active, viable cankers serve as the primary source of inoculum in the nursery. Like many anthracnose fungi, *C. gloeosporioides* produces fruiting bodies (acervuli) on cankers, which release spores (conidia) in a gelatinous matrix under conditions of warm temperatures and prolonged periods of high relative humidity or free moisture. Spores are spread from plant to plant by splashing rain and irrigation water.

Infection occurs through wounds on twigs, stems, and the trunk. Any wound can serve as a point of ingress, but the pathogen gains entry primarily through fresh leaf scars created during the spring when older (i.e., two- to three-year-old) leaves senesce and fall off as new growth begins. On *C. sasanqua* and *C. oleifera* with leaf gall (caused by *Exobasidium camelliae*), the incidence of dieback is greater because galls provide points of entry for *G. cingulata.*

The pathogen survives primarily in cankers but also in apparently healthy vegetative buds. It can be spread over long distances on infected plants, moved from one nursery to another or into the landscape. Dieback can be a serious problem in coastal areas of the Southeast where temperatures and relative humidity are high throughout the growing season and rainfall is frequent. It also is a serious problem any time on camellias grown in greenhouses. The disease is not as serious in inland areas of the Southeast, and it usually is not a problem in regions with low relative humidity (e.g., California and other areas of the Pacific Coast).

Host Range

G. cingulata has a wide host range and causes diseases on many plant species. However, the strain that attacks camellia is host-specific – i.e., it does not attack other host plants. Over 20 species of *Camellia* have been tested, and all except tea (*C. sinensis*) are susceptible when artificially inoculated under controlled conditions.

Management

An integrated approach is the most effective way to manage this disease in the nursery. Cultural practices, sanitation, host resistance, and fungicides all have an important role in preventing occurrence and limiting the spread of dieback.

Cultural practices Because the pathogen is spread by splashing water, keep diseased plants well away from healthy plants. Orient rows and spread out plants as much as possible to allow adequate air circulation and exposure to sunlight, which will promote drying. Adjust irrigation practices to minimize the amount of time foliage and stems remain wet. Avoid injuries to plants that might create wounds, which could become infection sites. Take cuttings only from near the tops of healthy plants; do not take cuttings from diseased plants.

Sanitation Prune out diseased twigs, shoots, and cankers as soon as they are recognized. Prune at least six inches below any visible symptoms. Remove and discard severely affected plants. Destroy or compost diseased plant material. Disinfest pruning tools regularly (e.g., in a 10% chlorine bleach solution) when removing diseased shoots. Follow good sanitation practices when propagating new plants to avoid introducing pathogen inoculum.

Resistance Although the pathogen attacks all species of *Camellia* except *C. sinensis* when plants are inoculated under controlled conditions, species and cultivars vary considerably in susceptibility under field conditions. In general, *C. sasanqua* and *C.*

Dieback

Geographic occurrence Dieback is a constant threat in coastal areas of the Southeast (Virginia, North Carolina, South Carolina, Georgia, Florida, Alabama, Mississippi, Louisiana, and eastern Texas); it is less important in inland areas and usually is not a problem in the Pacific Coast region, particularly California, because of environmental conditions, but it always is a problem in greenhouse-grown camellias

Seasonal occurrence Primarily in the spring, but infection can occur at any time during the growing season when inoculum is present and environmental conditions are favorable

DISEASE FREQUENCY	DISEASE SEVERITY
5 annual	5 plants killed
[4]	**[4]**
3	3
2	2
1 rare	1 very little damage

CHEMICAL TREATMENT	CULTURAL PRACTICES
3 used every year	**[3 very important]**
[2]	2
1 not used	1 not important

SANITATION	RESISTANT CULTIVARS
[3 very important]	3 many cultivars
2	**[2]**
1 not important	1 no resistance

oleifera are more susceptible than *C. japonica. C. sasanqua* 'Cleopatra' is particularly susceptible, whereas *C. japonica* 'Charles S. Sargent' and 'Governor Mouton' are particularly resistant. *C. hiemalis* 'Kanjiro' and several other cultivars of this species also are highly resistant to dieback.

Chemical treatment If dieback occurs or has been a problem in the recent past, apply a fungicide during the spring when old leaves are shedding, to protect leaf scars from infection. One well-timed application usually is sufficient. Soak cuttings and scion pieces in a fungicide suspension for at least 30 minutes prior to grafting to avoid inadvertently introducing inoculum into propagation wounds.

Leaf Gall

Exobasidium camelliae

Leaf gall is a common but usually innocuous disease of camellia, primarily *C. sasanqua,* that is caused by the basidiomycete fungus *Exobasidium camelliae.* Published reports have mentioned two morphological forms of this species, *E. camelliae* on *C. japonica* and *E. camelliae* var. *gracilis* on *C. sasanqua,* but the existence of distinct forms is not well documented. Little research has been done on this fungus and the disease it causes.

Symptoms

As in most hosts of *Exobasidium* spp., infection of camellia by *E. camelliae* results in hypertrophy and hyperplasia of leaf tissue. The predominant symptom is the formation of off-colored, thick, fleshy leaves (hence the name of the disease) in early spring, shortly after budbreak. The galls are distinctive and obvious; they usually are an off-shade of green or pink to rose. Most leaves in a developing vegetative bud are affected, but occasionally only one or two leaves or parts of a few leaves are affected. The optimum time for symptom expression is April through May. During the growing season, the galls shrivel and turn black. Many of them fall off the plant, but some may remain attached throughout the summer. The symptoms appear dramatic, but the damage rarely is economically important, except in young plants with few shoots.

Epidemiology

The life cycle of *E. camelliae* and the epidemiology of leaf gall are not well documented or understood. Like flower blight, leaf gall has only one disease cycle per year. Galls develop on tender new leaves as shoots begin to grow in the spring. Infected shoots may start growing several days before healthy, uninfected shoots. Infected leaves develop abnormally as mycelium of the fungus grows through them; the number, size, and types of cells produced and the orientation of cells in the leaf all are affected. As the galls mature, several layers of cells on the underside of the leaf break away, exposing a white layer of basidiospores borne on basidia. Basidiospores are blown by the wind and may be spread by splashing water (e.g., rainfall or overhead irrigation) to developing shoots on the same plant or on nearby plants. It is not known when primary infection occurs. After dispersal, basidiospores may remain quiescent on bark or in bud scales until the next spring, or they may germinate and cause superficial infections in developing buds. These infections become latent until the following spring when the buds begin to grow. The disease tends to be more severe on plants that are heavily shaded and whose foliage stays wet for prolonged periods.

Host Range

E. camelliae appears to be host-specific, attacking only species of *Camellia.* It apparently is distinct both morphologically and pathologically from the species of *Exobasidium* that attack rhododendrons, azaleas, and other ericaceous plants. All species of *Camellia* appear to be susceptible to some degree. *C. sasanqua* is the most susceptible, and the disease is common on this species. Leaf gall occurs only occasionally on *C. japonica, C. oleifera,* and other species and hybrids.

Management

Leaf gall is a common disease in the landscape and usually causes little, if any, economic damage. However, in the nursery, it can cause significant damage if left unchecked; young plants with only one or several shoots can become severely misshapen or distorted. A combination of sanitation, cultural practices, and fungicide applications should manage this disease effectively.

Cultural practices Orient rows and space plants to maximize air flow around the foliage to promote drying. Schedule irrigation applications to avoid prolonged periods of leaf wetness,

Leaf Gall

Geographic occurrence Leaf gall occurs wherever camellias are grown, but it is not a problem in California and other places with low relative humidity

Seasonal occurrence Spring (April–May)

DISEASE FREQUENCY	DISEASE SEVERITY
5 annual	5 plants killed
[4]	4
3	3
2	**[2]**
1 rare	1 very little damage

CHEMICAL TREATMENT	CULTURAL PRACTICES
3 used every year	3 very important
[2]	**[2]**
1 not used	1 not important

SANITATION	RESISTANT CULTIVARS
[3] very important	3 many cultivars
2	**[2]**
1 not important	1 no resistance

particularly in the spring. Avoid growing plants where they receive heavy or extended periods of shade.

Sanitation Sanitation is the single most effective means of managing leaf gall in the nursery. Monitor plants closely during the spring when budbreak occurs, particularly if leaf gall was present the previous year. Concentrate monitoring in blocks of *C. sasanqua.* Identify, remove, and destroy galls *before* the lower leaf surface separates and spores are released. Initially galls may be difficult to recognize by inexperienced observers.

Host resistance *C. japonica* is considerably more resistant than *C. sasanqua.* On the basis of observations in the landscape, other species of camellia also appear to be relatively resistant, but this has not been studied to any extent. Cultivars of *C. sasanqua* appear to vary in resistance, but no effort has been made to measure or document differences in resistance among cultivars.

Chemical treatment If galls are detected after the leaves have separated and basidiospores have been released, protect new growth on plants with several fungicide applications during May and June. Thorough coverage is important for effective control. Little is known about chemical management of this disease.

Camellia Yellow Mottle

Camellia yellow mottle varicosavirus

Camellia yellow mottle is the current name of the virus disease previously called variegation, infectious variegation, color breaking, camellia leaf yellow mottle, and camellia yellow mottle leaf. It causes symptoms on both flowers and leaves but apparently causes little damage, if any, in affected plants.

Symptoms

Symptoms of this disease are very striking, affecting both flowers and leaves. Symptom expression can vary seasonally and is not stable from year to year, and symptoms occasionally disappear altogether. Therefore, infected plants do not always show symptoms.

In pink and red flowers, infection results in a loss of color, known as color breaking; the flowers are variegated instead of uniform in color. White flowers do not show this symptom. Variegated flowers are considered desirable by camellia enthusiasts, and it has been suggested that the virus be used to "create" new flower types. Variegation of flower color is not always due to virus infection, however. It is also caused by genetic mutation, in which case it usually is a stable, inheritable trait. The cause of variegation usually can be determined by the distribution and uniformity of variegated color in the petals of individual flowers. Variegation caused by virus infection usually is not uniform on individual petals and among petals, whereas that caused by mutation is uniform on all petals of a flower.

Other conspicuous symptoms of this disease are yellow flecks, spots and blotches of various sizes and shapes, ring spots, and an overall mottling on leaves. The severity of symptoms usually varies from leaf to leaf, from a few small yellow flecks or spots to large mottled areas (Plate 25). On severely affected plants, entire leaves may be pale yellow.

Epidemiology

Camellia yellow mottle varicosavirus is a rod-shaped RNA virus that apparently occurs systemically in the plant. Transmission of the virus and infection occur only through grafting. Attempts to transmit the virus mechanically in sap have been unsuccessful, and no vectors are known. In addition, neither contact between plants nor pollen can spread this virus among plants. Consequently, the virus is spread in the nursery primarily during vegetative propagation. However, naturally occurring root grafts between adjacent plants in the landscape occasionally may move the virus between plants. Cuttings taken from infected plants, either with or without symptoms, will be infected. Grafting an infected scion onto a healthy rootstock or grafting a healthy scion onto an infected rootstock will produce an infected plant. Symptoms may take months to develop after infection occurs. There is little evidence that virus infection has any adverse effect on the host plant. However, plants with severe mottling of the foliage may be more sensitive to sunburn.

Host Range

The host range of camellia yellow mottle varicosavirus has not been well studied, but it appears to be very limited. This virus is known to infect only *C. japonica, C. sasanqua,* and some hybrids, but other species of *Camellia* are believed to be susceptible.

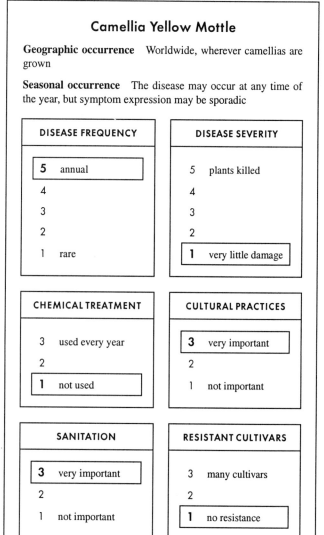

Camellia Yellow Mottle

Geographic occurrence Worldwide, wherever camellias are grown

Seasonal occurrence The disease may occur at any time of the year, but symptom expression may be sporadic

DISEASE FREQUENCY	
5	**annual**
4	
3	
2	
1	rare

DISEASE SEVERITY	
5	plants killed
4	
3	
2	
1	**very little damage**

CHEMICAL TREATMENT	
3	used every year
2	
1	**not used**

CULTURAL PRACTICES	
3	**very important**
2	
1	not important

SANITATION	
3	**very important**
2	
1	not important

RESISTANT CULTIVARS	
3	many cultivars
2	
1	**no resistance**

Management

Cultural practices are the only means of managing this disease. Periodically scout stock plants for symptoms, and then identify and mark infected plants. Beware that symptoms are not always evident. Diseased plants can be segregated or discarded if yellow mottle is deemed economically undesirable. Do not take cuttings for vegetative propagation or grafting from infected plants. No resistance is known among cultivars of *C. japonica* and *C. sasanqua,* and there are no chemicals to manage virus diseases.

Phytophthora Root Rot

Phytophthora cinnamomi

Phytophthora root rot is a serious and often fatal disease of camellias. The pathogen, *Phytophthora cinnamomi,* attacks many ornamental plant species and occurs commonly in the nursery. *C. japonica* is one of the most susceptible species of *Camellia.* In the nursery, cultivars of this species are sometimes grafted onto *C. sasanqua,* a more resistant species, to improve plant survival in the landscape. Refer to Chapter 14, "Phytophthora Root Rot and Dieback," for additional information on diseases caused by *Phytophthora* spp.

Symptoms

Symptoms of Phytophthora root rot of camellias are similar to those on other plants. Primary symptoms develop belowground, where infection occurs, well in advance of aboveground symptoms, but they usually go unnoticed unless plants are uprooted or removed from their containers so that roots are visible. Infected roots are discolored (e.g., various shades of brown) and rotted, feeder roots are lacking, and cankers occasionally develop on the root crown and lower stem. Although infection occurs belowground in the roots, the symptoms that first become obvious are aboveground, in the foliage. Leaves on infected plants initially turn a lighter shade of green (which can mimic mineral deficiency), become chlorotic, begin wilting, and eventually fall off. The plants become unthrifty and stunted, and they may die. This overall decline occurs slowly on older plants in the nursery as well as in the landscape, but on younger plants with limited root systems, particularly freshly rooted cuttings and new seedlings, decline and death may occur quickly.

Epidemiology

Disease development is favored by prolonged periods of saturated soil and moderate temperatures. The pathogen survives in soil and container mix as chlamydospores and, occasionally, oospores, both of which are thick-walled resting spores capable of persisting for years under most environmental conditions. *P. cinnamomi* often is present in container mix around potted plants and may be also in field soil at many nursery locations. When the soil or container mix is moist, the fungus produces sporangia, which release zoospores when the soil or mix is thoroughly wet or saturated. Zoospores are motile, swimming spores that move freely in water and can be moved over great distances in flowing water (e.g., surface water, streams, and rivers). These spores move to plant roots and initiate infection, usually on succulent feeder roots. Infection probably occurs during the spring and fall, when environmental conditions favor sporangium formation and zoospore release. However, aboveground symptoms often do not develop until mid- to late summer, when high temperatures increase the water requirements of plants, and plants with root rot cannot cope with the added stress. *P. cinnamomi* is spread around the nursery and among nurseries in or on anything that moves soil, container mix, or water – including shoes, containers, tools, equipment, and plant material.

Host Range

P. cinnamomi has an enormous host range, including over 1,000 species of plants in numerous genera and families. It attacks many different species of ornamental plants grown in nurseries and native plant species that grow naturally around nurseries. Other species of *Phytophthora* known to attack other ornamental crops and known to be present in nurseries have not been implicated in root rot of camellia.

Management

Phytophthora root rot can be managed effectively over the long term only with an integrated approach utilizing all available options – cultural practices, sanitation, host resistance, and fungicides.

Phytophthora Root Rot

Geographic occurrence Wherever camellias are grown

Seasonal occurrence Throughout the year, but particularly in spring, early summer, and fall when temperatures are most favorable for pathogen activity

DISEASE FREQUENCY	DISEASE SEVERITY
5 annual	**5 plants killed**
4	4
3	3
2	2
1 rare	1 very little damage

CHEMICAL TREATMENT	CULTURAL PRACTICES
3 used every year	**3 very important**
2	2
1 not used	1 not important

SANITATION	RESISTANT CULTIVARS
3 very important	**3 many cultivars**
2	2
1 not important	1 no resistance

Cultural practices Water management is the single most important strategy for managing Phytophthora root rot in the nursery. Every effort should be made to avoid overwatering, prevent standing water from collecting, and prevent saturation of the soil. Plants should be planted in a porous, well-drained container mix. Mixes containing a high percentage of composted or aged bark (75–100%) are used most frequently. Do not place containers directly on the soil surface where contamination can occur. The soil surface should be covered with a porous material (e.g., nursery cloth or gravel), to prevent roots from coming into contact with field soil and to prevent puddles and standing water. Avoid using plastics, which do not allow water to percolate through the covering and, therefore, encourage the collection of standing water. Nursery beds should be created and prepared to channel water away from the plants; raised or ridged beds frequently are used. Irrigation should be applied judiciously, based on the needs of the plants; apply only as much water as is necessary and only when it is necessary. Group plants in the nursery according to their water requirement so irrigation can be scheduled efficiently.

Sanitation Avoid introducing inoculum by following strict sanitation guidelines, particularly during propagation, when the plants are very susceptible. Always use fresh, uninfested container mix to pot or repot plants. Some composted bark media are naturally suppressive to many soilborne pathogens. Use new or disinfested pots and containers; avoid reusing pots that have not been cleaned and disinfested or sterilized. Take cuttings from plants free of disease symptoms and from high on the plant; disinfest cuttings prior to use. Use a suitable disinfestant regularly to disinfest tools and work surfaces. Treat any water that is to be recycled, recirculated, or reused in the nursery for irrigation (see Chapter 95, "Disease Management for Nurseries Using Recycling Irrigation Systems"). Do not use untreated recycled irrigation water during propagation or on extremely susceptible cultivars or species.

Host resistance Both *C. japonica* and *C. reticulata* are very susceptible to root rot, whereas *C. sasanqua* is resistant. It has been reported that *C. oleifera* also is resistant, but additional research is needed to confirm this. In some nurseries, particularly in the Southeast, *C. japonica* scions are grafted onto *C. sasanqua* rootstocks to produce plants resistant to root rot for use in the landscape. This practice adds significantly to the cost of these plants. Cultivars of *C. japonica* vary in degree of susceptibility to root rot; e.g., Debutante, Herme, Imura, and Pink Perfection are considered very susceptible.

Chemical treatment Several fungicides that are specifically active against *Phytophthora* spp. and other oomycete fungi are available. Most of these products do not eliminate or kill the pathogen in the soil or container mix but only prevent it from being active. Therefore, repeated applications are necessary when these products are used. Timely applications during periods of heavy rainfall and moderate temperatures, when infection is likely to occur, may be needed to manage root rot effectively in the nursery.

Cylindrocladium Black Rot

Cylindrocladium crotalariae

The fungus *Cylindrocladium crotalariae* causes a root and stem rot of camellias. The disease, known as Cylindrocladium black rot, was not recognized until 1987. It seems to be confined to nurseries, where it primarily affects rooted cuttings and young plants, although older plants also can become diseased. Disease development is rapid and frequently results in plant death. Refer to Chapter 11, "*Cylindrocladium* Diseases," for additional information on diseases caused by *Cylindrocladium* spp.

Symptoms

Cylindrocladium black rot usually affects rooted cuttings and plants, although three-year-old plants also have died after artificial inoculation. Affected plants initially become flaccid, wilt, and then die. Belowground, roots are discolored and rotted, and cankers may girdle the lower stem at the soil line. The entire process, from initial symptom expression to death, usually happens quickly; however, infection may have occurred months before. Sudden wilting and dying of cuttings or seedlings also is typical of dieback caused by *Glomerella cingulata,* but the two diseases can be differentiated by the color of the infected wood: wood affected by *C. crotalariae* is light red to pink, and that affected by *G. cingulata* is gray to brown.

Epidemiology

The epidemiology of Cylindrocladium black rot of has not been investigated. *C. crotalariae* is a relatively common soilborne plant-pathogenic fungus in areas with warmer climates, such as

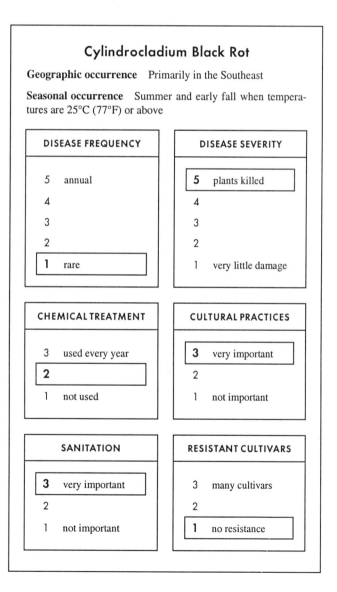

Cylindrocladium Black Rot

Geographic occurrence Primarily in the Southeast

Seasonal occurrence Summer and early fall when temperatures are 25°C (77°F) or above

DISEASE FREQUENCY	DISEASE SEVERITY
5 annual	**5 plants killed**
4	4
3	3
2	2
1 rare	1 very little damage

CHEMICAL TREATMENT	CULTURAL PRACTICES
3 used every year	**3 very important**
2	2
1 not used	1 not important

SANITATION	RESISTANT CULTIVARS
3 very important	3 many cultivars
2	2
1 not important	**1 no resistance**

the Southeast. It can survive in soil for many years as tiny microsclerotia but is sensitive to desiccation. Therefore, microsclerotia usually are not present in the top 1–2 inches of soil but occur deeper where soil moisture is more constant. The fungus also produces conidia and ascospores, but the role of these propagules in black rot of camellia has not been determined. Roots most likely are infected by mycelium growing from germinated microsclerotia. Infection, disease development, and plant death are favored by warm temperatures.

Host Range

C. crotalariae, like most species of *Cylindrocladium,* has a broad host range. It causes root and stem rots on a diverse array of plants, including alfalfa, clover, kiwi, peanuts, soybeans, and various species of *Vaccinium.*

Management

The best approach to managing black rot is to avoid introducing inoculum into the nursery by means of cultural practices and proper sanitation.

Cultural practices Avoid contaminating soilless container mixes with field soil or sand that may harbor *C. crotalariae.* Use gravel, nursery cloth, or some other material on nursery beds to prevent containers from contacting field soil. In particular, keep camellias away from soils where peanuts, alfalfa, soybeans, or clover have been grown.

Sanitation Disinfest tools and equipment that may have contacted soils infested with *C. crotalariae.* Remove and destroy diseased plants as soon as they are diagnosed so they do not supply inoculum that could infect other plants.

Resistance Cylindrocladium black rot has been observed in *C. japonica, C. sasanqua,* and *C. oleifera.* However, all species of *Camellia* are believed to be susceptible. Younger plants (e.g., rooted cuttings and seedlings) are more susceptible than older plants.

Chemical treatment If the disease occurs in the nursery, fungicide drenches applied around affected plants and plants adjacent to them may be useful in preventing the spread of the disease but probably will not eliminate the pathogen from soil or container mixes.

REFERENCES

Baxter, L. W., Jr., and Epps, W. M. 1957. Camellia flower blight. Am. Camellia Yearb. 1957:43–53.

Baxter, L. W., Jr., and Segars, S. B. 1988. Studies on Cylindrocladium black rot (CBR) of camellias. Am. Camellia Yearb. 1988:59–66.

Baxter, L. W., Jr., Epps, W. M., and Fagan, S. G. 1980. A 13-point program for the control of contagious camellia dieback caused by the fungus *Glomerella cingulata.* Camellia J. 35(2):22.

Baxter, L. W., Jr., Sprott, A. M., Segars, S. B., and Fagan, S. G. 1987. Black root rot and stem blight of camellia. Camellia J. 42(1):4–5.

Baxter, L. W., Jr., Fagan, S. G., and Segars, S. B. 1988. New suscepts of *Exobasidium camelliae,* the camellia leaf gall fungus. Plant Dis. 72:79.

Brunt, A. A., Crabtree, K., Dallwitz, M. J., Gibbs, A. J., Watson, L., and Zurcher, E. J., eds. 1996–. Plant Viruses Online: Descriptions and Lists from the VIDE Database. Version 16 January 1997. Australian National University, Canberra. http://biology.anu.edu.au/Groups/MES/vide/

Can, N. H., and Baxter, L. W., Jr. 1977. The status of our knowledge of root rot of camellias caused by *Phytophthora cinnamomi* in 1977. Am. Camellia Yearb. 1977:172–187.

Can, N. H., Baxter, L. W., Jr., and Fagan, S. G. 1978. The status of our knowledge in 1978 of twig blight, canker, and dieback of camellias caused by a strain of *Glomerella cingulata.* Am. Camellia Yearb. 1978: 75–91.

Dickens, J. S. W., and Cook, R. T. A. 1990. *Glomerella cingulata* on camellias. Plant Pathol. 38:75–85.

Hansen, H. N., and Thomas, H. E. 1940. Flower blight of camellias. Phytopathology 30:166–170.

Hirata, S. 1981. A new species of *Exobasidium* causing giant galls on *Camellia japonica.* Trans. Mycol. Soc. Jpn. 22:393–395.

Hiruki, C. 1985. A preliminary study on infectious variegation of camellia. Acta Hortic. 164:55–62.

Kohn, L. M., and Nagasawa, E. 1984. A taxonomic reassessment of *Sclerotinia camelliae* (= *Ciborinia camelliae* Kohn), with observations on flower blight of camellia in Japan. Trans. Mycol. Soc. Jpn. 25:149–161.

Marshall, R. E., and Cole, A. L. J. 1988. Infection of *Camellia* species by *Phytophthora cinnamomi.* Am. Camellia Yearb. 1988:104-109.

Milbrath, J. A., and McWhorter, F. P. 1946. Yellow mottle leaf, a virus disease of camellia. Am. Camellia Yearb. 1946:51–53.

Thomas, H. E., and Hansen, H. N. 1946. Camellia flower blight. Phytopathology 36:380–381.

Wolf, F. T., and Wolf, F. A. 1952. Pathology of camellia leaves infected by *Exobasidium camelliae* var. *gracilis* Shirai. Phytopathology 42:147–149.

CHAPTER 28

Jerry T. Walker • University of Georgia, Griffin

Cedrus Diseases

Geographic production USDA zones 5–9

Family Pinaceae

Genus *Cedrus*

Species			
	C. atlantica	Atlas cedar	zones 6–9
	C. deodara	deodar cedar	zones 7–8
	C. libani	cedar of Lebanon	zones 5–7

At least 14 woody trees have common names containing *cedar,* but botanically they are not in the genus *Cedrus,* although all are closely related conifers. Diseases of red cedar (*Juniperus* and *Thuja*), white cedar (*Chamaecyparis*), and Japanese cedar (*Cryptomeria*) are covered in other chapters.

Cedrus atlantica (Atlas cedar), *C. deodara* (deodar cedar), and *C. libani* (cedar of Lebanon) were introduced into the United States after the colonial period and prior to the Civil War. There are many forms or cultivars of deodar cedar in the South and the far West. This species may be injured by cold when planted at latitudes above that of central Tennessee. Other species are grown in the Northwest.

The principal fungal diseases of cedar are root rots and needle or twig blights.

Root Rots

Armillaria ostoyae **(syn. *A. mellea*)**
Cylindrocladium **sp.**
Heterobasidion annosum
Pestalotiopsis funera
Phytophthora cinnamomi
Pythium **sp.**
Rhizoctonia solani

Fungi causing root rots in *Cedrus* are *Phytophthora cinnamomi,* a *Pythium* sp., *Armillaria ostoyae* (syn. *A. mellea*), the damping-off fungus *Rhizoctonia solani,* and a *Cylindrocladium* sp. Except for *Armillaria,* these fungi destroy young root tissue, resulting in blackish or off-colored roots. Young seedlings may be affected soon after germination and during emergence in the seedbed. Once the seedlings straighten up, they apparently become more resistant. *Armillaria* causes a basal trunk rot and root deterioration in older trees.

Other fungi causing butt or root rots include *Heterobasidion annosum* and *Pestalotiopsis funera,* but these are encountered infrequently on *Cedrus* in the nursery, compared to the aforementioned root-rotting fungi.

Symptoms

Root rots are first noticed with the appearance of chlorotic needles that subsequently drop or wilting of twigs and branches without needle drop. Scraping away the bark will reveal a browning or discoloration of affected stems.

Root Rots

Geographic occurrence USDA zones 5–9

Seasonal occurrence All seasons

DISEASE FREQUENCY	DISEASE SEVERITY
5 annual	**[5] plants killed**
[4]	4
3	3
2	2
1 rare	1 very little damage

CHEMICAL TREATMENT	CULTURAL PRACTICES
3 used every year	**[3] very important**
[2]	2
1 not used	1 not important

SANITATION	RESISTANT CULTIVARS
[3] very important	3 many cultivars
2	2
1 not important	**[1] no resistance**

Management

Root rots occur in excessively wet soils, where good drainage is lacking, where irrigation water is taken from contaminated ponds, or where the soil or soil-media mixture has been contaminated. Using healthy mother stock is extremely important. Fungal isolations from diseased roots are necessary for correct diagnosis and pathogen identification. Selective media and commercial diagnostic kits are available for positive confirmation of the presence of some of these pathogens.

Many forms or cultivars of *Cedrus* spp. have not been tested for disease resistance.

Needle and Twig Blights

***Macrophoma* sp.**
***Melampsora cedri* (*Peridermium cedri*)**
***Phyllosticta* sp.**
***Sphaeropsis* sp. (syn. *Diplodia* sp.)**

More noticeable but perhaps of no less consequence than root rots, under nursery conditions, are needle blights of *Cedrus,* caused by several fungi. *Macrophoma* sp., *Sphaeropsis* (syn. *Diplodia*), and *Phyllosticta* all cause similar symptoms. *Melampsora cedri* causes a witches'-broom effect known as broom rust in deodar in the Himalayas.

Symptoms

Sphaeropsis infection begins as a shoot blight during shoot elongation in the spring, with purplish brown necrotic areas forming at the base of blighted twigs. The needles turn off-color and then brown, and they remain attached to their stems. Cankers producing copious resin are observed on older trees, and the death of branches creates a staghead appearance. Seed cones can be colonized by *S. sapinea,* resulting in seed rot or damping-off of seedlings.

Fruiting bodies (pycnidia) of the needle blight fungi are usually visible as black dots from which exuded spore masses can be observed under magnification. Spores are released at high relative humidity or if free moisture is present. They are disseminated by wind-driven rain and invade young, developing needles.

Dieback or broken tops on mature deodar cedars, as seen in the southern landscape, may be due to cold injury, insects (borer), lightning, or hurricanes.

Management

The impact of needle diseases can be devastating under the proper environmental conditions, but sometimes they can be much more severe under adverse environments or stressful conditions, including moisture stress, excessive nitrogen, or stress due to wounds, such as excessive pruning. The use of protective fungicides in nurseries, plantations, and ornamental plantings has become routine. Removal of diseased trees or pruning affected branches under dry conditions to reduce the inoculum in the nursery always is a good sanitary practice. Timing of irrigation to allow maximum drying of needles will reduce the wetness period required for the germination of conidia.

Other Problems

Cedars may be sensitive to (chlorosis of the needles) or injured by certain herbicides. Maintaining a record of herbicide applications may be helpful in explaining sudden problems with needles.

Needle and Twig Blights

Geographic occurrence USDA zones 5–9

Seasonal occurrence February to June

DISEASE FREQUENCY	DISEASE SEVERITY
5 annual	5 plants killed
4	**[4]**
[3]	3
2	2
1 rare	1 very little damage

CHEMICAL TREATMENT	CULTURAL PRACTICES
[3] used every year	**[3]** very important
2	2
1 not used	1 not important

SANITATION	RESISTANT CULTIVARS
3 very important	3 many cultivars
[2]	2
1 not important	**[1]** no resistance

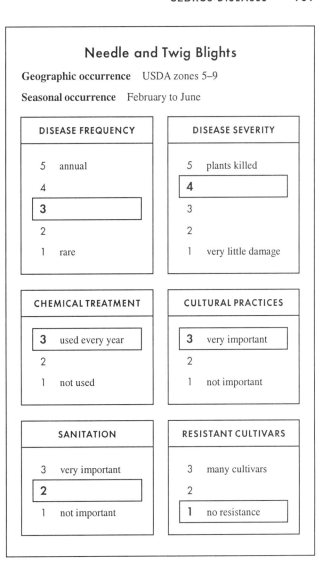

Young cedars in the nursery may show signs of sunscald or sunburn when one side of the tree is less vigorous than the other. This should not be confused with a disease; it is due to an environmental or abiotic factor, which the trees may outgrow.

A potential problem which nurseries may face in the future is wilt caused by the pinewood nematode, *Bursaphelenchus xylophilus.* This nematode, found in eastern Asia and the United States in exotic and native pines, has been reported in deodar and Atlas cedar (Sinclair et al., 1987). It is transmitted by a cerambycid beetle and causes extensive losses, because it builds up to extreme populations in a very few days. The pinewood nematode causes sudden death of ornamental spruce, Christmas trees, and forest pines.

REFERENCES

Dirr, M. A. 1998. Manual of Woody Landscape Plants: Their Identification, Ornamental Characteristics, Culture, Propagation and Uses. 5th ed. Stipes Publishing, Champaign, Ill.

Farr, D. F., Bills, G. F., Chamuris, G. P., and Rossman, A. Y. 1989. Fungi on Plants and Plant Products in the United States. American Phytopathological Society, St. Paul, Minn.

Hansen, E. M., and Lewis K. J., eds. 1997. Compendium of Conifer Diseases. American Phytopathological Society, St. Paul, Minn.

Lloyd, J., ed. 1997. Plant Health Care for Woody Ornamentals. International Society of Arboriculture, Savoy, Ill., and Cooperative Extension Service, College of Agricultural, Consumer, and Environmental Sciences, University of Illinois, Urbana-Champaign.

Sinclair, W. A., Lyon, H. H., and Johnson, W. T. 1987. Diseases of Trees and Shrubs. Comstock Publishing Associates, Cornell University Press, Ithaca, N.Y.

Stanosz, G. 1997. Sphaeropsis shoot blight and canker. Pages 42–43 in: Compendium of Conifer Diseases. E. M. Hansen and K. J. Lewis, eds. American Phytopathological Society, St. Paul, Minn.

CHAPTER 29

Elizabeth Hudgins ▪ Oklahoma State University, Stillwater

Judith O'Mara ▪ Kansas State University, Manhattan

Cotoneaster Diseases

Geographic production USDA zones 3–8

Family Rosaceae

Genus *Cotoneaster*

Species	*C. apiculatus*	cranberry cotoneaster
	C. dammeri	bearberry cotoneaster
	C. divaricatus	spreading cotoneaster
	C. horizontalis	rock cotoneaster
	C. lucidus	hedge cotoneaster
	C. multiflorus	many-flowered cotoneaster
	C. salicifolius	willowleaf cotoneaster

Fire Blight

Erwinia amylovora

Fire blight, caused by the bacterium *Erwinia amylovora,* is the most important disease of cotoneaster in nurseries and in the landscape. The severity of the disease is often increased by damp weather; the disease can develop within a week after infection if conditions are favorable. Significant losses can occur in cotoneaster, especially in locations with suitable climates or following a hailstorm, which inflicts wounds through which the pathogen can enter and infect the plant.

Symptoms

Fire blight is named for the typical scorched appearance of diseased tissues. Young shoots may start to wilt when infected and then quickly turn a scorched brown. In these cases, the dead tips may have a shepherd's-crook shape. The disease can continue to move into older tissues, where it causes significant injury and produces cankers.

Epidemiology

E. amylovora overwinters in cankers in infected woody tissues. In the spring, it multiplies and is spread primarily by insects; splashing rain can also carry the bacterium and is an important means of spread in nurseries. *E. amylovora* enters natural openings or wounds in the blossoms and leaves and begins to cause disease. Fire blight develops at temperatures of 59–90°F (15–32°C) in wet or humid weather, with optimal temperatures of 81–84°F (27–29°C) and relative humidity of 60% or greater.

Host Range

Fire blight occurs in many plants in the family Rosaceae, including apple (*Malus*), hawthorn (*Crataegus*), mountain ash (*Sorbus*), pear (*Pyrus*), pyracantha (*Pyracantha*), and quince (*Cydonia*), as well as cotoneaster.

Management

Cultural practices Grow plants in sunny locations, and avoid crowding of plants, to reduce humidity. Do not use overhead irri-

Fire Blight

Geographic occurrence All areas of the United States

Seasonal occurrence April to July, or when conditions are favorable for disease

DISEASE FREQUENCY	
5	**annual**
4	
3	
2	
1	rare

DISEASE SEVERITY	
5	plants killed
4	
3	
2	
1	very little damage

CHEMICAL TREATMENT	
3	used every year
2	
1	not used

CULTURAL PRACTICES	
3	**very important**
2	
1	not important

SANITATION	
3	**very important**
2	
1	not important

RESISTANT CULTIVARS	
3	many cultivars
2	
1	no resistance

gation, because it spreads the pathogen and provides the humid environment necessary for the development of the disease. Excessive nitrogen fertilizing and heavy pruning can increase the number of succulent shoots that are susceptible to disease.

Sanitation The best sanitation practice in the nursery is to remove infected plants. If this is not feasible, prune out infected branches 6–12 inches below the visibly diseased portion, to reduce the spread of the disease. To avoid spreading the disease when cutting the tissue, use a 70% alcohol or 10% household bleach solution to sterilize pruning shears between cuts, and do not prune when the foliage is wet. Burn or bury all removed plant material.

Resistance Most *Cotoneaster* species are susceptible to fire blight. *C. horizontalis* var. *perpusillus* has been reported to be especially susceptible. Species with reported resistance (less than 20% infection) include *C. adpressus* var. *praecox, C. amoenus, C. apiculatus, C. canadensis, C. dammeri* var. *radicans, C. microphyllus, C. nitens,* and *C. zabelii.* However, results may vary in different locations.

Chemical treatment Bactericides or antibiotics may help, but they are not likely to cure or prevent disease without cultural practices and sanitation. Conducive weather can significantly increase disease pressure, which may reduce the effectiveness of chemicals.

Entomosporium Leaf Spot

Diplocarpon mespili

(asexual stage, *Entomosporium mespili*)

Entomosporium leaf spot has been reported to occur in many hosts worldwide. It can occur in cotoneaster in production areas with prolonged cool, wet weather. In certain years, extensive defoliation can occur under conditions favorable for infection. The disease is caused by the fungus *Diplocarpon mespili* and its asexual stage, *Entomosporium mespili*.

Symptoms

During late spring or early summer, leaf symptoms initially develop on the lower portion of the plant. Leaf spots progress upward through the plant as long as the weather continues to favor secondary infection. Small, irregular, reddish brown spots develop on the upper and lower surfaces of leaves. Sometimes a yellow halo will develop around the margin of a leaf spot. Fruiting bodies of the pathogen (acervuli) appear as raised white bumps in the center of the lesion. The asexual state of the fungus, *E. mespili,* is easily identified microscopically by its spores (conidia) with insect-like appendages. Leaf yellowing and defoliation occur as leaf spots darken, enlarge, and run together.

Epidemiology

During the growing season, *E. mespili* is commonly found on leaf lesions. The fungus also overwinters in this spore stage on dead leaves and young twigs. In the northern part of the United States the fungus can overwinter in both the conidial and the sexual state on some hosts. Spring infections develop when spores are splashed from infected leaf litter to newly developing leaves. Prolonged cool and wet spring weather, with temperatures of 58–87°F (14–30°C), favors infection and progressive disease development. The optimum temperature for disease activity is 68°F (20°C).

Host Range

In addition to its occurrence in cotoneaster, Entomosporium leaf spot has been reported worldwide in other hosts, including apple, crabapple, firethorn, hawthorn, Indian hawthorn, Yedda hawthorn, loquat, mountain ash, photinia, flowering quince, serviceberry, stanvesia, and toyon.

Management

Fungicides may be necessary in areas with annual infections of Entomosporium leaf spot. Applications should begin at budbreak and continue as long as the weather favors disease activity. Cultural measures should include removal and destruction of diseased leaf litter in the fall and spring. Overhead irrigation should be avoided or done early in the morning to reduce the length of time leaves stay wet. Disease incidence can also be favored by poor air circulation and crowding of cotoneaster plantings. Disease-resistant cultivars have not been documented.

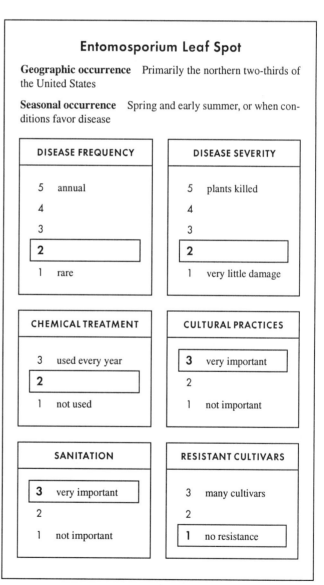

Entomosporium Leaf Spot

Geographic occurrence Primarily the northern two-thirds of the United States

Seasonal occurrence Spring and early summer, or when conditions favor disease

DISEASE FREQUENCY	DISEASE SEVERITY
5 annual	5 plants killed
4	4
3	3
2 (selected)	**2** (selected)
1 rare	1 very little damage

CHEMICAL TREATMENT	CULTURAL PRACTICES
3 used every year	**3 very important** (selected)
2 (selected)	2
1 not used	1 not important

SANITATION	RESISTANT CULTIVARS
3 very important (selected)	3 many cultivars
2	2
1 not important	**1 no resistance** (selected)

REFERENCES

Dirr, M. A. 1998. Manual of Woody Landscape Plants: Their Identification, Ornamental Characteristics, Culture, Propagation and Uses. 5th ed. Stipes Publishing, Champaign, Ill.

Dreistadt, S. H., Clark, J. K., and Flint, M. L. 1994. Pests of Landscape Trees and Shrubs: An Integrated Pest Management Guide. Publ. 3359. University of California, Division of Agriculture and Natural Resources.

Riffle, J. W., and Peterson, G. W. 1986. Diseases of trees in the Great Plains. U.S. Dep. Agric. For. Serv., Rocky Mountain For. Range Exp. Stn. (Ft. Collins, Colo.), Gen. Tech. Rep. RM-129.

Sinclair, W. A., Lyon, H. H., and Johnson, W. T. 1987. Diseases of Trees and Shrubs. Comstock Publishing Associates, Cornell University Press, Ithaca, N.Y.

Smith-Fiola, D. C. 1995. Pest Resistant Ornamental Plants. Rutgers Cooperative Extension, Rutgers, N.J.

Whitcomb, C. E. 1996. Know It and Grow It, III. Lacebark, Stillwater, Okla.

CHAPTER

30

Austin K. Hagan ▪ Auburn University, Auburn, Alabama

Crapemyrtle Diseases

Geographic production USDA zones 7–9

Family Lythraceae

Genus *Lagerstroemia*

Species *L. fauriei*
L. indica
L. indica × *fauriei*

Crapemyrtle (*Lagerstroemia indica*) is common in the United States, since its introduction in the mid-1700s from China, in landscapes and gardens across the South. Within the last three decades, the U.S. National Arboretum has released a number of *L. indica* × *fauriei* cultivars, characterized by brilliant fall color; handsome, smooth, exfoliating bark; lush, dark green foliage; large, showy flower clusters; and, in many, resistance to powdery mildew. The recently released cultivar *L. fauriei* 'Fantasy' shares nearly all of the desirable characteristics of the hybrid *L. indica* × *fauriei.* Crapemyrtle cultivars range in size from dwarf shrubs to small trees, and their flower color varies from white to pink, purple, or dark red. In field and container nurseries, the two predominate diseases of crapemyrtle are powdery mildew and Cercospora leaf spot.

Powdery Mildew

Erysiphe lagerstroemia

Powdery mildew is the most widely recognized and possibly the most commonly reported disease of crapemyrtle. It usually does not significantly threaten plant health, but heavily colonized foliage and flower buds are unattractive and may render mildew-damaged plants unsalable.

Symptoms

Powdery mildew is easily identified by the appearance of cottony white to buff-colored patches or colonies of the causal fungus, *Erysiphe lagerstroemia,* on new leaves, tender shoots, and the scales of flower buds. Heavily colonized leaves or shoots may turn nearly white and often curl upward or are twisted (Plate 26). In susceptible cultivars, tender shoots that are heavily colonized may wither and die. Heavily diseased plants may shed some leaves prematurely. Diseased flower buds may abort and fall to the ground. Powdery mildew typically appears a month or two after leaf-out and may continue to intensify into early summer.

Epidemiology

E. lagerstroemia usually overwinters as hyphae in dormant buds. In the spring, spores of the fungus are dispersed to new leaves and tender shoots by air currents. Weather patterns in the spring and early summer have a substantial impact on the severity of powdery mildew of crapemyrtle. Warm days and cool nights

Powdery Mildew

Geographic occurrence USDA zones 7–9

Seasonal occurrence April to October

DISEASE FREQUENCY	DISEASE SEVERITY
5 annual	5 plants killed
4	4
3	**3**
2	2
1 rare	1 very little damage

CHEMICAL TREATMENT	CULTURAL PRACTICES
3 used every year	3 very important
2	**2**
1 not used	1 not important

SANITATION	RESISTANT CULTIVARS
3 very important	**3 many cultivars**
2	2
1 not important	1 no resistance

along with extended periods of relatively dry weather reportedly favor rapid development and spread of the disease.

Host Range

The host range of *E. lagerstroemia* is limited to selections of *L. indica, L. fauriei,* and the hybrid cultivars.

Management

Strategies for controlling powdery mildew in container and field nurseries are largely limited to the production of disease-resistant cultivars and the use of protective fungicides.

Cultural practices Spacing plants to improve air circulation around the foliage may help slow disease spread.

Sanitation Except for the collection and disposal of fallen leaves, sanitation is of little value in controlling powdery mildew of crapemyrtle.

Resistance Substantial differences in the reaction of crapemyrtle cultivars to powdery mildew have been observed in field trials. The antique *L. indica* selections commonly found around old southern homesites typically are highly susceptible to the disease. Numerous crapemyrtle cultivars, particularly the hybrids released by the National Arboretum, are highly resistant to almost immune. The reactions of some commercially grown cultivars to powdery mildew are listed below.

Resistant cultivars include Acoma, Apalachee, Basham's Party Pink, Biloxi, Bourbon Street, Caddo, Catawba, Centennial Spirit, Cherokee, Choctaw, Comanche, Fantasy, Glendora White, Hope, Hopi, Miami, Muskogee, Natchez, Near East, Osage, Pecos, Sarah's Favorite, Seminole, Sioux, Tonto, Tuskegee, Tuscarora, Wichita, and Yuma.

Moderately susceptible cultivars include Hardy Lavender, Hardy White, Lipan, Majestic Beauty, Peppermint Lace, Potomac, Powhatan, Prairie Lace, William Toovey, Velma's Royal Delight, and Zumi.

Susceptible cultivars include Carolina Beauty, Christmastime, Country Red, Firebird, Gray's Red, New Orleans, New White, Orbin Adkin, Petite Plum, Pink Lace, Raspberry Sundae, Regal Red, Royalty, and Wonderful White.

Chemical treatment For susceptible cultivars, begin applications of protective fungicides shortly after leaf-out or as symptoms first appear on the foliage. Continue fungicide applications at intervals specified on the product label until the threat of further disease spread in early summer has passed. Shorten the time interval between applications when disease is severe. Production of disease-resistant cultivars largely eliminates the need for fungicide treatment programs.

Cercospora Leaf Spot

Cercospora lythracearum

Cercospora leaf spot often is not recognized as a damaging disease of crapemyrtle, but the extensive defoliation seen in late summer on susceptible cultivars greatly detracts from the beauty of their fall color display.

Symptoms

The disease first appears on mature leaves as circular to irregular brown spots, about 1/4 inch in diameter, with no halo (Plate 27). Diseased leaves may be twisted or distorted. As the spots get larger and more numerous, a portion to the entire leaf turns yellow to bright red and then falls to the ground. Typically, leaf spotting starts around the base of the plants and then quickly spreads upward through the plant canopy, until all but the youngest leaves near the shoot tips are lost. In field plantings, a year or two may pass after the establishment of a susceptible cultivar before severe leaf shed occurs. In resistant cultivars, symptoms are usually confined to the foliage in the center of the canopy, a foot or two above the soil line. Symptoms of Cercospora leaf spot usually do not appear before late June or early July, and premature leaf drop may begin as early as mid-August. By early September, susceptible cultivars may already have shed over 50% of their leaves as a result of the disease.

Epidemiology

The causal fungus, *Cercospora lythracearum,* is a resident on crapemyrtle in field and landscape plantings across the South. Fallen diseased leaves are probably a critical source of inoculum. Thousands of spores produced in each leaf spot are spread to healthy leaves by air currents. Frequent showers or heavy dews coupled with warm, cloudy weather appear to accelerate disease development.

Cercospora Leaf Spot

Geographic occurrence USDA zones 7–9

Seasonal occurrence June to September, with an optimum period in August

DISEASE FREQUENCY	DISEASE SEVERITY
5 annual	5 plants killed
4	4
3	**3**
2	2
1 rare	1 very little damage

CHEMICAL TREATMENT	CULTURAL PRACTICES
3 used every year	3 very important
2	**2**
1 not used	1 not important

SANITATION	RESISTANT CULTIVARS
3 very important	**3 many cultivars**
2	2
1 not important	1 no resistance

Host Range

The host range of *C. lythracearum* has not been defined, but most species in the genus *Cercospora* attack only a few closely related plant species. Cercospora leaf spot of crapemyrtle has been observed in cultivars of *L. indica, L. fauriei,* and *L. indica* × *fauriei.*

Management

As is the case with powdery mildew, control options are largely restricted to the production of disease-resistant cultivars, with some help from certain production and sanitation practices.

Cultural practices Fertilization, irrigation, and pruning are not known to specifically influence the severity of Cercospora leaf spot of crapemyrtle. However, maintaining recommended soil fertility will help maintain plant vigor, which may offset the detrimental effects of premature leaf drop. Drip or surface irrigation may be a preferable to overhead impact sprinklers for watering crapemyrtle. When an overhead sprinkler system is used, irrigate crapemyrtle either between 2 and 6 AM or shortly after noon. Overhead sprinkling in the late afternoon or early evening may intensify disease development.

Sanitation Collection and disposal of fallen leaves may slow the onset of the disease in the following year.

Resistance Significant differences in the susceptibility of crapemyrtle cultivars to Cercospora leaf spot have been observed in field trials. However, relatively few cultivars have good resistance to the disease.

Resistant cultivars include Apalachee, Basham's Party Pink, Caddo, Fantasy, Glendora White, Jet Stream, New White, Tonto, Tuscarora, Tuskegee, and Velma's Royal Delight.

Moderately resistant cultivars include Biloxi, Catawba, Centennial Spirit, Cherokee, Choctaw, Cotton Candy, Lipan, Miami, Muskogee, Osage, Pecos, Pink Lace, Regal Red, Sarah's Favorite, Twilight, and Victor.

Moderately susceptible cultivars include Byer's Red, Country Red, Hardy Lavender, Hardy White, Hopi, Majestic Beauty, Near East, New Orleans, Peppermint Lace, Powhatan, Raspberry Sundae, Seminole, Sioux, and Yuma.

Susceptible cultivars include Acoma, Carolina Beauty, Comanche, Orbin Adkin, Prairie Lace, and Wonderful White.

Chemical treatment If fungicides are applied for the control of Cercospora leaf spot, the first application should be made in mid-June to early July, when symptoms first appear on leaves around the base of the plant. Continue applications at the intervals specified on fungicide label until early September or until the threat of further disease spread has passed. Several fungicides registered for the control of powdery mildew are also registered for diseases incited by *Cercospora* in other woody trees and shrubs.

REFERENCES

Dirr, M. A. 1998. Manual of Woody Landscape Plants: Their Identification, Ornamental Characteristics, Culture, Propagation and Uses. 5th ed. Stipes Publishing, Champaign, Ill.

Hagan, A. K., and Mullen, J. M. 1997. Common foliage diseases of crape myrtle. Ala. Coop. Ext. Circ. ANR-1047.

Hagan, A. K., Gilliam, C. H., Keever, G. J. and Williams, J. D. 1997. Susceptibility of cultivars of crape myrtle to Cercospora leaf spot and powdery mildew. Biol. Cult. Tests Control Plant Dis. 12:57.

Holcomb, G. E. 1997. Reaction of crape myrtle cultivars to Cercospora leaf spot and powdery mildew, 1996. Biol. Cult. Tests Control Plant Dis. 12:56.

Windham, M. T., Witte, W. T., Sauve, R. J. and Flangan, P. C. 1995. Powdery mildew observations and growth of crapemyrtle in Tennessee. HortScience 30:813.

CHAPTER

31

Tom Creswell ▪ North Carolina State University, Raleigh

Robert L. Wick ▪ University of Massachusetts, Amherst

Cryptomeria Diseases

Geographic production USDA zones 5–9

Family Taxodiaceae

Genus *Cryptomeria*

Species *C. japonica*

Cryptomeria japonica (Japanese cryptomeria, Japanese cedar, or sugi) is the main timber tree species in most forestry plantations in Japan and is increasingly being planted as an ornamental worldwide. It forms conical trees with stout, erect branches, providing useful screening while young and potentially reaching 50–60 feet in height and 20–30 feet in width. It has had few significant diseases in plantings in the United States and has good cold-hardiness and adaptability. It is reported to be resistant to *Phytophthora cinnamomi* and to several isolates of *Seiridium cardinale* and *S. unicorne.* Its growth rate is medium.

Cercospora Blight (Cercospora Leaf Spot)

Cercosporidium sequoiae

(syns. *Asperisporium sequoiae, Cercospora cryptomeriae, Cercospora sequoiae, Cercospora thujina*)

Cercospora blight, caused by *Cercosporidium sequoiae,* was first reported in the United States in Virginia in 1985. It is not known to be of widespread importance; however, if the disease becomes a problem in nurseries, a dramatic increase in incidence could occur.

The pathogen was first described in Pennsylvania in 1887 on seedlings of *Sequoia gigantea* and later independently (as *Cercospora cryptomeriae*) in Japan, where it causes a widespread disease of *Cryptomeria japonica.* In Japanese nurseries the fungus kills seedlings within a season, and young trees may die within a few years. Older trees develop sunken cankers. Japanese researchers speculate that the fungus was introduced into Japan from the United States. The pathogen, as *Cercospora thujina,* has been reported to cause symptoms in *Thuja orientalis.*

Symptoms

In *Cryptomeria,* symptoms occur as stem cankers and brown to reddish brown needle blight (Plate 28). When disease pressure is high, the entire foliage and small branches become blighted. When conditions are conducive to sporulation, examination of the stems reveals fascicles of large conidiophores and conidia of the pathogen.

Epidemiology

C. sequoiae overwinters in infected needles or twigs, where it produces spores (conidia) that infect new growth in spring. Conidia are spread by wind-blown rain (since they are dry, they can be transported by the wind). Symptoms appear after about a three-

Cercospora Blight (Cercospora Leaf Spot)

Geographic occurrence USDA zones 5–9

Seasonal occurrence New infections begin in early spring; secondary spread to new growth occurs in summer

DISEASE FREQUENCY	
5	annual
4	
3	
2	
1	**rare**

DISEASE SEVERITY	
5	**plants killed**
4	
3	
2	
1	very little damage

CHEMICAL TREATMENT	
3	**used every year**
2	
1	not used

CULTURAL PRACTICES	
3	**very important**
2	
1	not important

SANITATION	
3	**very important**
2	
1	not important

RESISTANT CULTIVARS	
3	many cultivars
2	
1	**no resistance**

week incubation period. The fungus spreads secondarily to new growth during the summer. Twig lesions may spread to larger stems and establish perennial cankers. Susceptibility to the fungus declines with the age of the tree.

Host Range

The host range of *C. sequoiae* includes *Cupressus arizonica, C. lustianica, C. marcocarpa, C. sempervirens, Cryptomeria japonica, Juniperus virginiana, Sequoia sempervirens, Sequoiadendron gigantea, Taxodium distichum,* and *Thuja orientalis.*

Management

In nurseries, diseased plants should be removed and destroyed. Provide adequate spacing to decrease the duration of leaf wetness and lower humidity. Application of maneb with a spreader-sticker has been reported to provide protection from Cercospora blight.

Chloroscypha Needle Blight

Chloroscypha seaveri

Chloroscypha needle blight is common in cryptomeria in Japan, but the pathogen has been reported only on arborvitae in the United States, where it is regarded as a weak pathogen. Occurrence in Japan is sporadic, at six- to 10-year intervals. Disease severity is greater at high elevations, where cold injury is a problem. Twig death suppresses tree growth by one-third to one-half, but the trees are not killed. The pathogen is *Chloroscypha seaveri,* a fungus in the ascomycete group.

Chloroscypha Needle Blight

Geographic occurrence Reported in Japan on this host

Seasonal occurrence April to October

DISEASE FREQUENCY	DISEASE SEVERITY
5 annual	5 plants killed
4	4
3	**3**
2	2
1 rare	1 very little damage

CHEMICAL TREATMENT	CULTURAL PRACTICES
3 used every year	**3 very important**
2	2
1 not used	1 not important

SANITATION	RESISTANT CULTIVARS
3 very important	3 many cultivars
2	2
1 not important	**1 no resistance**

Symptoms

Symptoms appear in late spring. Gradual browning of old needles occurs, beginning in the lower crown and spreading to the entire crown except the top of the tree, often giving the entire tree the appearance of having been scorched. In severely infected trees most one-year-old needles and many small branches may be killed. New growth produced following the loss of old needles typically hides the symptoms during summer.

Epidemiology

Spores of *C. seaveri* are released in April and May in mild regions and in June in areas with heavy snow cover. Spores infect needles through stomates and may progress to branches. The fungus overwinters on dead twigs. It grows as a saprophyte on dead tissue and infects new tissue following stress, typically from cold injury.

Host Range

Chloroscypha needle blight has been reported in *Cryptomeria japonica* and a *Thuja* sp.

Management

Cultural practices for management of Chloroscypha needle blight include removal and destruction of blighted and fallen twigs, to reduce the spread of the disease. Avoid planting where cold-hardiness is marginal.

No resistant cultivars are available, and there is no chemical control for this disease.

Black Mold

Chalara thielavioides
(syn. *Chalaropsis thielavioides*)

Black mold causes graft failure and loss of cuttings in several hosts.

Symptoms

Recently taken cuttings and new graft unions are covered with a white mycelium resembling mildew, which darkens with age. The production of black chlamydospores of the fungal pathogen and heavy mycelial growth give affected areas a characteristic black, moldy appearance. The fungus prevents successful graft unions and inhibits callus formation and rooting of cuttings. It is reported to give off a sweet, fruity odor in incubation of rose tissue in a moist chamber.

Epidemiology

The pathogen is the soilborne fungus *Chalara thielavioides.* It may be spread by contaminated host tissue.

Host Range

Black mold has been reported in *Cryptomeria japonica, Juniperus* sp., *Rosa* sp., and *Thuja* sp.

Management

For sanitation, remove and destroy all infected plant tissues. Sterilize propagation tools, benches, and pots. Use sterilized propagation medium. Propagate only from disease-free stock.

No resistance is known in *Cryptomeria.*

No chemical products are specifically labeled for the control of black mold of *Cryptomeria,* but general-purpose fungicides, especially those effective against *Thielaviopsis,* may help reduce the spread of the pathogen.

Mushroom Root Rot
Armillaria sp.

Mushroom root rot, also called shoestring root rot and Armillaria root rot, is a common disease affecting many plants and having a wide geographic distribution. The pathogen is an *Armillaria* sp., a fungus that persists in decaying wood in soil and attacks roots and root collars. It may be strongly pathogenic but often attacks plants already weakened by stress.

Symptoms

Growth reduction, yellowing, and dieback of infected plants are common symptoms. Some large trees may survive infection for several years, but young plants may die quickly. Bark often flakes off easily at the soil, revealing a layer of white or cream-colored fungal mycelium between the bark and the wood. *Armillaria* also usually produces brown to black strands of tissue, from which the name *shoestring root rot* is derived. These strands, called rhizomorphs, are flattened under the bark but round in soil, leaf litter, and decayed wood. Light brown mushrooms often arise from decayed wood and in nearby soil in fall or early winter.

Epidemiology

The disease is restricted to field plantings, typically at recently cleared sites. Infection occurs when roots grow into soil containing old, decaying wood. It may also spread by movement of contaminated soil or by wind-blown spores of the fungus (if mushrooms are produced).

Host Range

The pathogen has a very wide host range, including hundreds of ornamental trees and shrubs.

Black Mold

Geographic occurrence USDA zones 5–9

Seasonal occurrence Whenever *Cryptomeria* is propagated

DISEASE FREQUENCY	DISEASE SEVERITY
5 annual	5 plants killed
4	4
3	3
2	**2**
1 rare	1 very little damage

CHEMICAL TREATMENT	CULTURAL PRACTICES
3 used every year	3 very important
2	2
1 not used	**1 not important**

SANITATION	RESISTANT CULTIVARS
3 very important	3 many cultivars
2	2
1 not important	**1 no resistance**

Mushroom Root Rot

Geographic occurrence USDA zones 5–9

Seasonal occurrence Long-term infection, not seasonal in appearance

DISEASE FREQUENCY	DISEASE SEVERITY
5 annual	5 plants killed
4	**4**
3	3
2	2
1 rare	1 very little damage

CHEMICAL TREATMENT	CULTURAL PRACTICES
3 used every year	**3 very important**
2	2
1 not used	1 not important

SANITATION	RESISTANT CULTIVARS
3 very important	3 many cultivars
2	2
1 not important	**1 no resistance**

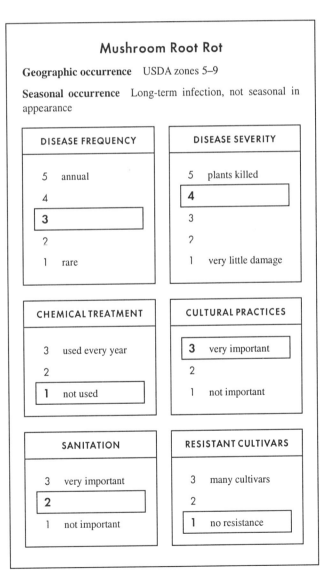

Phyllosticta Needle Blight

Geographic occurrence USDA zones 5–9

Seasonal occurrence New infections are initiated in spring with the beginning of new growth and may spread at any time during active growth

DISEASE FREQUENCY	DISEASE SEVERITY
5 annual	5 plants killed
4	4
3	**3**
2	2
1 rare	1 very little damage

CHEMICAL TREATMENT	CULTURAL PRACTICES
3 used every year	**3 very important**
2	2
1 not used	1 not important

SANITATION	RESISTANT CULTIVARS
3 very important	3 many cultivars
2	2
1 not important	**1 no resistance**

Management

Cultural practices Avoid planting susceptible hosts at infested sites.

Sanitation Avoid movement of infested soil to other sites. Remove infected roots from previous trees.

Resistance No resistance to mushroom root rot is known.

Chemical treatment No chemical treatment is known.

Phyllosticta Needle Blight

Phyllosticta cryptomeriae

Phyllosticta needle blight, caused by the fungus *Phyllosticta cryptomeriae,* is widespread in Japan, where *Cryptomeria* is native. It has been recognized in the United States only since 1978 but is likely to have been present much earlier. Most published reports on this disease are in Japanese and refer to the disease as the "red plague."

Symptoms

Individual needles die, and twig dieback subsequently occurs.

Epidemiology

Phyllosticta species are typically spread by splashing water, but in the ascomycetous perfect (sexual) stage they may be spread by wind.

Host Range

Phyllosticta needle blight has been reported in *Cryptomeria japonica.*

Management

Cultural practices Avoid sprinkler irrigation.

Sanitation Propagate only from clean stock.

Resistance No resistance to Phyllosticta needle blight is known.

Chemical treatment No chemical treatment is known.

REFERENCES

Hess, C. E., and Welch, D. S. 1954. *Chalaropsis thielavioides* Peyronel found on evergreen grafting stock. Plant Dis. Rep. 38:415–416.

Hodges, C. S. 1962. Comparison of four similar fungi from *Juniperus* and related conifers. Mycologia 54:62–69.

Hodges, C. S., and May, L. S. 1972. *Cercospora sequoiae* no Brasil. Fitopatologia 7(1–2):32–34.

Horst, R. K. 1983. Black mold. Page 17 in: Compendium of Rose Diseases. American Phytopathological Society, St. Paul, Minn.

Ito, K., Shibukawa, K., and Kobayashi, T. 1952. Etiological and pathological studies on the needle blight of *Cryptomeria japonica.* I. Morphology and pathogenicity of the fungi inhabiting the blighted needles. Gov. For. Exp. Stn. (Tokyo) Bull. 52.

Ito, K., Kobayashi, T., and Shibukawa, K. 1967. Etiological and pathological studies on the needle blight of *Cryptomeria japonica.* III. A comparison between *Cercospora cryptomeria* Shirai and *Cercospora sequoiae* Ellis & Everhart. Gov. For. Exp. Stn. (Tokyo) Bull. 204.

Kobayashi, T. 1976. Important forest diseases and their control measures in Japan. Pages 270–280 in: Plant Protection in Japan, 1976. Agriculture Asia, spec. issue 10. Association of Agricultural Relations in Asia, Tokyo.

Kubono, T. 1993. *Gloeosporidina cryptomeriae,* new species causing twig blight of Japanese cedar (*Cryptomeria japonica*). Trans. Mycol. Soc. Jpn. 34 (2):261–265.

Kubono, T. 1994. Symptom development of the twig blight of Japanese cedar caused *Gloeosporidina cryptomeriae.* J. Jpn. For. Soc. 76:52–58.

Kubono, T., and Hosoya, T. 1994. *Stromatinia cryptomeriae* sp. nov., the teleomorph of *Gloeosporidina cryptomeriae* causing twig blight of Japanese cedar. Mycoscience 35(3):279–285.

Sinclair, W. A., Lyon, H. H., and Johnson, W. T. 1987. Diseases of Trees and Shrubs. Comstock Publishing Associates, Cornell University Press, Ithaca, N.Y.

Tsugio, S. 1994. Studies on the occurrence of Chloroscypha needle blight of *Cryptomeria japonica* D. Don: Physiological and ecological characteristics and pathogenicity of the causal fungus. For. For. Prod. Res. Inst. Bull. 368, pp. 23–63.

Vegh, I., and Le Berre, A. 1981. Etude experimentale de la sensibilité de quelques cultivars de bruyères et de conifères d'ornement vis-a-vis du *Phytophthora cinnamomi* Rands. Phytopathol. Z. 103:301–305.

Wick, R. L., and Lambe, R. C. 1985. Occurrence of Cercospora blight on *Cryptomeria japonica* (L.F.) D. Don in the United States. J. Environ. Hortic. 3:18–19.

CHAPTER 32

Austin K. Hagan ▪ Auburn University, Auburn, Alabama

Daphne Diseases

Geographic production USDA zones 7–9

Family Thymelaeaceae

Genus *Daphne*

Species

D. × *burkwoodii*	Burkwood daphne
D. cneorum	rose daphne
D. mezereum	February daphne
D. odora	winter daphne

Winter daphne (*Daphne odora*) is a small evergreen shrub most noted for its production of fragrant, rosy purple blooms in late winter. This attractive, slow-growing import from China, which is best adapted to USDA hardiness zones 7–9, is the most common of the shrub-type daphne species in the nursery trade. Other available species include *D. cneorum* (rose daphne), *D.* × *burkwoodii* (Burkwood daphne), and *D. mezereum* (February daphne). Because of its limited production, the identity and impact of diseases, particularly leaf spot and blights, on the production of container-grown daphne is not well documented. Winter daphne, however, is very sensitive to several destructive diseases caused by soilborne fungi, and significant losses of container stock have been reported.

Southern Blight

Sclerotium rolfsii

The fungus causing southern blight, *Sclerotium rolfsii,* is an aggressive pathogen of numerous annual, perennial, and woody greenhouse and nursery crops. Southern blight outbreaks in container and field nurseries are limited largely to the hot, summer months in the southern United States and California. Under favorable weather conditions, disease development is rapid, and large blocks of plants may be killed before a disease outbreak is recognized and control practices are implemented. Typically, southern blight is much more prevalent on well-aerated sandy soils or potting medium than on poorly drained soil with a high clay or silt content.

Symptoms

In container stock, the most noticeable symptom of southern blight is sudden wilt or flagging of the foliage of plants randomly scattered across a block of containers. Wilted plants quickly die, and the foliage turns brown. Brown, girdling cankers usually develop on the main stem or root collar at or just below the surface of the soil or potting medium. When temperatures are high and the soil or potting medium is wet, the characteristic dense white, fan-like hyphal mat of *S. rolfsii* grows on stems just above the soil line, on the surface of the soil or potting medium, and on fallen leaves around the base of the plant. When the soil surface dries,

Southern Blight

Geographic occurrence USDA zones 7–9

Seasonal occurrence June to September, with an optimum period in July and August

DISEASE FREQUENCY

- 5 annual
- 4
- 3
- **[2]**
- 1 rare

DISEASE SEVERITY

- **[5] plants killed**
- 4
- 3
- 2
- 1 very little damage

CHEMICAL TREATMENT

- 3 used every year
- **[2]**
- 1 not used

CULTURAL PRACTICES

- **[3] very important**
- 2
- 1 not important

SANITATION

- **[3] very important**
- 2
- 1 not important

RESISTANT CULTIVARS

- 3 many cultivars
- 2
- **[1] no resistance**

the white hyphal mat may be found intertwined with roots, to a depth of 2 to 3 inches. Clusters of round, light tan to brown sclerotia, about the size of mustard seed, form on the surface of the white mat on the soil surface and dead stem tissues. Over time, the white mat disintegrates, but the diagnostic sclerotia remain.

Epidemiology

S. rolfsii overseasons in the soil or potting medium as sclerotia. Typically, sclerotia at or near the soil surface survive for several years, while those buried in several inches of soil die quickly. They are readily dispersed in bark or soil added to potting media, in soil or crop debris clinging to tools or tillage equipment, and in flowing water. Sclerotial germination is triggered by volatile organic compounds released from partially decayed crop resides. *S. rolfsii* often uses undecomposed leaves and other crop debris as a food base prior to attacking its host.

Southern blight is a hot- and humid-weather disease. Outbreaks in both container- and field-grown plants occur during the hottest months of the summer, usually when day and night temperatures exceed 30 and 21°C, respectively. At this time of year, frequent heavy showers or overwatering may accelerate disease development.

Host Range

Winter daphne is a host of *S. rolfsii.* The status of other daphne species is unknown. Overall, the host range of *S. rolfsii* encompasses over 500 species of plants in 100 families.

Management

Sanitation and cultural practices which either exclude the pathogen or create an environment unfavorable for its activity will greatly reduce the risk of a disease outbreak. When southern blight consistently causes significant losses of susceptible host plants, fungicide drenches are an option. For a more detailed discussion of recommended control practices, see Chapter 18, "Southern Blight."

Cultural practices Container beds must be crowned, covered with black plastic or weed cloth, and then topped with a layer of crushed rock or similar coarse material. To avoid over- or underwatering, block plant material by container size and water needs. Account for daily rainfall when scheduling irrigation. Fertilize according to need, and maintain the soil or medium pH at a slightly acidic level.

Sanitation Store potting medium components on an asphalt or a concrete pad. Do not add nonsterile soil or used potting media to fresh potting media. Avoid recycling containers unless they are carefully rinsed of potting media and disinfected. Clear all fallen leaves and other debris from container production areas. Immediately remove and discard diseased container and field stock offsite.

Resistance No information on resistance to southern blight in any cultivar of any daphne species is available.

Chemical treatment Fungicides will protect daphne from southern blight. For effective control, begin fungicide drenches before the onset of hot, humid weather in late spring or early summer. Repeat treatments at the interval listed on the product label until cooler weather in late summer to early fall.

For specific recommendations for chemical control of southern blight, contact your county extension agent or state specialist responsible for diseases of nursery crops.

Phytophthora Root Rot

Phytophthora cactorum
Phytophthora parasitica

Phytophthora root rot is a widespread and often destructive disease of numerous container and field-grown woody trees and shrubs. In the southern United States, winter daphne appears to be particularly susceptible to root rot incited by *Phytophthora parasitica.*

Symptoms

Sudden and permanent wilting of the top growth is quickly followed by plant death. Feeder roots colonized by either *P. cactorum* or *P. parasitica* are brittle and reddish brown to brown. Rotting of the feeder roots usually starts at or near the bottom of the root ball and usually continues until the root system is killed or the foliage collapses. If the root ball is removed from its container, areas of discolored, rotted roots can usually be easily seen.

Epidemiology

The causal fungi survive as resting structures (chlamydospores and oospores) and hyphae on the roots of diseased plants and in

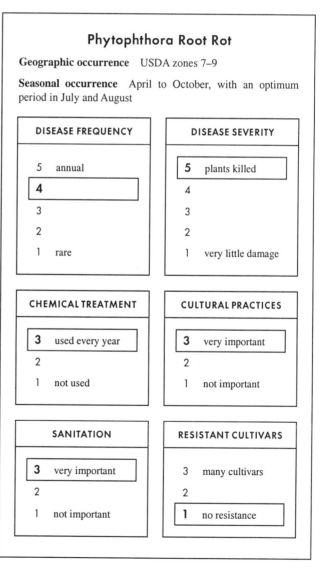

Phytophthora Root Rot

Geographic occurrence USDA zones 7–9

Seasonal occurrence April to October, with an optimum period in July and August

DISEASE FREQUENCY	DISEASE SEVERITY
5 annual	**[5] plants killed**
[4]	4
3	3
2	2
1 rare	1 very little damage

CHEMICAL TREATMENT	CULTURAL PRACTICES
[3] used every year	**[3] very important**
2	2
1 not used	1 not important

SANITATION	RESISTANT CULTIVARS
[3] very important	3 many cultivars
2	2
1 not important	**[1] no resistance**

crop debris and the soil below container beds. They are easily spread in contaminated potting media, diseased cuttings or liners, and splashing or flowing water. Recycled irrigation or runoff water is also a source of these fungi. The heaviest losses due to root rot usually occur on flood-prone production beds where water stands or flows around container stock. Over- or underwatering may accelerate the pace of disease development. Root rot fungi are generally active at temperatures which also favor plant growth. For further information on the epidemiology of the disease, refer to Chapter 14, "Phytophthora Root Rot and Dieback."

Host Range

The susceptibility of daphne species and cultivars to *P. cactorum* and *P. parasitica* has not been studied. Winter daphne is highly susceptible to attack by *P. parasitica*. February daphne (*D. mezereum*) and rose daphne (*D. cneorum*) are hosts of *P. cactorum*. The host ranges of both of these fungi also include numerous other woody shrubs and trees.

Management

Effective control of Phytophthora root rot can be obtained only by sanitation and cultural practices together with preventative fungicide treatments. For a detailed discussion of recommended control strategies, see Chapter 14.

Cultural practices Container beds must be crowned, covered with black plastic or a weed barrier, and then topped with crushed rock or a similar coarse material. Ponding of water around container stock must be avoided. To avoid over- or underwatering, block plant material by container size and water needs. Account for rainfall when scheduling irrigation. Fertilize according to need, maintain a slightly acid pH, and do not overfertilize with nitrogen. Use a bark-based medium with a high percolation rate and 20–30% air-filled pore space. Place rooted cuttings and liners on raised benches.

Sanitation Store potting medium components on an asphalt or a concrete pad. Do not add nonsterile soil or used potting media to fresh potting media. Avoid recycling containers unless they are carefully rinsed of old potting media and disinfected. Clear all fallen leaves, discarded media, and other debris from propagation and production areas. Periodically treat propagation benches and similar work areas with a surface disinfectant. Immediately discard any diseased cuttings and liners as well as container and field stock off-site. Collect cuttings for propagation only from blocks of healthy stock plants. Clean cutting knives or similar tools after each cut with denatured alcohol or a similar surface disinfectant. Root cuttings in a fresh soilless medium in new flats or containers on clean raised benches.

Chemical treatment A chemical control program is successful only when combined with good cultural practices and sanitation. Whenever possible, mix a granular or wettable powder fungicide into fresh potting medium prior to transplanting rooted cuttings, liners, or container stock. Depending on the fungicide incorporated, start a program of fungicide sprays or drenches one to three months after the plants are transplanted. If a fungicide is not incorporated into the medium, begin preventative fungicide treatments immediately after transplanting, and continue treatment at the intervals specified on the product label through the production cycle, until the crop is shipped.

Fungicides will not "cure" plants of Phytophthora root rot but will temporarily suppress further colonization of the root system and suppress the production of spores, which could splash to adjacent healthy plants.

For fungicide recommendations for the control of Phytophthora root rot, contact your county extension agent or state specialist responsible for diseases of nursery crops.

REFERENCES

Dirr, M. A. 1998. Manual of Woody Landscape Plants: Their Identification, Ornamental Characteristics, Culture, Propagation and Uses. 5th ed. Stipes Publishing, Champaign, Ill.

Hoitink, H. A. J., Benson, D. M., and Schmitthenner, A. F. 1986. Phytophthora root rot. Pages 4–8 in: Compendium of Rhododendron and Azalea Diseases. D. L. Coyier and M. K. Roane, eds. American Phytopathological Society, St. Paul, Minn.

McRitchie, J. J. 1983. Southern blight of ajuga. Fla. Dep. Agric. Consumer Serv. Plant Pathol. Circ. 252.

Sinclair, W. A., Lyon, H. H., and Johnson, W. T. 1987. Diseases of Trees and Shrubs. Comstock Publishing Associates, Cornell University Press, Ithaca, N.Y.

CHAPTER 33

Margery L. Daughtrey ▪ Cornell University, Riverhead, New York

Austin K. Hagan ▪ Auburn University, Auburn, Alabama

Dogwood Diseases

Geographic production USDA zones 5–9

Family Cornaceae

Genus *Cornus*

Species			
	C. alba	tartarian dogwood	zones 3–7
	C. alternifolia	pagoda dogwood	zones 3–7
	C. amomum	silky dogwood	zones 4–8
	C. asperifolia	roughleaf cornel	zones 4–9
	C. controversa	giant dogwood	zones 5–7
	C. florida	flowering dogwood	zones 5–9
	C. kousa	kousa dogwood	zones 5–8
	C. mas	corneliancherry dogwood	zones 4–8
	C. nuttallii	Pacific dogwood	zones 7–9
	C. obliqua	pale dogwood	zones 3–7
	C. racemosa	gray dogwood	zones 3b–8
	C. sanguinea	bloodtwig dogwood	zones 4–7
	C. sericea	redosier dogwood	zones 2–7

The flowering dogwood, *Cornus florida,* is one of the small flowering trees most widely grown by the U.S. nursery industry. The range of this hardy native tree extends from the Northeastern states south into Florida and northern Mexico. Showy white, pink, or red bracts, great fall color, and a broad range of adaptation across much of the eastern United States make flowering dogwood a favorite of landscape contractors and homeowners, despite several significant diseases and insect pests. In many regions of the United States, kousa dogwood (*C. kousa*), an introduction from Asia, along with a half-dozen recently released hybrids (*C. kousa* × *florida*) are attractive alternatives to the native flowering dogwood. Outside of their native range in the Pacific Northwest, the Pacific dogwood (*C. nuttallii*) and *C. nuttallii* × *florida* tend to be sensitive to winter injury and are not widely grown. A number of other native and introduced species (listed above) are occasionally available in the nursery trade.

Dogwood Anthracnose

Discula destructiva

Dogwood anthracnose, first noted in Pacific dogwood (*C. nuttallii*) in the Pacific Northwest and flowering dogwood (*C. florida*) in the Northeast during the 1970s, is believed to be exotic to North America. Currently, the range of this disease in flowering dogwood extends south from Connecticut and New York into Alabama and west into Illinois. In Northeastern and mid-Atlantic states, the disease occurs in flowering dogwood in woodlands and landscapes. In the South, damaging outbreaks and associated tree mortality have largely been confined to understory trees in forested areas, particularly at higher elevations in the Appalachian Mountains and the adjacent Piedmont region. Anthracnose has been found in flowering dogwood in nurseries, but disease incidence has typically been low and tree damage minor. Diseased nursery stock may, however, be responsible for expanding the range of the disease into landscapes in the South and Midwest.

Symptoms

Small, tan to brown spots with purple margins appear several weeks after flowering, randomly scattered across newly expanded leaves. The spots may expand into spreading brown blotches, which sometimes blight entire leaves (Plate 29). Shot holes may appear where the centers of the spots disintegrate. Blighted leaves are not shed in the fall; they remain on the diseased plants through the winter. Typically, spotting of the leaves begins in the lower canopy and progresses upward. Twig cankers form at the attachment point of blighted leaves. Numerous fruiting bodies (acervuli) of the causal fungus, *Discula destructiva,* appear on blighted leaves and dead twigs. Several years after symptoms are first seen, elliptical cankers form on the scaffold limbs and main trunk at the base of the blighted twigs. In severely diseased trees, numerous water sprouts (epicormic shoots) develop in early spring on the scaffold limbs and trunk. Over a period of three to four years, extensive leaf blighting and numerous trunk cankers may eventually kill trees.

Epidemiology

Trees in partial to full sun are far less prone to injury from dogwood anthracnose than those maintained under heavy shade. Extended periods of cloudy, wet weather shortly after leaf-out of flowering dogwood favor splash dispersal of *D. destructiva* and infection of leaves. Birds and insects are other possible vectors of spores (conidia) of the pathogen. Under favorable weather conditions, disease development may occur throughout the growing season. Overhead watering during late afternoon and early evening will encourage disease development, particularly in the partially resistant kousa dogwood. Drought increases the susceptibility of flowering dogwood.

Host Range

The host range of *D. destructiva* is limited to *Cornus* spp. Considerable differences in the susceptibility of dogwood taxa have been observed. Natural infection by *D. destructiva* has been recorded only in selections of flowering, Pacific, and kousa dogwoods.

Management

A defined anthracnose management program is required for certification and shipment of dogwood nursery stock into certain states. In a field or container nursery, effective control of the disease depends on good nursery sanitation and cultural practices and, when needed, preventive fungicide treatments.

Cultural practices Maintain field and container stock in full sun. Space out container material to speed evaporation of water from the leaves and improve air circulation. In areas where this disease is endemic, attempt to water all dogwood taxa in field and container plantings using a drip or similar surface irrigation system. When overhead irrigation is used, water shortly before dawn or around midday. Avoid watering in the late afternoon or early evening, particularly on cloudy days or several hours prior to rain.

Sanitation Collect wood for grafting only from disease-free dogwood. Discard diseased stock plants. For valuable specimen plants, prune out cankered branches in the spring, before budbreak.

Dogwood Anthracnose

Geographic occurrence USDA zones 5–8

Seasonal occurrence Late May to June, with some secondary infection throughout the growing season, under average yearly conditions in zone 7 (adjust for your hardiness zone)

DISEASE FREQUENCY	DISEASE SEVERITY
5 annual	5 plants killed
4	**4**
3	3
2	2
1 rare	1 very little damage

CHEMICAL TREATMENT	CULTURAL PRACTICES
3 used every year	**3 very important**
2	2
1 not used	1 not important

SANITATION	RESISTANT CULTIVARS
3 very important	3 many cultivars
2	**2**
1 not important	1 no resistance

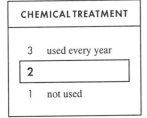

Resistance Until very recently, all commercially available selections and cultivars of flowering and Pacific dogwood have been highly susceptible to dogwood anthracnose. Several selections of disease-resistant flowering dogwood have been collected; the cultivar Appalachian Spring is the first of these to be made available to the nursery trade. Also susceptible are the redosier dogwood cultivars Flaviramea (yellow twig dogwood), Isanti, and Kelseyi; the giant dogwood cultivar Controversa; and the tartarian dogwood cultivar Elegantissma. The redosier dogwood cultivar Ruby and the corneliancherry, silky, and pagoda dogwoods have demonstrated good resistance to this disease. Resistance varies in cultivars of kousa and *C. kousa* × *florida,* as summarized below.

Resistant cultivars include the kousa cultivars Milky Way and Steeple and the hybrid cultivars Celestial, StarDust, and Stellar Pink.

Moderately resistant cultivars include the kousa cultivars Big Apple, China Girl, Elizabeth Lustgarten, Gay Head, Greensleeves, Julian, Milky Way Select, and Temple Jewel, and the hybrid cultivars Constellation and Ruth Ellen.

Susceptible cultivars include the kousa cultivars Autumn Rose, Chinensis, Moonbeam, and Wolf Eyes.

Chemical treatment Preventive fungicide treatments may be required as part of an anthracnose management program for the certification and shipment of dogwoods into selected states. To effectively protect the foliage, apply a recommended fungicide at 10- to 14-day intervals beginning at budbreak, until all new leaves mature. For at-risk trees, shorten the application interval to seven to 10 days during periods of frequent rain. Additional protective fungicide applications may be made during the remainder of the summer, as needed.

For fungicide recommendations and application guidelines for controlling dogwood anthracnose, contact your county extension agent or state specialist responsible for nursery crops.

Septoria Leaf Spot

Septoria cornicola
Septoria cornicola* var. *ampla
Septoria floridae

Outbreaks of Septoria leaf spot typically occur in late summer to early fall. The disease usually poses little threat to the health or the market value of field and container-grown flowering dogwood, and control measures are rarely justified. Septoria leaf spot is usually much more common in flowering dogwood, Siberian dogwood (*C. alba* cv. Siberica), and yellow twig dogwood (*C. sericea* cv. Flaviramea) in landscape plantings than in field or container nurseries.

Symptoms

Infection causes angular brown leaf spots, up to 1/4 inch in diameter, often with a broad purple border (Plate 30). Tiny, round pycnidia of the causal fungus can be seen on the upper surfaces of the lesions, with the aid of a hand lens. Heavily spotted leaves are often shed prematurely. Symptoms of this disease are almost indistinguishable from those of Cercospora leaf spot. Septoria leaf spot can be diagnosed in the field only if pycnidia are observed.

Epidemiology

Septoria leaf spot of dogwood is caused by *Septoria cornicola, S. cornicola* var. *ampla,* and *Septoria floridae.* These fungi overwinter as pycnidia embedded in fallen diseased leaves. Pycnidia produce spores (conidia), which are dispersed by splashing water to healthy leaves, particularly in the lower portion of the tree canopy. Extended periods of cloudy, wet weather in mid- to late summer favor the spread of the disease.

Host Range

Hosts of the three *Septoria* spp. include the pale, gray, pagoda, redosier, and flowering dogwood.

Management

In field and container nurseries, management programs designed to prevent outbreaks of spot anthracnose, powdery mildew, and dogwood anthracnose will also protect flowering dogwood from Septoria leaf spot.

Cultural practices Good production practices will prevent a reduction in tree vigor resulting from premature shedding of leaves. When using overhead irrigation, water shortly before dawn or around midday. Watering in the late afternoon or early evening, particularly on cloudy days or several hours prior to rain, may increase the risk of a disease outbreak.

Sanitation Remove diseased leaves and other crop debris from production ranges.

Septoria Leaf Spot

Geographic occurrence USDA zones 5–8

Seasonal occurrence Late June to October, under average yearly conditions in zone 7 (adjust for your hardiness zone)

DISEASE FREQUENCY	DISEASE SEVERITY
5 annual	5 plants killed
4	4
[3]	3
2	**[2]**
1 rare	1 very little damage

CHEMICAL TREATMENT	CULTURAL PRACTICES
3 used every year	3 very important
[2]	**[2]**
1 not used	1 not important

SANITATION	RESISTANT CULTIVARS
3 very important	3 many cultivars
[2]	**[2]**
1 not important	1 no resistance

Resistance Resistance to Septoria leaf spot among cultivars of flowering dogwood has not been studied. Of the other dogwood taxa, Siberian and yellow twig dogwoods appear to be the most susceptible.

Chemical treatment Typically, flowering dogwood rarely requires fungicide treatments to control Septoria leaf spot. As part of a regular nursery scouting program, examine the leaves for typical leaf spot symptoms in each block of flowering dogwood during the growing season. Should symptoms appear, a fungicide should be applied at 10- to 14-day intervals.

For fungicide recommendations and application guidelines, consult your county extension agent or state specialist responsible for nursery crops.

Powdery Mildew

Microsphaera pulchra
Phyllactinia guttata

Powdery mildew has emerged nationwide in recent years in both the nursery and the landscape as a common and sometimes damaging disease of flowering dogwood. Disease outbreaks have also been observed in cultivars of kousa and hybrid (*C. kousa* × *florida*) dogwoods. The damage to established flowering dogwood is largely cosmetic, but slowed shoot elongation, reduced vigor, and in some cases death of year-old seedlings have been attributed to severe disease outbreaks in field and container stock. Powdery mildew is equally damaging to dogwoods in full sun and in heavy shade.

Symptoms

The first signs of powdery mildew are often distinct white, granular-textured fungal colonies on the upper and lower surfaces of newly expanding leaves (Plate 31). At times, colony growth is so sparse that a hand lens is needed to find the thin coating of fungal hyphae on leaf surfaces. A reddish discoloration, curling, stunting, or scorching of the youngest leaves is often more noticeable than fungal colonies on leaves (Plate 32). In early fall, a few tiny, round, black fruiting bodies (cleistothecia) of the pathogen may appear on the colony surface.

Epidemiology

The powdery mildew fungi attacking dogwoods are *Microsphaera pulchra* and *Phyllactinia guttata.* Cleistothecia of these fungi are often scarce until late summer or early fall. Cleistothecia of *M. pulchra* formed on leaves during this period are the primary means of overwintering for this pathogen, and ascospores released in the spring are the primary inoculum for new infections. Typically, colony development is first seen on the new leaves in mid- to late spring. Conidia (*Oidium* sp. stage) of powdery mildew fungi are dispersed to healthy leaves by air currents. The effect of environmental factors on the severity of powdery mildew of dogwoods has not been determined.

Host Range

Nearly all cultivated selections and cultivars of flowering dogwood are susceptible to powdery mildew incited by *M. pulchra.* However, heavily colonized trees may be found adjacent to trees showing few signs of the causal fungus in native stands of flowering dogwood. The host range of *M. pulchra* also includes pagoda, gray, kousa, hybrid (*C. kousa* × *florida*), and redosier dogwoods.

Pagoda, Pacific, pale, gray, silky, and redosier dogwoods have been reported to be hosts of *P. guttata.*

Management

The most effective strategies for controlling powdery mildew in field- and container-grown dogwood are selecting disease-resistant species or cultivars for production and applying preventive fungicide treatments.

Cultural practices Spacing out container stock after the threat of cold damage has passed may help slow the spread and development of powdery mildew. Controlling weeds in field plantings of dogwood will improve air movement and thus may also slow disease development.

Sanitation With the exception of the removal of diseased leaves, sanitation practices are of limited value in controlling powdery mildew of dogwood.

Resistance Among commonly cultivated cultivars of flowering dogwood, Cherokee Brave has consistently shown excellent resistance to powdery mildew. Cultivars with partial resistance to powdery mildew and spot anthracnose are Cherokee Chief, Weaver's White, and Welch's Bay Beauty. Nearly all cultivars of *C. kousa* and *C. kousa* × *florida* are highly resistant to both powdery mildew and spot anthracnose. However, significant mildew outbreaks have been recorded in the hybrid cultivars Constellation, Ruth Ellen, and StarDust.

The reactions of selected cultivars of flowering dogwood (*C. florida*), kousa dogwood (*C. kousa*), and their hybrids to powdery mildew are summarized below.

Resistant cultivars, with little or no powdery mildew, include the flowering dogwood cultivar Cherokee Brave; the kousa cultivars Big Apple, China Girl, Gay Head, Greensleeves, Julian, Milky Way, Milky Way Select, National, Satomi, and Temple Jewel; and the hybrid cultivars Aurora, Galaxy, and Stellar Pink.

Moderately resistant cultivars, on which powdery mildew colonies are infrequent and restricted, include the flowering dogwood cultivars Barton White, Cherokee Chief, Cherokee Daybreak, Cherokee Princess, Cloud 9, Double White, Fragrant Cloud, Rainbow, Springtime, Weaver's White, Welch's Bay Beauty, and World's Fair; the kousa cultivars Elizabeth Lustgarten and Steeple; and the hybrid cultivars Constellation, Ruth Ellen, and StarDust.

Susceptible cultivars, on which powdery mildew colonies are frequent and spreading, include the flowering dogwood cultivars Autumn Gold, Cherokee Sunset, Dwarf White, First Lady, Junior Miss, Ozark Spring, Pink Beauty, Pink Flame, Purple Glory, Red Beauty, Rubra Pink, Stokes Pink, and Wonderberry.

Chemical treatment For susceptible cultivars, start fungicide applications when white fungal colonies first appear on the leaves. Continue fungicide treatment at the interval specified on the product label until the threat of disease spread is over. Shorten the interval between applications when foliage is heavily colonized. Production of disease-resistant cultivars will minimize the need for preventive fungicide treatment.

For fungicide recommendations and application guidelines, consult your county extension agent or state specialist responsible for nursery crops.

Powdery Mildew

Geographic occurrence USDA zones 5–8

Seasonal occurrence June to October, under average yearly conditions in zone 7 (adjust for your hardiness zone)

DISEASE FREQUENCY	DISEASE SEVERITY
[5] annual	5 plants killed
4	4
3	**[3]**
2	2
1 rare	1 very little damage

CHEMICAL TREATMENT	CULTURAL PRACTICES
3 used every year	3 very important
[2]	**[2]**
1 not used	1 not important

SANITATION	RESISTANT CULTIVARS
3 very important	3 many cultivars
[2]	**[2]**
1 not important	1 no resistance

Spot Anthracnose

Elsinoe corni

Spot anthracnose is a common and widespread disease of field- and container-grown flowering dogwood. Trees growing in partial to full sun are far more sensitive to the disease than trees growing in full shade. Leaves and bracts may be badly disfigured, but the disease does not have an adverse impact on tree health, although it may have a detrimental impact on market value.

Symptoms

The earliest symptoms of spot anthracnose are numerous small (1 to 3 mm in diameter), round, reddish purple to brown spots with dusty yellow to gray centers on unfurling bracts (Plate 33). In container and field nurseries, similar spots may appear at any time during the spring and summer on unfurling leaves, shoots, petioles, and fruit of actively growing flowering dogwood. The centers of spots may disintegrate, leaving shot holes in the leaves. Heavily spotted bracts and leaves are often twisted or deformed (Plate 34). Badly diseased bracts are often shed early.

In most dogwood taxa other than flowering dogwood, the only symptom is spotting of the bracts.

Spot anthracnose can be distinguished from dogwood anthracnose by the size of the leaf spots and the timing of the development of symptoms on leaves. Spot anthracnose lesions are considerably smaller and appear earlier than those of dogwood anthracnose.

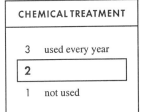

Epidemiology

The causal fungus, *Elsinoe corni,* overseasons as fruiting bodies in inconspicuous lesions (spots) on twigs. As the trees begin to bloom, conidia are spread by wind and splashing water to the unfurling bracts and leaves. Disease development is favored by extended periods of cloudy, wet weather. Frequent overhead watering along with favorable weather conditions may also contribute to the development of this disease in late spring and early summer on fast-growing nursery stock, particularly trees that have recently been pruned. Dry weather during bloom and leaf-out may sharply reduce disease severity.

Host Range

Flowering dogwood is the primarily target of the spot anthracnose pathogen. The disease has also been reported in small rough-leaf cornel (*C. asperifolia*). In Alabama, spot anthracnose has been noted in selections of *C. nuttallii × florida, C. kousa,* and *C. kousa × florida.*

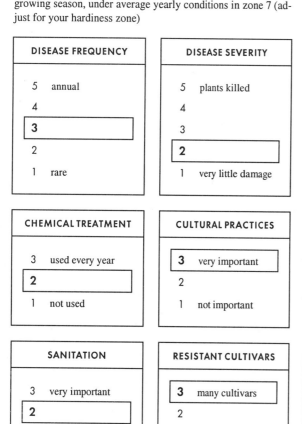

Dogwood Spot Anthracnose

Geographic occurrence USDA zones 5–8

Seasonal occurrence Late May to June, with some secondary infection of heavily pruned nursery stock throughout the growing season, under average yearly conditions in zone 7 (adjust for your hardiness zone)

DISEASE FREQUENCY	DISEASE SEVERITY
5 annual	5 plants killed
4	4
[3]	3
2	**[2]**
1 rare	1 very little damage

CHEMICAL TREATMENT	CULTURAL PRACTICES
3 used every year	**[3] very important**
[2]	2
1 not used	1 not important

SANITATION	RESISTANT CULTIVARS
3 very important	**[3] many cultivars**
[2]	2
1 not important	1 no resistance

Management

Strategies for preventing outbreaks of spot anthracnose in dogwood are largely limited to the production of resistant cultivars and preventive fungicide treatments.

Cultural practices Spacing out container stock, to speed the evaporation of free moisture from leaves and improve air circulation, may help to slow the spread of the disease. When using overhead irrigation, water shortly before dawn or around midday. Watering in the late afternoon or early evening, particularly on cloudy days or several hours prior to rain, may increase the risk of disease.

Sanitation Whenever possible, collect scion wood for grafting only from disease-free dogwood. Discard fallen diseased leaves and other crop debris collected from production ranges.

Resistance Production of cultivars of flowering dogwood resistant to spot anthracnose as well as powdery mildew is strongly recommended. Native flowering dogwoods differ in their sensitivity to spot anthracnose, but most are highly susceptible to it. The reactions of selected cultivars of flowering dogwood to spot anthracnose are summarized below.

Resistant cultivars, with little or no bract spotting, include Cherokee Chief, Cherokee Sunset, Weaver's White, and Welch's Bay Beauty.

Moderately resistant cultivars, with light spotting of bracts, include Cherokee Brave, Double White, Fragrant Cloud, Junior Miss, Red Beauty, and Rubra Pink.

Susceptible cultivars, with moderate spotting and some distortion of bracts, include First Lady, Pink Flame, Purple Glory, Wonderberry, and World's Fair.

Highly susceptible cultivars, with severe spotting and bract distortion, include Barton White, Cherokee Daybreak, Cherokee Princess, Cloud 9, Dwarf White, Ozark Spring, Pink Beauty, Rainbow, Springtime, and Stoke's Pink.

Cultivars of *C. kousa* and *C. kousa × florida* are almost all highly resistant to spot anthracnose.

C. nuttallii × florida 'Eddie's White Wonder' is as sensitive to spot anthracnose as most cultivars of flowering dogwood.

Chemical treatment To produce attractive, high-quality trees, particularly in susceptible cultivars of flowering dogwood, preventive fungicide treatments may be needed. For best results on nursery stock, begin fungicide applications at budbreak, and continue treatments at the interval specified on the product label until all new growth matures in late spring or early summer. For trees that are pruned during the summer, begin fungicide sprays after budbreak. During extended periods of cloudy, wet weather, shorten the time interval between fungicide applications.

A wide variety of fungicides have excellent activity against spot anthracnose. For fungicide recommendations and application guidelines for your area, consult your county extension agent or state specialist responsible for nursery crops.

Minor Fungal Leaf and Bract Diseases

Ascochyta leaf spot (*Ascochyta cornicola*)
Botrytis blight (*Botrytis cinerea*)
Cercospora leaf spot (*Cercospora cornicola*)

Several minor fungal diseases of flowering dogwood have been reported. Ascochyta leaf spot, Botrytis blight, and Cercospora leaf spot are among the more common of these diseases. In most cases, outbreaks of these and other minor leaf spots cause little more than cosmetic damage and are rare in container- and field-grown nursery stock. In Alabama, Botrytis blight and Cercospora leaf spot of flowering dogwood occur primarily in forests and landscape plantings.

Symptoms

Ascochyta leaf spot causes grayish brown, square to round spots, with a brown to red halo, to form on leaves. Damaged leaves eventually blacken and wither. Black fruiting bodies (pycnidia) of the causal fungus usually appear in the centers of the spots.

Botrytis blight is characterized by large, irregularly shaped, brown blotches on newly expanded bracts. Similar spreading lesions also develop where senescent bracts fall on the leaves.

Cercospora leaf spot causes angular to irregularly shaped, tan to brown spots, which closely resemble symptoms of Septoria leaf spot. After a rain or heavy dew, numerous dark tufts of conidia-producing structures may be seen with a hand lens on the upper surface of leaf spots. Heavily spotted leaves often turn yellow and prematurely drop. Generally, defoliation due to Cercospora leaf spot occurs in late summer or early fall and has little effect on tree vigor.

Minor Leaf and Bract Diseases

Geographic occurrence USDA zones 4–9

Seasonal occurrence

Ascochyta leaf spot Midsummer to early fall
Botrytis blight Early spring
Cercospora leaf spot Midsummer to early fall

DISEASE FREQUENCY	DISEASE SEVERITY
5 annual	5 plants killed
4	4
3	3
2 (selected)	**2** (selected)
1 rare	1 very little damage

CHEMICAL TREATMENT	CULTURAL PRACTICES
3 used every year	3 very important
2 (selected)	**2** (selected)
1 not used	1 not important

SANITATION	RESISTANT CULTIVARS
3 very important	3 many cultivars
2	2
1 not important (selected)	**1 no resistance** (selected)

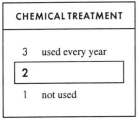

Epidemiology

The severity of leaf and bract diseases varies from year to year, depending on the amount and seasonal distribution of rainfall and the scheduling of overhead irrigation. Extended periods of cloudy, wet weather favor the development of these diseases in dogwood. Typically, Botrytis blight appears in early spring, during or shortly after bloom, while Cercospora leaf spot and Ascochyta leaf spot occur during hot, wet weather in mid- to late summer.

Host Range

Ascochyta leaf spot has been reported in flowering and red-osier dogwoods.

Botrytis blight has been reported in flowering, pagoda, and kousa dogwoods.

Cercospora leaf spot has been reported in cornelianchery and flowering dogwoods.

Management

These diseases are so sporadic in nursery stock that specific control programs are unnecessary.

Cultural practices To speed evaporation of free moisture from the leaves and to improve air circulation, space containers apart after the threat of a hard freeze has passed. When using an overhead irrigation system, water shortly before dawn or around midday. Watering in the late afternoon or early evening, particularly on cloudy days or several hours prior to rain, may increase the risk of disease.

Sanitation Discard fallen disease leaves and other crop debris collected from production ranges.

Resistance The susceptibility of cultivars of commonly cultivated dogwoods to these leaf and bract diseases has not been determined.

Chemical treatment Fungicide treatments for the control of spot anthracnose, dogwood anthracnose, and possibly powdery mildew should also control the leaf and bract diseases.

For fungicide recommendations and application guidelines for controlling leaf and bract diseases of dogwoods in your area, consult your county extension agent or state specialist responsible for nursery crops.

Dogwood Canker

Etiology unknown

Dogwood canker, a damaging disease of field-grown and native flowering dogwood, was first reported in Virginia in the late 1960s. Since then, it has been reported in several other southeastern states. Survey results indicate that the incidence of the disease in Tennessee field nurseries ranges from a few percent to nearly 60%. Dogwood canker was considered the most damaging disease of field-grown flowering dogwood prior to the appearance of dogwood anthracnose. Several fungi and nematodes have been suggested as possible causal agents, but none has been shown to cause the disease. Affected trees are often unmarketable.

Symptoms

Two distinct forms of canker have been recognized. On some flowering dogwoods, sunken cankers are formed, eventually girdle the scaffold limbs and trunk, and kill the top of the tree. Other trees display roughened, swollen areas of bark at two or more locations along the scaffold branches or trunk (Plate 35). These swollen cankers are often invaded by dogwood borer.

Epidemiology

Factors influencing the spread and development of dogwood canker are unknown.

Management

With the exception of sanitation practices, there are no controls for dogwood canker.

Sanitation Control of the disease involves the removal of cankered limbs or trees. Trees with noticeable cankers or swellings on the trunk or limbs should not be established in field plantings of flowering dogwood.

Phytophthora Leaf Blight

Phytophthora palmivora
Phytophthora parasitica

Phytophthora Root and Collar Rot

Phytophthora cactorum
Phytophthora cinnamomi

Sporadic outbreaks of Phytophthora leaf blight of container-grown flowering dogwood have been noted in Florida and Alabama. *Phytophthora parasitica* may also cause a root and crown rot of dogwood, as well as a number of other shrubs and trees.

Phytophthora root rot is a destructive disease of a wide variety of container- and field-grown trees and shrubs, including flowering dogwood. Damaging outbreaks are most likely to occur in container-grown trees, particularly in regions with heavy winter rains. Typically, flowering dogwood damaged by root and crown rot is unsalable. Refer to Chapter 14, "Phytophthora Root Rot and Dieback," for additional information.

Symptoms

Phytophthora leaf blight first appears as randomly scattered, gray green, water-soaked spots on both expanding and mature leaves. The spots rapidly enlarge into large, spreading brown blotches with gray green borders. Under conditions favorable for disease development, heavily damaged leaves wither and die. The symptoms of Phytophthora leaf blight are similar to those of dogwood anthracnose, but the regional distribution of the two diseases is somewhat different.

Phytophthora root and crown rot of field- and container-grown dogwood causes yellowing or reddening, bending or folding, and marginal necrosis of the leaves; premature defoliation; progressive limb dieback; slowed growth; sudden wilting of a limb or

Dogwood Canker

Geographic occurrence USDA zones 5–8

Seasonal occurrence Spring and summer

DISEASE FREQUENCY	DISEASE SEVERITY
5 annual	5 plants killed
4	4
3	**[3]**
[2]	2
1 rare	1 very little damage

CHEMICAL TREATMENT	CULTURAL PRACTICES
3 used every year	3 very important
2	2
[1] not used	**[1]** not important

SANITATION	RESISTANT CULTIVARS
[3] very important	3 many cultivars
2	2
1 not important	**[1]** no resistance

Phytophthora Leaf Blight and Phytophthora Root and Collar Rot

Geographic occurrence USDA zones 5–9

Seasonal occurrence Spring to early summer

DISEASE FREQUENCY	DISEASE SEVERITY
5 annual	5 plants killed
4	**[4]**
3	3
[2]	2
1 rare	1 very little damage

CHEMICAL TREATMENT	CULTURAL PRACTICES
[3] used every year	3 very important
2	2
1 not used	**[1]** not important

SANITATION	RESISTANT CULTIVARS
[3] very important	3 many cultivars
2	2
1 not important	**[1]** no resistance

an entire tree; and finally plant death. On field-grown flowering dogwood, cankers with cracked and sunken faces may develop on the root collar, at or just below the soil line. Often, a dark-colored fluid oozes from these cankers. Large, spreading, water-soaked cankers may also develop on the taproot of container-grown flowering dogwood, just below the medium surface. Diseased feeder roots are brittle and brown to reddish brown.

Epidemiology

Typically, *Phytophthora* spp. survive as resting structures (chlamydospores and oospores) in diseased roots, stems, and leaves and in crop debris, soil, and potting media. Motile zoospores of these fungi are spread to foliage and from pot to pot by splashing water or runoff flowing across field or container beds. Recycled irrigation water may also be a source of *Phytophthora* spp.

The development of Phytophthora leaf blight on flowering dogwood is favored by several days of warm, overcast, wet weather.

The heaviest losses due to root and collar rot often occur in flood-prone container beds and in poorly drained, compacted soils.

Overwatering often accelerates the development of both diseases.

Host Range

P. parasitica and *P. cinnamomi* are aggressive pathogens of numerous widely grown trees and shrubs. *P. cactorum* and *P. palmivora* have narrower host ranges.

In Alabama and Florida, Phytophthora leaf blight has been noted only in flowering dogwood.

Flowering and Pacific dogwoods are the primary targets of *P. cinnamomi* and *P. cactorum.*

Management

Strict sanitation, cultural practices, and preventative fungicide treatments are often required for the control of *Phytophthora* diseases of dogwood. A detailed discussion of recommended control practices can be found in Chapter 14, "Phytophthora Root Rot and Dieback."

Cultural practices Container beds must be crowned, covered with black plastic or weed cloth, and then topped with several inches of gravel or a similar coarse material. Drainage ditches must be of sufficient volume to handle heavy rains and prevent the ponding of water around container and field stock. A coarse aged or compost bark-based medium with a high percolation rate and 20–30% air-filled pore space will help suppress rot and collar rot. Fertilize according to need. When using overhead irrigation, water shortly before dawn or around midday. Watering in the late afternoon or early evening, particularly on cloudy days or several hours prior to rain, may greatly increase the risk of Phytophthora leaf blight.

Sanitation Store potting medium components on an asphalt or a concrete pad, and use clean hand tools, benches, and loaders and new containers to reduce the risk of disease outbreaks. Collect scion wood for grafting and cuttings for rooting only from disease-free container stock. Root cuttings in fresh potting media in new containers on well-drained raised benches. Diseased stock plants rarely recover and must be discarded, since they are a source of the causal fungi. Recycled irrigation water should be filtered, sterilized, or chlorinated before being reused on nursery stock.

Resistance The reactions of the commonly cultivated dogwood taxa to Phytophthora leaf blight and Phytophthora root rot have not been determined.

Chemical treatment Phytophthora leaf blight is not a common disease of container-grown flowering dogwood, and routine preventive fungicide treatments are not recommended. Preventive fungicide drenches for the control of Phytophthora root and crown rot should give some control of Phytophthora leaf blight. Scout blocks of container-grown flowering dogwood for typical leaf blight symptoms at bract fall at weekly intervals, and continue as long as the trees put on new leaves. Apply protective fungicides to the foliage at the first sign of disease, and continue treatments at the interval specified on the product label until conditions no longer favor disease development.

When combined with strict sanitation and good nursery management practices, preventive fungicides applied through the production cycle at the rates and intervals specified on the product labels will effectively control Phytophthora root and collar rot. Depending on the product chosen, fungicides may be incorporated into the soil or potting medium prior to planting, broadcast over field stock beds, applied as a drench on potting media or soil, or applied as a foliar spray. Preventive fungicide treatments are strongly recommended in nurseries where *Phytophthora* diseases have caused significant losses in other tree or shrub crops. Fungicides will not "cure" diseased plants or eradicate the causal fungi from their roots but may slow further symptom development.

For guidelines for chemical control of Phytophthora leaf blight and Phytophthora root and collar rot, contact your county extension agent or state specialist responsible for nursery crops.

Canker and Dieback

Botryodiplodia theobromae

Canker and dieback occurs sporadically in field- and container-grown flowering dogwood. The causal fungus, *Botryodiplodia theobromae,* is usually an aggressive pathogen only on trees and shrubs weakened by environmental stress, such as drought or low temperatures, or poor management practices.

Symptoms

Sudden wilting of the leaves on one or more limbs is one of the first symptoms of canker and dieback. The dead leaves usually remained attached to the limbs for part of the growing season. Large, girdling cankers are formed. The canker surface is slightly darker than the surrounding healthy tissues. Cankers often continue to expand down the scaffold branches to the trunk, thereby killing much of the top growth. Scattered clusters of black fruiting bodies (pycnidia) will develop on diseased limbs. Beneath the discolored bark, the underlying tissues turn brown. Flowering dogwood damaged by canker and dieback is usually unmarketable.

Epidemiology

Because of its wide host range and saprophytic capacity, *B. theobromae* is a resident of most native tree and shrub communities around container and field nurseries or has been introduced on diseased stock. Conidia of the fungus are dispersed by splashing water and possibly by pruning tools. Pruning wounds and other openings in the bark, such as frost cracks, are readily invaded by the fungus. Trees subjected to extreme cold or drought are much more readily invaded than healthy trees. Cankers are often much larger on stressed flowering dogwood than on unstressed trees.

Botryodiplodia Canker

Geographic occurrence USDA zones 6–9

Seasonal occurrence Spring and summer

DISEASE FREQUENCY	DISEASE SEVERITY
5 annual	5 plants killed
4	4
3	**[3]**
[2]	2
1 rare	1 very little damage

CHEMICAL TREATMENT	CULTURAL PRACTICES
3 used every year	**[3] very important**
2	2
[1] not used	1 not important

SANITATION	RESISTANT CULTIVARS
3 very important	3 many cultivars
[2]	**[2]**
1 not important	1 no resistance

Host Range

The host range of *B. theobromae* includes plants in nearly 100 genera. Redbud, rhododendron, rose, and sweet gum are among the many tree and shrub hosts of this fungus.

Management

Canker and dieback in both container- and field-grown flowering dogwoods can be avoided by following recommended management and sanitation practices.

Cultural practices Fertilize and irrigate container and field stock according to need. Follow other management practices which promote tree vigor.

Sanitation The causal fungus quickly invades wounded or freshly killed tissue, and therefore branches pruned from container and field stock should immediately be collected and discarded. Follow pruning practices which promote rapid healing of wounds. Sanitize pruning tools with denatured alcohol or a similar disinfectant.

Resistance Differences in the sensitivity of cultivars of flowering dogwood to canker and dieback have been noted. The cultivar Cherokee Red may be more sensitive to this disease than Barton White and Welch's Junior Miss. The reactions of most flowering dogwood cultivars and cultivars of other dogwoods are not known.

Chemical treatment The efficacy of fungicides against canker and dieback incited by *B. theobromae* has never been demonstrated.

REFERENCES

Alfieri, S. A., and El-Gholl, N. E. 1993. Phytophthora leaf blight of flowering dogwood, *Cornus florida* L. Fla. Dep. Agric. Consumer Serv. Plant Pathol. Circ. 368.

Brown, D. A., Windham, M. T., and Trigiano, R. N. 1996. Resistance to dogwood anthracnose among *Cornus* species. J. Arboric. 22:83–85.

Daughtrey, M. L., Hibben, C. R., Britton, K. O., Windham, M. T., and Redlin, S. C. 1996. Dogwood anthracnose: Understanding a disease new to North America. Plant Dis. 80:349–358.

Dirr, M. A. 1998. Manual of Woody Landscape Plants: Their Identification, Ornamental Characteristics, Culture, Propagation and Uses. 5th ed. Stipes Publishing, Champaign, Ill.

Gould, A. B., and Peterson, J. L. 1994. The effect of moisture and sunlight on the severity of dogwood anthracnose on street trees. J. Arboric. 20:75–78.

Hagan, A. K., Hardin, B., Gilliam, C. H., Keever, G. J., Williams, J. D., and Eakes, J. 1998. Susceptibility of cultivars of several dogwood taxa to powdery mildew and spot anthracnose. J. Environ. Hortic. 16:147–151.

Hagan, A. K., and Mullen, J. M. 1993. Dogwood anthracnose on dogwood in Alabama, its distribution and severity. Ala. Agric. Exp. Stn., Auburn Univ. Res. Rep. 8:34–35.

Hoitink, H. A. J., Benson, D. M., and Schmitthenner, A. F. 1986. Phytophthora root rot. Pages 4–8 in: Compendium of Rhododendron and Azalea Diseases. D. L. Coyier and M. K. Roane, eds. American Phytopathological Society, St. Paul, Minn.

Klein, L. A., Windham, M. T., and Trigiano, R. N. 1998. Natural occurrence of *Microsphaera pulchra* and *Phyllactinia guttata* on two *Cornus* species. Plant Dis. 82:383–385.

Lambe, R. C., and Wills, W. H. 1980. Current status of dogwood canker. Proc. Int. Plant Prot. Soc. 30:526–529.

Mmbaga, M. T. 2000. Winter survival and source of primary inoculum of powdery mildew of dogwood in Tennessee. Plant dis. 84:574–579.

Mullen, J. M., Gilliam, C. H., Hagan, A. K., and Morgan-Jones, G. 1991. Canker of dogwood caused by *Lasiodiplodia theobromae,* a disease influenced by drought stress or cultivar selection. Plant Dis. 75:886–889.

Neely, D. and Nolte, B. S. 1989. Septoria leaf spot on dogwoods. J. Arboric. 15:263–267.

Ranney, T. G., Grand, L. F., and Knighten, J. L. 1995. Susceptibility of cultivars and hybrids of kousa dogwood to dogwood anthracnose and powdery mildew. J. Arboric. 21:11–16.

Windham, M. T., and Montgomery, M. 1988. Spatial distribution of dogwood canker and comparison of cultural factors with disease incidence. Proc. South. Nurserymen's Assoc. Res. Conf. 33:165.

CHAPTER
34

Gail E. Ruhl ▪ Purdue University, West Lafayette, Indiana
Mary Francis Heimann, O.S.F. ▪ Madison, Wisconsin

Elaeagnus Diseases

Geographic production USDA zones 2–8

Family Elaeagnaceae

Genus *Elaeagnus*

Species

E. angustifolia	Russian olive	zones 2–7
E. commutata	silverberry	zones 2–5(6)
E. × *ebbingii*	. . .	zones 6–9
E. multiflora	cherry elaeagnus	zones 5–7
E. pungens	silverthorn, thorny elaeagnus	zones 6–9
E. umbellata	autumn olive	zones 3–8

Introduced from Asia and southern Europe, many species of *Elaeagnus* are of value for a number of reasons. Landscape architects frequently choose them because of their silvery foliage and silvery to red fruits, making them an attractive specimen tree. In northern states their winter-hardiness and salt tolerance are considered positive aspects. Adaptability to varied soils, drought tolerance, and an ability to do well in soils of elevated pH make *Elaeagnus* a tree of choice.

Despite these attributes, *Elaeagnus* spp. have been disappearing from the landscape in many parts of the country. Originally thought to be pest- and disease-free, *Elaeagnus* spp. have recently been subject to various fungal cankers and Verticillium wilt.

Cankers and Dieback

Numerous fungi

Several fungi, including species of *Botryosphaeria, Cytospora, Fusarium, Nectria, Phoma, Phomopsis, Phytophthora,* and *Tubercularia,* cause cankers and dieback of *Elaeagnus* spp.

Symptoms

Phomopsis arnoldiae (syn. *P. elaeagni*) causes cankers on the limbs and trunk. The first symptom is shriveled, gray foliage, which remains attached to the tree. Cankers form on limbs and can girdle stems. On smooth limbs cankered areas can take on a reddish brown, shiny appearance. In a few weeks, fruiting bodies (pycnidia) of the pathogen can be observed breaking through the bark. In some cases a gummy exudate is produced on the limbs or oozes out of the trunk. The cankers may not kill the entire tree, but they render it unacceptable as a landscape plant. Seedlings and saplings may wilt and die with no apparent canker formation.

Tubercularia ulmea also causes a canker and dieback of *Elaeagnus*. The cankers are subtle and often are not perceived until a branch dies, buds do not open in spring, or branches wilt as a result of girdling by cankers. On young smooth limbs cankers can cause a reddish brown discoloration of the bark and some drying of twig tissue. As in Phomopsis canker, a gummy exudate may become apparent. About three weeks after infection occurs, fungal

Cankers and Dieback

Geographic occurrence USDA zones 2–8

Seasonal occurrence Infection occurs from fall to spring, under average yearly conditions in the Midwest (adjust for your hardiness zone)

DISEASE FREQUENCY
5 annual
4
3
2
1 rare

DISEASE SEVERITY
5 plants killed
4
3
2
1 very little damage

CHEMICAL TREATMENT
3 used every year
2
1 not used

CULTURAL PRACTICES
3 very important
2
1 not important

SANITATION
3 very important
2
1 not important

RESISTANT CULTIVARS
3 many cultivars
2
1 no resistance

sporodochia push out from the cankered area. The sporodochia change color, turning from a light rust to brown to black. *T. ulmea* is a weak pathogen and apparently can enter plants only if they have been wounded. Whether or not this fungus is a variant of *T. vulgaris,* the asexual (imperfect) stage of *Nectria cinnabarina,* remains to be determined. A Nectria canker with a similar epidemiology is mentioned in the literature.

Botryosphaeria rhodina causes another canker disease of Russian olive and has killed many of these trees planted as windbreaks in the Great Plains. This fungus infects plants by entering wounds, and its aggressiveness is dependent on the degree of stress under which the host is growing. Gums or resins are frequently exuded from the edges of lesions on stems and trunks. The cankers are elongate, and girdling is due to several cankers side by side rather than one canker circumscribing the stem. Pycnidia of the imperfect state of the fungus, *Diplodia natalensis,* cause small, raised, pimple-like bumps as they push through the bark during moist weather. *B. obtusa* and *B. dothidea* are also reported to be pathogenic to Russian olive.

Epidemiology

Spores of these pathogens are disseminated by wind and rain, in soil, and on tools and equipment. They usually enter trees at wound sites, killing surrounding healthy tissues. New growth is often more susceptible than older tissue.

Host Range

Various fungal canker and dieback diseases have been reported in all of the species of *Elaeagnus* mentioned above. *E. angustifolia* appears to be the most susceptible to these diseases.

Management

Fungicides tend to be of no practical benefit for management of canker and dieback diseases. Cultural controls are the primary means of managing these diseases.

Cultural practices Since canker and dieback fungi primarily invade wounded tissue, it is of utmost importance to avoid wounding and unnecessary stress on trees. Practices that promote tree vitality are important for management of canker diseases. Inspect new trees carefully for symptoms of these diseases. Do not plant infected trees, since cankers are likely to continue to affect tree growth. Prune out dead branches on established trees, cutting well below the cankered areas, during dry weather, and destroy the cuttings. Disinfect pruning shears with rubbing alcohol or disinfectant between cuts. Remove and destroy dying trees.

Resistance Check for newly developed resistant varieties of Russian olive before planting this species.

Verticillium Wilt

Verticillium albo-atrum
Verticillium dahliae

Russian olive is one of many plants subject to wilt caused by either *Verticillium dahliae* or *V. albo-atrum.* These fungi are soilborne and survive for long periods in soil when no host is present.

Symptoms

Tissue necrosis results at points of infection and colonization. Various modes of pathogenesis have been reported: the pathogen plugs the vascular system; it produces a toxin or enzyme that kills tissue; it incites the host plant to produce toxins or enzymes; or it produces or causes the host to produce tyloses, which plug the vascular system. Whatever the mode, different parts of the host wilt and die. Wilting is frequently unilateral, with part of the tree dying each year. Various hosts display vascular discoloration. In *Elaeagnus* the discoloration tends toward brown, in contrast to the green to black discoloration associated with maples infected with *Verticillium.* Even if death does not occur, after a number of years the appearance of infected trees is usually no longer acceptable for landscape plantings.

Verticillium Wilt

Geographic occurrence USDA zones 2–8

DISEASE FREQUENCY	DISEASE SEVERITY
5 annual	5 plants killed
4	**[4]**
[3]	3
2	2
1 rare	1 very little damage

CHEMICAL TREATMENT	CULTURAL PRACTICES
3 used every year	**[3] very important**
2	2
[1] not used	1 not important

SANITATION	RESISTANT CULTIVARS
3 very important	3 many cultivars
2	**[2]**
[1] not important	1 no resistance

Epidemiology

V. dahliae survives in the soil in the form of microsclerotia, while *V. albo-atrum* survives in the form of resting hyphae. These structures germinate when they are stimulated by nutrients from nearby roots. The germination hyphae penetrate and infect roots of susceptible plants. They colonize root tissue at a relatively slow rate. However, once hyphae penetrate the system, conidia are produced, which move up the xylem tissue at a much faster rate. The conidia stop at the xylem vessel pit walls, germinate, and produce hyphae, which grow through the pit and then form more conidia, as the cycle repeats itself.

Foliar and Root Diseases

Geographic occurrence USDA zones 2–8 (the range of the host)

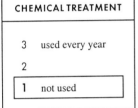

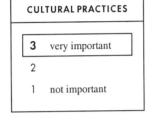

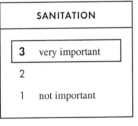

DISEASE FREQUENCY	DISEASE SEVERITY
5 annual	5 plants killed
4	4
3	3
2	**[2]**
[1] rare	1 very little damage

CHEMICAL TREATMENT	CULTURAL PRACTICES
3 used every year	**[3] very important**
2	2
[1] not used	1 not important

SANITATION	RESISTANT CULTIVARS
[3] very important	3 many cultivars
2	2
1 not important	**[1] no resistance**

Host Range

All *Elaeagnus* spp. are susceptible to Verticillium wilt.

Management

There is no cure for trees infected by *Verticillium.* Symptom progression may be delayed by maintaining tree vigor through proper watering, fertilization, and pruning. See Chapter 19, "Verticillium Wilt."

Other Foliar and Root Diseases

Numerous fungi

Several minor diseases of *Elaeagnus* have been reported, including fungal leaf spots, fungal root rots, powdery mildew, rust, (caused by *Puccinia*), and crown gall.

Leaf spot fungi documented to occur on *Elaeagnus* spp. include *Alternaria, Cercospora, Colletotrichum, Gloeosporium, Mycosphaerella, Phyllosticta,* and *Septoria.*

Fungi associated with major root rots of *Elaeagnus* spp. include *Armillaria, Fusarium, Phymatotrichopsis, Phytophthora, Pythium,* and *Rhizoctonia. E. angustifolia* is reported to be resistant to Armillaria root rot.

Host Range

Fungal leaf spot diseases have been reported in all of the *Elaeagnus* spp. mentioned above, rust in *E. angustifolia* and *E. commutata,* and powdery mildew in *E. commutata* only. Fungal root rot pathogens have been documented in *E. angustifolia* and *E. pungens.*

Management

Leaf spot diseases of *Elaeagnus* are rarely severe enough to warrant the use of protective fungicides. Cultural controls are the primary means of managing these diseases.

CHAPTER 35

Ned Tisserat ▪ Kansas State University, Manhattan
James Sherald ▪ National Park Service, Washington, D.C.
Gary Moorman ▪ Pennsylvania State University, University Park
Phillip Colbaugh ▪ Texas A&M Agricultural Experiment Station, Dallas

Elm Diseases

Geographic production USDA zones 2–9

Family Ulmaceae

Genus *Ulmus*

Species

U. alata	winged elm	zones 6–9
U. americana	American elm	zones 2–9
U. carpinifolia	smoothleaf elm	zones 3–6
U. crassifolia	cedar elm	
U. davidiana	David elm	
U. glabra	Scots elm	zones 4–6
U. lamellosa	golden bark elm	zones 5–9
U. lanciniata	Manchurian elm	zones 3–9
U. macrocarpa	large fruited elm	
U. parvifolia	Chinese elm, lacebark elm	zones 4–9
U. pumila	Siberian elm	zones 4–9
U. rubra	red elm, slippery elm	zones 2–9
U. wilsoniana	Wilson elm	zones 3–9

Elm Black Spot
Stegophora ulmea

Elm black spot, sometimes referred to as anthracnose, caused by the fungus *Stegophora ulmea,* is a common and widespread foliar disease of elms. It is generally considered to be a minor disease in mature trees, but it may cause significant defoliation and twig dieback of susceptible cultivars during nursery production.

Symptoms

Small yellow spots initially develop on the upper surface of leaves as they unfold in spring. On susceptible elms, the spots may rapidly expand and coalesce to blight entire leaves. Lesions may also girdle petioles and succulent shoot growth. Twig blighting in successive seasons may result in the formation of witches'-brooms. On more resistant elms, leaf spots remain smaller (0.5–3.0 mm in diameter) and eventually develop a slightly raised, black crust, or stroma, containing fruiting structures (acervuli) of *S. ulmea.* The spots may be surrounded by a white to light yellow halo. Repeated infections in early summer result in multiple spotting and premature defoliation.

Epidemiology

S. ulmea overwinters in dead leaves on the ground and in dormant buds. In spring, fruiting bodies (perithecia) of the fungus release spores (ascospores), which infect newly developing leaves and stems. The impact of splashing rain causes the release of secondary spores (conidia) from acervuli embedded in the stroma. Leaves are most susceptible to infection in spring and early summer. A third, noninfectious spore stage (sporidia) is formed later in summer.

Elm Black Spot

Geographic occurrence USDA zones 3–8

Seasonal occurrence Infection may occur from budbreak until leaves are fully developed; defoliation and other symptoms may continue to be apparent throughout the summer.

DISEASE FREQUENCY	DISEASE SEVERITY
5 annual	5 plants killed
4 (selected)	4
3	**3** (selected)
2	2
1 rare	1 very little damage

CHEMICAL TREATMENT	CULTURAL PRACTICES
3 used every year	3 very important
2 (selected)	**2** (selected)
1 not used	1 not important

SANITATION	RESISTANT CULTIVARS
3 very important	3 many cultivars
2 (selected)	**2** (selected)
1 not important	1 no resistance

Host Range

Most elm species are susceptible to black spot, although there is considerable variation in disease severity among cultivars of the same species. Certain cultivars of Chinese elm appear to be highly susceptible.

Management

Remove leaf debris and dead shoots from trees during winter to reduce overwintering inoculum of the fungus. Avoid close spacing and overhead irrigation of cuttings and young trees. Preventive fungicide applications may be required for susceptible elm cultivars. Make the first application at budbreak in spring, and continue at regular intervals until leaves are fully developed.

Canker Diseases

Botryosphaeria dothidea
***Botryosphaeria stevensii* (syn. *B. hypodermia*)**
Dothiorella ulmi
Nectria cinnabarina

Several fungi cause twig dieback and branch and trunk cankers on elms. The cankers are often associated with nursery practices that may injure bark tissue. Cankers are also more common on elm cultivars damaged by low temperatures.

Symptoms

Canker development is variable, depending on the fungal pathogen and host susceptibility. A few elm cankers are perennial, but many are annual and start at branch tips damaged by cold temperatures or at pruning cuts. Initially, infected bark becomes slightly sunken and develops a water-soaked appearance. The inner bark color changes from light tan to reddish brown to brownish black. Small-diameter twigs may be quickly girdled and killed. Branch and trunk cankers are often elliptical. Normally, there is a sharp demarcation between healthy and diseased tissue. As a canker ages, the dead bark dries, cracks, and pulls away from developing callus tissue at the canker margin. Small red to black fruiting structures of the pathogen may develop in diseased tissue. Eventually the dead bark may slough from the tree, exposing dead sapwood beneath.

Epidemiology

Most canker fungi are facultative parasites that survive on dead wood or in cankers on living trees. Infection is influenced by many factors, including tree age and vigor; environmental condi-

Canker Diseases

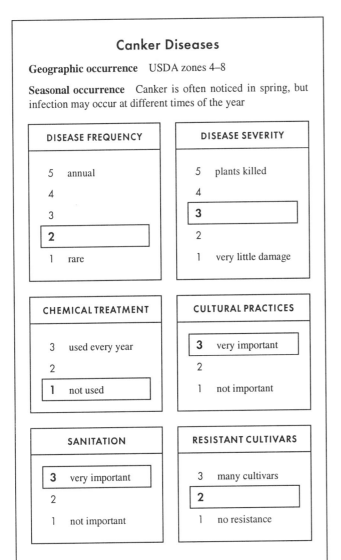

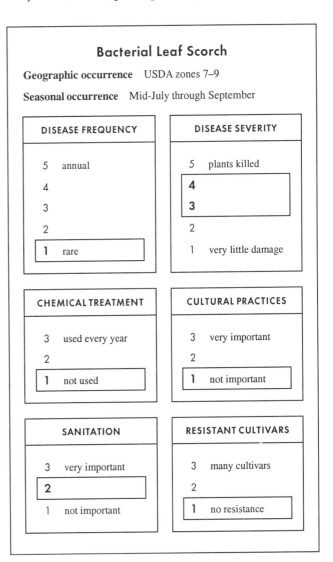

Geographic occurrence USDA zones 4–8

Seasonal occurrence Canker is often noticed in spring, but infection may occur at different times of the year

DISEASE FREQUENCY	DISEASE SEVERITY
5 annual	5 plants killed
4	4
3	**[3]**
[2]	2
1 rare	1 very little damage

CHEMICAL TREATMENT	CULTURAL PRACTICES
3 used every year	**[3 very important]**
2	2
[1 not used]	1 not important

SANITATION	RESISTANT CULTIVARS
[3 very important]	3 many cultivars
2	**[2]**
1 not important	1 no resistance

Bacterial Leaf Scorch

Geographic occurrence USDA zones 7–9

Seasonal occurrence Mid-July through September

DISEASE FREQUENCY	DISEASE SEVERITY
5 annual	5 plants killed
4	**[4**
3	**3]**
2	2
[1 rare]	1 very little damage

CHEMICAL TREATMENT	CULTURAL PRACTICES
3 used every year	3 very important
2	2
[1 not used]	**[1 not important]**

SANITATION	RESISTANT CULTIVARS
3 very important	3 many cultivars
[2]	2
1 not important	**[1 no resistance]**

tions, such as moisture stress and temperature fluctuations; and cultural practices that may damage bark tissue.

Host Range

All elm species are susceptible to canker.

Management

Avoid bark injury resulting from mowing or other management practices. Prune out all dead branches or those exhibiting cankers, and remove them. When pruning, do not leave stubs or make flush cuts. Protect the trunks of young, thin-barked elms from direct sunlight in late winter to prevent sunscald. Promote tree vigor by maintaining adequate soil moisture and fertilization. In northern climates, avoid propagation of cultivars that are sensitive to cold. Fungicides generally are not effective in preventing or suppressing canker development.

Bacterial Leaf Scorch

Xylella fastidiosa

Bacterial leaf scorch affects elms throughout the mid-Atlantic, southeastern, and Gulf Coast states but is rare in nurseries. It is a chronic systemic disorder caused by the xylem-inhabiting bacterium *Xylella fastidiosa,* which causes diseases of many woody and herbaceous hosts.

Symptoms

Leaf symptoms appear in midsummer as an undulating marginal necrosis. Necrotic tissue is often preceded by a chlorotic halo as it expands toward the midvein. The symptoms decrease in severity from the older leaves at the base of a branch to the younger leaves at the tip. Scorched leaves curl and abscise early, leaving tufts of unaffected or slightly affected leaves at the tips of bare branches. Symptoms usually reappear each year, affecting increasingly larger areas of the crown. Twig and branch dieback develops in trees that have been affected for several years.

The leaf necrosis of bacterial leaf scorch resembles symptoms of the later stages of Dutch elm disease. Unlike Dutch elm disease, bacterial leaf scorch does not cause wilt or vascular discoloration. Bacterial leaf scorch, however, may lead to Dutch elm disease. Elms chronically stressed by bacterial leaf scorch attract the European elm bark beetle, *Scolytus multistriatus,* which transmits *Ophiostoma ulmi,* the fungus causing Dutch elm disease.

Epidemiology

X. fastidiosa inhabits xylem tissue and is transmitted by xylem-feeding insects, principally sharpshooter leafhoppers in the subfamily Cicadellinae. The specific vector or vectors responsible for transmission in elm have not been confirmed.

X. fastidiosa has a wide host range, including many plant species that do not exhibit symptoms. Consequently, other hosts may harbor strains of *X. fastidiosa* which are pathogenic in elm.

Table 35.1 Susceptibility to elm yellows and elm leaf beetle in commonly propagated elms (*Ulmus* spp.) and hybrids with moderate to high levels of resistance to Dutch elm disease

Species and cultivar	Elm yellows[a]	Elm leaf beetle[a]	Comments
U. americana			
American Liberty	S	R	Multiclonal variety; juvenile trees may be susceptible to *Ophiostoma novo-ulmi*
Delaware	S	R	Originally designated Delaware #2
Independence	S	R	
New Harmony	S	R	
Valley Forge	S	R	
U. japonica	R	?	Primarily used as a parent in hybrid breeding programs
U. parvifolia	R	R	All of the many cultivars of this species have resistance to Dutch elm disease and good resistance to elm yellows and elm leaf beetle, but low temperatures in zones 4 and 5 may damage the trees
U. pumila	R	S	Resistant to Dutch elm disease and elm yellows, but highly susceptible to elm leaf beetle; this species should not be used in the landscape in most locations
U. wilsoniana			
Prospector	R	R	
Elm hybrids			
Accolade	R	R	*U. japonica* x *U. wilsoniana;* reported to be resistant to elm leaf miner
Cathedral	R	S?	*U. pumila* x *U. japonica;* reported to be resistant to Verticillium wilt
Charisma	R	?	Hybrid of Accolade and Vanguard
Danada	R	?	*U. japonica* x *U. wilsoniana*
Frontier	R	R	*U. carpinifolia* x *U. parvifolia;* some winter damage has been reported in the Great Plains
Homestead	R	S	Multispecies hybrid; probably not suitable where elm leaf beetle is present
New Horizon	R	S?	*U. japonica* x *U. pumila;* reported to be resistant to Verticillium wilt and elm leaf miner
Patriot	R	R	Hybrid of Urban and Prospector
Pioneer	R	S	*U. glabra* x *U. carpinifolia;* probably not suitable where elm leaf beetle is present
Regal	R?	?	Complex hybrid
Sapporo Autumn Gold	R	S	*U. pumila* x *U. japonica;* resistant to Verticillium wilt
Urban	R	S	Should not be used in areas where elm leaf beetle is present
Vanguard	R	S?	*U. japonica* x *U. pumila*

[a] R = resistant. S = susceptible.

Host Range

U. americana is the most common elm host of *X. fastidiosa*. *U. pumila* and *U. glabra* are also affected by bacterial leaf scorch. Other elm species are likely to be susceptible to the disease.

Management

Symptoms of bacterial leaf scorch are easily confused with those of other disorders. Suspected cases should be confirmed in the laboratory. Currently, there is no known cure. Affected nursery stock should be discarded.

Dutch Elm Disease

Ophiostoma ulmi
Ophiostoma novo-ulmi

The devastating effects of Dutch elm disease on native elms in North America resulted in decreased use of all elm species in the landscape for many years. The introduction of new cultivars of Asian elms and elm hybrids with resistance to Dutch elm disease has sparked a renewed interest in elms in the urban landscape.

It is unlikely that Dutch elm disease will be a problem in the nursery during the propagation of resistant elms. Therefore a detailed description of disease symptoms and management is not provided (stem symptoms are illustrated in Plate 36). However, managers should be familiar with elm cultivars and hybrids that are resistant to Dutch elm disease and should be aware of their susceptibility to other diseases and insect pests (Table 35.1). Elms susceptible to the elm leaf beetle should not be planted in many areas of the United States.

Elm Yellows

Phytoplasma

Elm yellows can be a serious disease in mature American elm and other native species, it is unlikely to be common in nurseries, because many of the newer elm hybrids in propagation are resistant. Nevertheless, some clones of American elms selected for a high degree of resistance to Dutch elm disease may be susceptible to elm yellows. Details on the relative susceptibility of elm species and hybrids to elm yellows are presented in Table 35.1.

Verticillium Wilt

Verticillium dahliae

Elm trees are susceptible to infection by *Verticillium dahliae*. Refer to Chapter 19, "Verticillium Wilt," for information on symptoms and management. Some new elm hybrids with resistance to Dutch elm disease are also reported to be resistant to Verticillium wilt (Table 35.1).

Damping-Off and Root Rots

Various soilborne fungi

Damping-off and root rots may be a significant problem in elm seedbeds or during rooting of cuttings taken from hybrid elms resistant to Dutch elm disease. Fungi associated with damping-off and root rot include *Chalara, Fusarium, Pythium,* and *Rhizoctonia*. See Chapter 12, "Damping-Off of Seeds and Seedlings and Cutting Rot," for information on management.

Leaf Blister

Taphrina ulmi

Pseudomonas Blight

Pseudomonas syringae

Powdery Mildew

***Microsphaera* spp.**
***Phyllactinia* spp.**
***Uncinula* spp.**

Several fungal and bacterial pathogens are capable of causing leaf spotting and defoliation of elms in the nursery.

Elm leaf blister, caused by the fungus *Taphrina ulmi,* results in the formation of slightly raised, thickened, yellow leaf spots soon after leaf emergence. However, the disease rarely causes extensive defoliation.

Pseudomonas leaf spot and blight, caused by the bacterium *Pseudomonas syringae,* affects a wide range of woody plants, including elm. For information on symptoms and management of the disease in the nursery, consult Chapter 16, "Diseases Caused by *Pseudomonas syringae*."

Powdery mildew, caused by species of the fungi *Microsphaera, Phyllactinia,* and *Uncinula,* has been widely reported in a number of elm species. In Texas, a high proportion of cedar elms showed slight to severe symptoms of the disease. A few trees showed no visible symptoms, suggesting that elm selections could be used to overcome susceptibility to powdery mildew. For information on symptoms and management, consult Chapter 15, "Powdery Mildew."

REFERENCES

Becker, H. 1996. New American elms restore stately trees. Agric. Res. 44: 4–8.

Colbaugh, P. F., Fields, S. L., and Simpson, B. J. 1981. Powdery mildew resistance among native populations of cedar elm. (Abstr.) Phytopathology 77:640.

Farr, D. F., Bills, G. F., Chamuris, G. P., and Rossman, A. Y. 1989. Fungi on Plants and Plant Products in the United States. American Phytopathological Society, St. Paul, Minn.

Hearon, S. H., Sherald, J. L., and Kostka, S. J. 1980. Association of xylem-limited bacteria with elm, sycamore, and oak leaf scorch. Can. J. Bot. 58:1986–1993.

Riffle, J. W., and Peterson, G. W. 1986. Diseases of trees in the Great Plains. U.S. Dep. Agric. For. Serv., Rocky Mountain For. Range Exp. Stn. (Ft. Collins, Colo.), Gen. Tech. Rep. RM-129.

Santamour, F. S., and Bentz, S. E. 1995. Updated checklist of elm (*Ulmus*) cultivars for use in North America. J. Arboriculture 21:122–131.

Sherald, J. L. 1992. Bacterial leaf scorch of landscape trees caused by *Xylella fastidiosa*. J. Arboric. 18:57–63.

Sherald, J. L. 1995. Leaf scorch of amenity trees. In: *Xylella fastidiosa* and Associated Diseases. A. B. Gould and J. L. Sherald, eds. Plant Diagnostician's Q. 16(3):119–123.

Sinclair, W. A., Lyon, H. H., and Johnson, W. T. 1987. Diseases of Trees and Shrubs. Comstock Publishing Associates, Cornell University Press, Ithaca, N.Y.

Smalley, E. B., and Guries, R. P. 1993. Breeding elms for resistance to Dutch elm disease. Annu. Rev. Phytopathol. 31:325–352.

Stipes, R. J., and Campana, R. J. 1981. Compendium of Elm Diseases. American Phytopathological Society, St. Paul, Minn.

CHAPTER 36

Jacqueline Mullen ▪ Auburn University, Auburn, Alabama
Larry W. Barnes ▪ Texas A&M University, College Station
Gary W. Simone ▪ University of Florida, Gainesville

English Ivy Diseases

Geographic production USDA zones 4–9

Family Araliaceae

Genus *Hedera*

Species *H. helix*

English ivy is one of the most well known ground covers in the United States and elsewhere. It performs well in shade or full sun and is thus widely planted where a ground cover 6 to 8 inches tall is desired. English ivy also can add a storied look to the facades of buildings after a few years' growth. Several diseases affect English ivy in the nursery, including anthracnose, bacterial leaf spot, and root rots caused by *Phytophthora* spp. and *Rhizoctonia solani.*

Anthracnose

Colletotrichum trichellum
Other *Colletotrichum* spp.

Anthracnose is a leaf spot, stem canker, and blight disease, first reported in 1928. Today it is a common disease of English ivy, reported in all sections of the United States. The pathogen most commonly reported as the cause of anthracnose of English ivy is the fungus currently called *Colletotrichum trichellum.* The name of this fungus has changed several times; early reports refer to it as *Vermicularia trichella, C. gloeosporioides* var. *hedera, C. hedericola,* and *Amerosporium trichellum.* Anthracnose is a problem in the landscape and nursery, but disease incidence is higher in the landscape.

Symptoms

The first symptoms of anthracnose are circular, necrotic areas, which usually develop into irregular, dry, brown to reddish brown or black spots about 3/4 inch (2 cm) in diameter (Plate 37). Spots may first appear at or near the margins of leaves. Reproductive bodies of the pathogen appear as black specks in leaf spots. With a low level of magnification (10×), black, needle-like fungal structures (setae) may be seen in the centers of spots on the upper surfaces of leaves. Lesions that develop on petioles and stems may result in dieback and defoliation.

Symptoms of anthracnose are sometimes confused with those of bacterial leaf spot. Culture isolations are sometimes necessary to determine which of these diseases is developing in affected plants.

Epidemiology

Anthracnose most often develops in nurseries where plants or cuttings are closely packed and watered overhead. Spores of the pathogen are spread by splashing water and wind-blown rain and can also be carried by insects.

Anthracnose

Geographic occurrence USDA zones 4–9

Seasonal occurrence Late spring through early fall

DISEASE FREQUENCY	DISEASE SEVERITY
[5] annual	5 plants killed
4	4
3	**[3]**
2	2
1 rare	1 very little damage

CHEMICAL TREATMENT	CULTURAL PRACTICES
[3] used every year	3 very important
2	**[2]**
1 not used	1 not important

SANITATION	RESISTANT CULTIVARS
3 very important	3 many cultivars
[2]	**[2]**
1 not important	1 no resistance

Host Range

C. trichellum is reported to occur on most cultivars of English ivy. Pierce and McCain (1983), testing the cultivars Crested, Glacier, Gold Dust, Needlepoint, and Shamrock, found that leaf spots developed on plants in all of these cultivars after inoculation with *C. trichellum*. In Crested, leaf spots developed in wounded and unwounded areas of leaves; in the other cultivars tested, leaf spots developed only where the leaves were wounded. None of the cultivars tested were immune to disease development. In all cultivars, large spots developed where leaves were wounded prior to inoculation. Leaf spots were significantly larger on Crested, Glacier, and Needlepoint than on Gold Dust and Shamrock.

Management

Anthracnose management requires a good sanitation program and cultural practices by which foliage is kept as dry as possible. Protective fungicide treatment is often necessary for complete disease control, once the disease has occurred in an area.

Cultural practices It is important to keep the foliage as dry as possible. Irrigation at the pot level or ground level is recommended. Overhead irrigation should be avoided. Greenhouse workers should not handle plants when the plants are wet, as this contact and movement will spread the disease. Insects and mites should be controlled, as they can carry spores from diseased to healthy foliage. Water early in the day, so foliage will be dry before evening. Avoid high humidity and high temperatures. Laboratory tests indicate that *C. trichellum* grows best at 20–25°C.

Sanitation Remove dead leaves, stems, and other plant debris frequently.

Resistance Osborne et al. (1984), testing several common cultivars of English ivy for susceptibility to anthracnose, found that Gold Dust and Green Feather exhibit a low level of disease after inoculation; Hahn Variegated and Perfection exhibit moderate disease levels; and Manda Crested is highly susceptible, developing extensive symptoms. Pierce and McCain (1983) found that Crested, Glacier, Gold Dust, Needlepoint, and Shamrock are susceptible to varying degrees, with Crested being the most severely diseased, Glacier and Needlepoint having intermediate reactions, and Gold Dust and Shamrock exhibiting the lowest level disease.

Chemical treatment Complete control of anthracnose in a nursery often requires protective fungicide treatments in addition to sanitation and cultural control. Fungicide treatments should be applied in accordance with directions on the product label.

Bacterial Leaf Spot

Xanthomonas campestris* pv. *hederae

Bacterial leaf spot is often considered the most common disease of English ivy. Nursery, landscape, and interiorscape ivy plantings may be affected by this disease, but it is most severe in nurseries and landscapes where humidity levels tend to be high. This disease occurs worldwide. Other members of the family Araliaceae, such as scheffleras and aralias, are also susceptible.

Symptoms

The initial symptoms are water-soaked (oily), green brown, brown, or black leaf spots, often with red brown or brown black centers (Plate 38). The spots may expand into lesions 2–10 mm in diameter, which are round or angular where they are bounded by leaf veins. They may have a yellow halo, 1–2 mm wide. Similar symptoms occur on both the upper surface and the corresponding lower surface of the leaf. Under moist conditions, an orange red ooze is sometimes exuded at the edges of lesions. Lesions that occur along veins may extend into the petioles, with consequent leaf death. Stems may be invaded by bacteria from petiole infections.

Older leaves are often the first to develop symptoms. Infection of immature leaves often causes leaf speckling and deformity. When succulent stems become infected, a soft dark brown decay may develop rapidly, cankers (lesions) form on the stems, and affected plants become unthrifty and dwarfed and have poor color. Twig tip dieback may occur.

Old lesions become cracked, often in their centers. Leaf margins are not water-soaked. In some cultivars, a red discoloration develops around old lesions in mature leaves. Small, dark brown, girdling cankers may develop on old plants.

Epidemiology

Bacterial leaf spot is caused by *Xanthomonas campestris* pv. *hederae*. The pathogen is spread by splashing water and by plant cultivation under wet conditions. Dense foliage, with plants packed closely together, facilitates disease spread. The bacterium survives in infected plant debris and in the soil. Infections occur

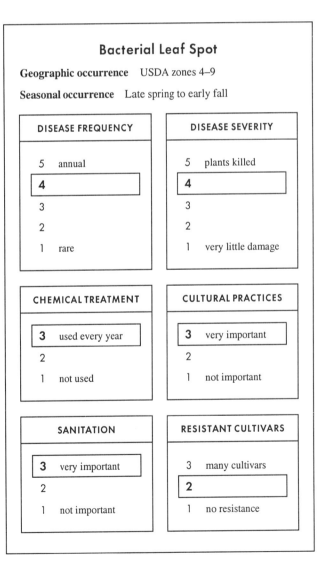

Bacterial Leaf Spot

Geographic occurrence USDA zones 4–9

Seasonal occurrence Late spring to early fall

DISEASE FREQUENCY	DISEASE SEVERITY
5 annual	5 plants killed
4	**4**
3	3
2	2
1 rare	1 very little damage

CHEMICAL TREATMENT	CULTURAL PRACTICES
3 used every year	**3 very important**
2	2
1 not used	1 not important

SANITATION	RESISTANT CULTIVARS
3 very important	3 many cultivars
2	**2**
1 not important	1 no resistance

most often where leaves are shaded and wet. Wet, poorly drained areas are conducive to disease development.

Host Range

X. campestris pv. *hederae* has been reported on many cultivars of English ivy. It has also been reported on Algerian ivy, *H. canariensis.*

Management

Pathogen-free stock plants must be maintained. Overhead watering should be kept to a minimum. Chemical treatments are not completely effective. The best control is achieved by planting resistant cultivars, minimizing overhead irrigation, managing the use of fertilizer, and applying chemical treatments.

Cultural practices Eliminate overhead watering, if possible, as splashing water will spread the bacterium. Keep workers out of the area when plants are wet. Control insects and mites, as they will spread the pathogen. Research has shown that application of fertilizer at higher than recommended rates may cause ivy susceptibility to decrease, but this practice is risky, because some disease agents are sometimes more aggressive when nitrogen is at high levels in plant tissues.

Sanitation Remove all diseased plant materials, including infected stock plants and plant debris. Clean benches and tools with a disinfectant solution.

Resistance Highly resistant cultivars have not been reported. Osborne and Chase (1985), in an inoculation study, found that the cultivars California, Eva, Gold Dust, Manda Crested, Perfection, and Sweet Heart were relatively resistant, with three to six lesions per plant; Brokamp and Hahn Variegated were most susceptible, with 16–22 lesions per plant; and Gold Heart, Green Feather, Ivalance, and Telecurl were intermediate in susceptibility, with eight to 12 lesions per plant.

Chemical treatment Traditionally, chemical control of bacterial leaf spots is difficult and often ineffective. However, reports indicate that up to 90% control of bacterial leaf spot is possible with the use of some fungicides.

Phytophthora Root Rot and Leaf Spot

Phytophthora cinnamomi
Phytophthora palmivora
Phytophthora parasitica

Phytophthora palmivora was first observed on English ivy (*H. helix* 'California') in 1974 in California, causing stem blight and leaf spot, together with internal vascular necrosis of stems and roots and a reduction of root systems of plants grown in clay pots in a greenhouse setting. It was reported in 1978 as a leaf spot and stem blight pathogen of ivy (no root involvement was noted) in a commercial nursery in Hawaii and subsequently as a leaf spot, stem blight, and root rot pathogen of ivy in Florida.

In 1989, *P. cinnamomi* was reported to cause root rot and stem and leaf blight of *H. helix* varieties in two Danish nurseries. In 1993, the disease was reported in potted ivy in Norway, and by 1997 it was widespread in pot plant nurseries in Denmark and Norway. Symptoms of damage caused by *P. cinnamomi* and by *P. palmivora* are similarly described.

Other *Phytophthora* spp. reported on ivy include *P. nicotianae* var. *parasitica* (syn. *P. parasitica*), causing root rot and stem blight in North Carolina, and unidentified species causing root rot in Florida and Alabama and stem blight and root rot in California.

Symptoms

Symptoms of infection by *P. palmivora* are poor plant growth and color. Basal leaves turn brown and curl downward. Root rot sometimes occurs, but leaf spot and stem rot usually are more common. Large, gray black, water-soaked spots form on leaves. An internal vascular necrosis of stems and roots was reported in *H. helix* 'California.'

Symptoms of leaf spot, stem blight, and root rot caused by *P. cinnamomi* are similar to those due to *P. palmivora,* as are symptoms of stem blight and root rot caused by *P. nicotianae* var. *parasitica* and an unidentified *Phytophthora* sp.

Epidemiology

Like most *Phytophthora* infections, these diseases are spread by the movement of the pathogen in water, soil, and infected roots and other Plant parts. *Phytophthora* diseases require prolonged periods of wet soil. Warm conditions (temperatures of 23–24°C) generally favor the development of *P. palmivora.* Losses due to disease caused by *P. cinnamomi* in Denmark were reported to be greatest at temperatures above 23°C. *P. parasitica* is generally reported to be most active at temperatures of 27–32°C, but different isolates have been observed growing at temperatures ranging from 5 to 40°C.

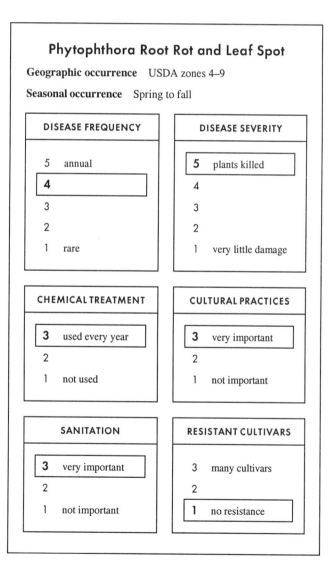

Phytophthora Root Rot and Leaf Spot

Geographic occurrence USDA zones 4–9

Seasonal occurrence Spring to fall

DISEASE FREQUENCY	DISEASE SEVERITY
5 annual	**[5 plants killed]**
[4]	4
3	3
2	2
1 rare	1 very little damage

CHEMICAL TREATMENT	CULTURAL PRACTICES
[3 used every year]	**[3 very important]**
2	2
1 not used	1 not important

SANITATION	RESISTANT CULTIVARS
[3 very important]	3 many cultivars
2	2
1 not important	**[1 no resistance]**

Host Range

The host range of *P. palmivora* includes *Dieffenbachia picta,* tropical plants (e.g., rubber, cacao, citrus, breadfruit, and papaya), and other plants, including many ornamentals. *P. nicotianae* var. *parasitica* and *P. cinnamomi* both have very large host ranges.

Management

General management includes using pathogen-free potting media, pathogen-free stock plants, and clean pots; applying minimal amounts of water; providing good drainage; and using media that drain well. Diseased plants should be removed from the area. Protective fungicides are often recommended for disease control in nurseries.

Cultural practices *Phytophthora* species require wet conditions for growth and infection. Care should be taken to prevent water from standing in plant areas or in potting media. Soil or media may need to be mixed with amendments to improve drainage in plant areas. Composted pine bark or hardwood bark is sometimes recommended as a potting mix component; these materials have good drainage properties and have been demonstrated to be protective against *Phytophthora* infection. Diseased plants should be removed from the area.

Sanitation Remove all damaged plants and associated media. Contaminated media must be treated with heat or fumigated before reuse. Clean pot benches. Test continuous ebb-and-flow watering systems when *Phytophthora* has been a problem. A pear bait technique used in North Carolina and some other states is effective in checking water sources for the presence of *Phytophthora* and related fungi.

Resistance Information is not available on the susceptibility or resistance of English ivy cultivars to *Phytophthora. H. helix* 'California' has been reported to be susceptible to *P. palmivora.*

Chemical treatment A fungicide drench may be applied as a protective treatment for the control of *Phytophthora* diseases of English ivy. Treatments should be applied in accordance with the directions on the product label. Fungicide treatments are most effective in combination with recommended sanitation and cultural practices.

Rhizoctonia Root Rot and Aerial Blight

Rhizoctonia solani

Rhizoctonia solani is a widespread pathogen of many crops, causing aerial blight and root rot. It has been reported to occur on English ivy in Connecticut, Florida, Hawaii, and North Carolina.

Symptoms

Aerial blight Brown, irregularly shaped lesions occur on all aboveground plant parts. The disease often develops initially on lower sections of the foliage, where humidity levels are highest, and it may be confined to this foliage or may be especially severe on it. The pathogen may form a red brown mycelial web in infected areas when the disease is severe.

Rhizoctonia root rot Brown, dried lesions develop on roots and may spread to involve the whole root system.

Epidemiology

Warm to hot, humid conditions favor the spread of these diseases. *R. solani* survives in plant debris and in soil. Aerial blight and root rot may spread where susceptible plants come in contact with the fungus in soil of moderate moisture. Generally, disease spread occurs by the movement of infected plants or infested soil, bringing them into contact with susceptible plants.

Host Range

The host range of *R. solani* is enormous. Subspecies groupings have been made for isolates of *Rhizoctonia* from some hosts, and this information can help to define the host preferences of the different subspecies, but this information is not available for isolates of *Rhizoctonia* from ivy.

Management

Sanitation, cultural practices, and application of protective fungicide sprays will effectively control aerial blight. Sanitation, cultural practices, crop rotations, and chemical practices are available for root rot control.

Cultural practices Always use disease-free potting media, pots, and plants. Choose a potting medium that drains well. Studies of the development of *Rhizoctonia* diseases on other plants have shown that plants heavily fertilized with nitrogen may develop more severe levels of disease than plants receiving a normal or recommended amount of nitrogen. Water at a time of day and by a method that will allow the foliage to remain as dry as possible.

Rhizoctonia Root Rot and Aerial Blight

Geographic occurrence USDA zones 4–9

Seasonal occurrence Late spring to early fall

DISEASE FREQUENCY	DISEASE SEVERITY
5 annual	5 plants killed
[4]	**[4]**
3	3
2	2
1 rare	1 very little damage

CHEMICAL TREATMENT	CULTURAL PRACTICES
[3 used every year]	**[3 very important]**
2	2
1 not used	1 not important

SANITATION	RESISTANT CULTIVARS
[3 very important]	3 many cultivars
2	2
1 not important	**[1 no resistance]**

Watering at ground level early in the day and spacing plants to reduce humidity levels will help.

Sanitation Remove all damaged plants and plant parts. Do not use soil or media associated with infected plants. Treat the soil or media with heat or fumigation before reuse with ivy or other plants.

Resistance Resistance to *R. solani* has not been reported in English ivy cultivars.

Chemical treatment Protective fungicide sprays are most effective for control of the aerial phase of the disease when coupled with sanitation and cultural practices. Fungicides available for control of root rot are protective in action. Rhizoctonia root rot can only be controlled when diseased plants are removed and the potting medium is removed or treated. Protective fungicides will help prevent a recurrence of the disease in the damaged area when sanitation and cultural control methods are practiced.

REFERENCES

Alfieri, S. A., Jr., Langdon, K. R., Kimbrough, J. W., El-Gholl, N. E., and Wehlburg, C. 1994. Diseases and disorders of plants in Florida. Fla. Dep. Agric. Consumer Serv. Bull. 14.

Chase, A. R. 1988. Compendium of Ornamental Foliage Plant Diseases. American Phytopathological Society, St. Paul, Minn.

Chase, A. R. 1989. Effect of fertilizer rate on susceptibility of *Hedera helix* to *Xanthomonas campestris* pv. *hederae,* 1988. Biol. Cult. Tests Control Plant Dis. 4:82.

Chase, A. R. 1997. Foliage Plant Diseases: Diagnosis and Control. American Phytopathological Society, St. Paul, Minn.

Farr, D. F., Bills, G. F., Chamuris, G. P., and Rossman, A. Y. 1989. Fungi on Plants and Plant Products in the United States. American Phytopathological Society, St. Paul, Minn.

Keim, R., Zentmeyer, G. A., and Klure, L. J. 1900. *Phytophthora palmivora* on ivy in California and its control with pyroxychlor. Plant Dis. Rep. 60:632–633.

Miller, J. W. 1970. Bacterial leaf spot of English ivy. Fla. Dep. Agric. Consumer Serv. Div. Plant Ind. Plant Pathol. Circ. 95.

Osborne, L. S., and Chase, A. R. 1985. Susceptibility of cultivars of English ivy to two-spotted spider mite and Xanthomonas leaf spot. HortScience 20:269–271.

Osborne, L. S., Chase, A. R., and Henley, R. W. 1984. English ivy. Pages 5–9 in: AREC-A Foliage Plant Res. Note RH-1984-H. University of Florida, Institute of Food and Agricultural Science, Agriculture Research Center, Apopka.

Pierce, L., and McCain, A. H. 1983. Colletotrichum leaf spot of English ivy: Chemical control and cultivar susceptibility. Calif. Plant Pathol. 62:1–4.

Ridings, W. H., and Alfieri, S. A., Jr. 1973. Colletotrichum leaf spot of English ivy. Fla. Div. Agric. Consumer Serv. Plant Pathol. Circ. 131.

Sinclair, W. A., Lyon, H. H., and Johnson, W. T. 1987. Diseases of Trees and Shrubs. Comstock Publishing Associates, Cornell University Press, Ithaca, N.Y.

Thinggaard, K. 1997. First report of *Phytophthora cinnamomi* root rot, stem, and leaf blight on ivy. Plant Dis. 81:960.

Uchida, J. Y., and Aragaki, M. 1978. Leaf and stem blight of ivy caused by *Phytophthora palmivora.* Plant Dis. Rep. 62:699–701.

CHAPTER 37

John W. Olive ▪ Ornamental Horticulture Substation, Mobile, Alabama

Robert L. Wick ▪ University of Massachusetts, Amherst

Euonymus Diseases

Geographic production USDA zones 3–9, depending on the species

Family Celastraceae

Genus *Euonymus*

Species *E. alatus*
E. americanus
E. fortunei
E. japonicus

Cultivars several

Euonymus is a small to medium-sized evergreen shrub grown for its dark green or variegated foliage.

Anthracnose

Colletotrichum gloeosporioides

Anthracnose of euonymus, caused by the fungus *Colletotrichum gloeosporioides,* can be devastating under certain environmental conditions. The disease is most often observed in *E. fortunei* and *E. japonicus.* It is a problem in the landscape as well as in container and field nurseries.

Symptoms

Small lesions, up to 3 mm in diameter, with dark borders and white or light tan to gray centers, form on both sides of leaves (Plate 39). In some cultivars, the area around the lesion turns reddish. The necrotic centers of lesions may fall out, leaving shot holes in the leaves. Stem cankers may be small and oval, like leaf spots, and tend to be raised and appear scabby. Larger cankers are elongate (several centimeters long) and may girdle the stem. On leaves and stems, spore tendrils of the pathogen may form, and fungal setae may be observed with a hand lens. Severe cases of anthracnose result in defoliation and stem dieback.

Epidemiology

Colletotrichum overwinters as mycelium or fruiting bodies (acervuli) in infected plant debris. Spores (conidia) are easily disseminated by rain or splashing irrigation water. The optimum temperature for spore germination and growth of the pathogen is 25–30°C (77–86°F). Disease severity is much greater after 24 hours of leaf wetness than after 18 hours of wetness. Symptoms develop on newly expanded leaves within 48 hours after spore deposition. Older leaves appear to be resistant to infection.

Management

A combination of cultural practices and chemical control is required to avoid epiphytotics of anthracnose. As in the control of most diseases, chemical treatment will not be effective without proper cultural measures. Sanitation is important but may be impractical in large nurseries. Complete sanitation would require re-

Anthracnose

Geographic occurrence USDA zones 3–9

Seasonal occurrence March to November during wet periods and whenever new growth is being produced

DISEASE FREQUENCY	DISEASE SEVERITY
5 annual	**5 plants killed**
4	**4**
3	3
2	2
1 rare	1 very little damage

CHEMICAL TREATMENT	CULTURAL PRACTICES
3 used every year	**3 very important**
2	2
1 not used	1 not important

SANITATION	RESISTANT CULTIVARS
3 very important	3 many cultivars
2	**2**
1 not important	1 no resistance

moving all plant debris from the ground and containers, removing infected leaves and stems from plants, and disposal of dead plants. Space plants to increase air circulation and decrease the duration of leaf wetness. Time watering to maximize leaf drying.

E. japonicus often does not thrive with repeated pruning, and this practice can increase disease severity. Fungicides have been effective in reducing disease severity. Applications should begin as new growth begins to emerge.

Powdery Mildew

Oidium sp.

(sexual stage, *Microsphaera penicillata*)

Powdery mildew caused by *Oidium* sp. is a common and often damaging fungal disease of euonymus. It seldom causes serious damage, but it reduces the market value of heavily infected plants. The sexual stage, *Microsphaera penicillata,* is seldom found on euonymus.

Symptoms

Powdery mildew first appears on leaves and tender stems as small, scattered white patches of fungal mycelia and spores. Typically, they are first seen on new growth. The mycelial patches can become very extensive on stems and leaves (Plate 40).

Epidemiology

The *Oidium* sp. causing powdery mildew of euonymus overwinters as mycelia and spores on infected leaves and stems. In early spring, airborne spores are spread to newly emerged leaves and succulent stems. Infection is favored by temperatures between 70 and 80°F and relative humidity between 85 and 100%. Free water on leaf surfaces will often suppress spore germination. Symptoms usually appear within one week of infection. In plants in cold-protection greenhouses, disease development can be very severe during warm, humid periods in the spring.

Host Range

The host range of the *Oidium* sp. causing powdery mildew of euonymus is not well known.

Management

Powdery mildew is controlled by a combination of sanitation, cultural practices, and fungicide treatment.

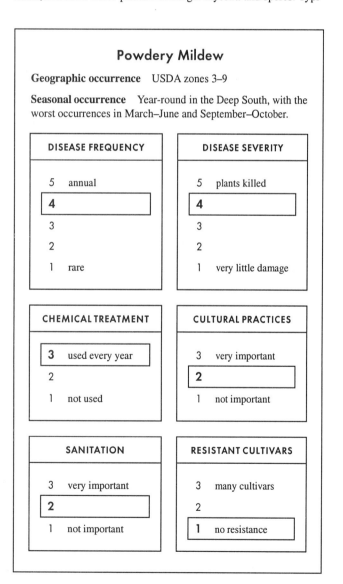

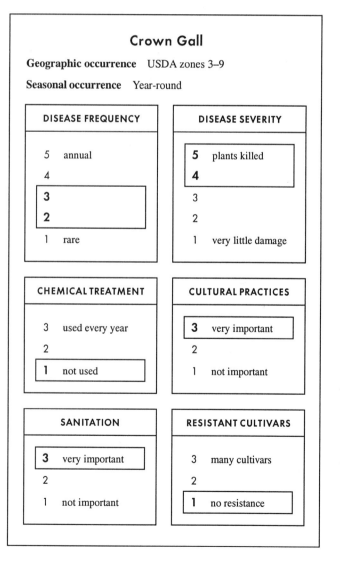

Powdery Mildew

Geographic occurrence USDA zones 3–9

Seasonal occurrence Year-round in the Deep South, with the worst occurrences in March–June and September–October.

DISEASE FREQUENCY	DISEASE SEVERITY
5 annual	5 plants killed
4 (boxed)	**4** (boxed)
3	3
2	2
1 rare	1 very little damage

CHEMICAL TREATMENT	CULTURAL PRACTICES
3 used every year (boxed)	3 very important
2	**2** (boxed)
1 not used	1 not important

SANITATION	RESISTANT CULTIVARS
3 very important	3 many cultivars
2 (boxed)	2
1 not important	**1 no resistance** (boxed)

Crown Gall

Geographic occurrence USDA zones 3–9

Seasonal occurrence Year-round

DISEASE FREQUENCY	DISEASE SEVERITY
5 annual	**5 plants killed** (boxed)
4	**4** (boxed)
3 (boxed)	3
2 (boxed)	2
1 rare	1 very little damage

CHEMICAL TREATMENT	CULTURAL PRACTICES
3 used every year	**3 very important** (boxed)
2	2
1 not used (boxed)	1 not important

SANITATION	RESISTANT CULTIVARS
3 very important (boxed)	3 many cultivars
2	2
1 not important	**1 no resistance** (boxed)

Cultural practices Disease severity and spore production can be reduced by pruning off all heavily infected plant parts. Improve air movement and reduce humidity around plants.

Chemical treatment Spray applications should begin when new growth starts or at the first sign of the disease. Continue fungicide applications at the interval specified on the product label until the threat of further disease spread ends.

Crown Gall

Agrobacterium tumefaciens

Crown gall, caused by *Agrobacterium tumefaciens,* can be a very serious disease of euonymus. Infected plants are unsalable.

Symptoms

In euonymus, the disease causes cream-colored galls, ranging from 1/2 inch to 3–4 inches in diameter, at or just below the surface of the medium (Plate 41). With age the galls become rough on the surface and turn brown. Infected plants may be stunted, lack normal green color, and die back during severe winters.

Epidemiology

A. tumefaciens is a soilborne bacterium and is systemic in some hosts. The biovar of *Agrobacterium* that infects euonymus is not known. In most woody plants, crown gall spreads during propagation or following wounding of the lower stem or roots.

Management

Crown gall of euonymus is controlled by avoidance. Do not purchase plants showing symptoms of the disease. All plants showing symptoms of crown gall should be dumped away from the nursery. Since the disease occurs in euonymus grown in nonsoil media in containers, the bacterium may be systemic in stock plants. Therefore, use only plants that are free of crown gall for propagation. Avoid planting in a field with a history of the disease.

REFERENCES

Lambe, R. C., and McRitchie, J. J. 1982. Euonymus diseases. Pages 55–56 in: Diseases of Woody Ornamental Plants and Their Control in Nurseries. R. K. Jones and R. C. Lambe, eds. N.C. Agric. Ext. Serv. Publ. AG-286.

Mahoney, M. 1979. Identification, etiology and control of *Euonymus fortunei* anthracnose caused by *Colletotrichum gloeosporioides.* M.S. thesis. University of Massachusetts, Amherst.

CHAPTER 38

Gary W. Simone · University of Florida, Gainesville

Fatsia and ×*Fatshedera* Diseases

Geographic production USDA zones 7–10

Family Araliaceae

Genus *Fatsia*
Species *F. japonica* zones 8–10

Genus ×*Fatshedera*
Species ×*F. lizei* zones 7–10

Fatsia (Japanese fatsia), a monotypic genus from Japan, is a shrubby evergreen used as an indoor foliage plant and also planted outdoors in areas with a mild climate. A cross of *F. japonica* 'Moseri' and *Hedera helix* var. *hibernica* resulted in the hybrid genus ×*Fatshedera* (aralia, or tree ivy). The hybrid resembles its ivy parent but lacks the holdfasts of a true ivy. *Fatsia* and ×*Fatshedera* are subject to many of the same root rots and leaf spots that affect *Hedera.*

Alternaria Leaf Spot

Alternaria panax
Other *Alternaria* spp.

Symptoms

The initial symptoms of Alternaria leaf spot of *Fatsia* are spots, 1–2 mm in diameter, which can enlarge into round to oblong lesions more than 2 cm in diameter, with a characteristic fine zonate pattern (Plate 43). Lesions may coalesce, causing necrosis of entire lobes of leaves. In a mature lesion, a dark zone forms in the center, where the fungal pathogen produces spores (conidia). The severity and speed of development of the disease in *Fatsia* suggest that it might better be called a blight than a leaf spot.

An Alternaria leaf spot of ×*Fatshedera* has been reported, but it has not been clearly defined. The pathogen may be *Alternaria alternata,* which causes a similar disease in ivy.

Epidemiology

Alternaria leaf spot of *Fatsia* is caused by *A. panax.* This fungus is infective at temperatures between 18 and 30°C but is most damaging between 24 and 27°C. The disease develops fast – lesions form in two to three days. Fungal spores produced in mature lesions are spread by air currents and splashing water. *A. panax* sporulates well whether leaves abscise or are held on the plant. It can survive on fallen infected plant debris.

The *Alternaria* sp. infecting ×*Fatshedera* is believed to be similar to *A. panax* in its persistence and spread.

Host Range

A. panax attacks only plants in the family Araliaceae. It has been reported in *Aralia, Dizygotheca, Fatsia, Panax, Polyscias, Schefflera arboricola, S. actinophylla,* and *Tupidanthus.*

Alternaria Leaf Spot

Geographic occurrence USDA zones 8–10

Seasonal occurrence Mid-spring to late summer

DISEASE FREQUENCY	DISEASE SEVERITY
5 annual	5 plants killed
[4]	**[4]**
3	3
2	2
1 rare	1 very little damage

CHEMICAL TREATMENT	CULTURAL PRACTICES
3 used every year	3 very important
[2]	**[2]**
1 not used	1 not important

SANITATION	RESISTANT CULTIVARS
[3] very important	3 many cultivars
2	2
1 not important	**[1] no resistance**

Management

Growth of *Fatsia* from seed can offer lower risk than vegetative propagation in areas where Alternaria leaf spot is established. Never propagate cuttings from diseased plants, as the propagation cycle is highly favorable for the reproduction and spread of *Alternaria* spp. Infected plants should be sanitized of diseased tissue, and fallen leaves should be collected and destroyed. Minimize overhead irrigation where the disease is active. Try to schedule irrigation for early morning or daylight hours, to disfavor the infection cycle. Application of protectant fungicide is important in arresting disease development. There are no data on the relative susceptibility of cultivars of *F. japonica*.

Management of Alternaria leaf spot of × *Fatshedera* follows a program similar to that for *Fatsia*.

Anthracnose

Colletotrichum gloeosporioides
(sexual stage, *Glomerella cingulata*)
Colletotrichum trichellum

Symptoms

In × *Fatshedera,* anthracnose begins as small, angular, chlorotic spots on leaves, which enlarge slowly to a diameter of 8–10 mm. As the lesions age, their centers turn necrotic. In *Fatsia,* which has thinner leaves, lesion development is more rapid, and the lesions are larger. The lesions are dark brown to black, with broad zonation. They may reach 2–3 cm in diameter and effectively kill a lobe of the leaf, much like lesions caused by *Alternaria panax.* Where lesions coalesce on *Fatsia,* necrotic tissue may detach itself from the leaf, creating a windowpane effect.

The pathogen can be observed in both its asexual stage, *Colletotrichum,* and its sexual stage, *Glomerella,* on infected leaves. The *Colletotrichum* stage is frequently observed as rings of dark, pincushion-like fruiting bodies (acervuli), which exude a pink to white spore mass amidst dark, sterile hyphae (setae). The *Glomerella* stage forms dark, globose perithecia in the centers of lesions, usually late in the calendar year.

Epidemiology

The *Colletotrichum* stage is the more important in disease spread. Spores (conidia) are spread by splashing water from infected tissue to adjacent plants. Because of the broad host range of *C. gloeosporioides,* it is almost impossible to isolate *Fatsia* and × *Fatshedera* from other hosts.

Host Range

C. gloeosporioides is predominantly the pathogen causing anthracnose of *Fatsia* and × *Fatshedera.* More than 500 plant species, including many grown in woody ornamental nurseries, have been reported to be hosts of this fungus.

C. trichellum invades *F. japonica* in Florida.

Management

Propagate *Fatsia* and × *Fatshedera* from disease-free stock, or grow *Fatsia* from seed, when possible. Avoid sunscald, which may occur when propagated liners are set outside in the nursery. Sunscalded tissue is highly susceptible to invasion by *Colletotrichum.* Infected tissue should be removed and destroyed prior to fungicide use. Minimize overhead irrigation. Protectant and systemic fungicides are available for nursery use and will aid in disease management. Resistance to *Colletotrichum* is not known in available cultivars.

If *C. trichellum* is present where *Hedera* is grown, physical separation of *Fatsia* and *Hedera* in the nursery (beyond splashing distance) is recommended. Management for the control of *C. trichellum* on *Fatsia* follows the program described above.

Bacterial Spot Diseases

Pseudomonas cichorii
Pseudomonas syringae* pv. *syringae
Xanthomonas campestris* pv. *hederae

Symptoms

Lesions due to infection by *Pseudomonas* spp. are initially sunken, water-soaked leaf spots about 1 mm in diameter. They expand and develop into black or tan, irregularly shaped spots, often elongating along veins. The lesions occur on wounded and unwounded leaves. Mature spots on *Fatsia* leaves appear papery tan.

Infection of *Fatsia* by *Xanthomonas* causes small, tan spots, less than 1 mm in diameter, often at the leaf margins (Plate 44). Infection at leaf emergence can cause leaf deformity.

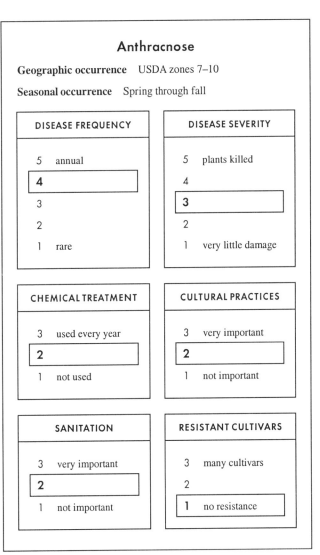

Anthracnose

Geographic occurrence USDA zones 7–10

Seasonal occurrence Spring through fall

DISEASE FREQUENCY	DISEASE SEVERITY
5 annual	5 plants killed
[4]	4
3	**[3]**
2	2
1 rare	1 very little damage

CHEMICAL TREATMENT	CULTURAL PRACTICES
3 used every year	3 very important
[2]	**[2]**
1 not used	1 not important

SANITATION	RESISTANT CULTIVARS
3 very important	3 many cultivars
[2]	2
1 not important	**[1] no resistance**

Host Range

Both *P. cichorii* and *X. campestris* pv. *hederae* have been reported in *Fatsia* and × *Fatshedera. P. syringae* pv. *syringae* also infects *Fatsia.*

P. cichorii has a broad host range among herbaceous floral and foliage species, woody plants, and vegetables. Common nursery hosts include *Acalypha, Codiaeum, Ficus, Hedera, Hibiscus, Jasminum, Magnolia, Platanus, Prunus, Rhaphiolepis,* and *Rhododendron* spp.

P. syringae pv. *syringae* also invades many ornamentals, fruits, and vegetables. Important nursery hosts include camphor tree, citrus, *Hedera, Hibiscus, Magnolia, Nerium, Platanus, Prunus, Syringa,* and *Viburnum* spp.

Epidemiology

The bacterial spot diseases of *Fatsia* and × *Fatshedera* are favored by periods of rainfall or overhead irrigation that enable bacteria to enter leaves through stomates or wounds. In addition, *P. cichorii* is an excellent leaf surface colonizer in the absence of disease symptoms. These bacteria are not systemic in their hosts.

Management

As *Fatsia* and × *Fatshedera* are grown under partial shade, overhead irrigation should be avoided, where possible. Space plants to ensure air circulation for rapid drying of the canopy if overhead irrigation is used. Sanitize severely infected plants prior to the use of a bactericide, such as copper hydroxide. Only disease-free cuttings should be used for propagation. There are no data on the interaction of plant fertility and disease severity in *Fatsia* and × *Fatshedera.* No cultivars are known to be more or less susceptible to these pathogens.

Wet Root Rots

Phytophthora nicotianae
Pythium splendens

Symptoms

Phytophthora nicotianae (syn. *P. parasitica*) causes a root and crown rot of *Fatsia* and × *Fatshedera* spp. Feeder roots are invaded under conditions of saturated soil moisture. The outer, cortical cylinders of invaded roots decay, leaving the feeder root sys-

Bacterial Spot Diseases

Geographic occurrence USDA zones 7–10

Seasonal occurrence Spring through fall, in wet periods

DISEASE FREQUENCY	DISEASE SEVERITY
5 annual	5 plants killed
4	4
3	**[3]**
[2]	2
1 rare	1 very little damage

CHEMICAL TREATMENT	CULTURAL PRACTICES
3 used every year	**[3] very important**
[2]	2
1 not used	1 not important

SANITATION	RESISTANT CULTIVARS
3 very important	3 many cultivars
[2]	2
1 not important	**[1] no resistance**

Wet Root Rots

Geographic occurrence USDA zones 7–10

Seasonal occurrence Late spring through mid-fall, in rainy periods

DISEASE FREQUENCY	DISEASE SEVERITY
5 annual	**[5] plants killed**
[4]	4
3	3
2	2
1 rare	1 very little damage

CHEMICAL TREATMENT	CULTURAL PRACTICES
3 used every year	**[3] very important**
[2]	2
1 not used	1 not important

SANITATION	RESISTANT CULTIVARS
3 very important	3 many cultivars
[2]	2
1 not important	**[1] no resistance**

tem discolored and mushy. Feeder root loss leads to progressive wilt followed by nighttime recovery. If the disease is unchecked, plants will wilt to death.

In *Fatsia*, if *P. nicotianae* is splashed upward into the canopy, it causes a rapid leaf blight. Infection results in a wet decay of leaves. Affected tissue is initially grayish green and then grayish black (Plate 45). An obvious layer of white sporangia is often present on the undersides of infected leaves.

Pythium splendens causes a seedling damping-off of *Fatsia* and a cutting rot of *Fatsia* and × *Fatshedera* in vegetative propagation. The pathogen invades feeder roots of container-grown plants, causing a soft, discolored decay of the cortex. Damage to the root system results in a progressive yellowing of lower leaves and slight wilting. The lower leaves desiccate, and the entire plant will be killed.

Host Range

The potential host range of *P. nicotianae* encompasses plants in more than 85 genera, including fruit, ornamentals, and vegetable species. Common nursery hosts include *Buxus, Carissa, Cornus, Hedera, Hibiscus, Ilex, Jasminum, Juniperus, Pinus, Prunus, Rhododendron,* and palms.

P. splendens also has a wide host range among crops and has been reported in plants belonging to more than 125 genera, including many woody species.

Epidemiology

Both pathogens are water molds and are highly dependent on saturated soil for vegetative survival and reproduction. Periods of warm to hot weather with frequent rains favor disease incidence. *P. nicotianae* is most aggressive at temperatures of 30–35°C. *P. splendens* can invade hosts at temperatures between 15 and 37°C, with an optimal temperature of 30°C. Both fungi persist as mycelia, oospores, and sporangia in soil and host debris.

Management

Irrigation management is of considerable importance in the production of *Fatsia* and × *Fatshedera* under shade. Do not overlap irrigation and natural rainfall. If aerial *Phytophthora* is present on *Fatsia*, affected plants should be sanitized and spaced apart prior to fungicide application. Foliar fungicide sprays should follow sanitation for best management of aerial blight. The root rots caused by these pathogens are better prevented than cured, because of the speed of root disease development in the nursery. Preventative spray and drench fungicides are available for use on *Fatsia* and × *Fatshedera*. There is no known resistance to or tolerance of *P. nicotianae* or *P. splendens* in either host.

REFERENCES

Alfieri, S. A., Jr., Langdon, K. R., Kimbrough, J. W., El-Gholl, N. E., and Wehlburg, C. 1994. Diseases and disorders of plants in Florida. Fla. Dep. Agric. Consumer Serv. Bull. 14.

Chase, A. R. 1984. *Xanthomonas campestris* pv. *hederae* caused leaf spot of five species of Araliaceae. Plant Pathol. 33:439–440.

Chase, A. R., and Brunk, D. D. 1984. Bacterial leaf blight incited by *Pseudomonas cichorii* in *Schefflera arboricola* and some related plants. Plant Dis. 68:73–74.

Farr, D. F., Bills, G. F., Chamuris, G. P., and Rossman, A. Y. 1989. Fungi on Plants and Plant Products in the United States. American Phytopathological Society, St. Paul, Minn.

Jones, J. B., Raju, B. C., and Englehard, A. W. 1984. Effects of temperature and leaf wetness on development of bacterial spot of geraniums and chrysanthemums incited by *Pseudomonas cichorii*. Plant Dis. 68:248–251.

CHAPTER 39

Gary A. Chastagner ▪ Washington State University, Puyallup

Fir Diseases

Geographic production USDA zones 1–9

Family Pinaceae

Genus *Abies*

Species

A. alba	European silver fir	zone 5
A. amabilis	Pacific silver fir	zone 6
A. balsamea	balsam fir	zone 3
A. balsamea var. *phanerolephis*	Canaan fir	zone 3
A. bracteata	bristlecone fir	zone 7
A. bornmuelleriana	Turkish fir	
A. cephalonica	Greek fir	zone 6
A. cilicica	Cilician fir	zone 6
A. concolor	white fir	zone 4
A. firma	momi fir	zone 6
A. fraseri	Fraser fir	zone 5
A. grandis	grand fir	zone 6
A. holophylla	Manchurian fir	zone 6
A. homolepis	Nikko fir	zone 5
A. koreana	Korean fir	zone 5
A. lasiocarpa	alpine fir, subalpine fir	zone 3
A. lasiocarpa var. *arizonica*	corkbark fir	zone 5
A. magnifica	red fir	zone 6
A. magnifica var. *shastensis*	Shasta fir	zone 6
A. nordmanniana	Nordmann fir	zone 5
A. numidica	Algerian fir	zone 7
A. pinsapo	Spanish fir	zone 7
A. procera	noble fir	zone 6
A. veitchii	Veitch fir	zone 4

There are about 40 species of fir, widely scattered throughout North and Central America, Europe, Asia, and northern Africa. Many of them are important timber species, some are grown as Christmas trees, and some are planted as ornamentals. There are many cultivars of some species (alpine, Fraser, grand, Korean, Nikko, noble, Nordmann, red, Spanish, and white fir), including prostrate, compact, contorted, yellow-foliaged, and blue-foliaged types (Dirr, 1998). They are often difficult to find in nurseries, but in the landscape they add texture, shapes, and colors unlike those of the more common species.

Firs are generally slow-growing, especially when they are planted outside their native habitats. They require moist, well-drained, acid soil and prefer cool, moist environments. Firs are not well adapted to areas with hot, dry summers and do not tolerate air pollution. Frost damage and winter hardiness can be a problem in some sites. Most firs are propagated from seed. Seedlings are produced in bare-root nurseries or greenhouse containerized nurseries (Arnott and Matthews, 1981). Specialized cultivars are typically grafted onto rootstocks.

Conifer seedlings are frequently affected by environmental stresses and diseases (Boyce, 1938; Hansen, 1997; James et al., 1991; Stein and Smith, 1991; Sutherland and Davis, 1991). Nursery site selection is an important factor in the management of disease problems. Sites with heavy, poorly drained soils have a much greater likelihood of damping-off and root rot than sites with deep, well-drained soils. Sites where seedlings will be exposed to temperature extremes and moisture stress will also have more problems than sites with good air drainage and irrigation equipment.

Several common and potentially destructive diseases of firs occur in nurseries: damping-off, Phytophthora root rot, felt blight, snow blights, and gray mold. Needle cast and rust diseases also occur in some nurseries.

Damping-Off

***Fusarium* spp.**
***Pythium* spp.**
***Rhizoctonia* spp.**

Damping-off is the rapid decay of young, succulent seedlings. Seeds or seedlings are killed prior to emergence (preemergence damping-off), or seedlings are killed shortly after emergence (postemergence damping-off). The disease is caused by several soilborne fungi. *Cylindrocladium, Fusarium, Phytophthora, Pythium,* and *Rhizoctonia* have been reported to cause damping-off of conifers in bare-root nurseries (Sutherland and Davis, 1991; Hansen, 1997). *Pythium* and *Fusarium* are the most common damping-off fungi in container nurseries and, in general, are probably the most common of these pathogens (McCain et al., 1989; Johnson et al., 1989; James et al., 1991; Sutherland and Davis, 1991; Hansen, 1997).

Damping-off is a much more serious problem in bare-root nurseries than in greenhouse container nurseries, unless growing media become contaminated or a pathogen is introduced on seeds or in irrigation water. Annual losses of 5–20% of seed and seedlings are not uncommon in bare-root nurseries, and losses can reach 60 to 80% when conditions are favorable for disease development (Sutherland and Davis, 1991; James et al., 1991).

Symptoms

Typical damping-off symptoms are failure of seedlings to emerge from the soil and rapid wilting of seedlings shortly after emergence. Seeds may be killed before they germinate. Older seedlings may be stunted, and their roots are poorly developed. Identification of the causal agent usually requires examination of seedlings for the presence of diagnostic fungal spores or isolation of the pathogen in culture. In some situations more than one pathogen can be present. In addition to causing damping-off, some of these pathogens also cause root rot or hypocotyl rot of seedlings (Plate 46), in which symptom development is usually later in the growing season (James et al., 1991).

Epidemiology

The fungi that cause damping-off are common soilborne organisms. They survive for various periods of time as resting spores or mycelium in residual host tissues. With high levels of soil moisture, these spores germinate when stimulated by exudates from seeds and roots. *Pythium* produces motile spores (zoospores), which can readily spread in water and infect roots or other susceptible host tissue. Some damping-off pathogens, such as *Fusarium,* are also seedborne (James et al., 1991). If environmental and cultural conditions, such as cool soil and improper planting depth, slow the rate of germination and seedling emergence, damping-off is much more likely to be a problem. In greenhouse nurseries, the containers used to grow seedlings can easily become contaminated with damping-off pathogens.

Host Range

Many plants are susceptible to damping-off. More than 60 species of conifers, including many firs, are known to be affected (Boyce, 1938).

Management

A number of cultural practices can minimize damping-off: avoiding poorly drained soils; providing proper irrigation management; properly conditioning and handling of seed; rotating fir crops with nonsusceptible hosts, such as small grains, to reduce inoculum levels; and disinfecting planting containers (McCain et al., 1989; Hansen, 1997; Brown and Baxter, 1991; Boyce, 1938; Cordell et al., 1989). Growers can also fumigate soils to reduce inoculum levels before planting, treat or filter irrigation water to prevent the spread of pathogens in irrigation water, and treat seeds to prevent the introduction of seedborne inoculum (Axelrood, 1991; Landis and Campbell, 1991).

Damping-Off

Geographic occurrence Wherever seedlings are grown

Seasonal occurrence Early spring

DISEASE FREQUENCY
- **5 annual**
- **4**
- **3**
- 2
- 1 rare

DISEASE SEVERITY
- **5 plants killed**
- **4**
- 3
- 2
- 1 very little damage

CHEMICAL TREATMENT
- 3 used every year
- **2**
- 1 not used

CULTURAL PRACTICES
- **3 very important**
- 2
- 1 not important

SANITATION
- **3 very important**
- **2**
- 1 not important

RESISTANT CULTIVARS
- 3 many cultivars
- 2
- **1 no resistance**

Phytophthora Root Rot

Numerous *Phytophthora* spp.

Several *Phytophthora* species, including *P. cactorum, P. cambivora, P. cinnamomi, P. citricola, P. cryptogea, P. drechsleri, P. gonapodyides, P. megasperma,* and *P. pseudotsugae,* can cause serious root rot of fir in bare-root nurseries (Adams and Bielenin, 1988; Benson et al., 1976; Chastagner et al., 1995; Grand and Lapp, 1974; Hamm and Hansen, 1987, 1991; Kenerley and Bruck, 1981; Shew and Benson, 1982). Generally, Phytophthora root rot is limited to nurseries with poorly drained field soils (Kuhlman et al., 1989). The disease may also develop in container nurseries if seedlings are irrigated with contaminated water (McCain and Scharpf, 1986). Irrigation with contaminated water or splashing contaminated soil on foliage can also result in a shoot blight.

Symptoms

Aboveground symptoms of Phytophthora root rot of fir include chlorosis of foliage, branch flagging, stunting, and death of seedlings (Plates 47 and 48). Roots are typically discolored reddish brown in the cambium region (Plate 49), and the discoloration may extend up the stem.

Phytophthora species also cause a shoot blight of fir seedlings. The initial symptoms are a reddening of the needles at the tips of shoots during late spring and early summer. The wood beneath the bark is medium brown, with a distinct margin between the dead and healthy tissue. The infection can spread down shoots and branches and into the stem of the tree. Branches anywhere on the tree can be affected.

Epidemiology

Saturated soils favor the development of Phytophthora root rot. *Phytophthora* species produce fruiting bodies (sporangia) that release swimming zoospores. These spores are attracted to nearby elongating or wounded roots, where they encyst, germinate, and infect the root. Sporangia require nearly saturated soil for germination and dispersal of zoospores. Root infection is progressive and eventually leads to aboveground foliar symptoms.

Shoot blight can occur when growers use contaminated water for overhead irrigation or when overhead irrigation water splashes infested soil onto the lower branches of trees. Some *Phytophthora* species produce resting spores in infected roots, enabling the fungi to survive for extended periods in the soil. *Phytophthora* spores can spread by surface runoff or irrigation water. See Chapter 14, "Phytophthora Root Rot and Dieback."

Host Range

The high susceptibility of many firs to Phytophthora root rot limits the areas where they can be grown. Balsam, Fraser, grand, noble, red, and Shasta firs are among the more susceptible species, while European silver, Japanese, momi, Nordmann, and Turkish firs tend to be much less susceptible (Benson et al., 1998a,b; Cooley et al., 1988; Chastagner et al., 1995). There is also considerable variation in the ability of the different *Phytophthora* species to cause disease. In general, *P. cinnamomi, P. cryptogea,* and *P. citricola* tend to be more aggressive than the other species.

Phytophthora Root Rot

Geographic occurrence Wherever firs are grown in heavy, poorly drained soil

Seasonal occurrence April to October, with an optimum period from May through July

DISEASE FREQUENCY	DISEASE SEVERITY
5 annual	**5** plants killed
4	**4**
3	**3**
2	2
1 rare	1 very little damage

CHEMICAL TREATMENT	CULTURAL PRACTICES
3 used every year	**3** very important
2	2
1 not used	1 not important

SANITATION	RESISTANT CULTIVARS
3 very important	3 many cultivars
2	2
1 not important	**1** no resistance

Management

Site selection is the most important cultural factor affecting the management of Phytophthora root rot in nurseries. Highly susceptible firs should not be planted on sites with heavy, poorly drained soils. Growers can alleviate high soil moisture conditions that favor disease development by installing drain tiles. Avoiding the use of overhead irrigation, particularly if it is contaminated with *Phytophthora* inoculum, will prevent shoot blight. Soil fumigation to control Phytophthora root rot is a common practice in bare-root nurseries. Selective fungicides registered for use in nurseries can suppress disease development.

Gray Mold (Botrytis Blight)

Botrytis cinerea

Gray mold, or Botrytis blight, caused by *Botrytis cinerea,* is a serious disease of conifer seedlings, resulting in damping-off or seedling blight (Srago and McCain, 1989). Under certain conditions, the disease also occurs as a shoot blight of larger nursery trees. Gray mold is a much more serious problem in greenhouse container nurseries than in bare-root nurseries (Sutherland and Davis, 1991). Seedling trees held in cold storage can also be damaged by the disease.

Symptoms

The initial symptoms of gray mold are water-soaked, dark, discolored spots on newly developing shoots (Plates 50 and 51). Tan to brown lesions develop on shoots and stems, and shoots wither and die if they are girdled by lesions. Under humid conditions, a web of gray brown mycelium with prolific clusters of diagnostic gray spores develops on diseased tissue. In storage the disease can spread through a bundle of seedlings as the pathogen grows over the surface of the needles.

Symptoms of gray mold can be confused with those of root rot and chemical injury.

Epidemiology

B. cinerea can survive for extended periods as sclerotia or as a saprophyte, colonizing dead and dying plant material. Spores of the fungus are spread to susceptible host tissues by wind or splashing water. Rapid changes in humidity can trigger the release of spores into the atmosphere. Infection is favored by cool, wet conditions. High humidity, plant stress due to frost or freeze injury, dense plantings, and the an abundance of juvenile tissue present in nurseries are highly conducive to gray mold. The disease can also damage seedlings in storage. See Chapter 9, "Botrytis Blight (Gray Mold)."

Host Range

B. cinerea attacks hundreds of species of woody and herbaceous plants. Many conifers, including firs, are susceptible.

Management

Growers should remove and destroy crop residues in greenhouses and nurseries where *Botrytis* can survive. To control gray mold, it is important to minimize periods of wetness of plant surfaces, keep humidity at a low level through proper ventilation, and avoid tight placement of seedlings. Avoid stresses and avoid producing overly succulent plants through proper temperature and fertility management. A number of fungicides can be used to protect seedlings from gray mold. Growers should alternate or tank-

Gray Mold

Geographic occurrence Wherever conifers are grown

Seasonal occurrence April to October, with an optimum period from May through June

DISEASE FREQUENCY
5 annual
4
3
2
1 rare

DISEASE SEVERITY
5 plants killed
4
3
2
1 very little damage

CHEMICAL TREATMENT
3 used every year
2
1 not used

CULTURAL PRACTICES
3 very important
2
1 not important

SANITATION
3 very important
2
1 not important

RESISTANT CULTIVARS
3 many cultivars
2
1 no resistance

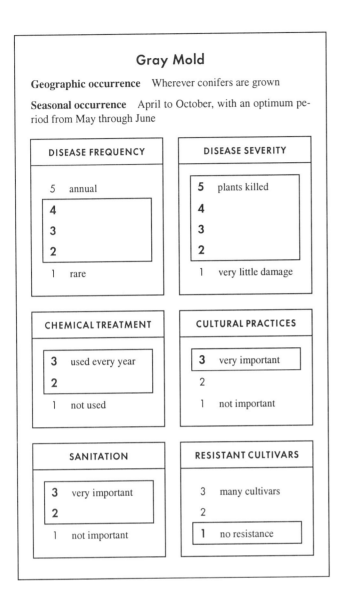

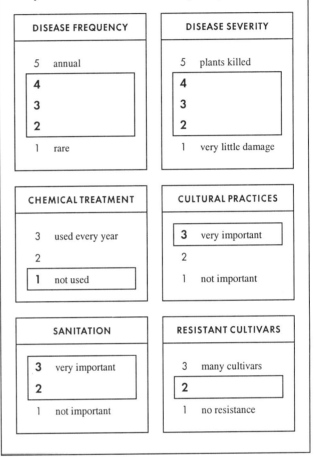

mix fungicides with different modes of action, to minimize the development of resistance.

Felt Blight

Herpotrichia juniperi

Snow Blight

Lophophacidium hyperboreum
Phacidium abietis
Sarcotrochila balsameae

Felt blight and snow blight affect conifer foliage covered by deep snow for an extended period (Sinclair et al., 1987; Skilling, 1989). Several fungi cause these diseases. The fungi that generally attack firs are *Herpotrichia juniperi*, causing felt blight, and *Lophophacidium hyperboreum, Phacidium abietis,* and *Sarcotrochila balsameae,* causing snow blight. *P. abietis* is the most common of the snow blight fungi.

Symptoms

Damage from snow blight and felt blight becomes evident as the snow melts in the spring. Damaged seedlings are commonly in patches, and the damage is usually more severe where the snow persists for an extended time.

On seedlings attacked by a snow blight pathogen, needles are covered with white mycelium of the fungus when the snow first melts. The needles are initially yellow but turn red to reddish brown and eventually gray. They remain attached to the branches.

On seedlings attacked by *H. juniperi,* a brown, feltlike growth of mycelium covers the surface of needles and tends to persist through the growing season. Groups of dead, brown, felt-covered needles bound together with mycelium fall from dead twigs.

Snow Blight

Geographic occurrence Northern regions where snow persists for an extended time in the spring

Seasonal occurrence Winter through fall; infection and colonization of needles occur primarily under snow; symptom development occurs from snowmelt through early summer

DISEASE FREQUENCY
5 annual
4
3
2
1 rare

DISEASE SEVERITY
5 plants killed
4
3
2
1 very little damage

CHEMICAL TREATMENT
3 used every year
2
1 not used

CULTURAL PRACTICES
3 very important
2
1 not important

SANITATION
3 very important
2
1 not important

RESISTANT CULTIVARS
3 many cultivars
2
1 no resistance

Epidemiology

The felt blight and snow blight pathogens survive in dead needles and twigs as mycelium and immature fruiting structures (Stone, 1997). The felt blight fungus, *H. juniperi,* can resume growth during the winter and spring, causing a perennial blight. It forms fruiting bodies (ascocarps) that mature in the summer, following initial symptom development. The snow blight fungi form small, brown to black fruiting structures on the undersides of diseased needles during late summer. Spores are spread from these fruiting bodies to nearby needles during moist weather from late summer until winter. Some of these pathogens also produce tiny, drought-resistant microsclerotia, which germinate on needles as soon as they are covered with snow; needles of all age classes are susceptible. Snow blight fungi can also spread from needle to needle as mycelium under the snow.

Host Range

H. juniperi has been reported on alpine, balsam, grand, noble, red, European silver, Pacific silver, and white firs and a number of other conifers. *P. abietis* is common on alpine, balsam, grand, Pacific silver, and white firs and also other conifers. Usually, conifers grown from seed from northern sources are not damaged as much as those grown from seed from sources farther south.

Management

Selecting nursery sites where snow does not persist for an extended time will reduce the potential development of felt blight and snow blight. Growers should remove the lower branches of susceptible trees near the nursery, to reduce levels of inoculum, which can be blown into the nursery. Susceptible firs should not be grown in areas where snow persists for an extended period of time. Removal of infected seedlings and apparently healthy ones in the vicinity of diseased trees will help reduce the spread of these diseases. Potassium applied to correct a deficiency of this nutrient will also help limit the development of these diseases.

REFERENCES

Adams, G. C., Jr., and Bielenin, A. 1988. First report of *Phytophthora cactorum* and *P. citricola.* Plant Dis. 72:79.

Arnott, J. T., and Matthews, R. G. 1981. Nursery production of true firs in British Columbia. Pages 195–201 in: Proc. Biol. Management True Fir Pacific Northwest Symp. C. D. Oliver and R. M. Kenady, eds.

Axelrood, P. E. 1991. Biological control of plant pathogens: principles and strategies. Pages 127–132 in: Diseases and insects in forest nurseries. Proc. IUFRO Working Party S2.07-09. For. Can. Pac. Yukon Reg. Info. Rep. BC-X-331. J. R. Sutherland and S. G. Glover, eds.

Benson, D. M., Grand, L. F., and Suggs, E. G. 1976. Root rot of Fraser fir caused by *Phytophthora drechsleri.* Plant Dis. Rep. 60:238–240.

Benson, D. M., Hinesley, L. E., Frampton, J., and Parker, K. C. 1998a. Evaluation of six *Abies* spp. to Phytophthora root rot caused by *Phytophthora cinnamomi.* Biol. Cult. Tests Control Plant Dis. 13:57.

Benson, D. M., Hinesley, L. E., and Parker, K. C. 1998b. Evaluation of Canaan seed sources for resistance to Phytophthora root rot caused by *Phytophthora cinnamomi.* Biol. Cult. Tests Control Plant Dis. 13:58.

Boyce, J. S. 1938. Forest Pathology. McGraw-Hill, New York.

Brown, B. N., and Baxter, A. G. M. 1991. Nursery hygiene in concept and practice. Pages 133–140 in: Diseases and insects in forest nurseries. Proc. IUFRO Working Party S2.07-09. For. Can. Pac. Yukon Reg. Info. Rep. BC-X-331. J. R. Sutherland and S. G. Glover, eds.

Chastagner, G. A., Hamm, P. B., and Riley, K. L. 1995. Symptoms and *Phytophthora* spp. associated with root rot and stem canker of noble fir Christmas trees in the Pacific Northwest. Plant Dis. 79:290–293.

Cooley, S. J., Hamm, P. B., and Hansen, E. M. 1988. Susceptibility of Northwest conifers to Phytophthora root rot. Tree Planters' Notes 40:15–18.

Cordell, C. E., Kelly, W. D., and Smith, R. S. 1989. Integrated nursery pest management. Pages 5–13 in: Forest nursery pests. U.S. For. Serv. Agric. Handb. 680. C. E. Cordell, R. L. Anderson, W. H. Hoffard, T. D. Landis, R. S. Smith, Jr., and H. V. Toko, eds.

Dirr, M. A. 1998. Manual of Woody Landscape Plants: Their Identification, Ornamental Characteristics, Culture, Propagation and Uses. 5th ed. Stipes Publishing, Champaign, Ill.

Grand, L. F., and Lapp, N. A. 1974. *Phytophthora cinnamomi* root rot of Fraser fir in North Carolina. Plant Dis. Rep. 58:318–320.

Hamm, P. B., and Hansen, E. M. 1982. Pathogenicity of *Phytophthora* spp. to Northwest conifers. Eur. J. For. Pathol. 12:167–174.

Hamm, P. B., and Hansen, E. M. 1987. Identification of *Phytophthora* spp. known to attack conifers in the Pacific Northwest. Northwest Sci. 61: 103–109.

Hamm, P. B., and Hansen, E. M. 1991. The isolation and identification of *Phytophthora* species causing damage in bare-root conifer nurseries. Pages 169–179 in: Diseases and insects in forest nurseries. Proc. IUFRO Working Party S2.07-09. For. Can. Pac. Yukon Reg. Info. Rep. BC-X-331. J. R. Sutherland and S. G. Glover, eds.

Hamm, P. B., Campbell, S. J., and Hansen, E. M. 1990. Growing healthy seedlings: Identification and management of pests in Northwest forest nurseries. Oregon State Univ. For. Res. Lab. Spec. Publ. 19.

Hansen, E. 1997. Diseases in special settings: Forest tree nurseries. Page 88 in: Compendium of Conifer Diseases. E. M. Hansen and K. J. Lewis, eds. American Phytopathological Society, St. Paul, Minn.

James, R. L., Dumroese, R. K., and Wenny, D. L. 1991. Fusarium diseases of conifer seedlings. Pages 181–190 in: Diseases and insects in forest nurseries. Proc. IUFRO Working Party S2.07-09. For. Can. Pac. Yukon Reg. Info. Rep. BC-X-331. J. R. Sutherland and S. G. Glover, eds.

Johnson, D. W., LaMadeleine, L. A., and Blomberg, W. J. 1989. Fusarium root rot. Pages 40–42 in: Forest nursery pests. U.S. For. Serv. Agric. Handb. 680. C. E. Cordell, R. L. Anderson, W. H. Hoffard, T. D. Landis, R. S. Smith, Jr., and H. V. Toko, eds.

Kenerley, C., and Bruck, R. I. 1981. Phytophthora root rot of Balsam fir and Norway spruce in North Carolina. Plant Dis. 65:614–615.

Kuhlman, E. G., and Hendrix, F. F., Jr. 1963. Phytophthora root rot of Fraser fir. Plant Dis. Rep. 47:552–553.

Kuhlman, E. G., Grand, L. F., and Hansen, E. M. 1989. Phytophthora root rot of conifers. Pages 60–61 in: Forest nursery pests. U.S. For. Serv. Agric. Handb. 680. C. E. Cordell, R. L. Anderson, W. H. Hoffard, T. D. Landis, R. S. Smith, Jr., and H. V. Toko, eds.

Landis, T. D., and Campbell, S. J. 1991. Soil fumigation in bareroot tree nurseries. Pages 191–205 in: Diseases and insects in forest nurseries. Proc. IUFRO Working Party S2.07-09. For. Can. Pac. Yukon Reg. Info. Rep. BC-X-331. J. R. Sutherland and S. G. Glover, eds.

McCain, A. H., and Scharpf, R. F. 1986. Phytophthora shoot blight and canker disease of *Abies* spp. Plant Dis. 70:1036–1037.

McCain, A. H., Sauve, R. J., and Kaufmann, B. W. 1989. Pythium root rot. Pages 124–125 in: Forest nursery pests. U.S. For. Serv. Agric. Handb. 680. C. E. Cordell, R. L. Anderson, W. H. Hoffard, T. D. Landis, R. S. Smith, Jr., and H. V. Toko, eds.

Shew, H. D., and Benson, D. M. 1982. Fraser fir root rot induced by *Phytophthora citricola.* Plant Dis. 65:688–689.

Sinclair, W. A., Lyon, H. H., and Johnson, W. T. 1987. Diseases of Trees and Shrubs. Comstock Publishing Associates, Cornell University Press, Ithaca, N.Y.

Skilling, D. D. 1989. Snow blight of conifers. Pages 73–74 in: Forest nursery pests. U.S. For. Serv. Agric. Handb. 680. C. E. Cordell, R. L. Anderson, W. H. Hoffard, T. D. Landis, R. S. Smith, Jr., and H. V. Toko, eds.

Srago, M. D., and McCain. A. H. 1989. Gray mold. Pages 45–46 in: Forest nursery pests. U.S. For. Serv. Agric. Handb. 680. C. E. Cordell, R. L. Anderson, W. H. Hoffard, T. D. Landis, R. S. Smith, Jr., and H. V. Toko, eds.

Stein, J. D., and Smith, R. S., Jr. 1991. Diseases and insects in United States forest nurseries. Pages 109–115 in: Diseases and insects in forest nurseries. Proc. IUFRO Working Party S2.07-09. For. Can. Pac. Yukon Reg. Info. Rep. BC-X-331. J. R. Sutherland and S. G. Glover, eds.

Stone, J. 1997. Felt blights and snow blights. Pages 63–64 in: Compendium of Conifer Diseases. E. M. Hansen and K. J. Lewis, eds. American Phytopathological Society, St. Paul, Minn.

Sutherland, J. R., and Davis, C. 1991. Disease and insects in forest nurseries in Canada. Pages 25–32 in: Diseases and insects in forest nurseries. Proc. IUFRO Working Party S2.07-09. For. Can. Pac. Yukon Reg. Info. Rep. BC-X-331. J. R. Sutherland and S. G. Glover, eds.

CHAPTER 40

D. Michael Benson · North Carolina State University, Raleigh

Flowering Crabapple Diseases

Geographic production USDA zones 3–9

Family Rosaceae

Genus *Malus*

Species

M. angustifolia	southern crabapple
M. baccata	Siberian crabapple
M. coronaria	wild sweet crabapple
M. floribunda	Japanese flowering crabapple
M. halliana var. *parkmanii*	
M. hupehensis	tea crabapple
M. ioensis	prairie crabapple
M. prunifolia	plumleaf crabapple
M. pumila	'Niedzwetzkyana'
M. sargentii	Sargent crabapple
M. sieboldii	
M. sikkimensis	Sikkim crabapple
M. spectabilis	Chinese crabapple
M. yunnanensis	

Cultivars more than 500 types among these species and hybrids

There are many diverse crabapple cultivars, with a spectrum of shapes and flower color from white to pink to red. Crabapples are popular landscape trees. They make a splendid show of color in the spring, and the foliage remains attractive during the summer. A tremendous amount of disease evaluation has been done over the past two decades to promote disease-resistant cultivars. Programs initiated by Lester Nichols at Pennsylvania State University led to the National Crabapple Evaluation Program, which was carried out in many states in the East and Midwest and as far south as Alabama and as far west as Oklahoma. State-by-state results of disease evaluations have been published, and growers should take advantage of this information in selecting crabapples for production. Crabapples were evaluated for four major foliar diseases in the program: apple scab, fire blight, powdery mildew, and rust. High levels of resistance to all four diseases are available, as outlined below. All cultivars that are susceptible to apple scab and fire blight should be discarded.

Apple Scab

Venturia inaequalis

Apple scab and fire blight are the two most destructive diseases of crabapples. Susceptible cultivars should be avoided (see Table 40.1).

Symptoms and Epidemiology

The apple scab fungus, *Venturia inaequalis,* causes black lesions on foliage (Plate 52), fruit, and stems. The disease develops during wet weather in the late spring, summer, and fall. Initial inoculum consists of ascospores of the fungus, produced on fallen

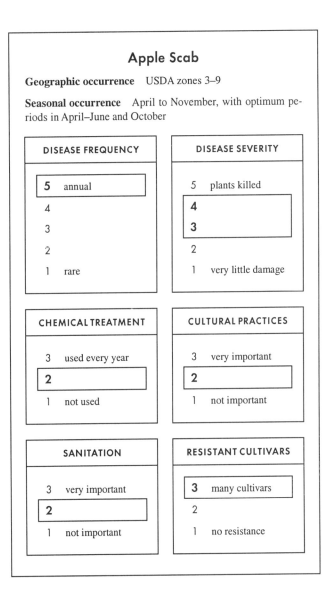

Apple Scab

Geographic occurrence USDA zones 3–9

Seasonal occurrence April to November, with optimum periods in April–June and October

DISEASE FREQUENCY	DISEASE SEVERITY
5 annual	5 plants killed
4	**4**
3	**3**
2	2
1 rare	1 very little damage

CHEMICAL TREATMENT	CULTURAL PRACTICES
3 used every year	3 very important
2	**2**
1 not used	1 not important

SANITATION	RESISTANT CULTIVARS
3 very important	**3 many cultivars**
2	2
1 not important	1 no resistance

Table 40.1 Resistance to apple scab in flowering crabapples (*Malus* spp. and cultivars) in Raleigh, North Carolina, 1988–1992[a]

Resistant species and cultivars	
Adams	Ralph Shay
M. baccata 'Jackii'	Red Baron
Beverly	Red Jade
Bob White	Red Jewel
Candied Apple	Red Splendor
Centurion	Robinson
Christmas Holly	*M. sargentii*
Dolgo	Selkirk
Donald Wyman	Sentinel
M. floribunda	Silver Moon
Harvest Gold	Snowdrift
M. henningii	Strawberry Parfait
Indian Summer	Sugar Tyme
Jewelberry	Tea Flowering
Liset	Velvet Pillar
Mary Potter	White Angel
Molten Lava	White Cascade
Ormiston Roy	Winter Gold
Prairie Fire	*M. yunnanensis* var. *veichii*
Profusion	*M. zumi* 'Calocarpa'
Moderately resistant cultivars	
David	Prof. Sprenger
Indian Magic	Royalty
Susceptible cultivars	
Hopa	
Radiant	
Ruby Luster	

[a] Some cultivars have higher or lower levels of disease and thus may be rated differently in other locations. For instance, apple scab has been found to be much more severe in these cultivars in Ohio (Draper et al., 1996).

Table 40.2 Resistance to fire blight in flowering crabapples (*Malus* spp. and cultivars) in Raleigh, North Carolina, 1988–1992[a]

Resistant species and cultivars	
Candied Apple	Robinson
Christmas Holly	Royalty
Liset	*M. sargentii*
Prairie Fire	*M. yunnanensis* var. *veichii*
Radiant	*M. zumi* 'Calocarpa'
Moderately resistant species and cultivars	
Adams	Prof. Sprenger
M. baccata 'Jackii'	Profusion
Beverly	Red Baron
Bob White	Red Jewel
Centurion	Red Splendor
David	Ruby Luster
Dolgo	Selkirk
Donald Wyman	Sentinel
M. floribunda	Strawberry Parfait
M. henningii	Sugar Tyme
Hopa	Tea Flowering
Indian Summer	Velvet Pillar
Jewelberry	White Angel
Molten Lava	White Cascade
Ormiston Roy	Winter Gold
Susceptible cultivars	
Harvest Gold	Red Jade
Indian Magic	Silver Moon
Mary Potter	Snowdrift
Ralph Shay	

[a] Some cultivars have more or less disease and thus may been rated differently in other locations. In evaluations conducted in Alabama, only *M. baccata* 'Jackii' was free of fire blight over a two-year period (Hagan et al., 1995a,b).

leaves that overwinter on the nursery floor. Subsequently, spores produced in lesions and dispersed to foliage by splashing rain serve as secondary inoculum. Severely infected trees defoliate by midsummer, leaving unsightly bare branches.

Management

Cultural practices to eliminate fallen leaf debris during the fall and winter, together with fungicide sprays at budbreak and throughout the growing season can suppress disease development. Highly susceptible cultivars are difficult to grow disease-free in the nursery or landscape. Crabapples resistant to apple scab are listed in Table 40.1.

Fire Blight

Erwinia amylovora

Like apple scab, fire blight can kill young trees and severely damage older ones, so highly susceptible cultivars should be avoided.

Symptoms

The initial symptoms of fire blight are a blighting of flowers and expanding twigs and leaves, which may appear black to brown (Plate 53). In seasons with frequent rain and cool weather during blooming and production of new growth, extensive flagging of diseased branches can develop. In highly susceptible cultivars, 3 to 6 feet of the terminal branch may be killed in a single season.

Epidemiology

Fire blight is caused by the bacterium *Erwinia amylovora,* which overwinters in cankers on twigs and stems and oozes from the cankers in early spring during cool, wet weather. The bacterial ooze is splashed or carried by insects, particularly honeybees, to opening flowers, expanding leaves, and fruit where infection occurs through natural openings. When the bacterium enters a stem from an infected petiole or peduncle, it can move through the vascular system, resulting in extensive flagging and dieback of branches.

Management

Overfertilization encourages succulent growth that increases susceptibility of trees. Bactericide sprays may give some control in the nursery. Crabapples resistant to fire blight are listed in Table 40.2.

Powdery Mildew

***Podosphaera* spp.**

Powdery mildew in crabapples results in loss of aesthetic value more than damage to the plants. Cultivars that are susceptible to powdery mildew but resistant to apple scab and fire blight can perform well in the nursery and landscape. The powdery mildew fungi *Podosphaera clandestina, P. leucotricha,* and an unidentified *Podosphaera* sp. have been reported on crabapples

Symptoms

Whitish colonies of the powdery mildew pathogen develop on susceptible cultivars of crabapple. The colonies are superficial, with only a small portion of the mildew pathogen actually infecting host tissues. The pathogen does little real damage to crabapples, but the whitish mildew colonies may be unsightly on very susceptible cultivars. When infection occurs early in the spring, the last two or three leaves to emerge may be distorted and disfigured.

Epidemiology

The powdery mildew fungi overwinter on infected buds and shoots as mycelium and cleistothecia, which are formed from the mycelium in mid- to late summer as the fungus undergoes sexual reproduction. Cleistothecia turn black and appear as small grains, like specks of pepper, against the background of the white mycelium. In the spring, cleistothecia release windborne ascospores, which cause leaf infections, resulting in the development of the white mycelium on leaves and stems. New spores (conidia) arise from the white mycelium shortly thereafter and are transported by the wind to new foliage. Germination of conidia and infection are favored by high humidity, but not free water. Conidia are produced daily on mildew colonies, so they serve as an important source of inoculum during the growing season.

Management

In the nursery, powdery mildew can be controlled with fungicide sprays. Crabapples resistant to powdery mildew are listed in Table 40.3.

Rust

Gymnosporangium spp.

Rust diseases of crabapple, like powdery mildew, cause little real damage. Heavy infection can result in numerous leaf spots, which reduce the aesthetic value of the plant. The rust fungi *Gymnosporangium clavipes, G. globosum, G. juniperi-virginianae,* and other *Gymnosporangium* spp. have been reported on crabapples.

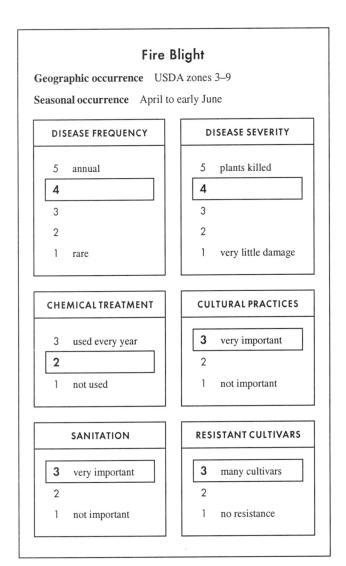

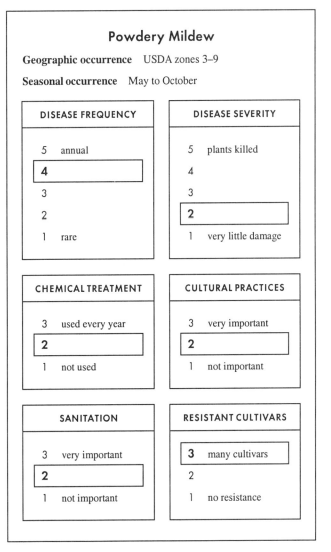

Fire Blight

Geographic occurrence USDA zones 3–9

Seasonal occurrence April to early June

DISEASE FREQUENCY: 5 annual; **4**; 3; 2; 1 rare

DISEASE SEVERITY: 5 plants killed; **4**; 3; 2; 1 very little damage

CHEMICAL TREATMENT: 3 used every year; **2**; 1 not used

CULTURAL PRACTICES: **3 very important**; 2; 1 not important

SANITATION: **3 very important**; 2; 1 not important

RESISTANT CULTIVARS: **3 many cultivars**; 2; 1 no resistance

Powdery Mildew

Geographic occurrence USDA zones 3–9

Seasonal occurrence May to October

DISEASE FREQUENCY: 5 annual; **4**; 3; 2; 1 rare

DISEASE SEVERITY: 5 plants killed; 4; 3; **2**; 1 very little damage

CHEMICAL TREATMENT: 3 used every year; **2**; 1 not used

CULTURAL PRACTICES: 3 very important; **2**; 1 not important

SANITATION: 3 very important; **2**; 1 not important

RESISTANT CULTIVARS: **3 many cultivars**; 2; 1 no resistance

Table 40.3 Resistance to powdery mildew in flowering crabapples (*Malus* spp. and cultivars) in Raleigh, North Carolina, 1988–1992[a]

Resistant species and cultivars	
M. baccata 'Jackii'	Ralph Shay
Beverly	Red Baron
Bob White	Red Jade
Candied Apple	Red Splendor
Christmas Holly	Robinson
David	Royalty
Dolgo	Ruby Luster
Donald Wyman	*M. sargentii*
M. floribunda	Selkirk
Harvest Gold	Silver Moon
M. henningii	Snowdrift
Hopa	Strawberry Parfait
Indian Magic	Sugar Tyme
Indian Summer	Tea Flowering
Ormiston Roy	Velvet Pillar
Prairie Fire	White Angel
Profusion	*M. zumi* 'Calocarpa'
Radiant	
Moderately resistant cultivars	
Centurion	Red Jewel
Jewelberry	Sentinel
Liset	White Cascade
Mary Potter	Winter Gold
Prof. Sprenger	
Susceptible species and cultivars	
Adams	
Molten Lava	
M. yunnanensis var. *veichii*	

[a] Some cultivars have more or less disease and thus may been rated differently in other locations.

Symptoms

Initial infections appear as yellow leaf spots (Plate 54). Infected fruit becomes swollen and irregular. Defoliation can occur in heavy infections but is rare.

Epidemiology

Like many rust fungi, the crabapple rust pathogens require two different hosts for the completion of their life cycles, in certain stages living on crabapple and in other stages living on juniper, the alternate host. On junipers, these fungi produce teliospores, which give rise to basidiospores, which in turn are carried by the wind to crabapple leaves in early spring, as the new growth develops. Normally basidiospores infect leaves from the upper surface, and yellow leaf spots develop. Fungal hyphae then grow through the leaf to produce aeciospores in pustules on the underside of the leaf or on the surface of the fruit. Aeciospores are windborne and infect juniper, thus completing the life cycle of the fungi.

Management

Maintaining a distance of 500 feet between junipers and crabapples in the nursery or landscape can limit the infection between hosts. Fungicide sprays on crabapple foliage at leaf break can protect the foliage of susceptible cultivars. Crabapples resistant to rust are listed in Table 40.4.

Table 40.4 Resistance to rust in flowering crabapples (*Malus* spp. and cultivars) in Raleigh, North Carolina, 1988–1992

Resistant species and cultivars	
Adams	Red Splendor
M. baccata 'Jackii'	Robinson
Beverly	Royalty
Candied Apple	Ruby Luster
Centurion	*M. sargentii*
Dolgo	Selkirk
Hopa	Sentinel
Indian Magic	Silver Moon
Liset	*M. tschonoskii*
Mary Potter	Velvet Pillar
Molten Lava	Winter Gold
Prof. Sprenger	*M. yunnanensis* var. *veichii*
Red Baron	*M. zumi* 'Calocarpa'
Red Jade	
Moderately resistant species and cultivars	
Bob White	Profusion
Christmas Holly	Ralph Shay
David	Red Jewel
Donald Wyman	Snowdrift
Harvest Gold	Strawberry Parfait
M. henningii	Sugar Tyme
Indian Summer	Tea Flowering
Jewelberry	White Angel
Prairie Fire	White Cascade
Susceptible species and cultivars	
M. floribunda	
Ormiston Roy	
Radiant	

[a] Some cultivars are more or less resistant in other locations.

Frogeye Leaf Spot (Black Rot)
Botryosphaeria obtusa

Infection of crabapple leaves by the fungus *Botryosphaeria obtusa* after the leaves have unfolded causes circular leaf spots with irregular purple margins. The spots eventually become brown, with definite margins. Secondary enlargements at different points around the margin of some leaf spots give rise to a "frogeye" pattern. In the leaf spots, *B. obtusa* produces spores (conidia), which are dispersed by splashing rain to other leaves, fruit, and damaged branches. Leaf spot is normally more severe in years following cold injury or severe fire blight, because the pathogen colonizes injured or diseased branches, forming cankers in which it sporulates. The fungus can also infect crabapple fruit, resulting in mummified fruit. Infected fruit and stem cankers serve as sources of overwintering inoculum for leaf infection the following year. Since the fungus is opportunistic following damage to crabapple branches, pruning branches damaged by cold injury or fire blight to eliminate cankers and potential infection sites can reduce initial inoculum.

Hartman et al. (1988, 1989, 1990, 1991) evaluated crabapples for resistance to frogeye leaf spot, as part of the National Crabapple Evaluation Program; their results are summarized in Table 40.5. Most cultivars evaluated were ranked immune (having no infection during the evaluation period) or resistant to *B. obtusa* (having spots on less than 10% of leaves).

Rust

Geographic occurrence USDA zones 3–9

Seasonal occurrence April to May

DISEASE FREQUENCY	DISEASE SEVERITY
5 annual	5 plants killed
4	4
3	**3**
2	**2**
1 rare	1 very little damage

CHEMICAL TREATMENT	CULTURAL PRACTICES
3 used every year	**3 very important**
2	2
1 not used	1 not important

SANITATION	RESISTANT CULTIVARS
3 very important	**3 many cultivars**
2	2
1 not important	1 no resistance

Frogeye Leaf Spot

Geographic occurrence USDA zones 3–9

Seasonal occurrence May to October

DISEASE FREQUENCY	DISEASE SEVERITY
5 annual	5 plants killed
4	4
3	3
2	**2**
1 rare	1 very little damage

CHEMICAL TREATMENT	CULTURAL PRACTICES
3 used every year	**3 very important**
2	2
1 not used	1 not important

SANITATION	RESISTANT CULTIVARS
3 very important	**3 many cultivars**
2	2
1 not important	1 no resistance

Fall Aesthetics

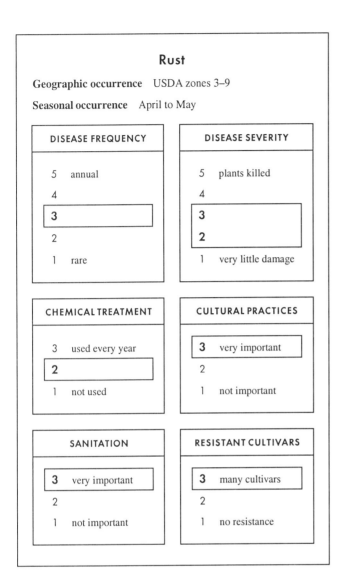

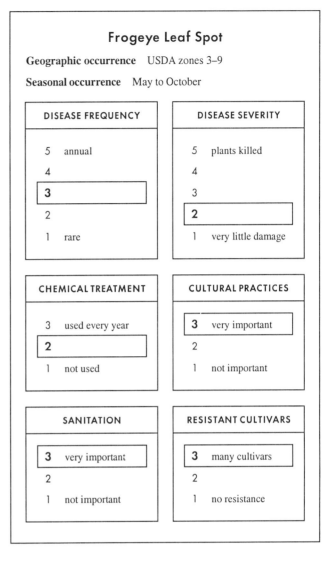

In addition to evaluations of flowering crabapples for disease resistance, fall aesthetic evaluations can provide nurserymen, horticulturists, and landscape architects with a guide to the overall appearance of trees at the end of the growing season. Tree habit, including foliage color, fruit retention, form, and disease resistance, is an indicator of overall aesthetics (Table 40.6). Guides for the selection of crabapple cultivars are available (Ranney et al., 1993). Many of the available crabapples have been described and their resistance to various diseases noted by Dirr (1998).

REFERENCES

Benson, D. M., and Daughtry, B. I. 1993. Disease response of selected crabapples in North Carolina, 1988–1992. Biol. Cult. Tests Control Plant Dis. 8:127.

Benson, D. M., Daughtry, B. I., and Jones, R. K. 1991. Crabapple cultivar response to foliar diseases, 1989, 1990. Biol. Cult. Tests Control Plant Dis. 6:106.

den Boer, J. H., and Green, T. L. 1995. Crabapple disease and aesthetic ratings. *Malus:* Int. Ornamental Crabapple Bull. 9(2):8–56.

Dirr, M. A. 1998. Manual of Woody Landscape Plants: Their Identification, Ornamental Characteristics, Culture, Propagation and Uses. 5th ed. Stipes Publishing, Champaign, Ill.

Draper, E. A., Chatfield, J. A., Cochran, K. C., Bristol, P. W., and Tubesing, C. E. 1996. Evaluation of crabapples for apple scab at Secrest Arboretum in Wooster, Ohio: 1995. Pages 22–25 in: Ornamental Plants – 1996: A Summary of Research. Ohio Agric. Res. Dev. Ctr. Spec. Circ. 152.

Hagan, A. K., Tilt, K. Williams, D., and Akridge, J. R. 1993. Susceptibility of crabapple cultivars to fireblight in Alabama. Proc. South. Nurserymen's Assoc. Res. Conf. 38:203–205.

Hagan, A. K., Tilt, K., and Akridge, J. R. 1995a. Reaction of crabapple cultivars to fireblight and apple scab in Alabama, 1994. Biol. Cult. Tests Control Plant Dis. 10:63.

Hagan, A. K., Tilt, K. Williams, D. and Akridge, J. R. 1995b. Fireblight-resistant crabapple cultivars. Alabama Agric. Exp Sta. Res. Rep. 10:13.

Hartman, J. R., Clinton, B., and McNeil, R. 1988. Reactions of flowering crabapple cultivars to foliar diseases, 1987. Biol. Cult. Tests Control Plant Dis. 3:81.

Hartman, J. R., Clinton, B., and McNeil, R. 1989. Reactions of flowering crabapple cultivars to foliar diseases, 1988. Biol. Cult. Tests Control Plant Dis. 4:79.

Hartman, J. R., Kennedy, B. S., and McNeil, R. 1990. Reactions of flowering crabapple cultivars to foliar diseases, 1989. Biol. Cult. Tests Control Plant Dis. 5:91.

Table 40.5 Resistance to frogeye leaf spot in flowering crabapples (*Malus* spp. and cultivars) in Lexington, Kentucky, 1987–1990[a,b]

Immune species and cultivars	
M. floribunda	Silver Moon
Liset	Winter Gold
Molten Lava	
Resistant species and cultivars	
M. baccata 'Jackii'	Profusion
Beverly	Radiant
Bob White	Red Jade
Candied Apple	Red Jewel
Christmas Holly	Red Splendor
David	Robinson
Donald Wyman	Royalty
Harvest Gold	*M. sargentii*
M. henningii	Sentinel
Hopa	Snowdrift
M. hupehensis	Strawberry Parfait
Indian Magic	Sugar Tyme
Jewelberry	Velvet Pillar
Mary Potter	White Cascade
Ormiston Roy	*M. zumi* 'Calocarpa'
Susceptible species and cultivars	
Dolgo	Selkirk
Prof. Sprenger	*M. tschonoskii*
Ralph Shay	White Angel
Red Baron	*M. yunnanensis* var. *veichii*
Ruby Luster	

[a] Data from Hartman et al. (1988, 1989, 1990, 1991).
[b] Crabapples were rated on a scale of 0–4, where 0 = no disease; 1 = up to 10% of leaves with spots; 2 = 10–25% of leaves infected; 3 = 25–75% of leaves infected; and 4 = more than 75% of leaves infected. Crabapples rated immune had no disease. Those rated resistant had an average four-year rating of 0.08–0.65. Those rated susceptible had average ratings of over 0.85.

Table 40.6 Fall aesthetic ratings of 48 cultivars of crabapples (*Malus* spp. and cultivars) in Raleigh, North Carolina, 1988–1992[a]

Cultivar	Rating	Cultivar	Rating
Harvest Gold	0.9	Hopa	2.7
Velvet Pillar	1.7	Selkirk	2.7
Profusion	1.8	Silver Moon	2.7
Sugar Tyme	1.8	*M. floribunda*	2.8
Snowdrift	1.9	Sentinel	2.9
M. zumi 'Calocarpa'	1.9	Prof. Sprenger	2.9
Donald Wyman	2.0	Christmas Holly	2.9
Dolgo	2.2	Tea Flowering	3.0
Red Splendor	2.2	Jewelberry	3.0
Winter Gold	2.2	Red Baron	3.1
Prairie Fire	2.3	Strawberry Parfait	3.1
Royalty	2.3	Liset	3.1
Centurion	2.3	Indian Magic	3.3
David	2.3	Ruby Luster	3.3
M. baccata 'Jackii'	2.4	Mary Potter	3.5
Beverly	2.4	Molten Lava	3.5
Robinson	2.4	White Cascade	3.5
Red Jewel	2.5	Ormiston Roy	3.6
Adams	2.5	Candied Apple	3.8
Indian Summer	2.5	White Angel	3.9
Ralph Shay	2.6	*M. sargentii*	4.0
M. henningii	2.7	Red Jade	4.3
Radiant	2.7	*M. tschonoskii*	4.4
Bob White	2.7	*M. yunnanensis* var. *veichii*	4.9

[a] Fall aesthetic rating: 0 = perfect tree; 1 = highly ornamental in foliage color, fruit, and form; 2 = ornamental, acceptable but less than perfect; 3 = barely ornamental; 4 = undesirable (severe disease, messy fruit drop, etc.); 5 = unacceptable, tree should be replaced.

Hartman, J. R., Doney, J. and McNeil, R. 1991. Reactions of flowering crabapple cultivars to spring frost and to foliar diseases, 1990. Biol. Cult. Tests Control Plant Dis. 6:107.

Ranney, T. G., Benson, D. M., and Powell, M. A. 1993. Superior crabapple trees for the landscape. N.C. Coop. Ext. Serv. Hortic. Info. Leafl. 613.

Smith, E. M. 1979. A 10-yr evaluation of flowering crabapple susceptibility to apple scab in Ohio. Pages 36–39 in: Ornamental plants – 1979: A summary of research. Ohio Agric. Res. Dev. Ctr. Res. Circ. 246.

CHAPTER
41

Austin K. Hagan ▪ Auburn University, Auburn, Alabama

Flowering Pear Diseases

Geographic production USDA zones 5–9

Family Rosaceae

Genus *Pyrus*

Species *P. calleryana*
P. fauriei

Flowering pear, *Pyrus calleryana,* was introduced into the United States from China in the early 20th century. The first improved cultivar, Bradford, was released by the U.S. National Arboretum in 1963 and has been widely used in street and landscape plantings in the mid-Atlantic and southeastern United States. Since then, approximately 12 additional selections of flowering pear have been marketed by the nursery industry. Flowering pears are most noted for their spectacular early spring floral display and brilliant fall foliage and for their tolerance of relatively poor growing conditions often encountered in the urban landscape. Because the tree produces numerous branches at acute angles, some selections of flowering pear, particularly mature specimens of Bradford, are often broken apart by high winds, ice, or wet, heavy snow. With the exception of fire blight, most selections of flowering pear are subject to relatively few diseases.

Fire Blight
Erwinia amylovora

Fire blight is the most common and destructive disease of flowering pear and many other members of the rose family (Rosaceae). In the United States, it occurs wherever flowering pears are produced or established in landscape plantings. Under conditions favorable for disease development, some selections of flowering pears may be badly disfigured and in some cases succumb to fire blight.

Symptoms

Fire blight severity varies according to the sensitivity of different selections of flowering pear to fire blight and the weather during bloom. In susceptible selections, diseased blooms first become water-soaked, then shrivel, and quickly turn brown. The infection usually progresses down through the peduncle and flower spur into the twig. Blighted peduncles, immature fruit, and spurs also shrivel and turn brown to almost black. Mummified fruit often remains on the tree for several months. Shortly after the blossom blight phase, leaf blight and twig dieback occur. Blighted leaves on diseased twigs curl downward, turn black, and remain on the tree. The tips of blighted twigs droop, in the diagnostic shepherd's-crook symptom. Symptoms are often more severe in older flowering pear trees than in younger trees.

Brown to black, slightly sunken cankers often form on twigs, scaffold branches, and trunks at the base of blighted spurs or shoots. The face or surface of the canker is usually smooth, but cracks often develop along its outside edge. Reddish brown streaks can be seen in the sapwood underneath the edges of discolored cankers.

In resistant selections, such as Bradford, the damage is often limited to blackening of a few scattered leaves or modest dieback of numerous shoot tips, but it does not disfigure the tree. In susceptible selections, extensive dieback of a sizable portion of the tree canopy may occur.

Epidemiology

Fire blight is caused by the bacterium *Erwinia amylovora.* The pathogen overseasons in tissues surrounding twig and branch cankers. During wet, humid weather in the early spring, beads of a sticky, amber-colored fluid containing millions of bacteria oozes from blighted spurs and cankers. *E. amylovora* is spread primarily by insects, particularly honeybees, and splashing rain to flower buds, blooms, leaves, and tender shoots of flowering pear. Pear blooms are invaded through the nectaries and pistils, where *E. amylovora* rapidly reproduces and then colonizes adjacent healthy tissues. Natural openings on the surfaces of the leaves and succulent tissues are other sites of infection for the pathogen.

Extended periods of warm, cloudy, and humid weather, particularly during bloom, greatly increase the risk of a fire blight outbreak in flowering pear. Temperatures between 75 and 81°F are most favorable for disease development. Other weather events, such as frost, hail, or high winds, create wounds through which the causal bacterium can enter the plant. Severe pruning and excessive nitrogen fertilization, which promote the production of soft, succulent shoots, greatly increase the vulnerability to fire blight.

Host Range

All selections of flowering pear are subject to fire blight, but significant differences in the susceptibility of selections have been noted.

Management

An effective management program combines the production and establishment of selections resistant to fire blight, sound cultural practices, good sanitation, and pesticides, when needed.

Cultural practices To maintain a moderate rate of shoot growth in container- and field-grown trees, apply recommended rates of a nursery-grade slow-release fertilizer. When fertilizer containing a water-soluble source of nitrogen is used, delay application until after petal fall, make several light applications during the growing season, and avoid applying high rates of nitrogen to trees in late summer or fall. In field-grown trees, avoid cultivating for weed control after midsummer. If possible, install a surface or drip irrigation system around field- and container-grown trees. When using an overhead irrigation system, avoid watering in the late afternoon or early evening.

Sanitation Pruning blighted twigs and cankered branches helps limit the spread of fire blight. Preferably, pears should be inspected for cankers and then be pruned in late winter, during cold, dry weather. At this time of the year, make pruning cuts 6 to 8 inches below the edge of the canker. During the spring and early summer, field and container stock should be regularly inspected for fire blight symptoms. At this time of the year, prune diseased twigs and branches about 12 to 14 inches below the margin of the canker. To avoid accidentally spreading the causal bacterium, pruning tools must be sanitized with denatured alcohol or a similar surface disinfectant after each cut.

Resistance Significant differences in susceptibility to fire blight have been noted in selections of flowering pear. Susceptibility, however, is greatly influenced by weather patterns during and shortly after bloom. Of the common selections of flowering pear, Bradford is moderately resistant to fire blight; Capital, Cleveland Select, Earlyred, and Whitehouse are moderately susceptible; and Aristocrat, Autumn Blaze, Fauriei (*P. fauriei*), and Redspire are susceptible.

Chemical treatment Timing of bactericide applications is critical for effective control of fire blight. Begin application of a recommended bactericide several days before the first flower buds open, and continue spraying every four or five days through petal fall. When postbloom weather favors continued disease development in susceptible selections in a container or field nursery, additional applications may be made at weekly intervals for another month. Repeated use of an antibiotic may result in the development of resistant strains of *E. amylovora,* which would not be controlled by the treatment.

Biocontrol agents, such as BlightBan, are being labeled and tested for fire blight control.

Fire Blight

Geographic occurrence USDA zones 5–9

Seasonal occurrence March to May, with an optimum period in April

DISEASE FREQUENCY	DISEASE SEVERITY
[5] annual	5 plants killed
4	4
3	**[3]**
2	2
1 rare	1 very little damage

CHEMICAL TREATMENT	CULTURAL PRACTICES
3 used every year	3 very important
[2]	**[2]**
1 not used	1 not important

SANITATION	RESISTANT CULTIVARS
[3] very important	**[3] many cultivars**
2	2
1 not important	1 no resistance

Bacterial Blossom Blast

Pseudomonas syringae* pv. *syringae

Bacterial blossom blast is a damaging disease of fruiting pears in the Pacific Northwest. It has been reported in flowering pear (*P. calleryana*) in Oregon, but its importance there and in the remainder of the United States is not well documented. The pathogen is the bacterium *Pseudomonas syringae* pv. *syringae,* which incites foliar diseases of other woody landscape plants nationwide.

Symptoms

The symptoms of bacterial blossom blast may be confused with those of fire blight. The early symptoms are a browning or blackening of portions or entire flowers, pedicels, and fruit clusters. Raised tan to brown blisters or cankers may appear on year-old twigs at the base of blighted spurs. As a blister dries, the paper-like periderm peels off the surface of the canker. A shoot dieback, similar to that associated with fire blight, may occur where the cankers girdle year-old twigs.

Epidemiology

The causal bacterium overseasons in perennial cankers and in buds and leaf scars of host trees. It is spread to blossoms by splashing water and insects and enters blossoms and leaves through natural openings as the trees break dormancy. Cool, wet weather during bloom favors the development of blossom blast. A hard frost or freeze just prior to bloom will greatly increase the severity of the disease.

Host Range

P. syringae pv. *syringae* causes blossom blight and dieback of a wide range of woody landscape plants and fruit trees. In addition to flowering pear, common woody landscape hosts include flowering dogwood; forsythia; golden rain-tree; hibiscus; lilac; Japanese, Norway, and red maples; live oak; oleander; and black poplar.

Management

No program for the management of blossom blast of fruiting or flowering pear has been developed.

Cultural practices and sanitation The cultural and sanitation programs described for the control of fire blight may help reduce the severity of blossom blast of flowering pear.

Resistance No information on the susceptibility of flowering pear selections to blossom blast is available.

Chemical treatment Bactericide applications have proved ineffective in controlling blossom blast of fruiting pear. Their effectiveness against the disease in flowering pear has not been studied but is presumed to be poor.

Entomosporium Leaf Spot

Entomosporium mespili

Anthracnose

Colletotrichum gloeosporioides

Two leaf spot diseases, anthracnose and Entomosporium leaf spot, occur sporadically in container- and field-grown flowering pear. Isolated severe outbreaks of either disease can cause extensive defoliation, which then greatly reduces tree vigor, beauty, and value.

Symptoms

Entomosporium leaf spot of pear is characterized by numerous reddish to purple pinpoints on young leaves; the lesions enlarge to form circular, brown spots, sometimes with a yellow halo (Plate 55). On heavily diseased leaves, spots may merge to form large, irregular blotches. Several tiny, black fruiting bodies of the causal fungus, *Entomosporium mespili,* develop in the center of each leaf spot. Severe disease outbreaks often result in early and heavy leaf drop, particularly in the lower portion of the tree canopy. Although young leaves are quickly colonized by *E. mespili,* the mature leaves are highly resistant to infection.

Symptoms of anthracnose are similar to those of Entomosporium leaf spot. Spots first appear on leaves as tiny red flecks and quickly enlarge into brown blotches. Heavily spotted leaves abscise and fall to the ground. Concentric rings (in a target spot pattern, or bull's-eye) of black fruiting bodies of the causal fungus, *Colletotrichum gloeosporioides,* can be seen in the blotches.

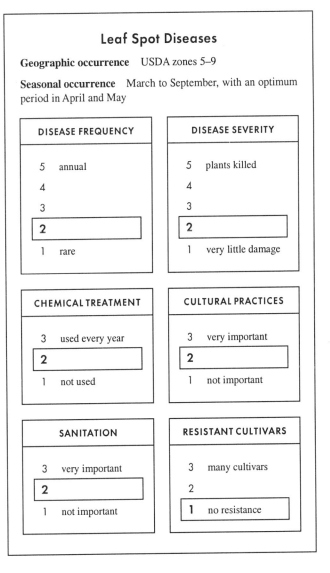

Bacterial Blossom Blast

Geographic occurrence USDA zones 6–9

Seasonal occurrence March to July, with an optimum period in April and May

DISEASE FREQUENCY
- 5 annual
- 4
- 3
- 2
- **1 rare**

DISEASE SEVERITY
- 5 plants killed
- 4
- 3
- **2**
- 1 very little damage

CHEMICAL TREATMENT
- 3 used every year
- 2
- **1 not used**

CULTURAL PRACTICES
- 3 very important
- **2**
- 1 not important

SANITATION
- 3 very important
- **2**
- 1 not important

RESISTANT CULTIVARS
- 3 many cultivars
- 2
- **1 no resistance**

Leaf Spot Diseases

Geographic occurrence USDA zones 5–9

Seasonal occurrence March to September, with an optimum period in April and May

DISEASE FREQUENCY
- 5 annual
- 4
- 3
- **2**
- 1 rare

DISEASE SEVERITY
- 5 plants killed
- 4
- 3
- **2**
- 1 very little damage

CHEMICAL TREATMENT
- 3 used every year
- **2**
- 1 not used

CULTURAL PRACTICES
- 3 very important
- **2**
- 1 not important

SANITATION
- 3 very important
- **2**
- 1 not important

RESISTANT CULTIVARS
- 3 many cultivars
- 2
- **1 no resistance**

Epidemiology

Both fungi survive in small, inconspicuous cankers on one-year-old cankers, and their spores are dispersed by splashing water. The development of anthracnose and Entomosporium leaf spots is favored by several days of warm to hot, cloudy, wet weather. Extended periods of dry weather will suppress the spread and development of both diseases. The anthracnose fungus may be an especially aggressive pathogen on plants weakened by drought, poor management, or improper fertilization.

Host Range

Entomosporium leaf spot has been reported in most members of the apple subfamily (Pomodidae). The most common hosts of *E. mespili* are hawthorn, Indian hawthorn, loquat, and photinia.

The host range of *C. gloeosporioides* includes many widely grown woody trees and shrubs and some perennial and annual flowers. Notable hosts include aucuba, azalea, camellia, English ivy, holly, hydrangea, privet, and southern magnolia.

Management

Effective management of foliar diseases involves the propagation of disease-free cuttings, good cultural practices, sanitation, timely fungicide applications, and production of disease-resistant selections, when available.

Cultural practices For both field- and container-grown pear, surface or drip irrigation is preferred to watering with overhead sprinklers. When an overhead sprinkler system is used, water between 2 and 6 AM or at midday. Avoid overhead watering late in the afternoon or early evening, particularly on a cloudy day or several hours prior to rain. When disease development is most likely during the growing season, be sure that stock containers have been spaced sufficiently to speed the evaporation of water from the foliage and interfere with the movement of the pathogens.

Sanitation Collect cuttings for propagation from disease-free plants only. Discard diseased cuttings. Fallen diseased leaves, which are an important source of inoculum of both *E. mespili* and *C. gloeosporioides,* should be collected from production ranges and discarded, if possible. Avoid establishing or producing blocks of field- or container-grown flowering pear near woody shrubs or trees damaged by leaf spot or anthracnose.

Resistance Differences in the susceptibility of flowering pear selections to Entomosporium leaf spot and anthracnose are not known. However, seedlings of *P. calleryana* have a high level of resistance to Entomosporium leaf spot.

Chemical treatment Timely application of fungicides can effectively control Entomosporium leaf spot and anthracnose of flowering pear. Generally, the occurrence of these diseases in field- and container-grown flowering pear is so sporadic that preventative spray programs are needed only where outbreaks have occurred in a nearby block of trees in past years. Begin fungicide applications at or shortly after budbreak, and continue sprays until shoot growth has stopped and the leaves have matured. New growth on field- and container-grown pear should be regularly inspected for symptoms of leaf spot and anthracnose throughout the spring and early summer. Should either disease appear, immediately begin an intensive fungicide spray program to prevent further spotting of the leaves and defoliation. Once either disease has become established in a block of flowering pear, preventative fungicide sprays will be required each year until the trees are marketed.

Botryosphaeria Canker (Black Rot)

Botryosphaeria dothidea

Botryosphaeria canker (Bot canker, or black rot) has been reported sporadically in flowering pear in the southeastern United States. The causal fungus, *Botryosphaeria dothidea,* has a wide host range, which includes many woody trees and shrubs. It is generally considered an aggressive pathogen only on trees weakened by environmental stress or poor management. Botryosphaeria canker is rare in nursery-grown flowering pear. Other fungi, such as a *Coniothyrium* sp. and *Nectria cinnabarina,* cause similar canker and dieback diseases of flowering pear.

Symptoms

Cankers first appear as elongated dark brown to maroon blotches in the bark along the lateral branches and trunk (Plate 56). Leaves on damaged branches turn yellow and often drop pre-

Botryosphaeria Canker

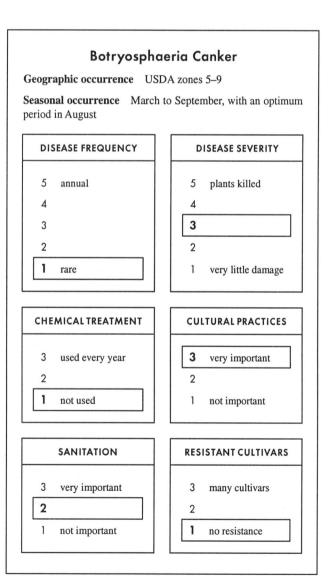

Geographic occurrence USDA zones 5–9

Seasonal occurrence March to September, with an optimum period in August

DISEASE FREQUENCY	DISEASE SEVERITY
5 annual	5 plants killed
4	4
3	**[3]**
2	2
[1 rare]	1 very little damage

CHEMICAL TREATMENT	CULTURAL PRACTICES
3 used every year	**[3 very important]**
2	2
[1 not used]	1 not important

SANITATION	RESISTANT CULTIVARS
3 very important	3 many cultivars
[2]	2
1 not important	**[1 no resistance]**

maturely. The paper-like periderm on the canker surface often peels away from its edge. Numerous pimple-like, black pycnidia of *B. dothidea* may be seen on the surface of the brown, slightly sunken cankers. In flowering pear, browning of the tissues around the canker may extend into the heartwood. Dieback and leaf shed may be so severe that some diseased trees are no longer marketable.

Epidemiology

With its wide host range and saprophytic nature, *B. dothidea* is probably a resident of most woody plant communities around most container and field nurseries or has been introduced into nurseries in diseased plant material. Spores (conidia) of the fungus ooze from fruiting bodies (pycnidia) and are spread by splashing water and possibly by pruning tools. Hyphae from germinating conidia readily invade plants through pruning wounds or other openings in the bark, such as frost cracks and lenticels. For some trees, several months may pass before symptoms are seen. Tissues of trees damaged by cold or drought are much more quickly invaded by *B. dothidea* than those of healthy trees. As a result, cankers on stressed trees are usually much larger than those on unstressed trees.

Host Range

The host range of *B. dothidea* includes plants in nearly 100 genera. Flowering dogwood, redbud, rhododendron, rose, and sweet gum are among the more popular woody tree and shrub hosts of this fungus.

Management

Outbreaks of canker and dieback in container- and field-grown flowering pear can largely be avoided by following good production and nursery sanitation practices.

Cultural practices Adjusting the fertility and pH of potting medium or field soil according to the results of soil fertility assay, watering according to plant needs, timely repotting of container material and other practices which promote plant vigor will greatly reduce the risk of an outbreak of canker or dieback disease induced by *B. dothidea*.

Sanitation Since the causal fungus quickly invades freshly killed plant tissue, immediately collect and discard all branches and twigs pruned from container and field stock. Follow pruning practices which promote rapid healing of wounds. Clean pruning tools with denatured alcohol or a similar disinfectant.

Resistance Given the opportunistic nature of the causal fungus and the role of stress in predisposing plants to infection, resistance to Bot canker is unlikely to occur.

Chemical treatment The activity of fungicides against diseases incited by *B. dothidea* has never been demonstrated in landscape trees and shrubs.

REFERENCES

Bell, R. L., and van der Zwet, T. 1988. Susceptibility of *Pyrus* germplasm to Fabraea leaf spot. Acta Hortic. 224:229–236.

Canfield, M. L., Baca, S., and Moore, L. W. 1986. Isolation of *Pseudomonas syringae* from 40 cultivars of diseased woody plants with tip dieback in Pacific Northwest nurseries. Plant Dis. 70:647–650.

Dirr, M. A. 1998. Manual of Woody Landscape Plants: Their Identification, Ornamental Characteristics, Culture, Propagation and Uses. 5th ed. Stipes Publishing, Champaign, Ill.

Fare, D. C., Gilliam, C. H., and Ponder, H. G. 1991. Fireblight susceptibility, growth and other characteristics in ornamental pears in Alabama. J. Arboric. 17:257–260.

Klarman, W. L. 1968. Coniothyrium canker on ornamental pear. Plant Dis. Rep. 52:792.

Kokalis-Burelle, N., Hagan, A. K., Gazaway, W. S., and Sikora, E. 1997. Fire blight on fruit trees and woody ornamentals. Ala. Coop. Ext. Serv. Circ. ANR-542.

Mansvelt, E. L., and Hattingh, M. J. 1990. Bacterial blossom blast. Page 64 in: Compendium of Apple and Pear Diseases. A. L. Jones and H. S. Aldwinckle, eds. American Phytopathological Society, St. Paul, Minn.

McNiel, R. E., Hartman, J. R., and Dunwell, W. C. 1986. Relative susceptibility of flowering pear (*Pyrus calleryana*) to fire blight (*Erwinia amylovora*) during the severe 1986 epiphytotic in Kentucky. Proc. South. Nurserymen's Assoc. Res. Conf. 31:156–159.

Mullins, J. M., Hagan, A. K., and Morgan-Jones, G. 1985. Bradford pear (*Pyrus calleryana*), a new host of *Botryosphaeria dothidea*. Plant Dis. 69:726.

Sinclair, W. A., Lyon, H. H., and Johnson, W. T. 1987. Diseases of Trees and Shrubs. Comstock Publishing Associates, Cornell University Press, Ithaca, N.Y.

Sutton, T. B. 1990. Bitter rot. Pages 15–16 in: Compendium of Apple and Pear Diseases. A. L. Jones and H. S. Aldwinckle, eds. American Phytopathological Society, St. Paul, Minn.

CHAPTER 42

Margery L. Daughtrey ▪ Cornell University, Riverhead, New York
Sharon von Broembsen ▪ Oklahoma State University, Stillwater

Forsythia Diseases

Geographic production USDA zones 4–8

Family Oleaceae

Genus *Forsythia*

Species		
F. europaea	Albanian forsythia	zone 5
F. × *intermedia*	border forsythia	zones 6–8(9)
F. ovata	early forsythia	zones 4–7
F. suspensa var. *sieboldii*	weeping forsythia	zones 5–8
F. viridissima	greenstem forsythia	zones 5–8

Sclerotinia Canker
Sclerotinia sclerotiorum

Individual branches of forsythia are affected by Sclerotinia canker in wet springs or when plants are frequently watered by overhead irrigation.

Symptoms

Scattered branches wilt suddenly. Under close examination, a sunken canker is evident on each affected branch. The canker is easiest to see at the margin between the diseased and healthy tissue. Signs of the fungal pathogen, *Sclerotinia sclerotiorum,* are often visible in this area: bits of white, fluffy mycelium and black sclerotia, 1/8–1/4 inch in diameter, may be evident, sometimes associated with matted-down petal tissue.

Epidemiology

Presumably ascospores released from apothecia (formed on sclerotia from the previous season) of the pathogen are the source of inoculum for spring infections. Cankers are often located where flowers are attached to the stem, suggesting that infection occurs during or just after flowering. The disease occurs in cool, moist regions worldwide.

Host Range

The pathogen has a wide host range, including *Camellia, Euonymus, Ficus, Hibiscus, Malus, Paeonia, Prunus, Schefflera,* and *Syringa,* in addition to *Forsythia.*

Management

Prune out affected branches, and bury or dispose of them at a distance from the nursery. Sclerotinia canker is usually not serious enough to merit fungicide treatment. Be careful to water early in the day when plants are in flower, and allow good air movement between plants.

Sclerotinia Canker

Geographic occurrence USDA zones 6–8

Seasonal occurrence Mid-April to May, under average yearly conditions in zone 7 (adjust for your hardiness zone)

DISEASE FREQUENCY	DISEASE SEVERITY
5 annual	5 plants killed
4 (boxed)	4
3	**3** (boxed)
2	**2** (boxed)
1 rare	**1** very little damage (boxed)

CHEMICAL TREATMENT	CULTURAL PRACTICES
3 used every year	**3** very important (boxed)
2	2
1 not used (boxed)	1 not important

SANITATION	RESISTANT CULTIVARS
3 very important (boxed)	3 many cultivars
2	2
1 not important	**1** no resistance (boxed)

Stem Gall

***Phomopsis* sp.**

Phomopsis stem gall is a common disease of forsythia. It is often mistaken for crown gall, a bacterial disease which it resembles, but crown gall does not typically cause galls above the root crown of this host.

Another fungal stem gall, caused by *Nectriella pironii,* occurs in forsythia and other shrubs in Florida, whereas Phomopsis gall is more common in the northern United States.

Symptoms

Galls the color of stem tissue form along the stems of infected forsythia. The galls occur anywhere along the stem and take the form of a cluster of nodules pressed tightly together. Multiple galls may cause some dieback.

Epidemiology

The disease is presumably spread by propagation from infected plants.

Host Range

It is not known whether the unidentified *Phomopsis* sp. affecting forsythia is able to cause galls on other species as well. Some other woody species also develop galls believed to be due to infection by unidentified *Phomopsis* spp., including azaleas, highbush blueberry, American elm, hickories, winter jasmine, maples, oaks, common privet, and cranberrybush viburnum.

Management

Avoid propagating from infested plant material. Prune out affected branches on salable stock.

Phytophthora Root and Crown Rot and Dieback

Phytophthora cinnamomi
Phytophthora parasitica

Forsythia is not particularly susceptible to *Phytophthora* spp., but losses from Phytophthora root and crown rot and dieback occur when conditions are favorable for these diseases.

Symptoms

Plants with root rot often are stunted and chlorotic and gradually die back or else wilt suddenly. Examination of underground

Stem Gall

Geographic occurrence USDA zones 6–8

Seasonal occurrence Unknown

DISEASE FREQUENCY	DISEASE SEVERITY
[5 annual]	5 plants killed
4	4
3	3
2	2
1 rare	**[1 very little damage]**

CHEMICAL TREATMENT	CULTURAL PRACTICES
3 used every year	**[3 very important]**
2	2
[1 not used]	1 not important

SANITATION	RESISTANT CULTIVARS
3 very important	3 many cultivars
2	2
[1 not important]	**[1 no resistance]**

Phytophthora Root and Crown Rot

Geographic occurrence USDA zones 4–8

Seasonal occurrence Throughout the growing season, under average yearly conditions in zone 7 (adjust for your hardiness zone)

DISEASE FREQUENCY	DISEASE SEVERITY
5 annual	5 plants killed
4	**[4]**
3	3
[2]	2
1 rare	1 very little damage

CHEMICAL TREATMENT	CULTURAL PRACTICES
3 used every year	**[3 very important]**
[2]	2
1 not used	1 not important

SANITATION	RESISTANT CULTIVARS
[3 very important]	3 many cultivars
2	2
1 not important	**[1 no resistance]**

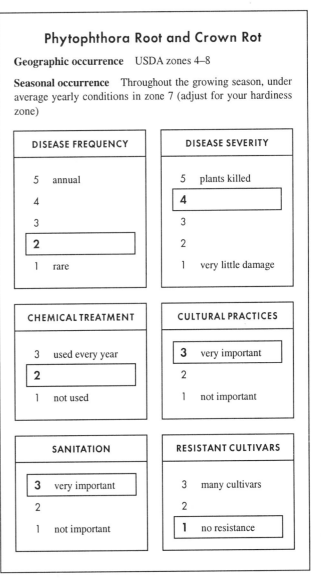

parts discloses decaying roots and sometimes extensive red to brown discoloration of crowns. Infection by *P. parasitica* occasionally causes shoot lesions and dieback.

Epidemiology

Root rot is favored by excessively wet conditions. Shoot infections result from inoculum splashed upward from soil or from application of recycled irrigation water containing spores of *Phytophthora* spp.

Host Range

P. parasitica and *P. cinnamomi* are common pathogens of a wide range of woody ornamentals grown in nurseries (see Chapter 14, "Phytophthora Root Rot and Dieback").

Management

See Chapter 14.

Pseudomonas Blight

Pseudomonas syringae* pv. *syringae

The bacterium *Pseudomonas syringae* pv. *syringae,* first isolated from lilac, attacks many hosts, causing a variety of symptoms, including leaf spots, blossom blight, shoot blight, gummosis, canker, and dieback.

Symptoms

Pseudomonas blight of forsythia is mainly a shoot tip blight and dieback.

Epidemiology

The bacterium overwinters on plant surfaces and in buds and twig cankers. It infects new tip growth in spring, being spread primarily by splashing water.

Host Range

Important hosts of *P. syringae* pv. *syringae* include lilac, citrus, and stone fruits and pome fruits, including ornamental forms, such as flowering almond and flowering pear. It also occurs on flowering and Himalayan ash; trembling aspen; basswood; flowering dogwood; golden-chain tree; hibiscus; saucer magnolia; Japanese, Norway, and red maples; live oak; oleander; Monterey pine; Carolina and black poplars; and willow.

Management

Pseudomonas blight is very difficult to control but can be suppressed by copper fungicides and bactericides applied at the onset of disease and at regular intervals thereafter. Management strategies should be employed to prevent a buildup of copper resistance.

Pseudomonas Blight

Geographic occurrence USDA zones 5–8

Seasonal occurrence April to September, with an optimum period in June–August, under average yearly conditions in zone 7 (adjust for your hardiness zone)

DISEASE FREQUENCY	DISEASE SEVERITY
5 annual	5 plants killed
[4]	**[4]**
3	3
2	2
1 rare	1 very little damage

CHEMICAL TREATMENT	CULTURAL PRACTICES
3 used every year	3 very important
[2]	**[2]**
1 not used	1 not important

SANITATION	RESISTANT CULTIVARS
3 very important	3 many cultivars
[2]	2
1 not important	**[1 no resistance]**

REFERENCE

Sinclair, W. A., Lyon, H. H., and Johnson, W. T. 1987. Diseases of Trees and Shrubs. Comstock Publishing Associates, Cornell University Press, Ithaca, N.Y.

CHAPTER 43

A. R. Chase ▪ Chase Research Gardens, Mt. Aukum, California
Robert L. Wick ▪ University of Massachusetts, Amherst

Gardenia Diseases

Geographic production USDA zones 8–10

Family Rubiaceae

Genus *Gardenia*

Species *G. jasminoides*

Cultivars Fortuniana
Radicans

Gardenia is a small to large shrub grown for its dark green, glossy evergreen foliage and highly fragrant white flowers.

Myrothecium Leaf Spot and Petiole Rot

Myrothecium roridum

Myrothecium leaf spot and petiole rot was first described in greenhouse-grown gardenias destined for a potted flowering crop. Since that time it has been found in most places where gardenias are grown, but rarely in landscape plantings.

Symptoms

Myrothecium leaf spot most frequently appears on wounded areas of leaves, such as the tips and the main veins if they are broken during handling of cuttings. the disease is especially prevalent in rooting cuttings under high humidity and moderately high temperature (25–30°C). The leaf spots are watery and nearly always contain black and white fungal fruiting bodies in concentric rings. These form near the outer edges of spots on the undersides of leaves.

Host Range

M. roridum infects a wide range of plants, including pansy, snapdragon, tomato, violet, and many foliage plants. Isolates of the pathogen are not host-specific.

Management

Cultural practices Newly planted cuttings are especially susceptible to this disease since they are frequently rooted under mist. Avoid wounding leaves, and keep the foliage as dry as possible. Under most conditions, proper cultural practices are sufficient for disease control.

Chemical treatment Several fungicides have been shown to be somewhat effective in controlling Myrothecium leaf spot in a number of ornamental crops.

Phytophthora Stem Rot and Leaf Spot

Phytophthora parasitica

Phytophthora stem rot and leaf spot is common in gardenia in the southeastern United States, where the pathogen, *Phytophthora*

Myrothecium Leaf Spot and Petiole Rot

Geographic occurrence USDA zones 9 and 10; reported in Pennsylvania, probably in the greenhouse

Seasonal occurrence Year-round in greenhouses and in Arizona

DISEASE FREQUENCY	
5	annual
4	
3	
2	
1	rare

DISEASE SEVERITY	
5	plants killed
4	
3	
2	
1	very little damage

CHEMICAL TREATMENT	
3	used every year
2	
1	not used

CULTURAL PRACTICES	
3	very important
2	
1	not important

SANITATION	
3	**very important**
2	
1	not important

RESISTANT CULTIVARS	
3	many cultivars
2	
1	**no resistance**

parasitica, attacks many other woody and herbaceous ornamentals.

Symptoms

This disease occurs primarily in plants grown in or on the ground. Infection results in leaf spots that are initially small and water-soaked, with irregular margins. They may become tan and papery under dry conditions, and their centers may fall out under wet conditions. Stem rot usually begins at the soil line, where the stem becomes soft and watery, and lower leaves turn yellow. Eventually, the area becomes sunken, and a cavity may form, resulting in lodging of the stem.

Epidemiology

Phytophthora is most active in wet, poorly drained soils. The relatively high temperatures in the summer are also conducive to Phytophthora root rot (Plate 58). Splashing irrigation water can spread the pathogen to leaves, resulting in aerial blight.

Host Range

All gardenias and many other subtropical shrubs are susceptible to *P. parasitica.*

Management

Because of similarities between Phytophthora stem rot and leaf spot and several other diseases, diagnosis must be confirmed by a diagnostic laboratory before optimum control strategies can be chosen. *Phytophthora* can become a resident in nursery soils, where it can then develop into a chronic problem for susceptible plants. If it becomes is a persistent problem in the nursery, consider using a protective fungicide. However, do not use a systemic fungicide regularly, or the pathogen may develop resistance to it.

See Chapter 14, "Phytophthora Root Rot and Dieback."

Cultural practices Growing plants away from the soil, which is the natural source of infection, is the best way to avoid this disease. Place containers on several inches of gravel to avoid contamination by splashing rain and by direct contact with soil. Use a well-drained growing medium with composted bark. Use pathogen-free cuttings and new flats and potting media to produce healthy plants.

Chemical treatment Fungicides active against oomycete fungi provide control of Phytophthora stem rot and leaf spot.

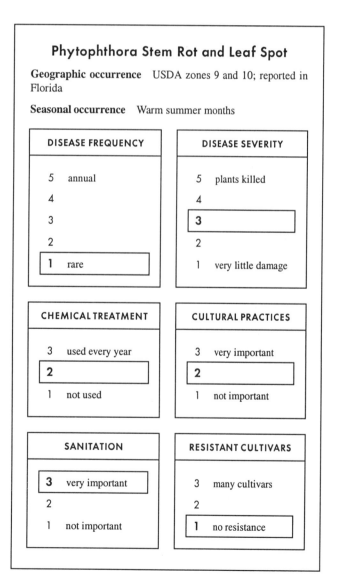

Phytophthora Stem Rot and Leaf Spot

Geographic occurrence USDA zones 9 and 10; reported in Florida

Seasonal occurrence Warm summer months

DISEASE FREQUENCY	DISEASE SEVERITY
5 annual	5 plants killed
4	4
3	**[3]**
2	2
[1] rare	1 very little damage

CHEMICAL TREATMENT	CULTURAL PRACTICES
3 used every year	3 very important
[2]	**[2]**
1 not used	1 not important

SANITATION	RESISTANT CULTIVARS
[3] very important	3 many cultivars
2	2
1 not important	**[1] no resistance**

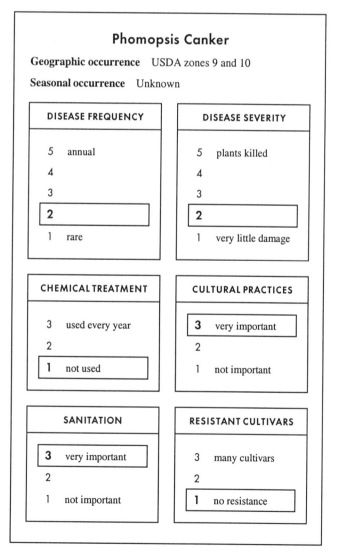

Phomopsis Canker

Geographic occurrence USDA zones 9 and 10

Seasonal occurrence Unknown

DISEASE FREQUENCY	DISEASE SEVERITY
5 annual	5 plants killed
4	4
3	3
[2]	**[2]**
1 rare	1 very little damage

CHEMICAL TREATMENT	CULTURAL PRACTICES
3 used every year	**[3] very important**
2	2
[1] not used	1 not important

SANITATION	RESISTANT CULTIVARS
[3] very important	3 many cultivars
2	2
1 not important	**[1] no resistance**

Phomopsis Canker

Phomopsis gardeniae
(sexual stage, *Diaporthe gardeniae*)

Phomopsis canker was described in 1894 in England and has since been reported in most places where gardenias are grown. The disease is caused by the fungus *Phomopsis gardeniae* (sexual stage, *Diaporthe gardeniae*), which is characterized as an opportunistic pathogen, infecting only hosts that have been wounded. Phomopsis canker of gardenia is generally thought to be a serious disease, however.

Symptoms

Cankers form at the sites of wounds on cuttings. Foliage wilts on a portion of the plant above the infected site. The entire plant may wilt when the infection occurs at the main stem. Stem cankers are circular to oblong and appear slightly browner than healthy areas of the stems. They can enlarge to as much as 5 cm in diameter and can girdle stems. The cankers are rough-textured and may have longitudinal cracks. Abscission of flower buds before they open is sometimes the first sign of an advanced stem canker infection.

Management

Fungicide sprays are sometimes effective against Phomopsis canker.

Rhizoctonia Aerial Blight

Rhizoctonia solani

Rhizoctonia aerial blight of gardenias has not been fully described, but the disease is common in many subtropical ornamentals. Symptoms, diagnosis, and control are similar for most of these woody crops.

Symptoms

Rhizoctonia aerial blight occurs primarily during the summer months. Severe disease can occur in less than a week, so plants should be checked carefully and frequently. Brown, irregularly shaped spots form anywhere in the foliage, but most commonly in the crown of the plant, which stays wet for long periods. Sometimes the first spots form near the top of plant, obscuring the source of inoculum, which is usually the soil. The spots spread rapidly, and the entire plant becomes covered with the brown, weblike mycelium of the pathogen, *Rhizoctonia solani*.

Rhizoctonia Aerial Blight

Geographic occurrence USDA zones 9 and 10, mainly in the southeastern United States

Seasonal occurrence Summer

DISEASE FREQUENCY	DISEASE SEVERITY
5 annual	5 plants killed
4	4
3	**[3]**
[2]	2
1 rare	1 very little damage

CHEMICAL TREATMENT	CULTURAL PRACTICES
3 used every year	3 very important
[2]	**[2]**
1 not used	1 not important

SANITATION	RESISTANT CULTIVARS
[3 very important]	3 many cultivars
2	2
1 not important	**[1 no resistance]**

Xanthomonas Leaf Spot

Geographic occurrence USDA zones 9 and 10, mainly in Florida

Seasonal occurrence Summer

DISEASE FREQUENCY	DISEASE SEVERITY
5 annual	5 plants killed
4	4
3	3
2	2
[1 rare]	**[1 very little damage]**

CHEMICAL TREATMENT	CULTURAL PRACTICES
3 used every year	3 very important
2	**[2]**
[1 not used]	1 not important

SANITATION	RESISTANT CULTIVARS
3 very important	3 many cultivars
[2]	2
1 not important	**[1 no resistance]**

Management

Because of similarities between Rhizoctonia aerial blight and several other diseases, the cause of symptoms must confirmed by a diagnostic laboratory before an optimum control strategy can be chosen.

Cultural practices Grow plants away from the soil, which is the natural source of infection, to avoid this disease. Use only pathogen-free stock for propagation of gardenia.

Chemical treatment Since *R. solani* inhabits the soil, both the roots and the foliage of plants must be treated to provide optimal disease control. Several fungicides give effective control of *Rhizoctonia* on ornamentals.

Xanthomonas Leaf Spot

Xanthomonas campestris* pv. *maculifolii-gardeniae

Xanthomonas leaf spot, caused by the bacterium *Xanthomonas campestris* pv. *maculifolii-gardeniae,* was reported in 1941 and described in an outbreak in California in 1946. It is more common (although still rare) in Florida and other southeastern states today.

Symptoms

Pinpoint-sized, water-soaked spots start on immature leaves. They rapidly turn black and often have a wide yellow margin. The spots enlarge and may be roughly circular or angular, if they expand until they reach a leaf vein. Abscission of severely infected leaves may occur. Unexpanded flower buds are also susceptible to the pathogen.

Host Range

Different cultivars vary in susceptibility, but all appear to be hosts of *X. campestris* pv. *maculifolii-gardeniae.* Some related tropical woody ornamentals, such as *Ixora* spp., are also hosts of this bacterium.

Management

Cultural practices Xanthomonas leaf spot is most severe in warm to hot weather, and plants should be scouted for symptoms especially when the temperature is between 22 and 28°C. Use of pathogen-free plants, roguing symptomatic plants, and minimizing exposure to overhead irrigation and rainfall are critical cultural controls.

Chemical treatment Chemical control of Xanthomonas leaf spot is not feasible. Most plants are infected by inoculum spread by rain or overhead irrigation, and under these conditions the benefits of bactericides are minimal at best.

REFERENCES

Buddin, W., and Wakefield, E. M. 1937. A stem-canker disease of gardenias. Gard. Chron. 101:226–227.

Farr, D. F., Bills, G. F., Chamuris, G. P., and Rossman, A. Y. 1989. Fungi on Plants and Plant Products in the United States. American Phytopathological Society, St. Paul, Minn.

Pirone, P. P. 1940. Diseases of the gardenia. N.J. Agric. Exp. Stn. Bull. 679.

Wick, R. L., Haviland, P., and Tattar, T. 1994. New hosts and associations for *Phytophthora* species. (Abstr.) Phytopathology 84:549.

CHAPTER 44

Ronald K. Jones ▪ North Carolina State University, Raleigh

Steven N. Jeffers ▪ Clemson University, Clemson, South Carolina

Ginkgo Diseases

Geographic production: USDA zones 3–9

Family Ginkgoaceae

Genus *Ginkgo*

Species *G. biloba*

Cultivars several

Ginkgo (maidenhair tree) is a medium-sized to large deciduous tree with a straight trunk, pyramidal shape, and highly variable width. It is considered a very primitive tree because the leaves have an open-ended, fan-type vein pattern, which lacks a midvein. Ginkgo trees are considered relatively disease-free and do not have many serious disease problems.

Phytophthora Root Rot

Phytophthora cinnamomi

Phytophthora root rot of ginkgo seedlings, caused by the fungus *Phytophthora cinnamomi,* was recently confirmed in South Carolina.

Symptoms

Symptoms of Phytophthora root rot of ginkgo are typical of the disease in other hosts. Aboveground, seedlings become chlorotic and turn yellow, wilt, and eventually die. Belowground, the root system lacks feeder roots and has a "rat-tail" appearance. The taproot becomes necrotic and turns dark brown to black, and cortical tissues slough off.

Epidemiology

Phytophthora root rot has been observed only in seedlings of ginkgo. Diseases caused by *P. cinnamomi* commonly occur in heavy, poorly drained soils or where irrigation or rainfall is excessive. Experiments conducted in a greenhouse suggest that periods of prolonged soil saturation are not needed for the disease to occur and that the pathogen is aggressive on plants receiving normal watering.

For more information, see Chapter 14, "Phytophthora Root Rot and Dieback."

Management

Cultural practices Cultural management is most important in avoiding Phytophthora root rot. Avoid soils where *P. cinnamomi* is present. Manage soil water content to avoid periods of prolonged saturation – e.g., avoid low-lying areas where water tends to accumulate, improve drainage, do not overirrigate, and grow seedlings in raised beds.

Resistance The plant may become more resistant to the pathogen as it matures and becomes woody. Because this disease

Phytophthora Root Rot

Geographic occurrence USDA zones 3–9

Seasonal occurrence Periods when soil temperature is above 50°F

DISEASE FREQUENCY	DISEASE SEVERITY
5 annual	**5 plants killed**
4	4
3	3
2	2
1 rare	1 very little damage

CHEMICAL TREATMENT	CULTURAL PRACTICES
3 used every year	**3 very important**
2	2
1 not used	1 not important

SANITATION	RESISTANT CULTIVARS
3 very important	3 many cultivars
2	2
1 not important	**1 no resistance**

has only recently been identified in ginkgo, resistance in cultivars and selections has not been investigated. Variations in the susceptibility of different ginkgo selections is likely.

Chemical treatment Chemicals will reduce the incidence of Phytophthora root rot but will not provide complete control or eliminate the pathogen altogether. It is very difficult to avoid this disease in a susceptible host growing in infested soil – particularly if soil water content is not managed effectively. Fumigate seedbeds prior to planting, and then drench the beds and seedlings with fungicides active against *Phytophthora* spp. before symptoms occur.

Biological control Commercially available products for biological control are not recommended for *Phytophthora* diseases and probably will not provide adequate control of them.

CHAPTER
45

Ned A. Tisserat ▪ Kansas State University, Manhattan
Judith O'Mara ▪ Kansas State University, Manhattan

Hawthorn Diseases

Geographic production USDA zones 3–8

Family Rosaceae

Genus *Crataegus*

Species	*C. crusgalli*	cockspur hawthorn	zones 3–7
	C. laevigata	English hawthorn	zones 4–7
	C. mollis	downy hawthorn	zones 3–6
	C. monogyna	singleseed hawthorn	zones 4–7
	C. phaenopyrum	Washington hawthorn	zones 3–8
	C. viridis	green hawthorn	zones 3–7

Entomosporium Leaf Spot

Diplocarpon mespili

(asexual stage, *Entomosporium mespili*)

Entomosporium leaf spot or leaf blight can cause extensive defoliation of plantings of English hawthorn, particularly the cultivar Paulii. Most other hawthorn species are moderately resistant to the disease. In production areas with extended cool, wet weather during late spring and early summer, hawthorn plantings can become almost completely defoliated by early July. Repeated annual infections can reduce growth and weaken the vigor of the trees.

Symptoms

Leaf symptoms develop in early summer, starting as small, irregular, reddish brown spots on the upper and lower surfaces of leaves. The area bordering the leaf lesions may initially remain green, creating a green island effect. Leaf spots range from 2 to 5 mm in diameter and can reach 1 cm. Raised bumps appear in the centers of the lesions, as the fungal pathogen forms fruiting bodies (acervuli). The asexual stage of the fungus, *Entomosporium mespili,* produces distinctive spores (conidia). The appendages of the cells may look like microscopic insects.

As the disease progresses, the tissue between leaf spots turns yellow. Elongated lesions can occur on leaf veins and tender shoot growth. Coalescing leaf spots may blight the entire leaf, resulting in rapid defoliation. If the weather remains cool and wet through the early part of the summer, susceptible hawthorn trees can lose most of their foliage.

Epidemiology

The pathogen overwinters in its asexual (conidial) stage on dead leaves and young twigs. In the northern part of the United States it can overwinter in both the conidial stage and the sexual (perfect) stage. It is not known how much the perfect stage of the fungus contributes to new infection in the spring. Infection starts in the spring when spores are splashed to developing leaves, and it continues through the growing season as long as the weather remains wet and cool (58–87°F). The optimum temperature for disease development is 68°F. Different strains of the pathogen may be associated with particular hosts.

Entomosporium Leaf Spot

Geographic occurrence USDA zones 3–8

Seasonal occurrence May through September

DISEASE FREQUENCY	DISEASE SEVERITY
5 annual	5 plants killed
4	4
[3]	**[3]**
2	?
1 rare	1 very little damage

CHEMICAL TREATMENT	CULTURAL PRACTICES
3 used every year	**[3] very important**
[2]	2
1 not used	1 not important

SANITATION	RESISTANT CULTIVARS
[3] very important	**[3] many cultivars**
2	2
1 not important	1 no resistance

Host Range

English hawthorn, including the cultivars Paulii and Crimson Cloud, is very susceptible to Entomosporium leaf spot. Most other commercial hawthorns (cultivars of cockspur, Washington, downy, singleseed, and green hawthorn) appear to be resistant. Entomosporium leaf spot has been reported worldwide as a disease of other plants, including apple, crabapple, mountain ash, cotoneaster, firethorn, Indian hawthorn, Yedda hawthorn, loquat, photinia, flowering quinces, serviceberry, stanvesia, and toyon.

Management

In production areas where the weather favors the annual occurrence of Entomosporium leaf spot of English hawthorn, disease management involves both cultural practices and chemical treatment. Rake and destroy leaf litter in the fall. Replace or remove mulches, which may contain diseased leaf debris, prior to budbreak. Avoid overhead irrigation, to help reduce leaf wetness. Begin fungicide applications at budbreak, and continue at regular intervals through the summer.

Hawthorn Rust

Gymnosporangium globosum

Quince Rust

Gymnosporangium clavipes

The fungi that cause hawthorn rust and quince rust spend part of their life cycles on rosaceous hosts, such as hawthorn, flowering crabapple, and apple, and the other part on junipers (the alternate host). These diseases can have a significant impact on hawthorn production in the nursery, especially in locations where a large population of native junipers is nearby. Hawthorn rust causes leaf spotting and premature defoliation, whereas quince rust causes fruit distortion, twig galls, and branch dieback. The damage to juniper is usually minimal, but the disease may be injurious to susceptible cultivars (see Chapter 52, "Juniper Diseases").

Symptoms

Hawthorn rust Initial symptoms appear in mid-May to early June. Infection usually is limited to the leaves and rarely occurs on fruit or twigs. Small, pale yellow spots, approximately 6 mm in diameter, first develop on the upper surfaces of leaves. The spots are slightly raised and eventually turn a shiny yellow orange. Pinpoint-sized, black fruiting structures (pycnia) of the hawthorn rust fungus, *Gymnosporangium globosum,* form in the centers of the leaf spots. In midsummer, off-white to light orange, protruding cylindrical tubes (aecia), 3 mm long, form on the lower surface of the leaf. The aecia may be scattered or grouped in circles. Leaves with multiple spots drop prematurely, and heavily infected trees may be defoliated by late summer.

Quince rust The quince rust fungus, *G. clavipes,* primarily attacks fruit and succulent stem tissue. Leaf lesions are not common. Infected fruit is covered with protruding, off-white aecia of the fungus. The fruit eventually becomes necrotic, dries out, and drops from the tree. Petioles, thorns, and twigs begin to swell and turn orange soon after infection. Spindle-shaped galls, 5–10 cm long and covered with aecia, are eventually formed. By late summer, twig galls expand and girdle stems, resulting in branch dieback. The dead twigs may remain attached to the tree for several years. Occasionally, living rust galls persist on twigs for more than one season.

Epidemiology

G. globosum and *G. clavipes* have complex life cycles, similar to that of the cedar-apple rust fungus (see Chapter 40, "Flowering Crabapple Diseases," and Chapter 52, "Juniper Diseases"). Hawthorn becomes infected in spring, as basidiospores produced by these fungi on juniper are blown to hawthorn foliage and twigs. Infection of hawthorns may continue into early June, as long as basidiospores are released from active juniper galls and the weather is favorable. Aeciospores produced on hawthorn leaves in summer are blown by the wind to junipers, and infection of junipers results in perennial spheroid galls (in hawthorn rust) or spindle-shaped galls (in quince rust).

Host Range

English hawthorn is reported to be resistant to hawthorn rust, but all other species and cultivars appear to be susceptible. Other hosts of *G. globosum* include *Amelanchier, Juniperus, Malus, Pyrus,* and *Sorbus.*

Hawthorn Rust and Quince Rust

Geographic occurrence USDA zones 3–8

Seasonal occurrence Hawthorn infection occurs from bloom through mid-May to early June, and symptoms are apparent through the summer, under average yearly conditions in the Midwest (adjust for your hardiness zone).

DISEASE FREQUENCY	DISEASE SEVERITY
5 annual	5 plants killed
[4]	4
3	**[3]**
2	2
1 rare	1 very little damage

CHEMICAL TREATMENT	CULTURAL PRACTICES
3 used every year	3 very important
[2]	**[2]**
1 not used	1 not important

SANITATION	RESISTANT CULTIVARS
3 very important	3 many cultivars
[2]	**[2]**
1 not important	1 no resistance

Fire Blight

Geographic occurrence USDA zones 3–8

Seasonal occurrence From bloom through midsummer

DISEASE FREQUENCY	DISEASE SEVERITY
5 annual	5 plants killed
4	**[4]**
3	3
[2]	2
1 rare	1 very little damage

CHEMICAL TREATMENT	CULTURAL PRACTICES
3 used every year	**[3] very important**
[2]	2
1 not used	1 not important

SANITATION	RESISTANT CULTIVARS
[3] very important	3 many cultivars
2	**[2]**
1 not important	1 no resistance

There are no reports of commercially available hawthorn species or cultivars with resistance to quince rust. *Crataegus crus-galli* var. *inermis* and *C. phaenopyrum* are particularly susceptible. Other hosts of *G. clavipes* include *Amelanchier, Aronia, Chaenomeles, Cydonia, Juniperus, Malus, Mespilus, Photinia,* and *Pyrus.*

Management

Where possible, eliminate susceptible junipers within a 1- to 3-km radius of hawthorn production. Elimination of the alternate host disrupts the life cycles of the rust fungi. Unfortunately, eradication is often difficult to implement in urban areas and other locations where junipers are abundant. Nevertheless, try to locate hawthorns as far as possible from juniper plantings.

Preventive fungicide applications may be required in locations where hawthorn and quince rusts are perennial problems. The timing of fungicide applications is similar to that for cedar-apple rust of flowering crabapple. The first treatment of hawthorns should be applied when the orange telial galls on junipers become gelatinous (usually at the time of hawthorn bloom), and additional applications should be made at regular intervals to protect newly developing foliage. Fungicide treatment is not required after the juniper galls become dry and inactive, typically in mid-May to early June, in the central United States.

Fire Blight

Erwinia amylovora

Fire blight may be a serious problem in nursery production of many members of the rose family, including hawthorn. For information on symptoms, epidemiology, and management of fire blight, consult Chapter 40, "Flowering Crabapple Diseases."

REFERENCES

Pirone, P. P. 1978. Diseases and Pests of Ornamental Plants. 5th ed. John Wiley & Sons, New York.

Riffle, J. W., and Peterson, G. W. 1986. Diseases of trees in the Great Plains. U.S. Dep. Agric. For. Serv., Rocky Mountain For. Range Exp. Stn. (Ft. Collins, Colo.), Gen. Tech. Rep. RM-129.

Shurtleff, M., and Neely, R. D. 1987. Leaf spot or blight of hawthorn. Report on Plant Diseases. Coop. Ext. Serv., Dep. Plant Pathol., Univ. Ill. Urbana-Champaign, RPD 637.

Sinclair, W. A., Lyon, H. H., and Johnson, W. T. 1987. Diseases of Trees and Shrubs. Comstock Publishing Associates, Cornell University Press, Ithaca, N.Y.

CHAPTER 46

A. R. Chase ▪ Chase Research Gardens, Mt. Aukum, California

Hibiscus Diseases

Geographic production USDA zones 5–10

Family Malvaceae

Genus *Hibiscus*

Species *H. rosa-sinensis* rose-of-China
H. syriacus rose-of-Sharon, althea

Cultivars many

Hibiscus syriacus is a medium-sized evergreen flowering shrub grown in zones 9 and 10 and (as a houseplant) in other zones. *H. rosa-sinensis* is a large, flowering deciduous shrub grown in zones 5–10. This old plant is becoming increasingly popular and is hardy into zone 5.

Pseudocercospora Leaf Spot

Pseudocercospora hibiscina

Pseudocercospora leaf spot, caused by the fungus *Pseudocercospora hibiscina,* has been found primarily in Hawaii but has been reported in most other areas where hibiscus is grown. This disease can easily be confused with bacterial leaf spots.

Symptoms

Leaf spots caused by *P. hibiscina* are usually angular, bordered by leaf veins, and easily seen on both leaf surfaces.

Management

Avoid using stock plants with a history of Pseudocercospora leaf spot, even when they do not appear to be infected. Keep overhead irrigation to a minimum, and promote rapid leaf drying by irrigating early in the day.

Chemical treatment Various fungicides are effective in the control of Pseudocercospora leaf spot of many ornamentals. Be sure to direct the sprays to the lower leaf surfaces, in order to make contact with the spores of the pathogen and reduce disease spread.

Choanephora Blight

Choanephora cucurbitarum
Choanephora infundibulifera

Choanephora blight, caused by the fungi *Choanephora cucurbitarum* and *C. infundibulifera,* has been reported in old flowers from Florida and North Carolina, but it probably occurs wherever hibiscus is grown, as the pathogens are common organisms.

Symptoms

Spent flowers rot very rapidly. Black spores of the pathogen form in the dead flowers and leaves within a day of first symptom expression. Choanephora blight is most common under very hot conditions with high levels of light.

Pseudocercospora Leaf Spot

Geographic occurrence USDA zones 5–10

DISEASE FREQUENCY	DISEASE SEVERITY
5 annual	5 plants killed
4	4
[3]	**[3]**
2	2
1 rare	1 very little damage

CHEMICAL TREATMENT	CULTURAL PRACTICES
3 used every year	3 very important
[2]	**[2]**
1 not used	1 not important

SANITATION	RESISTANT CULTIVARS
3 very important	3 many cultivars
[2]	2
1 not important	**[1] no resistance**

Management

No fungicides are available for the control of *Choanephora* spp. on hibiscus. When the weather is hot and humid and the light level is high, plants should be examined daily and old flowers removed and discarded, to reduce the spread of spores to adjacent plants. *Choanephora* spp. are weak pathogens, which attack only old flowers. If an infected petal falls on a succulent leaf, leaf blight can occur under the blighted petal.

Root Rots

***Fusarium* spp.**
***Phytophthora* spp.**
***Pythium* spp.**
***Rhizoctonia* spp.**

Root rots are usually not a problem in hibiscus once the plants are established in the landscape. Rhizoctonia root rot has been reported in Texas, Florida, and California.

Symptoms

In containers, infected plants are small and unthrifty and can give the appearance of lacking fertilizer. Roots should be examined to determine their health. Rotted roots are mushy and wet and disintegrate when handled. They are usually brown to black. Since these symptoms can be caused by nematodes as well as pathogenic fungi, a laboratory diagnosis is required to determine the best course of treatment.

Management

Keep plants fertilized and irrigated for optimal growth. Extremes of either fertilizing or irrigation can weaken plants, making them more susceptible to disease. Root rots caused by *Pythium* or *Phytophthora* spp. are especially exacerbated in plants that are overwatered or grown in poorly drained potting media or soils that become waterlogged. Use only new or thoroughly cleaned pots and only new potting media. Never start new plants from cuttings showing signs of root loss, mushiness, or stem rot.

Chemical treatment Many fungicides are routinely used to combat root rot pathogens, but since they are not equally effective for all pathogens, the exact cause must be determined prior to starting a treatment program. When applying a soil drench, be sure to use a high-quality potting medium that allows rapid and complete penetration of water.

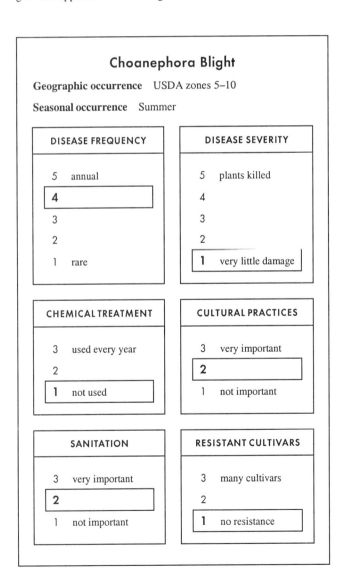

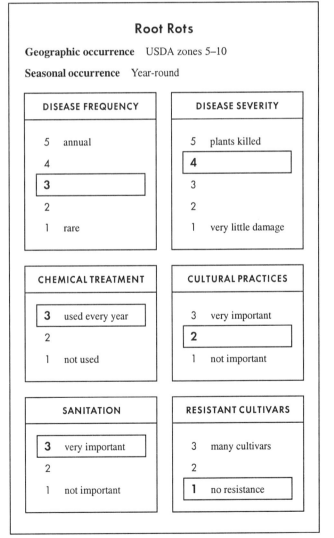

Choanephora Blight

Geographic occurrence USDA zones 5–10

Seasonal occurrence Summer

DISEASE FREQUENCY	DISEASE SEVERITY
5 annual	5 plants killed
[4]	4
3	3
2	2
1 rare	**[1] very little damage**

CHEMICAL TREATMENT	CULTURAL PRACTICES
3 used every year	3 very important
2	**[2]**
[1] not used	1 not important

SANITATION	RESISTANT CULTIVARS
3 very important	3 many cultivars
[2]	2
1 not important	**[1] no resistance**

Root Rots

Geographic occurrence USDA zones 5–10

Seasonal occurrence Year-round

DISEASE FREQUENCY	DISEASE SEVERITY
5 annual	5 plants killed
4	**[4]**
[3]	3
2	2
1 rare	1 very little damage

CHEMICAL TREATMENT	CULTURAL PRACTICES
[3] used every year	3 very important
2	**[2]**
1 not used	1 not important

SANITATION	RESISTANT CULTIVARS
[3] very important	3 many cultivars
2	2
1 not important	**[1] no resistance**

Rust

Kuehneola malvicola

Rust of hibiscus, caused by *Kuehneola malvicola,* is common throughout the southeastern states and in Central and South America and the West Indies.

Symptoms

Rust is easily recognized by the small, orange brown pustules which form on the undersides of leaves or, occasionally, on the upper surface of the leaf. The spots that first appear are white or light yellow when viewed from the upper surface. Premature defoliation can occur in severe infections, weakening the plants.

Host Range

K. malvicola causes rust of *Alcea, Hibiscus,* and *Malvaviscus.* The disease is most common in *H. syriacus* (rose-of-Sharon, or althea).

Management

Removal of dead, infected leaves is crucial to achieving control of rust. *K. malvicola* does not have an alternate host, so weed control is not as critical as it is for some other rust diseases of ornamentals. In the landscape, this may be sufficient for control, without chemical treatment.

Chemical treatment Rust diseases are inhibited by frequent rainfall or irrigation and high temperatures. Fungicides for rust control should be applied before symptoms develop in plantings with a history of the disease.

Bacterial Leaf Spots

Pseudomonas cichorii
Pseudomonas syringae* pv. *hibisci
Xanthomonas campestris* pv. *malvacearum

Bacterial leaf spots have been found primarily in Florida, although *Pseudomonas syringae* has been reported in California.

Symptoms

Foliar infection of hibiscus starts as pinpoint-sized, water-soaked areas, which rapidly enlarge (Plate 59). They tend to remain confined to areas between leaf veins and sometimes have a bright yellow border. Leaf drop is common in severe infections.

Rust

Geographic occurrence USDA zones 5–10

Seasonal occurrence Summer

DISEASE FREQUENCY
- 5 annual
- **4**
- 3
- 2
- 1 rare

DISEASE SEVERITY
- 5 plants killed
- 4
- 3
- **2**
- 1 very little damage

CHEMICAL TREATMENT
- 3 used every year
- 2
- **1 not used**

CULTURAL PRACTICES
- 3 very important
- 2
- **1 not important**

SANITATION
- 3 very important
- **2**
- 1 not important

RESISTANT CULTIVARS
- 3 many cultivars
- 2
- **1 no resistance**

Bacterial Leaf Spots

Geographic occurrence USDA zones 5–10

Seasonal occurrence Rainy periods when temperatures are between 60 and 92°F

DISEASE FREQUENCY
- 5 annual
- **4**
- 3
- 2
- 1 rare

DISEASE SEVERITY
- 5 plants killed
- 4
- **3**
- **2**
- **1 very little damage**

CHEMICAL TREATMENT
- 3 used every year
- **2**
- 1 not used

CULTURAL PRACTICES
- **3 very important**
- 2
- 1 not important

SANITATION
- **3 very important**
- 2
- 1 not important

RESISTANT CULTIVARS
- 3 many cultivars
- **2**
- 1 no resistance

Host Range

The bacteria causing leaf spot diseases of hibiscus are *P. cichorii, P. syringae* pv. *hibisci,* and *Xanthomonas campestris* pv. *malvacearum.* All hibiscus species tested have been found susceptible to these pathogens, although some cultivars are more resistant than others. Some cultivars show resistance to one bacterial leaf spot pathogen but high susceptibility to another. Unless the exact identity of the pathogen is determined, information on cultivar resistance is not useful.

Management

Minimize foliar wetting through irrigation or rainfall, to reduce spread of bacteria and their ability to infect. Bacterial leaf spot diseases are more of a problem during warm, wet periods of the year. Never use infected plants for cuttings, even when the cuttings appear healthy, since latent infections will continue to develop as the cuttings root.

Bacterial leaf spot diseases are significantly affected by temperature. Optimum temperatures for *P. cichorii* are 70–81°F; for *P. syringae,* 60–65°F; and for *X. campestris,* 75–92°F.

Xanthomonas leaf spot is most severe in plants receiving appropriate amounts of fertilizer. Plants receiving inadequate or excess fertilizer are less susceptible to the pathogen.

Chemical treatment Preventative applications of compounds containing copper are sometimes helpful but cannot control bacterial leaf spot diseases under most growing conditions. In hibiscus, Xanthomonas leaf spot appears harder to control with copper than either of the Pseudomonas leaf spots.

Phytophthora Leaf Spot

Phytophthora parasitica

Phytophthora leaf spot, caused by *Phytophthora parasitica,* has been found primarily in Florida. It has also been reported in Hawaii and Texas.

Phytophthora Leaf Spot

Geographic occurrence USDA zones 5–10

Seasonal occurrence Spring

DISEASE FREQUENCY	DISEASE SEVERITY
5 annual	5 plants killed
4	**4**
3	3
2	2
1 rare	1 very little damage

CHEMICAL TREATMENT	CULTURAL PRACTICES
3 used every year	3 very important
2	**2**
1 not used	1 not important

SANITATION	RESISTANT CULTIVARS
3 very important	3 many cultivars
2	**2**
1 not important	1 no resistance

Symptoms

The initial symptoms of foliar infection of hibiscus are water-soaked, black areas, which can rapidly enlarge. They are usually irregular in shape and extend across leaf veins as they enlarge. Leaf drop is common in severe infections.

Epidemiology

Phytophthora leaf spot is reported to be most severe at temperatures between 68 and 72°F.

Host Range

All hibiscus species tested have been found susceptible to *P. parasitica,* although some cultivars are more resistant than others. Painted Lady and Pink Versicolor are highly susceptible; Brilliant Red and Senorita are moderately susceptible; and American Beauty is slightly susceptible.

Management

Minimize foliar wetting through irrigation or rainfall to reduce spread of the pathogen and its ability to infect. Phytophthora leaf spot is more of a problem during the warm wet periods of the year. Never use infected plants for cuttings even when the cuttings appear healthy since latent infections will continue to develop as the cuttings root.

Chemical treatment Preventative applications of several fungicides have been moderately successful in controlling Phytophthora leaf spot. See Chapter 14, "Phytophthora Root Rot and Dieback."

CHAPTER 47

Jay W. Pscheidt ▪ Oregon State University, Corvallis
Robert L. Wick ▪ University of Massachusetts, Amherst
D. Michael Benson ▪ North Carolina State University, Raleigh

Holly Diseases

Geographic production USDA zones 3–10

Family Aquifoliaceae

Genus *Ilex*

Species	*I. aquifolium*	English holly	zones 6–9
	I. cornuta	Chinese holly	zones 7–9
	I. crenata	Japanese holly	zones 5–8
	I. glabra	inkberry	zones 4–9
	I. opaca	American holly	zones 5–9
	I. verticillata	common winterberry	zones 3–9
	I. vomitoria	yaupon holly	zones 7–10

There are many species and cultivars of holly, with a wide range of growth habits and forms. Hollies are utilized extensively in the landscape as foundation plants or specimen individuals, and they can be grown in a wide range of hardiness zones. Several diseases affect hollies in the nursery, including black root rot, Rhizoctonia web blight, nematode decline, Phytophthora blight, and leaf spot diseases.

Black Root Rot

Thielaviopsis basicola

Black root rot of Japanese holly was first reported in 1976 in Virginia. *Ilex crenata* cvs. Compacta, Convexa, Helleri, Highlander, Microphylla, and Rotundifolia are known to be susceptible. The disease is very destructive in liners, but infected landscape plants may remain symptomless for years.

Symptoms

The foliage of liners and young plants in containers may appear dull, off-color, and desiccated (Plate 60). Yellowing and defoliation occurs after most of the root system has been lost. Larger plants may not show symptoms, or they may grow slowly, lose color, or die back, depending on the extent of root rot. Infected feeder roots are dark brown to black (Plate 61). The dark brown to black lesions are the result of pseudoparenchymatous hyphae and chlamydospores of the pathogen.

Epidemiology

The pathogen, *Thielaviopsis basicola,* may be endemic in some soils, but it is likely that infected nursery plants are a major vehicle for its dissemination. *T. basicola* is commonly found in *Ilex* in nurseries. It has a wide host range, but only a few woody plants have been reported as hosts. The pathogen is soilborne and survives as chlamydospores for several years in the absence of a host.

Black Root Rot

Geographic occurrence USDA zones 4–9, possibly

Seasonal occurrence Most severe during the cooler months of the growing period, with an optimum period in April to May, under average yearly conditions in zone 8 (adjust for your hardiness zone)

DISEASE FREQUENCY	DISEASE SEVERITY
5 annual	**5 plants killed**
4	**4**
3	**3**
2	**2**
1 rare	1 very little damage

CHEMICAL TREATMENT	CULTURAL PRACTICES
3 used every year	**3 very important**
2	2
1 not used	1 not important

SANITATION	RESISTANT CULTIVARS
3 very important	3 many cultivars
2	2
1 not important	**1 no resistance**

Black root rot is more severe when the soil is cool and at pH 6–7. The optimum soil temperature for disease is 63–73°F. The effect of soil moisture on disease development is not known, but observations indicate that wetter soils are more conducive to the disease than drier soils.

Management

Avoid propagating hollies from plants that have bare ground under them, as splashing water could introduce spores to the foliage. Use only soilless mix for propagation, and follow the guidelines for avoiding cutting rot, as described in Chapter 12, "Damping-Off of Seeds and Seedlings and Cutting Rot." Liners received from another nursery should be examined for evidence of black root rot, and infected plants should not be accepted. Liners that develop disease should be discarded, and contaminated soil and plant debris should be removed. Avoid excessive lime, and maintain pH at 5.2 to 5.5. Fungicides will control the disease.

Rhizoctonia Web Blight

Rhizoctonia solani

Rhizoctonia web blight is common in container-grown Japanese hollies when plant canopies are crowded together (Plates 62 and 63). It develops in the summer, during periods of rainy, humid weather. Severe defoliation due to web blight may make plants unsalable.

Symptoms

Infected leaves abscise readily. Leaves fallen on the container surface may be the first symptom observed. Some abscised leaves are bound to the stem by mycelium of the pathogen, *Rhizoctonia solani.* Severely defoliated plants look leggy. Until tightly spaced plants are separated, the disease may go unnoticed.

Epidemiology

The disease is favored by rainy, humid weather in the summer. See Chapter 17, "Rhizoctonia Web Blight."

Host Range

Rhizoctonia web blight is common on Japanese hollies, azaleas, and cotoneasters.

Management

Spacing plant containers to provide adequate air movement around and in the plant canopy is a primary cultural practice. Fungicides can be used, but cultural practices should be the first line of defense against web blight.

Nematode Decline

***Criconemella xenoplax*, ring nematode**
***Meloidogyne arenaria*, root-knot nematode**
***Meloidogyne incognita*, southern root-knot nematode**
***Tylenchorhynchus claytoni*, stunt nematode**

Field-grown hollies, particularly Japanese holly, are very susceptible to root-knot nematodes (Plate 64). Nematodes are typically more damaging in light-textured, sandy soils with low moisture-holding capacity, but they also occur in loam and clay soils. Nematodes may cause severe foliar symptoms during periods of drought, when plants are already under stress due to the reduced availability of water.

Symptoms

The first aboveground symptom of infection by ring, root-knot, and stunt nematodes is generally chlorosis of leaves. In severe infections, leaves may turn necrotic and defoliate. However, in some cultivars that tolerate nematode infection, a general lack of new growth may be the only aboveground symptom; infected plants are smaller than plants without nematodes.

Belowground, however, ring and stunt nematodes cause lesions and necrosis on roots, and root-knot nematodes cause the formation of root galls.

Galls formed as a result of feeding by root-knot nematodes may be numerous on severely infected plants. The galls are small, initially not much larger than the diameter of the root, but grow as they develop, and can reach a half inch in diameter at multiple feeding sites with several females. The gall tissue tends to crush the water-conducting cells of the root system and therefore restricts the ability of the root system to take up water from the soil. The damage is exacerbated during dry weather.

Rhizoctonia Web Blight

Geographic occurrence Primarily in the Gulf Coast and southeastern states

Seasonal occurrence June to August during rainy, humid weather

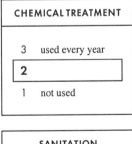

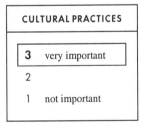

DISEASE FREQUENCY	DISEASE SEVERITY
5 annual	5 plants killed
4	4
3	3
[2]	**[2]**
1 rare	1 very little damage

CHEMICAL TREATMENT	CULTURAL PRACTICES
3 used every year	**[3] very important**
[2]	2
1 not used	1 not important

SANITATION	RESISTANT CULTIVARS
3 very important	3 many cultivars
[2]	2
1 not important	**[1] no resistance**

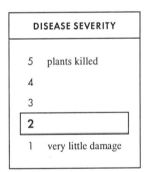

Table 47.1 Responses of selected holly cultivars to ring, root-knot, and stunt nematodes

Cultivar	Response[a] to: Ring nematode	Root-knot nematode	Stunt nematode
Chinese holly			
Burfordi	T	T	T
Rotunda	S	S	S
Japanese holly			
Compacta	S	HS	T
Convexa	S	HS	T
Helleri	S	HS	S
Rotundifolia	S	HS	S
Yaupon holly			
Nana	T	T	T

[a] HS = highly susceptible (severe stunting, twig dieback, and death). S = susceptible (some stunting but usually acceptable growth). T = tolerant (no apparent damage to plants).

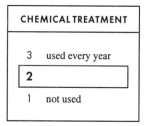

Nematode Diseases

Geographic occurrence Primarily in the Gulf Coast and southeastern states, Arizona, and California

Seasonal occurrence Nematodes are persistent on plants year-round once infection has occurred, but symptoms often develop during hot, dry periods

DISEASE FREQUENCY	DISEASE SEVERITY
5 annual	5 plants killed
4	**[4]**
[3]	3
2	2
1 rare	1 very little damage

CHEMICAL TREATMENT	CULTURAL PRACTICES
3 used every year	**[3]** very important
[2]	2
1 not used	1 not important

SANITATION	RESISTANT CULTIVARS
3 very important	3 many cultivars
[2]	**[2]**
1 not important	1 no resistance

Epidemiology

Ring and stunt nematodes lay eggs in soil around roots. Root-knot nematodes reproduce from eggs laid in egg masses attached to an encysted female on the root surface. Eggs are dispersed by mechanical means in cultivation and by the lateral movement of water in the soil profile. Ring and stunt nematode eggs develop into larvae, which begin feeding at the root surface. Root-knot nematode larvae begin feeding by burrowing into roots, between the root cells. Female root-knot nematodes eventually become nonmotile and develop pear-shaped bodies as they continue to feed at specialized giant cells formed at their feeding sites. Giant cells and the subsequent abnormal division of nearby cells create the galls observed on infected roots. See Chapter 13, "Nematode Diseases."

Host Range

Ring, root-knot, and stunt nematodes attack a wide variety of field, vegetable, and fruit crops as well as woody ornamentals. Japanese holly is the most susceptible among the hollies. Several cultivars of Chinese holly are tolerant or resistant (Table 47.1).

Management

Avoid fields known to be severely infested with nematodes. If infested fields must be used, preplant treatment with one of several broad-spectrum fumigants or nematicides may be necessary. Postplant nematicides are generally unavailable or ineffective. Selection of nematode-resistant hollies for field production or shifting susceptible cultivars to container production with soilless mixes may be the best alternative when ring, root-knot, or stunt nematodes are present in field soils.

Phytophthora Leaf and Twig Blight

Phytophthora ilicis

Phytophthora leaf and twig blight, caused by *Phytophthora ilicis,* has been a problem primarily in the Pacific Northwest, in English holly orchards used for cut greenery during the Christmas holidays.

Symptoms

The disease is characterized by purple black spots on leaves and berries, defoliation, and twig cankers. Leaf spotting frequently begins at the leaf margins and on the lower part of the tree in late fall. The disease progresses up the tree during the winter. Twigs die back, and cankers develop on larger stems later in winter. Infection may occur at a berry cluster, and the pathogen may girdle and kill the twig. Young plants infected in nursery beds may be defoliated and killed.

Epidemiology

The pathogen produces oospores in leaf spots and the cortex of dead twigs, and presumably it survives in this form from one season to the next. Pruning wounds and spine punctures, created by wind-whipped leaves, are ideal sites for new infections. Zoospores and sporangia can also infect through leaf scars. The disease develops in cool, rainy weather and subsides in the warm, dry summer. It may be very severe in dense orchards where air circulation is poor. In some orchards on the Oregon coast, the disease has caused almost complete defoliation and severe twig blighting, particularly on lower limbs. Defoliation is related to the production of ethylene by infected plant tissue.

Host Range

P. ilicis has been reported only on English holly (*I. aquifolium*).

Management

Before planting, select a site that permits good air drainage. Space and prune trees to permit good air circulation. Select cultivars that have resistance. In established orchards, practice good weed and bramble control to help air circulation. Remove and destroy fallen leaves and dead infected twigs during the summer.

Chemical control can be implemented, but make the first application before the fall rains begin. Additional applications after severe winter storms or after harvest may be useful. Fixed copper can also be used as a preventive postharvest dip. Dip tanks should have a copper concentration of 30 ppm. The amount of copper depends on the copper formulation, tank size, and pH of the water. Dip cuttings promptly after harvest, and change the tank water frequently.

Phytophthora Leaf and Twig Blight

Geographic occurrence Reported only in Oregon nurseries

Seasonal occurrence October to March, with an optimum period in November, under average yearly conditions in Oregon (adjust for your hardiness zone)

DISEASE FREQUENCY	DISEASE SEVERITY
5 annual	5 plants killed
4	**4**
3	**3**
2	2
1 rare	1 very little damage

CHEMICAL TREATMENT	CULTURAL PRACTICES
3 used every year	**3 very important**
2	2
1 not used	1 not important

SANITATION	RESISTANT CULTIVARS
3 very important	3 many cultivars
2	**2**
1 not important	1 no resistance

Scab

Sclerophoma sp.

Scab has been a sporadic problem in the Pacific Northwest, in English holly orchards used for cut greenery during the Christmas holidays.

Symptoms

Irregular, translucent to reddish black, swollen areas form on leaves of English holly. The symptoms appear mainly on the undersides of leaves, but in some cultivars they appear on both sides. The leaf spots are the result of cellular proliferation in the leaf mesophyll. The epidermal layers generally remain intact. The symptoms may resemble edema.

Epidemiology

The fungus causing scab, a *Sclerophoma* sp., does not sporulate readily in plant tissue under field or laboratory conditions, but it sporulates on culture plates after exposure to ultraviolet light. In inoculated plants, symptoms developed three months after inoculation. Healthy plants placed in the field in June developed scab by September.

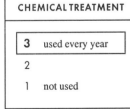

Scab

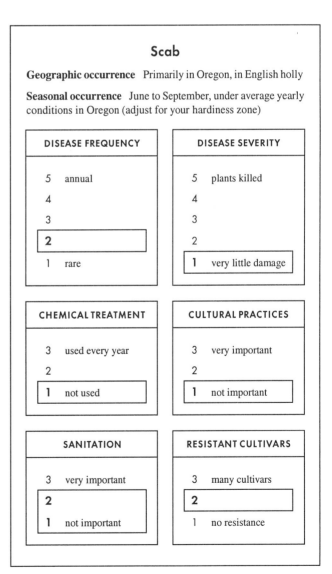

Geographic occurrence Primarily in Oregon, in English holly

Seasonal occurrence June to September, under average yearly conditions in Oregon (adjust for your hardiness zone)

DISEASE FREQUENCY	DISEASE SEVERITY
5 annual	5 plants killed
4	4
3	3
2	2
1 rare	**1 very little damage**

CHEMICAL TREATMENT	CULTURAL PRACTICES
3 used every year	3 very important
2	2
1 not used	**1 not important**

SANITATION	RESISTANT CULTIVARS
3 very important	3 many cultivars
2	**2**
1 not important	1 no resistance

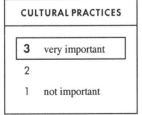

Host Range

The disease has been reported only on English holly (*I. aquifolium*).

Management

No management program for the control of scab is known. A broad-spectrum fungicide may be effective when applied to developing shoots in the spring.

Other Diseases

Cylindrocladium leaf spot, caused by *Cylindrocladium avesiculatum,* can cause defoliation of container-grown hollies under humid, wet conditions. Defoliation is sometimes followed by twig dieback. Several hollies, including *I. cornuta, I. crenata, I. opaca,* and *I. vomitoria,* are known to be susceptible to this disease. The initial symptoms are small, chlorotic leaf spots, which turn purplish black with a light green border. Disease management strategies include avoiding excess overhead irrigation, adopting drip irrigation for highly susceptible cultivars, and applying fungicide sprays when susceptible cultivars are grown under conditions that are highly favorable for the disease.

In *I. opaca* and *I. opaca* × *I. cassini* infected with *Sphaeropsis tumefaciens,* witches'-brooms and galls form on leaves, and foliage dies back. This fungal pathogen is transmitted on pruning tools and by cuttings taken from infected stock plants. Thus, sanitation of propagating tools is important in the control of the disease.

Field-grown Japanese holly can be infected by *Phytophthora cinnamomi* in soils that are particularly poorly drained. Feeder roots are infected, and eventually dark streaks extend up the crown and lower stem. Plants exhibit poor growth, and foliage may be chlorotic. See Chapter 14, "Phytophthora Root Rot and Dieback."

REFERENCES

Aycock, R., Barker, K. R., and Benson, D. M. 1976. Susceptibility of Japanese holly to *Criconemoides xenoplax, Tylenchorhynchus claytoni,* and certain other plant-parasitic nematodes. J. Nematol. 8:26–31.

Barker, K. R., and Benson, D. M. 1977. Japanese hollies : Intolerant host of *Meloidogyne arenaria* in microplots. J. Nematol. 9:330–334.

Barker, K. R., Benson, D. M., and Jones, R. K. 1979. Interaction of Burfordi, Rotunda, and Dwarf Yaupon hollies and *Aucuba* with selected plant parasitic nematodes. Plant Dis. Rep 63:113–116.

Benson, D. M., and Barker, K. R. 1982. Susceptibility of Japanese boxwood, dwarf gardenia, Compacta (Japanese) holly, Spiny Greek and Blue rug junipers, and nandina to four nematode species. Plant Dis. 66: 1176–1179.

Brown, W. M. 1970. A leaf spot disease of English holly caused by a species of *Sclerophoma.* Phytopathology 60:1144–1145.

Buddenhagen, I. W., and Young, R. A. 1957. A leaf and twig disease of English holly caused by *Phytophthora ilicis* n. sp. Phytopathology 47: 95–101.

Chase, A. R. 1991. Characterization of *Rhizoctonia* species isolated from ornamentals in Florida. Plant Dis. 75:234–238.

Frisina, T. A., and Benson, D. M. 1987. Characterization and pathogenicity of binucleate *Rhizoctonia* spp. from azaleas and other woody ornamental plants with web blight. Plant Dis. 71:977–981.

Lambe, R. C., and Wills, W. H. 1976. Thielaviopsis root-rot of Japanese holly. (Abstr.) Proc. Am. Phytopathol. Soc. 3:264.

Lambe, R. C., and Wills, W. H. 1977. Further incidence of black root rot of Japanese holly. (Abstr.) Proc. Am. Phytopathol. Soc. 4:217.

Lambe, R. C., Wills, W. H., and Bower, L. A. 1979. Susceptibility of some *Ilex* species to *Thielaviopsis basicola.* Southern Nurserymen's Res. J. 6:8–13.

Pscheidt, J. W., and Ocamb, C. M. 2001. Pacific Northwest Plant Disease Control Handbook. Oregon State University Extension Service, Corvallis.

Wick, R. L. 1987. Occurrence of *Thielaviopsis basicola* and phytopathogenic nematodes on healthy and declining landscape-grown *Ilex crenata* 'Helleri.' J. Environ. Hortic. 5:131–133.

CHAPTER
48

James L. Sherald ▪ National Park Service, Washington, D.C.

Honeylocust Diseases

Geographic production USDA zones 3–9

Family Fabaceae

Genus *Gleditsia*

Species	*G. japonica*	Japanese honeylocust
	G. tricanthos var. *inermis*	thornless honeylocust

Honeylocust has many desirable characteristics and is commonly used in street plantings, shopping malls, and home landscapes. The tree is so popular that, in the opinion of some, it is being planted to excess. As its production and use have increased, several diseases, some serious, have become apparent. The most serious problems are two canker diseases, caused by *Thyronectria austroamericana* and *Nectria cinnabarina.*

Thyronectria Canker

Thyronectria austroamericana

Nectria Canker

Nectria cinnabarina

Symptoms

Thyronectria canker *Thyronectria austroamericana* causes canker in honeylocust from Colorado to Massachusetts and south to the Gulf Coast. Thin, chlorotic foliage, early leaflet drop, failure of spring leaf development, and dieback are all early indications of infection by this fungus.

These symptoms are caused by the development of cankers, which are flattened, sunken, oval areas on the undersides of branches or along the trunk. Cankers are usually associated with wounds or branch stubs. A canker may start as small as a pinhead but can expand quickly, sometimes six times as fast longitudinally as in circumference. Elongated trunk cankers become so pronounced that they take on the appearance of lightning strikes. Early in development the canker surface is sunken, and the bark turns yellow orange. As the canker ages, the bark sloughs off, the color fades, and decay fungi invade exposed wood. Beneath the canker the sapwood develops a wine red or yellowish stain, and a gum exudate sometimes forms in the vicinity of the canker.

Callus develops along the margins of cankers. The formation of callus varies, but it is particularly prominent in vigorous trees. It generally does not form at the ends of cankers, where expansion is most rapid. Cankers can coalesce and girdle the trunk, killing the tree in a few years, or they may girdle branches or affect large sections of the trunk.

Fruiting bodies of the pathogen form a year or less after infection. The fruiting bodies, both sexual and asexual, are produced in compact masses of hyphae called stromata. Stromata are visible with the naked eye; they look like little cushions emerging through the lenticels in active cankers and on dead wood.

In *Gleditsia japonica* (Japanese honeylocust), *T. austroamericana* does not cause cankers, but rather causes a vascular wilt,

Canker Diseases

Geographic occurrence USDA zones 4–8

Seasonal occurrence Throughout the year

DISEASE FREQUENCY	DISEASE SEVERITY
[5 annual]	5 plants killed
4	**[4]**
3	3
2	2
1 rare	1 very little damage

CHEMICAL TREATMENT	CULTURAL PRACTICES
3 used every year	**[3 very important]**
2	2
[1 not used]	1 not important

SANITATION	RESISTANT CULTIVARS
[3 very important]	3 many cultivars
2	**[2]**
1 not important	1 no resistance

which quickly kills the tree. The acute virulence of *T. austroamericana* in *G. japonica,* in contrast to the more chronic canker pathogenesis in the native *G. tricanthos* var. *inermis,* suggests that the pathogen is native to the United States.

Nectria canker Honeylocust canker caused by *Nectria cinnabarina* has been reported in Minnesota, Michigan, Alabama, and South Carolina. Like Thyronectria canker, this disease can cause significant dieback and even death. Macroscopically, cankers caused by *Nectria* are very similar to those caused by *T. austroamericana.* They initially appear as sunken brown to reddish brown areas in the bark. The cankers are usually annual and, if the host is vigorous, callus tissue can cover the canker. However, in trees under stress, cankers will expand and coalesce, girdling branches and trunks. As in cankers caused by *T. austroamericana,* decay fungi often develop in exposed wood.

Nectria canker is also known as coral spot, with reference to the coral pink asexual fruiting structures (sporodochia) of the pathogen, which emerge from lenticels and cracks in cankers. Sporodochia are present throughout the year but are most prevalent during wet weather. As they age, they turn dark brown or black. Round, red sexual fruiting structures (perithecia), also develop in cankers and are visible with the naked eye.

Since the symptoms of Thyronectria canker and Nectria canker and the fruiting structures of the pathogens are macroscopically similar, these diseases can only be distinguished if the pathogens have been identified microscopically.

Epidemiology

T. austroamericana and *N. cinnabarina* are ubiquitous saprophytes and not particularly aggressive pathogens. Water, wind, and possibly pruning tools and other equipment spread spores of these fungi. They enter plants through fresh wounds, sunburned bark, pruning cuts, branch stubs, and other openings. Insects, particularly the wood-boring beetles (*Agrilus* spp.) have been suspected as vectors of *T. austroamericana,* but their role has not been demonstrated. Monocultures of closely planted trees in the nursery or in the landscape may encourage the spread of the pathogen.

Generally, canker expansion is promoted by stress caused by root damage, moisture and nutrient deficiencies, temperature extremes, and other stress factors. Rapid wound closure, a consequence of vigorous growth, reduces susceptibility and restricts canker expansion.

Many cultivars of *G. tricanthos* var. *inermis* are not particularly cold-hardy, and canker infection may increase susceptibility to cold injury.

Host Range

Thyronectria canker and Nectria canker affect cultivars of *G. tricanthos* var. *inermis.*

Management

Cultural practices and sanitation Both *T. austroamericana* and *N. cinnabarina* enter plants through wounds, and therefore care should be taken to minimize wounding in cultivation, handling, and other practices. Pruning should be performed when maximum closure can be expected but should be avoided in damp weather. Tools should be disinfected between prunings of different trees. Every effort should be made to maintain the plants in a healthy condition and avoid unnecessary stress. If infection does occur, vigorous trees will be able to contain the infection. Since both pathogens survive in dead tissue, infected plant material should be removed and destroyed.

Nursery field management practices may also affect canker infection. The infection rate for *N. cinnabarina* was found to be greater in fields where the soil is kept bare by herbicides or cultivation than in fields where trees were grown with cover crops. The higher incidence and severity in bare soil may be associated with root injury caused by cultivation or with winter injury due to delayed cold acclimation as a result of temperature extremes in bare soil.

Resistance Of the many cultivars of *G. tricanthos* var. *inermis,* only a few have been tested for resistance to *T. austroamericana,* and studies have not been consistent, possibly because of differences in location and methodology. Generally, of the cultivars tested, Sunburst is the most susceptible to *T. austroamericana.*

In a study conducted in Kentucky, the cultivars Rubylace, Shademaster, and Sunburst were found to be particularly susceptible, Imperial was variable in susceptibility, and Skyline and Trusdale were less susceptible. In another evaluation performed in Illinois, cankers formed most frequently on Sunburst, less on Moraine and Skyline, and least on Imperial, Shademaster, and Halka. In Colorado, Sunburst was found to be more susceptible than Imperial and Skyline.

Chemical treatment Fungicides are not recommended for management of canker diseases of honeylocust.

REFERENCES

Bedker, P. J., and Wingfield, M. J. 1983. Taxonomy of three canker-causing fungi of honey locust in the United States. Trans. Br. Mycol. Soc. 81:179–183.

Bedker, P. J., Blanchette, R. A., and French, D. W. 1982. *Nectria cinnabarina:* The cause of a canker disease of honey locust in Minnesota. Plant Dis. 66:1067–1070.

Calkins, J. B., and Swanson, B. T. 1997. Susceptibility of 'Skyline' honeylocust to cankers caused by *Nectria cinnabarina* influenced by nursery field management system. J. Environ. Hortic. 15:6–11.

Hudler, G. W., and Oshima, N. 1976. The occurrence and distribution of *Thyronectria austro-americana* on honeylocust in Colorado. Plant Dis. Rep. 60:920–922.

Jacobi, W. R. 1989. Resistance of honeylocust cultivars to *Thyronectria austro-americana.* Plant Dis. 73:805–807.

Neely, D., and Himelick, E. B. 1989. Susceptibility of honeylocust cultivars to Thyronectria canker. J. Arboric. 15:189–191.

Potter, D. A., and Hartman, J. R. 1993. Susceptibility of honeylocust cultivars to *Thyronectria austro-americana* and response of *Agrilus* borers and bagworms to infected and non-infected trees. J. Environ. Hortic. 11: 176–181.

Seeler, E. V., Jr. 1940. Two diseases of *Gleditsia* caused by a species of *Thyronectria.* J. Arnold Arbor. Harvard Univ. 21:405–427.

Sinclair, W. A., Lyon, H. H., and Johnson, W. T. 1987. Diseases of Trees and Shrubs. Comstock Publishing Associates, Cornell University Press, Ithaca, N.Y.

CHAPTER 49

Jean L. Williams-Woodward ▪ University of Georgia, Athens

Margery L. Daughtrey ▪ Cornell University, Riverhead, New York

Hydrangea Diseases

Geographic production USDA zones 3–9

Family Hydrangeaceae

Genus *Hydrangea*

Species

H. anomala subsp. *petiolaris*	climbing hydrangea	zones 4–7(8)
H. arborescens	smooth hydrangea	zones (3)4–9
H. macrophylla	bigleaf hydrangea	zones 6–9
H. paniculata	panicle hydrangea	zones 3–8
H. quercifolia	oakleaf hydrangea	zones 5–9

Cultivars many

Several species of *Hydrangea* are grown in container nurseries and greenhouses. These deciduous vines and shrubs are grown for their showy flowers. Numerous diseases of *Hydrangea* have been reported, but none are devastating in nursery production.

Fungal Leaf Spots

Various fungi

Leaf spots caused by various fungi occur from time to time on hydrangea in the nursery and landscape, but these diseases are generally not serious.

Symptoms

Fungal leaf spots are rounded or angular, dead, brown spots or blotches on leaves, often with a dark-colored (red, purple, brown, or black) border surrounding a dead center. Spore structures produced by fungal pathogens are sometimes visible on the upper and lower surfaces of leaves, usually appearing as small, dark bumps or dots on dead tissue.

In gray mold disease, caused by *Botrytis cinerea,* a fuzz of grayish brown fungal sporulation will grow out of dead leaf or stem tissue under humid conditions.

Fungal leaf spot diseases can be mistaken for powdery mildew, which can also cause dark blotches on leaves. However, strands of mycelium of the powdery mildew fungus growing on the surface of the tissue will be visible on close inspection with a hand lens.

Epidemiology

In nursery production of hydrangeas, leaf spot diseases become troublesome only when plants are exposed to frequent rainfall or overhead irrigation.

Host Range

Fungi causing leaf spots on hydrangea include *Ascochyta, Botrytis, Cercospora, Colletotrichum, Corynespora, Phyllosticta,* and *Septoria* spp.

B. cinerea has a broad host range, but many of the other leaf spot fungi affecting hydrangea are specific to that host. *Cercos-*

Fungal Leaf Spots

Geographic occurrence Wherever hydrangeas are grown

Seasonal occurrence May to October, under average yearly conditions in zone 7 (adjust for your hardiness zone)

DISEASE FREQUENCY	DISEASE SEVERITY
5 annual	5 plants killed
4	4
3	3
2 (selected)	2
1 rare	**1 very little damage** (selected)

CHEMICAL TREATMENT	CULTURAL PRACTICES
3 used every year	**3 very important** (selected)
2 (selected)	2
1 not used (selected)	1 not important

SANITATION	RESISTANT CULTIVARS
3 very important	3 many cultivars
2	2
1 not important (selected)	**1 no resistance** (selected)

pora hydrangeae has been reported on *H. arborescens* and *H. macrophylla; P. hydrangeae* on *H. anomala* var. *petiolaris, H. macrophylla,* and *H. paniculata;* and *S. hydrangeae* on *H. anomala* subsp. *petiolaris, H. arborescens,* and *H. paniculata.*

Management

Minimize leaf wetness periods by watering early in the day and allowing foliage to dry before nightfall. Provide adequate space between plants for good air circulation.

Powdery Mildew

Erysiphe polygoni
***Oidium* spp.**

Powdery mildew develops in the late spring, summer, and late fall when humidity is high and nights are cool. Plants produced in greenhouses or under shade are particularly vulnerable.

Symptoms

Mildew colonies initially develop on the lower surface of leaves, where they may at first escape notice. Opposite these areas, on the upper surface, yellow blotches develop in the leaf tissue. Under optimum conditions for disease development, purple brown blotches covered with white may be seen on the upper surfaces of leaves. Powdery mildew may affect inflorescences as well.

Epidemiology

See Chapter 15, "Powdery Mildew."

Host Range

Powdery mildew of hydrangea is caused by *Erysiphe polygoni* and *Oidium* spp. *E. polygoni* has been reported on many other hosts, but the host range of the strain that attacks hydrangea is unclear. Among hydrangeas, *H. anomala* subsp. *petiolaris, H. arborescens, H. macrophylla,* and *H. paniculata* are known hosts.

Management

See Chapter 15.

Powdery Mildew

Geographic occurrence Wherever hydrangeas are grown

Seasonal occurrence In greenhouses, year-round; outdoors, beginning in early summer, under average yearly conditions in zone 7 (adjust for your hardiness zone)

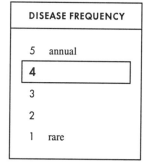

DISEASE FREQUENCY	DISEASE SEVERITY
5 annual	5 plants killed
[4]	4
3	**[3]**
2	2
1 rare	1 very little damage

CHEMICAL TREATMENT	CULTURAL PRACTICES
3 used every year	3 very important
[2]	**[2]**
1 not used	1 not important

SANITATION	RESISTANT CULTIVARS
3 very important	3 many cultivars
2	2
[1] not important	**[1] no resistance**

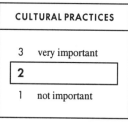

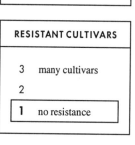

Bacterial Leaf Spot

Xanthomonas campestris

Bacterial leaf spot of hydrangeas, caused by a pathovar of *Xanthomonas campestris,* most commonly affects *H. arborescens, H. macrophylla,* and *H. quercifolia.* It is typically a problem under warm, wet conditions in late spring and summer. Not much is known about the disease.

Symptoms

On *H. quercifolia* (oakleaf hydrangea), angular purple spots develop first on the lower leaves and then spread upward in the plant. The spots are usually delineated by leaf veins. On *H. macrophylla* and *H. arborescens* cultivars, the spots are small, purple, and sometimes angular. As the leaf spots age, their centers dry, turn tan to brown, and sometimes drop out.

Epidemiology

In the nursery, bacterial leaf spot usually occurs in late spring and summer under humid, moderately wet conditions. Infection begins on the lower leaves and then spreads to adjacent leaves and plants as the pathogen is transported by splashing water. Small, water-soaked spots develop on leaves within three to five days after inoculation, with typical purple spots developing within seven days.

Host Range

Bacterial leaf spot of hydrangea is seen mostly in *H. arborescens, H. macrophylla,* and *H. quercifolia.*

Management

Avoid overhead irrigation, to reduce the spread of the bacterium. Promptly remove affected leaves, and discard severely infected plants. Space plants early in the season, to improve air circulation around them and avoid plant-to-plant spread. Copper-containing fungicides may provide some control when applied in late spring but will not be effective if plants are grown under conditions favoring the disease.

Hydrangea Virescence

Hydrangea virescence phytoplasmas

Phytoplasmas, previously known as mycoplasmalike organisms (MLOs), cause diseases with viruslike symptoms, the most famous of which is aster yellows. Virescence disease of hydrangea gets its name from the virescent (green) florets which occur on some cymes of affected plants.

Symptoms

Green florets are typical of the disease. Florets and leaves may also be stunted, and leaf veins may turn yellow. Leafy structures may grow out from the pistils of flowers. Symptoms in hydrangea stock may intensify over time. Three different forms of the virescence agent have been identified, which produce three different levels of symptom severity.

Epidemiology

Virescence is spread by grafting and other forms of asexual propagation. No insect vector has been identified, and the disease agent is not mechanically transmitted.

Host Range

It is not known whether the hydrangea virescence phytoplasmas are able to infect plants other than *H. macrophylla.*

Bacterial Leaf Spot

Geographic occurrence Mostly in the Southeast (USDA zones 7–9) and possibly other areas as well

Seasonal occurrence May to October, with a peak in May–July, under average yearly conditions in zone 7 (adjust for your hardiness zone)

DISEASE FREQUENCY	DISEASE SEVERITY
5 annual	5 plants killed
4	4
3	3
2	**2**
1 rare	**1 very little damage**

CHEMICAL TREATMENT	CULTURAL PRACTICES
3 used every year	**3 very important**
2	2
1 not used	1 not important

SANITATION	RESISTANT CULTIVARS
3 very important	3 many cultivars
2	2
1 not important	**1 no resistance**

Management

Avoid using infected plant material for propagation. Hydrangea virescence phytoplasmas can be eliminated from nuclear stock by heat treatment or meristem tip culture.

Hydrangea Viruses

Hydrangea ringspot potexvirus
Tobacco ringspot nepovirus
Tomato ringspot nepovirus
Tomato spotted wilt tospovirus

The name of a plant virus is usually derived from the name of the plant in which the virus was first found and the type of symptom produced in the plant. Although the names of the viruses infecting hydrangea may indicate wilt and spot symptoms, the type of symptom produced in hydrangea may be quite different. Viruses are fairly common in some hydrangea species, but virus diseases of hydrangeas are rare in nurseries.

Symptoms

Hydrangea ringspot potexvirus (HRSV) causes yellow or brown leaf spots or rings and leaf distortion.

Virescence

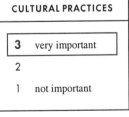

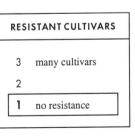

Geographic occurrence USDA zones 6–8

Seasonal occurrence Symptoms appear only at flowering in some cases

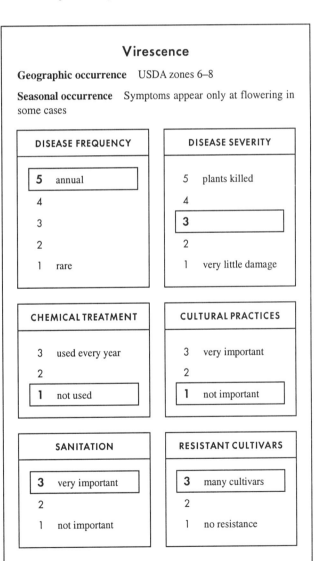

DISEASE FREQUENCY	DISEASE SEVERITY
5 annual	5 plants killed
4	4
3	**3**
2	2
1 rare	1 very little damage

CHEMICAL TREATMENT	CULTURAL PRACTICES
3 used every year	3 very important
2	2
1 not used	**1 not important**

SANITATION	RESISTANT CULTIVARS
3 very important	**3 many cultivars**
2	2
1 not important	1 no resistance

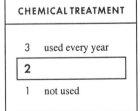

Hydrangea Viruses

Geographic occurrence Wherever hydrangeas are grown

Seasonal occurrence Variable, depending on the virus and the cultivar

DISEASE FREQUENCY	DISEASE SEVERITY
5 annual	5 plants killed
4	4
3	3
2	**2**
1 rare	1 very little damage

CHEMICAL TREATMENT	CULTURAL PRACTICES
3 used every year	3 very important
2	2
1 not used	**1 not important**

SANITATION	RESISTANT CULTIVARS
3 very important	**3 many cultivars**
2	2
1 not important	1 no resistance

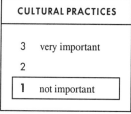

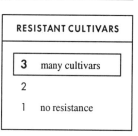

Tomato ringspot nepovirus (ToRSV) causes stunting, some leaf distortion, and leaf chlorosis with green blotches on leaves. Flower virescence can also occur.

Tobacco ringspot nepovirus (TRSV) has been identified in hydrangea, but no symptoms have been described in this host.

Tomato spotted wilt tospovirus (TSWV) has been recovered from hydrangeas affected with ring spots.

Epidemiology

HRSV may be transmitted in sap on cutting knives, but no insect vector is known.

TRSV and ToRSV are primarily vectored by nematodes.

TSWV is vectored by certain species of thrips.

Host Range

H. macrophylla is the only known host of HRSV.

Numerous woody and herbaceous species are hosts of TRSV and ToRSV.

TSWV has an extraordinarily wide host range.

Management

Purchase *H. macrophylla* stock free of HRSV. Growing hydrangeas in containers in soilless media will avoid problems with nematode-transmitted viruses if the stock is virus-free.

If TSWV were to become a significant problem in hydrangea, thrips control would be key to its management.

Weed control will help to eliminate potential reservoirs of ToRSV and TSWV.

Do not propagate plants with virus symptoms.

REFERENCES

Lawson, R. H. 1985. Hydrangea. Pages 259–273 in: Diseases of Floral Crops. Vol. 2. D. L. Strider, ed. Praeger Scientific, New York.

Uddin, W., and McCarter, S. M. 1996. First report of oakleaf hydrangea bacterial leaf spot caused by a pathovar of *Xanthomonas campestris*. Plant Dis. 80:599.

CHAPTER

50

Phillip Colbaugh ▪ Texas A&M Agricultural Experiment Station, Dallas

Austin Hagan ▪ Auburn University, Auburn, Alabama

Jerry Walker ▪ University of Georgia, Griffin

Larry Barnes ▪ Texas A&M University, College Station

Indian Hawthorn Diseases

Geographic production USDA zones 8–10

Family Rosaceae

Genus *Rhaphiolepis*

Species *R. × delacourii*
R. indica
R. umbellata

Indian hawthorn (India hawthorn) belongs to the rose family (Rosaceae), which includes its familiar botanical relatives loquat, pear, photinia, quince, and cotoneaster. The various cultivars of Indian hawthorn are mostly compact evergreen shrubs originating from two native species in Japan and southern China. Commercial cultivars produce attractive pink or white flowers. Plants in commercial production are grown outdoors in containers, in the warm climates of the Gulf Coast and southern California.

Entomosporium Leaf Spot

Entomosporium mespili

Entomosporium leaf spot, caused by the fungus *Entomosporium mespili,* is the predominant disease of Indian hawthorn in commercial container nurseries. It is most active during prolonged periods of cool, wet weather in the winter and spring. Severely diseased container-grown Indian hawthorn can be defoliated before midsummer, and the plants are rendered unsalable.

Symptoms

Leaf lesions arise as minute dots on newly developing leaves in the spring. The lesions enlarge to form gray brown, irregularly shaped spots with a red or brown border (Plate 66). The leaf spots mature as circular, brown lesions, 1/4 inch in diameter, with raised black fruiting bodies of the pathogen in the center of each spot. Leaf lesions can be few and scattered or can become so numerous that they coalesce to form large dead areas. Defoliation is heavy with the onset of severe outbreaks of the disease. Infected areas on older leaves frequently have a gray white cast due to the production of spore masses when the leaves are wet. Infection of older leaves in late spring may also appear as white "ghost spots" on leaves, where higher temperatures abort the progress of infection.

Epidemiology

The disease cycle is initiated early in winter to early spring by fungal spores overwintering on infected leaves or juvenile shoots. The blister-like fruiting bodies of the pathogen sporulate extensively as infected leaves become wet, producing white ribbons of spores that are spread to healthy leaves by splashing water. After the initiation of foliar disease in the spring, the disease can spread rapidly. High humidity, cool weather, crowded plants, and splashing water from rainfall or overhead irrigation provide an ideal

Entomosporium Leaf Spot

Geographic occurrence USDA zones 8–10

Seasonal occurrence Along the Gulf Coast, as early as November to December; in other areas, March through June

DISEASE FREQUENCY	DISEASE SEVERITY
5 annual (boxed)	5 plants killed
4	**4** (boxed)
3	3
2	2
1 rare	1 very little damage

CHEMICAL TREATMENT	CULTURAL PRACTICES
3 used every year (boxed)	3 very important
2	**2** (boxed)
1 not used	1 not important

SANITATION	RESISTANT CULTIVARS
3 very important	3 many cultivars
2 (boxed)	**2** (boxed)
1 not important	1 no resistance

environment for the spread of the disease. As daytime temperatures in late spring approach 85°F, disease progress is halted until cooler weather returns in the fall.

Host Range

E. mespili is pathogenic to several members of the family Rosaceae, including apple, cotoneaster, crabapple, hawthorn, loquat, medlar, mountain ash, pear, photinia, quince, and serviceberry. Strains of the fungus that attack only certain plant species may exist.

Management

If Entomosporium leaf spot has been a problem in container-grown plants, several fungicide treatments may be required to bring the disease under control. The disease is especially damaging to plants with newly developing foliage, so protective fungicide applications can be necessary until the new leaves have matured. Many nurseries attempt to trim off the infected foliage or prune back defoliated plants for resale, but unless this practice is coupled with a tightly scheduled fungicide spray, the disease will intensify.

Cultural practices Careful inspection of container-grown plants in early spring will be helpful in identifying early signs of leaf spot. Light pruning before budbreak in the spring can prevent the formation of extensive new growth requiring late spring and summer pruning. Scheduling overhead irrigation during the morning allows the foliage to dry rapidly after the watering cycle. Spacing of container plants is helpful in preventing the spread of the disease to healthy plants. Fungicides are generally not needed for field-grown landscape plants if the planting stock is disease-free and isolated from other sources of the pathogen.

Sanitation Growing plants from infected liners is not recommended. Take cuttings for propagation only from disease-free plants. Discard diseased cuttings and container stock. Do not place blocks of newly transplanted container stock near blocks of shrubs or trees damaged by leaf spot. Remove diseased leaves and other debris from production beds.

Resistance *R.* × *delacourii* (hybrida) and the cultivars Eleanor Tabor, Olivia, and Ovata have demonstrated resistance to Entomosporium leaf spot. Janice and Majestic Beauty are moderately resistant. Bay Breeze, Becky Lynn, Enchantress, Harbinger of Spring, Heather, Pinkie, Spring Rapture, Springtime, and White Enchantress are susceptible. Clara, Jack Evans, Rosalinda, and Snow White are moderately susceptible.

Chemical treatment Protective fungicide treatment should begin before the spread of the causal fungus and the onset of symptoms. In areas along the Gulf Coast, where pathogen activity often begins early, start fungicide sprays in late November to early December. In other areas, delay the first fungicide spray until February or early March. Preventive fungicide applications should continue at 10- to 14-day intervals until new shoot growth is finished. During periods of heavy rain, shorten the spray interval to seven to 10 days.

Phytophthora Aerial Blight

Phytophthora parasitica

Aerial blight caused by *Phytophthora* sp. have become more common in woody ornamental nurseries. The disease usually does not threaten the health of the plant, but sufficient cosmetic damage may occur to make plants unmarketable without a cleanup of the canopy.

Symptoms

Small, newly emerged leaves blacken, desiccate, and deteriorate. Terminal twig tissue commonly blackens an inch or two from the tip, but a well-defined canker is usually not apparent. Symptoms of Phytophthora aerial blight can be confused with those of bacterial fire blight.

Epidemiology

Phytophthora aerial blight is usually most common and severe where recycled irrigation water is used in overhead irrigation. Pathogen-infested soil or gravel below the plants can also serve as a source of inoculum. Plant-to-plant spread by irrigation or rainwater can intensify disease development. Tender, young foliage or foliage injured from pruning or shearing appears to be most susceptible to infection.

Host Range

All *Rhaphiolepis* species appear to be susceptible to aerial blight caused by *P. parasitica,* as are numerous other woody plants.

Phytophthora Aerial Blight

Geographic occurrence Gulf Coast

Seasonal occurrence Mid-spring through late fall

DISEASE FREQUENCY	DISEASE SEVERITY[a]
5 annual	5 plants killed
4	4
3	3
2	**2**
1 rare	**1** very little damage

[a] The disease causes cosmetic damage, primarily requiring labor to repair

CHEMICAL TREATMENT	CULTURAL PRACTICES
3 used every year	**3** very important
2	2
1 not used	1 not important

SANITATION	RESISTANT CULTIVARS
3 very important	3 many cultivars
2	2
1 not important	**1** no resistance

Management

Efforts to eliminate or reduce *Phytophthora* from irrigation water should be considered. Make sure plants are on a well-crowned gravel bed to minimize splash dispersal of the pathogen from the soil onto foliage. Protective fungicides should be applied at the first sign of symptoms, immediately following shearing operations, and during periods of flushes of new growth. Phytophthora aerial blight is usually not a problem when new growth is not being produced or when pruning has not been done recently, so fungicide application should not be needed at these times.

Cultural practices Tube or drip irrigation minimizes direct contact between the pathogen and susceptible host tissue, and it should be considered where feasible. Space plants to encourage good air circulation and to facilitate rapid drying of foliage. Schedule fungicide application following shearing to protect wounded tissue from infection.

Sanitation Remove infected terminals and other plant debris from the production area.

Resistance Information on resistance is not available, but most common cultivars are likely to be susceptible to infection under the conditions described above.

Phytophthora Root Rot

Phytophthora cinnamomi
Phytophthora parasitica

Phytophthora root rots are damaging to a wide variety of container-grown trees and shrubs. Outbreaks of root rot are periodically observed in container-grown Indian hawthorn. Poor nursery sanitation and inadequate drainage or periodic flooding of production ranges are often associated with outbreaks of the disease.

Symptoms

Typical root rot symptoms are yellowing (chlorosis) of leaves, slowed shoot growth, limb dieback, and plant death. The fine feeder or fibrous roots colonized by *Phytophthora* spp. are brittle and brown to reddish brown. Discolored roots can usually be observed by removing the root ball from the container and inspecting the root system.

Epidemiology

Phytophthora spp. survive as resting structures (chlamydospores and oospores) in decaying roots, plant debris, potting media, and soil around production beds. The surviving spores are easily spread through blocks of liners and container stock in contaminated containers, raw soil in potting media, and flowing or splashing water. Zoospores, released in saturated infested soil or potting media, are readily spread through container beds by runoff and recycled irrigation water taken from holding ponds. The heaviest losses usually occur in flood-prone beds where water stands around container stock for an extended period. Over- or underwatering and poor drainage or compaction of potting media often accelerate the development of root rot. For additional information concerning the epidemiology of this disease, refer to Chapter 14, "Phytophthora Root Rot and Dieback."

Host Range

The host ranges of *P. parasitica* and *P. cinnamomi* include a wide array of container-grown trees and shrubs. For a listing of trees and shrubs susceptible to these pathogens, see Farr et al. (1989).

Management

Adoption of strict sanitation and cultural controls along with preventative fungicide treatments will prevent significant plant loss due to Phytophthora root rot.

Cultural practices Place flats of rooted cuttings and liners on raised benches or crowned gravel-covered beds. To avoid ponding of water around container stock, beds must be crowned, covered with black plastic or similar material, and then covered with a layer of gravel or other coarse rock. Drainage ditches must be deep enough to handle flooding rains. Block container stock by container size and water needs. Avoid damaging roots with high rates of water-soluble nitrogen fertilizers. Use a coarse bark medium with a high percolation rate and 20–30% air-filled pore space.

Sanitation Store potting medium components on asphalt or concrete pads. Do not store bark or containers on bare soil. Rinse and disinfect recycled containers. Clean and disinfect propagation benches. Take cuttings from disease-free container stock. Root cuttings in a clean soilless potting medium. Discard diseased container stock.

Resistance Resistance to Phytophthora root rot among Indian hawthorn varieties is not known.

Chemical treatment Since destructive outbreaks of root rot of Indian hawthorn are unusual, good sanitation and cultural practices should prove effective for the control of this disease. Blocks of container stock should be routinely inspected for symptomatic plants. Should diseased plants be found, immediately begin preventive fungicide drenches or foliar sprays at the intervals listed on the fungicide label. Where Phytophthora root rot causes chronic losses, apply a fungicide throughout the production cycle. Fungicides should be used to protect roots from *Phytophthora* infection as well as suppress the onset of symptoms. For fungicide recommendations, contact your county extension office.

Fire Blight

Erwinia amylovora

Fire blight affects a number of members of the apple subfamily (Pomodidae) and occasionally occurs in Indian hawthorn. It is particularly damaging to susceptible Indian hawthorn cultivars during bloom when the weather is wet and mild.

Symptoms

Early symptoms of fire blight of Indian hawthorn include sudden blighting and death of bloom clusters and spurs. The causal bacterium, *Erwinia amylovora,* often spreads down through spurs and into twigs. Blighted bloom clusters, spurs, and terminal leaves on twigs turn brown and die. In susceptible cultivars, the infection will continue to spread down to the scaffold branches until a large portion of the plant dies. The progress of the disease slows markedly in woody twigs and branches.

Epidemiology

E. amylovora survives in cankers on twigs or infected branches. During wet, humid spring weather, a thick, amber-colored fluid containing millions of bacteria oozes from the cankers. The bacterium can be spread by insects or splashing water to flower buds or

Fire Blight

Geographic occurrence USDA zones 8 and 9

Seasonal occurrence March to April

DISEASE FREQUENCY	DISEASE SEVERITY
5 annual	5 plants killed
[4]	**[4]**
3	3
2	2
1 rare	1 very little damage

CHEMICAL TREATMENT	CULTURAL PRACTICES
[3] used every year	**[3] very important**
2	2
1 not used	1 not important

SANITATION	RESISTANT CULTIVARS
[3] very important	**[3] many cultivars**
2	2
1 not important	1 no resistance

wounds on leaves and tender shoots. Extended periods of warm, cloudy, wet weather during bloom favors the development of fire blight. Severe pruning and excessive nitrogen fertilization can greatly increase disease severity.

Host Range

Fire blight is commonly reported in certain cultivars of cotoneaster, crabapple, firethorn, flowering pear, hawthorn, loquat, photinia, and several other members of the apple subfamily, in addition to Indian hawthorn.

Management

A combination of nursery sanitation, proper fertilization and pruning, selection of disease-resistant cultivars, and timely applications of a bactericide will largely prevent outbreaks of fire blight in container nurseries.

Cultural practices Fertilize with a slow-release fertilizer. When using a water-soluble fertilizer, avoid applying it before petal fall. Make frequent, light applications of fertilizer during the growing season, and do not apply it at high rates in late summer or early fall. Avoid heavy pruning when the plants are in bloom. When using overhead irrigation, avoid watering in late afternoon or early evening, particularly during bloom.

Sanitation Discard container stock damaged by fire blight. If damaged plants are saved, prune out the perennial cankers during the winter. Cuts should be made 6–8 inches below the visible canker margin. Do not take cuttings from stock that has been damaged by fire blight. Clean pruning knives and shears with denatured alcohol or a commercial disinfectant after each cut.

Resistance Significant differences in the susceptibility of Indian hawthorn cultivars to fire blight have been recorded. Susceptibility is greatly influenced by weather patterns during and shortly after bloom. Highly resistant cultivars include Bay Breeze, Clara, Dwarf Yedda, Eleanor Tabor, Enchantress, Harbinger of Spring, Heather, Indian Princess, Majestic Beauty, Pinkie, Rosalinda, Snow White, Spring Rapture, Springtime, and White Enchantress. Susceptible cultivars include Jack Evans and Janice.

Chemical treatment The timing of bactericide applications is critical for effective control of fire blight in susceptible cultivars. Begin applications as the blooms begin to show color, and continue spraying at four- or five-day intervals until petal fall. If postbloom weather patterns favor continued disease development in susceptible cultivars, additional applications may be made at weekly intervals.

REFERENCES

Dirr, M. A. 1998. Manual of Woody Landscape Plants: Their Identification, Ornamental Characteristics, Culture, Propagation and Uses. 5th ed. Stipes Publishing, Champaign, Ill.

Farr, D. F., Bills, G. F., Chamuris, G. P., and Rossman, A. Y. 1989. Fungi in Plants and Plant Products in the United States. American Phytopathological Society, St. Paul, Minn.

Raabe, R. D., and Hansen, H. N. 1955. Entomosporium leaf spot of *Rhaphiolepis.* Phytopathology 45:55.

Sinclair, W. A., Lyon, H. H., and Johnson, W. T. 1987. Diseases of Trees and Shrubs. Comstock Publishing Associates, Cornell University Press, Ithaca, N.Y.

Tilt, K., Hagan, A. K., Williams, J. D., and Akridge, J. R. 1997. Indian hawthorn: Cultivar selection is important. Proc. South. Nursery Assoc. Res. Conf. 42:523–526.

CHAPTER 51

A. R. Chase ▪ Chase Research Gardens, Mt. Aukum, California

Ixora Diseases

Geographic production USDA zones 9 and 10

Family Rubiaceae

Genus *Ixora*

Species *I. coccinea*

Cultivars numerous

Ixora, also called flame-of-the-woods and jungle geranium, is a member of the madder family. This plant is hardy in only the warmest USDA zones in the United States (southern Florida and southern California). It prefers partial shade and moist soil, reaching heights of 6 feet in the landscape. Dwarf cultivars, which grow only 2 to 3 feet tall, are available. In the nursery, *I. coccinea* is affected by bacterial leaf spot and root-knot nematodes, and it suffers iron deficiency in highly limed soils. With the notable exceptions of these three disorders, ixora is relatively problem-free.

Bacterial Leaf Spot

Xanthomonas campestris* pv. *maculifoliigardeniae

Symptoms

The first symptoms of foliar infection by the bacterial leaf spot pathogen, *Xanthomonas campestris* pv. *maculifoliigardeniae,* are pinpoint-sized water-soaked areas, which can rapidly enlarge. They tend to remain confined to the areas between leaf veins and sometimes have a bright yellow border. Leaf distortion is common when immature leaves are infected; they pucker and fail to expand normally. In severe infections leaf drop is common.

Host Range

All ixora cultivars tested have been found susceptible to the pathogen. Gardenias also are sometimes infected with the same pathogen. The disease is more common in ixora, despite having been first reported in gardenia.

Management

Minimize foliar wetting through irrigation or rainfall, to reduce the spread of the bacterium and its ability to infect. This disease is more of a problem during warm, wet periods of the year. Never use infected plants for cuttings, even when the cuttings appear healthy, since latent infections simply continue to develop as the cuttings root. Bacterial leaf spot is most severe at temperatures between 75 and 95°F, and thus it is most common during the summer.

Chemical treatment Preventative applications of copper-containing compounds are sometimes helpful but cannot control the disease under most growing conditions. Cultural controls are superior and necessary even when chemicals are used.

Bacterial Leaf Spot

Geographic occurrence USDA zones 9 and 10

Seasonal occurrence Rainy and warm to hot

DISEASE FREQUENCY	DISEASE SEVERITY
5 annual	5 plants killed
[4]	4
3	**[3]**
2	**[2]**
1 rare	**[1]** very little damage

CHEMICAL TREATMENT	CULTURAL PRACTICES
3 used every year	**[3]** very important
[2]	2
1 not used	1 not important

SANITATION	RESISTANT CULTIVARS
[3] very important	3 many cultivars
2	**[2]**
1 not important	1 no resistance

Iron Deficiency and High Soil pH

Iron deficiency is the most common disorder of ixora. It occurs commonly in Florida.

Symptoms

Leaves turn yellow to varying degrees. Older leaves may simply have a yellow margin, while young, immature leaves may turn almost any shade of yellow and even white. Intermediate symptoms include yellowing of the tissues between leaf veins. The symptoms can be confused with those of severe manganese deficiency.

Management

Soils should be maintained at pH 5.5–6.5 when possible. Iron deficiency is usually not a problem in acid soils, but in heavily limed soils the disorder is common in sensitive crops, such as ixora, azalea, gardenia, and hibiscus.

Chemical treatment Soil amendments with iron chelates should follow pH adjustment with agricultural sulfur to lower the pH. Foliar iron treatments are sometimes effective but not preferred, since their benefits are not as long-term as those of soil treatment. Additionally, iron sprays may stain brick, concrete, and other surfaces surrounding affected plants.

Root-Knot Nematode

***Meloidogyne* sp.**

Root-knot nematodes are a greater problem in the landscape than in production, but they are sometimes present in nurseries.

Symptoms

Infected plants are stunted, off-color, and generally unthrifty. The characteristic swollen galls produced by the nematodes are obvious on roots. Infected roots frequently decay.

Management

Sanitation Root-knot nematodes must be avoided. Most ixora is grown in soilless mix in containers, and the medium must be kept free of soil. Adequate surface drainage must be provided, to avoid contamination with surface water, which can carry nematodes.

Chemical treatment No effective chemicals are registered for the control of root-knot nematodes in ixora. Methyl bromide can be used as a preplant treatment to fumigate soil, infested pots, and media.

Iron Deficiency

Geographic occurrence Florida

Seasonal occurrence Spring and summer

DISEASE FREQUENCY	DISEASE SEVERITY
5 annual	5 plants killed
[4]	4
3	**[3]**
2	**[2]**
1 rare	**[1] very little damage**

CHEMICAL TREATMENT	CULTURAL PRACTICES
[3 used every year]	**[3 very important]**
2	2
1 not used	1 not important

SANITATION	RESISTANT CULTIVARS
not applicable	not applicable

Root-Knot Nematode

Geographic occurrence USDA zones 9 and 10

Seasonal occurrence Periods when the soil temperature is above 50°F

DISEASE FREQUENCY	DISEASE SEVERITY
[5 annual]	**[5 plants killed**
4	**4**
3	**3]**
2	2
1 rare	1 very little damage

CHEMICAL TREATMENT	CULTURAL PRACTICES
3 used every year	3 very important
2	2
[1 not used]	**[1 not important]**

SANITATION	RESISTANT CULTIVARS
[3 very important]	3 many cultivars
2	2
1 not important	**[1 no resistance]**

CHAPTER 52

Ned A. Tisserat ▪ Kansas State University, Manhattan

D. Michael Benson ▪ North Carolina State University, Raleigh

Juniper Diseases

Geographic production USDA zones 2–9

Family Cupressaceae

Genus *Juniperus*

Species			
	J. chinensis	Chinese juniper	zones 3–9
	J. communis	common juniper	zones 2–6
	J. conferta	shore juniper	zones 6–8
	J. horizontalis	creeping juniper	zones 3–9
	J. sabina	savin juniper	zones 3–7
	J. scopulorum	Rocky Mountain juniper	zones 3–7
	J. virginiana	eastern red cedar	zones 2–9

Junipers are used extensively in landscape plantings throughout North America, because of their diverse growth characteristics and coloration, their tolerance for extreme and rapid temperature fluctuations, and their ability to grow in poor soils and at arid sites where other plants would fail. In fact, they may be the most widely used plants in the United States. Most junipers are considered low-maintenance plants in the landscape, because they are relatively free of major diseases and insect pests. Nevertheless, several diseases can cause significant damage to junipers in nursery production.

Phomopsis Tip Blight
Phomopsis juniperovora
Kabatina Tip Blight
Kabatina juniperi

Two fungal diseases, Phomopsis tip blight and Kabatina tip blight, are common in nursery production of junipers. Both diseases result in needle blighting and branch tip dieback. Damage to young seedlings or recently grafted trees can be extensive. The two diseases cause almost identical symptoms, but aspects of their development and control differ, and therefore it is important to distinguish between them.

Symptoms

Phomopsis tip blight New needle growth and succulent branch tips of junipers are damaged. Newly developing needles that are still yellow green are most susceptible to infection. Older, mature foliage is resistant, and therefore most blighting occurs on the terminal 4–6 inches of branches during the growing season. Affected foliage first turns dull red or brown (Plate 67) and finally ash gray. Small gray lesions often girdle branch tips and cause blighting of foliage beyond the diseased tissue. Small, black, spore-containing fruiting bodies of the pathogen develop in the lesions. These diagnostic fungal structures can be viewed with a hand lens. Repeated blighting in early summer can result in abnormal bunching (witches'-broom) and discoloration of foliage, stunting of young trees and shrubs, and in severe cases plant death.

Kabatina tip blight Symptoms of Kabatina tip blight appear in late winter or early spring, well before symptoms of Phomopsis tip blight. The terminal 2 to 6 inches of diseased branches turn dull green and then red or yellow (Plate 68). Small, ash gray to silver lesions dotted with black fruiting bodies of the pathogen are visible at the base of the discolored tissue. The brown, desiccated branch tips eventually drop from the tree in late spring and summer. Symptoms of Kabatina tip blight develop only in early spring, and the blighting does not continue through the summer. The blighting is restricted to branch tips and does not cause extensive branch dieback or tree death.

Epidemiology

Phomopsis tip blight The Phomopsis tip blight fungus, *Phomopsis juniperovora,* overwinters in infected plant tissue. Spores of the fungus are produced in flask-shaped fruiting structures (pycnidia) and are dispersed by rain or irrigation throughout the growing season. Infection can occur whenever young, succulent foliage is available and humidity is high. Most infections occur in spring and in fall, when the weather is favorable and susceptible tissue is present. Fewer infections occur in midsummer and winter.

Kabatina tip blight Kabatina tip blight fungus, *Kabatina juniperi,* also survives on infected plant debris. Spores of the fungus are produced in black, disk-shaped fruiting bodies (acervuli) and are released by rain or irrigation. The primary infection period is thought to be in autumn even though visible symptoms are not apparent until late winter or early spring. Infection of the foliage requires a wound and is often associated with small wounds on branch tips caused by insect feeding or mechanical damage.

Host Range

P. juniperovora has been reported on several trees in the cypress family (Cupressaceae), including juniper (*Juniperus* spp.), arborvitae (*Chamaecyparis*), white cedar (*Thuja*), and baldcypress (*Taxodium*).

K. juniperi has been reported on juniper, arborvitae, white cedar, and incense cedar (*Calocedrus*).

Management

Several practices can help reduce the severity of Phomopsis tip blight and Kabatina tip blight. Provide enough space between plants and rows to promote good air circulation. Maintain adequate fertility, but do not overfertilize. If overhead irrigation is used, water plants in the morning, so the foliage will dry as the day progresses. Do not irrigate in early evening. Prune out diseased branch tips during dry summer weather and destroy them, but avoid excessive pruning or shearing, especially in the spring and fall. This creates wounds for penetration by *Kabatina* and encourages succulent new growth, which is susceptible to Phomopsis blight.

Juniper varieties vary significantly in susceptibility to these diseases (Table 52.1). Use resistant varieties whenever possible.

Fungicide sprays are usually required to control Phomopsis tip blight in juniper seedling beds. Application of fungicides at seven- to 21-day intervals during rapid plant growth in the spring and again in the fall will give adequate control of this disease, but not Kabatina tip blight.

Kabatina blight infections occur in the fall. There are no fungicides labeled for control of this disease.

Phomopsis Tip Blight and Kabatina Tip Blight

Geographic occurrence USDA zones 3–8

Seasonal occurrence

Phomopsis tip blight May through October, with an optimum period in May–June

Kabatina tip blight Infection occurs in September to October; symptoms appear from March through May

DISEASE FREQUENCY	DISEASE SEVERITY
5 annual	5 plants killed
4	**4**
3	**3**
2	**2**
1 rare	1 very little damage

CHEMICAL TREATMENT	CULTURAL PRACTICES
3 used every year	**3** very important
2	2
1 not used	1 not important

SANITATION	RESISTANT CULTIVARS
3 very important	**3** many cultivars
2	2
1 not important	1 no resistance

Cercospora Needle Blight

Cercospora sequoiae var. *juniperi*

Cercospora needle blight is a potentially damaging disease of container- and field-grown junipers, especially Rocky Mountain juniper (*J. scopulorum*). Repeated infections over several years may result in defoliation and tree death. The disease is most severe in areas with high summer humidity and frequent rains.

Symptoms

Symptoms of Cercospora needle blight first appear on the inner branch needles (those located nearest the main tree trunk) and in the lower portion of the tree in late summer and fall. Needles turn dull brown or red and eventually drop. Small, fuzzy or hairy spore-bearing fungal structures, easily visible with a hand lens, form on the dead needles. Defoliation in succeeding years con-

Cercospora Needle Blight

Geographic occurrence USDA zones 3–8

Seasonal occurrence June to September, with an optimum period in June and July, under average yearly conditions in the Midwest (adjust for your hardiness zone)

DISEASE FREQUENCY	DISEASE SEVERITY
5 annual	5 plants killed
4	**4**
3	3
2	2
1 rare	1 very little damage

CHEMICAL TREATMENT	CULTURAL PRACTICES
3 used every year	**3** very important
2	2
1 not used	1 not important

SANITATION	RESISTANT CULTIVARS
3 very important	**3** many cultivars
2	2
1 not important	1 no resistance

tinues from the inner portion of the branch toward the tip, and from the bottom of the tree toward the top. Severely infected trees are open and spindly and may appear as if they had been scorched by fire.

Epidemiology

The Cercospora needle blight fungus, *Cercospora sequoiae* var. *juniperi,* overseasons on dead needles on the ground or attached to the tree. Spores of the fungus are dispersed throughout the growing season by rain or irrigation, but most infection occurs in early to midsummer. However, symptoms typically do not develop until late summer or early fall.

Host Range

Cercospora needle blight has been reported in species of *Juniperus* and *Cupressus.*

Management

Follow the cultural management practices detailed for Phomopsis tip blight. Cercospora needle blight may occur in several juniper species but is particularly severe in selections of Rocky Mountain juniper. This species should be avoided in areas where the disease is a chronic problem (e.g., the Great Plains region). Most selections of eastern red cedar and Chinese juniper have good resistance to this disease (Table 52.1).

Chemical control may be necessary to protect susceptible junipers during wet summers. Make two to three applications of a labeled fungicide at two- to four-week intervals beginning in early summer. Cover the foliage thoroughly, especially the lower two-thirds of the tree crown.

Cedar-Apple Rust

Gymnosporangium juniperi-virginianae

Cedar-Hawthorn Rust

Gymnosporangium globosum

Cedar-Quince Rust

Gymnosporangium clavipes

Cedar-apple rust and related rusts are the most striking and colorful diseases of juniper. The fungi causing these diseases live on rosaceous hosts, such as hawthorn, flowering crabapple, and apple, during a portion of their life cycles and live on junipers during the other portion. They can cause considerable damage to the crabapples and hawthorns as a result of premature defoliation and fruit distortion, but the effect on junipers is normally minimal.

Symptoms

Both cedar-apple rust and cedar-hawthorn rust produce woody, reddish brown, brain-shaped galls on twigs of juniper. The galls vary in size and shape but are usually 1/2 inch to 2 inches in diameter. They may be numerous on susceptible cultivars and eventually result in a dieback of branch tips. Galls begin to swell and produce orange, gelatinous tendrils in early spring (Plate 69). They usually remain active for six to eight weeks and then turn black and shrivel. Dried galls may remain attached to the tree for several months.

Cedar-quince rust produces perennial cigar-shaped galls on small twigs.

Epidemiology

The rust fungi infecting juniper have complex life cycles that involve multiple spore stages and two hosts (junipers and rosaceous trees and shrubs). Juniper infection usually occurs in midsummer by spores produced on crabapple, hawthorns, or other hosts and blown onto juniper foliage. Spore production and infection of junipers may continue for several months during wet weather. For more details on the life cycles of these rust fungi, see Chapter 40, "Flowering Crabapple."

Host Range

Rust diseases occur in all juniper species. Many selections of eastern red cedar and Rocky Mountain juniper are susceptible to cedar-apple rust, whereas Chinese junipers, as a group, are resistant.

Management

Rust damage to junipers is usually not serious, but the presence of hundreds of galls on susceptible varieties may result in

Cedar-Apple Rust and Related Rust Diseases

Geographic occurrence USDA zones 3–9

Seasonal occurrence Galls on juniper are active from April through May; infection of juniper occurs from July to September

DISEASE FREQUENCY	DISEASE SEVERITY
[5] annual	5 plants killed
4	4
3	**[3]**
2	2
1 rare	1 very little damage

CHEMICAL TREATMENT	CULTURAL PRACTICES
3 used every year	**[3] very important**
[2]	2
1 not used	1 not important

SANITATION	RESISTANT CULTIVARS
3 very important	**[3] many cultivars**
2	2
[1] not important	1 no resistance

Table 52.1 Relative resistance to various diseases in juniper selections (*Juniperus* spp.)[a,b]

	Cedar-apple rust	Cedar-hawthorn rust	Phomopsis tip blight	Kabatina tip blight	Cercospora needle blight	Botryosphaeria canker
J. chinensis						
Ames	0	0	L–M	0	0	0
Aureo-globosa	0	0	L	. . .[c]	. . .	. . .
Blue Point	0	. . .	. . .	M	0	0
Columnaris	0	0	S	. . .	. . .	. . .
Femina	0	0	L	. . .	. . .	. . .
Fortunei	0	0	. . .	. . .	. . .	. . .
Globosa	M	L	L	. . .	. . .	. . .
Hetzii	. . .	. . .	L	. . .	. . .	. . .
Hetzii Columnaris	0	0	M	0	0	0
Japonica	0	0	S	. . .	. . .	. . .
Keteleeri	0	0	L	0	0	0
Leeana	0	0	. . .	. . .	. . .	. . .
Maney	0	. . .	M	0	0	0
Mas	0	. . .	L	. . .	. . .	. . .
Mountbatten	0	. . .	L	0	0	0
Oblonga	0	0	. . .	. . .	. . .	. . .
Parsonsii	0	0	. . .	. . .	. . .	. . .
Pendula	0	0	S	. . .	. . .	. . .
Perfecta	0	. . .	. . .	0	0	. . .
Pfitzeriana	L	0	M	. . .	. . .	0*
Pfitzeriana Aurea	0	0	M	. . .	. . .	. . .
Pfitzeriana Compacta	0	0	L	. . .	. . .	. . .
Plumosa	0	. . .	M	. . .	. . .	. . .
Plumosa Aurea	0	0	. . .	. . .	. . .	. . .
Pyramidalis	0	0	L	. . .	. . .	. . .
Robusta Green	0	. . .	0	0	0	0
var. *sargentii*	L?	L?	L–M	. . .	. . .	. . .
Spartan	L	. . .	. . .	M	0	0
Variegata	0	0	M	. . .	. . .	. . .
Watereri	0	0	M	. . .	. . .	. . .
Wintergreen	0	. . .	M	0	0	0
Wren	0	. . .	. . .	L	0	0
J. communis						
Aurea	0	. . .	. . .	. . .	. . .	. . .
Depressa	0	0	0	. . .	. . .	. . .
Hibernica	0	0	. . .	. . .	. . .	. . .
Oblonga-pendula	0	. . .	0	. . .	. . .	. . .
J. conferta	0	0	0	. . .	. . .	. . .
J. horizontalis						
Admirabilis	0	0	M	. . .	. . .	. . .
Adpressa	0	0	S	. . .	. . .	. . .
Alpina	M	0	S	. . .	. . .	. . .
Argentea	0	0	S	. . .	. . .	. . .
Bar Harbor	. . .	. . .	S	. . .	. . .	. . .
Douglasii	L?	0	L	. . .	. . .	. . .
Eximius	0	0	S	. . .	. . .	. . .
Filicina	0	0	M	. . .	. . .	. . .
Glomerata	0	. . .	. . .	. . .	. . .	. . .
Livida	0	0	S	. . .	. . .	. . .
Petraea	0	0	S	. . .	. . .	. . .
Plumosa	L?	0	S	. . .	. . .	. . .
Procombens	0	0	L	. . .	. . .	. . .
Variegata	0	0	S	. . .	. . .	. . .
Wiltonii	0	0	M	. . .	. . .	. . .

(continued on next page)

[a] Data from Hartman et al. (1994, 1995), Himelick and Neely (1960), Schoeneweiss (1969), and Tisserat and Pair (1997).
[b] 0 = no disease; L = light disease; M = moderate disease; S = severe disease. Ratings were gathered during seasons favorable for disease development. An asterisk denotes ratings based on general field observations and not on trees in replicated plots.
[c] Not determined.

Table 52.1 (continued) Relative resistance to various diseases in juniper selections (*Juniperus* spp.)[a,b]

	Cedar-apple rust	Cedar-hawthorn rust	Phomopsis tip blight	Kabatina tip blight	Cercospora needle blight	Botryosphaeria canker
J. sabina						
Broadmoor	0	0	. . .	. . .	. . .	S*
Fastigiata	0	0	L	. . .	. . .	. . .
var. *tamariscifolia*	0	0	M–H	. . .	. . .	. . .
Variegata	0	. . .	. . .	. . .	. . .	. . .
J. scopulorum						
Blue Haven	M	. . .	L–S	S	M	S
Cologreen	M	. . .	. . .	L	M	S
Dewdrop	L	. . .	. . .	M	M	?
Gray Gleam	M	. . .	. . .	M	0?	0
Horizontalis	M	S	S	. . .	. . .	. . .
McFarland	M	. . .	. . .	L	S	M
Medora	0	. . .	. . .	M	L	0
Moffettii	M	0	L	M	M	0
Moonglow	0	. . .	. . .	M	S	S
Pathfinder	L	. . .	. . .	M	L	0
Pendula	S	0	S	. . .	. . .	. . .
Platinum	L	. . .	S	L	L	0
Silver Globe	L	. . .	. . .	L	L	0
Sky Rocket	L	. . .	. . .	S	S	S
Sparkling Skyrocket	M	. . .	. . .	S	S	S
Sutherland	S	. . .	. . .	L	L	?
Viridifolia	L?	. . .	. . .	. . .	. . .	. . .
Welchii	L	. . .	S	M	L	0
Wichita Blue	L	. . .	. . .	M	M	S
J. squamata						
Albo-variegata	0	0	M	. . .	. . .	. . .
Fargesii	. . .	. . .	0	. . .	. . .	. . .
Meyeri	. . .	. . .	L	. . .	. . .	. . .
Wilsonii	0	. . .	. . .	. . .	. . .	. . .
J. virginiana						
Admiral	M	. . .	. . .	M	0	0
Albospica	M?	M?	S	. . .	. . .	. . .
Aurea	0	. . .	. . .	. . .	. . .	. . .
Blue Mountain	0	. . .	. . .	0	0	0
Burkii	S	. . .	. . .	L	0	0
Canaertii	S	S	M–S	L	0	0
Chamberlaynii	S	L	M	. . .	. . .	. . .
Cinerascens	S	S	L	. . .	. . .	. . .
Elegantissima	S	S	M	. . .	. . .	. . .
Emerald Sentinel	M	. . .	. . .	L	0	0
Glauca	S	S	M	. . .	. . .	. . .
Globosa	L?	0	L	. . .	. . .	. . .
Henryii	S	. . .	. . .	L	0	0
Hillii Dundee	S	. . .	M–S	L	0	0
Hillspire (Cupressifolia)	0	0	M–S	L	0	0
Kosteri	0	0	. . .	. . .	. . .	. . .
Manhattan Blue	S	. . .	. . .	L	0	0
Nova	S	S	. . .	. . .	. . .	. . .
Oxford	M	. . .	. . .	M	0	0
Pendula	S	M	M	. . .	. . .	. . .
Pseudocupressus	0	0	. . .	. . .	. . .	. . .
Pyramidalis	L?	0	. . .	. . .	. . .	. . .
Pyramidiformis	S	S	M	. . .	. . .	. . .
Reptans	S	0	L	. . .	. . .	. . .
Schottii	S	0	. . .	. . .	. . .	. . .
Tripartita	0	0	0	. . .	. . .	. . .
Variegata	M	L	. . .	. . .	. . .	. . .
Venusta	0	. . .	. . .	. . .	. . .	. . .

some branch dieback and make the plants aesthetically unacceptable. Furthermore, the perennial galls of cedar-quince rust may result in significant problems in both juniper and the alternate host in the nursery and in the landscape after outplanting.

In areas where rust has been a problem, select resistant junipers (Table 52.1). Avoid growing susceptible rosaceous hosts, such as crabapple and hawthorn, adjacent to junipers.

Galls may be physically removed from junipers during winter if the incidence of galling is low. The galls are killed, once detached from the tree.

Alternatively, apply fungicides at two-week intervals from early July through August to prevent infection and gall formation on susceptible junipers. Fungicides are ineffective in controlling the gelatinous galls in the spring.

Botryosphaeria Canker
Botryosphaeria stevensii

Seiridium Canker
Seiridium unicorne

At least two canker diseases affect junipers: Botryosphaeria canker and Seiridium canker. Both cause distinct, sunken lesions in the bark and restrict the movement of water and nutrients in the tree, and they may ultimately lead to branch dieback and tree death.

Symptoms

Botryosphaeria canker A canker disease caused by the fungus *Botryosphaeria stevensii* has been reported in the Great Plains region and in Pennsylvania, but it is probably more widely distributed in the United States. Affected junipers develop elongated, flattened, often resinous cankers, which may occur anywhere on woody stems but are commonly located near branch crotches in the interior portion of the crown. Cankers are difficult to see, and it is often necessary to cut off a dead branch and carefully scrape away the outer bark to expose the chocolate brown, dead tissue in the canker. The surrounding healthy tissue will be pearl white. Small, black fruiting bodies of the fungus also develop in cankers, but they may be partially hidden by thin pieces of dead bark.

Occasionally, girdling stem cankers cause rapid death of the top one-third to one-half of the tree crown. More commonly, the disease kills branches throughout the crown and causes a gradual tree decline.

Botryosphaeria canker should not be confused with Kabatina tip blight and Phomopsis tip blight, which affect only foliage and succulent branch tips.

Seiridium canker Symptoms of Seiridium canker, caused by the fungus *Seiridium unicorne,* are similar to those of Botryosphaeria canker but are less common. Elongated, flattened cankers form on small branches and main stems. Bleeding and resin formation is usually associated with cankers. Multiple branch and stem cankers can coalesce, causing branch dieback and, in some cases, tree death.

Epidemiology

Both canker fungi survive in infected branch tissue. Spores of *B. stevensii* are produced in black fruiting structures and are dispersed by rain or irrigation, primarily in late May and June. Spores of *S. unicorne* are dispersed throughout the summer, but infection periods are not known.

Host Range

Botryosphaeria canker is primarily a problem in Rocky Mountain and savin junipers. Eastern red cedar and Chinese juniper are more resistant (Table 52.1). Seiridium canker occurs in oriental arborvitae; baldcypress; Arizona, Italian, and Leyland cypresses; and occasionally junipers.

Management

Avoid planting highly susceptible Rocky Mountain juniper cultivars. Remove cankered branches in winter or late spring. Do not prune or shear diseased junipers in May or June. Pruning and sanitation may not completely suppress canker development.

Seiridium canker tends to be associated with trees suffering from winter damage, drought, or other environmental stresses. Suppress canker development by irrigation and by protecting trees from winter desiccation. Prune cankered branches, and destroy them.

There are no chemical controls for canker diseases of junipers.

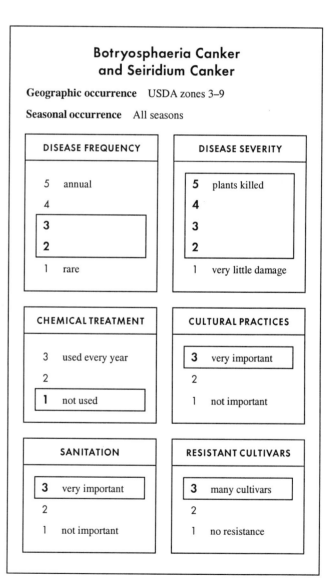

Botryosphaeria Canker and Seiridium Canker

Geographic occurrence USDA zones 3–9

Seasonal occurrence All seasons

DISEASE FREQUENCY	DISEASE SEVERITY
5 annual	**5 plants killed**
4	**4**
3	**3**
2	**2**
1 rare	1 very little damage

CHEMICAL TREATMENT	CULTURAL PRACTICES
3 used every year	**3 very important**
2	2
1 not used	1 not important

SANITATION	RESISTANT CULTIVARS
3 very important	**3 many cultivars**
2	2
1 not important	1 no resistance

Root Diseases

Phytophthora cinnamomi
Phytophthora cryptogea
Rhizoctonia solani

Several soilborne fungi can injure juniper roots. Root rot is most damaging in container media or soils with poor aeration and drainage. Laboratory analysis is normally necessary to for identification of the pathogens associated with rotted roots.

Symptoms

Aboveground symptoms of root injury include stunting and yellowing of foliage, branch dieback, low vigor, and sometimes rapid plant death. Roots will appear dark and discolored, in contrast to the whitish root tips and reddish brown secondary roots of healthy junipers. Damage is more common in prostrate and horizontal forms of juniper.

Root Diseases

Geographic occurrence

Phytophthora cinnamomi USDA zones 6–9
Phytophthora cryptogea USDA zones 7–9
Rhizoctonia solani USDA zones 3–10

Seasonal occurrence

Phytophthora cinnamomi* and *P. cryptogea May to September, with an optimum period in July

Rhizoctonia solani April to October, with an optimum period from June through September

DISEASE FREQUENCY	DISEASE SEVERITY
5 annual	**5 plants killed**
4	4
3	3
2	2
1 rare	1 very little damage

CHEMICAL TREATMENT	CULTURAL PRACTICES
3 used every year	**3 very important**
2	2
1 not used	1 not important

SANITATION	RESISTANT CULTIVARS
3 very important	**3 many cultivars**
2	2
1 not important	1 no resistance

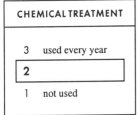

Epidemiology

Inoculum of *Rhizoctonia solani* consists of sclerotia and mycelium associated with organic debris in soil. *Phytophthora cinnamomi* and *P. cryptogea* survive as chlamydospores and sporangia in soil. These pathogens infect just behind the root tip and at the sites of lateral root emergence. Infection of root tissues results in foliar chlorosis, stunting, and necrosis. Small plants may be killed in one growing season; larger plants may decline over a period of several years.

Host Range

Most taxa of juniper in the nursery trade are susceptible to Phytophthora root rot. However, some cultivars are tolerant (Table 52.2). In North Carolina, *P. cinnamomi* severely affected top growth and root growth of rooted cuttings of *J. chinensis* 'Sargent's' and *J. procumbens* 'Nana' and cuttings of *J. horizontalis* 'Andorra' and 'Bar Harbor,' *J. chinensis* 'Parsonii,' and *J. conferta* but did not affect the growth of *J. horizontalis* 'Blue Rug' and 'Douglasii' (Kuske et al., 1981) (Table 52.2). In California, *P. cinnamomi* caused severe root rot in *J. chinensis* 'Gold Coast' and *J. sabina* var. *tamariscifolia,* whereas *P. cryptogea* was pathogenic only to *J. conferta* 'Blue Pacific,' unless plants were subjected to periodic flooding, under which conditions most species were susceptible (Standish et al., 1982).

Management

To avoid root rots, plant junipers in well-drained soils and away from locations where water has a tendency to puddle or

Table 52.2 Responses of juniper cultivars (*Juniperus* spp.) to *Phytophthora cinnamomi* and *P. cryptogea*[a]

	P. cinnamomi	*P. cryptogea*[b]
J. chinensis		
Gold Coast	Susceptible	Tolerant
Parsonii	Susceptible	. . .[c]
Pfitzer	Tolerant	. . .
Plumosa	Tolerant	Tolerant
Prostrata	Tolerant	Tolerant
Sargent's	Susceptible	. . .
J. conferta	Susceptible	. . .
Blue Pacific	Tolerant	Susceptible
J. horizontalis		
Andorra	Susceptible	. . .
Bar Harbor	Susceptible	Tolerant
Blue Rug	Tolerant	. . .
Douglasii	Tolerant	. . .
Prince of Wales	Tolerant	Tolerant
Winter Blue	Susceptible	. . .
J. procumbens		
Nana	Susceptible	. . .
J. sabina		
var. *tamariscifolia*	Susceptible	Tolerant
J. virginiana		
Prostrata	Tolerant	Tolerant

[a] Data from Kuske et al. (1981) and Standish et al. (1982)

[b] Most cultivars were tolerant of *P. cryptogea* under normal growing conditions, but many plants developed severe root rot when they were periodically flooded for 48 hours.

[c] Not tested.

Nematode Decline

Geographic occurrence USDA zones 3–10

Seasonal occurrence April to October, with an optimum period in May, under average yearly conditions in the southern United States (adjust for your hardiness zone)

DISEASE FREQUENCY	DISEASE SEVERITY
[5 annual]	5 plants killed
4	4
3	**[3]**
2	2
1 rare	1 very little damage

CHEMICAL TREATMENT	CULTURAL PRACTICES
3 used every year	**[3 very important]**
[2]	2
1 not used	1 not important

SANITATION	RESISTANT CULTIVARS
[3 very important]	3 many cultivars
2	**[2]**
1 not important	1 no resistance

stand for extended periods. Do not use nonporous, plastic mulches as a weed barrier under junipers. Plastic mulch inhibits water and air movement and is detrimental to root growth. Instead, select organic mulches and breathable fabrics for weed control in planting beds.

Selections of *J. horizontalis* and *J. sabina* appear to be particularly susceptible to root rots and should not be planted in heavy, poorly drained soils. Junipers with tolerance for Phytophthora root rot have been identified among the popular cultivars grown today and should be selected for regions such as the West Coast, the South, and the mid-Atlantic states, where Phytophthora root rot is common (Table 52.2).

Fungicide and nematicide soil drenches are sometimes used to control root rots in container nursery production but are not often used in field plantings.

Nematode Decline

Pratylenchus vulnus

Nematode decline of junipers is common in field-grown nursery stock but may go unnoticed until foliar symptoms develop during periods of drought. In the absence of drought, depression of normal plant growth may still be a significant factor in field production of junipers by the second growing season.

The effects of several nematodes on the growth and development of junipers have been studied, but only the lesion nematode *Pratylenchus vulnus* has been shown to cause severe stunting and poor growth in certain juniper taxa and cultivars under field conditions. The pathogen in a migratory endoparasitic nematode.

In greenhouse experiments, *J. horizontalis* 'Douglasii' and 'Plumosa' grown in soil infested with the root-knot nematode *Meloidogyne incognita* were significantly smaller than plants in nematode-free soil (Nemec and Struble, 1968). Top and root growth were affected, but no root galling was observed.

Symptoms

In *J. excelsa* var. *stricta* (spiny Greek juniper) and *J. horizontalis* 'Wiltonii' ('Blue Rug') grown in microplots of soil infested with *P. vulnus,* foliar symptoms by were not evident until 12–17 months after planting (Benson and Barker, 1982). The only symptom in spiny Greek juniper was a dwarfing of the plant canopy, whereas severe necrosis and dwarfing of foliage occurred in Blue Rug (Plate 70). Generally, root symptoms are not associated with *P. vulnus.*

Epidemiology

Parasitic nematodes such as *P. vulnus* live in the soil feeding in roots of susceptible plants. Nematodes reproduce by eggs that can survive for many years in soil in the absence of a suitable host. Crop rotation may be only partially successful in controlling nematode diseases, because of the survival of egg and juvenile stages between susceptible crops. Nematodes may be introduced into a field on infected planting material, in soil adhering to equipment coming from an infested area, and naturally by surface runoff from surrounding areas in which cultivated or native plants are infected.

Host Range

The host range of *P. vulnus* in juniper taxa has not been studied extensively. Only spiny Greek juniper and Blue Rug are known to be susceptible, but other taxa and cultivars may be as well.

Other nematode taxa have been associated with soil samples from the root zones of various junipers, but no damage thresholds have been established.

Management

Management of nematodes is difficult once plants have become infected, because there are few, if any, suitable nematicides for postplant application. Field production sites should be sampled for parasitic nematodes prior to establishing a planting. If damaging species of nematodes are detected in samples, preplant treatments should be used. Damage from nematodes during drought periods can be minimized with organic mulches and adequate irrigation.

REFERENCES

Benson, D. M., and Barker, K. R. 1982. Susceptibility of Japanese boxwood, dwarf gardenia, Compacta (Japanese) holly, spiny Greek and Blue Rug junipers, and nandina to four nematode species. Plant Dis. 66: 1176–1179.

Hartman, J. R., Doney, J., Johnson, M., Fountain, W., and McNiel, R.

1994. Evaluation of juniper cultivars for susceptibility to Phomopsis and Kabatina blights, 1993. Biol. Cult. Tests Control Plant Dis. 9:161.

Hartman, J. R., Doney, J., Johnson, M., Fountain, W., and McNiel, R. 1995. Evaluation of juniper cultivars for susceptibility to Kabatina blight, 1994. Biol. Cult. Tests Control Plant Dis. 10:68.

Himelick, E. B., and Neely, D. 1960. Juniper hosts of cedar-apple rust and cedar-hawthorn rust. Plant Dis. Rep. 44:109–112.

Kuske, C. R., Benson, D. M., and Jones, R. K. 1981. Susceptibility of juniper cultivars to root rot caused by *Phytophthora cinnamomi.* Pages 114–115: in Proc. South. Nurserymen's Assoc. Res. Conf., 26th.

Nemec, S., and Struble, F. B. 1968. Response of certain woody ornamental plants to *Meloidogyne incognita.* Phytopathology 58:1700–1703.

Ostrofsky, A., and Peterson, G. 1981. Etiological and cultural studies of *Kabatina juniperi.* Plant Dis. 65:908–910.

Peterson, G. W. 1981. Pine and juniper diseases of the Great Plains. U.S. For. Serv. Gen. Tech. Rep. RM-86.

Schoeneweiss, D. E. 1969. Susceptibility of evergreen hosts to the juniper blight fungus *Phomopsis juniperovora* under epidemic conditions. J. Am. Soc. Hortic. Sci. 94:609–611.

Standish, E. D., MacDonald, J. D., and Humphrey, W. A. 1982. Phytophthora root and crown rot of junipers in California. Plant Dis. 66:925–928.

Tisserat, N., and Pair, J. 1997. Susceptibility of selected juniper cultivars to cedar-apple rust, Kabatina tip blight, Cercospora needle blight and Botryosphaeria canker. J. Environ. Hortic. 15:160–163.

Tisserat, N., Rossman, A. Y., and Nus, A. 1988. A canker disease of Rocky Mountain juniper caused by *Botryosphaeria stevensii.* Plant Dis. 72: 699–701.

Tisserat, N., Nus, A., and Barnes, L. W. 1991. A canker disease of the Cupressaceae in Kansas and Texas caused by *Seiridium unicorne.* Plant Dis. 75:138–140.

CHAPTER 53

Margery L. Daughtrey ▪ Cornell University, Riverhead, New York

Leucothoe Diseases

Geographic production USDA zones 4–8

Family Ericaceae

Genus *Leucothoe*

Species	*L. axillaris*	coast leucothoe	zone 5
	L. fontanesiana (*catesbaei*)	drooping leucothoe, fetterbush	zones 4–6
	L. keiskei	Keisk's leucothoe	zone 5
	L. populifolia	Florida leucothoe	zones (6)7–9
	L. racemosa	sweetbells leucothoe	zones 5–9

Leucothoe is primarily a plant for shady areas with plenty of moisture. It blends nicely with rhododendrons and other shade-loving ornamentals. Diseases affecting leucothoe in the nursery are powdery mildew, Cylindrocladium leaf spot, and Phytophthora root rot.

Powdery Mildew
Microsphaera sp.

Powdery mildew diseases have the same symptoms in different hosts, but many of the fungi causing these diseases are host-specific. In leucothoe, powdery mildew can easily be misdiagnosed, because only in severe cases is the growth of the fungal pathogen thick enough to appear white.

Symptoms

Leaves infected with powdery mildew may first show pale, yellow green, irregularly shaped blotches. These areas enlarge to a diameter of about half an inch (1 cm) and turn purplish. They may coalesce until much of the leaf is affected. The purple color may be most vivid on the underside of the leaf. These discolored areas result from a reaction of the plant tissue to a thin coating of mycelium of the pathogen, which is not always easily visible to the naked eye. Heavy infection may lead to defoliation.

Epidemiology

It is not known whether the fungus causing leucothoe powdery mildew overwinters primarily as cleistothecia (sexual spore cases) or within the buds of the plant. Once the disease is encountered in the nursery, it is likely that inoculum will be available for new infections each year. Plants held too long in overwintering houses in the spring are prone to severe powdery mildew injury.

Host Range

L. axillaris is quite susceptible to a *Microsphaera* sp. causing powdery mildew, which probably also attacks other members of the family Ericaceae.

Powdery Mildew

Geographic occurrence USDA zones 4–8

Seasonal occurrence May to August

DISEASE FREQUENCY	DISEASE SEVERITY
5 annual	5 plants killed
4	4
[3]	3
2	**[2]**
1 rare	1 very little damage

CHEMICAL TREATMENT	CULTURAL PRACTICES
[3] used every year	3 very important
2	2
1 not used	**[1]** not important

SANITATION	RESISTANT CULTIVARS
3 very important	3 many cultivars
2	**[2]**
[1] not important	1 no resistance

Management

Carefully monitor plants to note the first onset of symptoms, which is the cue for beginning fungicide treatment. Protection of new growth prior to spring and early summer sales is most important. See Chapter 15, "Powdery Mildew," for information on control.

Cylindrocladium Leaf Spot and Stem Rot

***Cylindrocladium* spp.**

Three species of *Cylindrocladium* are known to cause leaf spot diseases of *L. axillaris: C. leucothoeae, C. avesiculatum,* and *C. scoparium,* which also causes stem canker, dieback, and wilt. A stem lesion and wilt disease caused by *C. theae* has also been reported.

Symptoms

Infection of *L. axillaris* by *C. avesiculatum* causes brown spots, which enlarge into zonate blotches, with alternating light and dark rings. Brown spots may also develop on shoots, which die back if the infection girdles the stems. In severe cases, leaf drop may be extensive.

Epidemiology

Under warm, humid conditions, leaf spots will be visible 48 hours after infection occurs. Of the three species known to cause leaf spots on leucothoe, *C. scoparium* is the most cosmopolitan. *C. avesiculatum* and *C. leucothoeae* are reported in the southeastern United States.

Host Range

Rainbow leucothoe is susceptible to *C. scoparium,* which has a wide host range and affects many woody ornamentals. Azaleas and yaupon holly are also susceptible to *C. avesiculatum.* Only leucothoe has been reported to be a host of *C. leucothoeae.*

Management

See Chapter 11, "Diseases Caused by *Cylindrocladium.*"

Cylindrocladium Leaf Spot and Stem Rot

Geographic occurrence USDA zones 5–8

Seasonal occurrence May to August

DISEASE FREQUENCY	DISEASE SEVERITY
5 annual	5 plants killed
4	**[4]**
3	3
[2]	2
1 rare	1 very little damage

CHEMICAL TREATMENT	CULTURAL PRACTICES
[3] used every year	**[3] very important**
2	2
1 not used	1 not important

SANITATION	RESISTANT CULTIVARS
[3] very important	3 many cultivars
2	**[2]**
1 not important	1 no resistance

REFERENCES

Gill, D. L. 1979. Powdery mildew and *Cylindrocladium* attack *Leucothoe axillaris.* Plant Dis. Rep. 63:358–359.

Grand, L. F. 1985. North Carolina plant disease index. North Carolina Agric. Res. Serv. Tech. Bull. 240:1–157.

Mims, F., Benson, D. M., and Jones, R. K. 1981. Susceptibility of leucothoe, hybrid rhododendron, and azalea to *Cylindrocladium scoparium* and *C. theae.* Plant Dis. 65:353–354.

CHAPTER 54

Jean L. Williams-Woodward ▪ University of Georgia, Athens
Alan S. Windham ▪ University of Tennessee, Nashville

Leyland Cypress Diseases

Geographic production USDA zones 6–10

Family Cupressaceae

Genus × *Cupressocyparis*

Species *C. leylandii*

Cultivars several

Production of Leyland cypress as an ornamental evergreen and Christmas tree is now in the hundreds of thousands in field and container nurseries in the Southeast. Its quick growth rate makes it suitable for screens, groupings, and hedges. Trees can easily grow 3 feet per year and reach heights of 60 to 70 feet under landscape conditions. However, Leyland cypress is not likely to be a long-term landscape plant (10 to 20 years), because of diseases and insect pests.

Seiridium Canker and Twig Dieback

Seiridium cardinale
Seiridium unicorne

Seiridium canker is the most important and damaging disease of Leyland cypress. Plants of all sizes and ages, from liners and 1-gallon material to mature trees, can be infected. The pathogen most commonly associated with cankers and twig dieback in the Southeast is the fungus *Seiridium unicorne. S. cardinale* has also been reported to cause cankers on cypress species.

Symptoms

Infection results in the formation of cankers on stems and branches and in branch axils, causing twig and branch dieback (Plate 71). The cankered areas are sunken, with raised margins, and may be discolored dark brown to purple. Resin oozes profusely from cracks in the bark in infected areas. The cambial tissue beneath oozing sites is discolored reddish to brown. On affected stems and branches, the foliage above the cankers may turn yellow, reddish brown, or gray.

In infection studies using stems and branches (5–13 mm in diameter), infection occurred within two weeks following inoculation with *S. unicorne.*

Instead of producing one large canker that girdles a branch or stem, the infection produces numerous thin, elongated cankers, which collectively interfere with the flow of water through a branch and eventually kill it. Infected trees appear thinly branched and often decline from the interior outward and from the lower part of the tree upward.

Branch tip and twig dieback is also associated with *Seiridium* infection. Infected tips turn yellow or rust-colored, and a gray discoloration appears at the point of infection.

Seiridium Canker

Geographic occurrence USDA zones 6–10; the disease has become much more common in the Southeast in recent years

Seasonal occurrence Year-round, with a peak from May to October, under average yearly conditions in zone 7 (adjust for your hardiness zone)

DISEASE FREQUENCY	DISEASE SEVERITY
5 annual	**5** plants killed
4	**4**
3	3
2	2
1 rare	1 very little damage

CHEMICAL TREATMENT	CULTURAL PRACTICES
3 used every year	**3** very important
2	2
1 not used	1 not important

SANITATION	RESISTANT CULTIVARS
3 very important	3 many cultivars
2	2
1 not important	**1** no resistance

Epidemiology

Environmental stress, particularly drought stress, favors infection. Spring freeze injury may predispose trees to infection. Small, black, pimple-like spore-producing structures of the pathogen develop in cankered areas. Spores from these structures are washed down the tree or splashed from branch to branch or tree to tree by rain or overhead irrigation. Pruning tools and insects also spread the fungus. New infections occur when spores lodge in bark cracks and wounds. Infection through lenticels may occur following hot, dry weather.

Host Range

Seiridium canker has been reported as a disease of numerous needled evergreens, including arborvitae, cypress (*Cupressus* spp.), and *Chamaecyparis* spp.

Management

Seiridium infection can be reduced by avoiding water stress and wounding of trees. Field-grown trees should be irrigated during periods of drought. Container-grown trees should be placed in overwintering houses to protect them from cold injury prior to a sudden drop in temperature or prolonged subfreezing weather. Do not take cuttings from infected plants. Promptly prune infected shoots and branches, and remove and destroy severely infected plants. Fungicides may provide some control in the nursery, especially when applied during periods of temperature and water stress, to prevent the fungus from infecting trees. Fungicides applied to infected trees have little effect.

Botryosphaeria Canker

Botryosphaeria dothidea
(asexual stage, *Fusicoccum* sp.)
Botryosphaeria ribis
(asexual stage, *Botryodiplodia* sp.)

Botryosphaeria canker affects numerous plants worldwide. The fungi that cause this disease, *Botryosphaeria dothidea* and *B. ribis,* attack plants predisposed to infection by environmental factors, such as freezing or drought, or by wounds. The disease occurs in nurseries and in landscape plantings.

Symptoms

Botryosphaeria canker causes branch dieback and sunken, sometimes resinous cankers along branches and main stems. The cankers are often centered on twig or branch stubs or at wound sites. They enlarge rapidly and can girdle the branch or stem, killing the tree. The cambial tissue beneath the canker is discolored dark brown, so that there is a clear distinction between healthy and infected, dead tissue. Often the first indication of the disease is the rusty reddish brown color of infected branches following infection.

Epidemiology

Infection is likely to occur in late spring and early summer during periods of wet, humid weather when spores of *Fusicoccum* sp. and *Botryodiplodia* sp. (the asexual stages of *B. dothidea* and *B. ribis,* respectively) are produced. Spores are splashed from branch to branch or tree to tree or washed down branches by rain or overhead irrigation. Wounded or stressed trees are most susceptible to Botryosphaeria canker.

Host Range

Botryosphaeria spp. and their asexual stages are nonspecialized weak pathogens that can infect numerous evergreen and deciduous trees and shrubs.

Management

Infected branches should be pruned below the cankered area immediately upon detection. Infection usually occurs at the sites of wounds, and therefore avoiding wounds and injuries will reduce infection. Environmental and cultural stresses predispose plants to infection and should be avoided. Provide adequate irrigation during extended periods of dry weather, and avoid heavy fertilization and severe pruning of larger plants. Fungicides may help reduce infection of stressed and wounded trees in nurseries, but they have little effect on trees already infected.

Botryosphaeria Canker

Geographic occurrence USDA zones 6–10, wherever Leyland cypress is grown

Seasonal occurrence Year-round, with a peak period in April–July, under average yearly conditions in zone 7 (adjust for your hardiness zone)

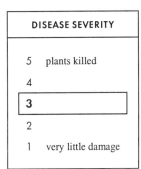

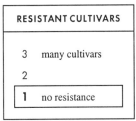

DISEASE FREQUENCY	DISEASE SEVERITY
5 annual	5 plants killed
4	4
3	**3**
2	2
1 rare	1 very little damage

CHEMICAL TREATMENT	CULTURAL PRACTICES
3 used every year	**3 very important**
2	**2**
1 not used	1 not important

SANITATION	RESISTANT CULTIVARS
3 very important	3 many cultivars
2	2
1 not important	**1 no resistance**

Cercospora Needle Blight

Asperisporium sequoiae (syn. *Cercospora sequoiae*)

Cercospora needle blight affects numerous plants in the families Cupressaceae and Taxodiaceae. Its distribution in Leyland cypress is not known, but it has been found in container, field, and Christmas tree nurseries in Alabama, Florida, Georgia, North Carolina, and South Carolina.

Symptoms

The disease causes a progressive browning and loss of foliage on one-year-old and older growth. Lower branches are affected first, and the disease spreads upward and outward. Needle drop is extensive. Reddish brown lesions are formed on one-year-old twigs. New growth from the current year is not affected. Infected needles and twigs turn brown and eventually gray. Dark, cushion-like fruiting structures of the fungal pathogen are produced on dead needles.

Epidemiology

Spores of the pathogen, *Asperisporium sequoiae* (syn. *Cercospora sequoiae*), are produced on dead needles in late spring and summer. Infection is likely to occur in summer, although infection of juniper has been reported to occur in the fall in the Great Plains. Spore dispersal and infection occur during extended periods of rainfall.

Host Range

Cercospora needle blight affects members of the family Cupressaceae, including *Chamaecyparis, Cupressus, Juniperus,* and *Thuja,* and members of the family Taxodiaceae, including *Cryptomeria, Sequoia, Sequoiadendron,* and *Taxodium.*

Management

No management program for the control of Cercospora needle blight of Leyland cypress is known at this time. However, pruning affected branches from trees and increasing air circulation around

Cercospora Needle Blight

Geographic occurrence USDA zones 6–10, wherever Leyland cypress is grown

Seasonal occurrence March–November, and especially May–October, under average yearly conditions in zone 7 (adjust for your hardiness zone)

DISEASE FREQUENCY	DISEASE SEVERITY
5 annual	5 plants killed
4	**4**
3	3
2	2
1 rare	1 very little damage

CHEMICAL TREATMENT	CULTURAL PRACTICES
3 used every year	**3** very important
2	**2**
1 not used	1 not important

SANITATION	RESISTANT CULTIVARS
3 very important	3 many cultivars
2	2
1 not important	**1** no resistance

Kabatina Blight

Geographic occurrence USDA zones 6–10, wherever Leyland cypress is grown

Seasonal occurrence March–June, with a peak period in May–June, under average yearly conditions in zone 7 (adjust for your hardiness zone)

DISEASE FREQUENCY	DISEASE SEVERITY
5 annual	5 plants killed
4	4
3	3
2	**2**
1 rare	1 very little damage

CHEMICAL TREATMENT	CULTURAL PRACTICES
3 used every year	3 very important
2	**2**
1 not used	1 not important

SANITATION	RESISTANT CULTIVARS
3 very important	3 many cultivars
2	2
1 not important	**1** no resistance

trees can reduce the spread of the disease. Avoid overhead irrigation. Fungicide applications beginning in early summer and continuing through the fall may help reduce infection of one-year-old and older growth.

Kabatina Blight

Kabatina juniperi
Kabatina thujae

Kabatina blight, a tip dieback of one-year-old growth, has been reported in juniper, arborvitae, and cypress. The disease is caused by the fungi *Kabatina juniper* and *K. thujae*.

Symptoms

During early spring infected shoots 6 to 8 inches long turn yellow brown and then reddish brown. Small cankers may be observed at the base of infected shoots. Small, black, spore-producing structures of the fungal pathogen may be observed at the base of infected shoots. Infected shoots remain on the tree or drop prematurely.

Epidemiology

Kabatina spp. are wound pathogens unable to infect healthy plant tissue. Wounds made by insects, freeze injury, or ice damage may serve as entry points for these fungi.

Host Range

Kabatina spp. infect juniper, arborvitae, and other cypresses.

Management

Infected shoots should be pruned below the site of infection.

REFERENCES

Hansen, M., and Lewis, K. J., eds. 1997. Compendium of Conifer Diseases. American Phytopathological Society, St. Paul, Minn.

Jones, R. K. 1993. Seiridium canker on Leyland cypress in North Carolina. Proc. South. Nurserymen's Assoc. Res. Conf. 38:220.

McCain, A. H. 1984. Cypress canker control with fungicides. J. Arboric. 10(7):212–214.

Sinclair, W. A., Lyon, H. H., and Johnson, W. T. 1987. Diseases of Trees and Shrubs. Comstock Publishing Associates, Cornell University Press, Ithaca, N.Y.

Strouts, R. G. 1973. Canker of cypress caused by *Coryneum cardinale*. Eur. J. For. Pathol. 3:13–24.

Tisserat, N. A., Nus, A., and Barnes, L. W. 1991. A canker disease of the Cupressaceae in Kansas and Texas caused by *Seiridium unicorne*. Plant Dis. 75:138–140.

Windham, A. S., Stebbins, T. C., and Windham, M. T. 1996. Canker and shoot blight diseases of Leyland cypress. Proc. South. Nurserymen's Assoc. Res. Conf. 41:178–179.

CHAPTER 55

Gary W. Simone · University of Florida, Gainesville

Ligustrum Diseases

Geographic production USDA zones 3–10

Family Oleaceae

Genus *Ligustrum*

Species

L. amurense	Amur privet
L. × *ibolium*	
L. japonicum	Japanese privet
L. lucidum	glossy privet, wax-leaf privet
L. obtusifolium	
L. ovalifolium	California privet
L. sinense	Chinese privet
L. vulgare	common privet, evergreen privet

Ligustrum is a familiar genus of over 50 species of deciduous and evergreen shrubs to short trees, commonly known as privets. They are used primarily as hedge plants and as specimen plants in the landscape. Common species in the nursery trade include common or evergreen privet (*L. vulgare*), California privet (*L. ovalifolium*), Amur privet (*L. amurense*), Japanese privet (*L. japonicum*), glossy or wax-leaf privet (*L. lucidum*), and Chinese privet (*L. sinense*). Privets are a large production item in nurseries, because of their wide use across the country. Privets are affected by powdery mildew and several other leaf diseases, and field-grown plants can be damaged by lesion and root-knot nematodes.

Anthracnose

Colletotrichum gloeosporioides
(sexual stage, *Glomerella cingulata*)

Symptoms

Anthracnose is a leaf spot disease affecting several *Ligustrum* spp. Lesions develop along the leaf margin and are often associated with wounds. Spots on *L. japonicum* and *L. lucidum* are round to oval, with tan centers, and dark fruiting bodies (acervuli) of the fungus can be seen scattered in the lesion centers. Anthracnose spots can span half a leaf, and during moist weather a pinkish white spore mass released by the acervuli will be evident.

Both the asexual stage of the pathogen, *Colletotrichum gloeosporioides,* and the sexual stage, *Glomerella cingulata,* can be observed in a canker and dieback complex in *Ligustrum* spp. The fungus invades branch or twig stubs and forms a canker, which gradually enlarges until the stem is girdled and its distal parts die. Leaves brown and are retained for some time. The dieback can affect more than 18 inches of growth at the tips of branches and twigs. Acervuli and black, globose perithecia of *Glomerella* can be observed in the bark in the centers of cankers or at points of infection.

Epidemiology

The pathogen is transported in splashing water and, in its asexual state, may also be carried by insects. It attacks stressed or

Anthracnose

Geographic occurrence USDA zones 4–10

Seasonal occurrence Late spring through fall

DISEASE FREQUENCY
- **5 annual**
- 4
- 3
- 2
- 1 rare

DISEASE SEVERITY
- 5 plants killed
- 4
- 3
- **2**
- 1 very little damage

CHEMICAL TREATMENT
- 3 used every year
- **2**
- 1 not used

CULTURAL PRACTICES
- 3 very important
- 2
- **1 not important**

SANITATION
- **3 very important**
- 2
- 1 not important

RESISTANT CULTIVARS
- **3 many cultivars**
- 2
- 1 no resistance

damaged tissue but can also directly penetrate leaf tissue. It also infects buds, soft twigs, and damaged twigs. An infection may be established and go dormant for weeks or months, until the environment or the host is more favorable for disease development. The pathogen is infective in both the sexual and the asexual stages, during periods of water splash.

Host Range

Leaf anthracnose has been reported in *L. lucidum, L. japonicum,* and *L. sinense* from New England to the Midwest, south to Texas, and east to Florida. The pathogen has a wide host range among woody and herbaceous plants in addition to *Ligustrum.*

Management

Use only healthy, disease-free stock for vegetative propagation or graft propagation. Stock should be sprayed with fungicides several times before cuttings are taken where the disease is severe. Systemic or protectant fungicides may be needed during rainy periods in the summer. Time fungicide applications for shoot emergence in the spring as well. Affected plants should be pruned below cankered, discolored wood. Prunings should be collected and destroyed. Cultivars of *L. vulgare* are the most susceptible to dieback. Cultivars of *L. amurense, L. ibota, L. obtusifolium* var. *regelianum,* and *L. ovalifolium* are more resistant.

Cercospora Leaf Spot

Cercospora adusta
Cercospora lilacis
Pseudocercospora ligustri

Three species of fungi cause Cercospora leaf spot diseases of *Ligustrum* (Plate 72): *Cercospora adusta, C. lilacis,* and *Pseudocercospora ligustri.* The symptoms of these diseases vary somewhat, depending on the pathogen.

Symptoms

Lesions caused by *C. adusta* are circular, 5–30 mm in diameter, and have a brown, sunken center, reddish brown margins, and a diffuse yellow halo. This pathogen causes two or three lesions per leaf, usually at the tip or margin, and primarily in the interior of the canopy.

C. lilacis produces small, light yellow areas of infection, which enlarge and change from lemon yellow to orange to brown. Mature lesions are circular to irregularly shaped, 5–30 mm in diameter, with a tan to dark brown center and reddish to purplish brown margins.

P. ligustri produces circular lesions, 5–15 mm in diameter, which are slightly zonate and sunken, with a tan to brown center, reddish purple borders, and a slight yellow halo. The lesions often coalesce. Lesions may be smaller on some *Ligustrum* spp. Heavy infection will cause leaf abscission, particularly in *L. japonicum.*

Epidemiology

In *L. lilacis,* leaf spot caused by *P. ligustri* develops in 10 to 14 days; another two to eight days is needed for reproduction of the pathogen from the lesion surface. *C. lilacis* reproduces on both leaf surfaces, while *P. ligustri* produces conidia primarily on the upper surface. The infection cycle of *C. adusta* is not known.

Spores (conidia) of these pathogens are airborne or transported in splashing water within the nursery. The pathogens survive in lesions in the canopy or on fallen leaves on the ground cloth or on the surface of the potting medium. There are no data on the relative susceptibility of *Ligustrum* spp. to Cercospora leaf spots.

Cercospora Leaf Spot

Geographic occurrence USDA zones 7–10

Seasonal occurrence Spring through fall

DISEASE FREQUENCY	DISEASE SEVERITY
[5 annual]	5 plants killed
4	4
3	**[3]**
2	2
1 rare	1 very little damage

CHEMICAL TREATMENT	CULTURAL PRACTICES
3 used every year	3 very important
[2]	**[2]**
1 not used	1 not important

SANITATION	RESISTANT CULTIVARS
3 very important	3 many cultivars
[2]	2
1 not important	**[1 no resistance]**

Host Range

C. adusta is a host-specific pathogen of *Ligustrum* and has been reported in *L. amurense, L.* × *ibolium, L. japonicum, L. lucidum,* and *L. ovalifolium* in the Southeast. It is perhaps the least common of the three species causing Cercospora leaf spot.

C. lilacis invades *Syringae* (lilac) as well as *L. japonicum, L. lucidum,* and *L. sinense.* It has been found only in Florida.

P. ligustri is specific to *Ligustrum* and has been reported in *L. japonicum, L. lucidum, L. ovalifolium, L. sinense,* and *L. vulgare* from Texas to Florida and north to North Carolina.

Corynespora Spot

Corynespora cassiicola

Symptoms

Corynespora spot lesions are initially tiny, circular, reddish spots. On variegated forms of *L. sinense,* they develop into light brown lesions with purple borders (Plate 73). On green cultivars, they develop into light brown lesions with dark brown margins surrounded by a bright yellow halo. Lesions coalesce and may cause leaf defoliation.

Epidemiology

Initial and secondary infections are caused by airborne or water-splashed spores (conidia) of the pathogen, *Corynespora cassiicola.* Where *L. sinense* is truly deciduous, the pathogen may survive on other hosts or on fallen infected leaves. In the Deep South, *L. sinense* does not entirely defoliate, and thus the fungus may survive on lesions in the canopy.

Host Range

C. cassiicola has a wide host range among ornamentals, fruits, and vegetables. It has been isolated from *L. sinense, L. lucidum,* and *L. japonicum* in the Southeast. The original research indicated host specificity of isolates derived from *Ligustrum.*

Management

Avoid excessive irrigation from overhead sources. Propagate from disease-free stock. Fungicides may be necessary to lessen defoliation during seasons favorable for disease development. Apply fungicides repeatedly according to instructions on the product label. All cultivars of *L. sinense* are susceptible. *L. lucidum* and *L. japonicum* are less vulnerable to the pathogen.

Edema

Water congestion

Symptoms

Edema develops primarily on the lower surface of leaves but may extend to petioles and soft new shoots. The first symptoms are small water blisters on the lower leaf surface. Long periods of favorable weather may produce greater leaf rupturing, which results in the formation of callus, appearing as light tan to rusty brown, raised warts on the lower leaf surface (Plate 74). Prolonged favorable weather may trigger the formation of small, chlorotic spots on the upper surface, directly over the rupture zone.

Epidemiology

Edema develops in midwinter and spring in the Deep South and somewhat later in areas farther north. Favorable conditions for disease development are high soil moisture; reduced air circula-

Corynespora Spot

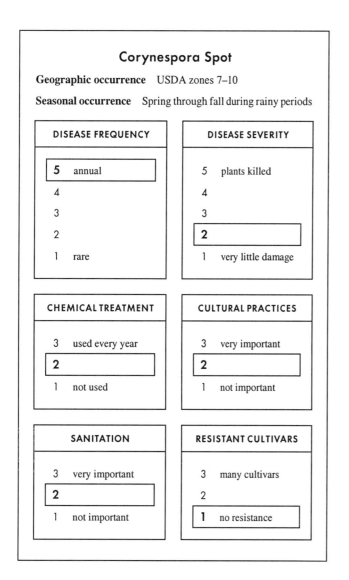

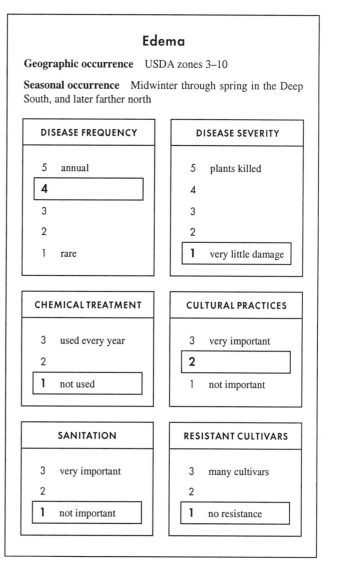

Geographic occurrence USDA zones 7–10

Seasonal occurrence Spring through fall during rainy periods

DISEASE FREQUENCY	DISEASE SEVERITY
5 annual	5 plants killed
4	4
3	3
2	**2**
1 rare	1 very little damage

CHEMICAL TREATMENT	CULTURAL PRACTICES
3 used every year	3 very important
2	**2**
1 not used	1 not important

SANITATION	RESISTANT CULTIVARS
3 very important	3 many cultivars
2	2
1 not important	**1 no resistance**

Edema

Geographic occurrence USDA zones 3–10

Seasonal occurrence Midwinter through spring in the Deep South, and later farther north

DISEASE FREQUENCY	DISEASE SEVERITY
5 annual	5 plants killed
4	4
3	3
2	2
1 rare	**1 very little damage**

CHEMICAL TREATMENT	CULTURAL PRACTICES
3 used every year	3 very important
2	**2**
1 not used	1 not important

SANITATION	RESISTANT CULTIVARS
3 very important	3 many cultivars
2	2
1 not important	**1 no resistance**

tion; cool, moist weather; and reduced solarization. Major storm systems often deliver the favorable conditions that trigger the development of edema overnight. Storm fronts decrease transpiration, because stomata close under low light and evaporation from media is retarded by reduced air circulation. The resultant internal water congestion in plants is expressed as edema.

Host Range

Edema develops in many herbaceous foliage and floral crops in greenhouse production. Woody ornamentals that sustain damage from edema include *Camellia, Eucalyptus, Fatsia,* × *Fatshedera, Hedera, Jasminum, Schefflera,* and *Taxus.*

Management

As edema is noncontagious, few efforts are made to manage the injury in container and field production of woody plants. Avoid overlapping automated irrigation with rainfall in the early season. Space plants properly to enhance air circulation and leaf drying. Greenhouse production of liners can be safeguarded during weather favorable for edema by the application of bottom heat and increased fan activity. Affected portions of the canopy are normally not removed, as one or two new growth flushes will mask the symptoms. No data on resistance to edema are available.

Nematode Damage

***Meloidogyne* spp., root-knot nematodes**
***Pratylenchus* spp., lesion nematodes**

Symptoms

Container-grown *Ligustrum* is normally not damaged by parasitic nematodes, except in unusual cases of contamination of media with nonsterile soil. Field-grown plants, however, may sustain root damage from lesion nematodes (*Pratylenchus* spp.), which feed externally, and root-knot nematodes (*Meloidogyne* spp.), which feed internally on roots. The damage is expressed as a gradual reduction in growth and vigor. Increments of growth are shorter and less frequent. The lower, innermost leaf canopy becomes chlorotic and prematurely senescent. More twigs and shoots become visible through the canopy as leaf loss progresses. Severely affected plants may exhibit nutrient deficiency symptoms due to root dysfunction and the loss of feeder roots.

Lesion nematodes are migratory parasites, causing numerous small, brown lesions on roots. Root symptoms are reduced size and overall discoloration due to the lesions. Damage resulting from feeding by these nematodes is often compounded by secondary invasion of wounded roots by fungi.

Adult female root-knot nematodes are sedentary parasites, causing galls to develop at their specialized feeding sites in the root tissue. Numerous galls of various sizes are formed on feeder roots, and the affected root system exhibits some discoloration.

Epidemiology

Field incidence of nematodes feeding on *Ligustrum* usually results from continued field production of susceptible woody hosts or ornamental crops at sites previously planted with agronomic or vegetable crops. Nematodes parasitize the fibrous feeder roots. Invasion by root-knot nematodes results in interference with water uptake and nutrient movement. Lesion nematode injury reduces the size of the feeder root system, limiting the uptake of water and nutrients and creating wounds for root-rotting fungi. These nematodes survive as adults, juveniles, and eggs in soil and root debris of host plants.

Host Range

M. arenaria, M. hapla, M. incognita, and *M. javanica* have been reported on privet species, including *L. japonicum, L. lucidum,* and *L. sinense.* Undoubtedly, other *Ligustrum* species are susceptible, but reports are poorly defined and not generally accessible.

Management

Nematode sampling should be done before *Ligustrum* is established in a field. If damaging populations of nematodes are present, soil fumigation may be needed before planting. Infested land should be rotated out of production of susceptible plants. Low disease severity in some fields may be mediated with contact granular nematicides. This treatment may restore plant quality to allow plants to be sold or at least used as stock plants for vegetative propagation.

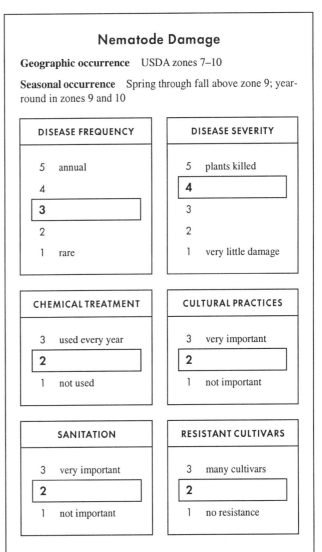

Nematode Damage

Geographic occurrence USDA zones 7–10

Seasonal occurrence Spring through fall above zone 9; year-round in zones 9 and 10

DISEASE FREQUENCY	DISEASE SEVERITY
5 annual	5 plants killed
4	**[4]**
[3]	3
2	2
1 rare	1 very little damage

CHEMICAL TREATMENT	CULTURAL PRACTICES
3 used every year	3 very important
[2]	**[2]**
1 not used	1 not important

SANITATION	RESISTANT CULTIVARS
3 very important	3 many cultivars
[2]	**[2]**
1 not important	1 no resistance

Powdery Mildew

Geographic occurrence USDA zones 4–9

Seasonal occurrence Cool periods from midwinter to spring and in the fall

DISEASE FREQUENCY	DISEASE SEVERITY
5 annual	5 plants killed
4	4
3	**3**
2	2
1 rare	1 very little damage

CHEMICAL TREATMENT	CULTURAL PRACTICES
3 used every year	3 very important
2	**2**
1 not used	1 not important

SANITATION	RESISTANT CULTIVARS
3 very important	3 many cultivars
2	**2**
1 not important	1 no resistance

Powdery Mildew

***Microsphaera penicillata* (asexual stage, *Oidium* sp.)**

Symptoms

Young, soft leaves and stems are susceptible to powdery mildew. The characteristic powdery white mycelium of the pathogen develops primarily on the lower surfaces of leaves. Infected leaves are distorted in shape and leaf plane, and chlorotic spots often form on their upper surfaces. Very young leaves may abscise.

Epidemiology

The pathogen survives in its asexual stage, *Oidium,* as spores (conidia) in the Deep South and likely as both conidia and ascospores in fruiting bodies (cleistothecia) on plant tissue in areas north of zone 9. Cool periods in spring and fall favor infection of newly emerging growth. The sexual stage of the fungus, *Microsphaera penicillata,* reproduces primarily on the lower surface of leaves but can reproduce on the upper surface as well. It does not need wet leaf surfaces for infection. Conidia are airborne in the nursery.

Host Range

M. penicillata has been reported on approximately 60 hosts, including such nursery crops as *Acer, Betula, Ceanothus, Cornus, Euonymus, Hydrangea, Ilex, Kalmia, Leucothoe, Lonicera, Magnolia, Quercus, Rhododendron, Spiraea, Syringa, Ulmus,* and *Viburnum.* It has been reported in *L. amurense, L. japonicum,* and *L. vulgare* in the Midwest and the mid-Atlantic and southern states.

Management

Avoid partially shaded areas for the production of *Ligustrum.* Monitor other susceptible hosts to detect the first disease incidence. Fungicides may be needed during cool, dry spring or fall periods to minimize leaf damage. *L. lucidum* is considered resistant to *M. penicillata.* The susceptibility of other *Ligustrum* species is unreported. Sanitation is important in reducing the survival of the fungus.

REFERENCES

Alfieri, S. A., Jr., and Stokes, D. E. 1971. Interaction of *Macrophomina phaseolina* and *Meloidogyne javanica* on *Ligustrum japonicum.* Phytopathology 61:1297–1298.

Bernard, E. L., and Witte, W. T. 1987. Parasitism of woody ornamentals by *Meloidogyne hapla.* Ann. Appl. Nematol. 1:41–45.

Farr, D. F., Bills, G. F., Chamuris, G. P., and Rossman, A. Y. 1989. Fungi on Plants and Plant Products in the United States. American Phytopathological Society, St. Paul, Minn.

Gill, D. L., and Sobers, E. K. 1974. Control of *Cercospora* sp. leaf spot of *Ligustrum japonicum.* Plant Dis. Rep. 58:1015–1017.

Holcomb, G. E. 1976. Powdery mildew, a new disease of *Ligustrum japonicum.* Plant Dis. Rep. 60:346–347.

Miller, J. W., and Alfieri, S. A., Jr. 1974. Leaf spot of *Ligustrum sinense* caused by *Corynespora cassiicola* and its control. Phytopathology 64: 255–256.

Mix, A. J. 1925. Anthracnose of European privet. Phytopathology 15:261–272.

Schoulties, C., and McRitchie, J. J. 1981. Edema. Fla. Dep. Agric. Consumer Serv. Plant Pathol. Circ. 215.

Sinclair, W. A., Lyon, H. H., and Johnson, W. T. 1987. Diseases of Trees and Shrubs. Comstock Publishing Associates, Cornell University Press, Ithaca, N.Y.

Sobers, E. K. 1964. Cercospora diseases of *Ligustrum.* Proc. Fla. State Hortic. Soc. 77:486–489.

CHAPTER 56

Jay W. Pscheidt ▪ Oregon State University, Corvallis
Gary W. Moorman ▪ Pennsylvania State University, University Park

Lilac Diseases

Geographic production USDA zones 2–7

Family Oleaceae

Genus *Syringa*

Species

S. meyeri	Meyer lilac	zones 3–7
S. reticulata	Japanese tree lilac	zones 3–7
S. villosa	late lilac	zones 2–7
S. vulgaris	common lilac	zones 3–7

Lilacs are valued for their fragrant flowers in a range of colors, from white to blue to magenta. They are very cold-hardy, but they are susceptible to a number of pathogens in the nursery and landscape. In the nursery, *Pseudomonas, Phytophthora,* and *Cylindrocladium* can threaten production, and a powdery mildew fungus affects foliage quality and aesthetics.

Shoot Blight

Phytophthora cactorum
Phytophthora syringae

Shoot blight is a problem when plants are "heeled in" (placed in sawdust piles) after bare-root stock has been dug prior to shipping to market. Shoot blight also can occur in plants in production.

Symptoms

Phytophthora shoot blight, like bacterial blight, causes shoot dieback in the spring; however, it also affects older wood. Shoots are often killed back to the ground. Shoots with *Phytophthora* infections have been described as darker brown or very black when compared with shoots affected by bacterial blight. Root sprouts that form under bushes may also be killed back to the ground.

Epidemiology

P. syringae is active during the dormant season during cool weather, while *P. cactorum* is active during warmer weather in the spring. See Chapter 14, "Phytophthora Root Rot and Dieback."

Host Range

P. syringae is mostly a problem on crabapple and pear stock; field-grown rhododendrons are also susceptible during the winter in the Pacific Northwest. On the other hand, *P. cactorum* has a wide host range.

Management

Prevent Phytophthora shoot blight by not covering nursery stock above the roots with sawdust. Plant resistant cultivars. Use minimal nitrogen fertility, for good but not excessive growth. Remove diseased limbs well below the blighted tissue. Fungicides have been recommended but not extensively tested for control of this disease.

Shoot Blight

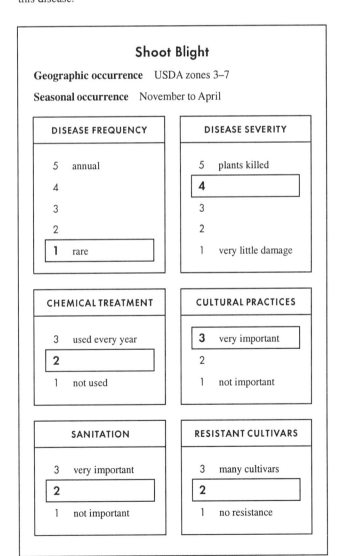

Geographic occurrence USDA zones 3–7

Seasonal occurrence November to April

DISEASE FREQUENCY	DISEASE SEVERITY
5 annual	5 plants killed
4	**4** (selected)
3	3
2	2
1 rare (selected)	1 very little damage

CHEMICAL TREATMENT	CULTURAL PRACTICES
3 used every year	**3 very important** (selected)
2 (selected)	2
1 not used	1 not important

SANITATION	RESISTANT CULTIVARS
3 very important	3 many cultivars
2 (selected)	**2** (selected)
1 not important	1 no resistance

Bacterial Blight

Pseudomonas syringae* pv. *syringae

Bacterial blight can be a limiting factor preventing successful nursery production. The pathogen, *Pseudomonas syringae* pv. *syringae,* also causes bacterial blight of blueberry, cherry, maple, pear, and many other woody plants.

Symptoms

The disease starts as brown spots, sometimes surrounded with a yellow halo, on stems and leaves of young shoots as they develop in early spring. The spots turn black and grow rapidly, especially during rainy periods. Further disease development depends on the age of the part attacked. On a young stem, the infection spreads around the stem and girdles it, so the shoot bends over at the lesion, and the parts above it wither and die. On a mature stem, spots usually enlarge along the stem, causing leaf death only within the infected area.

Infected young leaves blacken rapidly, with the discoloration starting near the margin and progressing in a wedge-shaped pattern down to the petiole. Eventually the entire leaf dies. On older leaves, spots enlarge slowly. Sometimes, several spots run together, and the leaf may crinkle at the edge or along the midvein.

Flower clusters may also be infected and rapidly blighted and blackened. Infected buds may die before opening in the spring.

Epidemiology

P. syringae pv. *syringae* overwinters on diseased twigs or as an epiphyte on healthy wood. Factors that weaken or injure plants predispose them to bacterial blight. Such factors include wounds, both accidental and from pruning or budding; frost damage; extremely low or high soil pH; poor or improper nutrition; and infection by other pathogens. Sources of the bacterium include old cankers, healthy buds, low populations on plants (with or without cankers), leaf surfaces, nearby weeds and grasses, and even soil. The pathogen is spread by wind, rain, insects, tools, and infected nursery stock. Like most bacterial diseases, bacterial blight of lilac is favored by mild, moist weather.

Two common genetic traits increase the ability of the bacterium to cause disease. Most strains produce a powerful plant toxin, syringomycin, which destroys plant tissues as the bacterium multiplies in wounds. *P. syringae* pv. *syringae* also produces a protein that acts as an ice nucleus, increasing frost injury, and the bacterium easily colonizes the wounded tissue.

Host Range

Most cultivars of *Syringa vulgaris* are susceptible, but some have been observed with less disease when planted in a garden, including Edith Cavell, Glory, Ludwig Spaeth, and Pink Elizabeth (Ludwig Spaeth is highly susceptible in intense nursery production systems). Some species of *Syringa* have shown resistance in western Washington, including *S. josikaea, S. komarowii, S. microphylla, S. pekinensis,* and *S. reflexa.*

Management

Maintain adequate spacing between plants, and prune to provide good air circulation in the canopy. Prune out and burn all affected tissues immediately. Plant resistant species or cultivars. Do not overfertilize young plants. In spring, protect plants from rain and frost with plastic hoop houses or similar structures. This treatment has been as good as the best chemical method. Spray bactericides before fall rains and again before budbreak in the spring. Bacteria with resistance to copper and antibiotic products have been detected in many nurseries.

Bacterial Blight

Geographic occurrence USDA zones 3–7

Seasonal occurrence March to April, under average yearly conditions in Oregon (adjust for your hardiness zone)

DISEASE FREQUENCY	DISEASE SEVERITY
5 annual	5 plants killed
4	4
3	**3**
2	2
1 rare	1 very little damage

CHEMICAL TREATMENT	CULTURAL PRACTICES
3 used every year	**3 very important**
2	2
1 not used	1 not important

SANITATION	RESISTANT CULTIVARS
3 very important	3 many cultivars
2	**2**
1 not important	1 no resistance

Powdery Mildew

Microsphaera syringae

Powdery mildew of lilac, caused by the fungus *Microsphaera syringae,* generally does not occur until the latter part of the growing season and does little damage.

Symptoms

A grayish white fungal growth develops extensively on leaves. New leaves may become distorted.

Host Range

The pathogen attacks ash and privet as well as lilac. Cultivars of *S. vulgaris* are very susceptible, but some species are resistant.

Powdery Mildew

Geographic occurrence USDA zones 3–7

Seasonal occurrence August to September, under average yearly conditions in Oregon (adjust for your hardiness zone)

DISEASE FREQUENCY	DISEASE SEVERITY
5 annual	5 plants killed
4	4
3	3
2	**2**
1 rare	1 very little damage

CHEMICAL TREATMENT	CULTURAL PRACTICES
3 used every year	3 very important
2	**2**
1 not used	1 not important

SANITATION	RESISTANT CULTIVARS
3 very important	3 many cultivars
2	**2**
1 not important	1 no resistance

Cylindrocladium Root Rot, Cutting Rot, and Blight

Geographic occurrence USDA zones 3–7

Seasonal occurrence Primarily during propagation

DISEASE FREQUENCY	DISEASE SEVERITY
5 annual	5 plants killed
4	**4**
3	3
2	2
1 rare	1 very little damage

CHEMICAL TREATMENT	CULTURAL PRACTICES
3 used every year	**3 very important**
2	2
1 not used	1 not important

SANITATION	RESISTANT CULTIVARS
3 very important	3 many cultivars
2	2
1 not important	**1 no resistance**

Management

Space plants or prune branches for good air circulation. Do not plant in shade. Use chemical control just before symptoms are expected to occur or when the first symptoms appear.

Cylindrocladium Root Rot, Cutting Rot, and Blight

Cylindrocladium floridanum
Cylindrocladium scoparium

Epidemiology

Cylindrocladium can be a major pathogen of young plants and cuttings in the nursery, especially during propagation if it is warm (25–30°C, or 75–85°F) and moist. Species of *Cylindrocladium* are not specific to lilac and attack over 100 different ornamental crops. The fungus competes and survives in soil very well, often in association with the roots of infected but symptomless plants. It can remain dormant in the form of microsclerotia and later produces spores. The fungus can be moved about in the nursery in microsclerotia-infested soil and on infected plants being transplanted. Spores splashed by rain or overhead irrigation spread from plant to plant. Brown fungal growths, sometimes covered with white tufts of spores, are signs of *Cylindrocladium.* They can be seen on infected stems, usually at or just above the soil line, and sometimes on infected leaves.

Symptoms

Root and cutting rot Dark brown to black lesions are scattered on the roots. Affected plants may be stunted and yellowed. As the severity of root rot increases, plants wilt and die. If the bases of cuttings are rotted, roots often form just above the rotted area. The fungus eventually spreads upward.

Blight Under hot, wet conditions, *Cylindrocladium* can invade leaves and girdle stems, causing a rapid blighting of the tissue.

Management

Propagation should be done in a pasteurized medium in sterile containers. Cuttings should be taken only from healthy stock plants. Infected stock plants and other infected plants should be destroyed immediately. Overhead irrigation should not be used on stock plants. Fungicides can be applied to protect healthy plants. It

Witches'-Broom

Geographic occurrence USDA zones 3–7

Seasonal occurrence The pathogen is systemic in the host after infection; witches'-brooms develop during shoot growth

DISEASE FREQUENCY	DISEASE SEVERITY
5 annual	5 plants killed
4	4
3	3
2	**2** (boxed)
1 rare (boxed)	1 very little damage

CHEMICAL TREATMENT	CULTURAL PRACTICES
3 used every year	**3 very important** (boxed)
2	2
1 not used (boxed)	1 not important

SANITATION	RESISTANT CULTIVARS
3 very important (boxed)	3 many cultivars
2	2
1 not important	**1 no resistance** (boxed)

has been reported that some populations of *Cylindrocladium* have developed resistance to fungicides in the benzimidazole class.

Witches'-Broom

Phytoplasma

A phytoplasma has been implicated in causing excessive branching and witches'-broom of lilac. See Chapter 8, "Phytoplasmas," to learn more about the pathogen.

Symptoms

At the tops of plants, many slender shoots produce small leaves, which may be one-quarter the normal size. The leaves may be yellow and contorted. The brooms do not properly develop winter dormancy and are killed. Growth elsewhere on the plant resumes in the spring, sometimes before the normal time. No flowering occurs on brooms. The entire plant dies within a few years of infection.

Table 56.1 Other pathogens reported to have been found on lilacs and the diseases associated with them

Pathogen	Disease
Fungi	
Alternaria	Leaf spot
Armillaria	Root rot
Ascochyta	Blight
Botrytis	Flower blight
Botryosphaeria	Canker
Cercospora	Leaf spot
Cladosporium	Leaf blotch
Colletotrichum	Leaf spot
Heterosporium	Blight
Phyllosticta	Leaf spot
Rhizoctonia	Stem blight
Sclerotinia	Stem blight
Sclerotium	Stem blight
Thielaviopsis	Root rot
Verticillium	Vascular wilt
Bacterium	
Agrobacterium	Crown gall
Phytoplasma	
Phytoplasma	Witches'-broom
Nematode	
Xiphinema (dagger nematode)	

Epidemiology

Leafhoppers spread phytoplasmas from plant to plant. The pathogen is also spread to plants grafted with infected cuttings.

Management

Destroy infected plants.

Other Pathogens of Lilac

Many fungi and a few other bacteria and nematodes are reported in various host indices and state reports to have been found on lilacs (Table 56.1). However, there are few descriptions of symptoms caused by these pathogens in lilac.

REFERENCES

Chester, K. S. 1932. A comparative study of three *Phytophthora* diseases of lilac and of their pathogens. J. Arnold Arbor. Harvard Univ. 13:232–268.

Farr, D. F., Bills, G. F., Chamuris, G. P., and Rossman, A. Y. 1989. Fungi on Plants and Plant Products in the United States. American Phytopathological Society, St. Paul, Minn.

Pscheidt, J. W., and Ocamb, C. M. 2001. Pacific Northwest Plant Disease Control Handbook. Oregon State University Extension Service, Corvallis.

Scheck, H. J., and Pscheidt, J. W. 1998. Effect of copper bactericides on copper-resistant and -sensitive strains of *Pseudomonas syringae* pv. *syringae*. Plant Dis. 82:397–406.

Sinclair, W. A., Lyon, H. H., and Johnson, W. T. 1987. Diseases of Trees and Shrubs. Comstock Publishing Associates, Cornell University Press, Ithaca, N.Y.

CHAPTER
57

Julie Beale ▪ University of Kentucky, Lexington

Linden Diseases

Geographic production USDA zones 2–8 (some species are less hardy)

Family Tiliaceae

Genus *Tilia*

Species		
	T. americana	American linden, basswood
	T. cordata	littleleaf linden
	T. tomentosa	silver linden
	T. vulgaris	common linden, European linden

Lindens grow 10 to 15 feet in five to 10 years, to become large trees in the landscape. *Tilia cordata, T. tomentosa,* and *T. vulgaris* are preferred for their landscape value. Anthracnose, canker, and powdery mildew are common diseases of lindens in the nursery.

Anthracnose

Glomerella cingulata
Gnomonia tiliae

Anthracnose of lindens is a foliar disease, similar to anthracnose of other shade trees, such as oaks and elms. It is not typically a serious disease. European lindens are occasionally affected.

Symptoms

Leaf spots become visible in spring in wet weather, which favors infection. Spots occur along the veins and at leaf tips, typically coalescing to form irregular blotches with black bands separating healthy and dead tissues. Severe infections may result in defoliation. Infection of twigs is also possible.

Epidemiology

Fruiting bodies of the fungal pathogen overwinter in fallen leaves and debris. Spores are released and carried by wind or splashing rain in spring to newly emerging leaves, where they cause infections. Twigs may also become infected. Cool, wet spring weather allows germination of spores and favors disease development.

Host Range

European lindens are most often affected by anthracnose.

Management

Cultural practices Cultural practices and sanitation are normally sufficient to reduce the incidence of anthracnose. Space plants to improve air circulation. Avoid overhead watering in spring, when infections occur.

Sanitation Prune out infected twigs. Rake up and destroy fallen leaves to reduce inoculum. Nurseries should be able to keep conditions sanitary and prevent outbreaks of this disease.

Resistance No resistant cultivars are available.

Chemical treatment Fungicides are typically not necessary

Anthracnose

Geographic occurrence USDA zones 2–8

Seasonal occurrence April to May

DISEASE FREQUENCY	DISEASE SEVERITY
5 annual	5 plants killed
4	4
[3]	3
2	2
1 rare	**[1 very little damage]**

CHEMICAL TREATMENT	CULTURAL PRACTICES
3 used every year	3 very important
[2]	**[2]**
[1 not used]	1 not important

SANITATION	RESISTANT CULTIVARS
3 very important	3 many cultivars
[2]	2
1 not important	**[1 no resistance]**

unless a severe outbreak occurs in the nursery. When they are used, fungicides are applied at budbreak, when leaves are half-expanded, and when they are fully expanded.

Leaf Spot

Cercospora microspora
(sexual stage, *Mycosphaerella microsora*)

Leaf spot of lindens is occasionally a problem in nurseries.

Symptoms

Brown circular spots with dark brown borders form on leaves, sometimes becoming quite numerous and coalescing. Severe infections may result in defoliation. Young trees are most seriously affected.

Epidemiology

Fruiting structures of the fungal pathogen in fallen leaves and debris release spores under warm, wet conditions, resulting in new infections.

Host Range

All species of linden are susceptible to this disease.

Leaf Spot

Geographic occurrence USDA zones 2–8

Seasonal occurrence May to July

DISEASE FREQUENCY	DISEASE SEVERITY
5 annual	5 plants killed
4	4
3	3
2	2
1 rare	**1 very little damage**

CHEMICAL TREATMENT	CULTURAL PRACTICES
3 used every year	3 very important
2	**2**
1 not used	1 not important

SANITATION	RESISTANT CULTIVARS
3 very important	3 many cultivars
2	2
1 not important	**1 no resistance**

Management

Cultural practices Space plants to improve air circulation.

Sanitation Clean up and destroy fallen leaves and debris, which serve as a source of inoculum.

Resistance There are no resistant cultivars.

Chemical treatment Chemical controls are often not needed. However, protectant fungicide applications in spring should prevent leaf spot.

Nectria Canker

Nectria cinnabarina

Nectria canker is the most common of several fungal canker diseases affecting linden. All of these diseases have similar effects on the tree and are controlled by the same management practices.

Symptoms

Cankers on small twigs are slightly sunken, with bark discolored brown to dark brown. Cankers are usually associated with wounds or dead branch stubs and may girdle twigs, causing dieback. On older branches with corky bark, cankers are not easily detectable until fruiting structures of the fungal pathogen, *Nectria cinnabarina,* erupt through the bark. The fruiting structures produced in spring and summer are orange to pink, cushion-like

Nectria Canker

Geographic occurrence USDA zones 2–8

DISEASE FREQUENCY	DISEASE SEVERITY
5 annual	5 plants killed
4	**4**
3	3
2	2
1 rare	1 very little damage

CHEMICAL TREATMENT	CULTURAL PRACTICES
3 used every year	3 very important
2	**2**
1 not used	1 not important

SANITATION	RESISTANT CULTIVARS
3 very important	3 many cultivars
2	2
1 not important	**1 no resistance**

Powdery Mildew

Geographic occurrence USDA zones 2–8

Seasonal occurrence May to June

DISEASE FREQUENCY	DISEASE SEVERITY
5 annual	5 plants killed
4	4
3	3
2	2
1 rare	**1 very little damage**

CHEMICAL TREATMENT	CULTURAL PRACTICES
3 used every year	3 very important
2	2
1 not used	**1 not important**

SANITATION	RESISTANT CULTIVARS
3 very important	3 many cultivars
2	2
1 not important	**1 no resistance**

sporodochia; the fungus later forms clusters of dark orange to red, pimple-like perithecia. The fruiting bodies are easily visible without a hand lens.

Epidemiology

N. cinnabarina attacks trees weakened by freezing, drought, injuries, or other stresses. Spores from fruiting bodies invade dead buds, twigs, and wounds throughout the year during periods of wet weather and moderate temperature (21–26°C), although infections can occur under much cooler conditions if adequate moisture is available.

Host Range

All lindens are susceptible to Nectria canker, as are many other trees and shrubs.

Management

Cultural practices Avoid stresses that will predispose trees to infection. Protect nursery stock from freeze injury. Avoid drought stress, injury, and stress from cultural practices such as pruning or transplanting at inappropriate times (e.g., summer or early fall) and excessive root pruning. Plant lindens in a favorable climate, noting that hardiness varies between species.

Sanitation Prune out and remove cankered twigs and branches. Keep the area free of debris and fallen infected twigs.

Resistance There are no resistant cultivars.

Chemical treatment There are no chemical treatments.

Powdery Mildew

Microsphaera penicillata
Phyllactinia guttata
Uncinula clintonii

Lindens are quite susceptible to powdery mildew, although they are rarely damaged by it. See Chapter 15, "Powdery Mildew."

Pseudomonas Blight

Pseudomonas syringae

Verticillium Wilt

Verticillium dahliae

Pseudomonas blight, caused by the bacterium *Pseudomonas syringae,* and Verticillium wilt, caused by the fungus *Verticillium dahliae,* occasionally affect linden in nurseries in the Pacific Northwest. See Chapter 16, "Diseases Caused by *Pseudomonas syringae,*" and Chapter 19, "Verticillium Wilt."

REFERENCES

Dirr, M. A. 1998. Manual of Woody Landscape Plants: Their Identification, Ornamental Characteristics, Culture, Propagation and Uses. 5th ed. Stipes Publishing, Champaign, Ill.

Farr, D. F., Bills, G. F., Chamuris, G. P., and Rossman, A. Y. 1989. Fungi on Plants and Plant Products in the United States. American Phytopathological Society, St. Paul, Minn.

Pirone, P. P. 1978. Diseases and Pests of Ornamental Plants. 5th ed. John Wiley & Sons, New York.

Sinclair, W. A., Lyon, H. H., and Johnson, W. T. 1987. Diseases of Trees and Shrubs. Comstock Publishing Associates, Cornell University Press, Ithaca, N.Y.

CHAPTER 58

Judith O'Mara ▪ Kansas State University, Manhattan

Elizabeth Hudgins ▪ Oklahoma State University, Stillwater

Lonicera (Honeysuckle) Diseases

Geographic production USDA zones 2–8

Family Caprifoliaceae

Genus *Lonicera*

Species

L. alpigena	Alps honeysuckle
L. × *bella*	belle honeysuckle
L. × *brownii*	Brown's honeysuckle
L. caerulea	bearberry honeysuckle
L. fragrantissima	winter honeysuckle
L. × *heckrottii*	goldflame honeysuckle
L. japonica	*Japanese honeysuckle
L. korolkowii	blueleaf honeysuckle
L. maackii	*Amur honeysuckle
L. maximowiczii var. *sachalinensis*	Sakhalin honeysuckle
L. morrowii	Morrow honeysuckle
L. nitida	boxleaf honeysuckle
L. pileata	privet honeysuckle
L. sempervirens	trumpet honeysuckle
L. tatarica	*Tatarian honeysuckle
L. xylosteum	European fly honeysuckle

*The species marked with an asterisk are invasive exotic pests in parts of the northeastern and the southeastern United States.

The honeysuckles are an enormous group of species and cultivars in shrub and vine forms. Relatively easy to grow from seeds and cuttings, honeysuckles may be attacked by leaf blight and powdery mildew in the nursery.

Insolibasidium Leaf Blight

Insolibasidium deformans

(asexual stage, *Glomopsis lonicerae*)

Insolibasidium leaf blight is a potentially destructive foliage disease of many honeysuckle species in the central Great Plains, the eastern United States, the Pacific Northwest, and Canada. This fungal disease can cause extensive leaf browning, defoliation, and a reduction in plant growth. Leaf blight is primarily a spring disease; however, cool, humid growing conditions and the development of new leaves favor the disease at any point in the growing season.

Symptoms

A week or so after becoming infected, new leaves start showing symptoms. Leaves initially exhibit interveinal yellowing, followed by browning of the affected area. Leaf veins remain green for a period of time but eventually turn brown. Infected leaves frequently develop crinkling, cupping, rolling, or twisting. A thin, white layer of fungal structures (basidia) and sexual spores (basidiospores) develop on the lower surfaces of leaves. On severely

Insolibasidium Leaf Blight

Geographic occurrence The northern two-thirds of the United States

Seasonal occurrence Spring or early summer; under cool, wet conditions throughout the growing season

DISEASE FREQUENCY
- 5 annual
- 4
- **[3]**
- 2
- 1 rare

DISEASE SEVERITY
- 5 plants killed
- 4
- **[3]**
- 2
- 1 very little damage

CHEMICAL TREATMENT
- 3 used every year
- **[2]**
- 1 not used

CULTURAL PRACTICES
- **[3]** very important
- 2
- 1 not important

SANITATION
- **[3]** very important
- 2
- 1 not important

RESISTANT CULTIVARS
- 3 many cultivars
- **[2]**
- 1 no resistance

blighted leaves, the conidial stage will develop later in the summer along the margins of the basidial lesions. Coalescing lesions produce blighted leaves, which dry and defoliate prematurely. Repeated infections can reduce plant vigor and plant growth. In production areas with extended cool, wet weather, a shoot blight symptom can develop, even though the pathogen does not infect stem tissue.

Epidemiology

The leaf blight pathogen produces both sexual spores (basidiospores) and asexual spores (conidia) during its life cycle. The fungus overwinters in its sexual state, *Insolibasidium deformans,* on diseased leaf debris. In the spring, basidiospores splash to new leaves, causing primary infections, which result in thin, white lesions on the undersides of the leaves. Masses of powdery white conidia develop in the lesions. The conidial stage is not thought to contribute to disease development, however. Basidiospores released from the lesions cause secondary infections throughout the growing season, as long as the weather remains cool (41–82°F, or 5–28°C) and wet. Optimum conditions for infection are temperatures in the range of 57–82°F (14–28°C) and recent rainfall. The disease has been noted in production nurseries in the northern and central Great Plains as well as the eastern United States and the Pacific Northwest.

Host Range

Insolibasidium leaf blight has been reported in most honeysuckle species and varieties. Iowa nurseries have noted resistance in *L. dioica, L. gracilipes,* and *L. sempervirens.*

Management

Production areas with annual periods of prolonged cool, wet weather are more likely to sustain damage from this disease. Under these growing conditions, it is important to avoid establishing long-term honeysuckle plantings near nursery plantings. Permanent plantings have the potential to act as a source of disease inoculum for nursery production. High-density seedbeds should be avoided. Heavy plantings reduce air circulation, increase humidity, and favor disease development. Likewise, irrigation should be scheduled early in the day, to reduce leaf wetness. Overwintering mulches and leaf litter should be removed prior to budbreak, to reduce the level of inoculum. In production areas with high disease incidence, fungicide applications should be initiated as young leaves emerge, and treatment should be continued throughout the growing season.

Powdery Mildew

***Microsphaera caprifoliacearum* (syn. *M. penicillata*)**
Microsphaera lonicerae* var. *ehrenbergii

Powdery mildew of honeysuckle is caused by fungi in the genus *Microsphaera,* which infect leaf surfaces and succulent tissues. The disease does not kill plants outright, but it reduces growth and may cause leaves to turn yellow and defoliate.

Symptoms

A white or grayish, powdery fungal growth develops on the upper leaf surfaces. The disease may begin with small, circular mats of mycelium, but over time the fungal growth may cover the entire leaf surface. Later in the season, small, black fruiting bodies (cleistothecia) may be evident in the mycelium on leaves. The disease tends to be worse in parts of the plant that are shaded or have dense foliage, where the humidity is highest.

Epidemiology

Microsphaera overwinters as cleistothecia on infected leaves. Spores released from the cleistothecia begin infections late in the spring. The fungus produces asexual spores on newly diseased areas of leaves, and these spores are spread by air movement. Powdery mildew spores require at least 95% humidity in order to infect and cause disease; however, they cannot germinate in free water.

M. lonicerae is found on *L. tatarica* and is believed to have been introduced into the United States with this honeysuckle species.

Host Range

It is suspected that the *Microsphaera* spp. causing powdery mildew of honeysuckle are limited to *Lonicera* spp., but this has not been confirmed.

Management

Cultural practices Irrigate the crop in the morning, to provide enough time for the foliage to dry during the day. Grow

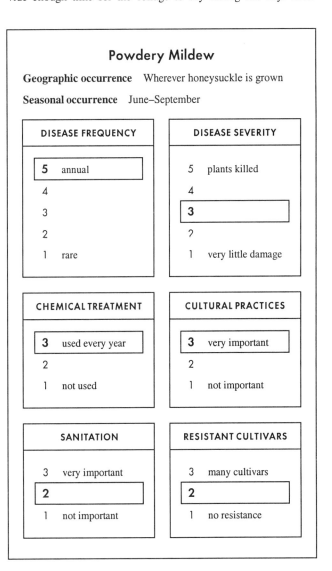

Powdery Mildew

Geographic occurrence Wherever honeysuckle is grown

Seasonal occurrence June–September

DISEASE FREQUENCY	DISEASE SEVERITY
5 annual (boxed)	5 plants killed
4	4
3	**3** (boxed)
2	2
1 rare	1 very little damage

CHEMICAL TREATMENT	CULTURAL PRACTICES
3 used every year (boxed)	**3 very important** (boxed)
2	2
1 not used	1 not important

SANITATION	RESISTANT CULTIVARS
3 very important	3 many cultivars
2 (boxed)	**2** (boxed)
1 not important	1 no resistance

plants in full sun, with good spacing, to increase air circulation and reduce humidity.

Sanitation Clean up leaf litter during the dormant season, to help remove sources of inoculum.

Resistance Resistance to powdery mildew has not been documented.

Chemical treatment Several fungicides are available for powdery mildew control.

REFERENCES

Braun, U. 1984. A short survey of genus *Microsphaera* in North America. Nova Hedwigia 39:211–243.

Dirr, M. A. 1998. Manual of Woody Landscape Plants: Their Identification, Ornamental Characteristics, Culture, Propagation and Uses. 5th ed. Stipes Publishing, Champaign, Ill.

Farr, D. F., Bills, G. F., Chamuris, G. P., and Rossman, A. Y. 1989. Fungi on Plants and Plant Products in the United States. American Phytopathological Society, St. Paul, Minn.

Hutchinson, T. F., and Vankat, J. L. 1997. Invasibility and effect of Amur honeysuckle in southwestern Ohio forests. Conserv. Biol. 11:1117–1124.

Riffle, J. W., and Peterson, G. W. 1986. Diseases of trees in the Great Plains. U.S. Dep. Agric. For. Serv., Rocky Mountain For. Range Exp. Stn. (Ft. Collins, Colo.), Gen. Tech. Rep. RM-129.

Sinclair, W. A., Lyon, H. H., and Johnson, W. T. 1987. Diseases of Trees and Shrubs. Comstock Publishing Associates, Cornell University Press, Ithaca, N.Y.

Woods, K. D. 1993. Effects of invasion by *Lonicera tatarica* L. on herbs and tree seedlings in four New England forests. Am. Midl. Nat. 130: 62–74.

CHAPTER 59

Austin K. Hagan ▪ Auburn University, Auburn, Alabama

Magnolia Diseases

Geographic production USDA zones 4–9

Family Magnoliaceae

Genus *Magnolia*

Species		
M. grandiflora	southern magnolia	zones 7–9
M. × *soulangiana*	saucer magnolia	zones 4–9
M. stellata	star magnolia	zones 4–9

The genus *Magnolia* encompasses a diverse group of native and introduced broadleaf flowering trees. Southern magnolia (*M. grandiflora*), known for its pyramidal crown, lustrous evergreen foliage, and large, white flowers, is widely used as a specimen tree in landscapes throughout the southern United States. The floral display of mature saucer magnolia (*M.* × *soulangiana*) and star magnolia (*M. stellata*) in late winter and early spring is spectacular; these are small trees, used as accent or specimen plants. The blooms of saucer and star magnolias are very sensitive to hard freezes and frost.

Other magnolia species occasionally seen in the nursery trade are cucumbertree magnolia (*M. acuminata*), lily magnolia (*M. liliiflora*), Loebner magnolia (*M.* × *loebneri*), bigleaf magnolia (*M. macrophylla*), sweet bay magnolia (*M. virginiana*), Gresham hybrids, and other hybrids.

Overall, foliar and soilborne diseases have relatively little impact on the production of container- and field-grown magnolia. Of the three major species, the saucer magnolia is considered the most disease-prone.

Bacterial Blight

Pseudomonas cichorii
Pseudomonas syringae

Bacterial leaf spot caused by *Pseudomonas syringae* is a common and sometimes damaging disease of container-grown saucer magnolia. A similar leaf spot of southern magnolia, caused by *P. cichorii,* is occasionally a problem in container and field stock.

Symptoms

Small, discrete, dark brown, water-soaked spots, often surrounded by a yellow halo, appear on newly unfolded leaves of southern and saucer magnolias (Plate 75). When lesion expansion is limited by leaf veins, spots on the leaves of saucer magnolia become angular to irregular in shape. Spots on southern magnolia may coalesce into large, irregular areas of dead tissue. As diseased leaves continue to enlarge, large holes may develop where dead tissues have split and then disintegrated (Plate 76). Distortion of young leaves and premature leaf drop may also occur in severely diseased southern magnolia.

Bacterial Blight

Geographic occurrence USDA zones 4–9

Seasonal occurrence March to July, with an optimum period in April and May

DISEASE FREQUENCY	DISEASE SEVERITY
5 annual	5 plants killed
4	4
3 (boxed)	3
2	**2** (boxed)
1 rare	1 very little damage

CHEMICAL TREATMENT	CULTURAL PRACTICES
3 used every year	3 very important
2 (boxed)	**2** (boxed)
1 not used	1 not important

SANITATION	RESISTANT CULTIVARS
3 very important	3 many cultivars
2	2
1 not important (boxed)	**1 no resistance** (boxed)

Epidemiology

Typically, bacterial plant pathogens survive in perennial cankers and in buds and leaf scars of their hosts. When trees break dormancy, the bacteria begin to reproduce and are spread to newly unfolding leaves by splashing water. Overhead watering contributes to disease development. Mild, wet weather for an extended period, starting at leaf-out and continuing through leaf maturity, favors severe disease outbreaks in container- and field-grown magnolia.

Host Range

The host range of *P. syringae* among members of the family Magnoliaceae has not been established, but it is an aggressive pathogen of saucer magnolia. The host range of this bacterium includes many trees and shrubs.

Among members of the magnolia family, southern magnolia is the primary host of *P. cichorii.* Leaf spot symptoms have also been recorded in artificially inoculated bigleaf, saucer, and umbrella magnolias.

Management

The general cultural practices and sanitation used to control leaf spot diseases of trees and shrubs should provide some protection from bacterial leaf spot.

Cultural practices and sanitation A ground-level irrigation system will slow the spread and reduce the severity of bacterial leaf spot in southern and saucer magnolia. Space out container stock just prior to leaf-out, to speed the evaporation of free moisture from the foliage and slow disease development. When possible, collect and discard fallen diseased leaves.

Resistance No information is available concerning reactions to bacterial leaf spot in cultivars of any magnolia taxa.

Chemical treatment. Weekly applications of selected copper-based bactericides or fungicides will reduce the severity of bacterial leaf spot in saucer and southern magnolias. Begin applications at budbreak, and continue spraying at weekly intervals until shoot growth stops. For pesticide recommendations for your state or region, contact your county extension agent or specialist responsible for diseases of nursery crops.

Anthracnose

Colletotrichum gloeosporioides

Phyllosticta Leaf Spot

Phyllosticta magnoliae

Numerous fungus-incited leaf spot diseases of members of the magnolia family, particularly southern magnolia, have been reported. Two of the more common are anthracnose, caused by *Colletotrichum gloeosporioides,* and Phyllosticta leaf spot, caused by *Phyllosticta magnoliae.* Neither disease seriously threatens the health of well-maintained magnolias, but they can detract from the beauty of the tree. Anthracnose and Phyllosticta leaf spot generally occur more frequently in mature landscape trees than in nursery stock.

Other foliar diseases of members of the magnolia family are Alternaria spot, algal leaf spot, angular leaf spot (Cercosporidium leaf spot), Coniothyrium leaf spot, Cylindrocladium leaf spot, Guignardia leaf spot, Phoma leaf spot, Phomopsis leaf spot, scab, Septoria leaf spot, and spot anthracnose.

Symptoms

Anthracnose is characterized by angular brown spots, up to 1/8 inch in diameter, surrounded by a yellow halo (Plate 77). Black, blister-like fruiting bodies (acervuli) of *C. gloeosporioides* develop on the upper surface of the spot and can easily be seen with a hand lens. Later, a distinctive pink spore mass oozes from the black fruiting bodies.

Phyllosticta leaf spot first appears as tiny purple to black spots on the upper surfaces of leaves. As the spots increase in size (the diameter may exceed 3/4 inch), their centers turn off-white, while their borders remain purple to almost black, and they may be surrounded by a faint yellow halo. Black, pinhead-sized fruiting bodies (pycnidia) of *P. magnoliae* appear in the centers of larger leaf spots.

Epidemiology

Typically, the fungi causing leaf spot diseases overseason on their host trees as hyphae or spores in fruiting bodies embedded in host tissues. Spores of both *C. gloeosporioides* and *P. magnoliae* are spread to new leaves and elongating shoots by splashing water. Extended periods of warm, cloudy, humid, wet

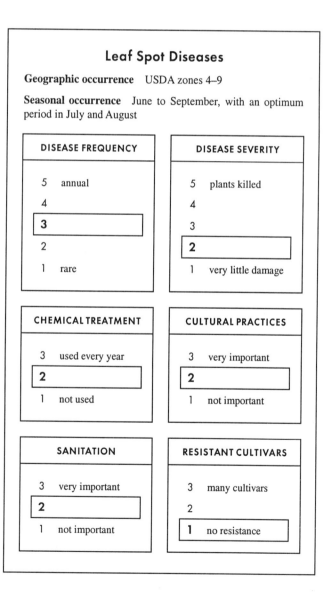

Leaf Spot Diseases

Geographic occurrence USDA zones 4–9

Seasonal occurrence June to September, with an optimum period in July and August

DISEASE FREQUENCY

5 annual
4
3
2
1 rare

DISEASE SEVERITY

5 plants killed
4
3
2
1 very little damage

CHEMICAL TREATMENT

3 used every year
2
1 not used

CULTURAL PRACTICES

3 very important
2
1 not important

SANITATION

3 very important
2
1 not important

RESISTANT CULTIVARS

3 many cultivars
2
1 no resistance

weather during the growing season favor disease development and increase symptom severity.

Host Range

Anthracnose is found most often in southern magnolia but has also been reported in saucer and star magnolias. Phyllosticta leaf spot occurs primarily in southern and sweet bay magnolias.

Management

Both anthracnose and Phyllosticta leaf spot occur sporadically, and therefore blocks of magnolia should be routinely inspected for typical leaf spot symptoms during the growing season. When symptoms are seen, a combination of timely fungicide applications and sanitation practices should prevent disease spread.

Sanitation and cultural practices Collect cuttings for propagation from disease-free trees. Discard diseased cuttings. When possible, install a ground-level irrigation system; overhead irrigation favors the spread of disease and can increase disease severity. Space out container stock just prior to budbreak, to speed the evaporation of free moisture from the foliage and slow disease development. When possible, collect and remove fallen diseased leaves from production ranges.

Resistance The susceptibility of cultivars of magnolia taxa to anthracnose and Phyllosticta leaf spot has not been reported.

Chemical treatment Preventative fungicides should be applied only on diseased trees. Inspect production blocks of magnolia for leaf spot symptoms regularly from early spring through midsummer. Should symptoms appear, immediately begin fungicide treatment. On trees damaged by either disease the previous year, begin fungicide treatment as the leaves begin to unfurl, and repeat applications at the interval specified on the fungicide label until shoot elongation stops. For diseased magnolias, continue the preventative fungicide program each year until the trees are sold. For fungicide recommendations, contact your county extension agent or state specialist responsible for diseases of nursery crops.

Powdery Mildew

Microsphaera penicillata

Powdery mildew occurs primarily in deciduous magnolia taxa, such as saucer and star magnolias, but it has also been reported in southern and lily magnolias. Typically, the disease causes no permanent injury, but affected foliage may be unsightly, and diseased trees unsalable.

Symptoms

Powdery mildew diseases are easily identified by the appearance of white to buff-colored, cottony colonies of the causal fungus, *Microsphaera penicillata,* on leaves and sometimes on tender shoots. A white mat of fungal growth may cover the upper surface of a heavily colonized leaf. When tissues under a fungal colony die, circular to irregular brown blotches may appear. On heavily diseased trees, some leaves may be shed prematurely.

Epidemiology

Powdery mildew fungi overwinter as hyphae in dormant buds or as spores in fruiting bodies (cleistothecia) on fallen diseased leaves. During the spring to early summer, spores are spread by air currents to new, expanding leaves and tender shoots. Warm days and cool nights along with several days of relatively dry weather favor the development and spread of the disease. For further information concerning the epidemiology of this disease, refer to Chapter 15, "Powdery Mildew."

Host Range

Among the Magnoliaceae, powdery mildew occurs primarily on saucer and star magnolias. Lily and southern magnolias are hosts of *M. penicillata,* but the disease rarely occurs in either tree.

Management

Control options for powdery mildew of magnolia are limited to good production practices and, when needed, protective fungicide treatments. A more detailed discussion of recommended control procedures for powdery mildew of nursery crops is presented in Chapter 15.

Sanitation and cultural practices Space out container stock just prior to budbreak, to improve air circulation around the leaves. Whenever possible, collect and remove fallen diseased leaves from production ranges.

Resistance The reactions of cultivars of mildew-susceptible taxa of magnolia have not been reported.

Chemical treatment Because powdery mildew only occurs sporadically in magnolia, preventative fungicide treatment is unnecessary. The leaves of magnolia in production blocks should be

Powdery Mildew

Geographic occurrence USDA zones 4–9

Seasonal occurrence April to September, with an optimum period in July

DISEASE FREQUENCY	DISEASE SEVERITY
5 annual	5 plants killed
4	4
3	3
[2]	**[2]**
1 rare	1 very little damage

CHEMICAL TREATMENT	CULTURAL PRACTICES
3 used every year	3 very important
[2]	2
1 not used	**[1] not important**

SANITATION	RESISTANT CULTIVARS
3 very important	3 many cultivars
2	2
[1] not important	**[1] no resistance**

regularly inspected for typical symptoms of powdery mildew from early spring through midsummer. Should mildew appear, immediately spray the affected block of trees with a labeled fungicide. Repeat fungicide applications at the rate and interval specified on the product label until the threat of further disease spread is over. For fungicide recommendations, contact your county extension agent or state specialist responsible for diseases of nursery crops.

Phytophthora Root Rot

Phytophthora cinnamomi

Phytophthora root rot is a destructive disease of numerous container- and field-grown woody shrubs and trees. Outbreaks of the disease have occasionally been observed in container-grown southern and saucer magnolias. Badly diseased trees are unsalable. See Chapter 14, "Phytophthora Root Rot and Dieback."

Symptoms

Typical symptoms of root rot include yellowing of leaves, slowed shoot growth, premature leaf shed, limb dieback, sudden wilting of leaves, and finally plant death. Feeder roots colonized by *P. cinnamomi* are brittle and turn brown to reddish brown. Browning of the feeder roots usually begins near the base of the root ball and often continues until the entire root system is destroyed. Areas of discolored, rotted roots can usually be seen by removing the root ball from the container.

Epidemiology

The causal fungus, *Phytophthora cinnamomi,* survives as resting structures (chlamydospores) and hyphae in roots of diseased plants, crop debris, and soil under container beds. It is easily spread in contaminated potting media and by splashing or flowing water. Recycled irrigation or runoff water has also been cited as a source of inoculum of the fungus. The heaviest losses to root rot often occur in flood-prone production beds where water is allowed to stand around container stock. Over- or underwatering may also accelerate the pace of root rot development. The root rot fungus is generally active at temperatures which also favor plant growth. For further information on the epidemiology of this disease, refer to Chapter 14.

Host Range

In the magnolia family, southern and saucer magnolias are the main hosts of *P. cinnamomi.* Other important nursery crops that are hosts of this fungus include numerous Kurume azaleas, flowering dogwood, many junipers, and rhododendron.

Management

Sanitation and cultural practices which minimize the risk of introduction of the pathogen and avoid conditions which favor disease development, along with fungicide treatments, are the keys to preventing outbreaks of Phytophthora root rot in container-grown magnolia. A detailed discussion of the recommended control practices can be found in Chapter 14.

Cultural practices Container production beds must be crowned and then covered with black plastic or weed cloth along with a layer of gravel or similar coarse material. The ponding of water around container stock must be prevented. To avoid over- or underwatering, block plant material by container size and water needs. Account for daily rainfall when scheduling irrigation, particularly for the deciduous saucer magnolia. Fertilize according to need, and avoid burning roots with water-soluble nitrogen fertilizers applied at high rates. Use a bark-based medium having a high percolation rate and 20 to 30% air-filled pore space.

Sanitation Store potting medium components on an asphalt or a concrete pad. Avoid recycling containers unless they are carefully rinsed and disinfected. Clear all debris, and disinfect propagation benches. Collect cuttings for propagation from disease-free stock plants. Root magnolia cuttings or germinate seed in fresh soilless medium in new flats or containers on clean raised benches.

Resistance The reactions of cultivars of southern and saucer magnolias and other magnolia taxa to Phytophthora root rot have not been reported.

Chemical treatment A chemical control program is successful only when combined with good production and sanitation practices. Prior to transplanting liners or container stock, mix a granular or selected wettable powder fungicide into the potting medium. Depending on the fungicide added to the medium, begin fungicide sprays or drenches one to three months later. If a fungicide has not been incorporated into the medium, begin preventative fungicide treatments immediately after liners or container stock are transplanted, and reapply the fungicide at the interval specified on the product label until the crop is shipped. Fungicides

Phytophthora Root Rot

Geographic occurrence USDA zones 6–9

Seasonal occurrence April to October, with an optimum period in July and August

DISEASE FREQUENCY	DISEASE SEVERITY
5 annual	5 plants killed
4	**4** (boxed)
3	3
2 (boxed)	2
1 rare	1 very little damage

CHEMICAL TREATMENT	CULTURAL PRACTICES
3 used every year	**3 very important** (boxed)
2 (boxed)	2
1 not used	1 not important

SANITATION	RESISTANT CULTIVARS
3 very important (boxed)	3 many cultivars
2	2
1 not important	**1 no resistance** (boxed)

will not "cure" plants of Phytophthora root rot. For fungicide recommendations for the control of this disease, contact your county extension agent or state specialist responsible for diseases of nursery crops.

REFERENCES

Dirr, M. A. 1998. Manual of Woody Landscape Plants: Their Identification, Ornamental Characteristics, Culture, Propagation and Uses. 5th ed. Stipes Publishing, Champaign, Ill.

Goff, W. D., Shumack, R. L., Tilt, K. M., and Hagan, A. K. 1996. Fungicide sprays affect leaf condition and tree appearance of southern magnolia. J. Arboric. 22:201–205.

Miller, J. W. 1976. Bacterial leaf spot on saucer magnolia. Fla. Dep. Agric. Consumer Serv. Plant Pathol. Circ. 163.

Mullen, J. M., and Cobb, G. S. 1984. Leaf spot of southern magnolia caused by *Pseudomonas cichorii.* Plant Dis. 68:1013–1015.

Mullen, J. M., Cobb, P. P., and Shumack, R. L. 1983. Management program developed for Alabama magnolia orchard. Ala. Agric. Exp. Stn. Auburn Univ. Res. Rep. Ser. 1:15–16.

CHAPTER

60

Mark Gleason ▪ Iowa State University, Ames

John Hartman ▪ University of Kentucky, Lexington

Maple Diseases

Geographic production USDA zones 2–9

Family Aceraceae

Genus *Acer*

Species

A. buergeranum	trident maple
A. campestre	hedge maple
A. ginnala	Amur maple
A. griseum	paperbark maple
A. macrophyllum	big leaf maple
A. miyabei	Miyabe maple
A. negundo	box elder
A. palmatum	Japanese maple
A. pensylvanicum	striped maple
A. platanoides	Norway maple
A. pseudoplatanus	sycamore maple
A. rubrum	red maple
A. saccharinum	silver maple
A. saccharum	sugar maple
A. saccharum subsp. *nigrum*	black maple
A. spicatum	mountain maple
A. tataricum	Tatarian maple

Maples are among the most widely grown ornamental plants. The many species of *Acer* grown in nurseries range from trees to shrubs, and their diseases are similarly diverse. Verticillium wilt is the most destructive disease of ornamental maples, but foliar diseases and cankers are more common in many species.

Verticillium Wilt

Verticillium dahliae

Verticillium wilt is the most destructive disease of maples. Further information about the disease is presented in Chapter 19, "Verticillium Wilt."

Symptoms

Symptoms are variable and can be confused with effects of herbicide injury, environmental stresses such as drought, or mechanical damage. Chronic symptoms include small, yellow foliage, leaf scorch (marginal browning), slow growth, abnormally heavy seed crops, and dieback of shoots and branches. Acute symptoms include abnormal red or yellow color between leaf veins, browning and curling up (wilting) of leaves, defoliation, and branch dieback. Symptoms can develop at any time during the growing season but are most likely to appear in July and August. Some trees show long-term, sublethal symptoms, while others die within a few months after symptoms appear. Symptoms often appear first on one side of the tree or on one branch. The course of the disease is difficult to predict; symptoms worsen rapidly in some cases, whereas in others they may abate for months or years.

Verticillium Wilt

Geographic occurrence USDA zones 2–9

Seasonal occurrence Year-round

DISEASE FREQUENCY	DISEASE SEVERITY
5 annual	**[5] plants killed**
4	4
[3]	3
2	2
1 rare	1 very little damage

CHEMICAL TREATMENT	CULTURAL PRACTICES
3 used every year	3 very important
2	**[2]**
[1] not used	1 not important

SANITATION	RESISTANT CULTIVARS
3 very important	3 many cultivars
[2]	**[2]**
1 not important	1 no resistance

Streaking of the vascular tissue or wood may accompany external symptoms, especially in the later stages of infection (Plate 79). Greenish brown streaks, parallel to the long axis of the trunk or branch, appear in the outer sapwood in new infections. Streaking of older (inner) wood may accompany older infections.

Epidemiology

See Chapter 19.

Host Range

All maples are susceptible to Verticillium wilt.

Management

The presence of vascular streaking, along with typical foliar symptoms, is fairly diagnostic for Verticillium wilt of maple. But because symptoms are variable and can be confused with other problems, a laboratory test should be performed to confirm the diagnosis. See Chapter 19 for additional information.

Cultural practices, sanitation, and chemical treatment See Chapter 19.

Resistance Maples are rated as generally susceptible to Verticillium wilt. However, the degree of susceptibility varies among species and among cultivars of the same species. For example, Norway maple cultivars Columnare Compacta, Jade Glenn, and Parkway are rated tolerant or resistant; Cleveland, Crimson King, Globosum, Greenlace, and Royal Red are susceptible; and Emerald Queen, Schwedleri, Silver Variegated, Summershade, and Superform are intermediate.

Canker Diseases

Nectria Canker (Coral Spot)

Nectria cinnabarina
(asexual stage, *Tubercularia vulgaris*)
Nectria galligena
(asexual stage, *Cylindrocarpon mali*)

Nectria spp. are common fungi on many species of woody plants. They invade trees through wounds and injuries caused by winter cold, drought, sunscald, leaf scars, cracks in the crotches of twigs or branches, or senescence of the lower branches.

Symptoms

N. cinnabarina causes girdling cankers on twigs, limbs, and sometimes trunks. Affected parts eventually die. The cankers are slightly sunken areas and are generally associated with wounds. They become easier to see when fungal fruiting bodies appear in them in large numbers (Plate 80), either the coral-colored perithecia of *N. cinnabarina* or the pinkish sporodochia of *Tubercularia vulgaris,* the asexual stage of the fungus.

Symptoms caused by *N. galligena* can resemble those of *N. cinnabarina.* On a branch or trunk that is not girdled, however, a "target canker," with raised, concentric ridges of callus tissue, can develop year after year. Swellings at the sites of target cankers can become weak points at which limbs break in the wind or after snow or ice storms.

Epidemiology

The pathogen infects trees through wounds and natural openings, including leaf scars, cracks in branch crotches, sunscalded areas, hail injuries, and pruning wounds. Cankers are most likely to appear on trees that have already been weakened by other stresses, such as root pruning, drought, or freeze damage. During wet weather, fruiting bodies of *Nectria* exude spores, which are spread by splashing rain or irrigation.

Host Range

All maples and many other woody species are susceptible to Nectria canker.

Management

Cultural practices Prune during dry weather near the end of the dormant period. Pruning at the wrong time of year increases the risk of canker development. Avoid wounding the trees. Maintain tree vigor through proper fertilizing and watering. Avoid severe root pruning. Avoid leaving pruning stubs when removing the rootstock shoot; make a sharp, clean cut which will heal promptly.

Sanitation Remove and destroy affected twigs and limbs. Burn or bury the pruned limbs promptly, so that they will not serve as sources of inoculum.

Resistance Resistance to Nectria canker in maples has not been identified.

Chemical treatment A fungicide application after pruning

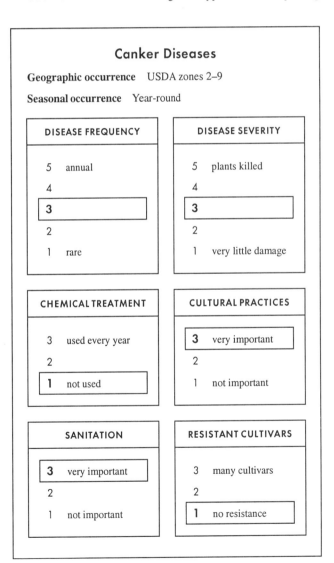

Canker Diseases

Geographic occurrence USDA zones 2–9

Seasonal occurrence Year-round

DISEASE FREQUENCY	DISEASE SEVERITY
5 annual	5 plants killed
4	4
[3]	**[3]**
2	2
1 rare	1 very little damage

CHEMICAL TREATMENT	CULTURAL PRACTICES
3 used every year	**[3] very important**
2	2
[1] not used	1 not important

SANITATION	RESISTANT CULTIVARS
[3] very important	3 many cultivars
2	2
1 not important	**[1] no resistance**

but before a rain or irrigation will protect wounds against infection. There are no specific fungicide registrations for control of Nectria canker of maple, however. Many of the fungicides recommended for anthracnose control will help to protect trees from Nectria canker.

Eutypella Canker

Eutypella parasitica

Symptoms

The fungus *Eutypella parasitica* causes a target canker of maples. Cankers typically appear on young trees, within 10 feet above the ground. During the first few years, Eutypella cankers are flattened or slightly sunken areas covered by bark, usually with a branch stub (the site of infection) at the center. A good diagnostic feature is the presence of white masses of fungus, called mycelial fans, under the bark at the edges of expanding cankers. Older cankers may have a rough exposed surface of decaying wood surrounded by a swollen ridge of callus. Affected trees become swollen at the site of the canker. Masses of small, black fruiting bodies (perithecia) form on bark in the centers of cankers that are at least six to eight years old.

Epidemiology

Cankers on trunks result from fungal invasion of branch stubs or wounds. Fruiting bodies of *E. parasitica* release ascospores during rain or irrigation at moderate temperatures, and the spores are dispersed by wind.

Host Range

Black, Norway, red, silver, sugar, and sycamore maples and box elder are susceptible to Eutypella canker.

Management

The same practices described for Nectria canker are likely to be helpful in the control of Eutypella canker.

Valsa Canker (Cytospora Canker)

Valsa ambiens
(asexual stage, *Cytospora leucosperma*)
Valsa ceratosperma
(asexual stage, *Cytospora sacculus*)

Valsa canker, also known as Cytospora canker, is a disease of girdling cankers on limbs. It is most common in eastern North America.

Symptoms

Cankers girdle twigs or branches, resulting in wilting and death of the affected limbs. Wilted leaves may stay attached to the limbs. Flags (limbs with brown, wilted leaves attached) develop when limbs are girdled during the growing season. Limb dieback is often sporadic and scattered, with only one or a few affected limbs per tree, but can be more common following a period of stress on the tree (e.g., drought or freeze injury). A canker may or may not be apparent on the outer bark at the site of infection (Plate 81). Beneath the bark, however, a dark greenish black line separating the healthy and the diseased sapwood is usually visible. White to grayish brown fruiting bodies (stromata) of the pathogen appear in cankers.

Epidemiology

The pathogens invade trees through branch stubs and other wounds. As in most canker diseases, the development and spread of Valsa canker are favored by predisposing stresses on the tree. Spores of *Cytospora* are exuded from fruiting bodies during rainfall or irrigation and are spread by water.

Host Range

Box elder; Norway, red, sugar, and silver maples; and ash are susceptible to Valsa canker.

Management

See the recommendations for management of Nectria canker.

Cryptosporiopsis Canker

***Cryptosporiopsis* sp.**

Cryptosporiopsis canker is common in some years in nurseries and landscapes in the northeastern United States.

Symptoms

The fungal pathogen, a *Cryptosporiopsis* sp., invades trunks and lower branches through oviposition wounds made by the narrow-winged tree cricket, *Oecanthus angustipennis.* Infection causes annual cankers (lasting only one year), which are long and narrow (up to 8 inches long and less than 1 inch wide) (Plate 82). Streaks of discolored sapwood can extend as much as 1–2 feet above a canker. In the spring, a dark brown fluid often bleeds from the oviposition hole at the center of the canker. Small trees and limbs may be girdled by cankers, and larger stems may be severely disfigured by multiple cankers.

Epidemiology

Infection occurs in the fall, when egg-laying crickets scrape fragments of outer bark, containing the fungus, into oviposition holes in the bark. Cankers develop mainly during the dormant season, and bleeding from them is conspicuous the following spring. Vigorous trees produce rolls of callus at canker margins and may contain cankers within one year. Secondary fungi, especially *Valsa ambiens,* may colonize the outer margins of cankers and cause some canker expansion. The occurrence of the disease varies greatly from year to year.

Host Range

Red maple is susceptible to Cryptosporiopsis canker.

Management

Remove and destroy infected limbs and trees. Fertilize and irrigate to ensure maximum tree vigor.

Foliar Diseases

Anthracnose

***Discula* spp.**
Kabatiella apocrypta

Anthracnose fungi produce a range of symptoms on maple leaves (Plate 83). The long-term effects of anthracnose on tree health are generally minimal, but foliar browning may be conspicuous in wet years.

Symptoms

The various anthracnose fungi cause somewhat different symptoms. *Kabatiella apocrypta* occurs on many species of maple throughout North America. Symptoms appear shortly after rainy periods in late spring through late summer. Irregularly shaped, randomly arrayed lesions develop on leaves. Lesions merge and kill large areas of the leaf. The color of the lesions varies with the host, from reddish brown on sugar maple to light tan on Japanese maple (Plate 84). Severe infection leads to premature defoliation.

One form of *Discula* causes elongated lesions along leaf veins in sugar and striped maples in the northeastern United States. Leaf tissue near affected veins turns yellow or red and then brown, and the discoloration extends to the margins of the leaf. These symptoms become most conspicuous in late summer.

Another form of *Discula* causes rounded, reddish brown spots with tan centers on leaves of sugar maple. The spots are usually centered on a vein.

Epidemiology

In growing seasons with persistently wet weather, several cycles of infection can occur. Spores of the pathogens are spread by splashing water.

Host Range

Japanese, mountain, Norway, red, silver, sugar, and Tatarian maples and box elder are susceptible to *K. apocrypta.*

Sugar maple and striped maple are susceptible to *Discula* spp.

Management

Damage due to anthracnose is usually minimal. However, contact fungicides can be applied as a preventative treatment in situations in which leaf blighting is intolerable.

Tar Spot

Rhytisma acerinum
Rhytisma punctatum

Tar spots are conspicuous lesions on leaves and are easy to recognize. The disease is caused by two fungal species: *Rhytisma acerinum,* which occurs nationwide but is more common in the eastern United States, and *R. punctatum,* which occurs nationwide to the north of the southernmost tier of states but is more common in the western United States. Control measures are usually not needed.

Symptoms

Symptoms of early infections appear in late spring or early summer, after leaves have reached full size. Small, water-soaked spots develop on leaves and turn light green, yellow, or brown (Plate 85). During mid- to late summer, the tar spot fungi form raised, black structures (stroma) within the spots on the upper surface of the leaf. *R. acerinum* forms relatively large stromata (a few millimeters to several centimeters in diameter) in leaf lesions. *R. punctatum* produces numerous small (1 mm in diameter), dark stroma in each lesion (tar spot caused by *R. punctatum* is also called small tar spot or speckled leaf spot). On the lower surface of the leaf, the tissue beneath a tar spot turns brown to yellow. Leaf tissue surrounding lesions caused by *R. punctatum* may stay green even after the rest of the leaf turns brown in fall, creating a "green island" effect. Leaves may fall prematurely in severe infection.

Epidemiology

Tar spot fungi overwinter in stroma on fallen leaves. Infection occurs as spores are ejected from these decayed leaves during rainy periods in the spring.

Host Range

Amur, bigleaf, hedge, mountain, Norway, red, Rocky Mountain, silver, sugar, and sycamore maples and box elder are susceptible to *R. acerinum.*

Bigleaf, mountain, red, Rocky Mountain, silver, striped, sugar, sycamore, and vine maples and box elder are susceptible to *R. punctatum.*

Management

Damage due to tar spot is usually minimal. When a high level of control is necessary, spray a labeled fungicide when buds are opening, and repeat two more times at 10-day intervals if the season is wet.

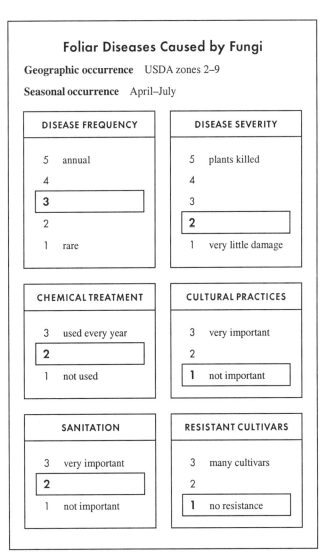

Foliar Diseases Caused by Fungi

Geographic occurrence USDA zones 2–9

Seasonal occurrence April–July

DISEASE FREQUENCY	DISEASE SEVERITY
5 annual	5 plants killed
4	4
3	3
2	**2**
1 rare	1 very little damage

CHEMICAL TREATMENT	CULTURAL PRACTICES
3 used every year	3 very important
2	2
1 not used	**1 not important**

SANITATION	RESISTANT CULTIVARS
3 very important	3 many cultivars
2	2
1 not important	**1 no resistance**

Phyllosticta Leaf Spot

Phyllosticta minima
Phyllosticta negundinis

Leaf spot diseases caused by *Phyllosticta* spp. occur primarily on lower leaves and usually have only a minimal impact on plant health. However, *P. minima* occasionally becomes severe enough to cause partial defoliation of red maple. Phyllosticta leaf spot diseases occur from the Great Plains to the East Coast.

Symptoms

P. minima causes eye spot, or purple-bordered leaf spot. The spots are irregularly round and usually less than 5 mm across. They are brown at first, changing to tan in the center and dark reddish or purple at the border (Plate 86). Small, black fruiting bodies (pycnidia) of the pathogen are often visible in lesions on the upper side of the leaf, typically arranged in a circle. In severe infection, lesions may grow together to form large, irregularly shaped areas of diseased tissue.

P. negundinis produces rounded leaf spots on box elder, up to 8 mm in diameter, which have yellowish brown margins and thin, translucent, pale yellow centers. The fragile tissue in the center of the lesion often falls out, leaving a ragged hole.

Epidemiology

The pathogens overwinter in lesions on fallen leaves. The following spring, spores are dispersed from pycnidia to new foliage by splashing rain or irrigation.

Host Range

Amur, hedge, Japanese, mountain, red, silver, sugar, sycamore, and Tatarian maples are susceptible to *P. minima.*

Box elder is susceptible to *P. negundinis.*

Management

Damage due to Phyllosticta leaf spot is seldom severe enough to warrant control measures. If a high level of control is required, a labeled fungicide could be applied at intervals of 10–14 days during wet periods in the spring.

Powdery Mildew

Phyllactinia guttata

See Chapter 15, "Powdery Mildew," for further details.

Symptoms

The powdery mildew fungus *Phyllactinia guttata* forms a whitish fungal mat, closely attached to the undersides of mature leaves, often with small (less than 1 mm in diameter), spherical, black fruiting bodies (cleistothecia) attached.

Epidemiology

The fungus overwinters in cleistothecia in fallen leaves. Spores are released in the spring and are transported to new leaves by air currents. Symptoms usually appear during the second half of the growing season, after leaves are mature. *P. guttata* also attacks many other hardwood species.

Host Range

All maples are susceptible to powdery mildew.

Management

Systemic fungicides labeled for powdery mildew control can be used to control the disease when it first appears.

Bacterial Leaf Spot and Dieback

Pseudomonas syringae* pv. *syringae

Bacterial leaf spot and dieback of maple is caused by *Pseudomonas syringae* pv. *syringae,* which also causes bacterial blights of lilac, fruit trees, and many woody ornamentals.

Symptoms

Symptoms include leaf spots, vein blackening, and tip dieback of Japanese, Norway, and red maples. The leaf spots vary in size, from pinpoints to 1/4 inch in diameter. Most spots begin as water-soaked areas, and they may have chlorotic halos. Spots may coalesce, destroying leaves or young seedlings. One-year-old twigs may turn black during the dormant season, die over the winter, and turn ash gray. Shoots may leaf out in the spring and then die back.

Epidemiology

P. syringae pv. *syringae* overwinters on infected plant parts or on the surfaces of healthy tissue. It is spread in windblown rain,

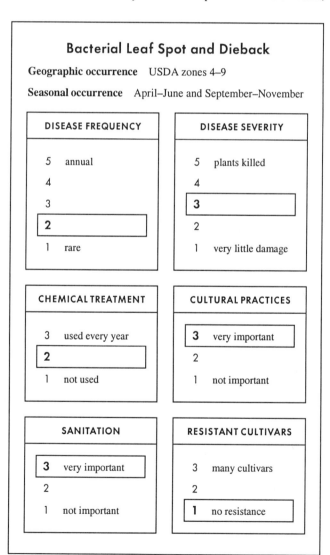

Bacterial Leaf Spot and Dieback

Geographic occurrence USDA zones 4–9

Seasonal occurrence April–June and September–November

DISEASE FREQUENCY	DISEASE SEVERITY
5 annual	5 plants killed
4	4
3	**[3]**
[2]	2
1 rare	1 very little damage

CHEMICAL TREATMENT	CULTURAL PRACTICES
3 used every year	**[3] very important**
[2]	2
1 not used	1 not important

SANITATION	RESISTANT CULTIVARS
[3] very important	3 many cultivars
2	2
1 not important	**[1] no resistance**

by insects, and on pruning tools. The bacterium enters plants through wounds or natural openings. Wound infection during bud grafting may interfere with bud-take. Frost damage, high levels of nitrogen fertilization, and heavy rains favor invasion by the pathogen.

Host Range

Most maple species except sugar maple are susceptible. Some Japanese maple cultivars, such as Sango Kaku and Oshi Beni, are highly susceptible.

Management

Cultural practices Handle young plants carefully to avoid wounding. If possible, perform operations that cause wounding, including pruning and budding, in dry weather. Protect plants from rain and frost, if feasible, in early spring and late fall. Maintain adequate spacing for good air circulation.

Sanitation Remove and destroy dead twigs and fallen leaves.

Resistance Plant resistant cultivars, or avoid highly susceptible cultivars.

Chemical treatment In areas with a high risk of outbreaks of bacterial leaf spot and dieback, apply a labeled antibacterial product, such as a fixed copper product, in fall to protect wounds and leaf scars, and in spring to protect new growth. Chemical applications are ineffective without the cultural controls described above.

CHAPTER 61

Margaret R. Williamson ▪ Clemson University, Clemson, South Carolina

James H. Blake ▪ Clemson University, Clemson, South Carolina

Mountain Laurel Diseases

Geographic production USDA zones 4–8

Family Ericaceae

Genus *Kalmia*

Species *K. latifolia*

Cultivars many

Mountain laurel is an evergreen shrub valued for its floral display. The most common disease of mountain laurel is Cercospora leaf spot.

Cercospora Leaf Spot

Cercospora kalmiae
Cercospora sparsa

Symptoms

Circular to irregularly shaped spots appear on new growth in mid- to late summer. Lesions form on both leaf surfaces and are initially medium to dark brown. As the disease progresses, a dark brown to purplish border develops, and the center fades to gray on the upper surface of the leaf. Mature leaf spots are 5–10 mm across. Severe infections can stunt plants and suppress flowering (Peterson et al., 1976).

Epidemiology

Cercospora leaf spot is most severe during exceptionally moist seasons and in shaded, enclosed areas. Most infections begin on newly expanded foliage just before and during bloom, with lesions appearing in late summer (Sinclair et al., 1987). Patches of thickened, black fungal tissue (stromata) develop in the gray centers of spots on the upper surface of the leaf. These stromata produce spores (conidia) on pincushion-like conidiophores, which can be seen under magnification with a hand lens. The conidia cause new infections. They are most likely spread by splashing water and possibly by air currents. Even heavily infected leaves remain on the plant through the winter, so they are probably the source of inoculum for new infections in spring.

Host Range

K. angustifolia and *K. latifolia* are susceptible to Cercospora leaf spot.

Management

Both cultural and chemical controls can be used to manage this disease. No resistant cultivars have been selected, but the severity of leaf spot diseases varies greatly in seedling populations. The cultivars Quinnipiac and Goodrich are especially susceptible (Dirr, 1998).

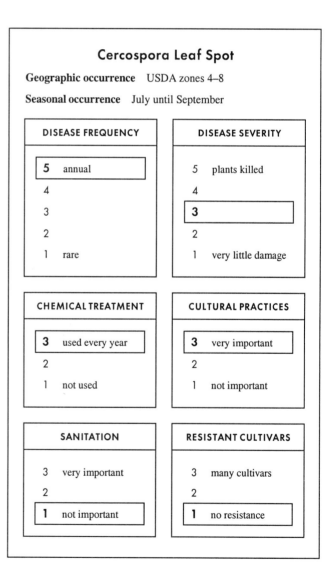

Cercospora Leaf Spot

Geographic occurrence USDA zones 4–8

Seasonal occurrence July until September

DISEASE FREQUENCY	DISEASE SEVERITY
[5] annual	5 plants killed
4	4
3	**[3]**
2	2
1 rare	1 very little damage

CHEMICAL TREATMENT	CULTURAL PRACTICES
[3] used every year	**[3] very important**
2	2
1 not used	1 not important

SANITATION	RESISTANT CULTIVARS
3 very important	3 many cultivars
2	2
[1] not important	**[1] no resistance**

Cultural practices Increase plant spacing. Improve air circulation and increase the amount of light penetration, where possible.

Chemical treatment Timing of applications is important. Spraying five times per season, starting at budbreak and continuing every two weeks, will give nearly complete control of the disease. However, adequate control can be achieved with only two sprays, the first applied at the emergence of new leaves and the second at flowering (Peterson et al., 1976).

Other Leaf Spot Diseases

Various fungi

Other fungi reported to cause leaf spot diseases of mountain laurel include a *Colletotrichum* sp., *Physalospora kalmiae, Phyllosticta kalmicola, Septoria angustifolia, S. kalmicola,* and *Venturia kalmiae* (Jaynes, 1997). Symptoms of diseases caused by these fungi are often indistinguishable. In general, small, yellow, brown, or reddish spots develop and enlarge to varying degrees. Concentric zones of dead or dying tissue may occur within the spots. Blighting may occur as spots enlarge and coalesce. In mild infections, only the aesthetic quality of the foliage is affected. Severe infection can result in defoliation and stem dieback.

Like Cercospora leaf spot, infections caused by other leaf spot fungi are favored by high humidity, poor air circulation, and wet foliage.

Fungicides to control leaf spot diseases are best applied prior to the onset of symptoms. If foliar diseases have occurred in the past, apply a protective fungicide when new growth is half-developed and then again when mature size is attained.

Phytophthora Root Rot

Phytophthora spp.

Phytophthora root rot is a common disease of mountain laurel and other crops in the family Ericaceae. Light infection causes slightly stunted leaves or minor root necrosis, and such symptoms often go unnoticed. As severity increases, the disease leads to noticeable stunting of leaves and plants, interveinal chlorosis and marginal necrosis of leaves, wilting, shoot dieback, and plant death. Root systems are greatly reduced, and roots are discolored. When roots are gently pulled, the outer root tissue often sloughs off.

Phytophthora spp. can also infect the crown. Infection is indicated by a reddish brown discoloration beneath the bark.

See Chapter 14, "Phytophthora Root Rot and Dieback," for information on epidemiology and management.

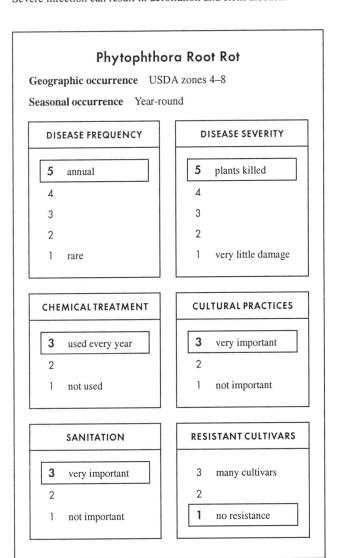

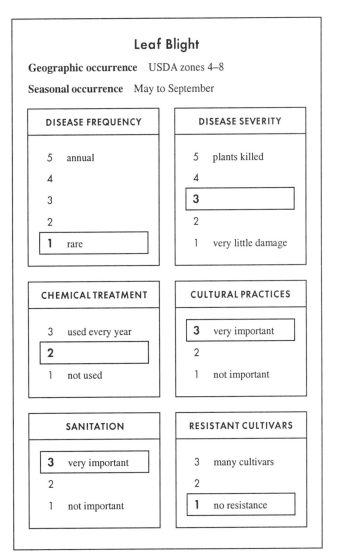

Phytophthora Root Rot

Geographic occurrence USDA zones 4–8

Seasonal occurrence Year-round

DISEASE FREQUENCY	DISEASE SEVERITY
5 annual	**5 plants killed**
4	4
3	3
2	2
1 rare	1 very little damage

CHEMICAL TREATMENT	CULTURAL PRACTICES
3 used every year	**3 very important**
2	2
1 not used	1 not important

SANITATION	RESISTANT CULTIVARS
3 very important	3 many cultivars
2	2
1 not important	**1 no resistance**

Leaf Blight

Geographic occurrence USDA zones 4–8

Seasonal occurrence May to September

DISEASE FREQUENCY	DISEASE SEVERITY
5 annual	5 plants killed
4	4
3	**3**
2	2
1 rare	1 very little damage

CHEMICAL TREATMENT	CULTURAL PRACTICES
3 used every year	**3 very important**
2	2
1 not used	1 not important

SANITATION	RESISTANT CULTIVARS
3 very important	3 many cultivars
2	2
1 not important	**1 no resistance**

Leaf Blight

Diaporthe kalmiae
(asexual stage, *Phomopsis kalmiae*)

Symptoms

Leaf blight lesions are initially small, brown spots, which enlarge gradually, becoming irregularly shaped and encompassing large areas of leaf tissue. The brown, necrotic tissue is darkest near the lesion margins, with a reddish zone separating the dead tissue from live tissue. Leaf tips and margins often become blighted. When leaf margins are affected, the leaves become cupped. Such severely infected leaves are often shed prematurely. Prior to leaf drop, the fungus may also move into petioles and from there into twigs and stems, causing dieback (Wolf and Cavaliere, 1965).

Epidemiology

Wounds allow the pathogen to enter plant tissue. Moist, humid conditions are conducive to disease development and spread. The fungus appears to overwinter in dead leaves, which are probably the source of inoculum for new infections in spring.

Host Range

K. latifolia is susceptible to leaf blight.

Management

Cultural practices Avoid wounding. Increase plant spacing and improve air circulation and light penetration when possible. Remove and destroy fallen leaves.

Chemical treatment If the pathogen has been a problem in the past, apply protective fungicides after pruning.

Flower Blight

Ovulinia azaleae

Flower blight is uncommon in *Kalmia.* The disease is most destructive in the humid southeastern states. It affects only floral tissue and most often occurs in mountain laurel grown in proximity to infected azaleas that bloom concurrently. To limit the likelihood of infection, separate mountain laurel from such late-blooming azaleas. See the section on petal blight in Chapter 23, "Azalea Diseases," for more information.

Flower Blight

Geographic occurrence USDA zones 4–8

Seasonal occurrence April to June, depending on the locality

DISEASE FREQUENCY	DISEASE SEVERITY
5 annual	5 plants killed
4	4
3	3
2	2
[1 rare]	**[1 very little damage]**

CHEMICAL TREATMENT	CULTURAL PRACTICES
3 used every year	**[3 very important]**
[2]	2
1 not used	1 not important

SANITATION	RESISTANT CULTIVARS
[3 very important]	3 many cultivars
2	2
1 not important	**[1 no resistance]**

Leaf and Flower Gall

Geographic occurrence USDA zones 4–8

Seasonal occurrence March to May, depending on the locality

DISEASE FREQUENCY	DISEASE SEVERITY
5 annual	5 plants killed
4	4
3	3
2	2
[1 rare]	**[1 very little damage]**

CHEMICAL TREATMENT	CULTURAL PRACTICES
3 used every year	**[3 very important]**
[2]	2
1 not used	1 not important

SANITATION	RESISTANT CULTIVARS
[3 very important]	3 many cultivars
2	2
1 not important	**[1 no resistance]**

Leaf and Flower Gall

Exobasidium vaccinii

Many ericaceous plants are affected by leaf and flower gall, and mountain laurel is susceptible, but reported incidences are rare. Only new growth is affected. See the section on leaf and flower gall in Chapter 23, "Azalea Diseases," for more information.

Powdery Mildew

Microsphaera penicillata
Microsphaera vaccinii

Mountain laurel is occasionally infected with powdery mildew. Patches of white, powdery growth become evident on the leaves in summer, and infection continues into the fall. For more information, see Chapter 15, "Powdery Mildew."

Necrotic Ringspot

Necrotic ringspot is a rare disease of *Kalmia.* It also occurs in hybrid rhododendron species of a certain parentage. The causal agent is presumed to be a virus in the potato virus X group (Coyier et al., 1977). No insect vectors have been identified, but viruses in this group are generally not insect-vectored. As far as is known, the virus is spread primarily by vegetative propagation and grafting.

Necrotic, reddish brown ring spots develop on two-year-old leaves as new growth emerges. Some leaves turn entirely red and drop off. In *Kalmia,* affected plants experience little or no loss of vigor. As in rhododendron, necrotic ringspot of mountain laurel seems to be associated with certain cultivars, such as Olympic Fire (Jaynes, 1997).

REFERENCES

Coyier, D. L., Stace-Smith, R., Allen, T. C., and Leung, E. 1977. Viruslike particles associated with a rhododendron necrotic ringspot disease. Phytopathology 67:1090–1095.

Dirr, M. A. 1998. Manual of Woody Landscape Plants: Their Identification, Ornamental Characteristics, Culture, Propagation and Uses. 5th ed. Stipes Publishing, Champaign, Ill.

Enlows, E. M. 1918. A leafblight of *Kalmia latifolia.* J. Agric. Res. 13: 199–212.

Farr, D. F., Bills, G. F., Chamuris, G. P., and Rossman, A. Y. 1989. Fungi on Plants and Plant Products in the United States. American Phytopathological Society, St. Paul, Minn.

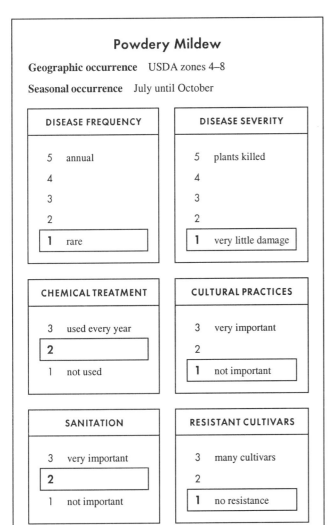

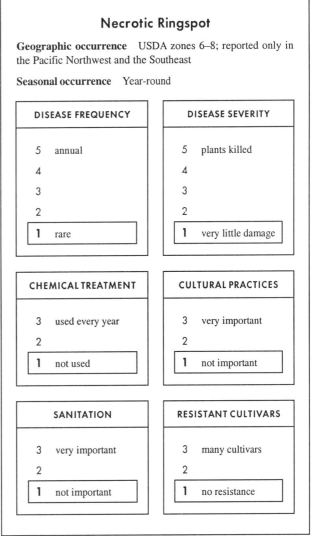

Powdery Mildew

Geographic occurrence USDA zones 4–8

Seasonal occurrence July until October

DISEASE FREQUENCY	DISEASE SEVERITY
5 annual	5 plants killed
4	4
3	3
2	2
[1 rare]	**[1 very little damage]**

CHEMICAL TREATMENT	CULTURAL PRACTICES
3 used every year	3 very important
[2]	2
1 not used	**[1 not important]**

SANITATION	RESISTANT CULTIVARS
3 very important	3 many cultivars
[2]	2
1 not important	**[1 no resistance]**

Necrotic Ringspot

Geographic occurrence USDA zones 6–8; reported only in the Pacific Northwest and the Southeast

Seasonal occurrence Year-round

DISEASE FREQUENCY	DISEASE SEVERITY
5 annual	5 plants killed
4	4
3	3
2	2
[1 rare]	**[1 very little damage]**

CHEMICAL TREATMENT	CULTURAL PRACTICES
3 used every year	3 very important
2	2
[1 not used]	**[1 not important]**

SANITATION	RESISTANT CULTIVARS
3 very important	3 many cultivars
2	2
[1 not important]	**[1 no resistance]**

Jaynes, R. A. 1997. *Kalmia,* Mountain Laurel and Related Species. Timber Press, Portland, Ore.

Peterson, J. L., Davis, S. H., Jr., and Judd, R. W., Jr. 1976. Effect of fungicide and application timing on Cercospora leaf spot of mountain laurel. Plant Dis. Rep. 60:138–140.

Sinclair, W. A., Lyon, H. H., and Johnson, W. T. 1987. Diseases of Trees and Shrubs. Comstock Publishing Associates, Cornell University Press, Ithaca, N.Y.

Wolf, F. A., and Cavaliere, A. R. 1965. Two new species of leafblight fungi on *Kalmia latifolia.* Mycologia 57:576–582.

CHAPTER 62

Gary W. Simone ▪ University of Florida, Gainesville

Nandina Diseases

Geographic production USDA zones 7–10

Family Berberidaceae

Genus *Nandina*

Species *N. domestica* nandina, heavenly bamboo

Nandina is a monotypic genus with a natural range extending from India through eastern Asia. This woody ornamental was introduced into western Europe in 1804 and into the Western Hemisphere some time later. *N. domestica* grows to 6–8 feet tall, although shorter cultivars have been developed. Standard cultivars include Firepower; dwarf cultivars include Harbor Dwarf, Nana purpurea, and Compacta. Nandina can be propagated from seed, but it does not breed true, and thus named cultivars are vegetatively propagated. *N. domestica* is root-hardy in zone 6. A number of diseases are important in nursery production of nandina.

Anthracnose

Glomerella cingulata
(asexual stage, *Colletotrichum gloeosporioides*)

Symptoms

Anthracnose infection often occurs in stressed leaf tissue, damaged by temperature extremes, phytotoxicity, senescence, or mechanical impact. Lesions expand into large, round to oval, brown spots around the point of damage or at the leaf tip (Plate 87). Small, dark specks of the pathogen in its reproductive stage, *Glomerella cingulata,* are often visible in concentric bands in leaf lesions. During moist weather, a pinkish spore matrix may be visible, extruded from each reproductive body.

Epidemiology

Infections may remain dormant for weeks prior to active disease development. Leaf spots expand, quickly covering high percentages of leaf tissue. Reproduction of the pathogen in lesions is rapid and consistent for long periods. Spores (conidia) of the pathogen are splashed by rain to adjacent leaves or plants. The fungus can persist within infected tissue on or below the plant.

Host Range

All cultivars of nandina are susceptible to *G. cingulata* (asexual stage, *Colletotrichum gloeosporioides*), a common foliar pathogen with a host range encompassing hundreds of plant species, including ornamentals as well as other crops.

Management

Time overhead irrigation for periods after dew formation or early during daylight. Severely damaged plants should be placed on a daytime water cycle to encourage fast drying of the canopy and discourage disease development. When disease severity war-

Anthracnose

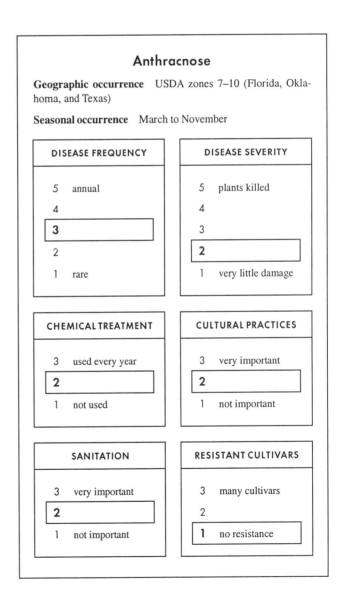

Geographic occurrence USDA zones 7–10 (Florida, Oklahoma, and Texas)

Seasonal occurrence March to November

DISEASE FREQUENCY	DISEASE SEVERITY
5 annual	5 plants killed
4	4
[3]	3
2	**[2]**
1 rare	1 very little damage

CHEMICAL TREATMENT	CULTURAL PRACTICES
3 used every year	3 very important
[2]	**[2]**
1 not used	1 not important

SANITATION	RESISTANT CULTIVARS
3 very important	3 many cultivars
[2]	2
1 not important	**[1] no resistance**

rants, shear infected tissue and remove prunings from production areas. Follow with applications of a protectant or a systemic fungicide to slow the spread of the disease. Sanitize stock of infected tissue before propagation of cuttings.

Cercospora Leaf Spot

Pseudocercospora nandinae

Symptoms

Cercospora leaf spot of nandina is caused by the fungus *Pseudocercospora nandinae.* Round, red to reddish brown spots develop on the upper surfaces of newly infected leaves. The lesions enlarge into round to irregular blotches ranging from 0.5 to 1.0 cm in diameter, with abundant red pigmentation on the upper leaf surface (Plate 88). The margins of the leaf spots are not distinct. Large areas of leaf tissue may be affected as lesions coalesce. The lower surface of the leaf looks gray to brown, because of abundant reproductive structures of the pathogen.

Epidemiology

P. nandinae reproduces exclusively on the lower surface of leaves. Spores (conidia) are splashed in water or are carried on the air to adjacent leaves. Disease severity is high in areas with overhead irrigation and in seasons with warm to hot, rainy weather.

Cercospora Leaf Spot

Geographic occurrence USDA zones 8–10 (Alabama, Florida, Oklahoma, and South Carolina)

Seasonal occurrence March to November

DISEASE FREQUENCY	DISEASE SEVERITY
5 annual	5 plants killed
4	4
[3]	**[3]**
2	2
1 rare	1 very little damage

CHEMICAL TREATMENT	CULTURAL PRACTICES
3 used every year	**[3] very important**
[2]	2
1 not used	1 not important

SANITATION	RESISTANT CULTIVARS
3 very important	3 many cultivars
[2]	2
1 not important	**[1] no resistance**

Host Range

All cultivars of nandina are believed to be susceptible. The pathogen is host-specific, attacking only plants in the genus *Nandina.*

Management

Examine exposure to light and the scheduling of irrigation in the management of Cercospora leaf spot. Disease severity is often higher in nursery production of nandina under partial shade than in production under higher levels of light. Time overhead irrigation for periods after dew formation or early during daylight. Space plants to ensure rapid drying of the canopy after irrigation. Where disease severity is high, both protectant and systemic fungicides are available. Apply foliar fungicides to both leaf surfaces.

Mosaic

Cucumber mosaic virus
Nandina mosaic virus
Nandina stem pitting virus

Symptoms

Mosaic may be due to one or more of three viruses affecting nandina (Plate 89). Although each virus has its own symptoms in this host, the frequency of multiple infections tends to make visual diagnosis difficult, because of new symptom combinations. Reliance on crown divisions or tip cuttings for propagation of named cultivars easily explains the increase and persistence of these viruses in nursery stock.

Cucumber mosaic virus (CMV) causes a wine red discoloration of new spring growth. Affected plants exhibit "shoestringed" leaves, which often cup downward. Flower buds on infected plants may be pink. There may be fewer buds than normal and fewer berries set as a result.

In contrast to CMV, Nandina mosaic virus (NaMV) causes a mild mosaic that does not persist in the plant. Reports of stunting and leaf distortion are inconsistent with diseases caused by this virus.

Nandina stem pitting virus (NSPV) causes leaf distortion, mosaic, leaf reddening, upward leaf roll, and stem pitting, but little if any stunting.

Epidemiology

None of these viruses is known to be seed-transmitted in *N. domestica.* The persistence and spread of these viruses within nurseries is tied to vegetative propagation. Only CMV is likely to be transmitted by aphid vectors, because of the wide host range of the virus and the numerous species of aphid vector.

Host Range

CMV has a known host range of more than 60 plant species in more than nine families, including agronomic, ornamental, vegetable, and weed species.

NaMV affects at least 10 species of plants in more than five families.

NSPV has the most limited host range, attacking plants in the fewer than three families, including a few crop and weed species.

Mosaic

Geographic occurrence USDA zones 7–10

Seasonal occurrence March to November

DISEASE FREQUENCY	DISEASE SEVERITY
5 annual	5 plants killed
[4]	**[4]**
3	3
2	2
1 rare	1 very little damage

CHEMICAL TREATMENT	CULTURAL PRACTICES
3 used every year	3 very important
2	**[2]**
[1] not used	1 not important

SANITATION	RESISTANT CULTIVARS
[3] very important	3 many cultivars
2	2
1 not important	**[1] no resistance**

Root Rot

Geographic occurrence USDA zones 7–10

Seasonal occurrence Warm to hot, rainy periods

DISEASE FREQUENCY	DISEASE SEVERITY
5 annual	**[5] plants killed**
[4]	4
3	3
2	2
1 rare	1 very little damage

CHEMICAL TREATMENT	CULTURAL PRACTICES
3 used every year	**[3] very important**
[2]	2
1 not used	1 not important

SANITATION	RESISTANT CULTIVARS
3 very important	3 many cultivars
[2]	2
1 not important	**[1] no resistance**

Management

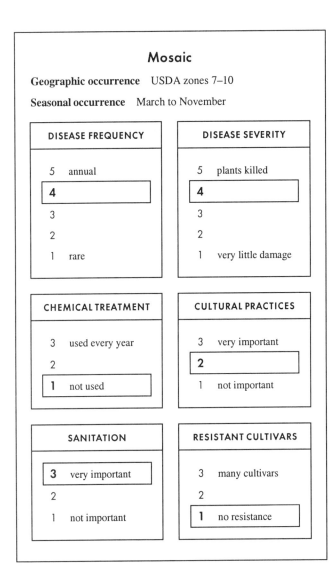

Vegetatively propagated stock should be routinely and rigorously scouted for the virus symptoms described above. Leaf reddening is a normal plant response to increased light intensity and a wound response to nonviral pathogens. Plants exhibiting virus symptoms should be rogued and destroyed in advance of propagation. Healthy-appearing stock should be used for propagation. Nurseries growing nandina from seed do not face a risk of virus in propagation. A nursery seeking to eliminate viruses from stock should consider tissue culture for production virus-free plantlets. There is a risk of reentry of CMV into clean stock if it is transmitted by winged aphids from outside the nursery.

Root Rot

Phytophthora nicotianae
Pythium myriotylum
Pythium spinosum
Pythium splendens

Symptoms

Plants affected by root rot produce fewer stems than normal, and their vertical growth is reduced. Lower leaves become chlorotic and then necrotic, and they abscise. New growth on one or more stems in a container may develop pronounced symptoms of minor element deficiency, resembling iron or manganese deficiency, even though the pH of the medium and nutrients are in acceptable production ranges. Examination of feeder roots will reveal root discoloration, soft decay, and a generally reduced volume of roots.

Epidemiology

Root rots can be related to propagation or production cycles in the nursery. Infested media, water sources, containers, and trays and excessive mist cycles can introduce root rot pathogens or trigger disease onset in seedlings, cuttings, and liners. The same factors can also operate in the production area, in addition to the risks posed by surface water runoff, excessive rainfall, and production stresses associated with fertilization, high temperatures in media, and inadequate pore space in media. These conditions favor initial colonization by root pathogens. The pathogens are spread by splashing water, media that drain onto beds, or the movement of infested soil.

Host Range

All cultivars of *N. domestica* are susceptible to the various fungi responsible for wet root rot diseases. In addition, these fungi all have broad host ranges among floral, foliage, and woody ornamental hosts in the nursery.

Management

The first step in management of root diseases is to identify the pathogen or pathogens involved. This allows critical use of cultural and chemical control options. For fungi causing wet root rot, manipulate water, media, and container use to prevent pathogen introduction. Avoid random contamination of liners by nonsterile media, soil, or water. Choose only healthy stock for crown divisions, to avoid a second cycle of root rot. Carefully manage watering cycles and fertilization to minimize stress on plants. Various fungicides are available for use as foliar sprays and drenches for the control of root rot pathogens. Always assess root rot severity as higher than what is indicated by aboveground symptoms.

Web Blight

Rhizoctonia solani

Symptoms

Web blight infection begins on the lower, inner leaves close to the soil line. Brown lesions of various sizes and shapes form on leaves and may encompass entire leaves (Plate 90). Infected leaves are held on and are often tied together by hyphae of the fungal pathogen, *Rhizoctonia solani.* Infection moves upward and outward in the plant. Leaf defoliation to death may occur in a matter of one to two weeks if the weather is favorable for disease development.

Epidemiology

The warm, moist summer period favors web blight. The pathogen is splashed by rain or irrigation from the soil surface upward into the lower canopy. It spreads from plant to plant in splashing water or where plants are in direct contact. The pathogen persists as a saprophyte in media and as hyphae and sclerotia in and on dead tissue. It can also survive in the remnant media in old containers.

Host Range

All cultivars of nandina are susceptible to *R. solani.* This fungus has a host range encompassing hundreds of plants, including many woody ornamentals, such as azalea, holly, pittosporum, and some junipers.

Management

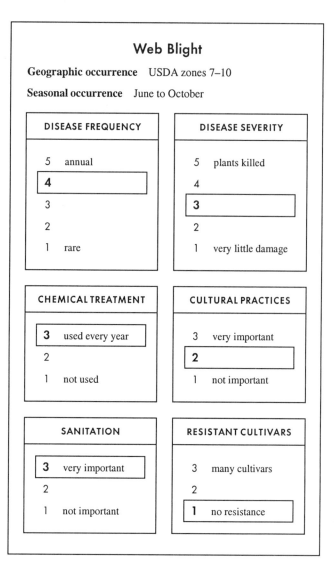

For cultivars that are vegetatively propagated, a routine scouting and fungicide spray program should be started at least four weeks before cuttings are taken. Take cuttings from elevated portions of the plant rather than close to the soil line. Use pathogen-free media and trays in propagation, and elevate cuttings and finished liners to keep them off nonsterile surfaces. In production beds, containers should be spaced to facilitate fast drying of the canopy after irrigation. In periods favorable for web blight development, fungicides are often required to halt the spread of the disease. Apply products with sufficient volume and pressure to penetrate the entire canopy, and direct some fungicide at the media surface as well. Where web blight is severe, try to remove necrotic tissue by blowing or raking leaves free from canopies and media surfaces prior to fungicide application.

Web Blight

Geographic occurrence USDA zones 7–10

Seasonal occurrence June to October

DISEASE FREQUENCY	DISEASE SEVERITY
5 annual	5 plants killed
[4]	4
3	**[3]**
2	2
1 rare	1 very little damage

CHEMICAL TREATMENT	CULTURAL PRACTICES
[3] used every year	3 very important
2	**[2]**
1 not used	1 not important

SANITATION	RESISTANT CULTIVARS
[3] very important	3 many cultivars
2	2
1 not important	**[1] no resistance**

REFERENCES

Ahmed, N. A., Christie, S. R., and Zettler, F. W. 1983. Identification and partial characterization of a closterovirus infecting *Nandina domestica.* Phytopathology 73:470–475.

Barnett, O. W., and Baxter, L. W. 1973. Cucumber mosaic virus on *Nandina domestica* in South Carolina. Plant Dis. Rep. 57:917–920.

McKenzie, E. H. C., and Dingley, J. M. 1996. New plant disease records in New Zealand: Miscellaneous fungal pathogens III. N.Z. J. Bot. 34:263–272.

Moreno, P., Attathom, S. and Weathers, L. G. 1976. Identification, transmission and partial purification of a potexvirus causing a disease of nandina plants in California. Proc. Am. Phytopathol. Soc. 3:319.

Sinclair, W. A., Lyon, H. H., and Johnson, W. T. 1987. Diseases of Trees and Shrubs. Comstock Publishing Associates, Cornell University Press, Ithaca, N.Y.

Zettler, F. W., Hiebert, E., Marciel-Zambolim, E., Abo El-Nil, M. M., and Christie, R. G. 1980. A potexvirus infecting *Nandina domestica* 'Harbor Dwarf.' Acta Hortic. 110:71–80.

CHAPTER
63

Gary W. Simone ▪ University of Florida, Gainesville

Oleander Diseases

Geographic production USDA zones 8–11

Family Apocynaceae

Genus *Nerium*

Species *N. oleander*

Oleander is a member of a small genus, with two species native to the region between the Mediterranean and Japan. *N. oleander* is the species in cultivation throughout the United States, on account of its salt tolerance and drought tolerance. It is available in standard and dwarf cultivars with a wide range of flower colors. Oleander is sensitive to many diseases in nursery production, including bacterial knot in the West, web blight and witches'-broom in the Southeast, and several dieback and leaf spot diseases throughout the United States.

Bacterial Knot

Pseudomonas syringae* pv. *savastanoi

Symptoms

Infection by the bacterium *Pseudomonas syringae* pv. *savastanoi* causes the formation of galls on stems (Plate 91), leaves, veins, flowers, and seed pods. No galls form belowground. The galls are rough in texture, round to irregularly shaped, and 1/8 to 1 inch in diameter. They are originally creamy white and then darken almost to black with time. Galls result from a stimulation of plant cell division and cell enlargement and are filled with pockets of bacteria. Infection of the soft stem results in cankers with raised edges and deep fissures. Infection of leaves, midribs, or lateral veins causes deformation and curled leaf margins. Infection of the inflorescence or seed pod causes distortion of the cyme and pod.

Epidemiology

The rainy period of winter to spring in the Pacific Northwest and spring in southern California and parts of the Southwest provides weather favoring bacterial knot. The occurrence of the disease in the Southeast coincides with the summer rainy period. *P. syringae* pv. *savastanoi* is active between 18 and 33°C and most damaging between 22 and 25°C. It survives epiphytically on bark and leaves. Late-season infection may not be expressed until the spring flush in the Northwest, while symptoms may develop in 10–14 days after infection in the Southeast. Bacteria ooze out of galls or cankers and are moved from plant to plant in splashing water and on pruners. Infection occurs at sites of pruning wounds or injuries, leaf scars, stomates, or lenticels. In oleander, bacterial cells move through laticifers a distance of 25 cm above and 20 cm below the point of infection.

Host Range

P. syringae pv. *savastanoi* has been reported in plants in the family Oleaceae worldwide, from such genera as *Chionanthus,*

Bacterial Knot

Geographic occurrence USDA zones 7–10

Seasonal occurrence Rainy seasons

DISEASE FREQUENCY	DISEASE SEVERITY
5 annual	5 plants killed
4	**4**
3	3
2	2
1 rare	1 very little damage

CHEMICAL TREATMENT	CULTURAL PRACTICES
3 used every year	**3 very important**
2	2
1 not used	1 not important

SANITATION	RESISTANT CULTIVARS
3 very important	3 many cultivars
2	2
1 not important	**1 no resistance**

Forestiera, Forsythia, Fraxinus, Jasminum, Ligustrum, Olea, Osmanthus, and *Syringa.* In the United States, distribution is limited to plants in the families Oleaceae (*Olea europaea*) and Apocynaceae (*N. oleander*). There is some evidence of host specificity among strains of *P. syringae* pv. *savastanoi;* primarily oleander strains can invade oleander, while oleander strains can also infect *Olea* sp., but not frequently.

Management

Inspect incoming stock for symptoms of bacterial knot. Isolate infected plants and prune out damage. Pruning tools should be disinfested in a suitable product between pruning one plant and pruning the next. Symptomatic plants should not be left under overhead irrigation and must be pruned or destroyed prior to the onset of the rainy season. Nurseries growing both *Nerium* and *Olea* spp. should physically separate them. Post-pruning use of copper-based fungicides or streptomycin sulfate can aid in disease management. See the product label and contact your county or state extension service for legal restrictions and details of use.

Dieback

Various fungi

Several fungi have been reported to cause dieback of oleander, including *Botryosphaeria dothidea, B. obtusa,* an unidentified *Botryodiplodia* sp., and species of *Dothiorella, Diaporthe, Phoma,* and *Phomopsis.*

Symptoms

New growth flushes turn chlorotic and wilt for increasingly longer periods as the disease progresses. Affected twigs and shoots brown out, often retaining their leaves for some time (Plate 92). The fungi causing dieback diseases cause a dark sapwood discoloration, which marks the upward or downward movement of the pathogen internally in the plant. Affected plants develop one or more dead shoots in a random pattern. Dieback fungi reproduce on the bark surface, and their globose, brown to black fruiting structures (perithecia or pycnidia) can be observed with a 10–15× hand lens.

Epidemiology

Dieback develops following such plant stresses as cold damage, wind whipping, or pruning. The pathogens readily invade wounds and can penetrate growth cracks and lenticels as well. Inoculum for oleander infection is most likely produced on other woody hosts in the nursery.

Ascospores are liberated from perithecia of *Botryosphaeria* spp. and *Diaporthe* sp. embedded in bark at sites of old infections. These spores are usually airborne or transported by splashing water in spring through midsummer.

The other dieback fungi reproduce as conidia from pycnidia embedded in the bark of infected twigs and shoots. Conidia depend on splashing rain or irrigation water for dispersal and are likely the primary initiators of dieback.

All these fungi survive as hyphae and pycnidia or perithecia in diseased tissue and prunings.

Host Range

Both *B. dothidea* and *B. obtusa* attack more than 100 woody hosts. The reported incidence of *Botryodiplodia* spp. and *Dothiorella* sp. is such that their host ranges may be similar. The host ranges of the *Diaporthe, Phoma,* and *Phomopsis* spp. causing oleander dieback are unknown, as they have not been studied sufficiently to be speciated.

Management

Prune symptomatic plants back to clean wood. Collect and destroy prunings. Avoid pruning wet canopy where there is a risk of spreading dieback fungi on pruning tools. Follow pruning with application of a fungicide. Oleander has no known resistance to dieback fungi.

Leaf Spot Diseases

Phyllosticta nerii
Pseudocercospora neriella
Septoria oleandrina

Symptoms

Leaf spots vary in size, shape, and color, according to the pathogen.

Lesions caused by *Phyllosticta* are circular to oblong, tan spots. Globose, black fruiting bodies (pycnidia) of the pathogen are visible in the centers of lesions.

Dieback

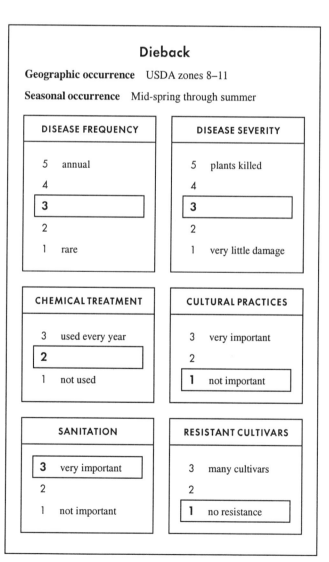

Geographic occurrence USDA zones 8–11

Seasonal occurrence Mid-spring through summer

DISEASE FREQUENCY	DISEASE SEVERITY
5 annual	5 plants killed
4	4
3	**3**
2	2
1 rare	1 very little damage

CHEMICAL TREATMENT	CULTURAL PRACTICES
3 used every year	3 very important
2	2
1 not used	**1 not important**

SANITATION	RESISTANT CULTIVARS
3 very important	3 many cultivars
2	2
1 not important	**1 no resistance**

Lesions caused by *Pseudocercospora* are chlorotic and angular to irregular in shape, with indefinite margins.

Lesions caused by *Septoria* are round to irregular in shape, with deep, wide maroon margins (Plate 93). Pycnidia of the pathogen are obvious as globose, black bodies embedded in the light tan centers of lesions. Where infection builds up, defoliation can occur.

Epidemiology

The fungi causing leaf spot diseases of oleander are introduced into nurseries in infected plant material, as they are host-specific and not subject to airborne spread. Infected plants or areas of fallen infected leaves serve to sustain the pathogens. Spores (conidia) ooze from pycnidia embedded in leaf spots and are splashed by rain or irrigation to susceptible growth flushes, where infection proceeds. Periods of cool weather from fall through spring favor the development of *Septoria* in southern latitudes, while growth flushes from spring into summer are more sensitive to *Phyllosticta* and *Pseudocercospora* spp.

Host Range

All three oleander pathogens appear to be host-specific in the United States. *Phyllosticta* has the broadest distribution, being reported in the Southwest, the Midwest, the Southeast, and the Northwest. *Septoria* has been reported from the Pacific Northwest to California and east to Florida. *Pseudocercospora* has been reported in Florida.

Management

Leaf spot diseases are generally not damaging, but hot spots of infestation can occur. Sanitize the most severely affected leaves, including those that have fallen to the container or ground cloth. Minimize overhead irrigation, and shift the timing of irrigation to early in the morning or midmorning. Coordinate the use of foliar protectant or systemic fungicides as needed to minimize disease on new growth flushes. Avoid propagation of diseased stock.

Root-Knot Nematodes

***Meloidogyne* spp.**

Symptoms

Field-grown oleanders are at risk of damage by root-knot nematodes (*Meloidogyne* spp.), which invade roots and induce the formation of root galls. As galling of the root system increases, affected plants exhibit reduced vigor, thinning of the older part of the canopy, and a reduction in shoot emergence.

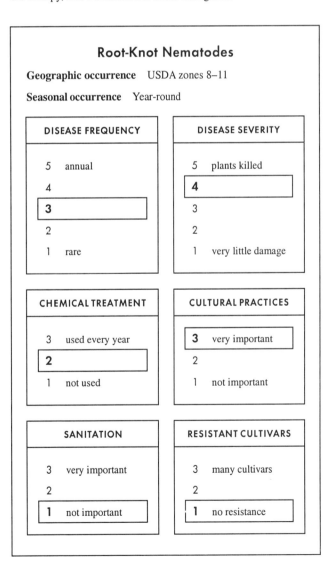

Leaf Spot Diseases

Geographic occurrence USDA zones 8–11

Seasonal occurrence Year-round, depending on the pathogen and the latitude

DISEASE FREQUENCY	DISEASE SEVERITY
5 annual	5 plants killed
4	4
3	3
2	**2**
1 rare	1 very little damage

CHEMICAL TREATMENT	CULTURAL PRACTICES
3 used every year	3 very important
2	**2**
1 not used	1 not important

SANITATION	RESISTANT CULTIVARS
3 very important	3 many cultivars
2	2
1 not important	**1 no resistance**

Root-Knot Nematodes

Geographic occurrence USDA zones 8–11

Seasonal occurrence Year-round

DISEASE FREQUENCY	DISEASE SEVERITY
5 annual	5 plants killed
4	**4**
3	3
2	2
1 rare	1 very little damage

CHEMICAL TREATMENT	CULTURAL PRACTICES
3 used every year	**3 very important**
2	2
1 not used	1 not important

SANITATION	RESISTANT CULTIVARS
3 very important	3 many cultivars
2	2
1 not important	**1 no resistance**

Epidemiology

Meloidogyne spp. survive in soil as eggs and larvae. Primary infection occurs at root tips and sites of lateral root emergence. Nematode larvae migrate within the root to the stele, where females encyst and form specialized feeding sites known as giant cells. These feeding sites enlarge and coalesce to form galls, which interfere with water uptake and the movement of nutrients in the plant. Oleander in container production is not prone to attack by root-knot nematodes unless the soilless medium is contaminated with soil.

Host Range

All cultivars of oleanders are subject to infection and damage by root-knot nematodes.

Management

Avoid planting a field before assessing the nematode population. Old cropland used for growing agronomic or vegetable crops is often unsuitable for oleander, because of its susceptibility to root-knot nematodes. Where land rotation or new land is unavailable, in-row fumigation may be needed. Granular contact nematicides will help suppress small populations of root-knot nematodes during production. Severely affected plants can be used as propagation stock but then should be rogued and destroyed.

Web Blight

Rhizoctonia solani

Symptoms

The first symptoms of web blight occur in the lower inner portion of the leaf canopy. Leaf lesions are round to irregular in shape and variable in size. They lack zonation or a distinctive border color. Infection may result in the death of the leaf tip, the leaf margins, or the entire leaf (Plate 94). A characteristic of web blight is the upward and outward development of the infection. Adjacent leaves are often webbed together with tan strands (hyphae) of the fungal pathogen, *Rhizoctonia solani,* and necrotic leaves are often held onto the plant by webs. Hyphae can be observed, by eye or with a hand lens, between leaves as they are pulled apart.

Leaf destruction is rapid in rainy periods, causing considerable death of the plant canopy. Often leaves are suspended by one or more strands within the canopy.

Epidemiology

R. solani survives as mycelium and often as dormant propagules (sclerotia) in media and diseased tissue. Warm to hot, rainy weather favors the spread of the fungus to the lower leaves from splashing soil and to other leaves in splashing water. *R. solani* actively spreads from leaf to leaf and from plant to plant through direct contact between leaves. Conditions are favorable for disease development from summer through fall. Only vegetative hyphae and sclerotia are involved in disease development; no sexual or asexual spore cycle is involved.

Host Range

The host range of *R. solani* is enormous, including both herbaceous and woody plants in probably more than 500 genera. Many common nursery crops are potentially susceptible to web blight.

Management

Avoid using cuttings harvested low in the plant canopy, in an effort to prevent the introduction of the pathogen into the propagation area. Set plants on ground beds with proper spacing, to enhance air circulation and avoid direct contact. Daytime irrigation cycles pose less risk of infection by *Rhizoctonia* than night cycles. Plants known to be susceptible, such as oleander, should receive preventative treatment with fungicides in advance of the rainy, warm production period. Infected plants should be cleared of infected canopy. Infected debris and prunings should be collected and destroyed prior to applications of fungicides. Always spray the surface of the medium as well when fungicides are applied.

Witches'-Broom

Sphaeropsis tumefaciens

Symptoms

Initial symptoms of witches'-broom include small blisters (1–6 mm in diameter) on stems, localized swelling or flattening of stem sections, and the stimulation of lateral buds. As the lateral buds grow, they form the witches'-brooms that are characteristic of the disease caused by *Sphaeropsis tumefaciens* in many hosts (Plate

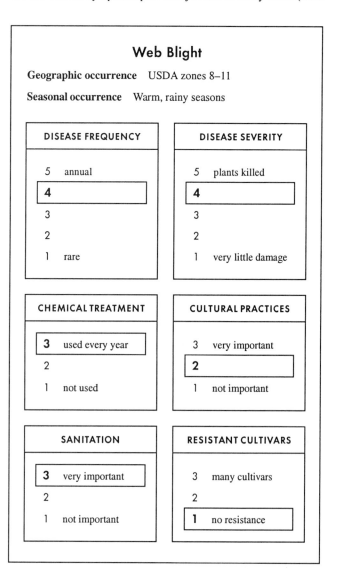

Web Blight

Geographic occurrence USDA zones 8–11

Seasonal occurrence Warm, rainy seasons

DISEASE FREQUENCY	DISEASE SEVERITY
5 annual	5 plants killed
4	**4**
3	3
2	2
1 rare	1 very little damage

CHEMICAL TREATMENT	CULTURAL PRACTICES
3 used every year	3 very important
2	**2**
1 not used	1 not important

SANITATION	RESISTANT CULTIVARS
3 very important	3 many cultivars
2	2
1 not important	**1 no resistance**

Witches'-Broom

Geographic occurrence USDA zones 8–11

Seasonal occurrence Year-long

DISEASE FREQUENCY	DISEASE SEVERITY
5 annual	5 plants killed
4	**[4]**
3	3
[2]	2
1 rare	1 very little damage

CHEMICAL TREATMENT	CULTURAL PRACTICES
3 used every year	3 very important
[2]	**[2]**
1 not used	1 not important

SANITATION	RESISTANT CULTIVARS
[3 very important]	3 many cultivars
2	2
1 not important	**[1 no resistance]**

95). Affected branches exhibit reduced growth and seldom initiate flowers. The witches'-brooms eventually die. The fungal pathogen produces visible spores (conidia) that ooze from fruiting bodies (pycnidia) embedded in the base of dead stems.

Epidemiology

Symptoms appear about four months after infection. *S. tumefaciens* invades actively growing terminal or lateral shoot tips. Splashing rain or irrigation water is needed to disperse the pathogen within the canopy and from plant to plant. Besides diseased oleanders, various woody hosts may serve as sources of initial inoculum. *S. tumefaciens* survives as mycelium and conidia in pycnidia in infected oleander tissue. Warm, rainy periods during active growth of the plants favor disease development.

Host Range

S. tumefaciens invades woody plants and trees in more than 30 genera, including *Carissa, Callistemon, Ilex,* and *Myrica,* in the nursery.

Management

Inspect stock prior to vegetative propagation cycles. Do not use cuttings from diseased plants. Inspect susceptible hosts in the nursery for symptoms of gall or witches'-broom. Severely prune infected material, 6–12 inches below a gall or broom. Collect and destroy prunings. Avoid pruning when the canopy is wet. Follow prunings with an application of a copper-based fungicide. The efficacy of preventative fungicides in controlling witches'-broom has not been determined.

It is questionable whether resistance to the pathogen is available. Growers believe that compact cultivars are resistant; however, experimentation does not support this hypothesis. The cultivars Hardy Pink and Isle of Capri develop symptoms more slowly than comparable cultivars, but they are still susceptible.

REFERENCES

Alfieri, S. A., Jr., Langdon, K. R., Kimbrough, J. W., El-Gholl, N. E., and Wehlburg, C. 1994. Diseases and disorders of plants in Florida. Fla. Dep. Agric. Consumer Serv. Bull. 14.

Atilano, R. 1981. Screening oleander cultivars for resistance to witches' Broom. Proc. Fla. State Hortic. Soc. 94:218–219.

Azad, H. R., and Cooksey, D. A. 1995. A semiselective medium for detecting epiphytic and systemic population of *Pseudomonas savastanoi* from oleander. Phytopathology 85:740–745.

Farr, D. F., Bills, G. F., Chamuris, G. P., and Rossman, A. Y. 1989. Fungi on Plants and Plant Products in the United States. American Phytopathological Society, St. Paul, Minn.

Jause, J. D. 1981. The bacterial disease of ash (*Fraxinus excelsior*) caused by *Pseudomonas syringae* subsp. *savastanoi* pv. *fraxini.* Eur. J. For. Pathol. 11:425–438.

Lavermicocca, P., and Surico, G. 1987. Presenza epifitica di *Pseudomonas syringae* pv. *savastanoi* e di altri batteri sull'olivo e sull'oleandro. Phytopathol. Medit. 26:136–141.

Ridings, W. H., and Marlatt, R. B. 1976. Sphaeropsis witches' broom of *Nerium oleander.* Proc. Fla. State Hortic. Soc. 89:302–303.

Smith, C. O. 1928. Oleander bacteriosis in California. Phytopathology 18: 503–521.

Varnow, L. 1983. Influenza della temperatura sulla moltiplicazione di *Pseudomonas syringae* pv. *savasteroi* (Smith) Young et al. Phytopathol. Medit. 22:39–40.

West, E. 1937. Witches' broom of oleander. Fla. Agric. Exp. Stn. Bull. 509.

CHAPTER 64

Gary W. Simone • University of Florida, Gainesville

Osmanthus Diseases

Geographic production USDA zones 6–9

Family Oleaceae

Genus *Osmanthus*

Species

O. americanus	devilwood
O. × fortunei	Fortune's osmanthus
O. fragrans	fragrant tea olive
O. heterophyllus	holly osmanthus

Osmanthus is a small genus of some 30–40 species of evergreen shrubs and small trees native to eastern Asia, Hawaii, New Caledonia, and North America. The most common species in production were introduced from Japan in the 1850s or earlier: *O. heterophyllus* (syn. *O. ilicifolius*), *O. fragrans,* and *O. × fortunei* (a hybrid of *O. heterophyllus* and *O. fragrans*). Devilwood (*O. americanus*) is native to North America and has some horticultural value. Anthracnose, Phyllosticta leaf spot, and dark mildews are the primary foliar diseases of *Osmanthus.* Root-knot nematodes can also affect field production of *Osmanthus* throughout its range.

Anthracnose

Colletotrichum gloeosporioides

Symptoms

Anthracnose is a leaf spot disease in which small, reddish brown spots form on leaves and enlarge into circular lesions, 4–8 mm in diameter, with a white center and a dark brown border (Plate 96). Considerable chlorosis develops around these spots on *O. fragrans.* Black, pincushion-like fruiting bodies (acervuli) of the pathogen, *Colletotrichum gloeosporioides,* form in the centers of lesions and can be observed with a hand lens. The acervuli ooze a cream-colored to pink spore mass during moist weather.

Epidemiology

C. gloeosporioides is introduced and spread primarily by the use of infected stock for propagation. It can invade stressed leaf tissue and remain dormant in plant tissue for a period of weeks to months. The high rate of asexual reproduction of the pathogen provides ample inoculum, which is spread by splashing rain or irrigation water in a planting of *Osmanthus* or between unrelated hosts of this pathogen in the nursery.

Host Range

Anthracnose has been reported in *O. fragrans* and *O. americanus* and in less commonly grown species. *C. gloeosporioides* can invade more than 500 plant species, including many common nursery species, such as aucuba, camellia, magnolia, and palms.

Management

Use only the most disease-free stock for vegetative propagation cycles where anthracnose is severe. Prune infected canopies, and remove the prunings. Avoid the use of overhead irrigation.

Anthracnose

Geographic occurrence USDA zones 6–9

Seasonal occurrence Late spring to early fall

DISEASE FREQUENCY	DISEASE SEVERITY
5 annual	5 plants killed
4	4
[3]	3
2	**[2]**
1 rare	1 very little damage

CHEMICAL TREATMENT	CULTURAL PRACTICES
3 used every year	3 very important
[2]	**[2]**
1 not used	1 not important

SANITATION	RESISTANT CULTIVARS
3 very important	3 many cultivars
[2]	2
1 not important	**[1] no resistance**

There is no information on resistant or tolerant species or cultivars of *Osmanthus.* Fungicide applications are useful to minimize leaf infection, especially in the months preceding the propagation cycle.

Dark Mildews

Numerous fungi

Dark mildews are caused by a large group of parasitic fungi (more than 1,500 species) that produce dark mat- or patchlike growth on leaves or soft shoots of susceptible hosts. Fungi reported to cause dark mildew of *Osmanthus* include *Asteridiella* spp., an *Asterina* sp., *Lembosia oleae, Meliola amphitricha, M. osmanthi,* and *M. osmanthina.* Dark mildew fungi do little real damage to their hosts, but they mimic sooty molds and hence visually affect plant quality.

Symptoms

Dark mildew fungi are initially superficial and then attach themselves to leaf or stem cuticle with anchoring structures, from which emerge filaments that penetrate plant epidermal cells and derive nutrients. The margins of these patches are highly dissected by radiating hyphae extending outward on the leaf. These fungi resemble the superficial and saprophytic sooty molds, which colonize the honeydew secretions of certain insects.

Epidemiology

The pathogens survive by sexual reproduction and the release of ascospores, which can infect adjacent leaves during moist weather in nonarid production zones.

Host Range

The dark mildew fungi are highly specialized parasites with narrow host ranges.

Meliola spp. have been reported on *O. americanus* in the South and on less commonly cultivated *Osmanthus* spp. in Hawaii.

Asterina spp. have been reported on *O. americanus* and *O. fragrans* in the Southeast.

Lembosia and *Asteridiella* have been reported on *O. americanus.*

Management

Where dark mildews develop, avoid tight spacing of plants, which prevents rapid leaf drying after rains and irrigation. Directing irrigation at the container surface will minimize periods of dis-

Dark Mildews

Geographic occurrence USDA zones 7–11

Seasonal occurrence Spring through early summer, coinciding with rainy periods

DISEASE FREQUENCY	DISEASE SEVERITY
5 annual	5 plants killed
4	4
3	3
[2]	2
1 rare	**[1] very little damage**

CHEMICAL TREATMENT	CULTURAL PRACTICES
3 used every year	3 very important
2	**[2]**
[1] not used	1 not important

SANITATION	RESISTANT CULTIVARS
3 very important	3 many cultivars
[2]	2
1 not important	**[1] no resistance**

Phyllosticta Leaf Spot

Geographic occurrence USDA zones 6–9

Seasonal occurrence April to August, during the rainy period

DISEASE FREQUENCY	DISEASE SEVERITY
5 annual	5 plants killed
4	4
3	3
[2]	**[2]**
1 rare	1 very little damage

CHEMICAL TREATMENT	CULTURAL PRACTICES
3 used every year	3 very important
[2]	**[2]**
1 not used	1 not important

SANITATION	RESISTANT CULTIVARS
3 very important	3 many cultivars
[2]	2
1 not important	**[1] no resistance**

ease development. Removal of infected, abscised leaves will reduce the surviving fungal population and decrease the amount of inoculum available in the next period of favorable wet weather. No resistant varieties of *Osmanthus* are available. Fungicides are not effective in controlling dark mildews. Plants in container production are more prone to the disease than plants in field production.

Phyllosticta Leaf Spot

Phyllosticta oleae
Phyllosticta osmanthi
Phyllosticta sinuosa
Phyllosticta terminalis

Symptoms

Phyllosticta spp. cause round to oblong leaf spots, usually with light tan centers and darker borders. Lesion size is variable, depending on the severity of infection and the coalescence of leaf spots. A random array or ring of brown to black, globose fruiting bodies (pycnidia) of the pathogen can typically be observed reproducing in lesion centers.

Epidemiology

Phyllosticta spp. invade host plants as asexual spores (conidia). It is suspected that these pathogens overseason as pycnidia in host tissue until a period of warm, wet weather stimulates spore release. Fungal spores are splashed by water, and thus infection is often correlated with periods of overhead irrigation or rain.

Host Range

The most widely distributed of the pathogens is *P. oleae,* which has been reported on *O. americanus, O. heterophyllus,* and *O.* × *fortunei* from the mid-Atlantic area southwest to Texas.

P. osmanthi and *P. sinuosa* have been reported on various *Osmanthus* spp. in the southeastern United States.

P. terminalis has been reported only in Florida on devilwood. This fungus has a wider host range, which includes *Forsythia, Ilex, Leucothoe,* and *Magnolia.*

Management

Infected leaves on or off the canopy should be collected and removed from production areas before the spring rainy period begins. Plants designated as stock material for the next propagation cycle should be well spaced to enhance leaf drying, especially if overhead irrigation is used. Only disease-free tips should be used for propagation, as the mist cycle favors the spread of the leaf spot fungi. Both protectant and systemic fungicides are available if disease severity warrants their use.

Root-Knot Nematode

***Meloidogyne* spp.**

Symptoms

Field-grown plants are at risk of infection by root-knot nematodes (*Meloidogyne* spp.), which invade roots and induce the formation of root galls. As root galls increase in number, affected plants exhibit slower growth and thinning of older parts of the canopy. Infected plants can be more sensitive to heat scald and cold damage.

Epidemiology

Meloidogyne spp. survive as eggs and larvae in soil, especially sandy or lighter coastal soil types. Infection occurs at root tips and sites of lateral root emergence. Nematode larvae migrate within roots to the stele, where females encyst and form specialized feeding sites known as giant cells. These feeding sites enlarge and coalesce to form galls, which interfere with water uptake and the movement of nutrients in the plant. *Osmanthus* in container production is not prone to attack by root-knot nematodes unless the soilless medium is contaminated with soil.

Host Range

Holly osmanthus and tea olive osmanthus have been reported to be hosts of root-knot nematodes. It is likely that *Osmanthus* spp. are susceptible to one or more species of *Meloidogyne.*

Management

Avoid planting susceptible species in nematode-infested soils. Where new land is unavailable for rotation, a preplant soil fumigant may be needed. Low levels of infection may be managed by application of a granular contact nematicide. Severely infected plants should be rogued and destroyed.

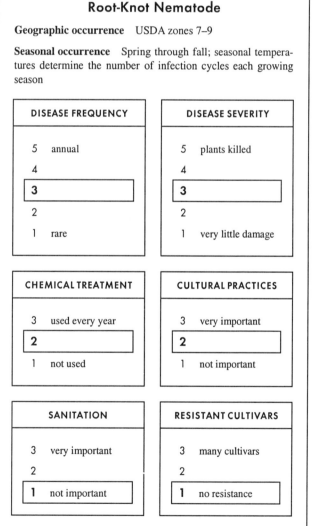

Root-Knot Nematode

Geographic occurrence USDA zones 7–9

Seasonal occurrence Spring through fall; seasonal temperatures determine the number of infection cycles each growing season

DISEASE FREQUENCY	DISEASE SEVERITY
5 annual	5 plants killed
4	4
[3]	**[3]**
2	2
1 rare	1 very little damage

CHEMICAL TREATMENT	CULTURAL PRACTICES
3 used every year	3 very important
[2]	**[2]**
1 not used	1 not important

SANITATION	RESISTANT CULTIVARS
3 very important	3 many cultivars
2	2
[1] not important	**[1] no resistance**

REFERENCES

Alfieri, S. A., Jr., Langdon, K. R., Kimbrough, J. W., El-Gholl, N. E., and Wehlburg, C. 1994. Diseases and disorders of plants in Florida. Fla. Dep. Agric. Consumer Serv. Bull. 14.

Farr, D. F., Bills, G. F., Chamuris, G. P., and Rossman, A. Y. 1989. Fungi on Plants and Plant Products in the United States. American Phytopathological Society, St. Paul, Minn.

Sinclair, W. A., Lyon, H. H., and Johnson, W. T. 1987. Diseases of Trees and Shrubs. Comstock Publishing Associates, Cornell University Press, Ithaca, N.Y.

CHAPTER
65

A. R. Chase ▪ Chase Research Gardens, Mt. Aukum, California

Palm Diseases

Geographic production USDA zones 8–10

Family Arecaceae

Genera numerous

Species numerous

Ornamental palms are an integral part of landscapes, both indoors and outdoors. Most palms are produced from seeds in shade or full sun. They are grown throughout the United States, mostly in tropical and subtropical regions, in Arizona, California, Florida, Georgia, Hawaii, Nevada, and Texas, and in coastal regions of other southern states. Palms are critical elements in landscape design and are especially important in interiorscapes of public facilities, such as hotels and malls. Many species and cultivars have diseases in common, since they are produced under similar conditions.

Algal Leaf Spot (Red Rust, Algal Rust)

Cephaleuros virescens

Algal leaf spot, also called red rust or algal rust, occurs only outdoors in moist tropical climates where rainfall and temperatures are relatively high. The disease is caused by *Cephaleuros virescens,* a green alga.

Symptoms

Young lesions appear as yellow pinpoint spots, primarily on the upper surface of leaves. They expand into crusty-appearing, gray green patches, which assume an orange cast during reproductive periods of the pathogen.

Epidemiology

Initial infection occurs toward the end of a humid rainy season, when rain disseminates the biflagellate zoospores of *C. virescens* from sporangiophores in old infections. Thalli of the alga penetrate leaves by mechanical force and grow chiefly along the horizontal plane between the leaf cuticle and the epidermal cell layers. Vegetative growth continues slowly for approximately eight to nine months, until rain or high humidity triggers a reproductive phase. Reproductive structures contain the pigment hematochrome, which is responsible for the orange color of the algal spots.

Host Range

Cephaleuros has been reported on *Bactris gasipaes, Butia capitata, Caryota* spp., *Cocos nucifera, Elaeis guineensis, Phoenix dactylifera, Sabal palmetto,* and *Trachycarpus fortunei.*

C. virescens has an extremely wide host range, including such economically important species as avocado, cacao, cashew nut, citrus, coffee, guava, litchi, mango, pepper, rubber, and tea.

Management

Cultural practices The incidence of algal spot on some nonpalm hosts seems to be correlated with either low plant vigor or

Algal Leaf Spot

Geographic occurrence Southeastern United States

Seasonal occurrence Hot, rainy periods

DISEASE FREQUENCY	DISEASE SEVERITY
5 annual	5 plants killed
4 (boxed)	4
3	3
2	2
1 rare	**1 very little damage** (boxed)

CHEMICAL TREATMENT	CULTURAL PRACTICES
3 used every year	3 very important
2	**2** (boxed)
1 not used (boxed)	1 not important

SANITATION	RESISTANT CULTIVARS
3 very important	3 many cultivars
2 (boxed)	2
1 not important	**1 no resistance** (boxed)

planting sites with high humidity and poor air circulation. Selective pruning and thinning of the plant canopy can increase air circulation and reduce incidence.

Chemical treatment Algal spot of some food crops has been successfully controlled by applications of fungicides timed for the end of the rainy season to prevent infection by zoospores.

Anthracnose (Colletotrichum Leaf Spot)

Colletotrichum gloeosporioides

Anthracnose, also called Colletotrichum leaf spot, is most common in small seedlings where high humidities favor disease development in the immature leaves. The disease is caused by the fungus *Colletotrichum gloeosporioides.*

Symptoms

Symptoms vary somewhat from host to host, from small, water-soaked speckles to large, necrotic and chlorotic lesions, which are circular to irregular in shape. The lesions are usually tan to black, with a bright yellow chlorotic halo, and generally coalesce as symptoms progress. On some hosts the lesions are angular, since they are confined to the area between leaf veins. On other hosts a frogeye spot may develop.

Anthracnose

Geographic occurrence Florida and Hawaii

Seasonal occurrence Most of the year

DISEASE FREQUENCY	DISEASE SEVERITY
5 annual	5 plants killed
[4]	4
3	3
2	2
1 rare	**[1 very little damage]**

CHEMICAL TREATMENT	CULTURAL PRACTICES
3 used every year	3 very important
[2]	**[2]**
1 not used	1 not important

SANITATION	RESISTANT CULTIVARS
[3 very important]	3 many cultivars
2	2
1 not important	**[1 no resistance]**

Host Range

Colletotrichum has been recovered from most palms grown in most palm-producing areas in the world.

Management

Cultural practices Any method which keeps foliage dry reduces the potential for infection and spread of the pathogen. Do not use overhead irrigation or expose plants to rainfall if possible. Improve rapid drying of the foliage by spacing plants adequately, and rogue severely infected seedling flats.

Chemical treatment Preventive applications of fungicides can greatly reduce the incidence of anthracnose.

Calonectria Leaf Spot (Cylindrocladium Leaf Spot)

Calonectria colhounii
Calonectria crotalariae
Calonectria theae

Calonectria leaf spot of palms has not been a widespread problem but has been reported in Florida and Hawaii. The disease is caused by several *Calonectria* spp. (asexual stage, *Cylindrocladium* spp.) and is sometimes referred to as Cylindrocladium leaf spot.

Symptoms

Leaf spots caused by *Calonectria* spp. are characteristically grayish brown, dark brown, or nearly black and circular to irregular in shape and sometimes have a dark border. Young spots are circular and often surrounded by a chlorotic band. Advanced stages of the disease are characterized by coalescence of leaf spots and chlorosis and necrosis of leaflet margins and tips.

Epidemiology

Spores are spread to healthy plants in splashing water, in contaminated potting media, and on tools. Potential spread is greatly increased with the production of ascospores, which are forcibly discharged from fruiting bodies (perithecia) of the fungus and are carried by air currents. Unlike conidia, ascospores do not require free moisture for their dissemination; however, the discharge of ascospores is induced by a rapid increase in relative humidity.

Host Range

Hosts of *Calonectria* spp. include *Chamaedorea elegans, Chrysalidocarpus lutescens, Howea belmoreana, Ptychosperma elegans,* and *Washingtonia robusta* (Table 65.1).

Management

Cultural practices Since a high level of moisture is required for disease development, moisture control is important in disease prevention. Any cultural practice that reduces free moisture on leaves (e.g., increased spacing among plants, covered greenhouses, drip irrigation, etc.) will reduce disease incidence and spread.

Sanitation Prompt removal and destruction of infected leaves, especially blighted leaves, before perithecia are formed is crucial to disease control. Reduction of inoculum sources greatly increases the cost-effectiveness of chemical treatments.

Chemical treatment Some fungicides are effective against the *Calonectria* spp. causing leaf spot diseases of palms.

Catacauma Leaf Spot (Tar Spot)

Catacauma mucosum
Catacauma sabal
Catacauma torrendiella

Catacauma leaf spot, or tar spot, is prevalent in Florida, Georgia, and Texas where humidities and temperatures are generally high.

Symptoms

Pinnae, rachides, and peduncles can become infected. Diamond-shaped lesions are formed, with the long axis of the lesion parallel with the venation of the leaflet. The lesions have a crusty, wartlike texture. A yellow halo surrounds the infected spot. Disease incidence can be quite high, such that large areas of tissue become necrotic.

Epidemiology

The incidence of Catacauma leaf spot seems to be higher in palms grown under less than optimal light levels. Planting sites with poor air circulation and overhead irrigation are likely to increase disease severity by stimulating the release and dissemination of spores under humid conditions favoring infection.

Host Range

Some of the palm hosts of *Catacauma* spp. are *Acoelorrhaphe wrightii, Butia capitata, Cocos nucifera, Livistona chinensis, Sabal causiarum, S. palmetto, Syagrus romanzoffiana,* and *Washingtonia robusta* (Table 65.1).

Management

Sanitation Little of the infection cycle of the tar spot pathogens is documented, and therefore the effects of sanitation and fungicide treatment are unknown. The removal of severely infected fronds will reduce the amount of inoculum left on the plant, but the critical timing for this activity is not known.

Chemical treatment The use of fungicides sometimes aids in the control of the disease.

Fusarium Wilt

Fusarium oxysporum

The Fusarium wilt pathogen, *Fusarium oxysporum,* has been shown to be transmitted during pruning operations by chain saws and handsaws, thus moving throughout a planting in California. In the mid-1990s Fusarium wilt was found in Florida.

Calonectria Leaf Spot

Geographic occurrence Florida and Hawaii

Seasonal occurrence Most of the year, under average yearly conditions in zones 9 and 10 (adjust for your hardiness zone)

DISEASE FREQUENCY	DISEASE SEVERITY
5 annual	5 plants killed
4	4
[3]	3
2	**[2]**
1 rare	1 very little damage

CHEMICAL TREATMENT	CULTURAL PRACTICES
[3] used every year	3 very important
2	**[2]**
1 not used	1 not important

SANITATION	RESISTANT CULTIVARS
3 very important	3 many cultivars
[2]	2
1 not important	**[1] no resistance**

Catacauma Leaf Spot

Geographic occurrence Florida, Georgia, and Texas

Seasonal occurrence Most months, under average yearly conditions in zone 9 (adjust for your hardiness zone)

DISEASE FREQUENCY	DISEASE SEVERITY
5 annual	5 plants killed
[4]	4
3	3
2	**[2]**
1 rare	1 very little damage

CHEMICAL TREATMENT	CULTURAL PRACTICES
3 used every year	3 very important
2	**[2]**
[1] not used	1 not important

SANITATION	RESISTANT CULTIVARS
3 very important	3 many cultivars
2	**[2]**
[1] not important	1 no resistance

Table 65.1 Some fungal diseases of ornamental palms

Scientific name	Common name	Diseases[a]
Acoelorrhaphe wrightii	Paurotis palm, Everglades palm	Ca, Ga, Gr, St
Archontophoenix alexandrae	Alexandra palm, king palm	He, Ph
Archontophoenix cunninghamiana	Piccabeen palm	Gl, He
Areca catechu	Betel nut palm	Ga, Th
Arenga pinnata	Sugar palm, areng palm, black-fiber palm	Gr
Bactris gasipaes	Peach palm	Ga
Borassus aethipium	. . .	St
Borassus flabellifer	Palmyra palm	Ph
Butia capitata	South American jelly palm, pindo palm	Ca, Ga, Gr, Ph, St
Caryota mitis	Burmese palm, clustered fishtail palm	He, St, Th
Caryota urens	. . .	St
Chamaedorea elegans	Parlor palm	Cl, Gl, He, Ph
Chamaedorea erumpens	Bamboo palm	Gl, Ph
Chamaedorea seifrizii	Bamboo palm, reed palm	Gl, He, Ph
Chamaerops humilis	European fan palm	Ga, Gr
Chrysalidocarpus cabadae	Cabada palm	Ga
Chrysalidocarpus lutescens	Areca palm, butterfly palm, golden palm	Cl, Ga, Gl, Gr, He, Ph, Ps, St
Coccothrinax argentata	Silver palm	Gr
Cocos nucifera	Coconut palm	Ca, Ga, Gr, He, St, Th
Howea belmoreana	Belmore palm	Cl, He
Howea forsteriana	Forster sentry palm, kentia palm	Cy, Gl, He, Ph, Ps, St
Licuala ramsayi	. . .	He, Ps
Livistona chinensis	Chinese fan palm	Ca, Ga, Ph, St
Phoenix canariensis	Canary Island date palm	Ga, Gl, Gr, Ph, St, Th
Phoenix dactylifera	Date palm	Gl, Gr, St, Th
Phoenix reclinata	Senegal date palm	Ga, St
Phoenix roebelenii	Dwarf date palm, pygmy date palm	Gr, He, St
Phoenix sylvestris	Wild date palm	Ga
Ptychosperma elegans	Alexander palm, solitaire palm	Cl, Ga, He
Ptychosperma macarthurii	Macarthur palm	Ga, Ph
Rhapis excelsa	Lady palm	Ps, St, Th
Rhapis subtilis	Thai dwarf lady palm	Ps
Roystonea elata	Florida royal palm	Ga, Gr, Ph, St, Th
Roystonea regia	Cuban royal palm	Ga, He, Ph, St
Sabal causiarum	Puerto Rican hat palm	Ca, Ga
Sabal palmetto	Cabbage palmetto	Ca, Ga, Gr, Ph, St, Th
Serenoa repens	Saw palmetto, scrub palmetto	Ga
Syagrus romanzoffiana	Queen palm	Ca, Ga, Gl, Gr, He, Ph, St, Th
Syagrus schizophylla	Arikury palm	Ga
Thrinax morrisii	Key thatch palm	St
Trachycarpus fortunei	Chinese windmill palm, windmill palm	Ph
Veitchia merrillii	Christmas palm, Merrill palm, Manila palm	St
Washingtonia filifera	California fan palm	Gl, Ph, Th
Washingtonia robusta	Mexican fan palm	Ca, Cl, Ga, Gl, Gr, He, Ph, St

[a] Ca = Catacauma tar spot. Cl = Calonectria leaf spot. Ga = Ganoderma butt rot. Gl = Gliocladium stem rot (pink rot). Gr = Graphiola leaf spot (false smut). He = Helminthosporium leaf spots. Ph = *Phytophthora* diseases. Ps = Pseudocercospora leaf spot. St = Stigmina leaf spot. Th = Thielaviopsis blight.

Symptoms

Fronds on one side of the tree may die more rapidly than the others, producing a lopsided appearance; fronds may die more rapidly than usual, from the lower fronds upward; or a ring of fronds may die, with green fronds below and above them. Spines and pinnae on one side of a frond may die progressively from the base to the tip, while the pinnae on the opposite side of the frond are still healthy. Pinnae may exhibit water soaking along the vascular elements. A dark brown streak often appears on the bottom of the rachis. Upon splitting of the rachis, the vascular elements will often be streaked dark brown, as in other Fusarium wilts, and fungal hyphae can be observed in cross sections of the vascular tissues.

Epidemiology

The pathogen is soil- and water-borne. In California it has been shown to be transmitted during pruning operations by chain saws and handsaws.

Host Range

Phoenix canariensis, P. reclinata, and seedlings of *P. dactylifera* are currently known to be susceptible to *F. oxysporum.*

Management

Sanitation Transmission of the pathogen can be prevented by using pruning saws that can be and are thoroughly sterilized between pruning trees. Sterilization can be accomplished by a 5-

minute dip in a 2.5% solution of sodium hypochlorite. Do not use chain saws on trees affected by Fusarium wilt, because of the difficulty of sterilizing these tools.

Ganoderma Butt Rot (Basal Stem Rot)

Ganoderma zonatum

Ganoderma butt rot, also called basal stem rot, caused by the fungus *Ganoderma zonatum* (syns. *G. applanatum, G. sulcatum, G. tumidum,* and *Polyporus lucidus* var. *zonatus*), has been reported in the southeastern United States (Alabama, Florida, Georgia, and South Carolina), throughout tropical Africa, and in Argentina. The disease is gaining in importance throughout the southeastern United States, because of the difficulty of early diagnosis and therefore poor chances for therapeutic treatments of any kind. The disease is seldom seen in nursery production.

Symptoms

The initial symptom of butt rot of palms is withering and drooping of older fronds. New growth slows, decreases in size, and develops a pale green to yellowish cast. As growth slows, several unopened spears may form in the crown. As the older canopy continues to die, younger leaves may exhibit nutrient deficiency symptoms, progressive periods of wilt, and tip necrosis. Finally, only one or more spears remain in the bud. Depending on the location of the infection in the tree, the head of the palm may fall off or the trunk collapse. Palm death may take as long as three to four years.

Host Range

Some of the palm hosts of *Ganoderma* are *Acoelorrhaphe wrightii, Areca catechu, Bactris gasipaes, Butia capitata, Chamaerops humilis, Chrysalidocarpus cabadae, C. lutescens, Cocos nucifera, Livistona chinensis, Phoenix canariensis, P. reclinata, P. sylvestris, Ptychosperma elegans, P. macarthurii, Roystonea elata, R. regia, Sabal causiarum, S. palmetto, Serenoa repens, Syagrus romanzoffiana, S. schizophylla,* and *Washingtonia robusta* (Table 65.1).

Management

Cultural practices Avoid establishment of *Ganoderma* in palms by planting them in sites where trunks will not be prone

Fusarium Wilt

Geographic occurrence California and Florida

Seasonal occurrence Most months, under average yearly conditions in zone 9 (adjust for your hardiness zone)

DISEASE FREQUENCY	DISEASE SEVERITY
5 annual	**5 plants killed**
4	4
3	3
2	2
1 rare	1 very little damage

CHEMICAL TREATMENT	CULTURAL PRACTICES
3 used every year	3 very important
2	2
1 not used	**1 not important**

SANITATION	RESISTANT CULTIVARS
3 very important	3 many cultivars
2	2
1 not important	**1 no resistance**

Ganoderma Butt Rot

Geographic occurrence Florida, South Carolina, Georgia, and Alabama

Seasonal occurrence Most months, under average yearly conditions in zone 9 (adjust for your hardiness zone)

DISEASE FREQUENCY	DISEASE SEVERITY
5 annual	**5 plants killed**
4	4
3	3
2	2
1 rare	1 very little damage

CHEMICAL TREATMENT	CULTURAL PRACTICES
3 used every year	3 very important
2	2
1 not used	**1 not important**

SANITATION	RESISTANT CULTIVARS
3 very important	3 many cultivars
2	2
1 not important	**1 no resistance**

to injury. Space palms adequately to prevent the spread of the pathogen within a site through belowground contact between root systems of infected and healthy palms. Exercise care during maintenance activities to avoid wounding trunks with mowing and trimming equipment. Remove dead palms immediately, and destroy them by burning or burial away from other palms.

Sanitation Sanitation is the major emphasis in butt rot treatment programs world wide. Promptly remove root systems, stumps, and trunks of dead palms in the landscape or nursery. This will prevent future airborne spread of the pathogen from basidiocarps formed on these trees. Root system removal will suppress the movement of the pathogen in the soil toward adjacent susceptible species. If a fallow period cannot be observed, the planting site should be sieved free of root fragments, filled with clean soil, and fumigated (if possible) prior to palm replacement. Currently, hand removal of reproductive brackets of *Ganoderma* is recommended until the stump can be removed.

Chemical treatment Fungicides have been examined for their possible use in controlling butt rot. Although some products have been effective in inhibiting the fungus in culture, field use of these products has failed. Stump treatments with creosote or copper-based fungicides have been ineffective.

Graphiola Leaf Spot (False Smut)

Graphiola congesta
Graphiola phoenicis
Graphiola thaxteri

Graphiola leaf spot, or false smut, is found worldwide in tropical to subtropical areas where rainfall is plentiful. *Graphiola phoenicis* is not very damaging in arid areas. *G. congesta* and *G. thaxteri* have been reported only in the southeastern United States.

Symptoms

The most obvious sign of false smut fungi is the wartlike, cup-shaped, black reproductive structures (sori) that rupture through both surfaces of leaves. Before the sori form, small yellow to brown spots develop on pinnae (on both surfaces), rachides, and leaf bases. These spots swell and eventually rupture as the reproductive sori emerge from below the leaf epidermis. Symptoms generally occur on two-year-old and older fronds, since the reproductive cycle of the pathogen is ten to eleven months. When lesions form at high densities, they can cause premature senescence of older fronds.

Host Range

Graphiola spp. have host ranges limited to plants in the family Arecaceae. Some hosts of *G. phoenicis* are *Acoelorrhaphe wrightii, Arenga pinnata, Butia capitata, Chamaerops humilis, Chrysalidocarpus lutescens, Coccothrinax argentata, Cocos nucifera, Phoenix canariensis, P. dactylifera, P. roebelenii, Roystonea elata, Sabal palmetto, Syagrus romanzoffiana,* and *Washingtonia robusta* (Table 65.1).

Management

Sanitation Removal of severely diseased fronds is usually an adequate control for false smut of ornamental palms.

Resistance Resistance to false smut is available in commercially grown date palm cultivars but is not known in other palm genera.

Chemical treatment Some fungicides have been successful in reducing the severity of false smut of date palms in India.

Helminthosporium Leaf Spot

Bipolaris cynodontis
Bipolaris incurvata
Bipolaris setariae
Exserohilum rostratum
Phaeotrichoconis crotalariae

Leaf spot diseases caused by the *Helminthosporium*-like fungi *Bipolaris, Exserohilum,* and *Phaeotrichoconis* are generally similar to each other and are frequently severe enough to drastically reduce the marketability of many ornamental palms. These diseases are common and require the same control measures.

Symptoms

Leaf spots begin as tiny water-soaked spots, which expand into circular to elliptical lesions and become brown, reddish brown, or dark brown to black. Lesions may be surrounded by a chlorotic, circular or spindle-shaped halo. Depending on the host species,

Graphiola Leaf Spot

Geographic occurrence Southeastern United States

Seasonal occurrence Most months, under average yearly conditions in zone 9 (adjust for your hardiness zone)

DISEASE FREQUENCY	DISEASE SEVERITY
5 annual	5 plants killed
4	4
[3]	3
2	**[2]**
1 rare	1 very little damage

CHEMICAL TREATMENT	CULTURAL PRACTICES
3 used every year	3 very important
2	**[2]**
[1] not used	1 not important

SANITATION	RESISTANT CULTIVARS
[3] very important	3 many cultivars
2	2
1 not important	**[1] no resistance**

the lesions may be slightly sunken, or eye spot symptoms may develop as lesions expand. In severe cases, young leaves are blighted by coalescing lesions, and fronds appear shredded. Severely blighted seedlings may be killed.

Epidemiology

Disease spread depends on the movement of conidia by wind or splashing water. Healthy plants are easily contaminated if diseased plants are brought into the greenhouse or placed on the same bench. Splashing water, such as that produced by overhead irrigation, is ideal for dislodging conidia in large numbers and dispersing them to healthy tissue.

Occurrence and Host Range

Helminthosporium leaf spots have been reported in most cultivated palms throughout the world. Some palm hosts of *Helminthosporium*-like fungi are *Archontophoenix alexandrae, A. cunninghamiana, Caryota mitis, Chamaedorea elegans, C. seifrizii, Chrysalidocarpus lutescens, Cocos nucifera, Howea belmoreana, H. forsteriana, Licuala ramsayi, Phoenix roebelenii, Ptychosperma elegans, Roystonea regia, Syagrus romanzoffiana,* and *Washingtonia robusta* (Table 65.1).

Helminthosporium Leaf Spot

Geographic occurrence Florida, California, and Hawaii

Seasonal occurrence Most months, under average yearly conditions in zone 10 (adjust for your hardiness zone)

DISEASE FREQUENCY	DISEASE SEVERITY
[5] annual	5 plants killed
4	4
3	**[3]**
2	2
1 rare	1 very little damage

CHEMICAL TREATMENT	CULTURAL PRACTICES
[3] used every year	3 very important
2	**[2]**
1 not used	1 not important

SANITATION	RESISTANT CULTIVARS
3 very important	3 many cultivars
[2]	2
1 not important	**[1] no resistance**

Management

Cultural practices Proper environmental considerations will greatly enhance the production of vigorous, disease-free plants. Since fungal growth, sporulation, germination, infection, and subsequent disease development are highly dependent on moisture and free water, any reduction in excess moisture will curb disease levels.

Sanitation Greenhouse sanitation is very important in the reduction of inoculum levels to minimize further disease development. *Bipolaris* spp. and *E. rostratum* tend to be nonspecialized among palms and nongramineous plants. Thus badly diseased palms could serve as a source of inoculum for infection of other crops in the vicinity, or conversely, other plants may serve as an inoculum source for infection of palm crops.

Chemical treatment Fungicides can provide excellent control of Helminthosporium leaf spots, but their efficacy is drastically reduced by heavy inoculum levels or stressful conditions. Producing areca palms in full sun creates conditions in which fungicides are usually not successful in controlling these diseases. In addition, some young palms (such as areca palms) are very sensitive to copper-based products and develop symptoms similar to those induced by the leaf spot pathogens.

Lethal Yellowing

Phytoplasma

In the United States lethal yellowing has been reported in the Rio Grande Valley in Texas, the lower east and west coasts of Florida, and the Florida Keys.

Symptoms

Lethal yellowing symptoms vary among palms. In coconut palms, the initial symptom is premature nut fall, or "shelling," of mature palms. Fallen nuts exhibit a dark brown to black, water-soaked zone at the stem end. Palms infected with lethal yellowing exhibit flowering stalks with black or dark brown male flowers on the tips. Flower necrosis is a definitive symptom of this disease. In general, one or more of the oldest, lower fronds turn yellow and then brown and hang parallel to the trunk. Immature fronds turn yellow, but remain upright, and then turn brown after bud death. Death of the bud occurs when half to two-thirds of the canopy has yellowed.

Epidemiology

Lethal yellowing is believed to be caused by a phytoplasma transmitted by the planthopper *Myndus crudus.*

Host Range

Some palms reported to be susceptible to lethal yellowing are listed in Table 65.2.

Management

Cultural practices In certain areas, quarantines have been imposed to control the importation of palms in genera and species known to be susceptible to lethal yellowing. As palms are purchased and planted, common sense dictates that a diversity of palms be used, to avoid massive losses of a single species due to this disease.

Sanitation Symptomatic palms in which more than 25% of the canopy is yellowed should be removed, as they are unlikely to respond to any treatment.

Resistance Table 65.2 lists some palms believed to be resistant to lethal yellowing.

Chemical treatment Therapeutic benefits have been demonstrated in symptomatic plants after injection with certain antibiotics.

Phytophthora Diseases

Phytophthora drechsleri
***Phytophthora nicotianae* (syn. *P. parasitica*)**
Phytophthora palmivora

Several *Phytophthora* spp. cause diseases of palms, including seedling blights and damping-off caused by *P. nicotianae* (syn. *P. parasitica*); trunk, crown, and root rots caused by *P. palmivora* and *P. drechsleri;* leaf spots, blights, and petiole rots caused by *P. nicotianae* and *P. palmivora;* and apical tip, bud, and heart rot caused by *P. palmivora.*

P. palmivora and *P. nicotianae* are distributed worldwide in tropical and warm temperate regions characterized by high rainfall.

Symptoms

P. nicotianae causes seedling blight of *Chrysalidocarpus lutescens.* The first symptoms are necrotic brown flecks, which expand to angular, irregularly shaped leaf spots, and finally into large, blighted areas, occasionally with yellow borders. Diseased tissue is initially grayish black to grayish green but turns tan to brown with the progression of the disease. Infection of emerging leaves spreads into unfurled leaves, eventually killing the terminal bud and the plant.

Bud decay of large palms is correlated with a two- to six-month period following a hurricane or severe storm. Symptoms begin as a pale green discoloration of the spear leaf and one or more of the newest leaves. Unfolding pinnae may exhibit dark brown lesions from infection that has occurred in the spear. The spear rots at the base and pulls out easily. Rapid disease development can result in crown death, while the lower canopy remains green for six to 12 months. Slow disease development will produce a progressive yellowing, browning, and collapse of fronds, from the lower canopy upward, until the crown is dead. Fronds often fall from the trunk, leaving the dead trunk naked.

Leaf spots begin as slightly grayish green, irregularly shaped or circular to elongate lesions, which turn brown to dark brown with tan centers. Lesions rapidly coalesce, forming blighted areas, which are dark brown to papery tan with a dark border. The expansion of leaf lesions is somewhat limited by the midrib.

Table 65.2 Susceptibility of palms to lethal yellowing disease

Palms known to be susceptible	Palms not known to be susceptible
Aiphanes lindeniana	*Acoelorrhaphe wrightii*
Allagoptera arenaria	*Arenga pinnata*
Arenga engleri	*Bactris gasipaes*
Borassus flabellifer	*Carpentaria acuminata*
Caryota mitis	*Chamaerops humilis*
Chrysalidocarpus cabadae	*Chrysalidocarpus lutescens*
Cocos nucifera	*Coccothrinax argentea*
Corypha elata	*Elaeis guineensis*
Corypha taliera	*Heterospathe elata*
Dictyosperma album	*Phoenix roebelenii*
Gaussia attenuata	*Ptychosperma macarthurii*
Howea belmoreana	*Rhapidophyllum hystrix*
Hyophorbe verschaffeltii	*Roystonea elata*
Latania sp.	*Roystonea hispaniola*
Livistona chinensis	*Roystonea regia*
Livistona rotundifolia	*Sabal causiarum*
Nannorrhops ritchiana	*Sabal palmetto*
Neodypsis decaryi	*Syagrus amara*
Phoenix canariensis	*Syagrus romanzoffiana*
Phoenix dactylifera	*Thrinax morrisii*
Phoenix reclinata	*Thrinax parviflora*
Phoenix sylvestris	*Washingtonia filifera*
Pritchardia affinis	*Washingtonia robusta*
Pritchardia pacifica	
Pritchardia remota	
Ravenea hildebrandtii	
Syagrus schizophylla	
Trachycarpus fortunei	
Veitchia arecina	
Veitchia merrillii	
Veitchia montgomeryana	

Lethal Yellowing

Geographic occurrence Florida

Seasonal occurrence Most months, under average yearly conditions in zone 10 (adjust for your hardiness zone)

DISEASE FREQUENCY	DISEASE SEVERITY
5 annual (boxed)	**5 plants killed** (boxed)
4	4
3	3
2	2
1 rare	1 very little damage

CHEMICAL TREATMENT	CULTURAL PRACTICES
3 used every year	3 very important
2 (boxed)	2
1 not used	**1 not important** (boxed)

SANITATION	RESISTANT CULTIVARS
3 very important (boxed)	3 many cultivars
2	**2** (boxed)
1 not important	1 no resistance

Host Range

Palms commonly affected by *Phytophthora* diseases include *Archontophoenix alexandrae, Borassus flabellifer, Butia capitata, Chamaedorea elegans, C. erumpens, C. seifrizii, Chrysalidocarpus lutescens, Howea forsteriana, Livistona chinensis, Phoenix canariensis, Ptychosperma macarthurii, Roystonea elata, R. regia, Sabal palmetto, Syagrus romanzoffiana, Trachycarpus fortunei, Washingtonia filifera,* and *W. robusta* (Table 65.1).

Management

Cultural practices Sources of clean plants or seeds are vital for disease avoidance. Seeds should never be gathered from the ground. Seeds from unknown sources should be surface-disinfested before planting, and then carefully monitored for symptoms. As with most fungal diseases, moisture control is all-important in crop management. Free water favors the development of *Phytophthora* in all stages of its life cycle, and it also favors and accelerates processes in the disease cycle.

Sanitation Once these diseases occur, either in the nursery or in the landscape, severely affected palms should be rogued and burned or buried. Early removal and destruction of infected plants will reduce inoculum levels and decrease the incidence of new infections.

Chemical treatment Curative efforts with foliar fungicides will be most effective in limiting disease incidence and spread.

Pink Rot (Gliocladium Blight)

Gliocladium vermoeseni

The pink rot pathogen, the fungus *Gliocladium vermoeseni,* is distributed worldwide and infects palms in greenhouses and landscapes. The disease is also known as Gliocladium blight.

Symptoms

Initially, dark brown, necrotic areas appear on the stems of infected plants, near the soil line or 0.7 to 1.0 m up the stem. On *Syagrus,* infections can occur on the trunk at any height. The spots are often associated with gummy exudates. Older fronds die prematurely, necrotic streaks appear from the rachis base, and pinnae turn chlorotic. The pathogen readily produces dusty masses of orange to pink spores (conidia), often in sporodochia, on this tissue. In severe infections, many stems die from girdling, giving

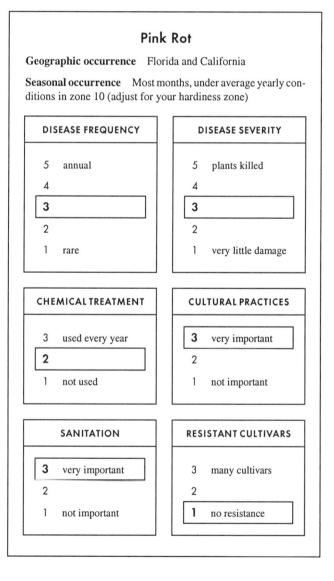

Phytophthora Diseases

Geographic occurrence Wherever palms are grown

Seasonal occurrence Most months, under average yearly conditions in zone 10 (adjust for your hardiness zone)

DISEASE FREQUENCY	DISEASE SEVERITY
5 annual	**5 plants killed**
4	4
3	3
2	2
1 rare	1 very little damage

CHEMICAL TREATMENT	CULTURAL PRACTICES
3 used every year	**3 very important**
2	2
1 not used	1 not important

SANITATION	RESISTANT CULTIVARS
3 very important	3 many cultivars
2	2
1 not important	**1 no resistance**

Pink Rot

Geographic occurrence Florida and California

Seasonal occurrence Most months, under average yearly conditions in zone 10 (adjust for your hardiness zone)

DISEASE FREQUENCY	DISEASE SEVERITY
5 annual	5 plants killed
4	4
3	**3**
2	2
1 rare	1 very little damage

CHEMICAL TREATMENT	CULTURAL PRACTICES
3 used every year	**3 very important**
2	2
1 not used	1 not important

SANITATION	RESISTANT CULTIVARS
3 very important	3 many cultivars
2	2
1 not important	**1 no resistance**

potted plants an open unsalable appearance. Removal of symptomatic fronds reveals stem infections, which are dark brown and irregularly shaped, sometimes with chlorotic margins.

Epidemiology

Pink rot is an invasive rot that affects bud tissues, petioles, leaf blades, and trunks. Often attacking trees under stress, the pathogen is believed to enter host tissue through wounds and areas damaged by such injuries as removal of leaves, sunburn, and freezing.

Host Range

Some of the palms affected by Gliocladium blight are *Archontophoenix cunninghamiana, Chamaedorea elegans, C. erumpens, C. seifrizii, Chrysalidocarpus lutescens, Howea forsteriana, Phoenix canariensis, P. dactylifera, Syagrus romanzoffiana, Washingtonia filifera,* and *W. robusta* (Table 65.1).

Management

Cultural practices In the landscape, maintain plant health and avoid wounds.

Sanitation Wounding facilitates infection, and therefore only completely dead leaves should be removed from palms with Gliocladium blight. Yellow leaves should be removed only when the temperature is at least 30°C, preferably 33°C, to reduce chances of infection.

Chemical treatment Fungicides have been found to provide excellent disease control in *Chamaedorea* palms during the summer in Florida. Transplanted specimen palms may benefit from prophylactic sprays of fungicides and stress-reducing measures. On *Syagrus romanzoffiana,* trunk surgery followed by application of protective fungicides has been reported to be effective.

Pseudocercospora Leaf Spot

***Pseudocercospora rhapisicola* (syn. *Cercospora rhapisicola*)**

Pseudocercospora is a notoriously slow-growing fungus, and the diseases it incites are also very slow to develop. The long incubation period makes it extremely difficult to ascertain whether symptomless leaves are free from infection. Pseudocercospora leaf spot has been reported in Japan and in Hawaii, California, and Florida.

Symptoms

Leaf spots begin as very tiny, faintly chlorotic or light green flecks. The primary lesions expand, the pathogen sporulates in them, and spores fall off to infect tissue immediately adjacent to the primary lesions. This results in larger spots surrounded by pinpoint satellite lesions in the later stages of disease development. Leaf spots eventually coalesce, becoming irregularly circular to elliptical. The infected areas are composed of slightly raised dark brown to reddish brown flecks and spots surrounded by chlorotic tissue, which gradually turns brown. Light infections produce a few groups of spots, while heavy infections produce leaves with mosaic patterns. Heavy infections result in premature leaf loss.

Epidemiology

Pseudocercospora leaf spot is a very slow-developing disease. A period of four weeks or more elapses between infection and the appearance of the initial symptoms.

Host Range

Some of the palms affected by Pseudocercospora leaf spot are *Chrysalidocarpus lutescens, Howea forsteriana, Licuala ramsayi, Rhapis excelsa,* and *R. subtilis* (Table 65.1).

Management

Sanitation Complete removal of diseased leaves followed by regular inspection and removal of newly diseased leaves will reduce inoculum levels. If sanitation is maintained and plants are kept under solid-cover greenhouses, chemical treatment should be unnecessary.

Stigmina Leaf Spot (Exosporium Leaf Spot)

***Stigmina palmivora* (syn. *Exosporium palmivorum*)**

Stigmina leaf spot, also called Exosporium leaf spot, is a foliar disease that develops during cool seasons in both hemispheres. It has been reported in Florida, Louisiana, Mississippi, and Texas.

Symptoms

Leaf spots develop first on lower, older leaves and appear as small, round, yellowish lesions. Mature lesions are tan to reddish

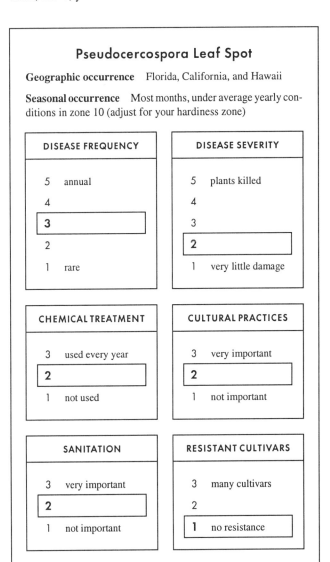

Pseudocercospora Leaf Spot

Geographic occurrence Florida, California, and Hawaii

Seasonal occurrence Most months, under average yearly conditions in zone 10 (adjust for your hardiness zone)

DISEASE FREQUENCY
5 annual
4
3
2
1 rare

DISEASE SEVERITY
5 plants killed
4
3
2
1 very little damage

CHEMICAL TREATMENT
3 used every year
2
1 not used

CULTURAL PRACTICES
3 very important
2
1 not important

SANITATION
3 very important
2
1 not important

RESISTANT CULTIVARS
3 many cultivars
2
1 no resistance

brown, with a dark brown to black, depressed center, and they are often surrounded by a diffuse yellow halo. Leaf spots may merge, killing large, irregularly shaped areas of tissue. Leaf pinnae and rachides can be infected.

Host Range

Some of the palm hosts of *Stigmina palmivora* are *Acoelorrhaphe wrightii, Borassus aethipium, Butia capitata, Caryota mitis, C. urens, Chrysalidocarpus lutescens, Cocos nucifera, Howea forsteriana, Livistona chinensis, Phoenix canariensis, P. dactylifera, P. reclinata, P. roebelenii, Rhapis excelsa, Roystonea elata, R. regia, Sabal palmetto, Syagrus romanzoffiana, Thrinax morrisii, Veitchia merrillii,* and *Washingtonia robusta* (Table 65.1).

Management

Cultural practices and sanitation The severity of Stigmina leaf spot is negligible except in the nursery. The disease may develop on the lower fronds of landscape palms planted in partial shade in areas with poor air circulation. Removal of infected fronds is quite effective controlling the disease. In nurseries with overhead irrigation and partial shade, disease severity may be high in containerized palms. Observations in the southeastern United States indicate that Stigmina leaf spot is most severe from late fall through spring. This period is typified by cool weather and frequent leaf wetness cycles.

Chemical treatment Fungicide efficacy trials with a related *Stigmina* sp. on stone fruits have indicated that some compounds are useful in reducing disease severity.

Stigmina Leaf Spot

Geographic occurrence Florida, Louisiana, Mississippi, and Texas

Seasonal occurrence Most months, under average yearly conditions in zone 9 (adjust for your hardiness zone)

DISEASE FREQUENCY	DISEASE SEVERITY
5 annual	5 plants killed
4	4
3	3
2	**2**
1 rare	1 very little damage

CHEMICAL TREATMENT	CULTURAL PRACTICES
3 used every year	3 very important
2	**2**
1 not used	1 not important

SANITATION	RESISTANT CULTIVARS
3 very important	3 many cultivars
2	2
1 not important	**1 no resistance**

Thielaviopsis Bud Rot

Ceratocystis paradoxa
(asexual stage, *Chalara paradoxa*)

Thielaviopsis bud rot (also known as stem bleeding, bitten leaf, black scorch, dry basal rot, and heart rot) is caused by the fungus *Ceratocystis paradoxa* (syns. *Ceratostomella paradoxa* and *Endoconidiophora paradoxa*) and its asexual stage, *Chalara paradoxa* (syn. *Thielaviopsis paradoxa*). *Chalara paradoxa* has been reported in most palm-producing areas of the world including Arizona, California, and Florida.

Symptoms

Bud rot and the associated bitten-leaf symptom start with blackish brown lesions on external and internal tissues. Buds die

Thielaviopsis Bud Rot

Geographic occurrence Arizona, California, and Florida

Seasonal occurrence Most months

DISEASE FREQUENCY	DISEASE SEVERITY
5 annual	5 plants killed
4	**4**
3	3
2	2
1 rare	1 very little damage

CHEMICAL TREATMENT	CULTURAL PRACTICES
3 used every year	3 very important
2	**2**
1 not used	1 not important

SANITATION	RESISTANT CULTIVARS
3 very important	3 many cultivars
2	2
1 not important	**1 no resistance**

or are damaged. New leaves may exhibit the bitten-leaf symptom as they emerge deformed, with reduced pinnae, and black, necrotic tips. In other situations, dark brown to black irregular lesions develop along the petiole and give the emerging frond a torched appearance. Pinnae are distorted during emergence or shorter than normal. The inflorescence may be invaded prior to the opening of the spathe. A soft, yellow decay develops in trunk tissue, darkening to black with age. As decay progresses, a reddish brown or rust-colored liquid bleeds from the point of invasion. This sap flow may extend several feet down the trunk, blackening the trunk as it dries.

Host Range

Palm hosts of *T. paradoxa* include *Areca catechu, Caryota mitis, Cocos nucifera, Phoenix canariensis, P. dactylifera, Rhapis excelsa, Roystonea elata, Sabal palmetto, Syagrus romanzoffiana,* and *Washingtonia filifera* (Table 65.1).

Management

Cultural practices and sanitation Avoid wounding palms, to limit disease incidence. Infected plants should be cut, removed, and destroyed. In some cases, localized infections can be excised through tree surgery followed by fungicide treatment.

Resistance Although some resistance to the pathogen has been documented in date palm, the range of disease reactions in ornamental palm species is not known.

Chemical treatment Early infection of bud, leaf, and root tissue can be effectively treated with fungicides.

REFERENCE

Chase, A. R., and Broschat, T. K. 1991. Diseases and Disorders of Ornamental Palms. American Phytopathological Society, St. Paul, Minn.

CHAPTER 66

Austin Hagan ▪ Auburn University, Auburn, Alabama
Larry W. Barnes ▪ Texas A&M University, College Station
Gary W. Simone ▪ University of Florida, Gainesville

Photinia Diseases

Geographic production USDA zones 4–9

Family Rosaceae

Genus *Photinia*

Species			
	P. ×*fraseri*	red-tip photinia	zones 8–9
	P. glabra	Japanese photinia	zone 9
	P. serrulata	Chinese photinia	zones 7–9
	P. villosa	oriental photinia	zones 4–7

Photinia is used extensively across the southeastern and northwestern United States as a screen and in foundation plantings around residences and commercial properties. Red-tip photinia (*P.* ×*fraseri*), with its brilliant red to maroon spring foliage, is by far the most widely planted of the four *Photinia* taxa. Of the other three, only Chinese photinia (*P. serrulata*) is available, in a limited supply, in the nursery trade. Japanese photinia (*P. glabra*) is sensitive to cold damage and is best adapted to zone 9. Oriental photinia (*P. villosa*) is the most cold-hardy of the photinias but apparently is highly sensitive to fire blight and is not widely planted. Foliar diseases, particularly Entomosporium leaf spot, fire blight, and bacterial leaf spot, have a significant impact on the beauty and health of photinia in production nurseries and landscape plantings.

Entomosporium Leaf Spot
Entomosporium mespili

Entomosporium leaf spot is a widespread and often destructive disease of photinia in both the nursery and the landscape. Of the *Photinia* taxa, red-tip photinia is the most sensitive to this disease and may be completely defoliated and badly weakened by it. Symptomatic red-tip photinia has little value in the nursery market. Because of difficulties in controlling this disease, many nurseries have dropped red-tip photinia from their production schedules. Occasional outbreaks of Entomosporium leaf spot also occur in Japanese photinia.

Symptoms

Tiny, circular, bright red spots on the upper and lower surfaces of young, expanding leaves are the earliest symptoms of Entomosporium leaf spot. The lesions enlarge to about 1/4 inch (1 cm) in diameter and have an ash brown to gray center with a distinctive dark red to maroon halo or margin (Plate 97). Large, irregular, purple to maroon blotches may also appear on heavily spotted leaves. Individual lesions may coalesce to form irregular brown areas of dead tissue. Tiny, black, slightly raised fruiting bodies (acervuli) of the fungal pathogen, *Entomosporium mespili,* are usually found in the center of each spot. Unlike the new growth, the mature leaves of photinia are resistant to infection. Lesions similar to those on the leaves often appear on leaf petioles and succulent shoots of badly diseased photinia. While light infections cause little more than cosmetic damage, severe infections usually result in early and severe leaf drop. In some cases, all but the youngest leaves near the shoot tips are prematurely lost. Severe defoliation increases sensitivity to stress induced by environmental factors or management practices.

Epidemiology

E. mespili survives in diseased leaves, shoots, and possibly fallen leaves. During periods of mild, humid weather, masses of spores (conidia) ooze from fruiting bodies on the lesion surface, beginning in late winter and continuing through much of the growing season. Along the Gulf Coast and adjacent areas, conidia may be produced sporadically through the winter. Conidia are spread to newly expanded leaves by wind and splashing water. Disease development is favored by several days of warm, cloudy, wet weather. Extended periods of dry weather during the spring may slow the spread of the disease. In container nurseries, year-round occurrence of Entomosporium leaf spot can be traced to a combination of continuous shoot growth due to heavy pruning, close plant spacing, and overhead irrigation.

Host Range

Entomosporium leaf spot has been reported as a disease of several shrubs and trees in the apple subfamily (Pomodidae), including *Cleyera,* flowering pear, hawthorn, Indian hawthorn, and loquat, as well as photinia.

Management

Effective management of Entomosporium leaf spot of photinia depends on propagating disease-free cuttings, intensive nursery sanitation, proper production practices, and a preventative fungicide treatment program.

Cultural practices and sanitation Outbreaks of Entomosporium leaf spot can usually be traced back to the propagation of diseased cuttings. Cuttings must be taken only from disease-free stock plants. Preventative fungicide applications must be made nearly year-round in order to ensure that stock plants remain disease-free. Discard diseased stock plants.

When using overhead irrigation systems, water late at night or at midday. Avoid watering blocks of photinia in the late afternoon or early evening, particularly on a cloudy day or several hours prior to a rain shower. Space out container material to speed evaporation of moisture from the foliage and improve air circulation.

Avoid frequent shearing or heavy pruning. Also avoid placing disease-free photinia near blocks of trees and shrubs damaged by Entomosporium leaf spot. Discard photinia defoliated by the disease. Collect fallen diseased leaves from production ranges whenever possible, and discard them.

Resistance Despite claims to the contrary, no cultivars or selections of red-tip photinia have demonstrated significant resistance to Entomosporium leaf spot. Japanese photinia is less susceptible than red-tip photinia but more susceptible than Chinese photinia. The reaction of oriental photinia is not well documented.

Chemical treatment Preventative fungicide treatments are required for production of leaf spot–free photinia. To prevent outbreaks of the disease in photinia, fungicide applications must be made throughout the growing season from the time cuttings root until finished container stock or ball-and-burlap stock is shipped to retail outlets. Along the Gulf Coast and adjacent areas, yearlong spray programs are usually required to protect photinia from leaf spot. Generally, preventative fungicide applications should be made during the growing season at 10- to 14-day intervals. Shorten the application interval to seven to 10 days during periods of frequent rain showers. During the winter, monthly applications should be sufficient to prevent disease spread. Do not water overhead immediately after a fungicide application. For fungicide recommendations, contact your county extension agent or state specialist responsible for diseases of nursery crops.

Entomosporium Leaf Spot

Geographic occurrence USDA zones 4–9

Seasonal occurrence Spring (April to May, in the southern United States)

DISEASE FREQUENCY	DISEASE SEVERITY
5 annual (selected)	5 plants killed
4	**4** (selected)
3	3
2	2
1 rare	1 very little damage

CHEMICAL TREATMENT	CULTURAL PRACTICES
3 used every year (selected)	3 very important
2	**2** (selected)
1 not used	1 not important

SANITATION	RESISTANT CULTIVARS
3 very important (selected)	3 many cultivars
2	2
1 not important	**1 no resistance** (selected)

Fire Blight

Erwinia amylovora

Fire blight is a destructive bacterial disease of numerous members of the apple subfamily (Pomodidae), including photinia, in both nurseries and landscape plantings. *Photinia* taxa differ significantly in their reaction to fire blight. In the United States, the disease is particularly damaging to susceptible plants wherever the weather is wet and mild during bloom in the spring. In container and field nurseries, however, serious damage to the widely grown red-tip photinia is rare.

Symptoms

Fire blight severity is influenced by the sensitivity of the *Photinia* species to the disease and the weather during bloom. Sudden wilting and downward curling of the succulent shoot tips is usually the first noticeable symptom of the disease in photinia. The leaves on blighted shoots usually turn reddish brown (Plate 98). In the resistant red-tip photinia, the damage is usually limited to blighting of a few scattered shoot tips. In susceptible photinia, blossom blight may be extensive, followed by the death of leaves and shoot tips.

Brown to black, slightly sunken cankers may form on susceptible photinia at the intersection of scaffold branches and the bases of blighted twigs and flower spurs. If the scaffold branches are girdled, large portions of the diseased shrub may die. The canker surface is usually smooth, but cracks often develop along its outer edge. Reddish brown streaks can be seen in the sapwood beneath the edge of each canker.

Epidemiology

The fire blight pathogen, *Erwinia amylovora,* overseasons in tissues surrounding twig and branch cankers. During mild, wet weather in late winter or early spring, beads of a sticky amber fluid, containing millions of bacteria, ooze from tissues around the margins of twig and branch cankers. The bacterium is spread from plant to plant by insects, particularly the honeybee, and in water splashing on flower buds, blooms, leaves, and tender shoots. Once in the flower nectaries, the bacterium quickly reproduces and colonizes adjacent tissues.

An extended period of warm, cloudy, wet weather during and shortly after bloom greatly increases the risk of a fire blight outbreak. Temperatures between 75 and 81°F are considered most favorable for disease development. Other weather events, such as frost, hail, or high winds, may also increase disease severity. Severe pruning and excessive nitrogen fertilization may increase plant vulnerability to fire blight.

Host Range

Photinia is just one of many genera in the apple subfamily that are susceptible to fire blight.

Management

For the resistant red-tip photinia, good sanitation and cultural practices are usually all that is necessary to control fire blight in the nursery. For susceptible photinia, pesticide treatments may be required, in addition to good sanitation and cultural practices, for production of disease-free plants.

Cultural practices Overfertilizing with nitrogen increases susceptibility to fire blight. To reduce the risk of disease, use a water-soluble nitrogen fertilizer in frequent applications at a low rate, or apply a slow-release fertilizer at the recommended rate. Avoid using a fertilizer containing water-soluble nitrogen at a high rate of application in early spring or late fall. When using an overhead irrigation system, do not water in the late afternoon or early evening.

Sanitation Generally, container stock damaged by fire blight should be discarded. Perennial cankers can be pruned out of damaged photinia during cold, dry winter weather. Cuts should be made 6 to 8 inches below the visible canker margin. Do not take cuttings from plants damaged by fire blight the previous spring. When taking cuttings or pruning damaged photinia, dip pruning tools between cuts in alcohol or a commercial germicide.

Fire Blight

Geographic occurrence Throughout the range of photinia production

Seasonal occurrence Spring, as new growth emerges

DISEASE FREQUENCY	DISEASE SEVERITY
5 annual	5 plants killed
4	4
3	3
[2]	**[2]**
1 rare	1 very little damage

CHEMICAL TREATMENT	CULTURAL PRACTICES
3 used every year	3 very important
[2]	**[2]**
1 not used	1 not important

SANITATION	RESISTANT CULTIVARS
[3 very important]	3 many cultivars
2	2
1 not important	**[1 no resistance]**

Resistance Red-tip and Japanese photinia are more resistant to fire blight than Chinese photinia. Oriental photinia may be highly sensitive to the disease. Among cultivars of red-tip photinia, no differences in resistance have been reported.

Chemical treatment Fire blight occurs sporadically in red-tip photinia, so that bactericide applications are unnecessary. For the susceptible Chinese and oriental photinia, bactericide treatments during and shortly after bloom may be needed for production of high-quality plants. Begin bactericide applications several days before the first flower buds open, and continue spraying at four- to five-day intervals through petal fall. When postbloom weather patterns favor disease development, applications should be continued at seven-day intervals for another month. For fungicide recommendations, contact your county extension agent or state specialist responsible for diseases of nursery crops.

Bacterial Leaf Spot

Xanthomonas campestris* pv. *vitians

Bacterial leaf spot has recently been recognized as a damaging disease of container-grown red-tip photinia. Occasional outbreaks have been reported in Texas, Florida, and possibly several other southeastern states. The pathogen is tentatively identified as *Xanthomonas campestris* pv. *vitians*. Little information is available concerning the biology of this bacterium, its dispersal, or conditions that favor disease development.

Symptoms

Bacterial leaf spot is characterized by dark purple, angular lesions, with no halo or margin, randomly scattered over the leaf surface. Young, rapidly growing leaves appear to be most susceptible to attack by the pathogen, as in Entomosporium leaf spot. On heavily diseased leaves, coalescing lesions lose their characteristic angular form and become more irregular in shape. On the underside of the leaf, tissues surrounding the angular lesions often appear water-soaked. Bacterial streaming from tissue sections taken from the water-soaked zone can be readily observed microscopically at low power.

Epidemiology

The pathogen is probably transmitted on cuttings taken from diseased red-tip photinia. Disease development is favored by warm weather in late spring and early summer, but significant spotting of the leaves may also occur in the fall. Frequent showers and overhead irrigation appear to intensify disease severity.

Host Range

Little information on the host range of *X. campestris* pv. *vitians* is available. Ornamental hosts include tropical foliage plants in the genera *Aglaonema* (Chinese evergreen) and *Syngonium* (nephthytis) and possibly other members of the family Araceae. A recent outbreak of bacterial leaf spot of lettuce in Florida was also attributed to this bacterium.

Management

Successful management of bacterial leaf spot on red-tip photinia depends on an aggressive sanitation program and protective applications of a bactericide.

Cultural practices Like most foliar bacterial pathogens, *X. campestris* pv. *vitians* is dispersed to nearby healthy plants by splashing water. To slow disease spread and speed the evaporation of moisture on the leaves, space out container material. If possible, overhead-irrigate red-tip photinia late at night or at midday. Watering in late afternoon or early evening, particularly on a cloudy day, may intensify damage to the foliage. Increased nitrogen fertility has been shown to help reduce *Xanthomonas* diseases in many ornamental crops.

Sanitation Outbreaks of bacterial leaf spot during liner propagation and production of red-tip photinia can often be traced to cuttings collected from diseased stock or production material. To avoid accidentally introducing the pathogen into blocks of production material, always take cuttings from spot-free red-tip photinia. Where the disease has previously caused serious losses, apply a recommended bactericide to stock plants at regular intervals during the summer and early fall. Avoid placing clean red-tip photinia near blocks of diseased production material. Discard diseased stock plants.

Resistance No information is available concerning the reactions of *Photinia* taxa to bacterial leaf spot. However, the disease has been reported only in red-tip photinia.

Chemical treatment Preventative applications of selected bactericides have been shown to reduce the severity of bacterial leaf spot of red-tip photinia. Where the disease has been previously diagnosed, apply a bactericide to rooted cuttings in propagation beds, liners, and container stock at 10- to 14-day intervals from mid- to late spring until early fall. Shorten the application interval to seven days during extended periods of warm, wet weather. For bactericide recommendations, contact your county extension agent or state specialist responsible for diseases of nursery crops.

Powdery Mildew

Sphaerotheca pannosa

Sporadic outbreaks of powdery mildew occur in Chinese photinia. The disease is little threat to plant health, but the market value of heavily mildewed stock may be greatly reduced. Red-tip photinia is apparently immune to powdery mildew.

Symptoms

Scattered, cottony, white to buff-colored colonies on the upper surfaces of leaves are the first signs of the photinia powdery mildew fungus, *Sphaerotheca pannosa.* In Chinese photinia, similar

Bacterial Leaf Spot

Geographic occurrence Texas, Florida and possibly other southeastern states

Seasonal occurrence Spring to fall, with an optimum period in late spring and early summer

DISEASE FREQUENCY
5 annual
4
3
[2]
1 rare

DISEASE SEVERITY
5 plants killed
4
[3]
2
1 very little damage

CHEMICAL TREATMENT
3 used every year
[2]
1 not used

CULTURAL PRACTICES
3 very important
[2]
1 not important

SANITATION
[3] very important
2
1 not important

RESISTANT CULTIVARS
3 many cultivars
2
[1] no resistance

Powdery Mildew

Geographic occurrence Throughout the range of photinia production

Seasonal occurrence Spring to fall, with an optimum period from early summer to midsummer

DISEASE FREQUENCY
5 annual
4
3
2
[1] rare

DISEASE SEVERITY
5 plants killed
4
3
[2]
1 very little damage

CHEMICAL TREATMENT
3 used every year
[2]
1 not used

CULTURAL PRACTICES
3 very important
[2]
1 not important

SANITATION
3 very important
2
[1] not important

RESISTANT CULTIVARS
[3] many cultivars
2
1 no resistance

colonies may also be seen on leaf petioles and new shoots. Heavily colonized leaves and shoots turn nearly white and may be twisted or distorted.

Epidemiology

S. pannosa overwinters as hyphae in dormant buds and possibly on leaf surfaces during mild winters. Spores (conidia) produced on diseased buds are spread to nearby healthy leaves by air currents. Warm days and cool nights during an extended period of dry but humid weather favor disease development.

Host Range

The host range of *S. pannosa* also includes rose and peach.

Management

Selection of disease-resistant cultivars or species and preventative applications of fungicides are the most widely employed strategies for controlling powdery mildew in nursery crops.

Cultural practices Spacing plants to improve air circulation around the foliage may help slow mildew development in susceptible cultivars.

Sanitation Fallen diseased leaves should be removed, but sanitation is otherwise of little value in controlling powdery mildew.

Resistance. Of commonly cultivated photinia taxa, only Chinese photinia is susceptible to powdery mildew. Japanese and red-tip photinia are highly resistant or immune to this disease.

Chemical treatment Powdery mildew occurs only very sporadically in Chinese photinia, so that preventative fungicide treatment is usually unnecessary. In the nursery, foliage of Chinese photinia should be inspected for colonies of *S. pannosa* in late spring and early summer. Should mildew appear, immediately apply a recommended fungicide to the affected block of plants. Repeat fungicide applications at the interval specified on the fungicide label until the threat of further disease spread is over. For fungicide recommendations, contact your county extension agent or state specialist responsible for diseases of nursery crops.

Root Rot

Phytophthora parasitica
Phytophthora palmivora
Pythium splendens

Root rots caused by *Phytophthora* spp. and *Pythium* spp. are common and often destructive diseases in a wide variety of container- and field-grown woody trees and shrubs. Serious losses related to root rots of several *Photinia* taxa, primarily red-tip and Japanese photinias, have occasionally been reported in container nurseries in Florida and along the Gulf Coast. Diseased photinia are generally unsalable. Poor nursery sanitation and inadequate drainage or periodic flooding of production ranges are typically associated with outbreaks of root rot in container stock. The developing practice of using recycled water in nursery production is likely to contribute to outbreaks of *Pythium* and *Phytophthora* diseases.

Symptoms

Typical symptoms of root rot of photinia include interveinal chlorosis, slowed shoot growth, limb dieback, wilting of foliage, and ultimately death of the plant. The interveinal chlorosis and slowed growth can be confused with symptoms of a nutritional disorder. Fine feeder roots colonized by root rot fungi are brittle and brown to reddish brown. Browning of the feeder roots often begins near the base of the root ball and usually continues until the entire root system is destroyed. Areas of discolored roots can often be observed by removing the root ball from the container.

Epidemiology

The causal fungi survive as resting structures (chlamydospores and oospores) in roots of diseased plants, crop debris, soil under container beds, and potting media. They are easily spread through blocks of liners or container stock in contaminated potting medium and by splashing or flowing water. Motile zoospores, produced during the growing season in saturated potting media, are also readily dispersed in runoff water flowing through container beds and in recycled irrigation water from holding ponds. The heaviest losses of plants usually occur in flood-prone container beds where water stands around container stock for extended periods of time. Over- or underwatering and poor drainage or compaction of potting media often accelerate root rot development. Typically, root rot fungi are active at temperatures which favor

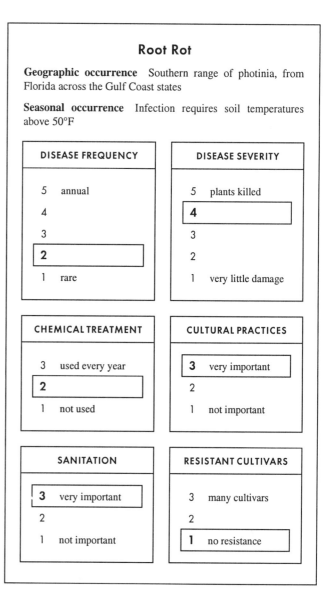

Root Rot

Geographic occurrence Southern range of photinia, from Florida across the Gulf Coast states

Seasonal occurrence Infection requires soil temperatures above 50°F

DISEASE FREQUENCY	DISEASE SEVERITY
5 annual	5 plants killed
4	**[4]**
3	3
[2]	2
1 rare	1 very little damage

CHEMICAL TREATMENT	CULTURAL PRACTICES
3 used every year	**[3] very important**
[2]	2
1 not used	1 not important

SANITATION	RESISTANT CULTIVARS
[3] very important	3 many cultivars
2	2
1 not important	**[1] no resistance**

plant growth. For additional information concerning the epidemiology of root rot diseases, refer to Chapter 14, "Phytophthora Root Rot and Dieback."

Host Range

Phytophthora parasitica, P. palmivora, and *Pythium splendens* are root rot pathogens attacking a wide array of container-grown trees and shrubs. For additional information concerning the host range of *Phytophthora* spp., refer to Chapter 14.

Management

Strict sanitation and cultural controls along with preventative fungicide treatments are generally required to prevent outbreaks of root rot in container-grown photinia. A more detailed discussion of recommended control practices can be found in Chapter 14.

Cultural practices Flats of rooted cuttings and liners should be placed on raised benches or on sloped, gravel-covered beds. Container beds must be crowned, covered with black plastic or weed cloth, and then topped with a layer of gravel or a similar coarse material. Ponding of water around container stock must be prevented. Drainage ditches must be deep enough to handle flooding rains. To avoid over- or underwatering container stock, block plant material by container size and water needs. Account for daily rainfall when scheduling irrigation. Fertilize according to need, and avoid damaging the roots with water-soluble nitrogen fertilizer applied at excessive rates. Use a coarse bark-based medium with a high percolation rate and 20 to 30% air-filled pore space.

Sanitation Store potting medium components on an asphalt or concrete pad. Take special care to avoid contamination of these materials by front-end loaders and other nursery vehicles and equipment. Do not place new containers on bare soil. Avoid recycling containers unless they are carefully rinsed and disinfested. Clear all debris and disinfest propagation benches. Collect cuttings for propagation from symptom-free container stock. Root cuttings in a fresh soilless potting medium in new flats or containers on clean raised benches. Discard diseased plant material.

Resistance The relative resistance of *Photinia* species and cultivars to Phytophthora and Pythium root rots is unknown. However, photinia taxa, including red-tip and Japanese photinias, are generally considered not nearly as susceptible to root rot diseases as selections of azalea, flowering dogwood, juniper, and rhododendron.

Chemical treatment. Root rot diseases of photinia are generally confined to a relatively small area, so preventative fungicide treatments are probably unnecessary outside of Florida and the Gulf Coast region. Within this area, blocks of red-tip and Japanese photinias should be regularly checked for symptoms of the disease. If root rot causes chronic losses in a nursery, then preventative fungicide treatments are recommended for future photinia crops. Otherwise, continue to scout, implement sanitation practices and cultural controls, but withhold fungicide treatment.

Fungicide treatment programs are usually effective only when combined in a management program with strict sanitation and good nursery management. A granular or selected wettable powder fungicide may be mixed into the potting medium prior to transplanting liners or stepping up container stock. If a soil-incorporated fungicide is not used, begin preventative fungicide drenches or foliar applications immediately after transplanting. In either case, repeat treatments at the intervals specified on the product label until the crop is shipped. Fungicides will not "cure" plants of root rot incited by *Phytophthora* or *Pythium* but will protect new roots from infection. For fungicide recommendations, contact your county extension agent or state specialist responsible for diseases of nursery crops.

REFERENCES

Alfieri, S. A. 1969. Entomosporium leaf spot of loquat. Fla. Dep. Agric. Consumer Serv. Circ. 82.

Alfieri, S. A., Jr., Langdon, K. R., Kimbrough, J. W., El-Gholl, N. E., and Wehlburg, C. 1994. Diseases and disorders of plants in Florida. Fla. Dep. Agric. Consumer Serv. Bull. 14.

Baudoin, A. B. A. M. 1986. Environmental conditions required for infection of photinia leaves by *Entomosporium mespili.* Plant Dis. 70:519–521.

Bowen, K. L., Hagan, A. K., and Foster, W. 1994. Application rates and spray intervals of ergosterol-biosynthesis inhibitor fungicides for control of Entomosporium leaf spot of photinia. Plant Dis. 78:578–581.

Hoitink, H. A. J., Benson, D. M., and Schmitthenner, A. F. 1986. Phytophthora root rot. Pages 4–8 in: Compendium of Rhododendron and Azalea Diseases. D. L. Coyier and M. K. Roane, eds. American Phytopathological Society, St. Paul, Minn.

Kokalis-Burelle, N., Hagan, A. K., Gazaway, W. S., and Sikora, E. 1997. Fire blight on fruit trees and woody ornamentals. Ala. Coop. Ext. Serv. Circ. ANR-542.

Lambe, R. C., and Ridings, W. H. 1979. Entomosporium leaf spot of photinia. Fla. Dep. Agric. Conumer Serv. Circ. 206.

Miller, J. W. 1997. *Xanthomonas* leaf spot of photinia. Fla. Dep. Agric. Consumer Serv. Circ. 383.

Pernezny, K., Raid, R. N., Stall, R. E., Hodge, N. C., and Collins, J. 1995. An outbreak of bacterial leaf spot of lettuce in Florida caused by *Xanthomonas campestris* pv. *vitians.* Plant Dis. 79:359–360.

Plakidas, A. G. 1957. New or unusual plant diseases in Louisiana. Plant Dis. Rep. 41:643–645.

CHAPTER

67

Jillanne R. Burns ▪ North Carolina State University, Raleigh

Pieris Diseases

Geographic production USDA zones 4–8

Family Ericaceae

Genus *Pieris*

Species *P. floribunda*
P. formosa
P. japonica

Dieback and Canker

***Botryosphaeria dothidea* (syns. *B. berengeriana, B. ribis*)**

Dieback and canker of *Pieris* spp. has been reported mostly in the eastern United States. However, given the wide and nonspecific host range of the pathogen, *Botryosphaeria dothidea* (syns. *B. berengeriana* and *B. ribis*), as well as its environmental versatility, the disease is probably more widespread.

Symptoms

Reddish brown to black, sunken cankers form on stems and are usually delimited by callus. Smaller twigs are usually girdled, while larger branches are able to wall off the pathogen within the cankers. As twig and stem cankers develop, the foliage becomes chlorotic and wilts. Affected foliage eventually turns necrotic, giving the plant a typical dieback appearance.

Epidemiology

The pathogen overwinters as fruiting bodies (pycnidia) in cankers. Pycnidia form throughout the year following the death of tissue in moist weather. They release spores (conidia), which are dispersed by splashing water and serve as the primary inoculum. In its sexual stage, the fungus forms other fruiting bodies (pseudothecia) in the cankers, and they release ascospores, which are dispersed by wind and water and provide secondary inoculum. Infection of plant tissue is limited unless the plant is under temperature stress or drought stress, when extensive colonization (infection) of stem tissue can occur.

Host Range

Dieback and canker has been reported in *P. japonica.*

Management

B. dothidea can be transmitted by pruning, so every effort should be made to maintain proper sanitation by disinfesting pruning tools. Environmental stress can predispose plants to infection by the pathogen. Mulching and watering during dry periods is an important part of preventing the disease. There are no resistant cultivars and no fungicides known to control *B. dothidea* on *Pieris* spp.

Dieback and Canker

Geographic occurrence USDA zones 4–8

Seasonal occurrence Most of the year, with an optimum period from April through July

DISEASE FREQUENCY	
5	**annual**
4	
3	
2	
1	rare

DISEASE SEVERITY	
5	**plants killed**
4	
3	
2	
1	very little damage

CHEMICAL TREATMENT	
3	used every year
2	
1	**not used**

CULTURAL PRACTICES	
3	**very important**
2	
1	not important

SANITATION	
3	**very important**
2	
1	not important

RESISTANT CULTIVARS	
3	many cultivars
2	
1	**no resistance**

Phytophthora Blight and Root Rot

***Phytophthora* spp.**

Phytophthora blight and root rot can be a devastating disease of *Pieris* spp., both in the nursery and in the landscape. *Phytophthora cinnamomi, P. citricola, P. citrophthora,* and *P. parasitica* have been reported on *Pieris* spp. *P. citrophthora* is considered the most damaging of these species to *Pieris japonica.*

Symptoms

Phytophthora spp. is distinguished from other organisms that cause dieback by usually attacking only healthy new growth and causing blight of young twigs and leaves. Mature wood is rarely affected. Blight caused by *Phytophthora* spp. begins as olivaceous leaf spots, which later turn brown and may be surrounded by a red margin. These leaves usually drop, harboring the fungus in the form of mycelium, sporangia, and oospores in the infected debris.

The initial symptoms of the root and crown rot phase of the disease are foliar chlorosis and stunting. In later stages, leaf epinasty, wilting, and dieback occur and can lead to the death of the plant.

Epidemiology

The disease occurs mainly during the spring and early summer and is favored by excess moisture and warm conditions. The lower stems and leaves are infected by zoospores and sporangia splashing from the soil and infected plant debris. Periods of cloudy weather following rain promote infection if plant surfaces remain wet. Leaf spots can expand rapidly. Once petioles are colonized, the pathogen can infect branches, causing dieback.

Host Range

Phytophthora blight and root rot has been reported in *P. floribunda* and *P. japonica.*

Management

Cultural practices and sanitation are effective in the management of Phytophthora blight and root rot. Soils should be well drained, with care taken to prevent overwatering and avoid late-afternoon irrigation. Fertilizer should be applied as needed; excessive fertilization increases susceptibility to this disease. Removing debris will reduce the amount of inoculum. Fungicides may be necessary to control the disease.

Phytophthora Blight and Root Rot

Geographic occurrence USDA zones 4–8

Seasonal occurrence April–September

DISEASE FREQUENCY	
5	annual
4	
3	
2	
1	rare

DISEASE SEVERITY	
5	plants killed
4	
3	
2	
1	very little damage

CHEMICAL TREATMENT	
3	used every year
2	
1	not used

CULTURAL PRACTICES	
3	very important
2	
1	not important

SANITATION	
3	very important
2	
1	not important

RESISTANT CULTIVARS	
3	many cultivars
2	
1	no resistance

REFERENCES

Agrios, G. N. 1997. Plant Pathology. 4th ed. Academic Press, San Diego, Calif.

Dirr, M. A. 1998. Manual of Woody Landscape Plants: Their Identification, Ornamental Characteristics, Culture, Propagation and Uses. 5th ed. Stipes Publishing, Champaign, Ill.

Erwin, D. C., and Ribeiro, O.K. 1996. Phytophthora Diseases Worldwide. American Phytopathological Society, St. Paul, Minn.

Gerlach, W. W. P., Hoitink, H. A. J., and Schmitthenner, A. F. 1976. *Phytophthora citrophthora* on *Pieris japonica:* Infection, sporulation and dissemination. Phytopathology 66:302–308.

Sinclair, W. A., Lyon, H. H., and Johnson, W. T. 1987. Diseases of Trees and Shrubs. Comstock Publishing Associates, Cornell University Press, Ithaca, N.Y.

CHAPTER 68

John Hartman ▪ University of Kentucky, Lexington

Charles Hodges ▪ North Carolina State University, Raleigh

Ed Barnard ▪ Florida Division of Forestry, Gainesville

Pine Diseases

Geographic production USDA zones 2–10

Family Pinaceae

Genus *Pinus*

Species

P. cembra	Swiss stone pine
P. densiflora	Japanese red pine
P. monophylla	pinyon pine
P. monticola	western white pine
P. mugo	mugo pine
P. nigra	Austrian pine
P. nigra var. *austriaca*	
P. palustris	longleaf pine
P. pinea	Italian stone pine
P. ponderosa	ponderosa pine
P. radiata	Monterey pine
P. strobus	eastern white pine
P. sylvestris	Scots pine
P. taeda	loblolly pine
P. thunbergiana	Japanese black pine
P. virginiana	Virginia pine
P. wallichiana	Himalayan white pine

Some pines other than ornamentals are mentioned in this chapter because they can serve as sources of inoculum if located near a nursery.

Cyclaneusma Needle Cast
Cyclaneusma minus

The fungus *Cyclaneusma minus* (syn. *Naemacyclus minor*) is believed to cause premature casting of needles of numerous species of pine, but environmental stress may be important in disease development. Cyclaneusma needle cast, also called Naemacyclus needle cast, causes early yellowing and needle loss, which weaken and degrade nursery trees.

Symptoms

The first symptoms are light green to yellow spots, which develop into brown bands on needles in late summer and fall on two-year-old and older needles. The needles turn yellow and then brown and are shed over the next six months. Fruiting bodies of the fungus develop in the brown bands or on other parts of the needles, appearing as elliptical swollen areas below the epidermis. When a fruiting body matures, the epidermis splits and exposes the waxy, white to yellowish spore-bearing surface.

The symptoms can be confused with symptoms of other fungal needle cast diseases (red band and Lophodermium, Lophodermella, and Elytroderma needle casts) and with damage due to environmental stresses, air pollution, and the pine needle sheath miner.

Cyclaneusma Needle Cast

Geographic occurrence USDA zones 4–9

Seasonal occurrence April to November

DISEASE FREQUENCY	
5	annual
4	
3	
2	
1	rare

DISEASE SEVERITY	
5	plants killed
4	
3	
2	
1	very little damage

CHEMICAL TREATMENT	
3	used every year
2	
1	not used

CULTURAL PRACTICES	
3	very important
2	
1	not important

SANITATION	
3	very important
2	
1	not important

RESISTANT CULTIVARS	
3	many cultivars
2	
1	no resistance

Epidemiology

Spores are forcibly discharged from fruiting bodies about 4 to 6 hours after the onset of rain, at any time during the growing season. When free moisture is present on needles, the spores germinate, and the fungus enters through stomata to infect current-year needles in June and one-year-old and older needles from April to November. Infection may occur at temperatures as low as 3°C. Ten to 15 months later, during the fall, infected needles begin to show symptoms, and fruiting bodies begin to form. Fruiting bodies mature and spores are first liberated after several weeks of warm weather in the spring; they continue to be released throughout the growing season. Spores may be liberated from fruiting bodies on both attached and fallen needles.

Infection and symptom development may vary with stress, genetic variation in the host, and environmental conditions.

Host Range

At least 15 species of pine are reportedly susceptible to *C. minus,* including Austrian, Jeffrey, lodgepole, mugo, ponderosa, and Scots pines.

Management

Cultural practices Control weeds and provide adequate tree spacing in field plantings to minimize periods of leaf wetness. Establish plantings with disease-free seedlings in locations well away from sources of inoculum. Avoid nutrient deficiencies and moisture stress.

Sanitation Remove and destroy highly susceptible cultivars.

Resistance Scots pine from northern European seed sources appear to be more resistant than those from Mediterranean sources.

Chemical treatment Infection periods occur throughout the growing season, and thus multiple season-long applications of fungicides, beginning at budbreak, are necessary to protect needles from infection. Even with partial control, heavy needle casting may still occur in the late fall.

Brown Spot Needle Blight

Mycosphaerella dearnessii
(asexual stage, *Lecanosticta acicola*)

Browning and early needle loss due to brown spot needle blight, caused by the fungus *Mycosphaerella dearnessii,* make infected pines unsalable as nursery stock.

Symptoms

Reddish brown, resin-soaked spots appear on needles in summer (Plate 99). Spotted needles turn yellow and then begin to turn brown, from the tips back, in late summer and fall. Black fruiting bodies of the pathogen appear flush with the surface of dry dead needles but protrude from the needle surface when wet. After spring growth, dead needles drop, leaving tufts of green growth on branch tips (Plate 100).

Other needle cast diseases, especially Dothistroma needle blight, have similar symptoms.

Epidemiology

Young needles are most susceptible. They become infected via stomata during warm, moist periods throughout the summer. In the South the fungus produces spores (conidia) on fruiting bodies formed on brown spots on living needles, and it produces ascospores from fruiting bodies formed on dead needles. In the North, only conidia are produced. Infected needles drop in the fall, and the fungus overwinters in the dead needles, emerging in late spring to begin the cycle again.

Host Range

Scots and longleaf pines are most affected, but many others pines are susceptible, including Austrian, Japanese black, loblolly, Monterey, mugo, ponderosa, Italian stone, Virginia, eastern white, and western white pines.

Management

Cultural practices Do not shear infected foliage during wet weather. Shear healthy plantations first, to avoid carrying spores from diseased to healthy plantations. Plant only disease-free nursery stock, and use resistant varieties of pines. Do not maintain windbreaks of susceptible pines near nursery plantings.

Sanitation Rake and remove fallen needles.

Resistance Certain longer-needled Scots pines from Central Europe are more resistant than others.

Chemical treatment Apply fungicides in late spring and summer to prevent infections.

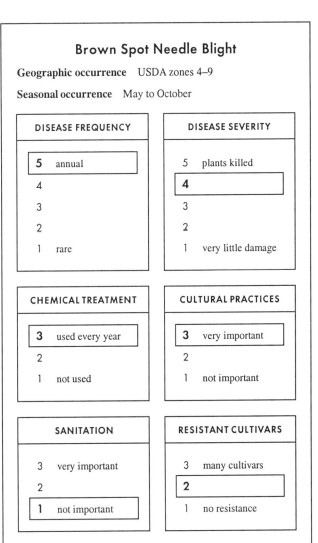

Brown Spot Needle Blight

Geographic occurrence USDA zones 4–9

Seasonal occurrence May to October

DISEASE FREQUENCY	DISEASE SEVERITY
5 annual	5 plants killed
4	**4**
3	3
2	2
1 rare	1 very little damage

CHEMICAL TREATMENT	CULTURAL PRACTICES
3 used every year	**3 very important**
2	2
1 not used	1 not important

SANITATION	RESISTANT CULTIVARS
3 very important	3 many cultivars
2	**2**
1 not important	1 no resistance

Dothistroma Needle Blight (Red Band)

Mycosphaerella pini

(asexual stage, *Dothistroma septospora*)

Dothistroma needle blight, or red band disease, affects more than 30 species and varieties of pine. The disease occurs worldwide, particularly in areas with persistent cool, moist weather. Needle blighting and premature drop can render nursery trees unmarketable, and when favorable weather prevails for several years, the disease may reach epidemic levels.

Symptoms

The first symptoms are yellow to tan spots on needles. The spots enlarge to form a red to brown band that encircles the needle (Plate 101). The needle dies beyond the infected band, and its tip turns tan to brown, while the base of the needle remains green. Needles in the lower canopy typically show symptoms first, and the disease progresses upward. Infected needles drop prematurely. After the needles die, black fruiting bodies of the pathogen rupture and burst through the epidermal tissues.

Symptoms of red band are similar to those of other fungal needle cast diseases (Lophodermium, Lophodermella, Elytroderma, and Cyclaneusma needle casts) and also resemble damage due to environmental stresses, air pollution, and the pine needle sheath miner.

Epidemiology

Spores of the fungal pathogen develop in fruiting bodies in the discolored red areas of infected needles and are spread primarily by splashing water and windblown rain throughout the year. The spores germinate in the presence of free moisture and enter needles through stomata. All needles except expanding and newly developed needles are susceptible. Infection takes three days, or maybe longer during cooler weather. Depending on the pine species, symptoms appear five to six months after infection. The fungus overwinters in infected needles.

Host Range

Austrian, Monterey, mugo, ponderosa, and Scots pines are the most susceptible, but the disease also occurs in numerous other species of pine.

Management

Cultural practices Improve tree spacing, control weeds, and remove the lowest whorl or two of branches to increase air circulation and sunlight penetration. Do not plant near infected windbreaks.

Sanitation In nursery plantations, dislodge diseased needles from the branches, then rake and destroy them.

Resistance Certain Austrian pines from Yugoslavian seed sources have shown resistance. Individuals of some species, such as Monterey and ponderosa pines, also exhibit resistance.

Chemical treatment Timely applications of fungicide have given effective control. Make two applications, one just prior to budbreak to protect the previous year's growth and a second after budbreak when considerable growth has occurred.

Dothistroma Needle Blight Disease

Geographic occurrence USDA zones 4–9

Seasonal occurrence September to November

DISEASE FREQUENCY	DISEASE SEVERITY
5 annual	5 plants killed
4	4
[3]	**[3]**
2	2
1 rare	1 very little damage

CHEMICAL TREATMENT	CULTURAL PRACTICES
3 used every year	3 very important
[2]	**[2]**
1 not used	1 not important

SANITATION	RESISTANT CULTIVARS
3 very important	3 many cultivars
2	**[2]**
[1] not important	1 no resistance

Lophodermium Needle Cast

Lophodermium seditiosum

Lophodermium needle cast can be a limiting factor in growing Scots pines where the environment is favorable for disease development.

Symptoms

Symptoms first appear on needles as small pale spots, which turn yellow and then reddish brown with yellow margins. As the spots enlarge, the needles appear mottled and eventually turn brown and die, changing from brown to straw-colored (Plate 102). Dead needles are usually cast prematurely in late winter and early spring, prior to new growth. In some situations, partially infected needles with areas of green tissue will remain on the tree. Large, elliptical, black fruiting bodies of the pathogen, up to 1/4 inch in diameter, develop in dead needles, just under the epidermis (Plate 103). When disease pressure has been high for several consecutive years, a majority of trees in a field planting will be totally brown prior to the production of new growth in the spring. Seedlings in nurseries can die when conditions favor disease development.

Lophodermium needle cast can be confused with other fungal needle cast diseases (red band and Cyclaneusma, Elytroderma, and Lophodermella needle casts). The symptoms also resemble damage from air pollution, drought stress, winter injury, and the pine needle sheath miner.

Epidemiology

Fruiting bodies of *Lophodermium seditiosum* on dead needles release spores in late summer and fall in moist weather, and infection takes place when windborne fungal spores germinate and penetrate directly into needle tissue. The disease progresses through the fall and winter, and in spring it destroys the xylem tissues, resulting in dead needles by spring and mature spores by fall.

Other species of *Lophodermium* with different life histories also infect pines.

Host Range

Two- and three-needle pines, such as Scots, Austrian, and red pines, are very susceptible to *L. seditiosum.* Other *Lophodermium* species attack two- and three-needle pines and occasionally five-needle pines.

Management

Cultural practices Increasing tree spacing and weed control will improve air circulation and sunlight penetration and thus help to create an environment less favorable for disease. Application of potassium may reduce losses due to *Lophodermium* in nurseries. Do not plant near infected windbreaks.

Sanitation In nursery plantations rake up and destroy infected fallen needles.

Resistance Scots, Austrian, and red pines are highly susceptible. However, long-needled Scots pines having some resistance are available.

Chemical treatment Fungicide applications at monthly intervals from August to October are needed to control infection. Where environmental conditions are favorable for disease development and multiple species of *Lophodermium* are present, as in the Pacific Northwest, monthly fungicide applications throughout the growing season may be necessary for production of acceptable trees.

Lophodermella Needle Cast

Lophodermella spp.

Five species of *Lophodermella* cause needle blight and premature needle drop of various species of pine: *L. arcuata, L. cerina, L. concolor, L. montivaga,* and *L. morbida.* Lophodermella needle cast is widespread and can impact nursery plantings. It can kill young trees after several years of high disease pressure.

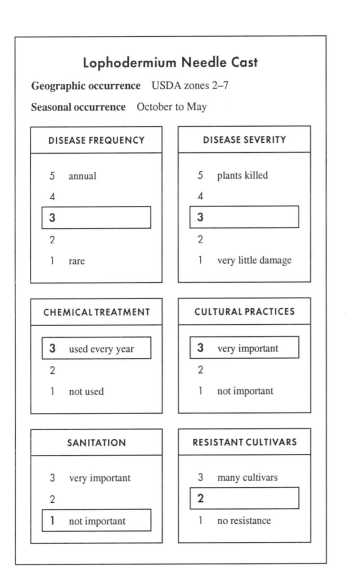

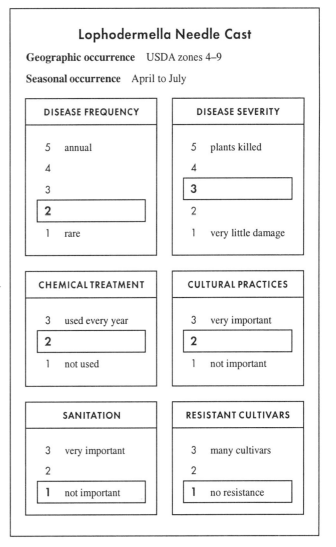

Lophodermium Needle Cast

Geographic occurrence USDA zones 2–7

Seasonal occurrence October to May

DISEASE FREQUENCY	DISEASE SEVERITY
5 annual	5 plants killed
4	4
3	**3**
2	2
1 rare	1 very little damage

CHEMICAL TREATMENT	CULTURAL PRACTICES
3 used every year	**3 very important**
2	2
1 not used	1 not important

SANITATION	RESISTANT CULTIVARS
3 very important	3 many cultivars
2	**2**
1 not important	1 no resistance

Lophodermella Needle Cast

Geographic occurrence USDA zones 4–9

Seasonal occurrence April to July

DISEASE FREQUENCY	DISEASE SEVERITY
5 annual	5 plants killed
4	4
3	**3**
2	2
1 rare	1 very little damage

CHEMICAL TREATMENT	CULTURAL PRACTICES
3 used every year	3 very important
2	**2**
1 not used	1 not important

SANITATION	RESISTANT CULTIVARS
3 very important	3 many cultivars
2	2
1 not important	**1 no resistance**

Symptoms

Second-year needles infected the previous summer typically begin to turn reddish brown in the spring, prior to budbreak. Needles may turn completely brown, or areas of green tissue may remain below the infected area. Gradually, infected tissue changes from reddish brown to straw color. Dark, elongated, elliptical fruiting bodies of the pathogen soon become evident on dead needles during the period of shoot elongation and needle growth (Plate 104). In some species the fruiting bodies are easily seen, although they form below the needle epidermis. In other species, the fruiting bodies are colorless and more difficult to find, especially prior to maturity. Diseased needles are cast following spore release. Badly affected trees may have only a one-year complement of needles at the end of a long bare branch.

Symptoms caused by *Lophodermella* are similar to those of other fungal needle cast diseases (red band and Cyclaneusma, Elytroderma, and Lophodermium needle casts) and resemble damage due to environmental stress, air pollution, and the pine needle sheath miner.

Epidemiology

Young and developing needles are susceptible to infection during periods of wet weather. Spores are released in the spring through a single longitudinal split in the center of a recently matured fruiting body. The spores are moved by splashing water and wind and adhere to young needles with the aid of a sticky sheath. In most cases, infected needles do not show symptoms until the following spring; however, necrotic tissues are sometimes evident by early summer or late fall.

Host Range

L. arcuata infects western white pine in western states.
L. cerina infects loblolly and Scots pine in southeastern states.
L. morbida affects ponderosa pine along the Pacific Coast.

Management

Cultural practices Practice weed control and provide proper tree spacing, to improve air movement and sunlight penetration.

Sanitation In valuable nursery plantings, remove diseased needles from trees and from the ground prior to budbreak in the spring.

Resistance Resistance to Lophodermella needle cast is not known.

Chemical treatment Apply fungicide soon after budbreak and about one month later, when new growth has ceased.

Elytroderma Needle Cast

Elytroderma deformans
(syn. *Hypoderma deformans*)

Elytroderma needle cast of two- and three-needle pines causes reddening and premature dropping of needles and a twig and branch infection that results in a perennial broom. The disease flourishes at low to moderately endemic levels throughout western North America, wherever pines are grown.

Symptoms

Infected needles turn reddish brown in spring and then gradually fade to tan or gray. The entire needle may be affected, but sometimes green areas remain. Long, narrow fruiting bodies of the pathogen appear in early spring as faint dark lines on the outer surface of the needle aligned with the needle axis. Twig infection results in brown, resinous lesions of the bark, phloem, and xylem tissues. The lesions become systemic and perennial, stimulating lateral buds and resulting in a witches'-broom. Many of the needles produced on broomed branches have the same red coloration as infected needles.

The symptoms can be confused with those of other diseases (red band and Cyclaneusma, Lophodermium, and Lophodermella needle casts) and with damage due to environmental stress and chemical injury. Dwarf mistletoe brooms could be confused with the brooms associated with Elytroderma needle cast, and some symptoms also could be confused with damage due to the pine needle sheath miner.

Epidemiology

Ascospores released from fruiting bodies of the pathogen, *Elytroderma deformans,* during periods of moist weather are readily dispersed by air currents. They need free moisture to germinate. It has been thought that infection occurs in the fall, during periods of wet weather and maximum spore release. However, it is possible that only young needles (less than one month old) are susceptible; in this case, infection would occur only in the spring,

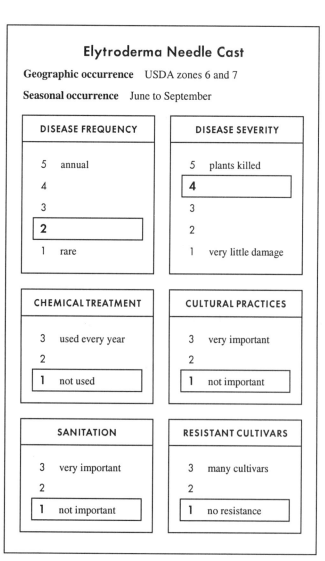

Elytroderma Needle Cast

Geographic occurrence USDA zones 6 and 7

Seasonal occurrence June to September

DISEASE FREQUENCY	DISEASE SEVERITY
5 annual	5 plants killed
4	**[4]**
3	3
[2]	2
1 rare	1 very little damage

CHEMICAL TREATMENT	CULTURAL PRACTICES
3 used every year	3 very important
2	2
[1] not used	**[1] not important**

SANITATION	RESISTANT CULTIVARS
3 very important	3 many cultivars
2	2
[1] not important	**[1] no resistance**

and the fungus would survive the winter as mycelium in infected twigs or in symptomless needles infected during the previous growing season.

Host Range

At least 10 different species of pine are reportedly susceptible to Elytroderma needle cast, including ponderosa and pinyon pines.

Management

Cultural practices Little information on the control of Elytroderma needle cast in nurseries is available.

Sanitation. Pruning out and destroying infected twigs and brooms may help.

Resistance Pine hosts resistant to *E. deformans* have not been identified.

Chemical treatment Fungicides have not been shown to be effective against this disease.

Botrytis Blight

Botrytis cinerea

Botrytis blight of pines, also called gray mold, is most commonly encountered as a damping-off or seedling blight. Under moist environmental conditions, the pathogen, *Botrytis cinerea,* occasionally incites a shoot blight of larger nursery trees. Seedlings held in cold storage can also be damaged by infection by this fungus.

Symptoms

Initial infection appears as water-soaked, dark, discolored spots on newly expanded needles and shoots. As the lesions enlarge, the diseased tissue turns tan to brown. If lesions enlarge enough to girdle expanding shoots, the shoots wither and die. Under humid conditions, a sparse web of gray brown mycelium develops on diseased plant tissues, and clusters of gray spores form on short stalks arising from the plant surface. When jostled, clouds of powdery spores can be seen floating from the diseased shoots and needles.

In nurseries, seedling blight could be confused with root rot or chemical injury.

Epidemiology

Release of spores may be triggered by rapid changes in humidity. Spores are spread by wind and by water splashing onto susceptible host tissue. During periods of wet, cool weather, they germinate and directly penetrate plant tissues. The fungus may continue to build up or maintain its population by colonizing dead and dying plant tissues. Plants weakened by frost or freeze injury are more susceptible to infection, even though the affected tissues may not be killed. Extended cool weather during shoot elongation prolongs the period when susceptible immature tissues are present and leads to greater disease development when sufficient moisture is present. High humidity, dense plantings, and abundant juvenile tissue in seedling nurseries create situations highly conducive to infection by *B. cinerea.*

Host Range

Botrytis blight occurs in hundreds of woody and herbaceous plants, in addition to pines.

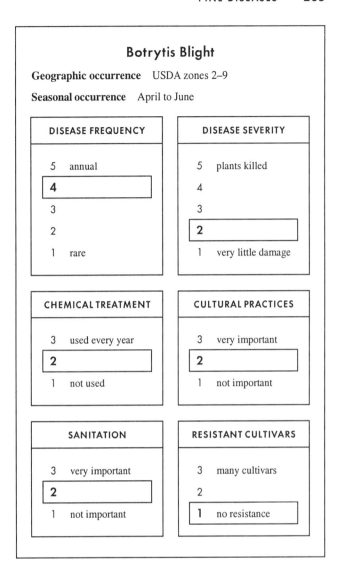

Botrytis Blight

Geographic occurrence USDA zones 2–9

Seasonal occurrence April to June

DISEASE FREQUENCY	
5	annual
4	
3	
2	
1	rare

DISEASE SEVERITY	
5	plants killed
4	
3	
2	
1	very little damage

CHEMICAL TREATMENT	
3	used every year
2	
1	not used

CULTURAL PRACTICES	
3	very important
2	
1	not important

SANITATION	
3	very important
2	
1	not important

RESISTANT CULTIVARS	
3	many cultivars
2	
1	**no resistance**

Management

Cultural practices Keep humidity at a low level by irrigation and ventilation management and by avoiding spacing seedlings tightly. Adjust temperature and fertility to avoid stressed or overly succulent plants. In field plantings improve air circulation to help reduce Botrytis blight.

Sanitation Remove and destroy crop residues in nurseries.

Resistance Resistant cultivars are not available.

Chemical treatment In the nursery, routinely apply protective fungicides when the disease is expected to occur. Fungicide treatment is generally not needed in field plantings. Manage fungicide applications to minimize the development of resistance to certain compounds in the pathogen.

Pine Needle Rust

Coleosporium asterum and other fungi

Needle rust is most common in young trees, up to sapling size. The disease slows tree growth and causes unsightly foliage. Together with insect pests and other agents that attack current-year foliage, needle rust may seriously damage or kill seedlings. There are many species of pine needle rust fungi, and each has its own host range. The discussion here pertains mostly to *Coleosporium asterum.*

Symptoms

On pine, chlorotic spots with orange pycnial droplets form on needles as the weather warms up in spring. These spots are followed by white, then orange blisters (aecia) erupting from needles in late spring (Plates 105 and 106). The aecia disappear by late summer, leaving yellow to brown spots or bands on the green needles.

In goldenrod, aster, and other alternate hosts, yellow rust pustules appear in early summer on the lower surfaces of leaves, and by late summer the pustules turn rusty brown. Heavily rusted leaves become blighted.

Epidemiology

C. asterum needs both pine and a herbaceous host, such as aster or goldenrod, to complete its one-year life cycle. Spores produced on pine are windborne and, when leaves are wet, infect goldenrod or aster in late spring. Urediniospores produced on may reinfect that host, increasing the level of inoculum. The pathogen forms fruiting bodies (telia) that produce teliospores and basidiospores in late summer, and the basidiospores infect pine needles when they are wet. The fungus overwinters as mycelium in pine needles or in alternate hosts and thus can survive two or more consecutive years of unfavorable environmental conditions.

Pine Needle Rust

Geographic occurrence USDA zones 2–9

Seasonal occurrence April to June

DISEASE FREQUENCY
5 annual
4
3
2
1 rare

DISEASE SEVERITY
5 plants killed
4
3
2
1 very little damage

CHEMICAL TREATMENT
3 used every year
2
1 not used

CULTURAL PRACTICES
3 very important
2
1 not important

SANITATION
3 very important
2
1 not important

RESISTANT CULTIVARS
3 many cultivars
2
1 no resistance

Host Range

Austrian, Japanese black, loblolly, longleaf, mugo, ponderosa, red, Scots, and Virginia pines are susceptible. Alternate hosts include aster, China aster, daisy fleabane, goldenrod, gumweed, and other members of the family Asteraceae.

Management

Cultural practices If practical, remove tall grass, weeds, goldenrod, and aster in and around the planting area. Avoid planting on humid sites north or west of a stand of tall trees, and avoid steep north- or west-facing slopes. If a nursery must be established on such a site, plant conifers that are resistant to pine needle rust. Plant pines next to 16- to 20-foot-tall windbreaks of resistant shrubs and trees that will separate the pines from alternate hosts of the rust fungus. Windbreaks intercept spores and alter air currents, so that fewer spores reach the pine needles.

Sanitation Mow goldenrods and asters before August. These plants are perennial, and mowing will be necessary each year until the trees are old enough for pine needle rust to have little or no impact on tree quality. Or apply a registered herbicide to kill goldenrods and asters within 1,000 feet of newly planted pine seedlings before August.

Resistance Susceptible species are mentioned above.

Chemical treatment Pines can be protected in late summer with fungicides.

Pine-Oak Gall Rusts:

Eastern Gall Rust

Cronartium quercuum* f. sp. *banksianae
Cronartium quercuum* f. sp. *echinatae
Cronartium quercuum* f. sp. *virginianae

Fusiform Rust

Cronartium quercuum* f. sp. *fusiforme

Eastern gall rust and fusiform rust, caused by several subspecies of *Cronartium quercuum,* are common in various pine species in the eastern United States. These diseases seldom kill seedlings, but infected seedlings usually die following field planting and must be discarded. Older plants may be killed or have unsightly galls, dead branches, or poor form.

Symptoms

Both diseases result in the formation of galls on the stems of seedlings. In eastern gall rust, the galls usually form during the year following infection; in fusiform rust, they usually form in late summer or fall on one-year-old seedlings. The two diseases can usually be distinguished by the shape of the galls, which are more or less globose in eastern gall rust (Plate 107) and spindle-shaped (fusiform) in fusiform rust (Plate 108). On older trees large perennial galls of the characteristic shape form on the trunk and branches. In the spring, masses of powdery, yellow orange spores (aeciospores) erupt from the galls.

The alternate hosts of *C. quercuum* are various species of *Quercus,* in which the fungus produces small, yellowish pustules (uredinia) and hairlike telia.

Epidemiology

Aeciospores of *C. quercuum,* produced on two- to four-year-old galls on pine in the spring, infect nearby oaks, in which the pathogen subsequently produces urediniospores, which reinfect oaks. Two to three weeks later, the fungus forms fruiting bodies (telia) producing basidiospores, which infect succulent needle or shoot tissue of pines. Cool, moist weather is most favorable for infection.

Host Range

Austrian, loblolly, mugo, ponderosa, Japanese red, Scots, Virginia, and other species of pines are susceptible to eastern gall rust. Some pathogenic strains of *C. quercuum,* which attack individual pine species or groups of species, have been segregated.

Fusiform rust is most often found in loblolly and longleaf pines, but other species are susceptible.

Alternate hosts of *C. quercuum* include more than 20 species of oaks, especially in the red and black oak groups.

Management

Cultural practices Establish field plantings with disease-free planting stock. In seedlings the disease may not become apparent until one year after infection, especially with eastern gall rust, so that nursery stocks can be infected but not show symptoms at the time of transplanting.

Sanitation Prune out branch galls when they are first observed, especially those near the main trunk.

Resistance Selections of loblolly pine resistant to fusiform rust have been developed for industrial plantations.

Chemical treatment Fungicides applied in spring and early summer are necessary to prevent seedling infection. Treatment of field plantings is not economical.

Pine-Pine Gall Rust: Western Gall Rust

Endocronartium harknessii

Western gall rust is found throughout the western United States, in most of Canada, and in the eastern United States as far south as Virginia. Unlike eastern gall and fusiform rusts, this disease often kills seedlings. Surviving infected seedlings usually die a few years after outplanting. Older plants may have numerous trunk and stem galls, which often kill limbs or tops and result in poor form.

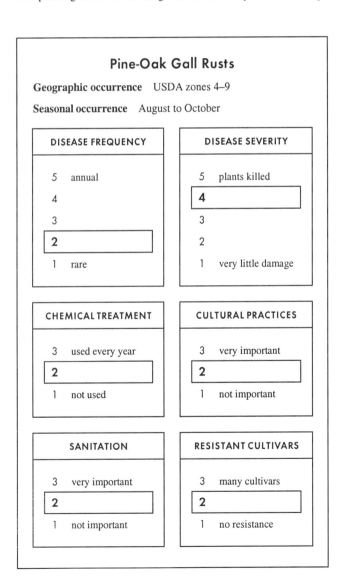

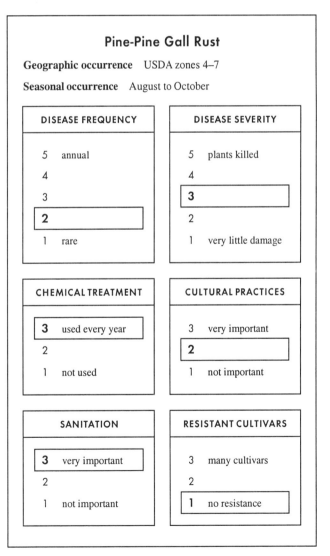

Pine-Oak Gall Rusts

Geographic occurrence USDA zones 4–9

Seasonal occurrence August to October

DISEASE FREQUENCY	DISEASE SEVERITY
5 annual	5 plants killed
4	**4**
3	3
2	2
1 rare	1 very little damage

CHEMICAL TREATMENT	CULTURAL PRACTICES
3 used every year	3 very important
2	**2**
1 not used	1 not important

SANITATION	RESISTANT CULTIVARS
3 very important	3 many cultivars
2	**2**
1 not important	1 no resistance

Pine-Pine Gall Rust

Geographic occurrence USDA zones 4–7

Seasonal occurrence August to October

DISEASE FREQUENCY	DISEASE SEVERITY
5 annual	5 plants killed
4	4
3	**3**
2	2
1 rare	1 very little damage

CHEMICAL TREATMENT	CULTURAL PRACTICES
3 used every year	3 very important
2	**2**
1 not used	1 not important

SANITATION	RESISTANT CULTIVARS
3 very important	3 many cultivars
2	2
1 not important	**1 no resistance**

Symptoms

In young seedlings, symptoms are usually not visible until the year after infection. Symptoms of western gall rust are very similar to those of eastern gall rust, and where the ranges of these two diseases overlap, it is difficult to distinguish them. The most recognizable symptoms are conspicuous spherical perennial stem galls, occurring mostly on branches but occasionally on trunks (Plate 109). New galls that form from August to October occur on succulent new growth and may be spindle-shaped. Branch galls may reach a diameter of 1 to 4 inches before they die; trunk galls can reach 8 to 12 inches in diameter. In the spring the galls are covered with masses of powdery, yellow orange aeciospores. Insect galls and burls are sometimes mistaken for gall rust symptoms.

Epidemiology

Endocronartium harknessii, the western gall rust fungus, does not require an alternate host for the completion of its life cycle, unlike the eastern gall and fusiform rust fungi. Spores of *E. harknessii* produced on pine can reinfect pine. Spores are produced in the spring and early summer (May–July) and are carried by wind to new shoots, where infection occurs. Wet, green plant surfaces are necessary for germination and infection. Infections on trunks (known as hip cankers) may remain active for many years, but most branch galls die within five to 10 years.

Host Range

Two- and three-needle pines are susceptible to western gall rust. Scots and Monterey pines are especially susceptible, but ponderosa, mugo, and Austrian pines are also infected.

Management

Cultural practices Establish field plantings with disease-free planting stock in areas not adjacent to infected pine. Since the disease does not become apparent until one year after infection, nursery stocks can be infected but not show symptoms at the time of transplanting. Disease-free areas for nursery production are essential.

Sanitation Prune out branch galls when they are first observed. Removing galls early, before they produce spores, is effective in intensively managed nurseries.

Resistance Resistant pines have not been identified.

Chemical treatment Systemic fungicides applied in spring have been shown to be effective.

White Pine Blister Rust

Cronartium ribicola

White pine blister rust is a common and destructive disease of five-needled pines in most areas of the United States. Infected pine seedlings seldom die in the nursery but do not survive field planting. In older plants, the disease kills branches or tops above girdling stem cankers and sometimes kills entire trees.

The pathogen, *Cronartium ribicola,* like many rust fungi, requires two unrelated host plants for the completion of its life cycle. Currants, gooseberries, and other *Ribes* spp. serve as the alternate hosts.

Symptoms

In seedlings, the first symptoms are yellowish or reddish spots on needles. Later, spindle-shaped swellings form on the stem and develop into sunken, necrotic cankers. On older plants, cankers are formed on trunks or branches (Plate 110). The necrotic area of the canker is surrounded by a band of yellowish green bark. Girdling cankers kill twigs or branches randomly throughout diseased trees, and girdling trunk cankers result in death of the tree distal to the canker. Copious resin often flows from cankers. In late spring yellow orange pustules (aecia) arise from beneath the living bark. The bark dies after the spores are released from the aecia, leaving remnants of the whitish wall which covered the spore masses.

In summer, diffuse yellow spots develop on the upper surface of infected leaves of *Ribes* spp., followed shortly by orange yellow pustules (uredinia). Brown, hairlike structures (telia) form in the uredinia in late summer and fall.

Epidemiology

Early in the growing season, aeciospores produced on active cankers on pine are spread by air currents to infect leaves of the alternate hosts, *Ribes* spp. Telia and teliospores form on the leaves of the alternate hosts in midsummer to early autumn. The teliospores germinate, producing basidiospores, which are dispersed on air currents to pines. The basidiospores enter pine needles through stomata during cool, moist weather when free moisture is present.

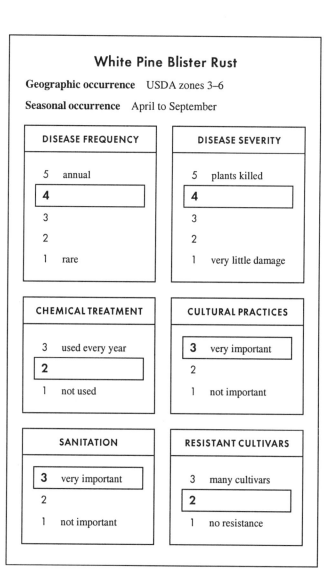

White Pine Blister Rust

Geographic occurrence USDA zones 3–6

Seasonal occurrence April to September

DISEASE FREQUENCY	DISEASE SEVERITY
5 annual	5 plants killed
[4]	**[4]**
3	3
2	2
1 rare	1 very little damage

CHEMICAL TREATMENT	CULTURAL PRACTICES
3 used every year	**[3] very important**
[2]	2
1 not used	1 not important

SANITATION	RESISTANT CULTIVARS
[3] very important	3 many cultivars
2	**[2]**
1 not important	1 no resistance

Host Range

Eastern and western white pines and other five-needle pines are susceptible to white pine blister rust. Alternate hosts are species of *Ribes,* including domestic currants and gooseberries and many native species, such as stink currant, white-stemmed gooseberry, black currant, and Sierra gooseberry.

Management

Cultural practices Prune out infected twigs and branches before the infection reaches the trunk of the tree.

Sanitation Remove all alternate hosts within 1,000 feet of the nursery.

Resistance Purchase rust-resistant stocks for production in high-risk areas.

Chemical practices Fungicide applications in late summer and early fall will protect seedlings from infection during periods of wet weather.

Sphaeropsis Tip Blight (Diplodia Tip Blight)

***Sphaeropsis sapinea* (syn. *Diplodia pinea*)**

Infection by the fungus *Sphaeropsis sapinea* (syn. *Diplodia pinea*) causes the formation of girdling cankers that kill current-year shoots on trees of all ages. Sphaeropsis tip blight, also called Diplodia tip blight, can kill nursery seedlings within the first year. It is especially damaging to trees growing under stress.

Symptoms

In seedlings, infected current-year shoots die and turn brown above girdling cankers. Affected shoots often become bent or curled. Infected tissue is resin-soaked. Black fruiting bodies (pycnidia) appear at the base of dead needles and shoot tissue (Plate 111). In older plants, the most common symptoms are bud and tip blights (Plate 112), but girdling cankers can also develop on branches and stems. The cankers are oblong, sunken, and often resin-soaked (Plate 113). *S. sapinea* produces a blue to black stain in the wood beneath the canker. Second-year cones may also be killed by the fungus, which then fruits heavily on the dead cones (Plate 114). Similar symptoms may be caused by drought and insect damage.

Epidemiology

Sphaeropsis overwinters in pine shoots, needles, bark, cones, or litter. Spores are released during wet weather from spring through fall, but most infection occurs in developing shoots in the spring. Trees that are stressed because of a poor site, drought, snow damage, or insect activity are very susceptible to Sphaeropsis tip blight. Wounds, such as those made by hail, shearing, or insects (for example, feeding injury due to the pine spittlebug), serve as entry points for the fungus.

Host Range

Austrian, ponderosa, and Scots pines are highly susceptible to Sphaeropsis tip blight, but loblolly, mugo, pinyon, eastern white, and other pines are also attacked.

Management

Cultural practices Do not plant trees next to windbreaks that are diseased. Examine windbreaks closely; although shoots may not be infected, cones may harbor the fungus. Avoid planting susceptible species, such as Austrian or red pine, on poor sites where they will be more vulnerable to both insect and fungal attack. Control insects that weaken trees and create entry points for *Sphaeropsis.* Do not shear infected trees during wet weather, because spores released then may be carried from tree to tree on shearing tools.

Sanitation Prune out infected twigs.

Resistance Resistant cultivars are not available.

Chemical treatment Apply fungicide every two weeks during bud swell and shoot elongation.

Sphaeropsis Tip Blight

Geographic occurrence USDA zones 3–9

Seasonal occurrence May to July

DISEASE FREQUENCY	DISEASE SEVERITY
5 annual	5 plants killed
[4]	**[4]**
3	3
2	2
1 rare	1 very little damage

CHEMICAL TREATMENT	CULTURAL PRACTICES
3 used every year	**[3] very important**
[2]	2
1 not used	1 not important

SANITATION	RESISTANT CULTIVARS
3 very important	3 many cultivars
[2]	2
1 not important	**[1] no resistance**

Pitch Canker

Fusarium circinatum

Pitch canker, named for the abundant resin exuded from infected tissues, is an important disease of pines in nurseries and plantations in the Southeast and California.

Symptoms

The most common symptom in seedling nurseries is stem, root collar, or taproot canker resulting in damping-off or sudden wilt and death. Infected xylem tissues are brown, necrotic, and impregnated with resin. In plantations, shoot dieback and limb cankers predominate. The first symptom of limb cankers is a reddish brown flagging; the needles later turn grayish brown, become matted in dried pitch, and often hang from killed shoots or branches for many months (Plate 115). Diseased seedlings in the field often occur in groups, where the pathogen has spread from an infection focus.

Epidemiology

Wounds made by insects, hail, wind, or machines are all suitable sites for infection by *Fusarium circinatum,* the pitch canker fungus. As invaded tissues die, the pathogen forms salmon-colored, cushion-like fruiting bodies (sporodochia), which produce large numbers of spores (conidia). Conidia are dispersed by air, water, and insects to nearby wounded tissue, where new infections can begin.

Host Range

Loblolly, longleaf, pitch, sand, shortleaf, slash, south Florida slash, table-mountain, Virginia, and eastern white pines are susceptible to pitch canker in the field.

Management

Cultural practices Avoid high soil fertility and water stress.

Sanitation Sanitation may be useful in preventing the spread of the disease to adjacent healthy pines.

Resistance Loblolly and eastern white pines are somewhat resistant. Genetically resistant planting stock of other pines may be available.

Chemical treatment Chemicals effective against pitch canker of pine are not available.

Procerum Root Disease (White Pine Rot Decline)

Leptographium procerum
(syn. *Verticicladiella procera*)

Procerum root disease, caused by the fungus *Leptographium procerum,* occurs in several *Pinus* spp. and other coniferous hosts throughout much of the United States, but it is primarily a problem in plantings of young eastern white pine in the Appalachian region. The disease is sometimes called white pine root decline. It is not known to occur in seedlings.

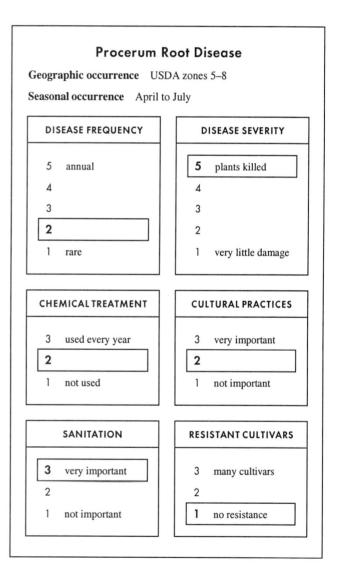

Pitch Canker

Geographic occurrence USDA zones 7–10

Seasonal occurrence October to March

DISEASE FREQUENCY	DISEASE SEVERITY
5 annual	**5 plants killed**
4	4
3	3
2	2
1 rare	1 very little damage

CHEMICAL TREATMENT	CULTURAL PRACTICES
3 used every year	**3 very important**
2	2
1 not used	1 not important

SANITATION	RESISTANT CULTIVARS
3 very important	3 many cultivars
2	**2**
1 not important	1 no resistance

Procerum Root Disease

Geographic occurrence USDA zones 5–8

Seasonal occurrence April to July

DISEASE FREQUENCY	DISEASE SEVERITY
5 annual	**5 plants killed**
4	4
3	3
2	2
1 rare	1 very little damage

CHEMICAL TREATMENT	CULTURAL PRACTICES
3 used every year	3 very important
2	**2**
1 not used	1 not important

SANITATION	RESISTANT CULTIVARS
3 very important	3 many cultivars
2	2
1 not important	**1 no resistance**

Symptoms

In young eastern white pine, the first apparent symptoms are delayed budbreak and a reduction in shoot elongation. These symptoms are followed by wilting and uniform browning of the needles on the entire tree (Plate 116). These symptoms usually occur in early spring but may occur at any time of the year. Dead needles may remain on the tree for a year or more. Resin is usually copiously exuded at the base of the tree, sometimes even before crown symptoms are evident (Plate 117). The wood of the lower stem and roots of infected trees is resin-soaked; dark, wedge-shaped streaks may be seen in a cross sections of the lower stem. Evidence of attack by weevils or bark beetles is usually present at the base of the tree.

Similar symptoms develop on other conifers affected by procerum root disease, except that visible wilting of the needles does not occur. On Scots pine, a basal restriction covered by a black crust may be present.

Epidemiology

L. procerum is spread primarily by weevils which feed on the lower stem and roots of susceptible hosts. Some bark beetles may also serve as vectors. These insects are usually attracted to trees under stress or trees with mechanical wounds near the base. The fungus produces sticky spores (conidia) on conidiophores in insect galleries or in wounds, where adult insects acquire the spores and transmit them to healthy trees.

Host Range

Austrian, red, Scots, Virginia, and eastern white pines are among the pine species most susceptible to procerum root disease.

Management

Cultural practices Avoid planting trees on sites where they may be subject to severe stress. Implement management practices such as proper spacing, fertilization, and weed control to keep plants growing vigorously.

Sanitation Stumps of recently felled pine should be removed or sprayed with insecticide before a field planting is established. Infected trees in planting areas should be cut, and the stumps should be either sprayed with insecticide or removed.

Resistance Resistant cultivars are not available.

Chemical treatment As a protective measure, spray the lower 3 feet of healthy trees and the soil for 1 to 2 feet around the base of the tree, with a registered insecticide before mid-March and again in September.

Phytophthora Root Rot

***Phytophthora* spp.**

See Chapter 14, "Phytophthora Root Rot and Dieback."

Charcoal Root Rot

Macrophomina phaseolina

Charcoal root rot is a serious disease of pine seedlings in nurseries in the South and the West. The causal fungus, *Macrophomina phaseolina,* has a wide host range.

Symptoms

Aboveground symptoms of infected seedlings include damping-off, stunted growth, off-color, wilt, reddish foliage, and seedling death. Infected root systems are blackened and necrotic and lack fine feeder roots. Infected cortical tissues of the root collar, taproot, and larger lateral roots become swollen, blackened, and cracked. Numerous tiny, black microsclerotia of *M. phaseolina* are readily visible in the cortical tissues and xylem of infected roots and root collars.

Epidemiology

M. phaseolina survives in soils for years in the form of highly durable microsclerotia. Stimulated by root exudates, microsclerotia germinate and infect nearby roots. After the fungus colonizes the roots, tissues die, and the fungus produces more microsclerotia. Charcoal root rot is favored by hot, dry soils. The pathogen is more aggressive on pines under heat or drought stress.

Host Range

Many pines are susceptible to charcoal rot.

Management

Cultural practices Provide irrigation to prevent dry soil. Reduce soil temperatures by watering and mulching.

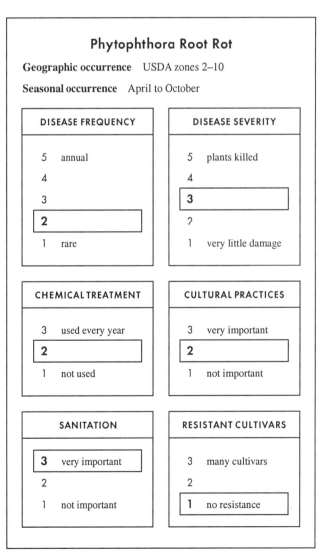

Phytophthora Root Rot

Geographic occurrence USDA zones 2–10

Seasonal occurrence April to October

DISEASE FREQUENCY	DISEASE SEVERITY
5 annual	5 plants killed
4	4
3	**[3]**
[2]	2
1 rare	1 very little damage

CHEMICAL TREATMENT	CULTURAL PRACTICES
3 used every year	3 very important
[2]	**[2]**
1 not used	1 not important

SANITATION	RESISTANT CULTIVARS
[3] very important	3 many cultivars
2	2
1 not important	**[1] no resistance**

Sanitation With its wide host range, *M. phaseolina* may be present in many soils, so sanitation measures may not be helpful.

Resistance Pines resistant to charcoal rot are not available.

Chemical treatment Soil fumigation with a mixture of methyl bromide and chloropicrin is effective against this disease. Soil drenches with certain fungicides have shown some effectiveness.

Cylindrocladium Root Rot and Stem Canker

Cylindrocladium scoparium

See Chapter 11, "Diseases Caused by *Cylindrocladium*."

Armillaria Root Rot (Shoestring Root Rot)

***Armillaria* spp.**

Armillaria root rot, also called shoestring root rot, is not a problem in pine seedlings, but it can cause severe losses in field plantings established on recently cleared forest land where the fungus is present. Death usually occurs when the root collar is girdled by the fungal pathogen. The disease is caused by several *Armillaria* spp.

Symptoms

Generally, young trees suddenly yellow, then turn reddish brown, and die, usually during the summer, when trees are subjected to moisture stress. A copious flow of resin may be present at the base of the tree, and white, fan-shaped mycelial mats are usually present beneath the bark at the root collar. The fungus forms dark brown to black rootlike structures (rhizomorphs, commonly called shoestrings), which can often be observed on or beneath the bark, especially on dead trees. Small clumps of honey-colored mushrooms sometimes form at the base of infected trees in the fall.

Epidemiology

Armillaria spp. can survive for extended periods (decades) in large infected stumps and roots. It spreads to adjacent plants by means of rhizomorphs or root contact. Decaying roots and stumps provide the fungus with the energy needed for it to spread and cause further infections. Spores produced by *Armillaria* mushrooms, when present, seldom cause direct infection of young trees.

Charcoal Root Rot

Geographic occurrence USDA zones 7–10

Seasonal occurrence July to October

DISEASE FREQUENCY	DISEASE SEVERITY
5 annual	5 plants killed
4	4
3	**3**
2	2
1 rare	1 very little damage

CHEMICAL TREATMENT	CULTURAL PRACTICES
3 used every year	3 very important
2	**2**
1 not used	1 not important

SANITATION	RESISTANT CULTIVARS
3 very important	3 many cultivars
2	2
1 not important	**1 no resistance**

Cylindrocladium Root Rot and Stem Canker

Geographic occurrence USDA zones 3–9

Seasonal occurrence May to September

DISEASE FREQUENCY	DISEASE SEVERITY
5 annual	5 plants killed
4	4
3	**3**
2	2
1 rare	1 very little damage

CHEMICAL TREATMENT	CULTURAL PRACTICES
3 used every year	3 very important
2	**2**
1 not used	1 not important

SANITATION	RESISTANT CULTIVARS
3 very important	3 many cultivars
2	2
1 not important	**1 no resistance**

Host Range

Armillaria spp. have wide host ranges among woody plants, including most species of pine.

Management

Cultural practices Do not establish field plantings on recently cleared forest land unless stumps and roots are removed.

Sanitation Remove and destroy root systems of trees in field plantings killed by *Armillaria.*

Resistance Pines resistant to Armillaria root rot are not available.

Chemical treatment Chemical controls have not proved effective.

Pinewood Nematode (Pine Wilt Nematode)

Bursaphelenchus xylophilus

When present in large numbers, the pinewood nematode (*Bursaphelenchus xylophilus*) can kill the pines it infests. This nematode, also called the pine wilt nematode, is more damaging during periods of drought or other stresses and is often found in trees that are dying from other causes. Disease due to the pinewood nematode is endemic throughout the eastern United States.

Symptoms

Pinewood nematode feeding causes yellowing and then browning of needles, affecting all needles on a branch, on one side of the tree, or throughout the tree, during the growing season. Brown needles remain on dead trees. There is little resin flow from wounds. Root rot diseases, wood borers, and bark beetles cause similar symptoms. Not all infected trees die, but those growing under stress are more likely to die.

Epidemiology

Pinewood nematodes are spread from infected dead pines in the spring by long-horned beetles in the family Cerambycidae (Plate 118). When infested beetles feed on shoots of healthy trees, they deposit nematodes in the feeding wounds. The nematodes develop systemically in the trees, feeding in the resin canals. They reproduce rapidly in the wood of infested trees during the summer, often killing the trees by fall. The insect vector uses dead and dying nematode-infested trees as breeding sites, and adult beetles are infested with nematodes when they emerge in spring.

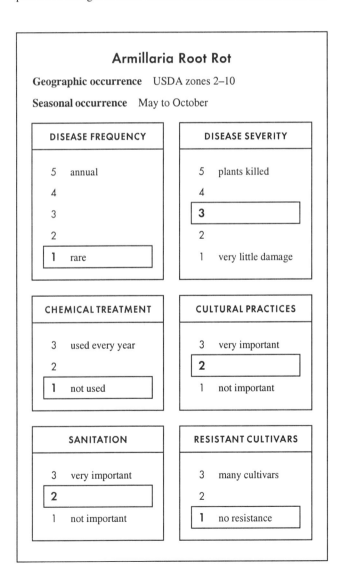

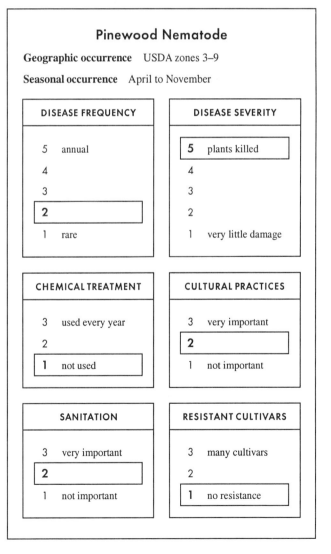

Armillaria Root Rot

Geographic occurrence USDA zones 2–10

Seasonal occurrence May to October

DISEASE FREQUENCY	DISEASE SEVERITY
5 annual	5 plants killed
4	4
3	**3**
2	2
1 rare	1 very little damage

CHEMICAL TREATMENT	CULTURAL PRACTICES
3 used every year	3 very important
2	**2**
1 not used	1 not important

SANITATION	RESISTANT CULTIVARS
3 very important	3 many cultivars
2	2
1 not important	**1 no resistance**

Pinewood Nematode

Geographic occurrence USDA zones 3–9

Seasonal occurrence April to November

DISEASE FREQUENCY	DISEASE SEVERITY
5 annual	**5 plants killed**
4	4
3	3
2	2
1 rare	1 very little damage

CHEMICAL TREATMENT	CULTURAL PRACTICES
3 used every year	3 very important
2	**2**
1 not used	1 not important

SANITATION	RESISTANT CULTIVARS
3 very important	3 many cultivars
2	2
1 not important	**1 no resistance**

Host Range

Many pines are susceptible, especially Scots and Japanese black pines.

Management

Cultural practices Avoid planting on dry sites.

Sanitation Destroy infested trees by burning or chipping before beetles emerge from them in the spring.

Resistance Resistant cultivars of species susceptible to the pinewood nematode are not available.

Chemical treatment Chemical control of the pinewood nematode is not practical.

Weather-Related Injuries

All pines are sensitive to some extreme weather, and weather-related injuries may cause problems for pines in nurseries. Some extreme weather events may occur simultaneously, such as drought, heat, and sunscald, and so injury symptoms may overlap.

Drought

All trees are subject to drought stress, but newly planted trees are the most sensitive. Wilting of new growth, top dieback, loss of interior needles, shortened needles, needle tip dieback, and death of trees are all possible symptoms of drought stress. Sometimes symptoms do not occur until the year after the drought. Older needles commonly turn yellow and are shed prematurely. Drought symptoms progress from the top of the tree down, and the damage may occur over several years. Root weevils, root rot diseases, and disturbance of roots by mammals can cause similar symptoms. Trees weakened by drought may be predisposed to other problems, such as insect feeding and canker-causing fungi.

Most nursery trees are grown with irrigation, so drought should be rare. Nevertheless, with each species having different moisture requirements and nurseries having different soil types, water management is important. Often growers do not identify drought as the primary problem because secondary problems caused by insects or diseases associated with drought damage are easier to see.

Sunscald

Sunscald results from high temperatures that kill the cambium in patches or sometimes along the length of the stem on the southern or southwestern side of young, thin-barked trees. Affected bark turns copper-colored to brown during the summer, eventually sloughing off. Branches above the damaged area may die. Winter desiccation damage also typically appears on the southern to southwestern side of the tree, on tissue above snow cover. Although the damage occurs during the winter, symptoms may not be evident until spring. Needles turn reddish brown, and the death of stem cambium near the top of the tree results in the death of shoots, needles, or buds.

Sunscald of the lower stem may result from basal pruning of young trees, particularly if they are grown in light-colored, reflective soils. Thinly branched trees on the southern edges of plantations or on south- or west-facing slopes may also be damaged. Once damaged, trees need many years for wounds to close.

Heat Injury

Nursery trees can be damaged by excessive heat. High temperatures during budbreak can damage emerging shoots, and groups of needles on a shoot will turn reddish brown. Needles emerging after an episode of heat usually appear normal. High temperatures later in the season can damage needles and shoots, causing the entire southern or southwestern side of the tree to appear reddened. Damage due to heat and drought can be confused with current-season needle necrosis, exhaust damage, and chemical injury.

Pines are also susceptible to groundline heat lesions, which develop on the stems of young seedlings in hot weather. Lesions first appear as superficial, white spots or streaks on the sides of stems facing the sunlight. In advanced stages, they appear as water-soaked areas, often associated with stem shrinkage, wrinkling, and constrictions of the hypocotyl at or near the groundline. Young seedlings may collapse. Older ones may gradually die from the girdling effect. This problem may be misdiagnosed as damping-off. For groundline heat lesions to occur on pine seedlings, soil temperatures must reach or exceed 52°C for perhaps 30 minutes. Trees in dark soils are more at risk than other trees.

Flooding

When trees are flooded, they may show drought symptoms, because the roots are unable to take up water for lack of oxygen. Flooding can also affect needle retention. Besides the obvious

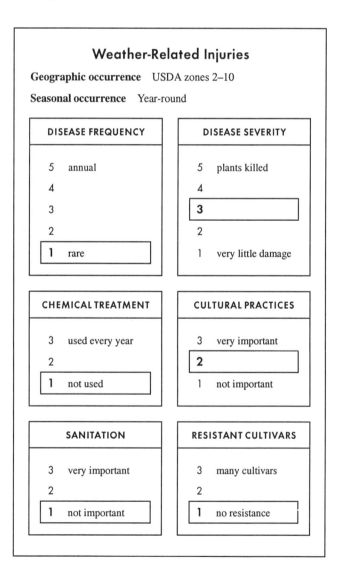

Weather-Related Injuries

Geographic occurrence USDA zones 2–10

Seasonal occurrence Year-round

DISEASE FREQUENCY	DISEASE SEVERITY
5 annual	5 plants killed
4	4
3	**[3]**
2	2
[1 rare]	1 very little damage

CHEMICAL TREATMENT	CULTURAL PRACTICES
3 used every year	3 very important
2	**[2]**
[1 not used]	1 not important

SANITATION	RESISTANT CULTIVARS
3 very important	3 many cultivars
2	2
[1 not important]	**[1 no resistance]**

damage, flooding can create conditions that favor Phytophthora root and crown rot, especially in newly planted trees.

Hail

Hailstones generally affect only succulent springtime shoots in the upper part of the tree. Small, depressed areas of injury appear in the bark on the upper side of twigs and branches. Hail increases occurrence and severity of Sphaeropsis tip blight in pine plantations.

Frost Injury

Frost damage occurs when temperatures drop well below freezing in the spring after new growth has appeared. The damage to foliage is usually evident within a few days. On very succulent growth, shoots may appear wilted, and needles may vary from a pale, water-soaked color to brown or red, depending on the degree of damage and the species involved. If budbreak is uniform in a nursery block, nearly all trees will be affected. When budbreak occurs at different times in different trees, some trees will show damage, while others escape injury.

Frost injury can be confused with Botrytis shoot blight, which tends to appear randomly and can be identified by the gray brown sporulation of the pathogen on dead tissue. Frost damage causes a darkened frost ring, observable under microscopic examination of cross sections of damaged shoots.

Winter Injury

Winter injury occurs as a direct result of exposure to abnormal cold, although the injury is uncommon in pines. Trees that are not yet fully winter-hardened are at risk in early winter, as are trees that break dormancy prematurely due to warm weather in late winter. Symptoms of winter injury are a sudden reddening and death of needles and the death of stems and buds in the entire upper portion of the tree. Sometimes only random branches are affected.

Winter desiccation occurs when trees continue to lose moisture through their needles but cannot supply new moisture to the tissues, because they cannot obtain water from frozen soil or are unable to move water through frozen stem tissues.

Management

Cultural practices With proper irrigation, drought should not be a problem in the nursery. New plantings should be established early enough in the season that the seedlings can develop root systems capable of surviving a summer or fall drought. In older plantings, little can be done other than to optimize soil moisture levels by conserving existing water, eliminating competing weeds and brush, planting drought-tolerant species, or providing supplemental water.

Groundline heat lesions can be prevented with adequate shade for young seedlings. As trees get older, branches generally provide sufficient shade to protect the bark from sunscald. Fall or winter basal pruning may help minimize the problem.

Prevent frost damage by avoiding planting sites in low areas or frost pockets where cold air collects. To prevent winter injury, avoid planting cold-sensitive species in open, windswept areas.

Heat damage and sunscald can be reduced on small seedlings with shade devices, such as shingles and cards.

Valuable nursery specimens can be protected from hail by netting.

Resistance Scots pine is fairly drought-tolerant.

Seed sources of trees that break bud early tend to be injured by frost more than those that break bud later. Frost and cold are likely to injure trees from exotic seed sources than locally adapted native species.

Soil-Related Abiotic Problems

All pines are subject to soil-related disorders. In the field, iron deficiency due to high soil pH and nitrogen deficiency are the most commonly observed mineral imbalances. In containers with artificial media, such imbalances are more likely to occur, but little is known of the symptoms that may result. Trees planted at poor sites or in poor growing media are more likely to show nutrient deficiency symptoms.

Mineral Nutrient Imbalances

The following symptoms may develop in pines as a result of mineral element deficiencies:

Nitrogen (N) deficiency Light green foliage; small, yellow needles; restricted terminal growth; browning and premature drop of older needles

Soil-Related Abiotic Problems

Geographic occurrence USDA zones 2 to 10

Seasonal occurrence Year-round

DISEASE FREQUENCY	DISEASE SEVERITY
5 annual	5 plants killed
4	4
3	3
[2]	**[2]**
1 rare	1 very little damage

CHEMICAL TREATMENT	CULTURAL PRACTICES
3 used every year	3 very important
[2]	**[2]**
1 not used	1 not important

SANITATION	RESISTANT CULTIVARS
3 very important	3 many cultivars
2	2
[1] not important	**[1] no resistance**

Phosphorus (P) deficiency Purple discoloration and death of older needles, under conditions of severe deficiency

Potassium (K) deficiency Progressive browning of needle tips, starting with the older foliage

Calcium (Ca) deficiency Death of terminal buds and some lateral buds

Magnesium (Mg) deficiency Browning of the tips of older needles, with a yellow margin adjacent to the green basal portion of the needle, the browning progressing until the entire needle is brown

Sulfur (S) deficiency Slight chlorosis of new foliage, while older needles remain green

Boron (B) deficiency Death of terminal buds; dark green foliage

Iron (Fe) deficiency Production of bright yellow new foliage, while older foliage remains green

Fertilizer Burn

Mineral element excesses resulting from misapplication of fertilizers can damage or kill trees. Newly developing needles may appear reddish brown or necrotic throughout the tree. Newly planted seedlings and small trees are especially sensitive to fertilizer burn. Too much fertilizer can damage or kill roots beneath the area where it is applied. In severe cases, where excess fertilizer is used on one side of the tree, needle damage will appear on branches progressing up the tree in a spiral pattern.

Soil Compaction

Soil compaction may be a problem in heavy-textured field soils, especially where heavy equipment is used. Poor cultivation practices may also lead to soil compaction. Compaction restricts air movement into the soil and reduces the amount of soil that a tree root system can utilize effectively. Thus the tree is likely to grow poorly and have poor color and may be predisposed to damage by insect pests.

Management

Cultural practices To reduce soil compaction, subsoiling when soils are relatively dry prior to planting can break apart hardpans or compacted areas. Using machinery equipped with flotation tires, minimizing travel through fields while soils are wet, breaking soil crusts with cultivation, and using living mulches are also helpful in reducing soil compaction.

Chemical treatment Applications of nitrogen, iron, or sulfur can improve tree color, quality, and growth rates. The greatest benefits have been observed at sites with shallow, infertile soils. Iron deficiency can also be overcome with sufficient amounts of sulfur to lower the soil pH.

Chemical Injury

All pines are susceptible to chemical injury. Air pollutants and chemical herbicides may reduce growth, cause early needle loss, increase vulnerability to insects and diseases, and even kill pines. The extent of damage caused by air pollutants and misapplied pesticides depends on the genetics of the tree, physiological state of the tree, needle age, chemical dose or pollutant concentration, time of exposure, and weather. New needles are most susceptible to air pollutants when elongating during early summer.

Air Pollutant Injury

Air pollutants such as ozone, fluorides, and sulfur dioxide may cause yellowing and premature needle shedding; yellow, red, or brown needle tips; or yellow stippling or banding of needles.

Herbicide Injury

Chemical damage from misapplication of pesticides generally has a definite pattern in a nursery block; i.e., it follows the pattern of the application equipment. In the case of drift from an application nearby, only the edge of the field may sustain damage. Where overlapping spray swaths lead to an overdose of a chemical, the symptoms are confined to one spray swath. Trees in skips in the spray pattern may avoid damage.

If misapplied chemicals are suspected, examine the spray application records to identify what was applied and when. Symptoms appear quickly during active growth, but at other times of the year they may take weeks or months to appear. Observation of the symptoms and their distribution, the presence or lack of weeds, and the pattern of damage may help in diagnosing damage from pesticide applications.

It is possible to "burn" new foliage with insecticides and fungicides, but most chemical injuries to pines in nurseries are due to

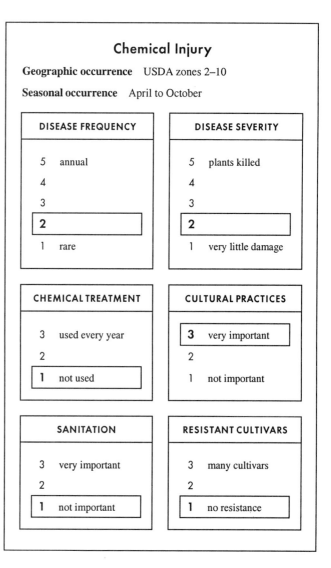

Chemical Injury

Geographic occurrence USDA zones 2–10

Seasonal occurrence April to October

DISEASE FREQUENCY	DISEASE SEVERITY
5 annual	5 plants killed
4	4
3	3
[2]	**[2]**
1 rare	1 very little damage

CHEMICAL TREATMENT	CULTURAL PRACTICES
3 used every year	**[3 very important]**
2	2
[1 not used]	1 not important

SANITATION	RESISTANT CULTIVARS
3 very important	3 many cultivars
2	2
[1 not important]	**[1 no resistance]**

herbicides. Injuries associated with some common herbicides are described below.

Atrazine Damage from atrazine is usually confined to chlorosis of the needle tips. Newly planted container seedlings may suffer more severe damage than established trees.

2,4-D Applied before budbreak, 2,4-D can stunt new growth, and during the growing season it will cause swelling of shoot tips and twisting and curling of new growth, resembling a growth hormone response.

Dichlobenil Needle tips yellow and may eventually turn reddish brown.

Glyphosate Applied prior to budbreak, glyphosate causes stunting of new growth and the production of short needles and pale green foliage. Applied after budbreak, it rapidly kills new needles and burns or desiccates foliage. Sprays targeted at weeds around the base of the tree can kill foliage near the ground.

Hexazinone Symptoms of hexazinone injury are similar to those of atrazine injury but are usually more severe. Affected needles are light yellow to pinkish and eventually fade to brown. If the tree does not die, subsequent needles are dark green and abnormally long. At extremely high rates of application, new foliage will desiccate quickly. Hexazinone is highly soluble in water and may accumulate and kill trees in low-lying areas of a field.

Triclopyr Applied prior to budbreak, triclopyr stunts new growth. Applied after budbreak, it rapidly kills succulent foliage, which may appear black.

Management

Cultural practices Most instances of chemical damage can be avoided by properly calibrating application equipment, avoiding spraying on windy days, and following the directions on the product label to determine the rate and timing of application and the species of trees to be sprayed.

Sanitation Prune off dead shoots and remove dead trees, so that insects and pathogens cannot build up on them.

Resistance Resistant cultivars are not available.

Chemical Application of chemicals to control chemical or air pollution damage is not practical.

Fall Needle Drop

Fall yellowing of inner foliage is a natural occurrence due to physiological causes and does not harm nursery pines. All pines are affected, but the disorder is especially noticeable in eastern white pine.

Yellowing and browning of the oldest foliage appears anywhere on the tree in September or October, and needles drop off in late fall. The symptoms may be similar to those of several needle cast diseases.

All pines shed their oldest needles each year. A healthy tree should have at least two years' growth of needles after the oldest needles drop off, except white pine, which may retain needles from the previous season only.

Disease management is not necessary.

Fall Needle Drop

Geographic occurrence USDA zones 2–10

Seasonal occurrence August to December

DISEASE FREQUENCY	DISEASE SEVERITY
5 annual	5 plants killed
4	4
3	3
2	2
1 rare	**1 very little damage**

CHEMICAL TREATMENT	CULTURAL PRACTICES
3 used every year	3 very important
2	2
1 not used	**1 not important**

SANITATION	RESISTANT CULTIVARS
3 very important	3 many cultivars
2	2
1 not important	**1 no resistance**

REFERENCES

Beynus, J. M., ed. 1983. Christmas tree pest manual. U.S. Dep. Agric. For. Serv., North Cent. For. Exp. Stn. (St. Paul, Minn.).

Chastagner, G. A., ed. 1997. Christmas Tree Diseases, Insects, and Disorders in the Pacific Northwest: Identification and Management. Washington State University Cooperative Extension.

Cordell, C. E., Anderson, R. L., Hoffard, W. H., Landis, T. D., Smith, R. S., and Toko, H. V., tech. coords. 1989. Forest nursery pests. U.S. Dep. Agric. For. Serv. Agric. Handb. 680.

Hansen, E. M, and Lewis, K. J., eds. 1997. Compendium of Conifer Diseases. American Phytopathological Society, St. Paul, Minn.

Sinclair, W. A., Lyon, H. H., and Johnson, W. T. 1987. Diseases of Trees and Shrubs. Comstock Publishing Associates, Cornell University Press, Ithaca, N.Y.

CHAPTER 69

Gary W. Simone ▪ University of Florida, Gainesville

Pittosporum Diseases

Geographic production USDA zones 8–11

Family Pittosporaceae

Genus *Pittosporum*

Species *P. crassifolium*
P. eugenioides
P. tenuifolium
P. tobira Japanese pittosporum

Pittosporum is a dicot genus of over 100 species of evergreen trees and shrubs native to tropical and subtropical regions of South Africa, China, Japan, Australia, New Zealand, Malaysia, and Hawaii. It is grown in the southern and Pacific Coast states as a specimen tree or shrub and as a hedge plant, and it is cultivated for cut foliage production.

Alternaria Leaf Spot

Alternaria tenuissima

Symptoms

Early symptoms of Alternaria leaf spot are obvious diffuse chlorotic areas, 1/32 to 1/8 inch (0.5–2 mm) in diameter, on the upper surface of leaves. The spots expand to become slightly depressed, circular lesions, ranging from 1/8 to 1/2 inch (3–12 mm) in diameter, with a tan center, dark margins, and a narrow chlorotic halo (Plate 119).

Epidemiology

Spores of the fungal pathogen, *Alternaria tenuissima,* are spread by air currents and splashing water. The fungus persists in leaf lesions on or below the plant canopy. Disease severity is often higher in partial shade, at high humidity, and under slow drying conditions.

Host Range

All cultivated *Pittosporum* spp. are susceptible to Alternaria leaf spot.

Management

Use an air blower to remove fallen infected leaves from blocks of plants, to reduce the amount of inoculum of the fungus prior to the spring flush period. Plants for nursery production should be grown under full sun, where possible. Fungicide applications should be timed for growth flush periods. Both protectant and systemic fungicides are available for use on *Pittosporum* spp. and are effective with repeated use.

Angular Leaf Spot

Cercospora pittospori

Symptoms

The initial symptoms of angular leaf spot are angular, chlorotic, yellow brown leaf spots on the upper surface of leaves. The spots expand and coalesce to form lesions 1/64 to 3/16 inch (1–5

Alternaria Leaf Spot

Geographic occurrence USDA zones 8–11

Seasonal occurrence March to November

DISEASE FREQUENCY	DISEASE SEVERITY
[5] annual	5 plants killed
4	4
3	**[3]**
2	2
1 rare	1 very little damage

CHEMICAL TREATMENT	CULTURAL PRACTICES
3 used every year	3 very important
[2]	**[2]**
1 not used	1 not important

SANITATION	RESISTANT CULTIVARS
[3] very important	3 many cultivars
2	2
1 not important	**[1] no resistance**

mm) in diameter or larger (Plate 120). Mature lesions are olive brown on the lower surface of the leaf, as a result of the production of spores by the fungal pathogen, *Cercospora pittospori*. Mature lesions seldom turn necrotic.

Epidemiology

C. pittospori sporulates primarily through stoma on the lower leaf surface. Hyphae of the fungus colonize the external leaf surface as well as the internal palisade and spongy mesophyll cell layers. The fungus survives as vegetative hyphae and spores (conidia) on retained infected leaves and fallen infected tissue. Spores are airborne and are carried by splashing water to spread within and among shrub canopies.

Host Range

All cultivated species and varieties of *Pittosporum* are susceptible to angular leaf spot. *C. pittospori* also invades *Ligustrum* spp.

Management

Use an air blower to remove fallen infected leaves from container and ground cloth surfaces. Time irrigations for periods after dew formation through late morning. Space plants out, to reduce leaf wetness periods. Both protectant and systemic fungicides will provide a degree of management if applied thoroughly to both leaf surfaces. Propagation should be conducted from production stock free of foliar disease.

Kutilakesa Gall and Dieback

Nectriella pironii
(asexual stage, *Kutilakesa pironii*)

Symptoms

Symptoms of Kutilakesa gall and dieback are round to oblong swellings with a rough, fissured surface on twigs and stems (Plate 121). The galls develop slowly, over a period of six to nine months, reaching sizes ranging from 3/16 inch to more than 1 inch (4 mm to more than 2.5 cm) in diameter. They often occur in series along stems and twigs. Affected twigs and stems slowly decline, resulting in a thinning of the distal canopy and a reduction in growth flushes. Galls can girdle entire twigs or stems, killing the distal portions of the plant.

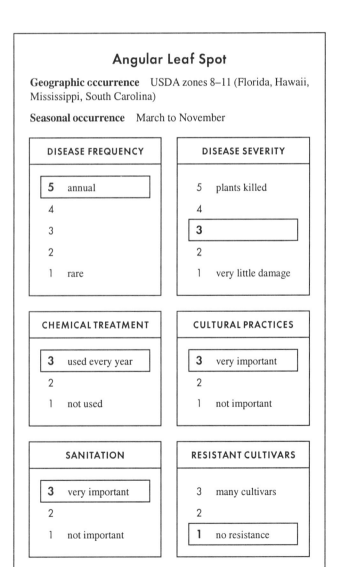

Angular Leaf Spot

Geographic occurrence USDA zones 8–11 (Florida, Hawaii, Mississippi, South Carolina)

Seasonal occurrence March to November

DISEASE FREQUENCY	DISEASE SEVERITY
5 annual	5 plants killed
4	4
3	**3**
2	2
1 rare	1 very little damage

CHEMICAL TREATMENT	CULTURAL PRACTICES
3 used every year	**3 very important**
2	2
1 not used	1 not important

SANITATION	RESISTANT CULTIVARS
3 very important	3 many cultivars
2	2
1 not important	**1 no resistance**

Kutilakesa Gall and Dieback

Geographic occurrence USDA zones 8–11 (Florida)

Seasonal occurrence March–November in zone 8; year-round in zones 9–11

DISEASE FREQUENCY	DISEASE SEVERITY
5 annual	**5 plants killed**
4	4
3	3
2	2
1 rare	1 very little damage

CHEMICAL TREATMENT	CULTURAL PRACTICES
3 used every year	3 very important
2	2
1 not used	**1 not important**

SANITATION	RESISTANT CULTIVARS
3 very important	3 many cultivars
2	2
1 not important	**1 no resistance**

Epidemiology

Both the asexual stage of the pathogen, *Kutilakesa pironii,* and the sexual stage, *Nectriella pironii,* are found on *Pittosporum.* The asexual stage is more common and more important in disease spread. Asexual spores (conidia produced on sporodochia) arise from superficial hyphae along gall fissures and appear as creamy yellow to orange spore masses. In the sexual stage, light orange fruiting bodies (perithecia) form in the bark of the galls. Spores of the fungus in both stages are spread by water splashing in the canopy and running down stems and twigs. This passive movement of spores results in the development of vertical series of galls on stems.

Host Range

All cultivated *Pittosporum* spp. are susceptible to *K. pironii.* Other reported woody hosts of the pathogen include *Bauhinia, Callistemon, Cercis, Clerodendrum, Codiaeum variegatum, Dizygotheca,* ×*Fatshedera lizei, Ficus, Gelsemium, Hedera, Hibiscus, Hydrangea, Jasminum, Lantana, Leucophyllum, Ligustrum, Lonicera, Mahonia, Nandina, Osmanthus, Parkinsonia, Polyscias, Psychotria, Rosa, Salix,* and *Ulmus* spp. The host range includes many herbaceous plants as well.

Management

Production stock should be carefully inspected for twig swellings or galls prior to propagation cycles. Infected stock should be dumped. Segregate propagation cycles on ground beds, and always isolate stock brought in from other nurseries, because of the slow development of the disease. Pruning is not effective in nursery stock, because numerous infections will be developing down the stem below existing galls. Cuttings should be taken when the canopy is thoroughly dry, to avoid spreading the pathogen during handling or on tools. Fungicides are not recommended.

Pink Limb Blight

Erythricium salmonicolor
(syn. *Corticium salmonicolor*)

Symptoms

The initial symptom of pink limb blight is tip dieback, as the fungal pathogen, *Erythricium salmonicolor,* girdles the water-conductive tissues in twigs and stems. The point of infection appears pinkish orange, and the fungus can be observed as a thin mantle of hyphae along declining shoots. Below the bark, the sapwood is discolored and brown. Leaves on affected plant parts turn brown and desiccated and eventually drop from the plant (Plate 122).

Epidemiology

Plants destined for vegetative propagation should be carefully inspected to prevent the movement of infected plant material into the next propagation cycle. The pinkish orange color of *E. salmonicolor* on the twig surface fades quickly, and young infections are often overlooked. Stock bought in from other nurseries should be isolated and monitored prior to use in propagation or distribution among other stock.

Host Range

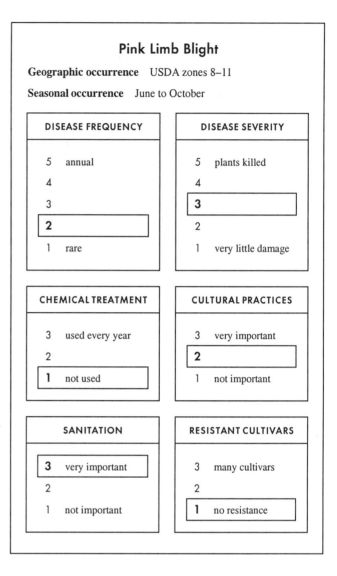

The host range of *E. salmonicolor* includes all species of *Pittosporum* and a wide range of common nursery crops, including *Buxus, Cercis, Euonymus, Ficus, Gardenia, Ilex, Illicium, Jasminum, Ligustrum, Liquidambar, Malus, Nerium, Photinia, Podocarpus, Pyrus, Quercus,* and *Ulmus* spp.

Management

Production stock should be scouted for pink limb blight. When signs and symptoms are observed, affected stock should be destroyed. Ground cloths should be raked clean, and the rakings should be destroyed, to prevent the overseasoning of the pathogen on nursery beds. Neither fungicides nor varietal selection offer useful solutions to nurseries at this time.

Pink Limb Blight

Geographic occurrence USDA zones 8–11

Seasonal occurrence June to October

DISEASE FREQUENCY	DISEASE SEVERITY
5 annual	5 plants killed
4	4
3	**[3]**
[2]	2
1 rare	1 very little damage

CHEMICAL TREATMENT	CULTURAL PRACTICES
3 used every year	3 very important
2	**[2]**
[1] not used	1 not important

SANITATION	RESISTANT CULTIVARS
[3] very important	3 many cultivars
2	2
1 not important	**[1] no resistance**

Root Knot

***Meloidogyne* spp., root-knot nematodes**

Symptoms

The secondary symptoms of root-knot nematode damage are usually noticed first, initially appearing as uneven plant growth in field-grown blocks of plants of the same age. After infection is established, the lower inner canopy declines, revealing more twig growth. Leaf size may be reduced, and leaves may appear off-color. Affected plants have shorter growth increments than healthy plants.

These secondary symptoms follow the development of root galls, the primary symptom. Long sections of thick roots may be variably swollen and distorted with galls, ranging from 1/16 to more than 3/4 inch (2 mm to more than 2 cm) in diameter. The number of feeder roots is reduced. Sections of affected root systems often become necrotic as a result of secondary invasion by soilborne pathogenic fungi following nematode invasion.

Epidemiology

Root-knot nematodes (*Meloidogyne* spp.) survive in soil as eggs and larvae. Initial infection occurs at root tips or sites of emergence of lateral roots. The nematode larvae migrate within roots to the conductive tissue cylinder (stele), where the females encyst and initiate specialized feeding sites, known as giant cells. The growth and enlargement of these feeding sites results in the formation of root galls. Galls interfere with the uptake of nutrients and water by the plant and the movement of water within the plant.

Root Knot

Geographic occurrence USDA zones 8–11

Seasonal occurrence March to November

DISEASE FREQUENCY	DISEASE SEVERITY
5 annual	5 plants killed
4	4
3	**[3]**
[2]	2
1 rare	1 very little damage

CHEMICAL TREATMENT	CULTURAL PRACTICES
3 used every year	**[3] very important**
[2]	2
1 not used	1 not important

SANITATION	RESISTANT CULTIVARS
[3] very important	3 many cultivars
2	2
1 not important	**[1] no resistance**

It is unusual to encounter root-knot nematode infection in container-grown plants unless infested field soil has been introduced into the production cycle.

Host Range

All cultivars of *P. tobira* are susceptible to root-knot nematodes, and other species are suspected to be vulnerable as well. *Meloidogyne* spp. infect a huge variety of ornamental plants.

Management

Field production sites should be assessed for root-knot nematode populations before plants are set. Select production sites with low levels of root-knot nematodes, rather than old agronomic or vegetable crop acreage, where nematode populations can be expected to be high. Where land choice or rotation is not an option, in-row fumigation may be needed. Granular contact nematicides will aid in the suppression of low populations of root-knot nematodes during the production cycle. Affected plants can be used as propagation stock but will be of low quality for marketing purposes.

Root Rots

***Phytophthora* spp.**
***Pythium* spp.**
Rhizoctonia solani

Symptoms

Root rots caused by *Phytophthora* spp., *Pythium* spp., and *Rhizoctonia solani* affect *Pittosporum* both in propagation and at production sites. Infected cuttings develop an off-color cast, resulting in lower leaf chlorosis and death. A soft decay develops at the base of the cutting, with dark discoloration at the end. Roots of infected liners may exhibit a dry brown to reddish brown root decay or a soft, mushy, grayish black decay, depending upon the pathogen.

Container-grown and field-grown plants exhibit primary symptoms in the root zone. Infection occurs mainly in the feeder root zone. Pruning of feeder roots is often a symptom of Rhizoctonia root rot, while a grayish black, mushy decay of feeder roots is often a symptom of Phytophthora root rot and Pythium root rot.

After a degree of root death has occurred, the lower innermost leaves exhibit marginal yellowing or necrosis progressing to leaf death. Leaves die from the lower canopy upward and from the inner canopy outward. During extremely wet weather, which favors disease development, invaded plants may wilt. Other symptoms of root rot include reduced leaf size, loss of color, and shorter growth flushes.

Epidemiology

The incidence of root diseases can be correlated with the use of contaminated media, infested containers, or (in the case of the water mold fungi, *Phytophthora* and *Pythium* spp.) contaminated water. Liners can be infected by being placed on nonsterile ground, by rapid changes in the groundwater level, or by surface water runoff. Production stresses, such as excessive fertilization, high media temperature, inadequate pore space in media, and excessive irrigation, can predispose root systems to invasion by soil pathogens. Once infection takes place, reproductive structures of the fungi can be transported by splashing water, drainage from potting media, or the movement of infested soil.

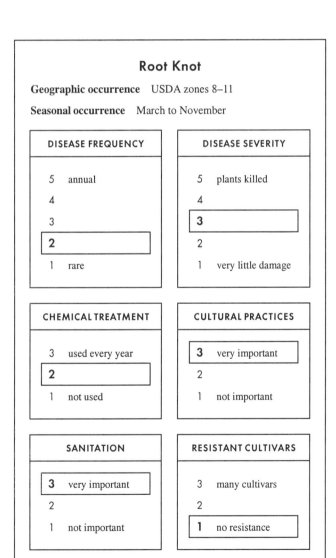

Host Range

No resistance to the soilborne fungi causing root rots is known to exist.

Different species of *Phytophthora* and *Pythium* have different host ranges among other woody plants.

Rhizoctonia solani has one of the largest host ranges of all plant pathogens, and thus it has great importance as a root rot pathogen of woody ornamentals. This fungus also causes of web blight of numerous plants, including *Pittosporum* spp. and many other woody ornamentals.

Management

Management of root rot depends on identification of the pathogen or pathogens causing the disease. Review plant production with regard to pathogen introduction, spread, and persistence. Manipulate water, media, and container use to prevent pathogen introduction. Avoid contamination of rooted liners with nonsterile soil or contaminated water. Carefully regulate irrigation and fertilization to avoid stress on plants in production. Various fungicides are available for root rot management, once the identity of the pathogen has been determined. In most cases, damage to the root system has already been sustained by the time foliar damage is evident.

Rough Bark Disease

Etiology unknown

Symptoms

Rough bark disease was initially believed to be caused by a virus, but this hypothesis remains to be proved. Early symptoms are a slight stem swelling and the concurrent development of rough-textured, scaly bark. Affected bark will loosen and drop from the plant in small pieces (Plate 123). Symptoms can appear on stems and twigs. Affected plants exhibit reduced vigor in the form or shorter growth flushes, general plant stunting, progressive defoliation of older parts of the canopy, and often dieback of affected parts due to stem or twig girdling. Chlorotic leaf spots with indefinite margins (much like leaf spots caused by *Cercospora*) and oak leaf and ringspot-like symptoms have been observed along the Pacific Coast but not in Florida.

Epidemiology

Until the causal agent is identified, rough bark disease remains undefined. Early work proved that the causal agent is graft-transmissible. The agent does not appear to be systemic within the host, as some cuttings taken from infected plants fail to develop

Root Rots

Geographic occurrence USDA zones 8–11

Seasonal occurrence March to November

DISEASE FREQUENCY
5 annual
4
3
2
1 rare

DISEASE SEVERITY
5 plants killed
4
3
2
1 very little damage

CHEMICAL TREATMENT
3 used every year
2
1 not used

CULTURAL PRACTICES
3 very important
2
1 not important

SANITATION
3 very important
2
1 not important

RESISTANT CULTIVARS
3 many cultivars
2
1 no resistance

Rough Bark Disease

Geographic occurrence USDA zones 8–11

Seasonal occurrence Year-round

DISEASE FREQUENCY
5 annual
4
3
2
1 rare

DISEASE SEVERITY
5 plants killed
4
3
2
1 very little damage

CHEMICAL TREATMENT
3 used every year
2
1 not used

CULTURAL PRACTICES
3 very important
2
1 not important

SANITATION
3 very important
2
1 not important

RESISTANT CULTIVARS
3 many cultivars
2
1 no resistance

symptoms. The pathogen does not appear to be seed-transmitted. Vegetative propagation seems to be the likely mechanism by which it is spread. Disease development takes as long as three to four months.

Host Range

Rough bark disease has been reported in *P. tobira, P. crassifolium,* and *P. viridiflorum.* The susceptibility of other species of *Pittosporum* is unknown.

Management

Inspect propagation stock for symptoms before taking cuttings. Rogue out any symptomatic plants and destroy them. Until the causal agent is identified, no specific management options are available.

Southern Blight

Sclerotium rolfsii

Symptom

The development of southern blight often mimics that of other root diseases. As affected plants decline, two signs of the fungal pathogen, *Sclerotium rolfsii,* become evident: coarse, white mycelium and masses of white to brown, seedlike sclerotia (the size of mustard seed). Mycelia and sclerotia form on the surface of the medium or soil at the stem base and may proceed some distance aboveground on stems, twigs, and leaves (Plates 124 and 125). Roots often exhibit significant decay.

Epidemiology

S. rolfsii survives as strands of hyphae and sclerotia on the lower stem and roots and in media and soil. Sclerotia persist for up to five to eight years in some soil types. Hot, wet weather favors infection by *S. rolfsii* and development of the fungus in root and stem tissue. The pathogen spreads from plant to plant as hyphae and sclerotia transported by splashing water or in infested soil, and it can physically spread through low, interlaced canopy between adjacent plants in production blocks.

Host Range

All species of *Pittosporum* are susceptible to *S. rolfsii,* which has a wide host range (see Chapter 18, "Southern Blight").

Management

Southern blight management revolves around a rigid program of prevention. Never take cuttings from low on the plant canopy.

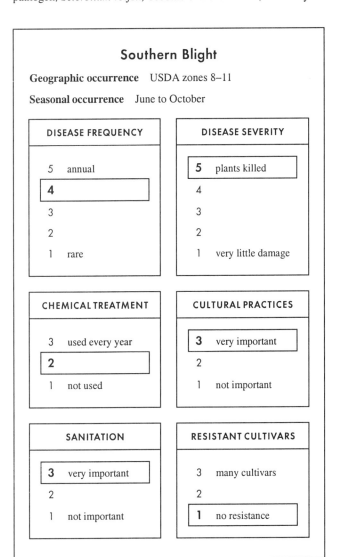

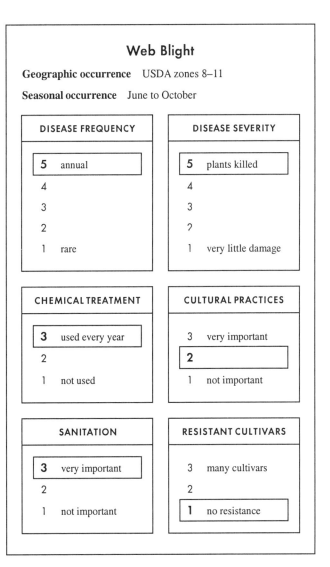

Southern Blight

Geographic occurrence USDA zones 8–11

Seasonal occurrence June to October

DISEASE FREQUENCY	DISEASE SEVERITY
5 annual	**5 plants killed**
4	4
3	3
2	2
1 rare	1 very little damage

CHEMICAL TREATMENT	CULTURAL PRACTICES
3 used every year	**3 very important**
2	2
1 not used	1 not important

SANITATION	RESISTANT CULTIVARS
3 very important	3 many cultivars
2	2
1 not important	**1 no resistance**

Web Blight

Geographic occurrence USDA zones 8–11

Seasonal occurrence June to October

DISEASE FREQUENCY	DISEASE SEVERITY
5 annual	**5 plants killed**
4	4
3	3
2	2
1 rare	1 very little damage

CHEMICAL TREATMENT	CULTURAL PRACTICES
3 used every year	3 very important
2	**2**
1 not used	1 not important

SANITATION	RESISTANT CULTIVARS
3 very important	3 many cultivars
2	2
1 not important	**1 no resistance**

Sterile trays and media should be used in the propagation cycle, and rooting trays should be raised to avoid touching nonsterile soil. Liners placed outside in the nursery should be placed on lengths of polyvinyl chloride (PVC) pipe to raise the cells above potentially contaminated soil or ground cloth. Infected plants should be rapidly removed from production beds and discarded. Drenches of protectant fungicides have some utility for the surrounding plants. Curative fungicides are not available.

Web Blight

Rhizoctonia solani

Symptoms

The initial symptoms of web blight commonly occur in the lower canopy near the soil line. Leaves exhibit irregularly shaped brown spots, which expand rapidly, causing rapid leaf death (Plate 126). Affected foliage is held onto the plant by the brown, weblike hyphae of the fungal pathogen, *Rhizoctonia solani,* giving the disease its name. Leaf defoliation and occasionally plant death can occur within several weeks. Many affected plants exhibit significant necrosis of the lower inner leaves but do not die.

Epidemiology

Web blight occurs in warm, moist summer periods. *R. solani* is spread to the lower canopy in water splashing from the surface of the medium or soil, and it spreads from plant to plant primarily in splashing water. The pathogen, in the form of hyphae or sclerotia, can persist as a saprophyte in media. It can also survive in contaminated media in used containers.

Host Range

R. solani infects hundreds of species of plants. All cultivated *Pittosporum* spp. appear to be susceptible to it.

Management

Production stock destined for propagation should be placed on a foliar fungicide program at least four weeks before cuttings are taken. All cuttings should be taken from elevated portions of the plant rather than close to the media surface. Pathogen-free media and sterile trays should be used in propagation, and plants should be propagated above nonsterile soil or ground cloth surfaces. Space container stock adequately to promote foliar drying after irrigation, to discourage physical movement of the pathogen across interlaced canopies. Fungicides are often required to halt foliar disease and prevent plant death. Fungicides should be applied thoroughly to both leaf surfaces and directed at the surface of the medium, where fallen infected leaves accumulate. Where disease is severe, bed and container surfaces should be blown or raked free of infected leaves before a fungicide is applied or new plants are reset.

REFERENCES

Alfieri, S. A., Jr. 1979. *Nectriella pironii* and its *Kutilakesa*-like anamorph, a parasite of ornamental shrubs. Mycologia 71:1178–1185.

Alfieri, S. A., Jr., and Seymour, C. P. 1970. Rough-bark of *Pittosporum tobira* Ait. in Florida. Proc. Fla. State Hortic. Soc. 83:438–440.

Chase, A. R. 1989. Alternaria leaf spot control in *Pittosporum.* Cut Foliage Grower 4(7/8):1–4.

Plakidas, A. G. 1940. Angular leaf spot of *Pittosporum.* Mycologia 32: 601–608.

Sinclair, W. A., Lyon, H. H., and Johnson, W. T. 1987. Diseases of Trees and Shrubs. Comstock Publishing Associates, Cornell University Press, Ithaca, N.Y.

Sobers, E. K. 1964. Alternaria leaf spot of *Pittosporum.* Phytopathology 54:478–480.

Thomas, H. E., and Baker, K. F. 1947. A rough-bark disease of *Pittosporum tobira.* Phytopathology 37:192–194.

CHAPTER 70

Gary W. Simone ▪ University of Florida, Gainesville

Podocarpus Diseases

Geographic production USDA zones 8–11

Family Podocarpaceae

Genus *Podocarpus*

Species

P. alpinus	
P. macrophyllus	southern yew, Japanese yew
P. nagi	nagi podocarpus
P. nivalis	
P. totara	

The genus *Podocarpus* consists of more than 75 species of primarily dioecious evergreen trees and some shrubs, originating in tropical and subtropical mountain habitat in the West Indies, South America, Asia, Australia, New Zealand, and Africa. Only a few species are commercially grown in the United States; *P. macrophyllus* (southern yew, or Japanese yew) and *P. nagi* (nagi podocarpus) are the most widely grown, from Florida to southern California. Conspicuous on the roots of species in this genera are abundant root nodules produced as a result of symbiotic association with bacteria that aid in nitrogen fixation by the plant. These nodules should not be mistaken for disease symptoms. Disease is not a significant factor in nursery production of *Podocarpus*.

Dieback

***Botryodiplodia* sp.**
***Botryosphaeria* sp.**
***Phoma* sp.**
***Phomopsis* sp.**

Symptoms

Liners infected by a dieback fungus may exhibit a progressive wilt to death as the pathogen girdles the stem, causing a reddish brown necrosis of parts of the plant distal to the infection (Plate 127). On larger plants, terminals and lateral shoots die after a growth flush. Dieback may continue downward into old wood and can kill to the soil line (Plate 128). In young plants, severe dieback can be confused with aboveground symptoms of root rot. Asexual reproduction of the pathogen can be observed at the point of invasion, where small, round, brown to black fruiting structures (pycnidia), visible with a hand lens, are formed on the browned epidermis.

Cold damage to unhardened new growth, mechanical damage from shearing activities, and sunscald can predispose *Podocarpus* spp. to dieback.

Epidemiology

The dieback fungi normally survive as fruiting bodies (pycnidia) in bark or as hyphae in invaded sapwood. Periods of frequent rainfall trigger the release of asexual spores (conidia) from pycnidia, and the spores are disseminated in splashing water.

Dieback

Geographic occurrence USDA zones 8 and 9

Seasonal occurrence March to October

DISEASE FREQUENCY	DISEASE SEVERITY
5 annual	5 plants killed
4	4
[3]	**[3]**
2	2
1 rare	1 very little damage

CHEMICAL TREATMENT	CULTURAL PRACTICES
3 used every year	**[3]** very important
2	2
[1] not used	1 not important

SANITATION	RESISTANT CULTIVARS
3 very important	3 many cultivars
[2]	2
1 not important	**[1]** no resistance

Host Range

Dieback of *Podocarpus* has been reported mostly in Japanese yew, the most common species in cultivation. It is expected that most other *Podocarpus* spp. would be susceptible to dieback pathogens, if the plants were under stress.

A *Botryodiplodia* sp., a *Botryosphaeria* sp., a *Phoma* sp., and a *Phomopsis* sp. have been reported to cause dieback of *Podocarpus.* Until the species causing the disease are identified, the host ranges of the pathogens are unknown.

Management

Closely examine stock prior to propagation cycles. Do not use diseased plants as a cutting source, because of the likelihood of latent infections or surface contamination of shoots. Severely affected plants should be physically separated from healthy plants and pruned down into the green wood. Prunings should be collected and destroyed. The common practice of planting several liners per container will usually mask the loss of one to dieback, but pruning will still be needed to sustain plant quality. Avoid late-season fertilization in your zone, which may result in freeze damage to unhardened growth flushes.

Dieback of *Podocarpus* is not managed with fungicide treatments or plant resistance.

Fusarium Wilt

Fusarium oxysporum

Symptoms

In young plants, infection by the fungus *Fusarium oxysporum* causes the lower leaves to develop a grayish green cast and then become necrotic. Plants in containers holding several liners and in field plantings may exhibit a one-sided dieback, in which only one liner in a clump is killed or a larger specimen exhibits symptoms on only one side. Vascular discoloration can be observed below the bark at or above the soil line for several inches. Symptoms are more pronounced in summer, when high temperatures and high water demand by the plant cause rapid foliar desiccation leading to plant death. Root systems are discolored and deteriorated.

Epidemiology

F. oxysporum is a soilborne pathogen, which survives as mycelia, conidia, and chlamydospores in infected soil and root debris. It invades feeder roots, most likely at points of emergence of lateral roots. This fungus is more likely to be present at damaging levels in field production than in container production. However, reuse of infested media or containers can introduce the pathogen into a new container production cycle.

Host Range

The biotype of *F. oxysporum* in *Podocarpus* is undetermined. The host range is assumed to be narrow, as is characteristic of formae speciales of *F. oxysporum.* No information on the host range of the pathogen is available.

Management

At the first diagnosis of wilt in a field nursery, rotate future plantings of *Podocarpus* into areas not infested with the pathogen. Do not harvest seed or cuttings from diseased trees, as the pathogen produces a microconidial stage capable of widespread internal distribution in the plant through the xylem tissue. Do not place rooted liners on nonsterile soil that may contain the pathogen. Avoid recycling media and containers in areas where wilt is known, unless they are sterilized by heat or chemical fumigation. Resistance to *F. oxysporum* is unknown in the genus *Podocarpus.* Fungicides are not effective for the management of vascular wilt diseases.

Root Rots

***Phytophthora* sp.**
Pythium irregulare

Symptoms

Root rot of *Podocarpus* is favored by warm to hot periods of frequent or excessive rainfall resulting in saturated soil moisture. The initial symptom is feeder root decay, as the pathogen invades and destroys the outer cylinder (cortex) of the root. The lower leaves become discolored and then necrotic, as the affected plant dies from the base upward. As the root system is destroyed, the lower crown tissues can be invaded, and a brownish black discoloration develops in the sapwood. One plant in a cluster may be killed at first. Dead plants retain their leaves.

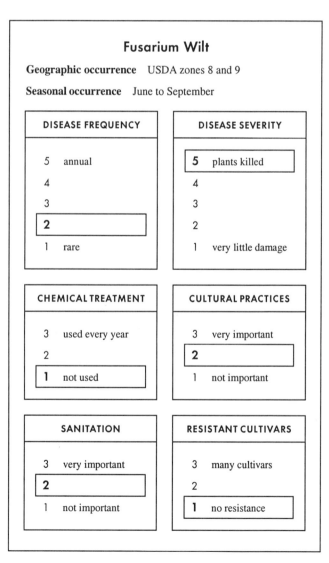

Fusarium Wilt

Geographic occurrence USDA zones 8 and 9

Seasonal occurrence June to September

DISEASE FREQUENCY	DISEASE SEVERITY
5 annual	**5 plants killed**
4	4
3	3
2	2
1 rare	1 very little damage

CHEMICAL TREATMENT	CULTURAL PRACTICES
3 used every year	3 very important
2	**2**
1 not used	1 not important

SANITATION	RESISTANT CULTIVARS
3 very important	3 many cultivars
2	2
1 not important	**1 no resistance**

Root Rots

Geographic occurrence USDA zone 8 and 9

Seasonal occurrence June to September

DISEASE FREQUENCY	DISEASE SEVERITY
5 annual	**5 plants killed**
4	4
3	3
2	2
1 rare	1 very little damage

CHEMICAL TREATMENT	CULTURAL PRACTICES
3 used every year	3 very important
2	**2**
1 not used	1 not important

SANITATION	RESISTANT CULTIVARS
3 very important	3 many cultivars
2	2
1 not important	**1 no resistance**

Epidcmiology

Pythium irregulare and a *Phytophthora* sp. have been reported to cause root rot diseases of *Podocarpus.* Frequent saturation of the soil favors infection of roots by zoospores of these fungi. Considerable deterioration of roots precedes symptoms of foliar decline. *P. irregulare* survives as mycelia, oogonia, and sporangia in soil and root tissue. The *Phytophthora* sp. causing root rot of *Podocarpus* is unidentified, but it probably survives in the form of mycelia and sporangia in soil or root debris.

Host Range

A *Phytophthora* sp. causing root rot has been reported in Japanese yew and other species.

Root rot of *Podocarpus* caused by *P. irregulare* has been reported only in Japanese yew. This pathogen has a diverse host range, including agronomic crops, grasses, floral crops, vegetables, and such woody hosts as *Buxus, Cornus, Dizygotheca, Juniper, Ligustrum, Pinus, Pittosporum,* and *Rhododendron.*

Management

Avoid exposure of rooted liners to surface runoff water. Keep liners raised above infested native soils or set them on ground cloth or a similar weed barrier. Monitor automated irrigation systems to avoid overlap with rainfall. Do not place container stock on unpitched beds or beds where puddles form. In field nurseries, plant *Podocarpus* in well-drained areas.

Resistance to root rot pathogens is unknown among cultivars and species of *Podocarpus.*

Preventative protectant or systemic fungicides can be used at the time of potting. Because of the slow development of foliar symptoms, in contrast to the more rapid root deterioration, curative fungicide treatments are not highly successful.

REFERENCES

Alfieri, S. A., Jr., Langdon, K. R., Klmbrough, J. W., El-Gholl, N. E., and Wehlburg, C. 1994. Diseases and disorders of plants in Florida. Fla. Dep. Agric. Consumer Serv. Bull. 14.

Farr, D. F., Bills, G. F., Chamuris, G. P., and Rossman, A. Y. 1989. Fungi on Plants and Plant Products in the United States. American Phytopathological Society, St. Paul, Minn.

Sinclair, W. A., Lyon, H. H., and Johnson, W. T. 1987. Diseases of Trees and Shrubs. Comstock Publishing Associates, Cornell University Press, Ithaca, N.Y.

CHAPTER

71

Gary A. Chastagner ▪ Washington State University, Puyallup

Poplar Diseases

Geographic production USDA zones 1–7

Family Salicaceae

Genus *Populus*

Section *Populus*

Species	*P. alba*	white poplar	zone 4
	P. canescens	gray poplar	zone 4
	P. grandidentata	bigtooth aspen	zone 4
	P. tremula	European aspen	zone 2
	P. tremuloides	trembling aspen	zone 1

Section *Tacamahaca*

Species	*P. angustifolia*	narrow-leaf cottonwood	zone 3
	P. balsamifera	balsam poplar	zone 2
	P. maximowiczii	Japanese poplar	zone 3
	P. simonii	Simon poplar	zone 2
	P. trichocarpa	black cottonwood	zone 5

Section *Aigeiros*

Species	*P. deltoides*	eastern cottonwood	zone 2
	P. fremontii	Fremont cottonwood	zone 7
	P. nigra	black poplar	zone 2

Poplars are grown for reforestation and conservation, as windbreaks, and to a lesser extent as landscape plants. Most poplars are propagated from hardwood cuttings, although some species, such as trembling aspen, are commonly propagated from seed. Taxonomists generally place species of poplars, cottonwood, and aspens in different sections (Eckenwalder, 1996; Dirr, 1998); most of the commonly grown species and cultivars are in *Populus* section *Populus* (aspens and white poplar), section *Tacamahaca* (balsam poplars), and section *Aigeiros* (cottonwoods and black poplar). Different taxonomists have recognized various numbers of species of *Populus,* ranging from 22 to 85; Eckenwalder (1996) recognized 29 species in six sections. In addition, many hybrids of poplars are grown and sold as named cultivars. One of the more common cultivars is Lombardy poplar, a hybrid of eastern cottonwood and black poplar. Descriptions of commonly grown species and cultivars of *Populus* are presented by Dirr (1998) and Bailey and Bailey (1976).

Poplars and aspens are subject to many diseases. Some of the more common ones are discussed below. An excellent source of additional information has been published by Ostry et al. (1989).

Leaf Rust

***Melampsora* spp.**

Leaf rusts are some of the most common and important diseases of poplar and occur wherever poplars are grown (Ostry et al., 1989; Shain and Filer, 1989; Ziller, 1974). In susceptible clones, rust causes significant reductions in growth, predisposes trees to winter injury, and has been associated with mortality of young plants. Of the fungi causing leaf rust of *Populus* in North America, the most important are *Melampsora albertensis, M. larici-populina, M. medusae* f. sp. *deltoidae, M. medusae* f. sp. *tremuloidae,* and *M. occidentalis* (Newcombe and Chastagner, 1993a,b; Shain and Filer, 1989). Some of these pathogens occur over a large geographic area, while others have a more limited distribution.

Symptoms

From late spring to midsummer, leaf rust fungi form small, bright yellow to orange pustules (uredinia) on the top or bottom surface or both surfaces of poplar leaves (Plate 129). An angular yellow area may form on the side of the leaf opposite a pustule. High infection levels may cause defoliation, lead to premature fall budbreak, predispose trees to other diseases and cold damage, and have been linked to mortality in young trees in stool beds (Newcombe et al., 1994). In autumn, black fruiting bodies (telia) develop in infected tissues.

Epidemiology

The *Melampsora* species that cause poplar leaf rust require two hosts for the completion of their life cycles. They alternate between poplar, on which the uredinial and telial states develop, and a coniferous host, on which the pycnial and aecial states develop.

These fungi overwinter as telia on dead poplar leaves. In the spring, the telia produce basidiospores, which infect current-season needles of nearby coniferous hosts (Douglas fir, larch, pine, and others). Pycnidia and yellow aecia form in approximately two weeks on infected needles, on which a chlorotic discoloration usually develops near the developing fruiting structures. Within a few weeks, the needle symptoms intensify, and affected needles turn necrotic and shrivel. Yellow orange aeciospores are dispersed by wind and infect nearby poplar leaves. After a couple of weeks, uredinia appear on the leaves in association with yellow leaf spots.

Urediniospore production is favored by humid or moist conditions. Urediniospores reinfect poplar throughout the summer and fall. Severely infected leaves can be covered with uredinia, and as a result they shrivel and fall prematurely. In late summer, brown, crustlike telia begin to form in place of the uredinia. Dormant telia are black and overwinter in fallen poplar leaves. They produce basidiospores in the spring, thus completing the life cycle of the fungus.

It is suspected that in areas with mild winters, some of these leaf rust fungi can survive in the uredinial host as uredinia or mycelium and thus are capable of perennial development in the absence of a coniferous host.

Host Range

Pinon (1992) and Newcombe (1996) reviewed the host specificity of *Melampsora* species in *Populus. Melampsora* species are thought to be specific to *Populus* sections *Aigeiros* and *Tacamahaca* or to *Populus* section *Populus.* Knowledge of host specificity is complicated by controversy over the taxonomy of the pathogen (Newcombe, 1996). Host resistance to leaf rust is under strong genetic control, and there are races of various leaf rust pathogens that vary in virulence patterns on poplar clones. Identification of various leaf rust pathogens and races is dependent on the characteristics of the uredinia and, in some cases, the telia; knowledge of the aecial host; virulence patterns on differential poplar clones; and molecular tests (Newcombe, 1996).

The *Melampsora* species reported in North America and their most common poplar hosts are *M. abietis-canadensis* (white and black poplars, trembling and bigtooth aspens, and eastern cottonwood), *M. albertensis* (trembling aspen), *M. larici-populina* (black cottonwood and its hybrids, black poplar, Lombardy poplar, and Japanese poplar), *M. medusae* f. sp. *deltoidae* (eastern cottonwood and its hybrids), *M. medusae* f. sp. *tremuloidae* (trembling aspen), *M. occidentalis* (black cottonwood and its hybrids), and *M. populnea* (white poplar).

Management

The use of resistant clones is the most important tool for the management of leaf rust of *Populus.* Susceptible poplars should not be planted in close proximity to alternate hosts. Plantings should be monitored to determine if there are shifts in the virulence of the rust population. Fungicides can be used to control disease development on susceptible clones in nurseries.

Leaf Rust

Geographic occurrence Wherever poplars are cultivated and in naturally occurring stands of poplar throughout its range

Seasonal occurrence Early summer through fall, with optimum symptom development in August–September

DISEASE FREQUENCY
5 annual
4
3
2
1 rare

DISEASE SEVERITY
5 plants killed
4
3
2
1 very little damage

CHEMICAL TREATMENT
3 used every year
2
1 not used

CULTURAL PRACTICES
3 very important
2
1 not important

SANITATION
3 very important
2
1 not important

RESISTANT CULTIVARS
3 many cultivars
2
1 no resistance

Venturia Leaf and Shoot Blight

Venturia spp. (asexual stages, _Pollaccia_ spp.)

Under moist conditions, Venturia leaf and shoot blight can be very damaging, especially to young trees, in the nursery and the landscape (Ostry et al., 1989; Anderson and Anderson, 1980; Dance, 1961). Infection causes premature defoliation and dieback of young shoots. The disease has caused extensive damage to older, established Lombardy poplars in areas of western Washington that have extended periods of precipitation in the spring.

Symptoms

Irregularly shaped, brown to black areas form on leaves in early spring (Plate 130). Necrotic areas expand rapidly down the midvein and petioles, killing whole leaves and shoots. Shoot dieback is characterized by black, brittle shoots that are curled to resemble a shepherd's crook (Plate 131). Shoot blight can distort the growth of small trees, resulting in poor tree form. Recently infected leaves commonly have an olive green appearance from sporulation of the pathogen on the leaf surface.

Epidemiology

Venturia species survive over the winter on blighted shoots in the tree. In the spring, primary infection is caused by windblown ascospores, which are released from pseudothecia on blighted shoots. Asexual spores (conidia) are then produced on infected leaf tissues, and they spread the disease during the growing season. Infections by *V. populina* can occur on both terminal and lateral shoots throughout the crown of the tree. In aspen, infections by *V. tremulae* tend to be limited to terminal shoots on small trees.

Host Range

The *Venturia* species reported on *Populus* are *V. populina* (asexual stage, *Pollaccia elegans*), *V. tremulae* var. *grandidentata* (asexual stage, *P. radiosa* var. *lethifera*), *V. tremulae* var. *populi-albae* (asexual stage, *P. tremulae* var. *populi-albae*), and *V. tremulae* var. *tremulae* (asexual stage, *P. radiosa* var. *radiosa*).

V. populina occurs primarily on species of cottonwood, black and balsam poplars, and their intersectional hybrids (Dance, 1961).

There are considerable differences of opinion regarding the taxonomic position of *V. tremulae* (syn. *V. macularis*) (Callan and Ring, 1994), and this species was recently separated into three varieties (Newcombe, 1996). *V. tremulae* var. *grandidentata* is the only variety of *V. tremulae* in North America; it occurs on trembling and bigtooth aspens, introduced white poplars, and some intrasectional hybrids. *V. tremulae* var. *populi-albae* causes necrotic spots on white poplars in Europe. *V. tremulae* var. *tremulae* occurs on European aspen and its hybrids and sometimes on white poplars in Europe.

Management

Pruning out infected shoots and avoiding overhead irrigation will help manage Venturia leaf and shoot blight. Little information is available regarding resistance to *Venturia* species. Repeated applications of fungicides during early shoot growth in the spring can protect foliage from infection until environmental conditions no longer favor disease development.

Marssonina Leaf Spots and Blights
Marssonina spp.

Marssonina leaf spots and blights are common diseases of poplars and aspens (Ostry et al., 1989; Ostry and Schipper, 1989; Sinclair et al., 1987). They occur wherever the host is grown, and under conditions favorable for disease development they cause premature defoliation and dieback and reduce the growth of trees. *Marssonina brunnea* f. sp. *brunnea* is probably the most common of the pathogens causing leaf spot and blight. Other species infecting *Populus* are *M. balsamiferae, M. brunnea* f. sp. *trepidae, M. castagnei,* and *M. populi.*

Symptoms

Small (1–5 mm in diameter), diffuse, orange brown spots with yellow margins form on leaves (Plate 132), and pustules form on young twigs. Fruiting structures (acervuli) initially appear as tiny blisters in the centers of spots. Under conditions favorable for disease, leaves may appear brown by midsummer, and trees become severely defoliated. During late summer and early fall, defoliated trees may resprout, increasing the potential for twig die-

Venturia Leaf and Shoot Blight

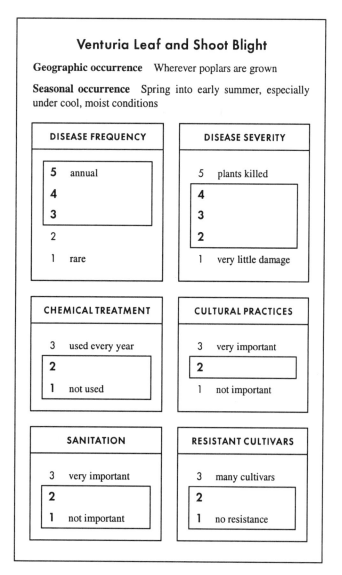

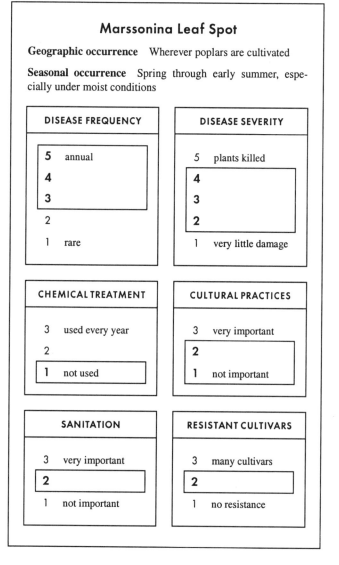

Geographic occurrence Wherever poplars are grown

Seasonal occurrence Spring into early summer, especially under cool, moist conditions

DISEASE FREQUENCY
5 annual
4
3
2
1 rare

DISEASE SEVERITY
5 plants killed
4
3
2
1 very little damage

CHEMICAL TREATMENT
3 used every year
2
1 not used

CULTURAL PRACTICES
3 very important
2
1 not important

SANITATION
3 very important
2
1 not important

RESISTANT CULTIVARS
3 many cultivars
2
1 no resistance

Marssonina Leaf Spot

Geographic occurrence Wherever poplars are cultivated

Seasonal occurrence Spring through early summer, especially under moist conditions

DISEASE FREQUENCY
5 annual
4
3
2
1 rare

DISEASE SEVERITY
5 plants killed
4
3
2
1 very little damage

CHEMICAL TREATMENT
3 used every year
2
1 not used

CULTURAL PRACTICES
3 very important
2
1 not important

SANITATION
3 very important
2
1 not important

RESISTANT CULTIVARS
3 many cultivars
2
1 no resistance

back, because of the lack of normal cold hardiness. Seedlings in nurseries can be killed.

Leaf spots caused by *M. brunnea* f. sp. *brunnea* typically are discrete, dark brown spots about 1 mm in diameter. Multiple infections result in leaf yellowing and browning followed by defoliation. Fruiting bodies appear in the centers of spots and are initially orange brown, turning gray white after sporulation.

Leaf spots caused by *M. castagnei* are dark brown to reddish brown, tend to have light-colored centers, and are about 3 mm in diameter. They often coalesce to form irregularly shaped blotches over the leaf surface. Fruiting structures are found mostly on the upper surface of the leaf.

Leaf spots caused by *M. populi* are initially bronze to chestnut-colored and 2–5 mm in diameter. They later coalesce into brownish black, vein-limited blotches.

It was previously thought that *Marssonina* species could be distinguished by the symptoms they cause. However, Spiers (1984) found that symptoms caused by different *Marssonina* species vary according to the host and environmental conditions, and this finding raises questions about the value of symptoms in distinguishing between species.

Epidemiology

The pathogens overwinter as stromata in fallen leaves and twig lesions (Ostry, 1987). In the spring, fungal fruiting bodies (acervuli) produce spores (conidia), which are dispersed by water and cause primary infection of new leaves. Initial symptoms appear after a couple of weeks. Fruiting bodies in leaf spots produce secondary conidia, which initiate repeated infection cycles during wet weather until defoliation occurs. In late summer and early fall, stromata develop in fallen leaves and twig lesions, completing the disease cycle.

M. brunnea is also seedborne and can infect seedlings grown from infested seed.

The sexual stages of three of the *Marssonina* species causing leaf spot and blight are species of *Drepanopeziza*. The sexual states are not common, and few studies have examined their role in disease development (Newcombe, 1996).

Host Range

Poplars and aspens vary considerably in their susceptibility to Marssonina leaf spot (Spiers, 1984; Ostry and McNabb, 1985). Some clones are extremely susceptible, while others are highly resistant. Isolates of *M. brunnea* f. sp. *brunnea* vary in aggressiveness, and young leaves are much more susceptible to infection than older leaves (Spiers, 1983).

Newcombe (1996) reviewed the taxonomy of *Marssonina* species and the hosts they attack. *M. brunnea* f. sp. *brunnea* is pathogenic on eastern cottonwood and its hybrids with black poplars (*P.* × *euramericana*). *M. brunnea* f. sp. *trepidae* is pathogenic on European and trembling aspens. *M. castagnei* typically occurs on white poplars, a few cottonwoods, and black and balsam poplars. *M. populi* attacks cottonwoods and black and balsam poplars. *M. balsamiferae* is found on balsam poplar.

Management

Removing or plowing under infected foliage, removal of infected plants, avoiding overhead irrigation, and the production of resistant clones will help control Marssonina leaf spot (Palmer et al., 1980).

Septoria Leaf Spot and Canker

Septoria musiva
(sexual stage, *Mycosphaerella populorum*)
Septoria populicola
(sexual stage, *Mycosphaerella populicola*)

Septoria populicola causes a severe leaf spot disease of poplars, and *S. musiva* causes leaf spots and a destructive canker disease (Filer, 1989; Filer et al., 1971; Moore et al., 1982; Newcombe et al., 1995; Thompson, 1941). The development of cankers weakens branches and main stems, and multiple cankers can kill trees. *S. musiva* has not been reported in western North America (Callan and Ring, 1994; Funk, 1985; Newcombe, 1996; Newcombe et al., 1995).

Symptoms

Symptoms vary considerably with the host. In susceptible hosts, initial symptoms consist of sunken, black flecks, which enlarge into rounded spots, usually less than 15 mm in diameter. Multiple infections result in dead blotches. Necrotic tissues even-

Septoria Leaf Spot and Canker

Geographic occurrence Septoria leaf spot occurs wherever susceptible poplars are cultivated; Septoria canker is not found in western North America

Seasonal occurrence Spring through summer, especially under moist conditions

DISEASE FREQUENCY	DISEASE SEVERITY
5 annual	**5** plants killed
4	**4**
3	**3**
2	**2**
1 rare	1 very little damage

CHEMICAL TREATMENT	CULTURAL PRACTICES
3 used every year	3 very important
2	**2**
1 not used	**1** not important

SANITATION	RESISTANT CULTIVARS
3 very important	3 many cultivars
2	**2**
1 not important	1 no resistance

tually turn brown, tan, or white and have a brown to black margin. In partially resistant hosts, the spots remain small and may have a silvery appearance. Infections are generally more prevalent on foliage in the lower portion of the tree. Masses of black pycnidia develop on blighted tissues.

Canker is associated with a high incidence of leaf spots. Most cankers form at the site of wounds or natural openings. The cankers initially have a water-soaked appearance and then turn black. Their centers turn tan as pycnidia develop in the cankers. They can enlarge to girdle stems in one or two growing seasons, depending on the susceptibility of the host and the site of infection (main stem or branch).

Epidemiology

S. populicola and *S. musiva* overwinter in infected leaves fallen to the ground (Thompson, 1941; Ostry, 1987). *S. musiva* can also overwinter in young stem cankers. The primary inoculum is ascospores, produced by pseudothecia, which develop on overwintered leaves on the ground. These spores are spread by wind and splashing water and, under moist conditions at mild temperatures, infect newly developing leaves. Symptoms appear about one to two weeks after infection. Fruiting bodies (pycnidia) are formed about two weeks later. Pink masses or tendrils of spores (conidia) produced by the pycnidia are spread by water to adjacent leaves and young stems, where infection takes place. Repeating cycles of infection occur during extended periods of moist conditions.

Host Range

S. musiva causes leaf spots on all native North American species of poplar and aspens (Ostry et al., 1989; Ostry and McNabb 1985). It also causes destructive cankers on eastern cottonwood; eastern cottonwood and black poplar hybrids (*P.* × *euramericana*); and hybrids with black cottonwood. *P.* × *euramericana* clones are more resistant to *S. musiva* than hybrids of black cottonwood (Ostry and McNabb, 1985; Strobl and Fraser, 1989).

S. populicola occurs on black, eastern, and narrow-leaf cottonwoods and balsam poplar.

Management

Removing or plowing under fallen leaves will help reduce the level of primary inoculum. Growers should also destroy all infected trees and use only canker-free propagating material. There is considerable variation in host resistance to Septoria leaf spot and canker. In areas where Septoria canker is a problem, growers should avoid planting black cottonwood and its hybrids.

Ink Spot

Ciborinia pseudobifrons
Ciborinia whetzelii

Ink spot occurs in the northern portions of North America and in the central Rocky Mountains (Sinclair et al., 1987; Ostry et al., 1989). The disease is caused by the fungi *Ciborinia whetzelii* and *C. pseudobifrons,* of which the former is more common (Groves and Bowerman, 1955; Baranyay and Hiratsuka, 1967). Disease outbreaks are often sporadic, but repeated defoliation can result in the death of young trees. High planting density favors severe outbreaks.

Symptoms

The disease is characterized by blighted foliage and the presence of prominent, dark brown to black sclerotia of the pathogen in the blighted tissue (Ostry, 1980). Ink spot is also referred to as shot hole, because of the holes that form in leaves after the sclerotia fall out, in mid- to late summer.

Epidemiology

Ink spot is a single-cycle disease. The pathogen overwinters as sclerotia on the ground. In the spring, sclerotia germinate and produce small, cuplike fruiting structures (apothecia). Ascospores form in the apothecia and are forcibly ejected from them and spread by the wind to nearby foliage. Under moist conditions, infection takes place, and the initial reddish brown blotches develop in about two to three weeks. The blotches expand, sometimes producing alternating concentric zones of light and dark color. Dead leaves are tan. In another two to three weeks, small, brown, circular to oval sclerotia begin to develop on blighted leaf tissue. The sclerotia turn black, giving rise to the symptom for which the disease is named. Sclerotia of *C. pseudobifrons* tend to form on leaf veins and petioles and remain attached to the leaf. In mid- to late summer, sclerotia of

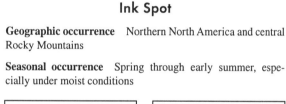

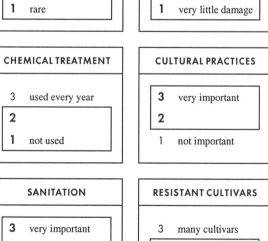

Ink Spot

Geographic occurrence Northern North America and central Rocky Mountains

Seasonal occurrence Spring through early summer, especially under moist conditions

DISEASE FREQUENCY	DISEASE SEVERITY
5 annual	5 plants killed
4	4
3	3
2	**2**
1 rare	**1** very little damage

CHEMICAL TREATMENT	CULTURAL PRACTICES
3 used every year	**3** very important
2	**2**
1 not used	1 not important

SANITATION	RESISTANT CULTIVARS
3 very important	3 many cultivars
2	**2**
1 not important	**1** no resistance

C. whetzelii drop out of the leaves and fall to the ground, leaving shot holes in the leaf (Funk, 1985).

Disease development is favored by the survival of large numbers of sclerotia, high plant density, low temperatures, and high relative humidity during sporulation and leaf expansion.

Host Range

Ink spot occurs primarily in trembling aspen and its hybrids. It also causes limited damage to bigtooth aspen, eastern cottonwood, and balsam and black poplars.

Management

Reducing the number of sclerotia and plant density and avoiding overhead irrigation help reduce disease pressure.

Blackstem

Cytospora chrysosperma **(sexual stage,** ***Valsa sordida*****)**
Dothichiza **spp.**
Phomopsis macrospora

Blackstem can cause extensive mortality of young cottonwood and hybrid poplar hardwood cuttings in nurseries and field plantings. Losses have been reported to range from 20 to 55% or more (Walla and Stack, 1980; Gray et al., 1965; Ostry et al., 1989; Filer, 1967). The pathogen most common associated with this disease is *Cytospora chrysosperma,* particularly in the more northern half of the United States (Filer, 1989). Other fungi reported to cause blackstem are *Dothichiza* spp. and *Phomopsis macrospora.*

Blackstem

Geographic occurrence Wherever poplars are cultivated

Seasonal occurrence Late fall through spring; most evident shortly after planting

DISEASE FREQUENCY	DISEASE SEVERITY
5 annual	**5** plants killed
4	**4**
3	**3**
2	2
1 rare	1 very little damage

CHEMICAL TREATMENT	CULTURAL PRACTICES
3 used every year	**3** very important
2	2
1 not used	1 not important

SANITATION	RESISTANT CULTIVARS
3 very important	3 many cultivars
2	**2**
1 not important	**1** no resistance

Symptoms

Characteristic symptoms of blackstem include slightly sunken black cankers with definite margins on the sides and ends of cuttings. Infected cuttings may fail to sprout after planting, or the new growth may wilt shortly after emergence, because of the lack of good root development in diseased cuttings.

Epidemiology

The fungi associated with blackstem are common on dead and dying poplars. There is some debate as to whether they are primary pathogens, secondary invaders, or endophytes (Newcombe, 1996). Cuttings taken from trees or stools that are under stress are much more likely to have problems with blackstem than cuttings obtained from healthy plants. Loss of internal bark moisture and improper storage can also predispose cuttings to the disease. Rain-splashed conidia and windblown ascospores infect cuttings through bud scales, leaf scars, and wounded or stressed bark during late fall or winter. Infections may not be evident at the time cuttings are collected or upon removal from storage, but symptoms develop rapidly after planting.

The symptoms associated with infection by the different blackstem pathogens are very similar, so positive identification of the causal agent is based on isolation of the pathogen or examination of spores produced by fruiting bodies that develop on cankered tissue.

Host Range

Blackstem occurs in hardwood cuttings of most cottonwood and hybrid poplars. Cuttings taken from clones exposed to environmental stresses and stress associated with premature defoliation caused by diseases such as leaf rust or Marssonina leaf spot are much more likely to develop blackstem than cuttings taken from healthy plants.

Management

A number of cultural practices can be effective in managing blackstem (Ostry and McNabb, 1982). Cuttings should be collected from healthy planting material only. Proper storage of cuttings until planting, usually at temperatures slightly below freezing, and prevention of moisture loss from cuttings will also help reduce the development of the disease. Dipping cuttings in fungicides has been effective in reducing the damage caused by blackstem (Walla and Stack, 1980).

Poplar Mosaic

Poplar mosaic virus

Several virus or viruslike diseases of aspens and poplars have been reported (Navratil, 1979; Ostry et al., 1989; Sinclair et al., 1987). Little is known about most of them. Poplar mosaic virus (PopMV), a carlavirus, has been reported in almost all commercially important cultivars of *P.* × *euramericana,* black and eastern cottonwoods, balsam and black poplars, and European and trembling aspens (Biddle and Tinsley, 1971; Cooper and Edwards, 1981). In highly susceptible hosts, PopMV can reduce growth and result in dieback of young trees. Older trees are more tolerant of the virus.

Symptoms

PopMV causes a variety of chlorotic spots and mosaic patterns on infected foliage. Various patterns of chlorosis, reddening, and necrosis may be evident on leaf veins. Necrotic spots may develop on main leaf veins and petioles, which can result in abnormal leaf development.

Epidemiology

Particles of PopMV are slightly flexuous rods, 670 to 685 nm long (Francki et al., 1987; Biddle and Tinsley, 1971). This virus is sap-transmissible and is known to occur naturally in poplar and aspen only. Symptoms are most evident on the oldest fully expanded leaves in late spring. High temperatures can mask foliar symptoms. The disease is spread by vegetative cuttings and grafting. Serological tests or sap inoculation of indicator plants can be used to confirm the presence of the virus (Van der Meer et al., 1980).

Management

Management of poplar mosaic depends on the use of virus-free propagation material.

Bacterial Blight

Pseudomonas syringae* pv. *syringae

Pseudomonas syringae causes bacterial blight of numerous economically important fruits, vegetables, and ornamental plants (Moore and Pscheidt, 1998). Bacterial blight of poplar seedlings or cuttings can be a limiting factor in nursery production.

Symptoms

The most obvious symptoms associated with bacterial blight are black leaf spots, blackened leaf veins and petioles resulting from systemic infection, and shoot tip dieback, in which dead, blackened twig tissue extends some distance back from the shoot tip. Cankers can also develop on stems and branches, especially on the south side of trees. The initial symptoms can be confused with those of Venturia leaf and shoot blight.

Poplar Mosaic

Geographic occurrence Wherever poplars are cultivated

Seasonal occurrence Late spring through summer; high temperatures can mask foliar symptoms

DISEASE FREQUENCY	DISEASE SEVERITY
5 annual	5 plants killed
4	4
3	3
2	**2**
1 rare	**1 very little damage**

CHEMICAL TREATMENT	CULTURAL PRACTICES
3 used every year	**3 very important**
2	2
1 not used	1 not important

SANITATION	RESISTANT CULTIVARS
3 very important	3 many cultivars
2	**2**
1 not important	**1 no resistance**

Bacterial Blight

Geographic occurrence Areas where nursery plants are exposed to temperatures near freezing in early spring

Seasonal occurrence Early spring, especially if plants are predisposed by exposure to cold

DISEASE FREQUENCY	DISEASE SEVERITY
5 annual	**5 plants killed**
4	**4**
3	**3**
2	2
1 rare	1 very little damage

CHEMICAL TREATMENT	CULTURAL PRACTICES
3 used every year	3 very important
2	**2**
1 not used	1 not important

SANITATION	RESISTANT CULTIVARS
3 very important	3 many cultivars
2	2
1 not important	**1 no resistance**

Epidemiology

P. syringae pv. *syringae* survives as an epiphyte on the surface of many different plants. Exposure of seedlings to temperatures at or below freezing and wounding can predispose them to severe infection. The pathogen can be moved from place to place by wind, rain, and insects; on contaminated hand tools; and on infested propagation material and nursery stock.

Host Range

The pathogen has a very wide host range. Damage to small seedlings is generally more severe than damage to plants obtained from large, dormant cuttings.

Management

Plants should be spaced far enough apart to promote good air circulation. Protecting plants from exposure to cold and other stress can significantly reduce disease pressure. Fixed copper fungicides have been useful in controlling bacterial blight of other crops, but in some regions strains of *P. syringae* pv. *syringae* have developed resistance to copper.

REFERENCES

Anderson, N. A., and Anderson, R. L. 1980. Leaf and shoot blight of aspen caused by *Venturia macularis* in northern Minnesota. Plant Dis. 64:558–559.

Bailey, L. H., and Bailey, E. Z. 1976. Hortus Third. Macmillan, New York.

Baranyay, J. A., and Hiratsuka, Y. 1967. Identification and distribution of *Ciborinia whetzelii* (Seaver) Seaver in western Canada. Can. J. Bot. 45:189–191.

Biddle, P. G., and Tinsley, T. W. 1971. Poplar mosaic virus. Descriptions of Plant Viruses, no. 75. Commonwealth Mycological Institute and Association of Applied Biologists, Kew, England.

Callan, B. E., and Ring, F. M. 1994. An annotated host fungus index for *Populus* in British Columbia. Can. For. Serv. FRDA Rep. 222.

Cooper, J. I., and Edwards, M. L. 1981. The distribution of poplar mosaic virus in hybrids poplars and virus detection by ELISA. Ann. Appl. Bot. 99:53–61.

Dance, B. W. 1961. Leaf and shoot blight of poplars (section *Tacamahaca* Spach) caused by *Venturia populina* (Vuill.) Fabric. Can. J. Bot. 99:53–61.

Dirr, M. A. 1998. Manual of Woody Landscape Plants: Their Identification, Ornamental Characteristics, Culture, Propagation and Uses. 5th ed. Stipes Publishing, Champaign, Ill.

Eckenwalder, J. E. 1996. Systematics and evolution of *Populus*. Pages 7–32 in: Biology of *Populus* and Its Implications for Management and Conservation. R. F. Stettler, H. D. Bradshaw, Jr., P. E. Hielman, and T. M. Hinckley, eds. NRC Research Press, National Research Council of Canada, Ottawa.

Filer, T. H, Jr. 1967. Pathogenicity of *Cytospora, Phomopsis,* and *Hypomyces* on *Populus deltoides*. Phytopathology 57:978–980.

Filer, T. H, Jr. 1989. Poplar cankers. Pages 101–102 in: Forest Nursery Pests. C. E. Cordell, R. L. Anderson, W. H. Hoffard, T. D. Landis, R. S. Smith, Jr., and H. V. Toko, eds. U.S. Dep. Agric. Agric. Handb. 680.

Filer, T. H., Jr., McCracken, F. I., Mohn, C. A., and Randall, W. K. 1971. Septoria canker on nursery stock of *Populus deltoides*. Plant Dis. Rep. 55:460–463.

Francki, R. I. B., Milne, R. G., and Hatta, T. 1987. Atlas of Plant Viruses. 2nd ed. Vol. 2. CRC Press, Boca Raton, Fla.

Funk, A. 1985. Foliar fungi of western trees. Can. For. Serv. Inf. Rep. BC-X-265.

Gray, L. E., Jokela, J. J., and Wycoff, H. B. 1965. Blackstem of cottonwood. Plant Dis. Rep. 49:867–868.

Groves, J. W., and Bowerman, C. A. 1955. The species of *Ciborinia* on *Populus*. Can. J. Bot. 33:577–590.

Moore, L. M, Ostry, M. E., Wilson, L. F., Morin, M. J., and McNabb, H. S., Jr. 1982. Impact of Septoria canker caused by *S. musiva* on nursery stock and first-year plantation coppice. Pages 44–50 in: Proc. North Am. Poplar Counc. Meet. J. Zavitkovski and E. A. Hansen, eds. Kansas State University, Manhattan.

Moore, L. W., and Pscheidt, J. W. 1998. Diseases caused by *Pseudomonas syringae*. Pages 25–27 in: Pacific Northwest Plant Disease Control Handbook. J. W. Pscheidt and C. M. Ocamb, eds. Oregon State University, Corvallis.

Navratil, S. 1979. Virus and virus-like diseases of poplar: Are they threatening diseases? Report 19 in: Poplar Research, Management, and Utilization in Canada. D. C. F. Fayle, A. Zsuffa, and H. W. Anderson, eds. Ont. Minist. Nat. Resource For. Res. Inf. Pap. 102.

Newcombe, G. 1996. The specificity of fungal pathogens of *Populus*. Pages 223–246 in: Biology of *Populus* and Its Implications for Management and Conservation. R. F. Stettler, H. D. Bradshaw, Jr., P. E. Hielman, and T. M. Hinckley, eds. NRC Research Press, National Research Council of Canada, Ottawa.

Newcombe, G., and Chastagner, G. A. 1993a. A leaf rust epidemic of hybrid poplar along the lower Columbia River caused by *Melampsora medusae*. Plant Dis. 77:528–531.

Newcombe, G., and Chastagner, G. A. 1993b. First report of the Eurasian poplar leaf rust fungus, *Melampsora larici-populina,* in North America. Plant Dis. 77:532–535.

Newcombe, G., Chastagner, G. A., Schuette, W., and Stanton, B. J. 1994. Mortality among hybrid poplar clones in a stool bed following leaf rust caused by *Melampsora medusae* f. sp. *deltoidae*. Can J. For. Res. 24:1984–1987.

Newcombe, G., Chastagner, G. A., Callan, B. E., and Ostry, M. E. 1995. An epidemic of Septoria leaf spot on *Populus trichocarpa* in the Pacific Northwest in 1993. Plant Dis. 79:212.

Ostry, M. E. 1980. How to identify ink spot of poplars. U.S. Dep. Agric. For. Serv., North Cent. For. Exp. Stn. (St. Paul, Minn.).

Ostry, M. E. 1987. Biology of *Septoria musiva* and *Marssonina brunnea* in hybrid *Populus* plantations and control of Septoria canker in nurseries. Eur. J. For. Pathol. 17:158–165.

Ostry, M. E., and McNabb, S. H., Jr. 1982. Preventing blackstem damage to *Populus* hardwood cuttings. Pages 36–43 in: Proc. North Am. Poplar Counc. Meet. J. Zavitkovski and E. A. Hansen, eds. Kansas State University, Manhattan.

Ostry, M. E., and McNabb, S. H., Jr. 1985. Susceptibility of *Populus* species and hybrids to disease in the north central United States. Plant Dis. 69:755–757.

Ostry, M. E., and Schipper, A. L., Jr. 1989. Marssonina blight. Pages 97–98 in: Forest Nursery Pests. C. E. Cordell, R. L. Anderson, W. H. Hoffard, T. D. Landis, R. S. Smith, Jr., and H. V. Toko, eds. U.S. Dep. Agric. Agric. Handb. 680.

Ostry, M. E., Wilson, L. F., McNabb, H. S., Jr., and Moore, L. M. 1989. A guide to insect, disease, and animal pests of poplars. U.S. Dep. Agric. Agric. Handb. 677.

Palmer, M. A., Ostry, M. E., and Schipper, A. L., Jr. 1980. How to identify and control Marssonina leaf spot of poplars. U.S. Dep. Agric. For. Serv., North Cent. For. Exp. Stn. (St. Paul, Minn.).

Pinon, J. 1992. Variability in the genus *Populus* in sensitivity to *Melampsora* rusts. Silvae Genet. 41:25–34.

Shain, L., and Filer, T. H., Jr. 1989. Cottonwood leaf rusts. Pages 90–91 in: Forest Nursery Pests. C. E Cordell, R. L Anderson, W. H. Hoffard, T. D. Landis, R. S. Smith, Jr., and H. V. Toko, eds. U.S. Dep. Agric. Agric. Handb. 680.

Sinclair, W. A., Lyon, H. H., and Johnson, W. T. 1987. Diseases of Trees and Shrubs. Comstock Publishing Associates, Cornell University Press, Ithaca, N.Y.

Spiers, A. G. 1983. Host range and pathogenicity studies of *Marssonina brunnea* to poplars. Eur. J. For. Pathol. 13:181–196.

Spiers, A. G. 1984. Comparative studies of host specificity and symptoms exhibited by poplars infected with *Marssonina brunnea, Marssonina castagnei,* and *Marssonina populi*. Eur. J. For. Pathol. 14:202–218.

Strobl, S., and Fraser, K. 1989. Incidence of Septoria canker of hybrid poplars in eastern Ontario. Can. Plant Dis. Surv. 69(2):109–112.

Thompson, G. E. 1941. Leaf spot diseases of poplars caused by *Septoria musiva* and *S. populicola.* Phytopathology 31:241–254

Van der Meer, F. A., Maat, D. Z., and Vink, J. 1980. Poplar mosaic virus: Purification, antiserum preparation, and detection in poplars with the enzyme-linked immunosorbent assay (ELISA) and with infectivity tests on *Nicotiana megalosiphon.* Neth. J. Plant Pathol. 86:99–110.

Walla, J. A, and Stack, R. W. 1980. Dip treatment for control of blackstem on *Populus* cuttings. Plant Dis. 64:1092–1095.

Ziller, W. G. 1974. The tree rusts of western Canada. Can. For. Serv. Publ. 1329.

CHAPTER 72

Jay W. Pscheidt ▪ Oregon State University, Corvallis
Ralph S. Byther ▪ Washington State University, Puyallup

Prunus Diseases

Geographic production USDA zones 2–10

Family Rosaceae

Genus *Prunus*

Species

P. americana	American plum	zones 3–8
P. armeniaca var. *mandshurica*	Manchurian apricot	zones 3(4)–8
P. avium	mazzard cherry	zones 3–8
P. besseyi	western sandcherry	zones 3–6
P. campanulata	Taiwan cherry	zones 7–9
P. caroliniana	Carolina cherrylaurel	zones 7–10
P. cerasifera	cherry plum	zones 5–8
P. glandulosa	dwarf flowering almond	zones 4–8
P. 'Hally Jolivette'		zones 5–7
P. laurocerasus	common cherrylaurel	zones 6–8
P. maackii	Amur chokecherry	zones 2–6
P. maritima	beach plum	zones 3–6
P. mume	Japanese apricot	zones 6–10
P. nigra	Canada plum	zones 2–5(6)
P. padus	European birdcherry	zones 3–6
P. pensylvanica	pin cherry, wild red cherry	zones 2–5(6)
P. persica	common peach	zones 5–9
P. sargentii	Sargent cherry	zones 4–7
P. serotina	black cherry	zones 3–9
P. serrula		zones 5–6
P. serrulata	Japanese flowering cherry	zones 5–6
P. subhirtella	Hilgan cherry	zones 4–8(9)
P. tenella	dwarf Russian almond	zones 2–6
P. tomentosa	Manchu cherry	zones 2–7
P. triloba var. *multiplex*	double-flowering plum	zones 3–6(7)
P. virginiana	common chokecherry	zones 2–6
P. × *yedoensis*	Yoshino cherry	zones 5–8

Prunus is a large genus, including apricots, almonds, cherries, peaches, and plums. The species listed above are grown mostly for their ornamental value in the landscape. Members of this group are valued for early-season flowering, while other ornamentals are still dormant, and for their unique foliage and bark. Several diseases afflict *Prunus* spp. in the nursery, including black knot, bacterial canker, brown rot, crown gall, leaf spots, powdery mildew, shot hole, Verticillium wilt, leaf curls, and viruses.

Black Knot

Apiosporina morbosa

Black knot occurs in at least 24 members of the genus *Prunus,* including ornamental, edible, and wild species, but it is most common and destructive in plums and prunes. The disease is distributed throughout North America but is most serious in the North Central and northeast states. Disease outbreaks are often observed adjacent to wild areas, neglected landscapes, or orchards where *Prunus* spp. are growing.

Symptoms

Rough, black, elongate, swollen galls form on woody twigs and branches of all sizes. During the first growing season following spring infection, branch tissues become swollen but remain relatively smooth, and symptoms are not easily detected until late summer and fall. The swelling continues the following year, causing the bark tissue to split, revealing an olive green fungal stroma. As the season progresses, the swollen tissues darken, become roughened, and form the elongated galls characteristic of black knot. Younger twigs may become severely bent at infection sites. Initially, galls seldom encircle the entire branch, but they continue to expand annually. Numerous infections on a branch or older galls will girdle the branch and cause subsequent dieback.

Epidemiology

Fruiting bodies of the fungal pathogen, *Apiosporina morbosa,* located on the surface of mature galls, eject ascospores into the air during wet spring weather. The spores infect new shoot growth and occasionally wounds. At the optimum temperature of 70°F (21°C), germination and infection can take place in 6 hours. This process can occur at temperatures as low as 43°F (6°C) but requires a considerably longer time. For the first several months, infections are undetectable. Eventually growth-regulating hormones are produced, increasing the number and size of plant cells and resulting in obvious swellings of the tissue in late summer and fall. After further enlargement during the next growing season, ascospores are produced in fruiting bodies on the surface of mature galls. Galls on large branches result from the invasion of the pathogen from adjacent smaller twigs.

Host Range

At least 24 species of *Prunus* have been reported to be susceptible to black knot. There is some evidence of specialized strains of *A. morbosa,* however, with more limited host ranges. Flowering almond, apricot, blackthorn, cherries (including bird, bitter, black, mahaleb, Nanking, pin, sand, western sand, and sour and sweet cherries), chokecherry, peach, and plums (including American, beach, Canada, common, Damson, Japanese, myrobalan, and Sierra plums) are reported to be susceptible hosts.

Management

Infected twigs and branches should be pruned out and destroyed. Cuts should be made 3 to 4 inches (8 to 10 cm) below the visible swelling. Since ascospores can be produced on detached galls, prunings must be removed and destroyed. *Prunus* spp. should be removed from wild areas adjacent to nursery plantings. Fungicide applications should be made in the spring during shoot elongation, to protect the new growth.

Bacterial Canker

Pseudomonas syringae* pv. *syringae

Bacterial canker is common west of the Cascades in the Pacific Northwest, because it is favored by cool, wet spring weather. It causes severe damage under environmental conditions favorable for disease development.

Symptoms

Infected buds may fail to open in spring. Small, greasy-looking spots appear on newly opened leaves. The spots turn dark brown and may eventually fall out, leaving shot holes. In wet weather, spots may expand and kill terminal shoots of susceptible cultivars. Shoots may appear blackened. Cankers may develop on branches. Leaves of branches may wilt during hot weather and turn brown. When infection becomes systemic, leaves of infected trees emerge smaller and yellower than leaves of healthy trees. Decline and death of the tree may follow.

Epidemiology

Trees are predisposed to bacterial canker by weakness and injury due to such factors as wounds, frost damage, early pruning, improper soil pH, poor nutrition, and infection by other pathogens, including *Verticillium, Nectria,* and nematodes. Old cankers, healthy buds, and systemically infected trees (with or without cankers) provide sources of inoculum of the bacterial canker pathogen, *Pseudomonas syringae* pv. *syringae.* It also survives as an

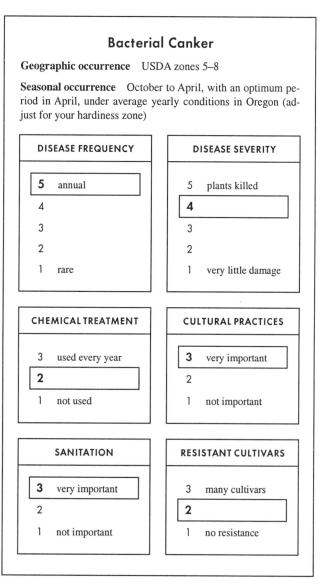

Black Knot

Geographic occurrence USDA zones 3–9

Seasonal occurrence New infections occur in spring, from March to May; knots are perennial

DISEASE FREQUENCY
5 annual
4
3
2
1 rare

DISEASE SEVERITY
5 plants killed
4
3
2
1 very little damage

CHEMICAL TREATMENT
3 used every year
2
1 not used

CULTURAL PRACTICES
3 very important
2
1 not important

SANITATION
3 very important
2
1 not important

RESISTANT CULTIVARS
3 many cultivars
2
1 no resistance

Bacterial Canker

Geographic occurrence USDA zones 5–8

Seasonal occurrence October to April, with an optimum period in April, under average yearly conditions in Oregon (adjust for your hardiness zone)

DISEASE FREQUENCY
5 annual
4
3
2
1 rare

DISEASE SEVERITY
5 plants killed
4
3
2
1 very little damage

CHEMICAL TREATMENT
3 used every year
2
1 not used

CULTURAL PRACTICES
3 very important
2
1 not important

SANITATION
3 very important
2
1 not important

RESISTANT CULTIVARS
3 many cultivars
2
1 no resistance

epiphyte on leaf surfaces, weeds, grasses, and even soil. The bacterium is spread by wind, rain, and insects and in infected budwood and infected nursery stock.

Host Range

P. sargentii 'Rancho' and *P.* × *yedoensis* 'Akebono' appear to have some resistance. The cherry cultivar Kwanzan appears to be resistant when plants are mature but not when they are young.

Management

If practical, destroy old seedling cherry trees growing in low areas; they may be a source of inoculum. Prune out cankered branches in summer or in dry weather late in the dormant season. Between cuts, disinfect pruning tools with 10% bleach or 70% rubbing alcohol or shellac thinner. Replace severely infected trees with resistant cultivars.

Copper products are typically applied in the Pacific Northwest before fall rains, during leaf fall, and late dormant just before buds open. Further applications of fixed copper at reduced rates during early shoot growth in the spring are recommended in British Columbia. Populations of bacteria resistant to copper products may compromise efforts to control bacterial canker with fixed copper products.

Brown Rot Blossom Blight

Monilinia fructicola
Monilinia laxa

The fungi *Monilinia fructicola* and *M. laxa* can incite a blossom blight, twig and branch dieback, and fruit rot of several *Prunus* spp., including many ornamental and fruit trees. The disease is a problem in most production areas.

Symptoms

Infected flower parts turn light brown and may develop areas of buff-colored spores of *M. fructicola* or gray spores of *M. laxa.* Infected petals may look water-soaked; this symptom can be mistaken for frost injury. Flowers generally collapse as the fungus invades through the pedicel. Infected flowers often adhere to twigs and spurs through harvest or even through winter. Depending on the fungus and the host, the infection may advance into twigs and spurs. Lesions may remain discrete or may girdle twigs, causing distal portions of the twigs to die. Profuse gumming may also occur in lesioned areas. Buff or gray spores may develop in sporodochia on necrotic twigs.

Epidemiology

The brown rot blossom blight fungi survive from year to year on infected twigs, branches, old flower parts, and mummified fruit. Spores (conidia) are produced on infected plant debris in the tree at temperatures above 40°F. Small, mushroom-like structures (apothecia) may be produced on fruit that drops to the ground. Wind and rain disperse spores (conidia and ascospores) to healthy blossoms in spring, to begin the infection process during wet weather. Infection by *M. laxa* occurs at temperatures above 55°F or, with 24 hours of wetness, at any temperature above freezing. Flowers can be blighted whenever floral tissue is exposed but are most susceptible at full bloom. More spores can be produced on floral tissue, initiating several more disease cycles during the spring.

Some infections may be symptomless until fruit begins to ripen. Ripening fruit is also highly susceptible to infection, and many more disease cycles can occur near harvest. Peach and nectarine fruit that fall to the ground because of lack of pollination, thinning, or overripeness can significantly increase the level of inoculum and the amount of fruit rot at harvest. Fruit infected in the orchard may not show symptoms until it is in storage or in transit. Nitrogen fertilization at high rates of application is also associated with increased levels of brown rot.

Host Range

Fruiting and ornamental cherries, almonds, nectarines, peaches, plums, and prunes are susceptible.

Management

Remove and destroy infected twigs and branches in summer. Use moderate amounts of nitrogen fertilizer. These practices must be supplemented by chemical control, especially in the wettest areas.

Apply fungicides during the bloom period at the early popcorn

Brown Rot Blossom Blight

Geographic occurrence USDA zones 4 to 8

Seasonal occurrence March to May, with an optimum period in April, under average yearly conditions in Oregon (adjust for your hardiness zone)

DISEASE FREQUENCY	DISEASE SEVERITY
5 annual	5 plants killed
4	4
3	3
2	**2**
1 rare	1 very little damage

CHEMICAL TREATMENT	CULTURAL PRACTICES
3 used every year	**3 very important**
2	2
1 not used	1 not important

SANITATION	RESISTANT CULTIVARS
3 very important	3 many cultivars
2	**2**
1 not important	1 no resistance

stage (red bud, pink bud, or green tip, depending on the crop), full bloom, and petal fall to control blossom blight.

Fungicides are also applied prior to harvest to control fruit rot. Tolerant strains of some fungi have become troublesome when the same fungicides are used exclusively in a spray schedule. To reduce the possibility of tolerance, alternate or tank-mix fungicides having different modes of action.

Crown Gall

Agrobacterium tumefaciens

The bacterium *Agrobacterium tumefaciens* has a wide host range, but plants more likely to have crown gall include apple, euonymus, poplar, rose, walnut, willow, and all *Prunus* spp.

Symptoms

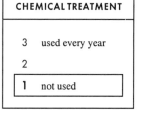

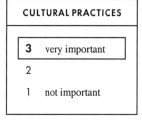

On young nursery trees, soft, spongy, or wart-like galls develop on the crown or roots. Galls on mature trees range from a fraction of an inch to several inches across. Galls on woody plants become hard and develop a rough, fissured surface as they age. Gall tissues are irregular and have no definite growth pattern. Young trees may die if their trunks are completely encircled by galls.

Crown Gall

Geographic occurrence USDA zones 3–9

Seasonal occurrence Year-round, with an optimum period when wounds occur in the spring, under average yearly conditions in Oregon (adjust for your hardiness zone)

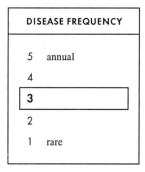

DISEASE FREQUENCY	DISEASE SEVERITY
5 annual	5 plants killed
4	4
3	**3**
2	2
1 rare	1 very little damage

CHEMICAL TREATMENT	CULTURAL PRACTICES
3 used every year	**3 very important**
2	2
1 not used	1 not important

SANITATION	RESISTANT CULTIVARS
3 very important	3 many cultivars
2	**2**
1 not important	1 no resistance

Epidemiology

A. tumefaciens can live for several years in the soil, often spreading from diseased nursery stock. It can also be moved by irrigation water and cultivation equipment. It enters wounds to begin the infection process and tumor formation.

Host Range

All *Prunus* spp., including Colt cherry rootstock, are reported to be quite susceptible. More than 600 plant species have been reported to be susceptible to *A. tumefaciens.*

Management

Plant only disease-free nursery stock after inspecting new trees to avoid planting any with gall symptoms. Do not plant nursery trees in soil with a recent history of crown gall. Use care in planting trees, avoiding injury to bark around the crown, because bacteria in the soil can enter the injured tissue. Plant in well-drained soils, when the soil temperature is below 50°F.

Preplant soil solarization has been effective for nursery stock in western Oregon. Solarize the soil by placing clear plastic on ground tilled with a rotary cultivator and irrigated nearly to field capacity, from mid-July to mid-September. This is more effective on sandy loam soil.

A. radiobacter strain 84 (see Chapter 98, "Biological Control") can be applied as a biological control to prevent infection but has no effect on latent (symptomless) infections and existing galls. A suspension of strain 84 may be used as a dip or spray. Thoroughly cover grafting wood, roots, and crowns, and spray until runoff.

Gummosis

Many causes

Mechanical injury, winter injury, insect damage, fungal diseases, and improper growing conditions can cause gumming, which often follows brown rot, shot hole (of peach and apricot), and twig infections. Bacterial canker, caused by *P. syringae,* can also produce a severe gumming of *Prunus* spp. However, gumming may also be spontaneous, especially in trees in which growth has been forced by too much water or nitrogen fertilizer (or both). Sweet cherry trees in wet or other unfavorable locations are particularly subject to gummosis.

Symptoms

In trees with gummosis, gum is exuded from buds, twigs, branches, or trunks. Pools or large deposits of gum collect beneath the bark at the crotch, on larger branches, or on the trunk. Gum eventually breaks through to the surface and runs down the bark.

Host Range

All *Prunus* spp. are subject to gummosis.

Management

Control insects and fungus diseases. In large cankers, cut away all dead tissue until a sound surface is exposed. Treat the wound with a reliable disinfectant. Follow cultural practices that produce firm, stocky, moderate growth rather than forced growth of soft wood. Prevent trunk injury when possible. Shielding or white-washing trunks can prevent winter injury.

Leaf Spot

Blumeriella jaapii **(syn. *Coccomyces hiemalis*)**

Losses due to leaf spot are associated with a weakening of the tree as a result of early summer defoliation.

Symptoms

Variably colored spots develop on the upper surface of leaves. The spots rapidly enlarge and turn brown or purple, and affected tissue dies from the center of the lesion outward. The lesions are round or irregularly shaped and occur over the entire leaf surface. Cream-colored spore masses of the pathogen appear on the lower leaf surface beneath the spots. Individual spots never become large, but numerous spots merge together to kill large areas of the leaf. After leaf spots have developed, affected leaves turn yellow and drop from the tree. The area adjacent to a spot may remain green while the rest of the leaf turns yellow. Diseased leaf tissue may separate from healthy tissue and drop out, giving the leaf a shot hole appearance.

Epidemiology

The fungal pathogen, *Blumeriella jaapii,* overwinters on fallen cherry leaves and in spring produces large numbers of spores on leaves infected during the previous year. The spores are spread by air currents and rain. In spring, with moisture, they invade the stomata of young leaves to initiate new infections. Once unfolded, leaves are susceptible throughout the growing season, but susceptibility decreases with age.

Host Range

Leaf spot has been reported as a disease of many hosts, including *P. caroliniana, P. laurocerasus, P. serrulata,* and *P. virginiana.*

Management

Rake up and destroy infected leaves. Apply a fungicide at petal fall, shuck fall, and two weeks later. Postharvest applications are

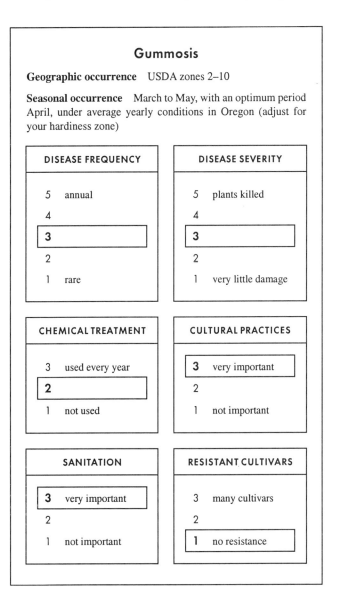

Gummosis

Geographic occurrence USDA zones 2–10

Seasonal occurrence March to May, with an optimum period April, under average yearly conditions in Oregon (adjust for your hardiness zone)

DISEASE FREQUENCY	DISEASE SEVERITY
5 annual	5 plants killed
4	4
[3]	**[3]**
2	2
1 rare	1 very little damage

CHEMICAL TREATMENT	CULTURAL PRACTICES
3 used every year	**[3] very important**
[2]	2
1 not used	1 not important

SANITATION	RESISTANT CULTIVARS
[3] very important	3 many cultivars
2	2
1 not important	**[1] no resistance**

Leaf Spot

Geographic occurrence USDA zones 3 to 8

Seasonal occurrence March to June, with an optimum period in May, under average yearly conditions in Oregon (adjust for your hardiness zone)

DISEASE FREQUENCY	DISEASE SEVERITY
5 annual	5 plants killed
[4]	4
3	**[3]**
2	2
1 rare	1 very little damage

CHEMICAL TREATMENT	CULTURAL PRACTICES
3 used every year	**[3] very important**
[2]	2
1 not used	1 not important

SANITATION	RESISTANT CULTIVARS
[3] very important	3 many cultivars
2	**[2]**
1 not important	1 no resistance

helpful in wet years. Rotate or tank-mix materials to prevent the development of resistant strains. A forecasting program from Michigan is available to help time applications. Tests in western Oregon show the program to be useful at higher temperatures.

Powdery Mildew

Podosphaera clandestina

Outbreaks of powdery mildew in *Prunus* spp., caused by the fungus *Podosphaera clandestina,* have been infrequent.

Symptoms

The first symptoms of powdery mildew on leaves are light green, circular lesions on either leaf surface. The first few infected leaves may be found on the leaves of suckers, branches close to the ground, or leaves on the main scaffold branches near tree crotches. A white, powdery growth develops in infected areas. Severe leaf infection can result in curling and blistering, and leaves become covered with the characteristic white, powdery growth. The pathogen forms small, black fruiting bodies (cleistothecia) on infected leaves as they age.

Epidemiology

P. clandestina overwinters as cleistothecia on dead leaves or trapped in bark crevices. In spring, as buds break, irrigation or rain releases ascospores, and wind spreads them to young leaves. With the right combination of temperature and leaf wetness duration, ascospores start the first colonies of the season. The colonies produce more (asexual) spores around shuck fall and continue the disease cycle. These secondary spores are favored by warm, humid conditions and the presence of dry leaves. The fungus can complete its life cycle in as little as two days under favorable conditions.

Host Range

All sweet and sour cherry cultivars are susceptible to powdery mildew, but the disease is most severe in Bing cherry. It has also been reported in *P. lyonii* and *P. virginiana.*

Management

Remove and destroy sucker shoots. Keep irrigation water off foliage by using low-angle irrigation nozzles. Follow cultural practices that promote good air circulation, such as pruning and judicious nitrogen management. Chemical control should begin at shuck fall; however, significant rain between budbreak and shuck fall warrants application as soon as possible.

Powdery Mildew

Geographic occurrence USDA zones 3–10

Seasonal occurrence May to June, with an optimum period in May, under average yearly conditions in Oregon (adjust for your hardiness zone)

DISEASE FREQUENCY	DISEASE SEVERITY
5 annual	5 plants killed
4	4
[3]	3
2	**[2]**
1 rare	1 very little damage

CHEMICAL TREATMENT	CULTURAL PRACTICES
3 used every year	**[3] very important**
[2]	2
1 not used	1 not important

SANITATION	RESISTANT CULTIVARS
[3] very important	3 many cultivars
2	**[2]**
1 not important	1 no resistance

Shot Hole

Many causes

Shot hole symptoms are commonly observed in *Prunus* spp. They are associated with leaf spots caused by the bacterium *Pseudomonas syringae* pv. *syringae* or any of several fungi, including a *Cercospora* sp., a *Blumeriella* sp., and *Wilsonomyces carpophilus* (the Coryneum blight pathogen); several earwigs; cherry mottle leaf virus; Prunus ringspot virus; prune dwarf virus; copper spray injury; and boron toxicity. When only advanced symptoms are present, it is not possible to definitely identify the cause.

Symptoms

Abscission layers develop around necrotic leaf spots, and the injured tissue drops away, leaving holes and tattered areas in the leaf (as though someone had fired a shotgun at the leaf – thus the name *shot hole*). After the tissues have dropped out, it is often difficult to determine specifically what caused the initial injury. Observations of early symptom development, signs, and symptoms on other areas of the plant may be helpful in making an accurate diagnosis.

Epidemiology

Cherrylaurel and sometimes other *Prunus* sp., including cherry and plums, commonly show shot hole symptoms resulting from cultural or environmental stress. Research has failed to identify what specific stress is responsible. Both container and field-grown laurel can develop symptoms. See also the information on specific biotic diseases that can result in shot hole symptoms for details concerning disease development.

Host Range

P. laurocerasus (cherrylaurels, including common or English laurel and the cultivars Otto Luyken and Zabeliana) are highly susceptible to shot hole.

Management

Avoid overhead irrigation. Remove and destroy fallen leaves. Do not plant near other flowering or fruiting *Prunus* spp. Chemical control can be used once the identity of the primary problem has been determined. No management practices have been shown to be beneficial in reducing physiological shot hole.

Verticillium Wilt

Verticillium dahliae

Verticillium wilt usually affects young trees coming into bearing, but it also occurs in older trees. It is caused by *Verticillium dahliae,* a soilborne fungus that can remain viable in soil for years. The disease affects numerous plants, including apricot, geranium, lambsquarters, maple, nightshade, peach, pepper, phlox, potato, raspberry, shepherdspurse, strawberry, and tomato. See Chapter 19, "Verticillium Wilt."

Symptoms

Some lower leaves yellow, and later so do higher leaves. Twigs and branches often wilt and die. Leaves may turn reddish orange. Spurs and twigs may die so rapidly that leaves remain attached. Leaves of current-season shoots and older wood may drop off or may be less numerous than on healthy trees, giving trees an open or bare appearance. A brown red discoloration of the sapwood of some diseased twigs and branches can be observed if they are cut open with a pocketknife.

Epidemiology

Microsclerotia of *V. dahliae* germinate and infect roots. The fungus grows through the vascular system, up into the trunk and branches. The cambium may die, resulting in an elongate canker, which can be colonized by other pathogens. Current-season sap-

Shot Hole

Geographic occurrence USDA zones 3–10

Seasonal occurrence April to June, with an optimum period in April, under average yearly conditions in Oregon (adjust for your hardiness zone)

DISEASE FREQUENCY	DISEASE SEVERITY
5 annual	5 plants killed
[4]	**[4]**
3	3
2	2
1 rare	1 very little damage

CHEMICAL TREATMENT	CULTURAL PRACTICES
3 used every year	**[3] very important**
[2]	2
1 not used	1 not important

SANITATION	RESISTANT CULTIVARS
[3] very important	3 many cultivars
2	2
1 not important	**[1] no resistance**

Verticillium Wilt

Geographic occurrence USDA zones 3–8

Seasonal occurrence June to August, with an optimum period June, under average yearly conditions in Oregon (adjust for your hardiness zone)

DISEASE FREQUENCY	DISEASE SEVERITY
5 annual	5 plants killed
4	4
3	**[3]**
[2]	2
1 rare	1 very little damage

CHEMICAL TREATMENT	CULTURAL PRACTICES
3 used every year	**[3] very important**
2	2
[1] not used	1 not important

SANITATION	RESISTANT CULTIVARS
[3] very important	3 many cultivars
2	2
1 not important	**[1] no resistance**

wood may not be infected, and symptoms may not reappear, or infection may occur without foliar symptoms. This may result in branch dieback or bud failure in spring. After diseased plant parts die, microsclerotia form and live for several years in soil. Many weeds are susceptible and can help the fungus survive and disperse. Disease incidence and severity can be greater in plants injured by parasitic nematodes.

Host Range

V. dahliae has been reported on *P. cerasifera* and *P. laurocerasus.*

Management

A preplant soil test for propagules of *V. dahliae* will aid in site selection. Do not plant in soils where *Verticillium*-susceptible crops have been grown previously. Control weeds. Avoid excessive irrigation, severe pruning, or other measures that promote succulent growth. Prune out dead branches. Trees have been shown to recover following proper cultural care.

Witches'-Broom

Geographic occurrence USDA zones 3–9

Seasonal occurrence April to May, with an optimum period in April, under average yearly conditions in Oregon (adjust for your hardiness zone)

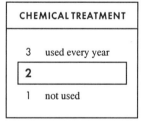

DISEASE FREQUENCY	DISEASE SEVERITY
5 annual	5 plants killed
4	4
3	3
2	**2**
1 rare	1 very little damage

CHEMICAL TREATMENT	CULTURAL PRACTICES
3 used every year	**3 very important**
2	2
1 not used	1 not important

SANITATION	RESISTANT CULTIVARS
3 very important	3 many cultivars
2	**2**
1 not important	1 no resistance

Witches'-Broom and Leaf Curl

Taphrina spp.

Several diseases of *Prunus* are caused by species of *Taphrina,* cause including peach leaf curl, caused by *T. deformans;* plum pockets, caused by *T. communis;* and witches'-broom, sometimes referred to as cherry leaf curl, caused by *T. wiesneri.*

Symptoms

Initially infected peach leaves are yellow to reddish, thickened, and crisp in texture. The highly distorted leaves eventually develop a dusty white coating of spores of the pathogen. Infected leaves may fall or turn brown and remain on the tree.

On cherries, large, broom-like tufts develop on branches. They are easily distinguished at blossom time, when they have few flowers, and they become leafy earlier than normal branches. At the base of the witches'-broom is a common stem, which may be much thicker than the branch to which it is attached. Long, slender branches grow from this stem. Affected leaves somewhat resemble peach leaves affected by peach leaf curl: they are thick and reddish, with a white growth of saclike structures (asci) of the pathogen on the undersurface.

Plum fruit infected by *T. communis* becomes greatly enlarged and distorted and fails to form a pit.

Epidemiology

Witches'-broom Ascospores from diseased cherry leaves fall on buds and, upon germinating, penetrate the branch, stimulating it to abnormal growth year after year. Once a cherry branch is infected, it will always remain so, and the leaves on the broom will be diseased every year.

Peach leaf curl Spores overwinter on bark, twigs, and old infected leaves. Infection occurs through bud scales in mid- to late winter, just as the buds begin to swell. The spores can remain dormant and survive on the surface of twigs and branches for several years.

Host Range

All *Prunus* spp. are susceptible to witches'-broom and leaf curl.

Management

Prune out the witches'-brooms. Trace the main stem of the broom to where it grows from a normal branch, and then cut the branch at least 12 inches below that point. Chemical control can be implemented at the delayed dormant stage of growth.

Diseases Caused by Viruses and Viruslike Agents

Numerous viruses and viruslike agents can cause problems in *Prunus.* Many of them are eliminated by intensive cleanup programs; however, reinfection can occur at any time. Some *Prunus* cultivars are selected for unusual horticultural properties that are lost once they go through the virus cleanup procedure. Flowering types may also be symptomless carriers of several important tree fruit viruses.

Symptoms

A wide variety of symptoms can occur, from deformed to off-color leaves, leaf chlorosis, veinbanding, leaf necrosis, shot hole, puckering, enations, ring spots, defoliation, stunted shoot growth

Virus and Viruslike Diseases

Geographic occurrence USDA zones 2–10

Seasonal occurrence March to July, with an optimum period April under average yearly conditions in Oregon (adjust for your hardiness zone)

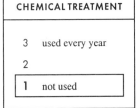

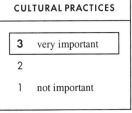

DISEASE FREQUENCY	DISEASE SEVERITY
5 annual	5 plants killed
4	4
3	3
2	**2**
1 rare	1 very little damage

CHEMICAL TREATMENT	CULTURAL PRACTICES
3 used every year	**3 very important**
2	2
1 not used	1 not important

SANITATION	RESISTANT CULTIVARS
3 very important	3 many cultivars
2	**2**
1 not important	1 no resistance

(including shortened internodes), reduced flowering, and many fruit symptoms. Symptom development may be dependent on environmental conditions. Day length and temperature often influence symptom development. For example, trees can show symptoms in the spring and early summer but show no leaf symptoms later in the growing season, even though they are still infected. Shock symptoms may develop in some trees when they are initially infected, but the trees may show few or greatly subdued symptoms in future years.

Symptoms of diseases caused by viruses and viruslike agents are easily confused with those of nutritional disorders, herbicide injury, and damage due to unfavorable environmental conditions. Many of the common virus diseases can be accurately identified by symptom development alone, but testing by a virus diagnostic laboratory is usually required for accurate diagnosis. Trees can be infected by more than one virus, making visual diagnosis even more difficult.

Epidemiology

All of the viruses attacking *Prunus* spp. can be transmitted from an infected plant to a healthy plant through grafting. Insects, mites, and nematodes can also transmit certain viruses, and some can be transmitted in pollen.

Host Range

All *Prunus* spp. are susceptible to virus and viruslike diseases.

Management

There is no effective treatment for virus-infected trees in nurseries or landscapes. Infected trees should be removed and destroyed. The urgency of removal depends on the virus or viruses and the vector that are involved.

Heat therapy is commonly used to produce virus-free propagation material. Buds taken from infected plants are grafted onto a seedling rootstock and grown for 50 to 70 days at a temperature of 100°F (38°C). Shoot tips are excised and cleft-grafted onto a succulent side branch of one-year-old seedling rootstock. The tree is grown at 78°F (25°C) in a humid chamber until it becomes adapted to ambient conditions. It continues to grow and is later tested for viruses. Budwood is then propagated from the virus-free tree for production.

Other Diseases

Prunus spp. may also be affected by Armillaria root rot, Cytospora canker, ring nematodes, root-lesion nematodes, Phytophthora root rot, replant disease, and silver leaf. These diseases may cause affect mature or cultivated fruit trees more than nursery stock, however.

REFERENCES

Anonymous. 1976. Virus diseases and noninfectious diseases of stone fruits in North America. U.S. Dep. Agric. Agric. Handb. 437.

Dirr, M. A. 1998. Manual of Woody Landscape Plants: Their Identification, Ornamental Characteristics, Culture, Propagation and Uses. 5th ed. Stipes Publishing, Champaign, Ill.

Grove, G. G. 1989. Powdery mildew of cherry. Wash. Coop. Ext. Plant Dis. Bull. EB 1539.

Ogawa, J. M., Zehr, E. I., Bird, G. W., Ritchie, D. F., Uriu, K., and Uyemoto, J. K. 1995. Compendium of Stone Fruit Diseases. American Phytopathological Society, St. Paul, Minn.

Pscheidt, J. W., and Ocamb, C. M. 2001. Pacific Northwest Plant Disease Control Handbook. Oregon State University Extension Service, Corvallis.

CHAPTER 73

Karen K. Rane ▪ Purdue University, West Lafayette, Indiana

Pyracantha Diseases

Geographic production USDA zones 5–9 (some hardy cultivars are produced in zone 5)

Family Rosaceae

Genus *Pyracantha*

Species *P. angustifolia*
P. atalantioides
P. coccinea
P. crenulata
P. fortuneana (formerly *P. crenato-serrata*)
P. koidzumii
P. rogersiana

Hybrids numerous

Pyracantha is a multipurpose ornamental used in the landscape for screens, foundation plantings, single specimen plants, and espaliers. As anyone who comes in contact with pyracantha soon learns, its common name *firethorn* was given not only for its sharp thorns but the lingering sting after contact. Pyracantha requires pruning to maintain its appeal in the landscape. It does well in soil between pH 5.5 and 7.5 and in areas with dry soils in summer. Only two diseases, fire blight and scab, are significant problems in pyracantha.

Fire Blight
Erwinia amylovora

Pyracantha is one of a number of members of the family Rosaceae that are susceptible to fire blight. For more information on this destructive bacterial disease, see Chapter 40, "Flowering Crabapple Diseases."

Symptoms

Infected new shoots and blossoms wilt and turn black in late spring. Blackened leaves and flowers remain attached to the stems, and the stems may curve downward in a shepherd's-crook shape. Cream-colored bacterial ooze may form on newly blighted shoots under moist conditions. Sunken cankers may develop on branches at the base of blighted shoots.

Epidemiology

The bacterial pathogen, *Erwinia amylovora,* overwinters in cankers from the previous year's infections. In the spring, bacteria multiply at the canker margins, forming an ooze, which may drip from the canker. The ooze is attractive to insects, such as bees, which may carry the pathogen to flowers or developing shoots. Bacterial ooze may also be spread from plant to plant during spring rainstorms. The pathogen enters the host through natural openings in flowers or developing stem tissue, causing discoloration of infected plant parts. Infections can also occur later in the

Fire Blight

Geographic occurrence USDA zones 6–9

Seasonal occurrence Primarily mid-spring through midsummer

DISEASE FREQUENCY	DISEASE SEVERITY
5 annual	**5** plants killed
4	**4**
3	**3**
2	2
1 rare	1 very little damage

CHEMICAL TREATMENT	CULTURAL PRACTICES
3 used every year	**3** very important
2	2
1 not used	1 not important

SANITATION	RESISTANT CULTIVARS
3 very important	3 many cultivars
2	**2**
1 not important	1 no resistance

growing season, when rainstorms spread the bacterium to wounds on leaves or stems. Disease spread is greatly dependent upon moisture – there is very little fire blight when the weather is dry in spring and early summer. Excessive nitrogen encourages the production of soft, succulent shoots, which are very susceptible to infection.

Host Range

Species and cultivars of *Pyracantha* vary in susceptibility to fire blight. Susceptible species include *P. angustifolia, P. koidzumii,* and *P. atalantioides. P. crenulata* is reported to be moderately susceptible, while *P. coccinea* var. *lalandii* and *P. fortuneana* are cited as being moderately resistant. *Pyracantha* hybrids Mohave, Navaho, Shawnee, Teton, and Yunan are reported to be resistant. Many plants in the pome fruit group of genera in the family Rosaceae are susceptible to fire blight: apple, crabapple, pear, mountain ash, etc.

Management

Severely infected plants should be removed. Plants with only a few infected branches can be rescued by careful pruning. Diseased branches should be pruned during the dormant season and in dry weather, to reduce the chances of spreading the bacterium during pruning. Infected twigs should be pruned at least 4 inches below the base of a visible canker during dormancy and at least 12 inches below the base of the canker if the shrubs are actively growing. Pruning tools should be disinfested between cuts. Bactericides and biocontrol agents may help protect new flowers and shoots from infection.

Scab

Spilocaea pyracanthae

Scab is a common fungal disease causing defoliation and fruit blemishes, which diminish the ornamental appeal of the host. *Spilocaea pyracanthae,* the fungus causing scab of pyracantha, is different from the apple scab pathogen, but the two diseases have many similarities in symptoms, epidemiology, and management.

Scab

Geographic occurrence USDA zones 5–9

Seasonal occurrence Throughout the growing season

DISEASE FREQUENCY	DISEASE SEVERITY
5 annual	5 plants killed
4	4
3	3
2	2
1 rare	1 very little damage

CHEMICAL TREATMENT	CULTURAL PRACTICES
3 used every year	3 very important
2	2
1 not used	1 not important

SANITATION	RESISTANT CULTIVARS
3 very important	3 many cultivars
2	2
1 not important	1 no resistance

Symptoms

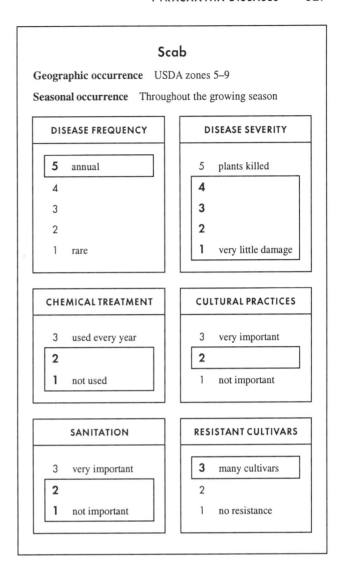

Olive green to black spots form on leaves, twigs, and fruit. Fungal growth in the spots can give them a velvety appearance. Heavily infected leaves are often yellow or reddish. Fruit can be completely covered with cracked, scabby, dark lesions. Infected leaves and fruit of highly susceptible cultivars may drop prematurely.

Epidemiology

S. pyracanthae overwinters in infected plant tissue on the host and possibly in infected plant debris on the ground. Spores from these tissues are splashed by rain to young, emerging leaves and flower buds in the spring. Infection requires a film of water on the plant surface. Secondary inoculum produced in the initial lesions results in the spread of the disease throughout the growing season. Mild summer temperatures and frequent rain favor disease development.

Host Range

Scab has been reported in all species of *Pyracantha,* but there is a wide range of susceptibility among cultivars. Plants considered resistant include *P. coccinea* cultivars Government Red and Prostrata; *P. koidzumii* cultivars Bella, Duval, and Santa Cruz Prostrata; *P. rogersiana* cultivar Flava; and *Pyracantha* hybrids Fiery Cascade, Mohave, Navaho, Rutgers, and Shawnee.

Management

The best means of managing scab is to plant resistant cultivars. Removal of infected shoots, fruits, and fallen leaves can aid in reducing inoculum that causes primary infections in the spring. Overhead irrigation can spread the pathogen to new foliage and fruit as well as provide films of water necessary for infection, so avoiding this method of irrigation will reduce disease incidence. The disease can be managed with fungicide sprays beginning when buds swell in the spring and repeated until dry, warm summer conditions occur.

REFERENCES

Dirr, M. A. 1998. Manual of Woody Landscape Plants: Their Identification, Ornamental Characteristics, Culture, Propagation and Uses. 5th ed. Stipes Publishing, Champaign, Ill.

Moorman, G. 1992. Scouting and Controlling Woody Ornamental Diseases in Landscapes and Nurseries. Pennsylvania State University, University Park.

Pirone, P. P. 1978. Diseases and Pests of Ornamental Plants. 5th ed. John Wiley & Sons, New York.

Pscheidt, J. W., and Ocamb, C. M. 2001. An Online Guide to Plant Disease Control. On-line publication. Oregon State University Extension Service, Corvallis.

Sinclair, W. A., Lyon, H. H., and Johnson, W. T. 1987. Diseases of Trees and Shrubs. Comstock Publishing Associates, Cornell University Press, Ithaca, N.Y.

CHAPTER 74

Karel A. Jacobs ▪ Morton Arboretum, Lisle, Illinois
D. Michael Benson ▪ North Carolina State University, Raleigh

Redbud Diseases

Geographic production USDA zones 4–9

Family Fabaceae

Genus *Cercis*

Species	*C. canadensis*	eastern redbud	zones 4–9
	C. chinensis	Chinese redbud	zones 6–9
	C. occidentalis	western redbud	zone 7 (Calif.)
	C. reniformis		zones 7–9
	C. siliquastrum	Judas-tree	zone 6

Eastern redbud is a native tree from New Jersey to northern Florida and west to Missouri, Texas, and northern Mexico. In cultivation since the mid-1600s, this floriferous landscape tree is adapted to most of the eastern United States as far north as Minnesota. Several cultivars have been selected, even some with white flowers. The most damaging disease of redbud is Botryosphaeria canker, but it is also susceptible to several other fungal diseases, including Verticillium wilt, Nectria canker, and Phytophthora root rot. Redbud has no important foliar diseases, although several leaf spot diseases may occur locally under environmental conditions favorable for disease development. Cold-hardiness can be a problem in trees grown in the north.

Botryosphaeria Canker

Botryosphaeria dothidea
(asexual stage, *Fusicoccum aesculi*)

Botryosphaeria canker of redbud was first reported in the mid-Atlantic states in 1939 (Watson, 1941). The disease is considered the greatest limitation on the growth of redbud in the landscape. It is less prevalent in nursery plants than landscape plants, but this may be due to a latent phase of the pathogen.

The fungus causing this disease was initially identified as *Botryosphaeria ribis,* but it is now considered to be *B. dothidea,* which also attacks apple, peach, and over 100 other woody plants in the United States (Brown and Britton, 1986; Sinclair et al., 1987). However, not all researchers agree that these two species are synonymous. Jacobs and Rehner (1998) found that the DNA of the redbud pathogen differs from that of strains isolated from peach, apple, and other hosts but is identical to that of strains isolated from *Melaleuca* in Florida and identified as *B. ribis.*

Although early research suggested that a single species infects numerous hosts (Punithalingam and Holliday, 1973), it appears more likely that a collection of fungi or a species complex is responsible for the numerous canker diseases attributed to *B. dothidea.*

Symptoms

The most obvious symptom of Botryosphaeria canker of redbud is a flag, or a branch on which the foliage dries out and turns orangish brown. Flags may occur on plants as young as one or two years old but are more typical in older plants in the landscape and nursery. Flagging usually begins in early summer. Earlier symptoms include a gradual loss of turgor and a faded green color in leaves. Close examination of a flagged branch or the juncture between a flagged branch and the trunk will reveal a sunken lesion, or canker, often with cracked bark and callused ridges (Plate 133). The bark and tissue underlying the lesion are discolored or dead.

Fruiting bodies of the fungus develop within cankers and, when viewed with a hand lens, appear as groups of black dots protruding through the bark. If a thin layer of the outer bark is scraped away, the fruiting bodies are cut open and appear white. This is diagnostic for cankers caused by *Botryosphaeria* and closely related fungi.

In young seedlings, cankers may girdle and kill the plants in one season. In older trees, branches are girdled one by one, resulting in weakened trees with deformed architecture (Plate 134).

Epidemiology

In general, Botryosphaeria canker is stress-related. The fungus is considered opportunistic and readily emerges on plants subjected to drought, cold injury, transplant, or other stresses (Cline, 1994; Crist and Schoeneweiss, 1975). The disease may also have a substantial latent phase and is probably resident in natural woodlands and forests, developing slowly as trees age. The fungus is spread by water (splashing rain and overhead irrigation) and, to a lesser extent, by wind and insects, which carry asexual spores (conidia) and sexual spores (ascospores) to healthy trees. Wounds and dying branches on otherwise healthy trees are common infection sites, but the fungus may also colonize natural plant openings, such as lenticels (Brown and Hendrix, 1981; Luttrell, 1950). Succulent green stems do not appear to be susceptible to infection, nor is the root system affected. Seeds can harbor the fungus, but the frequency of seed infection is apparently low and variable (Jacobs, 1996). *B. dothidea* remains viable indefinitely, perhaps several years, in the bark of diseased or dead trees, which serves as an important source of inoculum. If the bark is removed from infected branches, the fungus may die out (Drake, 1971).

Another *Botryosphaeria* species, *B. obtusa,* also occurs on redbud, but it did not cause substantial disease in greenhouse trials (K. A. Jacobs, unpublished data). Its role in the etiology of canker is unknown; it may become established after a branch has already been killed or colonized by *B. dothidea,* as is postulated to be its role in Botryosphaeria canker of apple (Brown and Britton, 1986).

Host Range

Cultivated redbuds (mainly *C. canadensis* and *C. chinensis*) are susceptible to canker wherever they are grown, but the disease is reportedly infrequent in nursery stock grown in the Midwest and the Pacific Northwest. The incidence of canker is greater in plants damaged by frost in these areas. *C. chinensis* is apparently less prone to canker than *C. canadensis* under natural conditions (Wester et al., 1950), but both species, as well a number of other *Cercis* species tested under greenhouse conditions, develop canker if artificially inoculated (Jacobs and Johnson, 1994). The disease has been observed in much of the natural range of *C. canadensis* in the eastern United States (Watson, 1941; Anonymous, 1960) as well as in forests and woodlands of Illinois and Indiana (K. A. Jacobs, unpublished data). It has been observed in naturally occurring *C. occidentalis* in California (Brooks and Ferrin, 1994). Nonetheless, a thorough survey of the disease in *Cercis* species throughout their natural distribution has not been conducted.

Botryosphaeria Canker

Geographic occurrence Wherever redbuds are cultivated, and in naturally occurring redbuds throughout their range; the exact distribution of the disease is not known

Seasonal occurrence May to October, with an optimum period for symptom development in July–August (mid- to late summer)

DISEASE FREQUENCY	DISEASE SEVERITY
5 annual	5 plants killed
4	**4**
3	**3**
2	2
1 rare	1 very little damage

CHEMICAL TREATMENT	CULTURAL PRACTICES
3 used every year	**3 very important**
2	2
1 not used	1 not important

SANITATION	RESISTANT CULTIVARS
3 very important	3 many cultivars
2	**2**
1 not important	**1 no resistance**

Management

Management of canker starts with planting clean seed. This is particularly important with redbud, as it is mostly seed-propagated. A simple seed dip in hot water or oil (131°F) for 10 minutes was found to eradicate both *B. dothidea* and *B. obtusa* from infected seed lots without reducing germinability (Jacobs, 1996).

To minimize the buildup of inoculum in dead and dying branches, infected branches and, if necessary, whole plants should be pruned and discarded as soon as canker is evident. Pruning cuts should be made several inches below the discolored wood underlying a canker. The discoloration can easily be seen if the bark is scraped away. Use clean shears when cutting into healthy tissue. Also, to minimize the likelihood of dispersing spores, avoid pruning during rainy periods, and do not use overhead irrigation.

Composting infected wood and bark will help eliminate potential inoculum as well as recycle green waste. Turn compost piles regularly, and make sure the pile temperature reaches at least 122°F (50°C).

Cultural practices that promote good tree health, such as planting redbuds at an appropriate site (with partial shade and good drainage) and providing adequate water and nutrients, will help to keep stresses in check and thereby help to prevent canker.

No immunity to canker has been found among *Cercis* spp. However, a few researchers are continuing to work toward developing a resistant redbud.

Several fungicides may provide protection against infection by *Botryosphaeria,* but chemical control is inefficient, because of the perennial nature of the disease and the potential for infection year-round. Canker of redbud can best be held in check by prevention, which includes maintaining tree health, planting under conditions appropriate for the *Cercis* species, and rapid removal of diseased plants.

Verticillium Wilt
Verticillium dahliae

Verticillium wilt can develop in field-grown redbud trees in nurseries in most eastern production areas. The fungal pathogen survives in soil for many years and infects root tips as the root system expands over time. *Verticillium albo-atrum* was originally thought to cause wilt in redbud and other ornamentals, but morphological and molecular characters indicate that the pathogen in ornamentals is primarily *V. dahliae* (Smith and Neely, 1979; Chen, 1994; Sinclair et al., 1987).

Symptoms

Initial symptoms of Verticillium wilt include fading and yellowing of leaves. On multiple-stemmed trees, leaves on only one stem may be affected, at least initially. Eventually, wilting follows, and leaves become necrotic. It is common for the leaf tips to die first. *V. dahliae* is a vascular wilt pathogen. It causes greenish brown streaks of discoloration in current-year wood just below the bark.

Epidemiology

V. dahliae survives in soil for many years as specialized structures of melanized cells called microsclerotia. When a root tip grows through the soil near a microsclerotium, root exudates stimulate the microsclerotium to germinate and penetrate the root. The pathogen then invades the vascular system of the tree, primarily the xylem, as it moves from the root into the stem, causing discoloration of the xylem as it goes. Eventually the pathogen can reach the leaf tissue. See Chapter 19, "Verticillium Wilt," for more information.

Host Range

Verticillium wilt occurs in many different field, fruit, vegetable and ornamentals crops and has been reported in eastern redbud (*C. canadensis*), Chinese redbud (*C. chinensis*), and Judas-tree (*C. siliquastrum*) in the field (Himelick, 1969; Carter, 1945; Ostasheva, 1982). Another Asian redbud, *C. yunnanensis,* was found to be susceptible to Verticillium wilt in greenhouse inoculation experiments (Jacobs et al., 1994).

Verticillium Wilt

Geographic occurrence USDA zones 3–8

Seasonal occurrence May to August, with an optimum period during hot weather (e.g., in July)

DISEASE FREQUENCY	DISEASE SEVERITY
5 annual	**5 plants killed**
4	4
3	3
2	2
1 rare	1 very little damage

CHEMICAL TREATMENT	CULTURAL PRACTICES
3 used every year	**3 very important**
2	2
1 not used	1 not important

SANITATION	RESISTANT CULTIVARS
3 very important	3 many cultivars
2	2
1 not important	**1 no resistance**

Management

Redbud produced in clean soilless mix in containers would avoid Verticillium wilt, since the pathogen survives in field soil. Field production sites with infested soil could limit the production of redbud and should be avoided. Fungicides are not effective against *V. dahliae* in the field, but fumigants (e.g., metam sodium) are useful in production agriculture systems and may benefit commercial operations with persistent Verticillium wilt. Soil heating (solarization) has been successful in reducing losses due to Verticillium wilt in perennial food crops (Hartz et al., 1993) and may be useful for nurseries in climates with extended periods of hot, sunny weather. Balanced fertilization and sufficient watering will help minimize the disease, as plants stressed by overfertilization or drought are more prone to infection (Neely, 1982; Caroselli, 1957).

Nectria Canker

Nectria galligena

(asexual stage, *Cylindrocarpon heteronemum*)

Perennial Nectria canker can infect redbud in container and field production. The cankers are small during the first year and may go unnoticed before the trees leave the nursery. Target-shaped cankers with concentric ridges of wood can develop on landscape trees.

Symptoms

Most cankers develop around branch stubs or in wounds and natural openings (lenticels) in young, smooth bark. They expand during the growing season and may gradually girdle large branches or the main trunk after several years. After the first year, successive ridges of callus tissue develop around the edge of the canker as the tree resumes active growth. Each year, during the dormant season, the callus ridge is killed by the fungus, and the next growing season the tree forms a new ridge, so that the canker acquires a target appearance after several years. As the canker enlarges, sections of foliage above the affected stem become chlorotic and then die. The target appearance of the cankers and the reddish fruiting bodies of the pathogen differentiate Nectria canker from the more common Botryosphaeria canker.

Epidemiology

Cankers on older trees serve as a source of inoculum for new infections. The asexual stage of the fungus, *Cylindrocarpon heteronema* (also known as *C. mali*), produces spores (conidia) in sporodochia, which emerge through the lenticels of infected stems. Conidia are dispersed in rainstorms or by overhead irrigation throughout the year in warm climates, but only during the growing season in colder climates. Reddish fruiting bodies (perithecia) of the fungus in its sexual stage, *Nectria galligena,* produce two-celled sexual spores (ascospores) in cankered tissue. These spores are dispersed by wind and rain, primarily during the dormant season.

Host Range

Nectria canker affects trees in more than 30 genera in addition to redbud, including dogwood, elm, hawthorn, holly, magnolia, maple, oak, and pear (Sinclair et al., 1987). Chinese redbud (*C. chinensis*) has not been reported to be a host.

Management

The most important method of controlling Nectria canker in nurseries is to avoid any cultural activity that would wound young stems and branches, especially during the active sporulation cycle in spring and autumn. Growers report that cold injury appears to predispose young redbuds to this and other canker diseases. Use local seed sources, and protect young stems at sites prone to early frost. Avoid late-afternoon irrigation that leaves stems wet overnight. Prune and destroy cankered tissue. Diseased wood can be composted and reused as long as temperatures in the compost pile exceed about 122°F (50°C) and the pile is turned regularly. Fungicides are not effective for control of Nectria canker.

Phytophthora Root Rot

Phytophthora cinnamomi

Other *Phytophthora* spp.

Phytophthora root rot may develop in redbuds produced in field soil for the ball-and-burlap trade if the soil is infested with *P. cinnamomi* or other *Phytophthora* spp. Generally, the disease is limited to redbuds produced in the southeastern United States, where these pathogens occur in field soils that are poorly drained. However, Phytophthora root rot may also develop in redbuds in container production.

Symptoms

Symptoms of Phytophthora root rot of redbud include chlorosis of foliage and wilting, followed by shoot dieback and collapse. Foliage on only one stem of a multiple-stemmed plant may show symptoms initially, and symptoms progress to the foliage on other stems over time. Roots are discolored, and the cortex sloughs off from the stele easily. Trees more than one year old may take two or more growing seasons to die. It may be difficult to differentiate this disease from Verticillium wilt, but the latter causes streaks in the wood and does not rot the base of the tree and its roots as extensively as Phytophthora root rot.

Epidemiology

Chlamydospores of *P. cinnamomi* survive in field soil for many years. When root tips of redbud grow near a chlamydospore, it germinates to form a sporangium, which releases swimming zoo-

Nectria Canker

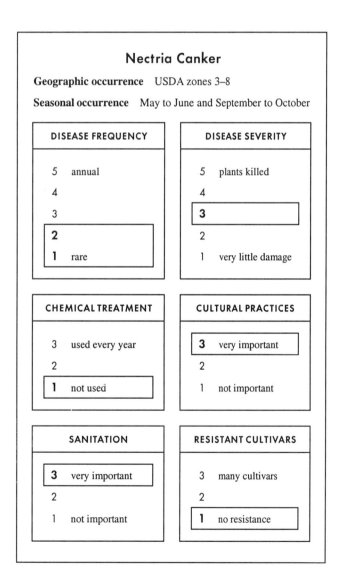

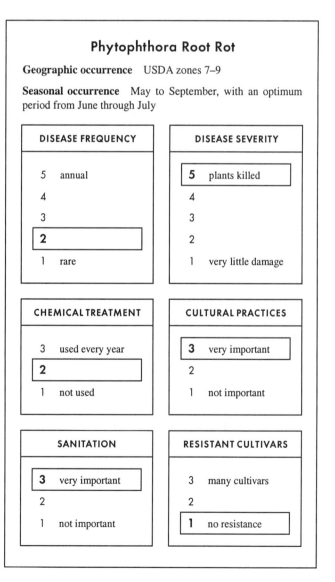

Geographic occurrence USDA zones 3–8

Seasonal occurrence May to June and September to October

DISEASE FREQUENCY
- 5 annual
- 4
- 3
- **2**
- **1 rare**

DISEASE SEVERITY
- 5 plants killed
- 4
- **3**
- 2
- 1 very little damage

CHEMICAL TREATMENT
- 3 used every year
- 2
- **1 not used**

CULTURAL PRACTICES
- **3 very important**
- 2
- 1 not important

SANITATION
- **3 very important**
- 2
- 1 not important

RESISTANT CULTIVARS
- 3 many cultivars
- 2
- **1 no resistance**

Phytophthora Root Rot

Geographic occurrence USDA zones 7–9

Seasonal occurrence May to September, with an optimum period from June through July

DISEASE FREQUENCY
- 5 annual
- 4
- 3
- **2**
- 1 rare

DISEASE SEVERITY
- **5 plants killed**
- 4
- 3
- 2
- 1 very little damage

CHEMICAL TREATMENT
- 3 used every year
- **2**
- 1 not used

CULTURAL PRACTICES
- **3 very important**
- 2
- 1 not important

SANITATION
- **3 very important**
- 2
- 1 not important

RESISTANT CULTIVARS
- 3 many cultivars
- 2
- **1 no resistance**

spores. These spores infect the plant just behind the root tip. The soil must be near saturation for germination to occur and for zoospores to be dispersed. Root infection is progressive and eventually leads to foliar chlorosis and wilting. See Chapter 14, "Phytophthora Root Rot and Dieback."

Host Range

More than 1,000 species of plants are susceptible to *P. cinnamomi,* including eastern redbud. However, Chinese redbud has not been reported as a host.

REFERENCES

Anonymous. 1960. Index of plant diseases in the United States. U.S. Dep. Agric. Agric. Handb. 165.

Brooks, F. E., and Ferrin, D. M. 1994. Branch dieback of southern California chaparral vegetation caused by *Botryosphaeria dothidea.* Phytopathology 84:78–83.

Brown, E. A., and Britton, K. O. 1986. *Botryosphaeria* diseases of apple and peach in the southeastern United States. Plant Dis. 70:480–484.

Brown, E. A., and Hendrix, F. F. 1981. Pathogenicity and histopathology of *Botryosphaeria dothidea* on apple stems. Phytopathology 71:375–379.

Caroselli, N. E. 1957. Verticillium wilt of maples. Univ. R.I. Agric. Exp. Stn. Bull. 335.

Carter, J. C. 1945. Isolations of *Verticillium* from trees and shrubs in Illinois, 1940–1945. Plant Dis. Rep. 29:95–96.

Chen, W. 1994. Vegetative compatibility groups of *Verticillium dahliae* from ornamental woody plants. Phytopathology 84:214–219.

Cline, W. O. 1994. Infection of cold-injured blueberry stems by *Botryosphaeria dothidea.* Plant Dis. 78:1010.

Crist, C. R., and Schoeneweiss, D. F. 1975. The influence of controlled stresses on susceptibility of European white birch stems to attack by *Botryosphaeria dothidea.* Phytopathology 65:369–373.

Drake, C. R. 1971. Source and longevity of apple fruit rot inoculum. Plant Dis. Rep. 55:122–126.

Goidanich, G. 1935. New cases of tracheomycosis caused by *Verticillium* in Italy. Boll. Stn. Patol. Veg. Rome 15:548–554. Cited in Himelick, 1969.

Hartz, T. K., DeVay, J. E., and Elmore, C. L. 1993. Solarization is an effective soil disinfestation technique for strawberry production. HortScience 28:104–106.

Himelick, E. B. 1969. Tree and shrub hosts of *Verticillium albo-atrum.* Ill. Nat. Hist. Surv. Biol. Notes, no. 66.

Jacobs, K. A. 1996. Presence of *Botryosphaeria* spp. in redbud seed and efficacy of seed treatments. (Abstr.) Phytopathology 86:S88.

Jacobs, K. A., and Johnson, G. R. 1994. Screening for disease resistance in redbud to Botryosphaeria canker. (Abstr.) U.S. Dep. Agric. Agric. Res. Serv. BARC Poster Day Proc., p. 19.

Jacobs, K. A., and Rehner, S. A. 1998. Comparison of cultural and morphological characters and ITS sequences in anamorphs of *Botryosphaeria* and related taxa. Mycologia 90:601–610.

Jacobs, K. A., Bentz, S. E., and Johnson, G. R. 1994. Screening three *Cercis* species for susceptibility to *Verticillium dahliae.* Plant Dis. 78: 925.

Luttrell, E. S. 1950. Botryosphaeria stem canker of elm. Plant Dis. Rep. 34:138–139.

Neely, D. 1982. Vascular wilt disease updates. J. Arboric. 8:285–288.

Ostasheva, N. A. 1982. Chinese *Cercis* is a potential carrier of the tracheomycosis of fruit trees. Byull. Gl. Bot. Sada 123:92–95.

Punithalingam, E., and Holliday, P. 1973. *Botryosphaeria ribis.* Descriptions of Pathogenic Fungi and Bacteria, no. 395. Commonwealth Mycological Institute, Kew, England.

Sinclair, W. A., Lyon, H. H., and Johnson, W. T. 1987. Diseases of Trees and Shrubs. Cornell University Press, Ithaca, N.Y.

Smith, L. D., and Neely, D. 1979. Relative susceptibility of tree species to *Verticillium dahliae.* Plant Dis. Rep. 63:328–332.

Watson, A. J. 1941. Studies of *Botryosphaeria ribis* on *Cercis* and *Benzoin.* Plant Dis. Rep. 25:29–31.

Wester, H. V., Davidson, R. W., and Fowler, M. E. 1950. Canker of linden and redbud. Plant Dis. Rep. 34:219–223.

CHAPTER 75

Margery L. Daughtrey ▪ Cornell University, Riverhead, New York

D. Michael Benson ▪ North Carolina State University, Raleigh

Rhododendron Diseases

Geographic production USDA zones 4–7 or 8

Family Ericaceae

Genus *Rhododendron*

Species	*R. carolinianum*	Carolina rhododendron	zones (4)5–8
	R. catawbiense	catawba rhododendron	zones 4–8
	R. maximum	rosebay rhododendron	zone 3
	R. mucronulatum	Korean rhododendron	zones 4–7

Hybrids many

Phytophthora Root Rot
Phytophthora spp.

Phytophthora root rot is a common disease of rhododendrons wherever they are grown. It occurs in container- and field-grown rhododendrons, particularly when soil drainage is poor. Losses occur in propagation as well. The disease is caused by numerous species of *Phytophthora,* of which *P. cinnamomi* is the most commonly found on rhododendron. Other species reported on rhododendron are *P. cactorum, P. cambivora, P. citricola, P. cryptogea, P. gonapodyides, P. heveae, P. lateralis, P. megasperma,* and *P. nicotianae* var. *parasitica.*

Symptoms

The foliage of infected cuttings develops a dull green cast, and the cuttings wilt and fail to initiate new shoots (Plate 135). The roots of cuttings with these symptoms are reddish brown rather than the normal healthy white.

Epidemiology

Frequent overhead watering, close plant spacing, and poor drainage contribute to the development of epidemics. Taking cuttings from diseased plants introduces the pathogen into propagation and thus leads to repeated cycles of infection in the nursery. Ordinarily *P. cinnamomi* does not survive winter conditions in the northeastern United States, but it may be introduced into a nursery annually on stock received from areas with milder climates.

Host Range

Many of the *Phytophthora* species affecting rhododendrons also infect other hosts. *P. cinnamomi, P. cactorum,* and *P. nicotianae* var. *parasitica,* in particular, are often found on rotted roots of other woody plants. Phytophthora root rots commonly occur in azalea, forsythia, heath, heather, juniper, leucothoe, and many other nursery crops.

Phytophthora Root Rot

Geographic occurrence USDA zones 5–8

Seasonal occurrence April to September, with an optimum period in June–August, under average yearly conditions in zone 7 (adjust for your hardiness zone)

DISEASE FREQUENCY	DISEASE SEVERITY
5 annual (selected)	**5 plants killed** (selected)
4	4
3	3
2	2
1 rare	1 very little damage

CHEMICAL TREATMENT	CULTURAL PRACTICES
3 used every year (selected)	**3 very important** (selected)
2	2
1 not used	1 not important

SANITATION	RESISTANT CULTIVARS
3 very important (selected)	3 many cultivars
2	**2** (selected)
1 not important	1 no resistance

Management

See Chapter 14, "Phytophthora Root Rot and Dieback," for comments relevant to the management of *Phytophthora* diseases in all susceptible crops.

Some resistant *Rhododendron* hybrids have been identified. Hoitink and Schmitthenner (1974) compared susceptibility to Phytophthora root rot in 336 hybrids and 198 species of rhododendron. The most resistant hybrids in their trial were Caroline, Professor Hugo de Vries, and Red Head. English Roseum was moderately resistant. One of the most susceptible rhododendrons was Purple Splendor. Twelve *Rhododendron* species with some resistance to *P. cinnamomi* were identified; *R. carolinianum* was one of the most susceptible.

Poorly drained soils allow disease development even in resistant rhododendrons. Composted hardwood bark media may be used to suppress root rot caused by *P. cinnamomi.*

Phytophthora Dieback

***Phytophthora* spp.**

Most *Phytophthora* spp. are better known as root pathogens, but under rainy, warm conditions several species can cause shoot dieback in rhododendrons. *P. cactorum, P. citricola, P. heveae, P. nicotianae* var. *parasitica,* and *P. syringae* have been reported to cause rhododendron dieback.

Dieback caused by *P. cactorum* was first reported in the 1930s and has become a significant disease in nurseries.

Soil surveys in the 1960s recorded *P. heveae* in forest stands in the Great Smoky Mountain National Park in Tennessee, but not until 1980 was this species described as a pathogen of rhododendron (Benson and Jones, 1980). No other hosts of *P. heveae* are known in the United States, but it is a blight and canker pathogen of avocado, Brazil nut trees, cocoa, kauri, and rubber in other parts of the world.

In the Pacific Northwest, *P. syringae* has been reported to cause stem dieback of field-grown rhododendron during the rainy, cold winter.

Symptoms

Brown blotches, irregular in outline, appear on infected leaves (Plate 136), or brown discolored areas develop along new shoots. Leaf lesions typically begin at the margin of the leaf undersurface, then extend through the leaf to the midrib, and next move through the petiole and into the stem. Sometimes the fungus grows along the stem and infects additional leaves.

Epidemiology

All of the Phytophthora dieback pathogens except *P. syringae* initiate infection during the growing season. New infections occur only on succulent current-season stems and leaves, but the fungus then slowly moves into older portions of the plant. In first-year plants the pathogen may move from a new leaf infection down to the soil line in seven days. Initial infections develop when zoospores are splashed from the soil to new foliage in early summer (Fig. 75.1). Symptoms occur within one or two days of infection. When infected tissue stays wet overnight, spores form and will be splashed by overhead irrigation or rain to nearby plants. The optimum temperature for sporulation and infection is 78–86°F (25–30°C), and infection may occur at temperatures as low as 50°F (10°C). Oospores may survive in debris in the soil for at least three years in the absence of a host, even when the soil temperature is near freezing.

In rhododendron leaves and stems infected with *P. syringae,* the disease develops during the winter. In the Pacific Northwest, winters are characterized by moderately low temperatures (40–50°F) and many rainy days. These conditions prove ideal for infection of leaves and stems of dormant rhododendrons in the field by *P. syringae.* The inoculum consists of sporangia or zoospores, which are splashed onto stems. Wounds may be needed for infection, along with low temperatures. Secondary dispersal of inoculum during the winter occurs from lesions formed on infected leaves and stems. The dieback can be extensive, but plants often form adventitious buds below infected stems, and these buds develop into new shoots at budbreak in the spring.

Host Range

Among nursery crops, rhododendron is the most common host of the Phytophthora dieback pathogens. *R. simsii, R. catawbiense,* an *Erica* sp., and a *Calluna* sp. have been affected by *P. citri-*

Phytophthora Dieback

Geographic occurrence USDA zones 5–8

Seasonal occurrence Infection by *Phytophthora syringae* occurs only in winter in mild climates; infection by other *Phytophthora* spp. occurs from June to August, with an optimum period in July, under average yearly conditions in zone 7 (adjust for your hardiness zone)

DISEASE FREQUENCY	
5	annual
4	
3	
2	
1	rare

DISEASE SEVERITY	
5	plants killed
4	
3	
2	
1	very little damage

CHEMICAL TREATMENT	
3	used every year
2	
1	not used

CULTURAL PRACTICES	
3	very important
2	
1	not important

SANITATION	
3	very important
2	
1	not important

RESISTANT CULTIVARS	
3	many cultivars
2	
1	no resistance

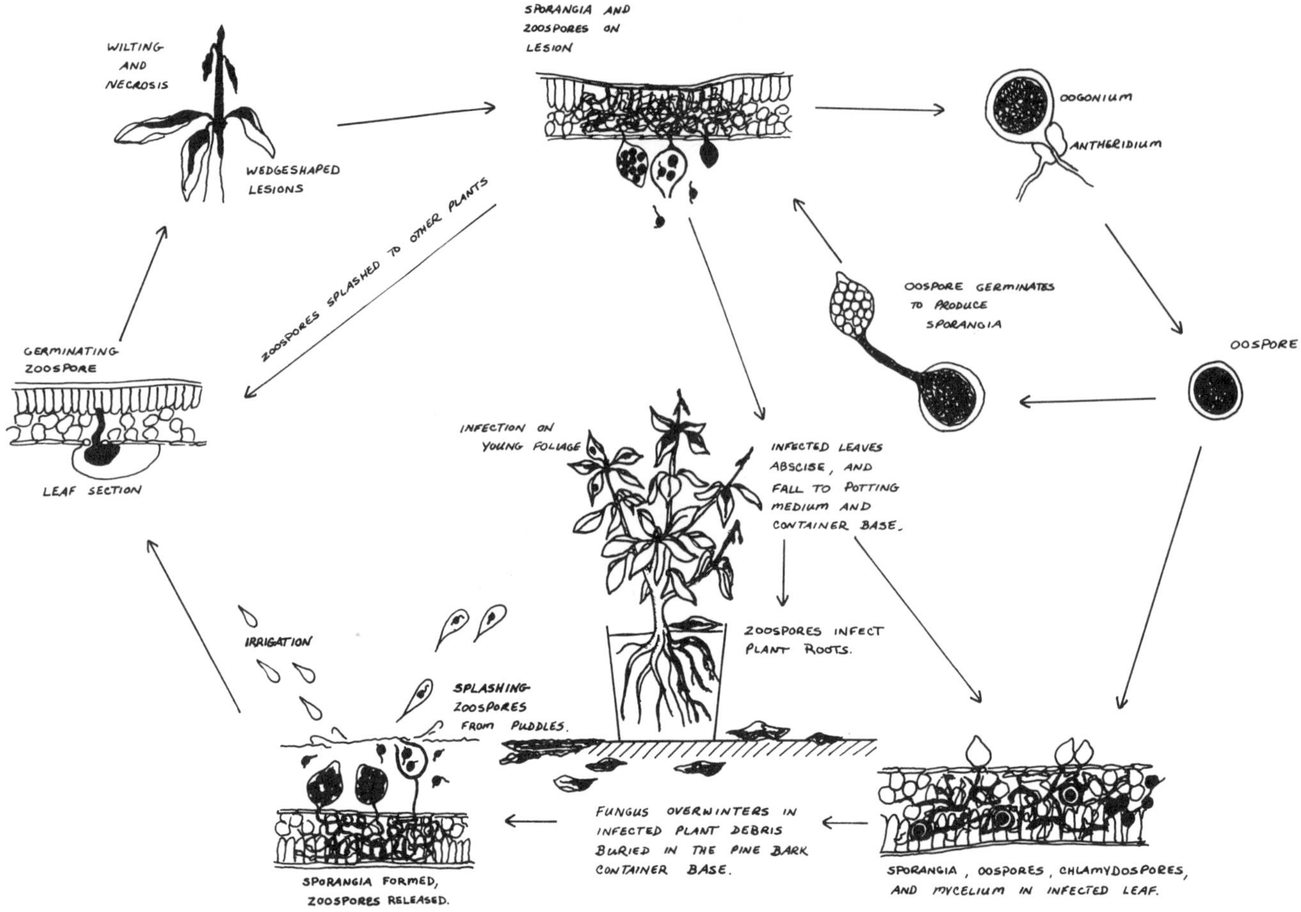

Fig. 75.1 Disease cycle of Phytophthora dieback of rhododendron. (After C. R. Kuske and D. M. Benson)

cola in nurseries in Germany. Dieback symptoms in *Calluna* and *E. gracilis* may be confused with those of anthracnose, caused by *Glomerella cingulata.* Numerous other woody plants are hosts of *P. citricola, P. cactorum,* and *P. nicotiana* var. *parasitica.* The herbaceous ornamentals vinca (*Catharanthus*) and petunia are commonly blighted by *P. nicotiana* var. *parasitica* in southern landscapes.

The susceptibility of rhododendrons to Phytophthora dieback varies from cultivar to cultivar. Under epidemic conditions in one nursery, Roseum Elegans and Roseum Pink were highly resistant, while Chionoides White and Scintillation were highly susceptible. *R. catawbiense* cultivars Album and Nova Zembla are very susceptible to *P. citricola.* Among azaleas, F. Scherrer is particularly susceptible to *P. citricola,* while Gloria, Stella Maria, and White Water are not likely to become infected in the field.

Management

Control of dieback depends on cultural practices for avoiding or minimizing the disease and application of fungicides to prevent infection. Avoid the use of recycled irrigation water, which may contain spores of *Phytophthora* spp. It may be necessary to develop a separate irrigation system for rhododendron and other susceptible crops if dieback becomes endemic in the nursery. See Chapter 95, "Disease Management for Nurseries Using Recycled Irrigation Systems," for information on appropriate water treatment options.

During early summer, when new growth has not yet hardened, growers should avoid late-afternoon irrigation, which would favor lesion expansion and spore formation in diseased tissue. When the foliage dries before nightfall, the production of new spores is greatly reduced. Irrigation is best done at midday, after dew on the foliage has dried, so that spores spread to new sites will dry out and die during the afternoon. All infected shoots should be pruned out twice weekly. Disinfect shears between cuts. Excess fertilizer, particularly high levels of nitrogen, will result in plants that are very succulent and hence more susceptible to dieback. Excess shade also increases disease incidence.

Fungicides are effective, if applied before infection occurs. Nurseries with a history of Phytophthora dieback of rhododendron will want to adjust irrigation practices and institute a preventive spray program during the summer. Grow only less susceptible cultivars in fields or container blocks contaminated with *Phytophthora.*

Botryosphaeria Dieback

Botryosphaeria dothidea
Botryosphaeria ribis

Botryosphaeria dieback is often misidentified as a *Phytophthora* disease. Although their symptoms are superficially similar, these diseases occur under different cultural and environmental conditions. Phytophthora dieback is more likely to occur in container nurseries and holding areas with overhead irrigation, while Botryosphaeria dieback is very common in the landscape but only rarely seen in nurseries.

Symptoms

Leaves on affected stems droop and roll inward along the midvein (Plate 137). The cause of the leaf symptoms is a reddish brown to black, sunken canker that girdles the stem. Cankers develop from infection at the point where a flower cluster was attached or at the site of natural wounds, such as leaf scars, or pruning wounds, up to two months after wounding. Stem discoloration is initially confined to the bark and later spreads to the wood and pith, generally moving downward along infected branches. The fungal pathogen, in the form of dark pycnidia, sporulates (often visibly) on dead stem tissue.

Epidemiology

Plants are more susceptible to invasion by *Botryosphaeria* during periods of drought stress. Dieback occurs primarily in field-grown or landscape plants, where drought stress is more likely than in container culture. Cankers caused by *Botryosphaeria* develop more slowly than those caused by *Phytophthora*. *Botryosphaeria* requires 135 days to colonize about 4 inches of stem, and it is usually restricted to the stem. *Phytophthora* spp. will usually spread from stems into the base of leaves, producing a wedge-shaped brown discoloration pointing down the midvein; the killed leaves remain attached to the dead stem. Canker symptoms caused by *Botryosphaeria* may develop at any time, particularly following drought, winter injury, or other stresses, whereas Phytophthora shoot blight ordinarily occurs during a growth flush.

Management

Cankers should be pruned out below the discolored portion. Treat pruning tools with disinfectant solution. Avoidance of drought stress is a critical cultural control.

No cultivars are known to be resistant to Botryosphaeria dieback, but Boursault, Cunningham's White, English Roseum, LeBar's Red, Roseum Elegans, and Roseum 2 were all observed to have less than 10% dieback in a study in which the cultivar Nova Zembla averaged 10–50% dieback.

Chemical treatments are not effective for the control of Botryosphaeria dieback.

Botryosphaeria Dieback

Geographic occurrence USDA zones 4–8

Seasonal occurrence March to November, under average yearly conditions in zone 7 (adjust for your hardiness zone), with optimum periods during droughts

DISEASE FREQUENCY	DISEASE SEVERITY
5 annual	5 plants killed
4	4
3	**3**
2	2
1 rare	1 very little damage

CHEMICAL TREATMENT	CULTURAL PRACTICES
3 used every year	**3 very important**
2	2
1 not used	1 not important

SANITATION	RESISTANT CULTIVARS
3 very important	3 many cultivars
2	2
1 not important	**1 no resistance**

Cylindrocladium Root Rot

Cylindrocladium scoparium
Cylindrocladium theae

Cylindrocladium root rot occasionally affects rhododendrons, but it much more common in azalea. It is primarily a problem in propagation.

Symptoms

The most common symptoms are root and stem rots. In propagation, dark brown to black lesions may develop on leaves, stems, and petioles. Necrotic flecks may also form on flower petals. Affected plants may wilt following invasion of the vascular system at the crown. Leaf tissues of affected azalea and heath (*Erica*) turn pale before yellowing or browning.

Epidemiology

The fungal pathogens, *Cylindrocladium scoparium* and *C. theae,* are highly competitive saprophytes. These fungi form thick-walled structures called microsclerotia, which facilitate their survival in the absence of a host plant. Warm conditions and excessive water favor disease development.

Host Range

Many woody plants, including *Abies, Calluna, Erica, Gaultheria, Ilex, Lagerstroemia, Leucothoe, Photinia, Picea, Pinus, Prunus, Rhododendron, Rosa, Syringa,* and *Tsuga,* are susceptible to root rot caused by *Cylindrocladium,* particularly if soil drainage is poor. Foliar infection is more unusual but sometimes occurs in azalea and rhododendron, especially those grown in greenhouses.

Management

Cylindrocladium root rot can be controlled with good water management, prompt roguing of diseased plants, and treatment with fungicide sprays. Take cuttings only from healthy plants. Dipping cuttings in a disinfectant may be necessary. Avoid excessive levels of nitrogen fertilization, overwatering, or drought, all of which may increase the susceptibility of the plants. Do not reuse flats or containers that may be contaminated by microsclerotia from a previous infected crop. Propagation beds should be steam-pasteurized before reuse.

Leaf Gall

Exobasidium vaccinii

Leaf gall, caused by the fungus *Exobasidium vaccinii,* is common in azalea but relatively rare in rhododendron. The folk name for this disease is *pinkster gall,* referring to the symptoms on a native azalea. Leaf gall occurs in native and landscape plants, as well as in production.

Symptoms

Thick, swollen galls develop on new leaves as they emerge. The galls may become large and convoluted. By early summer they develop a whitish coating of new spores of *E. vaccinii.*

Epidemiology

Spores produced on galls are liberated by air currents. Spores that land on newly formed buds of the next year's growth either infect the bud or remain dormant over the winter. In the spring, as the buds expand, the fungus produces growth regulator substances which stimulate gall development.

Management

The routine control of leaf gall is to pick off and bury galled tissue before it turns white. Since the disease causes little real damage, fungicide treatment is generally not necessary. If needed, fungicide sprays should be applied as the spore layer begins to develop on galled tissue, in order to protect the vegetative buds, and should continue at 10- to 14-day intervals until all galls have dried up and are no longer producing spores. Symptom reduction should be apparent in the spring of the following year.

Discula Leaf Spot

***Discula* sp.**

Brown areas on rhododendron leaves are usually symptoms of winter desiccation injury, overfertilization, or severe root injury from the feeding of grubs or weevils. Occasionally, brown areas on leaves are symptoms of contagious diseases caused by fungi. A

Cylindrocladium Root Rot

Geographic occurrence USDA zones 4–8

Seasonal occurrence March to October, with an optimum period in June–August, under average yearly conditions in zone 7 (adjust for your hardiness zone)

DISEASE FREQUENCY	DISEASE SEVERITY
5 annual	5 plants killed
4	**4**
3	3
2	2
1 rare	1 very little damage

CHEMICAL TREATMENT	CULTURAL PRACTICES
3 used every year	3 very important
2	**2**
1 not used	1 not important

SANITATION	RESISTANT CULTIVARS
3 very important	3 many cultivars
2	2
1 not important	**1 no resistance**

Leaf Gall

Geographic occurrence USDA zones 4–8

Seasonal occurrence April to June, with an optimum period in May, under average yearly conditions in zone 7 (adjust for your hardiness zone)

DISEASE FREQUENCY	DISEASE SEVERITY
5 annual	5 plants killed
4	4
3	3
2	**2**
1 rare	1 very little damage

CHEMICAL TREATMENT	CULTURAL PRACTICES
3 used every year	3 very important
2	**2**
1 not used	1 not important

SANITATION	RESISTANT CULTIVARS
3 very important	3 many cultivars
2	2
1 not important	**1 no resistance**

Discula sp. causes a leaf spot disease of rhododendron, and leaf spots caused by a *Phyllosticta* sp. are also observed occasionally. Leaf spot diseases are generally limited to a few cultivars, whereas stress-associated browning may be common to all cultivars.

Symptoms

Infection causes round, brown leaf spots with purple rims. The lesions contain rings of black dots, which are spore-bearing structures (acervuli) of the fungus. Spots develop on new foliage following spring infection. Badly stressed plants may be defoliated.

Epidemiology

An extended period of leaf wetness is necessary for infection. Spores produced on leaf lesions are spread by splashing water from irrigation or rainfall to infect nearby leaves.

Host Range

Other than rhododendron, hosts of the pathogen causing Discula leaf spot have not been identified. The cultivars Boule de Neige and Lee's Dark Purple are susceptible.

Discula Leaf Spot

Geographic occurrence USDA zones 6–8

Seasonal occurrence April to June, with an optimum period in May, under average yearly conditions in zone 7 (adjust for your hardiness zone)

DISEASE FREQUENCY	DISEASE SEVERITY
5 annual	5 plants killed
4	4
3	3
2	**[2]**
[1] rare	1 very little damage

CHEMICAL TREATMENT	CULTURAL PRACTICES
3 used every year	3 very important
[2]	**[2]**
1 not used	1 not important

SANITATION	RESISTANT CULTIVARS
3 very important	**[3] many cultivars**
[2]	2
1 not important	1 no resistance

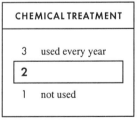

Management

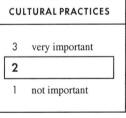

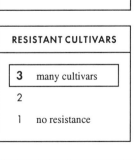

Maintain adequate nutrition. Avoid extensive periods of leaf wetness, which promote infection. Select less susceptible cultivars. Remove symptomatic leaves before the spring flush.

Ovulinia Petal Blight

Ovulinia azaleae

Ovulinia petal blight of rhododendrons occurs later in the growing season than the disease in azaleas and therefore may not be as damaging (Plate 138). Temperatures are higher and rainy weather is less frequent when rhododendrons are blooming than during azalea bloom. See Chapter 23, "Azalea Diseases," for more information.

Powdery Mildew

***Microsphaera azaleae* and other fungi**

Powdery mildew is common in azaleas and broad-leaved rhododendron species and hybrids in the Pacific Northwest and British Columbia. It is observed only rarely in rhododendron in other parts of the United States. Symptoms were detected on rhododendron cultivar Mary Belle and a few other Dexter hybrids for several consecutive years in Rhode Island, but the disease did not become permanently established there.

Azalea and rhododendron powdery mildew is caused by several fungi in the family Erysiphaceae. *Microsphaera azaleae* is the primary pathogen in the Pacific Northwest. In Britain, rhododendron powdery mildew is believed to be caused by two different fungi, one affecting plants in greenhouses and a second affecting plants growing outdoors.

The incidence of powdery mildew of rhododendron in both the United Kingdom and the Pacific Northwest has increased markedly in the past 10–15 years, causing sometimes dramatic injury in collections as well as commercial nurseries. Additionally, the host range appears to have expanded: broad-leaved evergreen rhododendrons are now sometimes seriously affected by powdery mildew, whereas in the past deciduous azaleas were the most common hosts.

Symptoms

The white mycelial growth typical of powdery mildew fungi on other crops is not always evident on rhododendrons, and symptoms can vary a great deal from cultivar to cultivar. When clearly recognizable powdery mildew colonies are not present, the fungal infection may be mistaken for a physiological disorder. On most azaleas and on some rhododendrons, such as Purple Splendor and Vulcan's Flame, white fungal colonies will be visible on both the upper and the lower surfaces of leaves. However, on many rhododendrons, it is more common for pale yellow spots (0.25 to 1 inch in diameter) with diffuse margins to develop on the upper surface of leaves. Purple to brown, circular patches with feathery margins develop on the undersurface, opposite the pale spots. Other symptoms that may appear on yellowing leaves are large, purple brown to black spots, dark purple brown vein discoloration, purple ring spots, and green patches. In some cultivars, such as Unique, colored spots appear where the fungus is visible on the undersurface of leaves, while the upper surface remains symptomless.

Defoliation may occur in some cases, as in severe infection of the cultivar Virginia Richards, *R. campylocarpum,* or *R. cinna-*

Powdery Mildew

Geographic occurrence USDA zones 6–8

Seasonal occurrence April to November, with an optimum period in early spring and fall, under average yearly conditions in zone 8 (adjust for your hardiness zone)

DISEASE FREQUENCY	DISEASE SEVERITY
5 annual (boxed)	5 plants killed
4	4
3	3
2	**2** (boxed)
1 rare	1 very little damage

CHEMICAL TREATMENT	CULTURAL PRACTICES
3 used every year	3 very important
2 (boxed)	**2** (boxed)
1 not used	1 not important

SANITATION	RESISTANT CULTIVARS
3 very important	**3 many cultivars** (boxed)
2	2
1 not important (boxed)	1 no resistance

barinum. Curious-looking yellow, red, or brown patterns develop on leaves as rhododendrons begin to defoliate, usually in fall or early spring in the Pacific Northwest. Some rhododendrons have discolored leaves but no significant defoliation.

Epidemiology

Rhododendron powdery mildew is favored by high humidity. The symptoms develop rapidly in late summer, when powdery mildew fungi spread in the form of asexually produced spores (conidia). Some colonies overwinter on the undersides of leaves. These fungi are also thought to overwinter in leaf buds. They also overwinter as cleistothecia (cases containing sexually produced spores), which are formed frequently on azalea but are comparatively rare on rhododendron. There is some circumstantial evidence that relatively low winter temperatures help to curb powdery mildew of rhododendron in field and landscape settings.

Host Range

Many rhododendron and azalea species and cultivars are susceptible to powdery mildew. Indumented species appear to be immune.

Management

Grow the less susceptible hybrid rhododendrons. Reduce relative humidity by spacing plants to allow for good air circulation. Water early in the day, to allow foliage to dry before nightfall. Scout for symptoms, so that chemical control may be initiated promptly if powdery mildew is observed. Grow plants in full sun or only light shade.

Tissue Proliferation

Unknown etiology

Tissue proliferation of rhododendrons is an abnormal growth of callus-like tissues and sometimes a proliferation of dwarfed shoots, usually at the root crown (Plate 139). Because it looks a great deal like crown gall disease, it has caused much concern. There is no evidence that tissue proliferation is contagious. It appears almost exclusively in micropropagated plants. See Chapter 100, "Tissue Culture of Woody Plants."

Many rhododendron cultivars and occasionally *Kalmia* and *Pieris* show symptoms of tissue proliferation. The rhododendron cultivar Montego is severely affected.

Various explanations of tissue proliferation have been proposed. One is that the proliferated tissue is a structure (called a lignotuber) formed naturally by some ericaceous plants. Another explanation is that epigenetic changes during tissue culture are responsible for the altered growth. Tissues which exhibit the characteristic symptoms appear to be overproducing cytokinins; researchers are currently exploring this hypothesis.

Plants exhibiting symptoms of tissue proliferation should not be used for vegetative propagation. Plants developed from cuttings of recently micropropagated rhododendrons with tissue proliferation are likely to be affected themselves.

REFERENCES

Backhaus, G. F. 1994. *Phytophthora citricola* (Sawada) – Cause of an important shoot rot of rhododendron and azalea. Acta Hortic. 364:145–154.

Backhaus, G. F. 1994. *Cylindrocladium scoparium* causing wilt disease in rhododendron and azalea. Acta Hortic. 364:163–165.

Basden, N., and Helfer, S. 1995. World survey of rhododendron powdery mildews. J. Am. Rhododendron Soc. 49(3):147–156.

Benson, D. M. 1980. Chemical control of rhododendron dieback caused by *Phytophthora heveae.* Plant Dis. 64:684–686.

Benson, D. M., and Hoitink, H. A. J. 1986. Phytophthora dieback. Pages 12–15 in: Compendium of Rhododendron and Azalea Diseases. D. L. Coyier and M. K. Roane, eds. American Phytopathological Society, St. Paul, Minn.

Benson, D. M., and Jones, R. K. 1980. Etiology of rhododendron dieback caused by four species of *Phytophthora.* Plant Dis. 64:687–691.

Benson, D. M., Daughtry, B. I., and Jones, R. K. 1990. Botryosphaeria dieback in hybrid rhododendron, 1986–1990. Biol. Cult. Tests Control Plant Dis. 6:108.

Brand, M. H., Ruan, Y., and Kiyomoto, R. 2000. Response of *Rhododendron* 'Montego' with tissue proliferation to cytokinin and auxin in vitro. HortScience 35:136–140.

Braun, U. 1984. A short survey of the genus *Microsphaera* in North America. Nova Hedwigia 39:211–243.

Helfer, S. 1994. Rhododendron powdery mildews. Acta Hortic. 364:155–159.

Hoitink, H. A. J., and Schmitthenner, A. F. 1969. Rhododendron wilt caused by *Phytophthora citricola.* Phytopathology 59:708–709.

Hoitink, H. A. J., and Schmitthenner, A. F. 1974. Resistance of rhododendron species and hybrids to Phytophthora root rot. Plant Dis. Rep. 58:650–653.

Hoitink, H. A. J., Benson, D. M., and Schmitthenner, A. F. 1986. Phytophthora root rot. Pages 4–8 in: Compendium of Azalea and Rhododendron Diseases. D. L. Coyier and M. K. Roane, eds. American Phytopathological Society, St. Paul, Minn.

LaMondia, J. L., Smith, V. L., and Rathier, T. M. 1997. Tissue proliferation in rhododendron: Association with disease and effect on plants in the landscape. HortScience 32:1001–1003.

Linderman, R. G. 1986. Phytophthora syringae blight. Pages 15–17 in: Compendium of Rhododendron and Azalea Diseases. D. L. Coyier and M. K. Roane, eds. American Phytopathological Society, St. Paul, Minn.

Linderman, R. G. 1986. Cylindrocladium blight and wilt. Pages 17–20 in: Compendium of Rhododendron and Azalea Diseases. D. L. Coyier and M. K. Roane, eds. American Phytopathological Society, St. Paul, Minn.

Pscheidt, J. W., and Ocamb, C. M. 2001. Pacific Northwest Plant Disease Control Handbook. Oregon State University Extension Service, Corvallis.

CHAPTER 76

George Philley ▪ Texas A&M University, Overton
Austin K. Hagen ▪ Auburn University, Auburn, Alabama
A. R. Chase ▪ Chase Research Gardens, Mt. Aukum, California

Rose Diseases

Geographic production USDA zones 7–9

Family Rosaceae

Genus *Rosa*

Species major hybrid groups include hybrid teas, grandifloras, polyanthas, floribundas, miniatures, and shrubs
R. multiflora (rootstock)
R. 'Manetti' (rootstock)
R. 'Dr. Huey' (rootstock)
R. 'Fortuniana' (rootstock)

Field-grown roses are produced primarily in California, Arizona, and Texas. The production cycle takes two years. In the western states the cultivar Dr. Huey is the favored rootstock, while thornless selections of *R. multiflora* are used in Texas. *R.* 'Fortuniana,' a cold-sensitive rootstock, is used in some production nurseries in the Southeast. Rootstocks are used because many cultivars do not perform well when rooted from cuttings. However, cultivars that root and grow consistently from cuttings are being grown in some areas, and all miniature roses are grown from rooted cuttings. Black spot and powdery mildew are two major diseases affecting roses in production nurseries. Regular fungicide applications are commonly made to maintain healthy foliage.

This chapter deals with the propagation of rose plants, and not greenhouse production of cut flowers. Greenhouses referred to in this chapter are for propagation or forcing plants in the spring.

Black Spot

***Marssonina rosae* (sexual stage, *Diplocarpon rosae*)**

Black spot is a major disease of roses. It can progress rapidly, resulting in heavy defoliation, which weakens plants by depleting carbohydrate reserves. The plants are then susceptible to winter injury, or they produce inferior foliage and blooms.

Symptoms

Round to irregularly shaped, light brown to black spots with feathery edges develop on infected leaves, which eventually become chlorotic (Plate 140). The chlorosis progresses until defoliation occurs. Defoliation is associated with fungal toxins and ethylene. One or a few infection sites on a leaf will cause it to fall. All aboveground parts of plants are susceptible to infection, but symptoms are seldom noticed except on leaves. Dark brown to black blotches may develop on current-season canes.

Epidemiology

Black spot is caused by the fungus *Marssonina rosae,* which overwinters in infected canes, leaves that do not fall, and fallen leaves. Infection and disease development are greatest when the relative humidity is above 85% and temperatures are 24–30°C

Black Spot

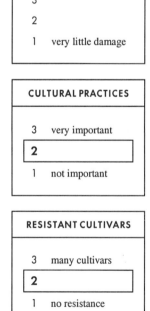

Geographic occurrence Wherever roses are grown

Seasonal occurrence All seasons, but incidence is lower under hot, dry conditions and the coldest months in northern climates

DISEASE FREQUENCY	DISEASE SEVERITY
5 annual (boxed)	**5 plants killed** (boxed)
4	4
3	3
2	2
1 rare	1 very little damage

CHEMICAL TREATMENT	CULTURAL PRACTICES
3 used every year (boxed)	3 very important
2	**2** (boxed)
1 not used	1 not important

SANITATION	RESISTANT CULTIVARS
3 very important	3 many cultivars
2 (boxed)	**2** (boxed)
1 not important	1 no resistance

(75–85°F). Spores (conidia) of *M. rosae* must be wetted for about 7 hours before they germinate and can infect a host. Symptoms appear in three to 16 days, and more conidia are formed 10–18 days after initial infection. Spores are spread primarily by splashing rain rather than by air currents. Young, developing leaves are most susceptible. Black spot is generally not a problem in dry areas or in greenhouses where climate controls limit leaf wetness. In temperate climates the disease is a threat for the entire growing season.

Host Range

Roses vary in susceptibility to black spot. The least resistant are hybrid teas, grandifloras, and miniature types. Cultivars of floribundas, shrub roses, and climbers are more resistant or tolerant.

Management

Cultural practices Avoid overhead watering, close spacing, and excessive shade.

Chemical treatment Preventing initial infection is critical in the control of black spot. Begin fungicide applications shortly after budbreak. Once the disease is established, it is difficult to bring under control. A surfactant in the fungicide spray solution will enhance coverage and should be added if suggested on the product label. Alternate unrelated chemicals to prevent the development of resistance in the fungal population. Plan on following the shortest recommended spray interval, which is usually seven days, for highly susceptible varieties growing in areas with moderate to heavy rainfall.

Powdery Mildew

Sphaerotheca pannosa

Powdery mildew is a common and often damaging disease of nearly all types of container- and bed-grown roses. It poses little real threat to plant health, but the market value of heavily mildewed container-grown roses may be reduced.

Symptoms

Powdery mildew first appears on leaves, tender shoots, and the peduncles and sepals of flower buds, as small, scattered, cottony white patches of mycelia and spores of the pathogen, *Sphaerotheca pannosa* (Plate 141) Typically, these patches are seen first on the underside and then on the upper surface of leaves. Localized white to buff-colored colonies may also develop on canes, particularly around thorns (Plate 142). Newly unfurled leaves are more susceptible to infection than mature leaves. Heavily colonized leaves, shoots, and flower buds turn almost completely white and may be curled or otherwise distorted. Symptom severity, particularly in container stock, depends on cultivar susceptibility, host growth stage, moisture levels, and local weather conditions.

Epidemiology

S. pannosa overseasons primarily as hyphae in dormant buds but also as colonies on canes and remaining leaves in areas with mild winters. In early spring, spores are spread by air currents to newly unfurled leaves and tender shoots. Infection is favored by temperatures between 70 and 80°F and relative humidity between 90 and 100%. Free water on leaf surfaces will often suppress disease development. Symptoms usually appear within one week of infection. In container-grown roses produced under plastic, extended periods of cloudy, warm, humid weather favor disease development. New infections rarely develop when daytime temperatures exceed 90°F.

Host Range

The host range of *S. pannosa* includes peach, photinia, and rose.

Management

Strategies for controlling powdery mildew in container- and field-grown roses are largely limited to the production of disease-resistant cultivars and treatment with protective fungicides.

Cultural practices Spacing out plants to increase air circulation around the foliage may help slow the spread of the disease. For container stock grown under plastic, venting the warm, humid air from the greenhouse in the evening and heating the incoming night air will sufficiently reduce the relative humidity to suppress disease development.

Resistance Considerable differences in the susceptibility of rose cultivars to powdery mildew have been noted. Certain cultivars are virtually immune to the disease, while others annually

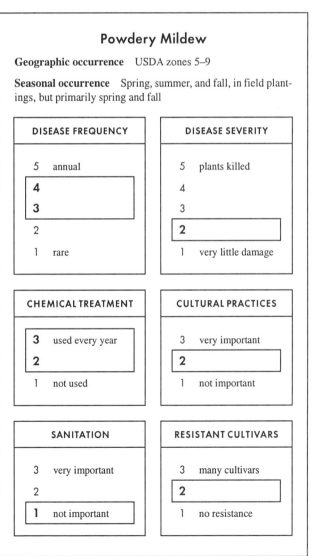

Powdery Mildew

Geographic occurrence USDA zones 5–9

Seasonal occurrence Spring, summer, and fall, in field plantings, but primarily spring and fall

DISEASE FREQUENCY	DISEASE SEVERITY
5 annual	5 plants killed
4	4
3	3
2	**2**
1 rare	1 very little damage

CHEMICAL TREATMENT	CULTURAL PRACTICES
3 used every year	3 very important
2	**2**
1 not used	1 not important

SANITATION	RESISTANT CULTIVARS
3 very important	3 many cultivars
2	**2**
1 not important	1 no resistance

suffer considerable damage. Because of the multitude of cultivars of rose in the nursery trade, however, no comprehensive listing of their reactions to powdery mildew has been published recently. Since there are different races of *S. pannosa,* regional variations in cultivar resistance to powdery mildew may occur.

Chemical treatment For susceptible cultivars, start applying protective fungicides shortly after leaf-out, or as symptoms first appear on foliage. Continue applications at the interval specified on the product label until the threat of further disease spread has passed or the until the stock is shipped. Shorten the interval between applications and use the full rate recommended on the label when disease is severe. Adding a wetting agent, if the label suggests one, may improve control. *S. pannosa* may have developed resistance to some fungicides.

Downy Mildew

Peronospora sparsa

Downy mildew of rose was first observed in England in 1862. The disease is erratic in expression, appearing in some years but not in others. It has been reported worldwide but is most common where the weather is commonly cold and wet.

Symptoms

Reddish black spots develop on leaves, petals, and stems of roses, well in advance of sporulation of the fungal pathogen, *Peronospora sparsa.* As the infection progresses, leaves become chlorotic (Plate 143). Major leaf veins often restrict fungal growth, so that lesions take on an angular shape as they enlarge (Plate 144). Sometimes the most obvious symptom is leaf burn, which may be confused with pesticide toxicity. When conditions are optimal for disease development, sporangiophores of the fungus emerge, primarily from the lower surface of leaves, and the characteristic downy coating appears. *P. sparsa* does not form these structures as abundantly as many other downy mildew fungi, and the disease may be hard to diagnose if characteristic fungal structures are not found, particularly in some cultivars. Heavy defoliation often occurs (Plate 145).

Epidemiology

The optimal temperature for rose downy mildew is 64°F, with relative humidity above 85%. *P. sparsa* appears to overwinter as dormant mycelia in canes. Although oospores have been observed, their role in the disease process has not been determined. Downy mildew develops rapidly under cool, wet conditions where plants are closely spaced. It usually occurs in late winter and early spring. The disease abates quickly when the weather is dry and warm.

Host Range

All types of roses, from potted miniatures to landscape and cut flowers, are variably susceptible to downy mildew. Cultivars probably vary in susceptibility, but the difficulty of working with *P. sparsa* makes cultivar trials almost impossible. The pathogen is reported to be the same fungus that causes downy mildew of *Rubus* species (blackberry, raspberry, etc.).

Management

In greenhouse production it is critical to keep the relative humidity below 85% to decrease the sporulation of the downy mildew fungus on infected plants and stop the germination of spores on uninfected plants. The relative humidity can be controlled by venting and raising the temperature at key times during the day. Improving the drying of wet leaves by means of fans is also recommended, but fans can cause problems with disease spread, since spores of *P. sparsa* are easily spread by air currents. Other methods (fungicide treatment or removal of infected plants) should be used to prevent the spread of spores by fans.

Chemical treatment Fungicides have been tested for control of downy mildew of roses throughout the world. For a fungicide to be effective, it must be applied before or as soon as the first symptoms develop. Downy mildew is an explosive disease and is difficult to bring under control after it has progressed. Some systemic fungicides provide excellent control when used alone, but resistance to them may develop in the pathogen. Tank mixing or alternating fungicides with different modes of action helps to prevent the development of resistant strains. Certain systemic fungicides have worked as soil drenches. When nonsystemic products are used, it is critical to apply them to the undersides of leaves, where infection and subsequent sporulation occur.

Downy Mildew

Geographic occurrence Entire United States

Seasonal occurrence Spring and fall, whenever the weather is rainy and cool

DISEASE FREQUENCY	DISEASE SEVERITY
5 annual	**5 plants killed**
4	4
3	3
2	2
1 rare	1 very little damage

CHEMICAL TREATMENT	CULTURAL PRACTICES
3 used every year	3 very important
2	**2**
1 not used	1 not important

SANITATION	RESISTANT CULTIVARS
3 very important	3 many cultivars
2	2
1 not important	**1 no resistance**

Rust

Phragmidium mucronatum

Rust occurs occasionally in the western United States under cool, wet conditions. It is rare in the central and eastern states.

Symptoms

Rust pustules are orange, red, or black, depending on the spore stage of the pathogen, *Phragmidium mucronatum.* Aeciospores (orange) appear first in the spring on the undersides of leaves. Urediospores (red) develop in pustules during the summer; they are the most common spores and are responsible for the rapid spread of the disease during the growing season. Teliospores (black) form in response to cool conditions and serve as the overwintering form of the fungus. Yellow spots appear on the upper surface of leaves, opposite pustules on the lower surface. In heavy infections, the pustules erupt on both surfaces. Heavy defoliation occurs if rust is not controlled. Canes and other plant parts are also susceptible.

Epidemiology

P. mucronatum survives on canes and fallen leaves. Infections that occur in spring and summer are favored by cool, wet weather.

Rust

Geographic occurrence Western United States; the disease is rare in other states

Seasonal occurrence Spring and fall

DISEASE FREQUENCY	DISEASE SEVERITY
5 annual	5 plants killed
4	**4**
3	**3**
2	2
1 rare	1 very little damage

CHEMICAL TREATMENT	CULTURAL PRACTICES
3 used every year	3 very important
2	**2**
1 not used	1 not important

SANITATION	RESISTANT CULTIVARS
3 very important	3 many cultivars
2	**2**
1 not important	1 no resistance

Free moisture on leaf surfaces is essential for spore germination and infection.

Host Range

Rose is the only known host of *P. mucronatum.* Varieties vary in susceptibility.

Management

Destroy fallen leaves and prune heavily in the spring, to reduce the level of inoculum. Begin fungicide applications at the first sign of rust or, where the disease is common, when new growth is produced.

Cylindrocladium Leaf and Stem Rot

Cylindrocladium scoparium

Cylindrocladium leaf and stem rot is primarily a disease of greenhouse-propagated roses, particularly miniature roses. The pathogen, *Cylindrocladium scoparium,* infects leaves and stems under wet, warm conditions and remains viable on decaying plant parts (Plate 146). Plants carrying the fungus may grow normally until a change in the environment triggers fungal growth. Disease development can be rapid. Sudden wilting is usually the first symptom. Affected stems are often girdled at or just below the surface of the medium (Plate 147).

Sanitation, propagation of clean cuttings, and application of fungicide sprays on stock plants are key elements in the control of this disease. For detailed information, see Chapter, 11, "Diseases Caused by *Cylindrocladium.*"

Common Canker

Coniothyrium fuckelii

Brown Canker

Cryptosporella umbrina

(asexual stage, *Phomopsis umbrina*)

Canker diseases generally affect plants weakened by foliar diseases, insect injury, cold damage, or other stresses. Pruning and grafting wounds are often points of entry for canker fungi. Cankers range from light brown to black.

Pruning infected wood and practicing good sanitation are usually effective in controlling canker disease in the nursery. Fungicide applications may be necessary if the disease is widespread.

Botrytis Blight (Gray Mold)

Botrytis cinerea

Botrytis blight, also known as gray mold, can cause extensive damage, particularly to buds and blooms, during the production and shipping of roses. Blighting of blooms and canes occasionally occurs in container- and field-grown roses following several days of mild, overcast, humid weather. Canes of bare-root roses held for an extended period in cold storage can also become infected. Cane infection may occur after newly emerging buds are damaged by late frost in the spring.

Symptoms

Buds, blooms, and peduncles are the main targets of the fungal pathogen, *Botrytis cinerea.* A cane dieback has also been reported (Plate 148). On the petals of buds and blooms, small, light-colored spots, sometimes with a maroon halo, are the earliest symptoms.

Under environmental conditions favorable for disease development, these fast-spreading spots enlarge into irregular, brown, often water-soaked blotches. Masses of gray brown spore clusters and mycelia of *B. cinerea,* easily seen with the naked eye, appear on the surface of blighted tissues.

Brown, slightly sunken cankers may extend down canes from blighted flowers or pruning cuts or are centered on lateral buds or wounds along the canes (Plate 149). Canker and dieback symptoms associated with Botrytis blight are similar in appearance to those caused by other fungal pathogens. Canes with advanced blight may turn light tan.

Mats of mycelia of *B. cinerea* usually develop on the canes of diseased bare-root roses in cold storage, and the blighted roses are typically either killed or are of such poor quality that they are unsalable.

Epidemiology

The onset of Botrytis blight is influenced by temperature, moisture, sanitation, and the growth stage and nutritional status of the host. Outbreaks of Botrytis blight develop much faster than other diseases, and under favorable conditions symptoms appear virtually overnight.

Botrytis Blight

Geographic occurrence USDA zones 5–9

Seasonal occurrence Spring, summer, and fall, in field plantings

DISEASE FREQUENCY	DISEASE SEVERITY
5 annual	5 plants killed
4	4
[3]	3
2	**[2]**
1 rare	1 very little damage

CHEMICAL TREATMENT	CULTURAL PRACTICES
3 used every year	3 very important
[2]	**[2]**
1 not used	1 not important

SANITATION	RESISTANT CULTIVARS
3 very important	3 many cultivars
[2]	2
1 not important	**[1]** no resistance

The optimum temperature for Botrytis blight is 65°F; and the disease can occur at temperatures ranging from 50 to 75°F. Frost or cold injury to canes and buds contributes to disease development. Bare-root stock held under very moist conditions may be destroyed at temperatures as low as 32°F.

B. cinerea can survive almost indefinitely in spent blooms, diseased canes, and other plant debris. Under overcast, humid conditions abundant fungal spores are produced on infected tissue. When the relative humidity decreases during the day, masses of spores are released into the air and then dispersed by air currents throughout the greenhouse and outdoors. Normal greenhouse maintenance activities, such as watering, harvesting of blooms, and shipping, also greatly increase the concentration of spores in the air. Often, healthy tissues are colonized only after the fungus has attacked adjacent dead or dying tissues or infected debris has fallen on otherwise healthy leaves, blooms, or buds. Temperatures well below the optimum for rapid plant growth may also facilitate the development of Botrytis blight in healthy host tissue.

Host Range

B. cinerea causes blossom blight and dieback of a wide variety of annual and perennial flowers and numerous woody trees and shrubs, including camellia, flowering dogwood, hydrangea, and rhododendron, as well as rose.

Management

Sound management practices, sanitation, and chemical control are required to prevent outbreaks of Botrytis blight of roses.

Cultural practices To prevent condensation of water on cooling foliage of greenhouse-grown roses and suppress disease development, ventilate the house in early evening, and heat the incoming air. In the greenhouse, use a drip irrigation system, to avoid wetting the foliage. If an overhead irrigation system or water breaker is used, stop watering early enough that the foliage dries before evening. To allow for some air circulation, space out container-grown roses so that the foliage does not touch. Install sufficient fans or a forced-air circulation system to eliminate pockets of humid, stagnant air. Avoid shipping damp plants, because they may rot in transit.

Calcium nutrition has a significant influence on the sensitivity of rosebuds and blooms to attack by *Botrytis.* The incidence of Botrytis blight in container stock can be significantly reduced if calcium nitrate, gypsum, or a similar calcium source is incorporated into the container medium or applied as a topdressing.

Sanitation Removal and disposal of spent blooms, blighted canes, dead plants, and other debris from production beds and storage coolers will help suppress disease development. The floors and walls of refrigerated coolers and other storage areas should periodically be cleaned with a surface disinfectant.

Resistance Under ideal conditions for disease development, cultivars and selections of all rose taxa are probably equally susceptible to Botrytis blight.

Chemical treatment Many fungicides used to control black spot or powdery mildew also provide some control of Botrytis blight. Some fungicides are specifically labeled for control of Botrytis blight of roses and other ornamentals. Fungicides may be applied to bare-root roses prior to refrigerated storage. Resistance management is important, since resistant fungal strains are known to develop after continual use of certain chemicals. Strategies have been developed to help maintain fungicide effectiveness and are usually outlined on product labels.

Crown Gall

Agrobacterium tumefaciens

Rose is one of many woody ornamentals susceptible to crown gall. The soilborne bacterium *Agrobacterium tumefaciens* enters through wounds, where a tumor-inducing plasmid DNA is transferred to host cells. Gall tissue develops from transformed cells.

High incidence of crown gall is often associated with cultural practices that increase wounding followed by exposure to bacteria in soil. A good way to prevent crown gall is to avoid injury to the roots and main stem and to avoid fields with a history of the disease. Cold storage of dug plants and rootstock cuttings, sanitation, and in certain cases propagation in soilless media have reduced crown gall incidence in nursery production. Biological control has shown positive results in some tests. For more information, see Chapter 10, "Crown Gall."

Nematodes

***Meloidogyne hapla*, root-knot nematode**
***Pratylenchus* spp., lesion nematodes**
***Xiphinema* spp., dagger nematodes**

Numerous plant-parasitic nematodes are known to feed on rose roots. Those most often found attacking roses are *Meloidogyne hapla* (root-knot nematode), *Pratylenchus* spp. (lesion nematodes), and *Xiphinema* spp. (dagger nematodes). Root-knot nematodes are a major problem in field production. Lesion nematodes are prevalent in cut rose production where plants are maintained several years. Nematode injury is subtle and often overlooked as a cause of poor plant growth.

Symptoms

No specific aboveground system is diagnostic for nematode feeding on roots. Because roots are injured, nematode feeding can result in poor overall growth and flowering, and nutrient deficiency symptoms may appear on foliage.

Root-knot and dagger nematodes cause swellings or galls on roots. Root-knot nematodes cause smaller and more numerous galls, and they also cause infected plants to produce excessive root branching (Plate 150). Only the northern root-knot nematode, *M. hapla,* has been reported on roses.

The swollen bodies of reproductive root-knot females can be observed in root galls with minimal magnification, but a trained observer with access to a microscope is often needed to confirm the presence of these and other nematodes.

Epidemiology

Most nematode injury occurs when roses are planted in sandy soil. As roots grow, they release chemicals that stimulate the hatching of nematode eggs and subsequent feeding. Nematodes molt several times before reaching maturity. For more information, see Chapter 6, "Plant-Parasitic Nematodes," and Chapter 13, "Nematode Diseases."

Host Range

Many plants are hosts of the nematodes that feed on roses.

Management

In field production the root-knot nematode causes the most damage and is the primary target of nematode control. Fumigation is necessary for field production on infested sites. Nematodes can be avoided in greenhouses and container production if clean plants are set in soilless growing media. If soil is added to a mix, it should ttle progress has been made in breeding acceptable nematode-resistant rootstocks. *R.* 'Manetti,' commonly used in greenhouse cut flower production, is resistant to root-knot nematodes but susceptible to lesion nematodes. Most roses, however, are produced on *R. multiflora, R.* 'Dr. Huey,' or *R.* 'Fortuniana' rootstocks, because they are better adapted to certain climatic conditions. All are considered susceptible to root-knot nematodes, but Fortuniana is the least susceptible. *R. multiflora* has a level of tolerance or resistance to lesion nematodes.

Viruses

Apple mosaic virus
Prunus necrotic ringspot virus
Other viruses

Apple mosaic virus (AMV) and Prunus necrotic ringspot virus (PNRSV) are found throughout the world and are the most common of all known viruses in roses. The disease caused by these

Nematodes

Geographic occurrence Wherever roses are grown

Seasonal occurrence Whenever soil is warm enough for root growth

DISEASE FREQUENCY	DISEASE SEVERITY
5 annual	5 plants killed
4	4
3	**3**
2	2
1 rare	1 very little damage

CHEMICAL TREATMENT	CULTURAL PRACTICES
3 used every year	**3 very important**
2	2
1 not used	1 not important

SANITATION	RESISTANT CULTIVARS
3 very important	3 many cultivars
2	**2**
1 not important	1 no resistance

Viruses

Geographic occurrence Wherever roses are grown

Seasonal occurrence All year, with an optimum period for symptom expression in spring

DISEASE FREQUENCY	DISEASE SEVERITY
5 annual	5 plants killed
4	4
3	3
2	**2**
1 rare	1 very little damage

CHEMICAL TREATMENT	CULTURAL PRACTICES
3 used every year	**3 very important**
2	2
1 not used	1 not important

SANITATION	RESISTANT CULTIVARS
3 very important	3 many cultivars
2	**2**
1 not important	1 no resistance

viruses, individually or together, was first called rose mosaic, but that name is less common now that AMV and PNRSV are known to be the causal agents.

Symptoms

Mosaic, ring spots, veinclearing, and line patterns appear on new growth, and AMV often causes leaves to have a golden appearance (Plates 151 and 152). However, there is no way to distinguish between these viruses by observing symptoms. Modern serological techniques have been developed for identification of specific viruses, but positive results are not always associated with symptoms. Virus symptoms subside in plant tissue during the hot summer, which could explain the negative results sometimes obtained when symptomatic tissue is tested.

The literature is confusing regarding the ill effects of rose mosaic. A few reports indicate a reduction in blooms and growth of infected plants, but many indicate no stunting or decrease in blooms. In most reports, the identity of the virus or viruses in the study was not confirmed or stated, and the disease was referred to as rose mosaic in a general sense. More thorough investigations are needed, now that specific viruses and virus combinations can be identified and their effects studied.

Epidemiology

There are no known vectors of PNRSV and AMV. Vegetative propagation is their only known means of transmission in rose. PNRSV can be transmitted in grains of pollen of some plants, and therefore transmission by this means is suspected in roses, but it is likely to occur at very low levels. Root grafts are another means of distribution. Natural spread, in whatever ways it may occur, is slow and inefficient compared to vegetative propagation by grafting.

Host Range

The hosts of PNRSV and AMV are primarily woody plants, including almond, apple, apricot, cherry, peach, and rose. Certain rose cultivars tend to show more symptoms than others.

Management

Propagating virus-free stock is the only means of control of virus diseases. Thermotherapy followed by micropropagation and virus indexing is the best method of cleaning up infected cultivars. This procedure consists of growing plants at high temperatures, so that new growth is free of virus particles. Shoot tips are excised, increased in number, and eventually budded to a highly susceptible indicator host as a final check for viruses. Rootstocks as well as scion varieties must be indexed. Virus indexing does not guarantee that all indexed material is virus-free, but it is the best means available for removing viruses from woody plants. The University of California at Davis has a virus indexing program in operation.

After growers acquire virus-indexed material, it has to be increased to commercial numbers. Systematic surveys of all propagation stock must be conducted annually, in an ongoing program to keep viruses out of propagation plant material. All plants showing symptoms of virus infection or testing positive in a virus assay must be rogued from the planting. If a virus is detected regularly in propagating stock, the entire block should be destroyed and replaced with new virus-indexed stock

Other Diseases

Other diseases occasionally observed in roses and their causal agents are

Cercospora leaf spot (*Cercospora rosicola*)
rose rosette (suspected virus)
southern blight (*Sclerotium rolfsii*)
Verticillium wilt (*Verticillium* spp.)

REFERENCES

Coyier, D. L. 1985. Roses. Pages 405–488 in: Diseases of Floral Crops. Vol. 2. D. L. Strider, ed. Praeger, New York.

Horst, R. K. 1983. Compendium of Rose Diseases. American Phytopathological Society, St. Paul, Minn.

Pirone, P. P. 1978. Diseases and Pests of Ornamental Plants. 5th ed. John Wiley & Sons, New York.

CHAPTER 77

Thomas J. Banko ▪ Virginia Polytechnic Institute and State University, Virginia Beach

Marcia A. Stefani ▪ Virginia Polytechnic Institute and State University, Virginia Beach

Sourwood Diseases

Geographic production USDA zones 5–9

Family Ericaceae

Genus *Oxydendrum*

Species *O. arboreum*

Sourwood is a small to medium-sized tree or large shrub native to the southeastern and central United States. Its natural range is east of the Mississippi River, throughout woodlands from southern Pennsylvania to northern Florida. It has been praised as an outstanding small landscape tree, and demand for it in the nursery trade has increased recently, although it is not yet in widespread use. Sourwood requires well-drained soil and will not tolerate wet, poorly drained soil. The most common diseases of sourwood are fungal leaf spots, of which the most important is Cercospora leaf spot, described below. Leaf spot diseases caused by *Mycosphaerella caroliniana, Cristulariella moricola,* and an *Alternaria* sp. have also been reported. Cankers and twig dieback also occasionally affect sourwood.

Cercospora Leaf Spot
Cercospora oxydendri

Cercospora leaf spot, caused by the fungus *Cercospora oxydendri,* is possibly the most common leaf disease of sourwood in the southern United States. Wild populations of sourwood have been observed with nearly 100% susceptibility.

Symptoms

Red to deep purple, circular to irregularly shaped lesions, 1–5 mm in diameter, form on leaves (Plate 153). The spots are often surrounded by a halo of reddened tissue radiating out over the leaf surface. Older lesions have papery gray centers surrounded by dark brown margins. The centers of lesions may drop out, leaving shot holes. Severe infection of young leaf tissue may cause extensive leaf distortion. Numerous spots may coalesce to form larger necrotic areas.

Epidemiology

Cercospora leaf spot is prevalent in warm, humid regions, such as the southeastern United States. *C. oxydendri* sporulates on both the upper and the lower surfaces of lesions, producing asexual spores (conidia), which are easily detached and may be blown long distances by the wind. The spores need water to germinate and penetrate the plant, but heavy dews seem to be sufficient for infection. *Cercospora* is favored by high temperatures and is therefore more damaging during the summer and in warm climates. It overwinters in dead leaves on the ground, which serve as a reservoir of inoculum for subsequent infections.

Cercospora Leaf Spot

Geographic occurrence USDA zones 5–9

Seasonal occurrence June to September, with an optimum period in July and August

DISEASE FREQUENCY	DISEASE SEVERITY
5 annual (selected)	5 plants killed
4	4
3	3
2	**2** (selected)
1 rare	1 very little damage

CHEMICAL TREATMENT	CULTURAL PRACTICES
3 used every year (selected)	3 very important
2	**2** (selected)
1 not used	1 not important

SANITATION	RESISTANT CULTIVARS
3 very important	3 many cultivars
2 (selected)	2
1 not important	**1 no resistance** (selected)

Host Range

C. oxydendri is a host-specific pathogen, occurring only on sourwood.

Management

Maintain good air movement around plants. Apply protectant fungicide treatments during warm, humid, or wet weather during or following leaf expansion.

Botryosphaeria Canker

Botryosphaeria obtusa

Botryosphaeria obtusa (syn. *Physalospora obtusa*) is a nonspecialized fungus which lives as a saprophyte in the bark of many woody plants. However, it is capable of pathogenicity if the living tissue of a susceptible host is under environmental stress or if the tissue has been wounded.

Symptoms

Lesions develop around wound sites or at the base of dead twigs. Newly formed lesions are dark red or purple brown; in time they may develop a tan margin. The lesions may involve only the external bark, or the fungus may spread into the cambium. Some cankers stop growing after the first year, but some continue to expand annually during the growing season, eventually reaching 30 cm or more in length. Canker growth is most rapid when trees are suffering from drought stress. Once the bark has been killed, it gradually cracks and falls away. This process takes a year or more. If a canker enlarges enough to surround a limb during the growing season, the leaves wilt or yellow prematurely and drop.

Botryosphaeria Canker

Geographic occurrence USDA zones 5–9

Seasonal occurrence March to September, with an optimum period in April and May

DISEASE FREQUENCY	DISEASE SEVERITY
5 annual	5 plants killed
4	**4**
3	**3**
2	2
1 rare	1 very little damage

CHEMICAL TREATMENT	CULTURAL PRACTICES
3 used every year	**3 very important**
2	2
1 not used	1 not important

SANITATION	RESISTANT CULTIVARS
3 very important	3 many cultivars
2	2
1 not important	**1 no resistance**

Epidemiology

Cool, wet spring weather triggers the release of spores (both conidia and ascospores) of *B. obtusa.* The pathogen is particularly favored by nighttime rains under cool conditions, with a constant temperature for 12–24 hours. A temperature of 20°C for a 24-hour period is optimal for infection.

B. obtusa overwinters as fruiting bodies (pycnidia and pseudothecia) in cankers and in the bark of dead branches and twigs. Pycnidia may form as early as the first season of canker growth. They eventually erupt through the bark, darkening and roughening the surface of the cankered area. Under cool, moist conditions, tendrils of brown conidia ooze from pycnidia. These spores are disseminated by splashing water and by insects. Pseudothecia develop in the second growing season and produce ascospores, which are disseminated by water and wind. Both conidia and ascospores may be released at any time during the growing season when conditions are favorable, but the greatest dissemination occurs during the spring.

Host Range

B. obtusa is a pathogen of a wide range of woody trees and shrubs, but it is best known in North America as a fruit tree pathogen. It causes black rot canker of apples and crabapples, grapevine dieback, canker and gummosis of peach, and fruit rot of citrus. Other woody hosts include arborvitae, ash, azaleas and rhododendrons, cotoneaster, crapemyrtle, dogwood, juniper, magnolia, maple, pine, rose, spirea, viburnum, wax myrtle, and willow, in addition to sourwood (Sinclair et al., 1987).

Management

Avoid or minimize stress caused by drought or extreme temperatures, such as freeze damage or sunscald of the bark. Prune out twigs or branches with lesions, to reduce inoculum production.

Nectria Canker

Nectria galligena

(asexual stage, *Cylindrocarpon heteronemum*)

Nectria canker is a widespread disease of many hardwood trees in the forests of the United States and Europe and occasionally occurs in sourwood. In the United States, it is more prevalent in the East than in the West. The disease is caused by the fungus *Nectria galligena,* which is most prevalent under cool, humid conditions. It causes distinctive perennial cankers in sourwood.

Symptoms

The earliest symptoms of Nectria canker are small, sunken areas on young smooth bark. The cankers are initially small and are often overlooked. Callus tissue forms at the edges of cankers, as the plant resists the spread of the pathogen. In the autumn, the fungus in its asexual stage (*Cylindrocarpon heteronemum*) forms inconspicuous cushiony tufts of white hyphae (sporodochia), which protrude through cracks near the edges of cankers. These hyphal mats in turn produce colorless asexual spores (conidia). Older cankers become increasingly obvious as the fungus continues to spread and new callus is formed, in a cycle that results in a definite target spot, with concentric rings of callus. If fungal growth is uniform, the concentric rings will be quite circular. However, if the pathogen encounters resistance in the host tissue, cankers will be irregular or elongated. During autumn through spring following the formation of sporodochia, small, flask-shaped, bright red to orange fruiting bodies (perithecia) form on fungal stromata in dead callus tissues. The perithecia contain colorless ascospores, the sexual spores of the pathogen.

Epidemiology

Asexual and sexual spores of the pathogen penetrate host tissue through cracks in the bark or wounds which expose the cambium. Wounds include leaf scars, cracks in the axils of twigs, and sunscald injury. Asexual spores (cylindrical macroconidia and ellipsoidal microconidia) are disseminated by splashing rain and wind year-round in warm climates and during the growing season in colder regions. The sexually produced ascospores are forcibly discharged from perithecia and dispersed by water and wind, primarily during the spring and fall, but they can be produced whenever there is adequate moisture and the air temperature is above freezing.

In order to form a canker, the infectious agent must reach cambial tissue. In an established canker, the fungus annually penetrates the callus tissue, so that canker growth resumes every dormant season. The pathogen produces indoleacetic acid, which may be responsible for the very visible concentric callus margins that characterize this disease.

Host Range

Nectria canker affects more than 60 species of deciduous trees in North America. *N. galligena* is a major pathogen of apples, birches, black walnut, maples, and pears.

Management

Inspect plants annually, after leaf drop, for the presence of small branches with cankers. Prune out and destroy any cankered areas.

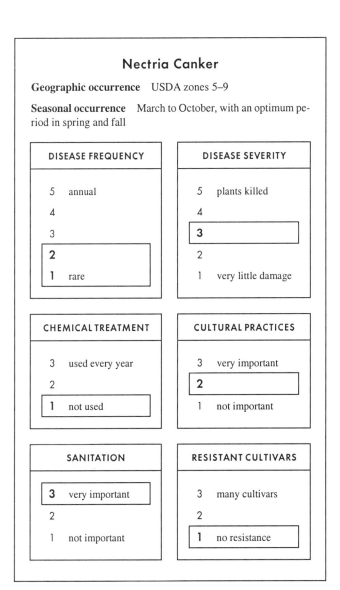

Nectria Canker

Geographic occurrence USDA zones 5–9

Seasonal occurrence March to October, with an optimum period in spring and fall

DISEASE FREQUENCY
- 5 annual
- 4
- 3
- **2**
- **1 rare**

DISEASE SEVERITY
- 5 plants killed
- 4
- **3**
- 2
- 1 very little damage

CHEMICAL TREATMENT
- 3 used every year
- 2
- **1 not used**

CULTURAL PRACTICES
- 3 very important
- **2**
- 1 not important

SANITATION
- **3 very important**
- 2
- 1 not important

RESISTANT CULTIVARS
- 3 many cultivars
- 2
- **1 no resistance**

Phomopsis Twig Blight

Geographic occurrence USDA zones 5–9

Seasonal occurrence March to June

DISEASE FREQUENCY
- 5 annual
- 4
- 3
- **2**
- **1 rare**

DISEASE SEVERITY
- 5 plants killed
- 4
- **3**
- 2
- 1 very little damage

CHEMICAL TREATMENT
- 3 used every year
- **2**
- 1 not used

CULTURAL PRACTICES
- **3 very important**
- 2
- 1 not important

SANITATION
- **3 very important**
- 2
- 1 not important

RESISTANT CULTIVARS
- 3 many cultivars
- 2
- **1 no resistance**

Phomopsis Twig Blight, Canker, and Dieback

***Phomopsis vaccinii* (sexual stage, *Diaporthe vaccinii*)**

Twig Blight

Dothiora wolfii

Phomopsis vaccinii was first identified as causing a twig blight and dieback of sourwood in Virginia in 1992. *Dothiora wolfii* (syn. *Sphaerulina polyspora*) has also been reported to cause a twig blight of sourwood.

Symptoms

The primary symptoms of these diseases are stem tip dieback and twig cankers. *P. vaccinii* penetrates host tissue at the tips of twigs and progresses toward the stem base. Twigs are eventually girdled and exhibit dieback symptoms. Lesions are formed on twig stems and eventually develop into cankers.

Epidemiology

Asexual spores (conidia) of *Phomopsis* are released from dark, nearly globose fruiting bodies (pycnidia), generally in the spring during rainy weather. Two types of conidia are produced. Alpha-conidia are one-celled, short, ovate or ellipsoidal spores and are capable of germinating readily to infect new tissue. Beta-conidia are one-celled, curved, filamentous spores which do not germinate readily. Both types are spread in splashed and windborne drops of water. *P. vaccinii* overwinters as mycelium and fruiting bodies in dead tissues, where it can remain viable for at least two years.

Host Range

P. vaccinii causes twig blight of blueberries (*Vaccinium* spp.) and fruit rot of cranberries (*V. macrocarpon*) as well as twig blight, canker, and dieback of sourwood.

Management

Avoid drought and other environmental stresses. Remove infected stems, pruning below the cankered areas and sanitizing shears between cuts. Prune in dry weather, when the plants are dry. Protectant fungicides may be helpful in a management strategy in nurseries, but frequent applications may be required.

Physiological Damage Due to Excess Water

Some plants are more or less tolerant of flooding or waterlogged soils, but sourwood is rated intolerant of excess water around the root system. Excess water may also be conducive to root rot organisms. Phytophthora root rot is a major pathogen of several species in the family Ericaceae, and although it apparently has not been reported in sourwood, it is suspected that sourwood could become infected under certain conditions.

Symptoms

Symptoms of excess water mimic those of other root problems: they may include nutrient deficiency symptoms, premature reddening of leaves, marginal necrosis of leaves, wilting, premature leaf abscission, twig dieback, and death of the entire plant.

Management

Locate sourwood plants in the nursery or in the landscape where there is no standing water. Use a well-drained growing medium with a high pine bark content.

REFERENCES

Blanchard, R. O., and Tattar, T. A. 1981. Field and Laboratory Guide to Tree Pathology. Academic Press, New York.

Coyier, D. L., and Roane, M. K., eds. 1986. Compendium of Rhododendron and Azalea Diseases. American Phytopathological Society, St. Paul, Minn.

Farr, D. F., Bills, G. F., Chamuris, G. P., and Rossman, A. Y. 1989. Fungi on Plants and Plant Products in the United States. American Phytopathological Society, St. Paul, Minn.

Grand, L. F. 1978. New hosts of *Cristulariella pyramidalis* in North Carolina. Plant Dis. Rep. 62:841–842.

Horst, R. K. 1979. Westcott's Plant Disease Handbook. 4th ed. Van Nostrand Reinhold, New York.

Likins, T. M. 1992. New host pathogen relationship. Va. Nurserymen's Assoc. Newsl., Mar.–Apr., p. 486.

Pirone, P. P. 1978. Diseases and Pests of Ornamental Plants. 5th ed. John Wiley & Sons, New York.

Sinclair, W. A., Lyon, H. H., and Johnson, W. T. 1987. Diseases of Trees and Shrubs. Comstock Publishing Associates, Cornell University Press, Ithaca, N.Y.

CHAPTER 78

Margaret R. Williamson • Clemson University, Clemson, South Carolina

Mark Windham • University of Tennessee, Nashville

Spirea Diseases

Geographic production USDA zones 4–8

Family Rosaceae

Genus *Spiraea*

Species White-flowered species
- *S. cantoniensis*
- *S.* × *cinerea*
- *S. nipponica*
- *S. prunifolia*
- *S. thunbergii*
- *S. trilobata*
- *S.* × *vanhouttei*

Pink-flowered or white-corymbed species
- *S. albiflora*
- *S. bullata*
- *S.* × *bumalda*
- *S. japonica*

Spiraea spp. are "old-fashioned" shrubs that have increased in popularity in recent years. The dwarf forms have become especially popular in mixed perennial beds and borders. Spireas possess good cold-hardiness and have relatively few diseases. To avoid pH-induced chlorosis, some species are best planted in acidic soils (pH 5.0–6.0).

Leaf Spot

***Cylindrosporium* spp.**
Phloeosporella filipendulae
(syn. *Cylindrosporium filipendulae*)

Leaf spot has been reported in a number of *Spiraea* spp., but most research has been conducted with *S.* × *vanhouttei* and *S. japonica.* Plant death due to leaf spot is unlikely.

Symptoms

Usually, leaf spot is first observed on foliage of the lower branches (Windham and Windham, 1988). The first symptoms are small, yellowish spots, which develop into small, reddish brown to purple lesions. The lesions may enlarge and coalesce. Early leaf drop is common, and severely affected plants may defoliate by midsummer (Ragan and Bernard, 1984; Windham and Windham, 1988).

Epidemiology

The leaf spot fungus *Phloeosporella filipendulae* (syn. *Cylindrosporium filipendulae*) overwinters in fallen spirea leaves, and initial inoculum is probably from this source (Williamson and Bernard, 1988). About a week after inoculation, *P. filipendulae* forms flesh-colored spore masses in fruiting bodies (acervuli) on both surfaces of the leaf, but more prevalently on the lower surface. The spores (conidia) are borne in a moist matrix and are probably disseminated by splashing water (Ragan and Bernard, 1984).

Leaf Spot

Geographic occurrence USDA zones 4–8

Seasonal occurrence April until September

DISEASE FREQUENCY	DISEASE SEVERITY
5 annual	5 plants killed
[4]	4
3	3
2	**[2]**
1 rare	1 very little damage

CHEMICAL TREATMENT	CULTURAL PRACTICES
3 used every year	**[3] very important**
[2]	2
1 not used	1 not important

SANITATION	RESISTANT CULTIVARS
[3] very important	3 many cultivars
2	2
1 not important	**[1] no resistance**

Host Range

Leaf spot has been reported in a number of *Spiraea* species (Alfieri et al., 1994; Farr et al., 1989; Ragan and Bernard, 1984), notably *S.* × *vanhouttei, S. japonica,* and *S. thunbergii.* The popular hybrid *S.* × *bumalda* is probably susceptible as well, since *S. japonica* is one of its parents.

Management

Both cultural and fungicidal controls can be used to manage leaf spot.

Cultural practices Since *P. filipendulae* overwinters in fallen leaves, they should be removed after leaf drop, to help prevent a buildup of inoculum causing new infections the following year. Avoid frequent overhead irrigation and increase plant spacing to help control the spread of the pathogen during the growing season.

Resistance Cultivars of *S. japonica* seem to exhibit a greater degree of tolerance to leaf spot (Ragan and Bernard, 1984), but no resistant cultivars have been selected.

Chemical treatment In fungicide trials, even the most effective chemicals had little effect on preventing initial infections, making leaf removal an important preventative measure. Several fungicides were effective in controlling disease spread, but applications must be started as soon as initial infection is noticed and should continue throughout the growing season (Windham and Windham, 1988).

Powdery Mildew

Microsphaera penicillata
***Oidium* sp.**
Podosphaera clandestina, P. tridactyla
Sphaerotheca fuliginea, S. macularis

Fungi in four different genera have been reported to cause powdery mildew of spirea: *Microsphaera penicillata,* an *Oidium* sp., *Podosphaera clandestina, P. tridactyla, Sphaerotheca fuliginea,* and *S. macularis.* However, the signs of these pathogens, the symptoms they cause, and the methods of controlling them are basically the same. Plant death due to powdery mildew is unlikely.

Symptoms

Powdery mildew causes distortion of young leaves, sometimes followed by premature defoliation. Signs of the powdery mildew pathogens are powdery fungal growth on leaves, which may be followed in the fall by small, dark, spherical fruiting bodies (cleistothecia).

Epidemiology

Little is known about the overwintering and dispersal of initial inoculum of the spirea powdery mildew fungi. They may overwinter as mycelium in vegetative buds or on leaves that persist during mild winters. Overwintering as cleistothecia is another possibility. Once initial infection has occurred, the white mycelium produces spores (conidia), which are dispersed by air currents to cause new infections.

Host Range

Most reports of the host ranges of the spirea powdery mildew fungi have listed *Spiraea* spp. without naming individual species. None of the widely cultivated species of *Spiraea* has been reported as a host, but odds are that most are susceptible to one of these pathogens.

Powdery Mildew

Geographic occurrence USDA zones 4–8 (the range of the host)

Seasonal occurrence April to September

DISEASE FREQUENCY	DISEASE SEVERITY
5 annual	5 plants killed
4	4
3	3
2	**2**
1 rare	1 very little damage

CHEMICAL TREATMENT	CULTURAL PRACTICES
3 used every year	3 very important
2	2
1 not used	**1 not important**

SANITATION	RESISTANT CULTIVARS
3 very important	3 many cultivars
2	2
1 not important	**1 no resistance**

Management

Specific cultural controls cannot be provided, since little is known about the epidemiology of these fungi on spirea. However, removing and destroying fallen leaves may be helpful in controlling initial spring infections.

Resistance No resistant cultivars are known.

Chemical treatment Fungicides will provide control during the growing season.

REFERENCES

Alfieri, S. A., Jr., Langdon, K. R., Kimbrough, J. W., El-Gholl, N. E., and Wehlburg, C. 1994. Diseases and disorders of plants in Florida. Fla. Dep. Agric. Consumer Serv. Bull. 14.

Farr, D. F., Bills, G. F., Chamuris, G. P., and Rossman, A. Y. 1989. Fungi on Plants and Plant Products in the United States. American Phytopathological Society, St. Paul, Minn.

Ragan, M. A., and Bernard, E. C. 1984. Etiology of spirea leaf spot. Proc. South. Nurserymen's Assoc. Res. Conf. Annu. Rep. 29:6–7.

Williamson, M. A., and Bernard, E. C. 1988. Life cycle of a new species of *Blumeriella* (Ascomycotina: Dermateaceae), a leaf-spot pathogen of spirea. Can. J. Bot. 66:2048–2054.

Windham, M. T., and Windham, A. S. 1988. Chemical control of spirea leaf spot. Tenn. Farm Home Sci. 146:16–17.

CHAPTER

79

John Hartman ▪ University of Kentucky, Lexington

Sycamore and Planetree Diseases

Geographic production USDA zones 4–9

Family Platanaceae

Genus *Platanus*

Species		
	P. × *acerifolia* (*P. orientalis* × *P. occidentalis*), or *P.* × *hybrida*	London planetree, plane
	P. occidentalis	American planetree, sycamore, buttonwood
	P. orientalis	oriental planetree

Anthracnose

Apiognomonia veneta
(asexual stage, *Discula platani*)

Anthracnose is the most common and important disease of sycamore. It kills twigs and blights foliage early in the season when the leaves are still small, so it can be confused with frost damage. Early defoliation is common during rainy seasons.

Symptoms

Leaf blight Irregular brown areas appear along the veins, midribs, and edges of infected leaves (Plate 154). Leaf blight causes premature defoliation and is most severe on the lower branches of the tree.

Shoot blight Shoot blight appears as a rapid death of newly developing shoots, following the entry of the pathogen into the succulent shoot tissues, or following the death of a cankered twig bearing the emerging shoot.

Twig canker Infected twigs and buds may be killed during the dormant season by cankers which form following infection in the previous season (Plate 155). The cankers are oval and brown and sometimes cause bark splitting, in contrast to the intact greenish healthy bark. Cankers kill twig terminals, so that lateral buds emerge to become branch leaders, causing the branches to change directions repeatedly in a zigzag fashion. Sometimes multiple twigs arise behind a dead leader, and the branch resembles a witches'-broom.

Epidemiology

Anthracnose of sycamore is caused by the fungus *Apiognomonia veneta,* which overwinters at nodes in twigs and by early spring forms cankers in which dark fruiting bodies (pycnidia) develop. It also overwinters in leaves, where it produces other fruiting structures (perithecia), which discharge airborne ascospores. As a source of primary inoculum, however, ascospores are less important than inoculum from overwintering cankers. Asexual spores (conidia) produced in cankers are dispersed by splashing rain, which carries them to emerging shoots and leaves, where they initiate new infections.

Anthracnose

Geographic occurrence USDA zones 4–9

Seasonal occurrence April and May

DISEASE FREQUENCY	DISEASE SEVERITY
5 annual	5 plants killed
4 (boxed)	**4** (boxed)
3	**3** (boxed)
2	2
1 rare	1 very little damage

CHEMICAL TREATMENT	CULTURAL PRACTICES
3 used every year	**3** very important (boxed)
2 (boxed)	2
1 not used	1 not important

SANITATION	RESISTANT CULTIVARS
3 very important (boxed)	3 many cultivars
2	**2** (boxed)
1 not important	1 no resistance

When the mean daily temperature during the two weeks following budbreak is below 13°C, shoot infections ensue, but in seasons with higher temperatures during that two-week period there is less shoot blight. If the mean temperature during these two weeks is above 16°C, shoot blight is absent, but leaf infections occur.

In spring, the asexual stage of the fungus, *Discula platani,* produces conidia in tiny, cream-colored fruiting structures (acervuli) on infected leaves. When leaves and twigs are moist, the acervuli release conidia, which are carried to healthy tissues by splashing rain and initiate secondary infections. Hyphae of the fungus also grow down leaf petioles and into the twig nodes, where the fungus overwinters.

Host Range

All planes are susceptible to anthracnose, but sycamore is much more susceptible than London and oriental planes.

Management

Cultural practices Maintain good spacing between plants and between rows, to improve the movement of air and the penetration of sunlight in the nursery. Overhead irrigation in spring must be avoided. Grow sycamore nursery stock away from fencerows where infected mature sycamores might be growing.

Sanitation Pruning out and destroying infected twigs and branches and raking up and destroying fallen leaves will help reduce overwintering inoculum.

Resistance Clonal London plane selections, such as Bloodgood, Columbia, and Liberty, are resistant to anthracnose.

Chemical treatment Fungicides applied two to three times beginning at budbreak may be used to prevent anthracnose infection.

Leaf Spots

Mycosphaerella platanifolia
Phyllosticta platani

There is some uncertainty about whether *Mycosphaerella platanifolia* and *Phyllosticta platani* cause different stages of the same disease or whether they cause separate diseases. The leaf spot associated with *M. platanifolia* is more severe than anthracnose in many southern states.

Symptoms

Phyllosticta leaf spot Circular to irregularly shaped lesions, 1–2 cm across, develop on sycamore leaves in late summer. The lesions are brown, with dark brown, indefinite margins. As lesions coalesce, leaves become blighted, turning yellow and then brown, and eventually become necrotic. Fruiting bodies (pycnidia) of *Phyllosticta* appear as tiny brown specks on the underside of lesions.

Mycosphaerella leaf spot Numerous small, brown spots, 1 mm across, become noticeable in early summer. Spots coalesce and cause leaf blighting. The asexual stage of the fungus, *Cercospora platanicola,* sporulates through stomata on both surfaces of the leaf.

Epidemiology

The leaf spot fungi overwinter on diseased fallen leaves and infect new foliage during moist weather. Secondary infections initiated by asexual spores (conidia) increase the number of spots on leaves during the growing season. The disease is more severe in rainy seasons.

Host Range

Sycamore is susceptible to *M. platanifolia* and *P. platani.*

Management

Cultural practices Space plants in the nursery to improve air movement and penetration of sunlight.

Sanitation Rake up and destroy fallen leaves in the nursery.

Resistance Resistance to leaf spot is not known in sycamore.

Chemical treatment Timely applications of a protectant fungicide in spring and early summer should prevent leaf spot.

Botryosphaeria Canker and Dieback

Botryosphaeria dothidea

Botryosphaeria canker and dieback of sycamores occurs occasionally in the nursery, causing minimal losses. The disease is caused by the fungus *Botryosphaeria dothidea.*

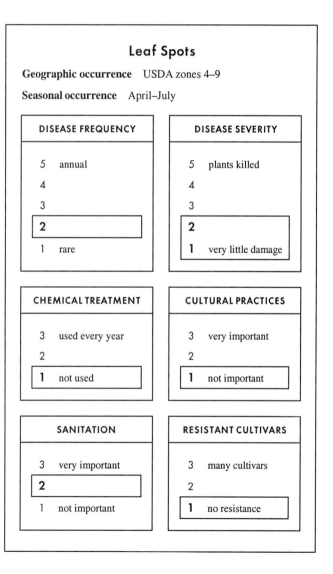

Leaf Spots

Geographic occurrence USDA zones 4–9

Seasonal occurrence April–July

DISEASE FREQUENCY	DISEASE SEVERITY
5 annual	5 plants killed
4	4
3	3
[2]	**[2]**
1 rare	**[1] very little damage**

CHEMICAL TREATMENT	CULTURAL PRACTICES
3 used every year	3 very important
2	2
[1] not used	**[1] not important**

SANITATION	RESISTANT CULTIVARS
3 very important	3 many cultivars
[2]	2
1 not important	**[1] no resistance**

Symptoms

Brown lesions occur in the bark of infected twigs and branches. In the lesions, tiny fruiting bodies (pycnidia) of *B. dothidea* are produced in stromata, which are visible to the unaided eye as black dots. When a canker girdles a twig or branch, leaves and shoots distal to the canker may decline and die. The symptoms can easily be confused with those of anthracnose.

Epidemiology

Spores (conidia and ascospores) of the pathogen, carried by splashing rain or pruning tools, initiate infection of twigs and branches, usually in spring and summer. The fungus may infect plants by invading wounds and natural openings, such as bark cracks and lenticels. It colonizes both phloem and xylem tissues, and cankers develop in a few days or weeks to several months. Susceptibility to Botryosphaeria canker is increased when sycamores are drought-stressed or injured by cold. Drought-stressed trees may regain their resistance to Botryosphaeria canker when they are watered.

Host Range

B. dothidea attacks a very wide range of woody plants, in addition to sycamore and planetree.

Botryosphaeria Canker and Dieback

Geographic occurrence USDA zones 4–9

Seasonal occurrence April–September

DISEASE FREQUENCY	DISEASE SEVERITY
5 annual	5 plants killed
4	4
3	3
2	**2**
1 rare	**1** very little damage

CHEMICAL TREATMENT	CULTURAL PRACTICES
3 used every year	**3** very important
2	2
1 not used	1 not important

SANITATION	RESISTANT CULTIVARS
3 very important	3 many cultivars
2	2
1 not important	**1** no resistance

Management

Cultural practices Because most nurseries are irrigated, drought stress is usually not a problem. Avoid other stresses, such as excessive root pruning.

Sanitation Prune out dead and cankered branches.

Resistance No cultivars are resistant.

Chemical treatment No chemical controls for canker and dieback of planes are available.

Canker Stain

Ceratocystis fimbriata* f. sp. *platani

Canker stain is an important disease of London plane; it also affects sycamore. The disease began killing London planes in the eastern United States in the 1920s and has now spread to cities and towns throughout the South and Midwest. It is also an important disease in Italy and southern France.

Symptoms

An elongated, brown or black discoloration appears on the smooth yellow or green bark of trunks and branches. This symp-

Canker Stain

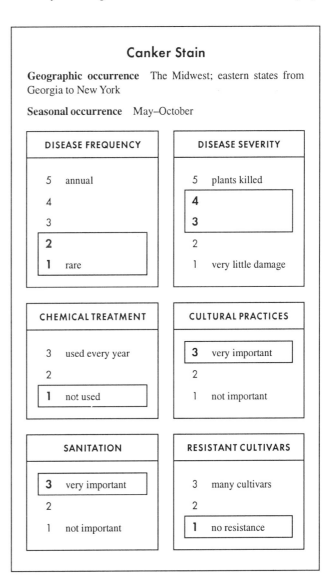

Geographic occurrence The Midwest; eastern states from Georgia to New York

Seasonal occurrence May–October

DISEASE FREQUENCY	DISEASE SEVERITY
5 annual	5 plants killed
4	**4**
3	**3**
2	2
1 rare	1 very little damage

CHEMICAL TREATMENT	CULTURAL PRACTICES
3 used every year	**3** very important
2	2
1 not used	1 not important

SANITATION	RESISTANT CULTIVARS
3 very important	3 many cultivars
2	2
1 not important	**1** no resistance

tom may not be observed at first on flaky bark. The discoloration is associated with slightly flattened cankers, a few centimeters wide and up to a meter or more in length. As the cankers expand, they may coalesce and girdle the trunk or a branch. Infected phloem and cambial tissues appear almost black, and the wood beneath is stained reddish brown or bluish black. The stained portion of the wood may form a wedge pattern extending radially to the pith. Older cankers shed the darkened, dead bark, exposing the wood, which dries and cracks. Cankers may spread both up and down the tree. Tree parts distal to a canker decline and die within a few months to a year or two after infection.

Epidemiology

Canker stain is caused by the fungus *Ceratocystis fimbriata* f. sp. *platani,* which sporulates abundantly on newly killed wood in wet weather from May to October, producing two kinds of endoconidia. It also forms long-beaked fruiting bodies (perithecia) bearing ascospores. The conidia and ascospores are sticky and are readily transmitted by pruning tools and tree maintenance equipment and, to a lesser extent, by splashing rain. They are also spread by insects, which are attracted by the fruity odors emitted by the fungus. Wounds are critical for the initiation of infection. During spring, summer, and fall, infection can occur in any fresh wound in living bark or sapwood. The fungus may begin to sporulate within a few days of infection, even though disease symptoms may not appear for several weeks.

Host Range

All planetrees are susceptible to canker stain, but London plane is the most susceptible.

Management

Cultural practices Avoid wounding trees at any time during the growing season. If nursery trees need pruning and shaping, these operations must be performed during the coldest part of the winter. Before pruning, be sure that the pruning tools are clean.

Sanitation Remove and destroy all infected planetrees from the nursery.

Resistance Resistant cultivars are not available.

Chemical treatment There are no chemical treatments for canker stain.

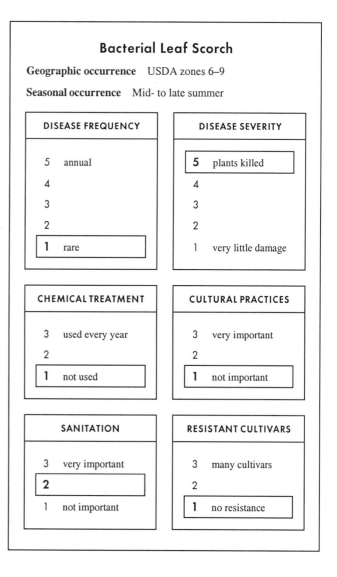

Powdery Mildew

Geographic occurrence USDA zones 4–9

Seasonal occurrence April–October

DISEASE FREQUENCY
5 annual
4
3
2
1 rare

DISEASE SEVERITY
5 plants killed
4
3
2
1 very little damage

CHEMICAL TREATMENT
3 used every year
2
1 not used

CULTURAL PRACTICES
3 very important
2
1 not important

SANITATION
3 very important
2
1 not important

RESISTANT CULTIVARS
3 many cultivars
2
1 no resistance

Bacterial Leaf Scorch

Geographic occurrence USDA zones 6–9

Seasonal occurrence Mid- to late summer

DISEASE FREQUENCY
5 annual
4
3
2
1 rare

DISEASE SEVERITY
5 plants killed
4
3
2
1 very little damage

CHEMICAL TREATMENT
3 used every year
2
1 not used

CULTURAL PRACTICES
3 very important
2
1 not important

SANITATION
3 very important
2
1 not important

RESISTANT CULTIVARS
3 many cultivars
2
1 no resistance

Powdery Mildew

Microsphaera platani
Phyllactinia guttata

The powdery mildew fungi *Microsphaera platani* and *Phyllactinia guttata* attack vigorous sprouts of sycamore in late summer. New growth in August may be distorted, stunted, and covered with fungal mycelium and spores (Plate 156), while older leaves nearby are unaffected. See Chapter 15, "Powdery Mildew."

Bacterial Leaf Scorch

Xylella fastidiosa

Bacterial leaf scorch occurs occasionally in landscape sycamores but is rare in the nursery. Nevertheless, an outbreak of the disease in a nursery would be disastrous, because infected trees moved from the nursery would not show symptoms until late in the first season after transplanting.

Symptoms

Marginal and interveinal leaf scorch appears in mid- to late summer. Scorched areas of leaves dry out and turn tan, with a brown margin separating the scorched and the green tissues (Plate 157). Affected leaves may roll upward. Early the next season, the leaves are green and appear normal, but by late summer the leaf scorch symptoms reappear.

Epidemiology

Xylella fastidiosa, the causal agent of bacterial leaf scorch, is thought to be transmitted from diseased to healthy trees by leafhoppers, treehoppers, and spittlebugs. The bacterium inhabits and grows in xylem tissues and, over time, may occlude the xylem and cause scorch. In the landscape, the disease appears to spread relatively slowly through a population of trees; the rate of disease spread in the nursery is not known.

Host Range

Elm, several maple species, mulberry, several oak species, and sweetgum are susceptible to *X. fastidiosa,* as well as sycamores.

Management

Little is known about the spread of bacterial leaf scorch in the nursery. No research-based recommendations for disease management are available. The suggestions offered below may or may not be effective.

Cultural practices Avoid establishing sycamores in the nursery near infected shade trees.

Sanitation Remove and destroy affected sycamores.

Resistance Resistance in sycamore is not known.

Chemical treatment There are no chemical treatments for bacterial leaf scorch in the nursery. Controlling insects may be of some benefit.

CHAPTER 80

Robert L. Wick ▪ University of Massachusetts, Amherst

Taxus Diseases

Geographic production USDA zones 2–7

Family Taxaceae

Genus *Taxus*

Species	*T. baccata*	English yew	zones 5–7
	T. baccata cv. Fastigiata		zones 5–7
	T. brevifolia	Pacific yew, western yew	
	T. canadensis	American yew, Canada yew, ground-hemlock	zones 2–6
	T. cuspidata	Japanese yew	zones 4–7
	T. cuspidata cv. Capitata		zones 4–7
	T. cuspidata cv. Nana		zones 4–7
	T. × *media* (*T. baccata* × *T. cuspidata*)		zones 4–7

The genus *Taxus* is relatively free of important diseases, with the exception of Phytophthora root rot. *Botryosphaeria, Dothiora, Herpotrichia, Mycosphaerella,* and *Phacidium* have been noted on twigs or leaves but have not been reported to cause problems in nursery stock. Rhizoctonia root rot and damping-off and Cylindrocladium root rot have been noted but are not common in container stock.

Seedling Blight, Root Rot, and Dieback

Phytophthora cinnamomi
Phytophthora citricola
Phytophthora citrophthora

Phytophthora species occur sporadically on *Taxus* in nursery production and in large, established landscape plantings. Few publications other than host indices mention *Phytophthora* diseases of *Taxus*. However, several *Phytophthora* species have wide host ranges, and *P. cinnamomi* in particular can cause significant damage to *Taxus*. *P. citrophthora* and *P. citricola* have been cultured from *Taxus* but are weakly pathogenic to it.

Symptoms

Infected seedlings develop symptoms remarkably similar to those of landscape plants (Plates 158–160). The first obvious symptom of root rot is a loss of normal green color. Dieback may occur, and seedlings may wilt. Roots appear decayed and develop a dry rot. Reddish brown to black streaking may be evident in larger roots and the crown.

Epidemiology

Phytophthora spp. are most active in wet, poorly drained soils. Relatively high temperatures in the summer are conducive to Phytophthora root rot. See Chapter 14, "Phytophthora Root Rot and Dieback."

Seedling Blight, Root Rot, and Dieback

Geographic occurrence USDA zones 4–7

Seasonal occurrence April to September, with an optimum period in July and August

DISEASE FREQUENCY	DISEASE SEVERITY
5 annual	**[5 plants killed]**
4	4
[3]	3
2	2
1 rare	1 very little damage

CHEMICAL TREATMENT	CULTURAL PRACTICES
3 used every year	**[3 very important]**
[2]	2
1 not used	1 not important

SANITATION	RESISTANT CULTIVARS
[3 very important]	3 many cultivars
2	2
1 not important	**[1 no resistance]**

Management

Phytophthora spp. can become resident in nursery soils and develop into a chronic problem for susceptible plants. Avoid introducing these pathogens into the nursery. Avoid using contaminated soil. Place containers on several inches of gravel to avoid contamination by rain splash and direct contact with soil. Use a growing medium with well-drained composted bark. If Phytophthora root rot is persistent in the nursery, consider using a protective fungicide.

REFERENCES

Farr, D. F., Bills, G. F., Chamuris, G. P., and Rossman, A. Y. 1989. Fungi on Plants and Plant Products in the United States. American Phytopathological Society, St. Paul, Minn.

Schreiber, L. R., and Green, R. J. 1959. Die-back and root rot disease of *Taxus* spp. in Indiana. Plant Dis. Rep. 43:814–817.

Wick, R. L., Haviland, P., and Tattar, T. 1994. New hosts and associations for *Phytophthora* species. (Abstr.) Phytopathology 84:549.

CHAPTER 81

Ronald K. Jones ▪ North Carolina State University, Raleigh

Ternstroemia Diseases

Geographic production USDA zones 7–10

Family Theaceae

Genus *Ternstroemia*

Species *T. gymnanthera*

Ternstroemia (known in the trade as *Cleyera japonica*) is a large, upright evergreen shrub. Mature leaves are dark glossy green, while new leaves are reddish. This plant has no damaging disease problems in the nursery.

Leaf Spot Diseases

Several fungi have been reported to cause leaf spot diseases of *Ternstroemia,* including a *Elsinoe leucospila,* and species of *Cercospora, Colletotrichum, Leptosphaeria,* and *Phyllosticta.* Leaf spots with dark red to purple borders and a light gray to white center form occasionally. These diseases seldom cause any significant damage to *Ternstroemia.*

Interveinal chlorosis sometimes occurs on new leaves during periods of rapid growth, but this is probably a nutritional problem rather than a disease symptom.

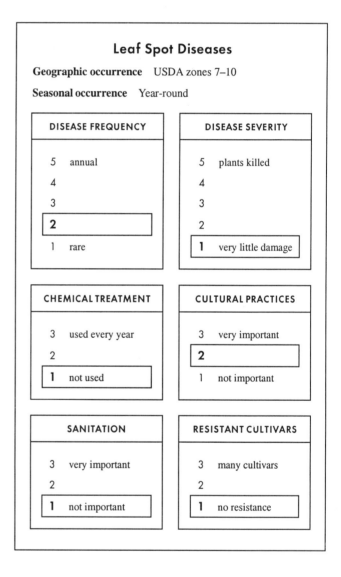

Leaf Spot Diseases

Geographic occurrence USDA zones 7–10

Seasonal occurrence Year-round

DISEASE FREQUENCY	DISEASE SEVERITY
5 annual	5 plants killed
4	4
3	3
2	2
1 rare	**1** very little damage

CHEMICAL TREATMENT	CULTURAL PRACTICES
3 used every year	3 very important
2	**2**
1 not used	1 not important

SANITATION	RESISTANT CULTIVARS
3 very important	3 many cultivars
2	2
1 not important	**1** no resistance

CHAPTER 82

Julie Beale ▪ University of Kentucky, Lexington

Tuliptree Diseases

Geographic production USDA zones 4–9

Family Magnoliaceae

Genus *Liriodendron*

Species *L. tulipifera* tuliptree, tulip magnolia, tulip poplar, whitewood, yellow poplar

The tuliptree, also known as tulip magnolia, tulip poplar, yellow poplar, and whitewood, is a relatively rapid-growing species, attaining a height of 150 feet or more in the landscape. It has beautiful flowers, but they form high in the tree and thus are often overlooked. Tuliptrees should be planted in large landscape areas rather than street plantings or small lots. Several tuliptree diseases are important in the nursery, including Botryosphaeria canker, anthracnose, and Verticillium wilt.

Botryosphaeria Canker and Dieback

Botryosphaeria dothidea

Botryosphaeria canker and dieback is the most destructive of several fungal canker diseases affecting stressed tuliptree. It is of limited occurrence in the nursery, where stressful conditions (particularly drought) can be minimized.

Symptoms

Cankers on small twigs are slightly sunken, and the bark is discolored. Cankers may be associated with wounds or dead branch stubs. A canker may girdle a twig, resulting in dieback of the twig from the point of infection to the tip. Black, raised fruiting bodies (pycnidia) of the fungal pathogen are visible in tissues of dead branch tips.

Epidemiology

The canker pathogen, *Botryosphaeria dothidea,* attacks trees weakened by freezing, drought, injuries, or other stresses. The pathogen produces sexual and asexual spores in cankered branch tips and in debris. The spores are moved by wind and splashing rain and invade dead buds, twigs, and wounds, usually in the spring and summer.

Host Range

Tuliptree and many other woody plants are susceptible to *B. dothidea.*

Management

Cultural practices Avoid stresses that will predispose trees to infection. Protect nursery stock from freeze injury. Avoid drought stress, injury, and stress from cultural practices such as pruning or transplanting at inappropriate times (e.g., summer or early fall) and excessive root pruning.

Botryosphaeria Canker and Dieback

Geographic occurrence USDA zones 4–9

Seasonal occurrence April to November

DISEASE FREQUENCY	DISEASE SEVERITY
5 annual	5 plants killed
4	**4** (selected)
3	3
2 (selected)	2
1 rare	1 very little damage

CHEMICAL TREATMENT	CULTURAL PRACTICES
3 used every year	3 very important
2	**2** (selected)
1 not used (selected)	1 not important

SANITATION	RESISTANT CULTIVARS
3 very important	3 many cultivars
2 (selected)	2
1 not important	**1 no resistance** (selected)

Sanitation Prune and remove cankered twigs and branches. Keep the area free of debris and fallen infected twigs.

Resistance No resistant cultivars are available.

Chemical treatment No chemical treatment for canker is available.

Anthracnose and Other Leaf Spot Diseases

Gloeosporium liriodendri

Other fungi

Anthracnose and other fungal leaf spot diseases occasionally occur on tuliptree but are rarely destructive.

Symptoms

Anthracnose causes irregularly shaped lesions on leaves, with necrotic centers and thin dark borders. The lesions may be concentrated along veins or leaf margins.

Other fungal leaf spot diseases may cause smaller, more circular lesions scattered on the leaf blade.

Epidemiology

Wet spring weather allows germination of fruiting bodies of the leaf spot fungi, which overwinter in fallen leaves and debris. Spores of *Gloeosporium liriodendri,* the tuliptree anthracnose pathogen, typically require cool conditions for germination and infection. Spores are released and carried by wind or splashing rain to leaves, where they cause infections. Secondary infections may occur later in the season, caused by spores produced in old lesions.

Host Range

All tuliptree cultivars are susceptible to leaf spot diseases.

Management

Cultural practices Cultural practices and sanitation are normally sufficient to reduce the incidence of anthracnose. Space plants to improve air circulation. Avoid overhead watering in spring when infections occur.

Sanitation Rake up and destroy fallen leaves to reduce inoculum. Maintaining sanitary conditions in the nursery should prevent outbreaks of this disease.

Resistance No resistant cultivars are available.

Chemical treatment Fungicides are typically not necessary, except in severe outbreaks in the nursery. Protectant fungicides, if needed, may be applied early, starting at budbreak.

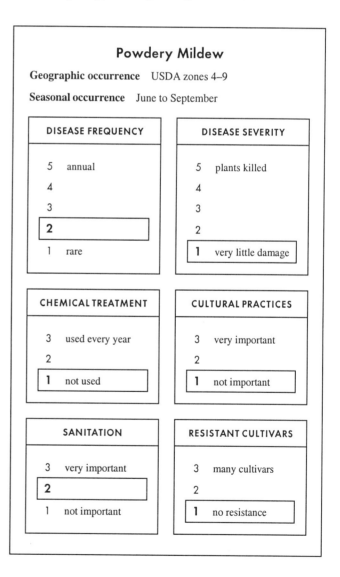

Anthracnose

Geographic occurrence USDA zones 4–9

Seasonal occurrence May to June

DISEASE FREQUENCY	DISEASE SEVERITY
5 annual	5 plants killed
4	4
[3]	3
2	**[2]**
1 rare	1 very little damage

CHEMICAL TREATMENT	CULTURAL PRACTICES
3 used every year	3 very important
2	**[2]**
[1] not used	1 not important

SANITATION	RESISTANT CULTIVARS
3 very important	3 many cultivars
[2]	2
1 not important	**[1] no resistance**

Powdery Mildew

Geographic occurrence USDA zones 4–9

Seasonal occurrence June to September

DISEASE FREQUENCY	DISEASE SEVERITY
5 annual	5 plants killed
4	4
3	3
[2]	2
1 rare	**[1] very little damage**

CHEMICAL TREATMENT	CULTURAL PRACTICES
3 used every year	3 very important
2	2
[1] not used	**[1] not important**

SANITATION	RESISTANT CULTIVARS
3 very important	3 many cultivars
[2]	2
1 not important	**[1] no resistance**

Powdery Mildew

Erysiphe polygoni
Phyllactinia guttata

Powdery mildew of tuliptree is typically not a serious disease, but some chemical treatment may be warranted for badly infected small trees. See Chapter 15, "Powdery Mildew."

Verticillium Wilt

Verticillium albo-atrum

Verticillium wilt is a common and destructive disease of tuliptree. See Chapter 19, "Verticillium Wilt."

Cylindrocladium Root and Stem Rot

Cylindrocladium scoparium

Cylindrocladium root and stem rot has been reported to cause a decline of large tuliptrees in the South. The disease has not been reported in nursery stock. See Chapter 11, "Diseases Caused by *Cylindrocladium*."

REFERENCE

Dirr, M. A. 1998. Manual of Woody Landscape Plants: Their Identification, Ornamental Characteristics, Culture, Propagation and Uses. 5th ed. Stipes Publishing, Champaign, Ill.

CHAPTER

83

Karen K. Rane ▪ Purdue University, West Lafayette, Indiana

Viburnum Diseases

Geographic production USDA zones 3–8

Family Caprifoliaceae

Genus *Viburnum*

Species

V. acerifolium	*V. opulus*
V. alnifolium	*V. plicatum* var. *tomentosum*
V. × *burkwoodii*	*V. prunifolium*
V. × *carlcephalum*	*V.* × *rhytidophylloides*
V. carlesii	*V. rhytidophyllum*
V. davidii	*V. sargentii*
V. dentatum	*V. setigerum*
V. dilatatum	*V. sieboldii*
V. lantana	*V. trilobum*
V. lentago	other species

Bacterial Leaf Spot (Bacterial Blight)

Pseudomonas syringae pv. *viburni*

Like other diseases caused by pathovars of *Pseudomonas syringae,* bacterial leaf spot of viburnum, also called bacterial blight, is especially severe in spring when the weather is cool and wet.

Symptoms

Leaf spots due to infection by *P. syringae* pv. *viburni* are initially water-soaked, irregularly shaped or angular lesions, which turn necrotic and dark brown with age. Bacterial ooze may be visible as a shiny exudate on the leaf surface. Early infection of developing leaves can result in leaf distortion as the leaves expand. Brown lesions may also appear on young shoots, and severe infection can result in dieback.

Epidemiology

P. syringae pv. *viburni* overwinters in infected bark tissues or on the surface of buds and twigs. Infection is restricted primarily to early spring, when the weather is cool and moist. See Chapter 16, "Diseases Caused by *Pseudomonas syringae,*" for additional information on epidemiology.

Host Range

Several species are susceptible to this disease. *V.* × *burkwoodii* and *V. carlesii* have been reported to be very susceptible. Cultivars reported to be resistant include *V.* × *burkwoodii* 'Mohawk,' *V.* × *carlcephalum* 'Cayuga,' *V. lantana* 'Mohican,' and *V. rhytidophyllum* 'Alleghany.'

Management

Removal of diseased shoots and leaves will help reduce the amount of inoculum for infection. Bacterial diseases are favored by long periods of leaf wetness, and therefore spacing and pruning plants to ensure maximum air circulation and promote drying of the foliage after rain or dew will reduce disease. Avoid high-nitrogen fertilizers, since they promote excessive growth of succu-

Bacterial Leaf Spot

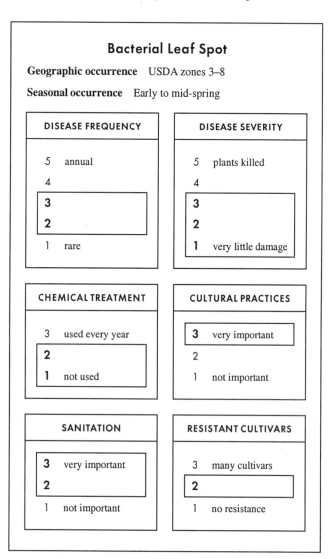

Geographic occurrence USDA zones 3–8

Seasonal occurrence Early to mid-spring

DISEASE FREQUENCY	DISEASE SEVERITY
5 annual	5 plants killed
4	4
3	**3**
2	**2**
1 rare	**1** very little damage

CHEMICAL TREATMENT	CULTURAL PRACTICES
3 used every year	**3** very important
2	2
1 not used	1 not important

SANITATION	RESISTANT CULTIVARS
3 very important	3 many cultivars
2	**2**
1 not important	1 no resistance

lent shoots, which are more susceptible to *Pseudomonas* infection. Copper products may help protect developing leaves and shoots from infection.

Downy Mildew
Plasmopara viburni

Downy mildew is common in viburnums throughout their natural range.

Symptoms

Angular lesions, bounded by leaf veins, form on infected leaves. The lesions are initially chlorotic and then necrotic. Symptoms develop first on the lowest leaves and progress up the plant. Individual lesions may coalesce, forming large blotches. Severely infected leaves will fall. White to tan growth of the fungal pathogen may be observed on the underside of lesions.

Epidemiology

The pathogen, *Plasmopara viburni,* overwinters in infected leaves on the ground and is splashed to lower leaves with spring rains. Leaf infection requires several hours of leaf wetness. Spores are produced on hyphae that emerge from the underside of leaf lesions. They can initiate new leaf spots when splashed to adjacent leaves, resulting in repeated cycles of infection. Downy mildew is favored by moist and cool to warm conditions. Disease spread is minimal during hot, dry summer weather.

Host Range

Downy mildew has been reported in *V. acerifolium, V. dentatum, V. opulus, V. trilobum,* and other *Viburnum* species.

Management

Remove and destroy all fallen infected leaves. Overhead irrigation will spread the pathogen and provide hours of leaf wetness needed for infection, and thus should be avoided. Spacing plants to promote air circulation will reduce disease incidence. Protectant fungicides may be used when weather conditions favor disease development.

Downy Mildew

Geographic occurrence USDA zones 3–8, primarily east of the Great Plains

Seasonal occurrence Spring through early summer, and later if the weather is cool and wet

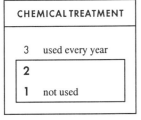

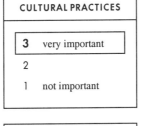

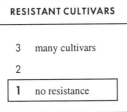

DISEASE FREQUENCY	
5	annual
4	
3	
2	
1	rare

DISEASE SEVERITY	
5	plants killed
4	
3	
2	
1	very little damage

CHEMICAL TREATMENT	
3	used every year
2	
1	not used

CULTURAL PRACTICES	
3	very important
2	
1	not important

SANITATION	
3	very important
2	
1	not important

RESISTANT CULTIVARS	
3	many cultivars
2	
1	no resistance

Botryosphaeria Canker
Botryosphaeria dothidea

Botryosphaeria canker is a common stem disease of viburnums under stress and can result in significant dieback. For more information on this disease see Chapter 74, "Redbud Diseases."

Symptoms

Cankers develop on stems and branches and may appear as areas of dark, discolored bark. Infected twigs die, while older stems and branches may show discrete cankers. Black fruiting structures of the fungal pathogen may give the canker surface a roughened appearance. Cankers may girdle stems and branches, resulting in the death of foliage beyond the infection point. Dark brown discoloration develops in the wood beneath the cankers.

Epidemiology

Botryosphaeria dothidea, the fungus causing canker, is an opportunistic pathogen, invading plants stressed by drought, freeze injury, or other adverse environmental conditions or characteristics of the planting site.

Host Range

Botryosphaeria canker has been reported to affect *V. acerifolium, V. prunifolium,* and other *Viburnum* species.

Management

Since viburnums are predisposed to Botryosphaeria canker by environmental stress, proper plant care (mulching and providing adequate moisture during dry periods) will help prevent infection. Pruning tools should be disinfested, since the fungus can be spread by pruning.

Powdery Mildew
Microsphaera penicillata

Powdery mildew is a common mid- to late-season disease of viburnums, particularly plants located at shady sites.

Symptoms

White, powdery growth of the fungal pathogen, *Microsphaera penicillata,* appears on both the upper and the lower surfaces of leaves. Severe infections may cause leaf deformity. In late sum-

Botryosphaeria Canker

Geographic occurrence USDA zones: 3–8

Seasonal occurrence Throughout the growing season

DISEASE FREQUENCY	DISEASE SEVERITY
5 annual	**5** plants killed
4	**4**
3	**3**
2	**2**
1 rare	1 very little damage

CHEMICAL TREATMENT	CULTURAL PRACTICES
3 used every year	**3** very important
2	2
1 not used	1 not important

SANITATION	RESISTANT CULTIVARS
3 very important	3 many cultivars
2	2
1 not important	**1** no resistance

Powdery Mildew

Geographic occurrence USDA zones 3–8

Seasonal occurrence Mid- to late summer through fall

DISEASE FREQUENCY	DISEASE SEVERITY
5 annual	5 plants killed
4	4
3	**3**
2	**2**
1 rare	1 very little damage

CHEMICAL TREATMENT	CULTURAL PRACTICES
3 used every year	**3** very important
2	2
1 not used	1 not important

SANITATION	RESISTANT CULTIVARS
3 very important	3 many cultivars
2	2
1 not important	**1** no resistance

mer, tiny black fruiting structures of the fungus appear as dots among the white mycelial growth. Refer to Chapter 15, "Powdery Mildew," for additional information.

Epidemiology

Powdery mildew is favored by humid periods with warm days and cool nights. Refer to Chapter 15.

Host Range

Several *Viburnum* species, including *V. acerifolium, V. lentago,* and *V. opulus,* are reported to be susceptible to powdery mildew. *V.* × *burkwoodii* 'Mohawk' and *V.* × *carlcephalum* 'Cayuga' have been reported to be resistant.

Management

Refer to Chapter 15 for management information. Fungicides are effective in controlling powdery mildew, but avoid sulfur fungicides, which can be phytotoxic.

REFERENCES

Dirr, M. A. 1998. Manual of Woody Landscape Plants: Their Identification, Ornamental Characteristics, Culture, Propagation and Uses. 5th ed. Stipes Publishing, Champaign, Ill.

Farr, D. F., Bills, G. F., Chamuris, G. P., and Rossman, A. Y. 1989. Fungi on Plants and Plant Products in the United States. American Phytopathological Society, St. Paul, Minn.

Moorman, G. 1997. Viburnum diseases. Pa. State Univ. Coop. Ext. Plant Dis. Facts.

Pirone, P. P. 1978. Diseases and Pests of Ornamental Plants. 5th ed. John Wiley & Sons, New York.

Pscheidt, J. W., and Ocamb, C. M. 1996–. An Online Guide to Plant Disease Control. On-line publication. Oregon State University Extension Service, Corvallis.

Sinclair, W. A., Lyon, H. H., and Johnson, W. T. 1987. Diseases of Trees and Shrubs. Comstock Publishing Associates, Cornell University Press, Ithaca, N.Y.

CHAPTER 84

Margaret R. Williamson ▪ Clemson University, Clemson, South Carolina
James H. Blake ▪ Clemson University, Clemson, South Carolina
Gary W. Simone ▪ University of Florida, Gainesville

Wax Myrtle (Bayberry) Diseases

Geographic production USDA zones 3–10

Family Myricaceae

Genus *Myrica*

Species			
	M. californica	California wax myrtle, California bayberry	zone 8
	M. cerifera	southern wax myrtle, southern bayberry	zones 7–10
	M. pensylvanica	northern bayberry	zones 3–7

Myrica spp. are extremely adaptable shrubs. They are often planted in coastal areas, because of their salt tolerance. They have few disease problems, but more are being discovered as the crop gains wider acceptance in the nursery industry.

Septoria Leaf Spot
Septoria hodgesii

Symptoms

Infection by the fungus *Septoria hodgesii* results in circular leaf spots, 1/16–1/8 inch in diameter. When numerous spots are formed, they often coalesce to form larger areas of necrotic tissue. The lesions, visible on both leaf surfaces, are brownish black and are surrounded by a purplish halo. Sparsely distributed black fruiting bodies (pycnidia) may be observed in lesions viewed at a magnification of 10–20×.

Epidemiology

Spores (conidia) are extruded in a moist matrix from pycnidia of *S. hodgesii* and are dispersed during wet weather. The spores are dispersed primarily by splashing water, but they can also be transported in wind-driven rain or irrigation water. Little is known about how *S. hodgesii* overwinters. Other *Septoria* spp. are known to overwinter as mycelium or pycnidia in fallen leaves. Since *M. cerifera* is evergreen, the fungus may overwinter in infected leaves on the plant. The disease can be detected year-round from Florida to South Carolina. In warm climates, the fungus appears to remain active throughout the year.

Host Range

Septoria leaf spot is probably the most common disease of *M. cerifera* in nurseries. It has not been reported in the other commonly cultivated *Myrica* spp. (Farr et al., 1989).

Management

Cultural and chemical controls can be used to manage Septoria leaf spot. Implement cultural controls described in Chapter 88, "Horticultural Practices for Reducing Disease Development," to reduce periods of leaf wetness and promote fast drying of foliage. Remove fallen infected leaves from container surfaces and ground

Septoria Leaf Spot

Geographic occurrence USDA zones 7–10

Seasonal occurrence Throughout the year in the warm climates where *M. cerifera* is grown

DISEASE FREQUENCY	DISEASE SEVERITY
5 annual	5 plants killed
4	4
[3]	3
2	**[2]**
1 rare	1 very little damage

CHEMICAL TREATMENT	CULTURAL PRACTICES
3 used every year	3 very important
[2]	**[2]**
1 not used	1 not important

SANITATION	RESISTANT CULTIVARS
3 very important	3 many cultivars
[2]	2
1 not important	**[1] no resistance**

cloths with a motorized blower. Apply recommended fungicides at the first sign of disease and then repeat according to the instructions on the product label, as needed for control.

Rust

Gymnosporangium ellisii

Rust of *Myrica* spp., caused by the fungus *Gymnosporangium ellisii,* occurs mainly along the East Coast, from Maine to Florida, where the pathogen's alternate host, southern white cedar (*Chamaecyparis thyoides*), grows wild. Southern white cedar is rather uncommon, as its swampy habitat is not widespread, but if cultivation of this tree becomes more common, the incidence of rust of *Myrica* will probably increase.

Symptoms

Orange to red spots form on leaves. The pathogen forms two types of fruiting structures in these lesions: tiny, yellow to black spermagonia, which develop first, on the upper surface of leaves, followed by the more conspicuous aecia, which form along midribs and veins on the lower surface of leaves and also on petioles and stems. Aecia are surrounded by light-colored tissue and produce orange to brown spores (aeciospores).

Rust

Geographic occurrence USDA zones 3–9

Seasonal occurrence Spring to late summer

DISEASE FREQUENCY	DISEASE SEVERITY
5 annual	5 plants killed
4	4
3	3
2	**[2]**
[1] rare	1 very little damage

CHEMICAL TREATMENT	CULTURAL PRACTICES
3 used every year	3 very important
[2]	**[2]**
1 not used	1 not important

SANITATION	RESISTANT CULTIVARS
3 very important	3 many cultivars
2	2
[1] not important	**[1] no resistance**

Epidemiology

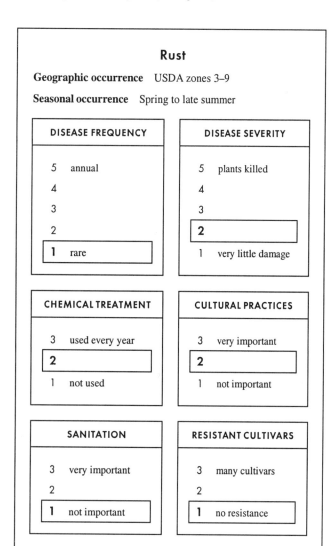

Wind-blown aeciospores infect white cedars in summer and fall, causing a witches'-broom the following year, as *G. ellisii* completes its life cycle. During winter, the pathogen is either shed with fallen leaves or dies out in *Myrica.* In spring and summer, the fungus produces basidiospores on white cedar, and these spores reinfect *Myrica.*

Host Range

Among the *Myrica* spp. important to the nursery industry, rust has been reported on *M. cerifera* (Farr et al., 1989).

Management

Where rust has not occurred in past years, there is no need to implement any control measures. Avoid digging *Myrica* from the wild where infected plants exist. In areas with a history of the disease, remove white cedars within a 1-mile radius of the nursery. If this is not possible, protect *Myrica* foliage with a recommended fungicide during periods of spore production on the alternate host. This generally occurs in spring, when buds are just beginning to break. Apply a fungicide according to the instructions on the product label for four to six weeks, depending on the frequency of rainfall. If both hosts are produced in the nursery, keep them widely separated, and apply protective fungicides during their respective infection periods.

Fusarium Wilt

Fusarium oxysporum

Fusarium wilt of *Myrica* has been reported only in Florida (Alfieri et al., 1994), However, the pathogen is ubiquitous, and the disease could probably occur anywhere *Myrica* spp. are grown.

Symptoms

Affected plants are generally unthrifty, have small, off-color leaves, and may wilt. Dieback of stems and branches may occur. The xylem in cut stems is discolored, either pale to dark brown or purplish. Disease symptoms are likely due to toxins and cellulose-degrading enzymes produced by the pathogen and physical plugging of the vascular system. Stem cankers may also be present. *F. oxysporum* can be isolated from stem cankers, but it is not known if these lesions are related to vascular infection.

Epidemiology

Little is known about Fusarium wilt of *Myrica,* but from the known biology of *Fusarium oxysporum,* it may be assumed that this soilborne fungus enters the plant through its roots and moves from there into the xylem and the vascular system. Mechanical injury to roots and possibly to stems may facilitate the entry of the pathogen. Feeding by plant-pathogenic nematodes may cause wounds that allow the fungus to enter plants. *F. oxysporum* survives in the soil as thick-walled spores (chlamydospores), which germinate when conditions are favorable and initiate infection.

Host Range

F. oxysporum has a wide host range among both herbaceous and woody plants. Many strains of this fungus exist worldwide, and many of them are plant- or genus-specific. Among *Myrica* spp., Fusarium wilt has been reported only in *M. cerifera,* but it is likely that the other species are susceptible. The host range of the strain of *F. oxysporum* causing Fusarium wilt of *M. cerifera* is unknown.

Management

Plants grown in field nurseries are at greater risk than those grown in containers, but *F. oxysporum* can invade through drainage holes and enter pots in splashing water. Do not procure specimens of *Myrica* from native areas where the fungus is endemic. Prepare the growing area properly to prevent entry of the pathogen. Sanitation measures, such as destroying infected plants and removing infested soil, should also be undertaken. In field nurseries, fumigation of infested soils may prove helpful.

Sphaeropsis Gall and Witches'-Broom

Sphaeropsis tumefaciens

Symptoms

Infection by the fungus *Sphaeropsis tumefaciens* triggers the stimulation of dormant lateral buds, which produce numerous shoots branching from a single node. In container-sized *Myrica,* witches'-brooms appear on twigs within the canopy. Continued internal growth of the fungus at the witches'-brooms often results in thinning of the leaves and eventual death of all leaves and shoots beyond the infection point. In field-grown, larger stock, galls form on the trunk. The galls are smooth and globose, ranging from 1 to 4 cm in diameter. Affected specimens exhibit low vigor and a thin canopy.

Epidemiology

S. tumefaciens exudes spores (conidia) from dark, globose fruiting bodies (pycnidia) embedded in the bark of invaded tissues. The spores are carried by splashing water from one branch to another and from plant to plant, and they may wash down the trunk during rain or irrigation. Shallow pruning of galls and witches'-brooms may leave the fungus in the pruning stub. During moist weather, spores may be moved during pruning or shearing, but there is insufficient evidence to prove that the disease is spread by pruning. *S. tumefaciens* enters twigs and branches through natural openings and pruning wounds. Symptom development is slow, often taking several months.

Host Range

S. tumefaciens has a wide host range among woody plants, including such common nursery crops as *Callistemon, Carissa, Citrus, Eucalyptus, Ficus, Ilex, Malus, Nerium, Pyrus,* and *Ulmus.* Among *Myrica* spp., it has been reported only on *M. cerifera* in Florida. The susceptibility of other *Myrica* spp. is unknown.

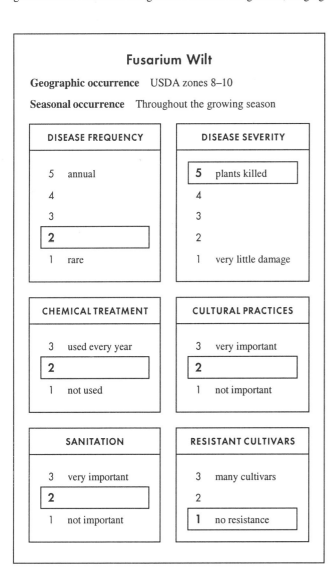

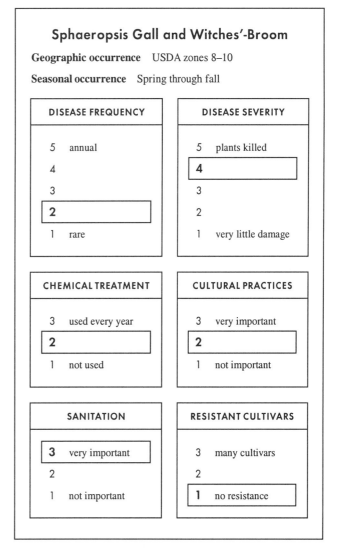

Fusarium Wilt

Geographic occurrence USDA zones 8–10

Seasonal occurrence Throughout the growing season

DISEASE FREQUENCY	DISEASE SEVERITY
5 annual	**5 plants killed**
4	4
3	3
2	2
1 rare	1 very little damage

CHEMICAL TREATMENT	CULTURAL PRACTICES
3 used every year	3 very important
2	**2**
1 not used	1 not important

SANITATION	RESISTANT CULTIVARS
3 very important	3 many cultivars
2	2
1 not important	**1 no resistance**

Sphaeropsis Gall and Witches'-Broom

Geographic occurrence USDA zones 8–10

Seasonal occurrence Spring through fall

DISEASE FREQUENCY	DISEASE SEVERITY
5 annual	5 plants killed
4	**4**
3	3
2	2
1 rare	1 very little damage

CHEMICAL TREATMENT	CULTURAL PRACTICES
3 used every year	3 very important
2	**2**
1 not used	1 not important

SANITATION	RESISTANT CULTIVARS
3 very important	3 many cultivars
2	2
1 not important	**1 no resistance**

Management

Inspect stock entering the nursery for galls or witches'-brooms, and segregate symptomatic plants pending a diagnosis. Infected stock may be severely pruned (more than 4–6 inches below symptomatic tissue) in an effort to eradicate the disease. Disinfest pruners or shears prior to pruning clean stock. Postpruning sprays of copper hydroxide may offer some insurance against the possible spread of the fungus. Severely infected plants should be rogued and destroyed. Resistant cultivars are not known.

REFERENCES

Alfieri, S. A., Jr., Langdon, K. R., Kimbrough, J. W., El-Gholl, N. E., and Wehlburg, C. 1994. Diseases and disorders of plants in Florida. Fla. Dep. Agric. Consumer Serv. Bull. 14.

Barnard, E. L., and Ash, E. C., III. 1991. Fusarium wilt of wax myrtle. Fla. Dep. Agric. Consumer Serv., Div. Plant Ind., Plant Pathol. Circ. 350.

Dodge, B. O. 1934. Witches'-brooms on southern white cedar. J. N.Y. Bot. Gard. 35:41–45.

Eleuterius, L. 1971. Notes on the occurrence of *Gymnosporangium ellisii* in Mississippi. Bull. Torrey Bot. Club 98:281–282.

Farr, D. F., Bills, G. F., Chamuris, G. P., and Rossman, A. Y. 1989. Fungi on Plants and Plant Products in the United States. American Phytopathological Society, St. Paul, Minn.

Schubert, T. S. 1986. Septoria leaf spot of wax myrtle. Fla. Dep. Agric. Consumer Serv., Div. Plant Ind., Plant Pathol. Circ. 282.

Sinclair, W. A., Lyon, H. H., and Johnson, W. T. 1987. Diseases of Trees and Shrubs. Comstock Publishing Associates, Cornell University Press, Ithaca, N.Y.

Tattar, T. A. 1978. Diseases of Shade Trees. Academic Press, New York.

CHAPTER

85

C. C. Powell, Jr. ▪ Plant Health Advisory Services, Worthington, Ohio

An Introduction to the Management of Infectious Plant Diseases in the Nursery

The Importance of Integrated Disease Management

This chapter presents an overview of general practices of infectious disease management in nurseries. One thing must be kept in mind to be successful at plant disease control: You do not manage a disease by relying on any one particular practice. Diseases are managed, prevented, or controlled by groups of practices loosely held together under the concepts of integrated health management, or holistic plant health management, an approach emphasized in this chapter and throughout this book. For different diseases of different crops in different situations, various practices take on primary or secondary importance. Attention to the entire group of applicable management practices, however, will always be needed for success.

Why Do Diseases Occur in Ornamental Plant Nurseries?

Cultivated plants are usually more susceptible to disease than their wild relatives, partly because large numbers of the same cultivated plant are often grown close together in pure stands. Under these conditions, disease-causing organisms (pathogens) often become established and, once established, can spread rapidly.

In addition, many of our valuable ornamental plants are highly susceptible to disease and would have difficulty surviving in undisturbed nature. Ornamental varieties are commonly bred and selected to increase or improve their horticultural characters, without regard to disease susceptibility.

Cultivation and crop husbandry – whether in the production of geraniums, ornamental cabbages, or oak trees – are quite different from conditions in nature and tend to create environmental conditions that favor disease development. For example, heavy fertilization can produce soft growth, causing plants to grow rapidly and produce a good potential profit for the grower, but also possibly weakening the plants and increasing their susceptibility to disease.

What Causes Plant Diseases?

A plant can become diseased when it is continuously disturbed by a factor or factors in its environment. Often these disturbing factors are living, infectious pathogens, including bacteria, fungi, viruses, phytoplasmas, nematodes, and parasitic seed plants. When the disturbances produce noticeable effects, called symptoms, the plant is considered diseased. Knowledge of normal growth habits, characteristics of the variety (cultivar), and the normal variability of plants of the same species is required in order to recognize abnormal or diseased conditions. As the ornamental industry is constantly introducing new plants and quickly moving them to nurseries across the United States and other countries, plants may encounter pathogens that are not present in their area of origin, and thus new diseases occur.

Infectious organisms are part of the environment, whenever and wherever plants exist. Surely, we practice sanitation to limit their numbers, but we cannot achieve total "sterility" of our nurseries and fields. We must manage the crop environment so that plant pathogens cannot develop and flourish, and this management involves both the microenvironment (e.g., the leaf or root surface) and the macroenvironment (e.g., the potting medium and its chemistry).

Integrated Plant Disease Management

Integrated plant disease management should really be called *plant health management.* It is the most important way to proceed and succeed in growing plants profitably in a nursery. It is not complicated or difficult but must be incorporated into the daily management program of the nursery. It is based on the concept that plants should not be chronically disturbed by elements in their environment, since plants under stress are subject to more infection by pathogens.

The specifics of integrated disease management are covered in more detail in other chapters. Suffice it to say that the reduction of plant stress involves an awareness of environmental disturbances, which cannot be avoided entirely in the nursery. The secret of crop health maintenance and ongoing disease management is in recognizing when stress-causing disturbances have taken place and promptly developing altered management strategies to cope with them.

Trying to set up and maintain a balanced environment for each and every type of plant in the nursery is one way to avoid stressing the plants. A single ornamental plant nursery usually produces many kinds of plants. Thus, it is important for the grower to recognize the value of various plant care practices in the nursery and their contribution to the good or poor health of the plants. A set of good general practices can usually be determined to satisfy most of the needs of the plants and allow them to remain disease-free most of the time.

Beyond Stress Management

When dozens of different types of plants are grown in a nursery, it is impossible to grow each one in its precisely balanced, optimum environment. Furthermore, some things cannot be completely controlled, such as amounts of sunlight and rainfall, which are the environmental factors that may give you the most trouble. Problems can arise as a result of mistakes made by new workers, equipment breakdowns, or changes in fertilizers and growing media. A bumper sticker with the message "Stress Happens" might be a good one for nursery managers! However, alert management can minimize the extent of the damage.

Pathogen Management

When stress happens and environmental conditions favor the growth and spread of the pathogens, plant pathogens are primed to strike. We must turn our attention to methods of dealing with pathogens. Successful pathogen management is based on accurate diagnosis of the cause, thorough knowledge of the pathogen and its disease cycle, an understanding of the effects of environmental factors on the host and pathogen, and an ability to evaluate the cost and the practicality of possible management actions.

The most important step in controlling plant diseases is choosing the best methods for a given situation. Control strategies for a particular disease of a certain host may not be the best strategies for another disease of the same or a different plant. Furthermore, the implementation of several control measures (integrated control) is necessary to provide a high level of disease management. For example, to manage fungal diseases in the nursery, irrigation is often adjusted along with implementation of a fungicide treatment program.

Basic Methods of Infectious Disease Control in Nurseries

Exclusion

Disease-causing organisms should be excluded from your nursery whenever possible. Purchasing pathogen-free plants, cuttings, or seed will be at the top of any plant pathologist's list of things you must do. So what else is new? The problem is that you cannot always be 100% certain you have done this, even if you pay extra for symptom-free, "indexed," or "cultured" material. Do the best you can, and buy from your most trusted sources.

Sanitation in and around the nursery – cleaning and disinfecting potting benches, soil bins, head houses, growing benches in propagation areas, tools, and equipment – is an excellent pathogen exclusion measure. Disinfecting tools and equipment is effective in preventing the spread of many pathogens. Liners or rooted cutting crops are grown on raised benches to keep soilborne and root rot pathogens away from the plants.

Exclusion also involves preventing the spread of pathogens, and various practices will help to accomplish this. Limit the number of people handling stock plants. Take cuttings from the healthiest plants. Avoid splashing water from plant to plant whenever possible. Use potting medium that is free of plant pathogens. Treat irrigation water to eradicate root rot pathogens, if necessary. Control weeds that harbor insects carrying pathogens or offer survival for pathogens between crops. Exclusion may be the most important method in the management of infectious plant diseases.

Protection

Plants can be protected from pathogens by uniform and timely applications of pesticides (fungicides, bactericides, and nematicides). Diseases caused by fungi are most commonly treated in this manner. There are many fungicides that are useful for treating specific diseases, but the recent trend is to use broad-spectrum products or a combination of products for general fungal disease protection in the nursery, when necessary. Managing leaf and flower diseases with a combination of fungicides is currently popular, and a combination of products is commonly used to control roots rots.

It is important to apply fungicides on the plant surfaces or in the soil where infection can take place, before it takes place. Most fungicides are protectants (preventives) and have limited curative or eradicant activity. Knowing when the pathogen is most likely to be active determines when to apply the fungicide.

Many commonly recommended cultural practices provide protection against infectious plant diseases. These practices generally involve altering the air and soil environments to make them less favorable for infection and the development, reproduction, and spread of pathogens. Well-drained and well-aerated potting medium is commonly used by nursery growers to manage root-rotting fungi. Proper dehumidification of greenhouses will aid greatly in the management of *Botrytis* and powdery mildews.

Prevention

Prevention of conditions favoring infection and disease development involves many of the same practices as protection and stress management. It never ceases to amaze me how tough plants are if they are not stressed. In addition to stress management, we can do a lot to prevent infection by keeping leaf surfaces dry. Never water plants late in the day, and dehumidify greenhouses at the end of the day to prevent dew from forming on the leaves. This is best done by venting, then heating, in short cycles as the sun sets.

To prevent root rots, avoid overwatering and persistent wetness of the growing medium. Also avoid excessive dryness between waterings. Use well-aerated growing media, and do not allow a buildup of soluble salts. Excess wetness and dryness, insufficient aeration, and accumulations of soluble salts injure roots and favor root-rotting pathogens.

Eradication

Plant pathogens can be eradicated by pesticides, but only rarely. Some products can eradicate powdery mildew pathogens, partly because these fungi remain mostly outside their hosts. Insects that spread pathogens can be controlled by pesticides, and insect control thus provides a form of eradication as well as plant protection.

The removal and destruction of diseased plants or plant parts is a common practice to help control fungal, bacterial, nematode, and viral diseases. Plants that should be destroyed may be part of the crop you are growing, or they may be weeds or other less important hosts growing in and around the nursery.

Another eradication method is crop rotation, which "starves" soil-invading pathogens, such as nematodes.

It is impossible to achieve total eradication of a plant pathogen that is well established in a nursery and has adapted to the local environment. Eradicative measures must usually be practiced routinely. Nevertheless, for many diseases, eradication strategies are the only ones that are effective.

In the End, Diseases Are Managed

As you read this book, note how the recommended control practices fit in the general disease management categories outlined here. If you develop a general understanding of the ways in which diseases are managed in the nursery, then the management of a particular disease of a specific plant in a particular situation will be easier to understand.

The disease triangle, discussed in Chapter 2 (see Fig. 2.1), can also help growers to understand integrated disease management. The same three factors – the host, the pathogen, and the environment – are the basis for disease development and integrated disease management. All three must be managed.

CHAPTER 86

Ronald K. Jones ▪ North Carolina State University, Raleigh

Gary W. Simone ▪ University of Florida, Gainesville

Sharon L. von Broembsen ▪ Oklahoma State University, Stillwater

Ethyl Dutky ▪ University of Maryland, College Station

Integrated Disease Management

Integrated disease management means deploying a wide variety of disease management strategies to prevent disease or keep it at the lowest possible level. There is no silver bullet – no one chemical or management practice that controls all plant diseases in nursery production. However, by deploying a wide variety of disease management strategies growers will accomplish several purposes:

- reduce the overall incidence and severity of diseases
- minimize the probability of a sudden explosive outbreak of a disease
- improve the quality of plant material
- improve nursery efficiency and increase profit
- minimize reliance on a strategy based on chemical treatment
- minimize the development of chemical-resistant pathogens in the nursery

An integrated disease management program is not cheap or easy to carry out every day, year-round. An effective program requires a major time commitment by the management team, an excellent working knowledge of the cultural needs of the crops grown and of the diseases of those crops, a strong commitment to produce the best-quality plants possible, a strong commitment to the total program, and continuous communication of the program to the entire work force.

The program should pay for itself in fewer plant losses from disease and better overall quality that will result in more salable high-quality plants that sell for a premium price. This program comprises a series of small actions, each one serving a critical part of the program that should not be left out. Integrated disease management is a part of an integrated pest management (IPM) program that includes control of insects and weeds as well as diseases.

An integrated disease management program for a nursery is a group of practices conducted in three main phases:

Preventative programs General practices, such as sanitation, cultural practices, and observation of regulatory requirements, have some effect on all diseases of the crops grown in the nursery. These practices are carried out every day throughout the nursery, whether or not there is evidence of disease. They are discussed in more detail in other chapters.

Scouting and monitoring programs Scouting and monitoring are conducted throughout the year but with increased emphasis and frequency during the growing season. These practices help growers stay in touch with what is happening in the nursery to identify problems at the very earliest moment so that corrective action can be implemented.

Corrective actions Disease management measures must be implemented at the first evidence of a problem. If a problem is detected early and the cause is correctly identified, effective corrective action aimed at the specific cause will minimize damage and losses. A timely integrated management program costs less than corrective knee-jerk reactions implemented after huge numbers of plants have been severely damaged.

Integrated disease management strategies include

- sanitation
- regulatory control
- chemical treatment
- planting resistant cultivars
- cultural practices
- biological control
- exclusion of pathogens

These practices are the topics of several chapters and are discussed throughout the book.

Sanitation

Sanitation deals with the general cleanliness and pathogen-free condition of the nursery operation. It is aimed at:

- reducing buildup of pathogens on dead or dying plant parts
- reducing carryover of pathogens from one set of plants to the next set
- reducing carryover and spread of pathogens on tools, benches, pots, hoses, etc.
- reducing contamination of healthy plants with infested soil or water
- reducing contamination of clean pots, media, etc.

A sanitation program is the very foundation of an effective integrated disease management program, and its value cannot be underestimated. It is aimed at general disease control, i.e., all diseases of all crops, by reducing the overall inoculum level in the nursery. Sanitation is discussed more completely in Chapter 87, "Sanitation: Plant Health from Start to Finish."

Monitoring for Diseases

Entomologists set out traps, sometimes with black lights or sex attractants, to monitor or measure the population levels of various insects. These numbers are used to determine when insect populations are increasing or decreasing. With established known thresholds for specific pests, recommendations concerning whether or not to treat can be made. It is not easy to monitor pathogen levels in nurseries. Few monitoring protocols have been developed for pathogens on ornamental plants. *Phytophthora* spp. can be monitored in irrigation water (this is discussed in more detail in Chapter 95, "Disease Management for Nurseries Using Recycling Irrigation Systems").

Plants in the nursery are an excellent monitoring device, or natural living bait, for diseases. Diseases must be detected at the very first appearance of symptoms through a regular, systematic scouting program. By the time symptoms become obvious, a pathogenic fungus is often already producing spores to cause additional infections.

Disease levels in target crops should be monitored and recorded during regular scouting. Any disease can be monitored in this way, but it is easier to monitor visible aboveground symptoms of diseases such as leaf spot. For example, if the incidence and severity of leaf spot is recorded weekly, disease incidence and severity can be compared month to month or year to year. Such records can be used to compare the relative susceptibility or resistance of cultivars, the effectiveness of a control program, the effect of weather, or the severity potential of the disease. This information can then be used to determine whether and precisely when to start a spray program and determine which chemicals have been the most effective.

Such records become more useful as the monitoring program is continued over time. The data can be used in producing a disease development calendar for a particular nursery. A chemical control program may need to be implemented 10–14 days before the first symptoms are observed. Applications of a protectant fungicide should begin before the onset of disease, as predicted by the disease development calendar, while applications of a systemic curative fungicide can begin as soon as the first symptoms are detected. Existing symptoms of disease cannot be erased, so fast action is a must.

Nematodes are common in most agricultural soils in which field-grown nursery stock is planted. The nematode species present and their populations can be determined by sampling the soil prior to planting. Soil samples can be submitted to many universities, state departments of agriculture, and private laboratories. for nematode analysis. Growers can make informed decisions based on a laboratory report on whether to plant a susceptible or a resistant crop, fumigate the field, or not plant nursery stock in the field. Since most fields are planted for multiple years, nematode levels must be essentially zero at planting if the crop is susceptible to the nematode species present. Nematodes have been virtually eliminated in the nursery industry through the use of soilless media in the production of container-grown plants.

Pest Thresholds

The term *threshold* is widely used in IPM programs. The threshold is the level or population of a pest that causes enough plant damage (yield or quality reduction) to justify the cost of the treatment to reduce or keep the population below that level. Thresholds are well established for many insect pests in field crops. Control measures (for example, pesticide applications) are not taken until the pest population reaches or exceeds the threshold.

Threshold levels are also well established for many nematode species in numerous annual crops. For many annual crops, pest control is necessary for only three to four months in order to achieve enough growth to produce an adequate yield. In contrast, ornamental plants may remain in nurseries for one to three years, and woody plants are grown in field plantings for one to five years. Even a very small population of a damaging nematode species can multiply to a high level (above the threshold) in one to five years. For field production of woody plants that are susceptible to nematodes, the threshold is zero: the soil must be free of damaging species, or the field must be treated with a fumigant prior to planting. Several nematode species are the subject of regulatory concern and have a threshold of zero.

Thresholds have not been established for any diseases of woody ornamentals. Most growers attempt to maintain a zero threshold for all diseases of all crops. A zero threshold may be unnecessary for some diseases with low severity potential and no regulatory concern. However, the goal, should be a level as close to zero as possible and economical.

The threshold for a disease, as determined by the following factors, must be established by each grower:

- the reputation the nursery desires to establish or maintain with customers
- the expectation of customers or markets
- the species of plants produced
- regulations of the states in which the producer and the markets are located
- the type of disease or the potential severity of the disease (for example, leaf spot versus root rot)

Scouting for Diseases

Scouting is a critical part of integrated disease management in any nursery. Scouting must be done regularly by a trained employee. Records should be kept and organized so that changes in disease incidence and severity can be observed at the earliest stage. Scouting can be either general or specific, but a combination of these two approaches is best.

A general scouting plan can focus on many different species of plants and include observations on all aspects of plant health, e.g., leaf color, rate of growth, stage of growth, and size of leaves. This type of scouting can identify the general vigor and growth of the plants. Atypical leaf color and size may indicate a nutrient or pH problem or root rot. Pour-through tests with a soluble salt meter or pH meter can confirm or rule out nutritional problems. If the nutritional status and pH are shown to be within acceptable levels, root rot may be developing. Samples of roots should be collected and sent to a diagnostic laboratory or plant clinic for analysis so that an appropriate treatment strategy can be implemented as soon as possible.

A specific scouting program focuses on specific crops or diseases. Such a program can be expensive and time consuming and must be knowledge-based. If a nursery has not had a scouting program, a general scouting program should be started at once. It may not be practical to start a specific disease scouting program for every crop grown and every disease of each crop. The value of each crop to the total nursery operation, the frequency of problems with specific crops, or the frequency of specific diseases should determine which disease problem or problems to target in a specific scouting program. These decisions should be based on the history of crop production at the nursery or a change in projected future production, such as adding new crops to the production mix. As experience is gained with the specific scouting program, additional crops and diseases should be added until the entire nursery is included in the general and specific program. Implementing a big scouting program is expensive and can be difficult to justify for small nurseries, but with time and experience it should pay dividends.

A block of plants can be scouted several different ways. A scout should walk through the block of plants following a zigzag, W-shaped, or X-shaped pattern (Fig. 86.1), observing plants while walking. Plants should be examined, both tops and roots, at regular intervals throughout the block. The scout should stop at selected intervals to carefully examine plants for particular problems or diseases, particularly diseases that could be occurring according to a disease development calendar. Plants in low areas should always be checked. Notes should be kept on each plant and block for every week. A form should be developed to carry in the field to standardize the information collected and minimize the time required.

Once a crop is selected for a specific scouting program, the first decision is to select diseases to monitor and the time of year when they occur. A specific scouting program must be based on the diseases affecting the crop; the symptoms of each disease, particularly the earliest symptoms; the time of year when each disease is actively developing and spreading in the nursery; and the time when the pathogen naturally goes dormant and treatment becomes ineffective. This information is presented in the disease summaries in the chapters on individual crops (Chapters 20–84). A specific scouting program may be too difficult for certain crops with multiple diseases; these crops should be incorporated into the program later. Select one or two diseases of the target crop that cause economic loss and have been recurring problems in your nursery, as determined by historical records or memory (i.e., what you have seen at your nursery).

How to Develop a Specific Scouting Program

As an example, suppose that azalea is the target crop for which a specific scouting program is to be developed. There are more than 20 pathogens known to cause diseases of azalea. Make a list of the problems and diseases that historically have occurred at the nursery. For example, assume that there have been six chronic problems in the nursery: (1) cold injury, (2) improper nutrition, (3) Phytophthora root rot, (4) Ovulinia petal blight, (5) Rhizoctonia web blight, and (6) Exobasidium leaf gall. Different nurseries may make different lists, because there can be a lot of variation in problems and diseases from one nursery to another.

To implement a specific program, the azalea crop should be scouted weekly, starting just before budbreak in the spring and continuing until fall. The scout should develop a scouting calendar showing when these selected problems and diseases are likely to occur (Fig. 86.2) based on the plant disease development calendar for the zone in which your nursery is located. With a scouting calendar, the scout knows what to examine plants for each week. The scout should know or review the symptoms and signs of each problem or disease selected for scouting.

1. Cold injury Has cold injury been a consistent problem at your nursery? Have you improved winter protection efforts? Azaleas are injured by rapid changes from mild to very cold conditions. In scouting for cold injury following a sudden cold snap, examine the least hardy cultivars for your hardiness zone. Check for split bark on the branches or the main trunk and burned foliage on new growth. If cold damage to stems or main trunks is observed, a course of corrective action must be determined. If damage is confined to the branches, it can be pruned out. If the lower main stem is damaged, plants may need to be dumped.

If cold injury has been a consistent problem for azaleas at the nursery, does it occur in certain cultivars and does it occur consistently every year or is it a very sporadic problem? If damage con-

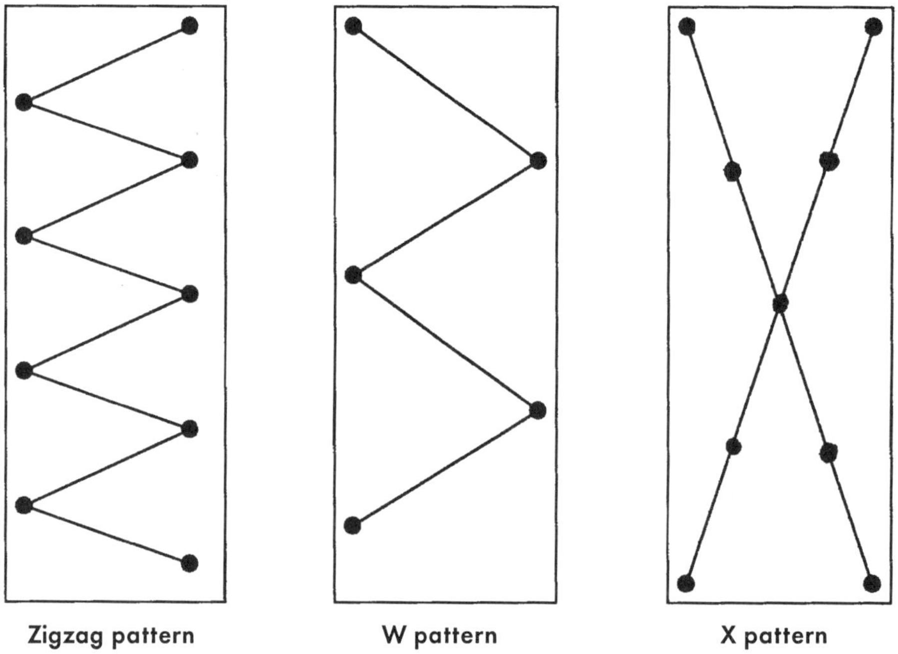

Fig. 86.1 Scouting patterns for a block of plants.

sistently occurs in a certain cultivar, it may be most economical to stop growing that cultivar or to provide better winter protection, if it is a very important cultivar. It may also be important to check the late-summer or fall fertilization program, because azaleas should have time to harden off before cold weather. The grower should consider reducing the late-summer or fall nitrogen application.

2. Improper nutrition Various problems can occur in azalea as a result of improper nutrition, either too high or too low, and the symptoms of nutritional problems can mimic disease symptoms. Low nutrition causes symptoms such as chlorosis and slow growth, while high nutrition can cause very similar symptoms. Root rot also causes slow growth and chlorosis on azalea. Extremes in nutrition can also cause root injury and can increase the susceptibility of a species or cultivar to Phytophthora root rot.

Soluble salt levels and pH should be monitored weekly by the pour-through technique. This can help determine irrigation scheduling to minimize high levels of soluble salts and to determine when the fertility level is getting too low and a topdressing of additional fertilizer is needed.

3. Phytophthora root rot If Phytophthora root rot has been a problem at the nursery, it should be a part of the scouting program. As part of the weekly scouting of azaleas, plants should be removed from the containers and the root systems examined to help identify periods of root growth and the ratio of live roots to dead roots. If roots are dark reddish brown and pull apart easily, Phytophthora root rot may be present. Careful examination of roots may detect the disease before aboveground symptoms become apparent. If dark, discolored roots are observed, collect samples of roots and potting medium from several pots, place the samples in a plastic bag (labeled with the name of the genus, species, and cultivar; the location of the block; the date, etc.), and submit them to a diagnostic laboratory for testing. If the analysis is negative for *Phytophthora,* other problems, such as cold injury, excessive dryness, or excess soluble salts, should be suspected.

A positive analysis for Phytophthora root rot requires devising and implementing a management program. Review the irrigation program and media drainage properties. It may be necessary to start drenching containers with an appropriate fungicide.

4. Ovulinia petal blight If petal blight is selected for specific management, the following questions should be asked:

Does the disease occur in your area? yes no
Has petal blight been a problem at the nursery? yes no
What cultivars should be scouted?
early-season midseason late-season
When does petal blight occur? late April through early June (see the disease development calendar)
What are the symptoms of petal blight? See Chapter 23, "Azalea Diseases"

Scout problem cultivars as they come into flower. It is best to scout early in the morning. Check for wet, water-soaked spots on petals. If petal spots are present, is the tissue firm and dry or does the blighted tissue macerate easily between thumb and finger with little pressure and is the tissue soft and greasy? If the tissue is soft, watery, or greasy, place the petal tissue in a plastic bag and send it to a diagnostic laboratory. Blighted petals can also be held in a plastic bag at room temperature. If it is petal blight, black, round, flat, or curved sclerotia, the resting stage of the fungus, will form in petal tissue in about three to six weeks. This will be too late to implement a management program. A diagnostic laboratory can confirm petal blight in just a couple of days.

If the report or field observations are negative for petal blight, continue scouting later-blooming cultivars. The amount of rainfall and cool conditions greatly influence the development and severity of petal blight. Early-season cultivars may escape the disease, while mid- to late-season cultivars are severely affected.

If the field test or the diagnostic laboratory analysis is positive, a management program should be devised and implemented. If plants are sold during bloom, the damage may be severe enough to justify a fungicide application. If petal blight is developing on a midseason cultivar, a management program must be devised and implemented on all later-blooming cultivars. This pathogen also infects the flowers of rhododendron and several other plants in the family Ericaceae. If it is present in azalea, it would be wise to scout rhododendron for petal blight. The symptoms, damage, and life cycle of the pathogen are the same. If petal blight symptoms are observed on azalea and rhododendron, a management program

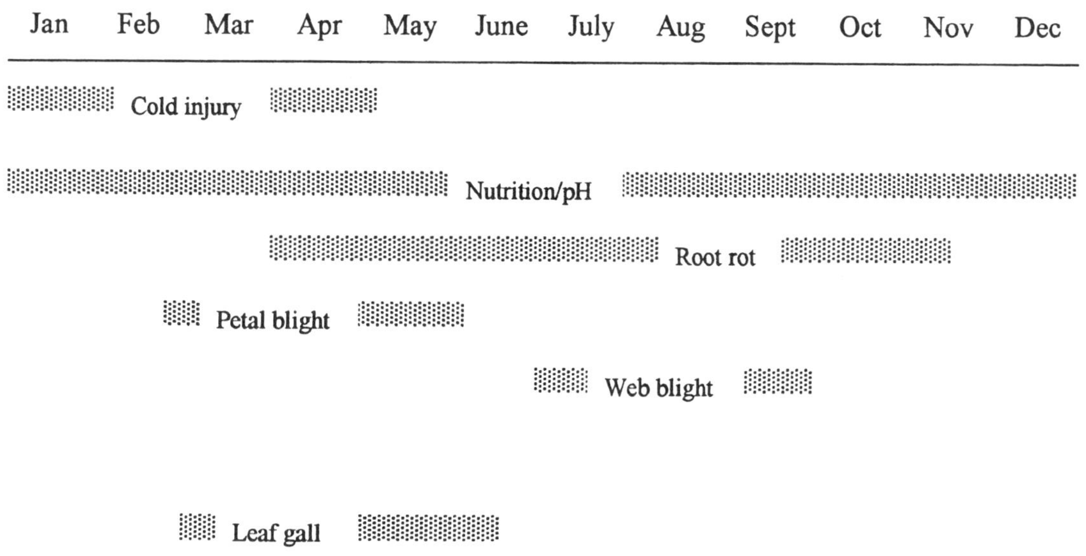

Fig. 86.2 Example of an annual scouting calendar for six problems and diseases affecting azalea.

should be implemented for both crops. Several fungicides are labeled and effective for this disease. Check the section on disease management of petal blight in Chapter 23, "Azalea Diseases."

5. Rhizoctonia web blight If web blight is selected for specific management, the following questions should be asked:

Does web blight occur in your area? yes no
Has web blight been a problem historically? yes no
Does web blight occur on the cultivars you grow? yes no
When does web blight develop? July, August, and early September (see the disease development calendar)
What are the symptoms of web blight? See Chapter 23, "Azalea Diseases"

If web blight occurs in your area and has repeatedly occurred at your nursery, it should be scouted for. Early detection and appropriate management practices will prevent further development.

Web blight usually first appears in mid-July, and the scouting program should begin July 1 (about two weeks before the first symptoms are detected). Plants within the block, rather than along the edge of the block, should be examined. The foliage of plants inside the block will stay wet longer than that of plants on the edge of the block. The scouting program should target cultivars with a thick, dense canopy, such as Gumpo azalea, and blocks where the plant canopies are touching.

Pick up plants, and examine for dead leaves on the inner and lower part of the canopy. An excellent field diagnostic feature of web blight disease is the webbing of dead leaves to the stems. Through careful examination of the plant with the unaided eye or a hand lens, the silvery brown strands of Rhizoctonia can be seen webbing the leaves to the stems. If dead leaves are observed in the plant canopy but the silvery fungal strands cannot be observed, place a large, clear plastic bag over the plant and place it at room temperature out of direct sunlight. The silvery fungal strands should be very visible in 24 to 48 hours. If you are still not certain, send the sample to a diagnostic laboratory. If the scouting indicates that no web blight is developing, make no management changes but continue to scout until mid-September.

If web blight is detected, check other cultivars of azalea to determine how widespread the disease may be. Other plants, such as Helleri holly (see Chapter 47, "Holly Diseases") are also susceptible to web blight (see Chapter 17, "Rhizoctonia Web Blight"). Scout for web blight symptoms in these crops also. Devise a management program and implement it. If the disease has a spotty distribution when first detected, the most economical way to stop further disease development is to treat the symptomatic plants. Increase the spacing between plants if they are crowded to obtain better drying in the plant canopy. If this plan is implemented, continue to scout carefully to detect other outbreaks or spread of the disease. If spread occurs after the first management strategy was implemented, fungicide spray applications over the entire block of plants may be warranted. An appropriate fungicide should be applied every seven to 10 days or at intervals recommended on the product label until mid-September, when the pathogen ceases activity.

6. Exobasidium leaf gall The potential severity of Exobasidium leaf gall of azalea is almost zero. The disease is of no economic importance and should be ignored. It tends to be more common in the landscape than in nurseries. Growers may want to stop producing highly susceptible cultivars.

After one or two years of scouting these six problems, the program should be expanded if it has been successful.

Case Study 1: Disease management in a total IPM program in a large, highly diversified nursery

Greenleaf Nursery Company, in Park Hill, Oklahoma, is a large and highly diversified woody ornamental nursery, occupying 570 acres and producing more than 10,000 liners and 8,500,000 finished plants per year, including 70 cultivars of conifers, 570 cultivars of broadleaf evergreen and deciduous shrubs, and 145 cultivars of shade and flowering trees. Disease management is totally integrated into an IPM program managed by a pest management supervisor. Scouting is done every day, year-round (with some allowances for weather), by four scouts. All plant material in greenhouses, overwintering houses, and production blocks is inspected at least once each week. Scouts are trained to detect 38 categories of pest and cultural problems (Table 86.1) and to gauge changes in the severity of these problems.

Scouting data for each production block and plant material are entered into hand-held computers in the field and uploaded to a central computer by 1:30 PM each day. The pest management supervisor uses a printout of this information to visit problem spots and formulate a plan of corrective action, taking into account many factors, including the history of the problem, its severity in the current season and previous seasons, corrective actions that have already been taken and their efficacy, and the current options for corrective action. The supervisor's decisions are then communicated to one of four division supervisors, if the problem is cultural, or the pest management foreman, if pest control is needed. The foreman is given a written directive for applications of pesticides, and a duplicate is retained by the supervisor.

The foreman coordinates the application of pesticides by a team of pesticide technicians and communicates with division supervisors to make sure that pest control actions integrate with production and sales activities. Every time a pesticide is applied, the foreman enters all pertinent information into a hand-held computer, and the information is then uploaded as a permanent record. The computer supplies the reentry interval for each pesticide, which is posted using modified traffic markers and flagging. The division supervisors are also informed, so that workers will not be sent to these areas. An internal radio system allows for easy

Table 86.1 Diseases and problems routinely scouted in a large, highly diversified nursery

1. Leaf spot	20. Insect, unknown
2. Mildew	21. Whitefly
3. Rust	22. Mealybug
4. Canker of stem or trunk	23. Spittlebug
5. Phomopsis blight	24. Lacebug
6. Root rot	25. Nutrition problem
7. Scab	26. Deer
8. Gummosis	27. Rodent
9. Crown gall	28. Rabbit
10. Virus	29. Weeds
11. Unknown, weakness	30. Spray damage
12. Aphid	31. Broken limbs
13. Spider mite	32. Wilt
14. Worm	33. Thrips
15. Leaf miner	34. Psyllids
16. Beetle	35. Wooly apple aphid
17. Scale	36. Southern blight
18. Leafhopper	37. Fire blight
19. Borer	38. Bleeding necrosis

communication among all involved at each stage of the process. Major problems are discussed daily at early-morning meetings in which all nursery managers, from supervisor to vice presidents, participate.

Case Study 2: Management of bacterial diseases of hibiscus in a Florida nursery

Production of hibiscus (*Hibiscus rosa-sinensis*) in the Gulf Coast states can be challenging because of the incidence of diseases caused by the bacterial pathogens *Xanthomonas campestris* pv. *malvacearum, Pseudomonas cichorii,* and *P. syringae,* which can invade and seriously disfigure the leaf canopy. These pathogens are particularly damaging at different temperature ranges: *P. syringae* at 15–18°C (60–65°F), *P. cichorii* at 21–27° (70–81°F), and *X. campestris* pv. *malvacearum* at 24–33°C (75–92°F). Disease severity increases with the use of overhead irrigation or the occurrence of seasonal rainy periods. Often infections result from a complex or two or more of these pathogens. A number of cultivars that vary in susceptibility to these bacterial pathogens are in demand from consumers. The yellow cultivar Hula Girl is particularly susceptible to *X. campestris* pv. *malvacearum* during the summer rainy months, preventing quality production.

A progressive nursery in Florida tried to meet the constant demand for high-quality hibiscus grown in 1-gallon pots and for standards in 10-inch containers through the implementation of an IPM program. This crop can be marketed whenever it is in color. Two objectives of this IPM effort were established: (1) to propagate at a seasonal period to escape conditions favorable for disease and (2) to grow out the crop under environmental conditions that maintained the pathogen-free status of the propagated liners.

Hibiscus production stock was placed under routine scouting for earliest disease detection and subsequent eradication. The stock bought outside the nursery was routinely segregated to exclude the introduction of bacterial pathogens to existing nursery stock. The initial stock was treated with copper-based fungicide in a spray program to manage existing disease and minimize disease spread.

The propagation cycles were moved from winter and summer periods to early spring (April). The cool, moist winter period favors diseases caused by *Pseudomonas* spp., while during the summer propagation period, plants are at high risk to infection by *X. campestris* pv. *malvacearum.* Propagation under mist was carefully controlled to minimize leaf wetness, with 4 seconds of mist every 16 minutes. Foliage was trimmed to reduce transpiration and potential surface area for infection.

Rooted liners were shifted from mist propagation houses to specially constructed poly houses with solid ceilings and drop-down sides screened with 22% shade cloth. The new production houses eliminated contact with natural rainfall during the hot summer, while still providing air circulation. During the winter, the houses offered frost protection, and condensation problems that could trigger outbreaks of infection by *Pseudomonas* could be managed. The 1-gallon production stock was watered by overhead irrigation at midday, eliminating nighttime periods of leaf wetness. Hibiscus standards were irrigated from 14- to 18-inch risers among rows, so that the upper canopy remained dry. The outer rows of plants, along the perimeters of the house, were subject to water splashed from the roadway and were thus treated with copper-based fungicide as needed.

The combination of exclusion, eradication, protection, and environmental avoidance improved the production of high-quality, disease-free hibiscus when implemented as a result of a sustained scouting program. Even such susceptible cultivars as Hula Girl could be reliably produced to meet consumer demand.

History and IPM Records

Keeping an organized record system of pest problems and the success or failure of management strategies can be of great value in guiding future IPM efforts in your nursery. Historical records can be examined to discern patterns of problem occurrence by year, by month, by crop, and by location in the nursery. Observations should be recorded on pesticide effectiveness or pesticide injury, rates used, combinations of pesticides, adjuvants, and nutrients. Careful review of historical records may reveal a pattern that can lead to a change in cultural practices, pesticide used, or timing of a pesticide application that may result in a significant quality improvement and more than pay for the cost of the IPM program.

When problems or diseases are detected by the scouts, take photographs for the records. These can be referred to in future years to compare current symptoms with what has been observed in the past. When new problems or diseases are observed and samples are sent to a diagnostic laboratory, be sure to include the diagnostic report and photographs with the scouting report for future reference.

Dumped plants should be another part of the historical record, including why were they dumped. Plants consistently dumped because of one disease should be counted to calculate loss estimates. Management strategies for that disease should be reviewed, new strategies examined, and cost-benefit ratios developed that will be helpful in making disease management decisions for the future. If disease distribution is mapped within the nursery, it may be possible to pinpoint a consistent pattern of disease development within the nursery. If so, the area can be analyzed for problems and factors that can be associated with disease occurrence, such as low areas, excess shade, and faulty irrigation equipment.

Severity Potential

Severity potential is the capacity for a pathogen to cause damage to a plant if no specific management strategies are implemented and conditions remain favorable for disease development. Some pathogens have unlimited potential to severely damage or ultimately kill a highly susceptible plant. Phytophthora root rot of rhododendron is an example of a disease with a very high severity potential; it frequently kills rhododendron plants. If you are growing rhododendron and have a history of Phytophthora root rot, you must practice a preventative management program. There is no cure for Phytophthora root rot in a highly susceptible host.

Many other diseases have very limited potential to cause significant real damage to a susceptible host. Real damage means a significant reduction in the growth or survival of the host. The aesthetic value of the plant must also be considered. This is a quality measurement and does not necessarily relate to the growth rate of the host or its survivability. Exobasidium leaf gall of azalea is an example of a disease that causes no real damage to the host. This disease has a very low severity potential. The pathogen is active for only four to six weeks each year. Visible evidence of the disease almost completely disappears two to three weeks later. For two to three weeks when the disease is highly visible, the apparent quality of infected plants may be reduced, but a buyer with knowledge of the limited potential of leaf gall should not be concerned. If the azaleas are not going to be sold during the time the disease is highly visible, the disease is of little importance and management strategies are unnecessary.

Leaf spot diseases are common on many hosts and vary widely in their severity potential. Black spot of most hybrid tea roses has the potential to cause severe defoliation requiring a preventative management strategy. Entomosporium leaf spot causes severe damage to red tip Photinia and several other crops in nursery production. Highly effective management practices have to be carried out to the point that it is often questionable whether it is economical to grow the plant. Entomosporium leaf spot has a very high severity potential and can cause severe damage on highly susceptible cultivars of Indian hawthorn, but it has a very low severity potential and causes little or no real damage on highly resistant cultivars. The cost of producing the highly susceptible cultivars is significantly more than that of the resistant cultivars because of the cost of the specific management strategies, i.e., spraying the susceptible cultivar numerous times each year to produce an acceptable plant. The fungus produces large quantities of spores on the highly susceptible cultivars and increases the disease pressure on the slightly resistant and moderately resistant cultivars. The susceptible cultivar also demands a costly management program in the landscape. A good management strategy for Entomosporium leaf spot of Indian hawthorn is to grow only resistant cultivars. This is a good decision for both the nursery and the customer.

To practice very limited management on diseases with little or no potential for causing severe damage to the host and accepting any aesthetic damage, the grower must correctly identify the disease, know its severity potential, and have a good understanding of the regulations in both the producing state and the receiving state. The knowledgeable customer will ignore these cosmetic diseases that have a low severity potential.

When to Scout

Most pathogens actively sporulate or otherwise produce propagules that infect plants at a specific time of the year. This active period is often determined by temperature, moisture, rainfall, humidity, and the growth stage of the host. *Phytophthora cinnamomi* is active whenever the temperature of the soil or medium is above 50–55°F. The fungi that cause black spot of rose and Entomosporium leaf spot of red-tip photinia are active about 10 months of the year in USDA zone 8. Most pathogens, however, are active for a shorter, more well-defined period. *Ovulinia azaleae,* the petal blight pathogen, is active only when mid- to late-season azaleas are in bloom.

The time to scout for these diseases is when the pathogen is active, or better still, just starting to be active in your hardiness zone. Visible symptoms may be present for months after the active period of the pathogen. *Erwinia amylovora,* the fire blight pathogen, is active for only four to six weeks in the spring, but blighted shoots are still visible the following fall or winter.

Scouting for a specific disease should be conducted when the pathogen is active. After the pathogen has gone dormant for the growing season, scouting can cease until the pathogen starts to become active during the next growing season. The disease development calendar in Chapter 2 (for central North Carolina, in USDA hardiness zone 8) shows the active periods of numerous pathogens, and the disease summaries in Chapters 20–84 also state the active periods of pathogens. The activity of pathogens may vary from one hardiness zone to another, and each nursery must adjust the disease development calendar for the zone in which it is located. Historical records of disease occurrence at the nursery are very helpful in developing a calendar for the diseases that need to be scouted.

The period of activity of a pathogen can also vary from year to year, mainly in response to temperature. With an early or late spring, a disease may occur a week earlier or a week later than normal. An early or late spring can also delay or speed up seasonal plant growth or flowering. Forsythia may bloom in the last week of February in an early spring or in mid-March in a late spring. Recording the dates of flowering of plants that bloom at a specific time of year, such as pear, apple, azalea cultivars, or forsythia, makes an excellent biological calendar, which can help in zeroing in on seasonal variations in plant growth and disease development. For example, a leaf spot disease may show up three weeks after budbreak, even if budbreak occurs earlier or later in one year than in another.

The average date of the last frost can also be used to determine whether seasonal plant development is ahead of or behind that noted in the disease development calendar in Chapter 2. If the last frost is two weeks later than in zone 8, the disease development calendar should be shifted two weeks later. Plant hardiness zones can be used to adapt the disease development calendar to your location. In zone 8, stages of plant growth and disease development occur one week earlier in the spring and one week later in the fall than in zone 7. In zones 9 and 10, many diseases are active during the entire growing season.

Geographical Distribution of Diseases

Another important consideration in determining what to scout for is whether a disease occurs in your area or whether plants were recently shipped from an area where the disease occurs. In the descriptions of diseases in this book, the geographical distribution of the pathogen is presented, if it is known. If a nursery is separated by more than one hardiness zone from the area where the pathogen is known to occur, the scouting program should not focus on this pathogen unless the plants were recently purchased from a nursery located in the hardiness zone where the pathogen occurs.

Regulatory Control

State departments of agriculture have plant inspection agencies that inspect nursery stock coming into the state and being shipped out of the state. Federal agencies inspect plant material coming into the United States. The objective of these inspections is to prevent or reduce the amount of infected plant material being shipped between regions. For more information on this subject, see Chapter 101, "Regulatory Control."

This type of regulatory control is one form of exclusion, which is a means of keeping pathogens out of the nursery. Regulatory restrictions may keep pathogens out of states or regions where they do not naturally occur. Pathogens that are excluded from your region or nursery are pathogens for which you do not have to develop and implement expensive disease management programs. This is a very valuable service to the nursery industry. Purchasing disease-free starter plants may keep the plants free of certain diseases for the entire production cycle.

Exclusion

Exclusion consists of steps taken to keep a pathogen out of a nursery. In addition to the benefits of regulatory exclusion, individual nurseries must purchase disease-free propagating stock, plants to grow on, and plants for resale. Cuttings taken from stock or landscape plants from the nursery should be as free of pathogens as possible. Starting with healthy plants has two benefits: it prevents the introduction of pathogens into the nursery, and it avoids placing diseased plants in the nursery. It is very expensive and often impossible to cure diseased plants. They may always remain stunted, off-color, or of low quality, even after excessive inputs, such as extra fertilizer, extra pruning, and extra pesticide applications. Pathogens introduced into a nursery on diseased liners can affect future crops, increasing future costs of production.

Resistance

Resistance is the natural capacity of a genus, species, or cultivar to resist disease. Some plants or cultivars are just too susceptible to diseases to be of economic value as ornamentals. Why grow highly susceptible plants? They are an inferior product for customers, and the cost of production is higher, because of the increased cost of disease management.

Furthermore, highly susceptible plants support a higher level of disease than less susceptible plants. Pathogens produce abundant quantities of propagules (e.g., spores) in highly susceptible plants, increasing the disease pressure on moderately resistant to resistant plants. Reducing the number of highly susceptible plants reduces the inoculum level (e.g., the number of spores), thus lowering disease pressure on moderately resistant to resistant plants. In addition, reducing the inoculum level decreases the level of disease and thus the need to apply pesticides, and this lowers the risk of creating pesticide-resistant pathogens. Fungicides are more effective when disease pressure is low than when an epidemic is raging.

Eliminating highly susceptible species and cultivars does the whole industry a favor. If customers demand a highly susceptible species or cultivar, let them purchase it elsewhere.

This topic is covered in more depth in Chapter 96, "Disease Resistance."

Biological Control

Biological control of plant diseases is developing slowly. Some products are available for biological control of root rot, primarily of seedlings. Biological control may reduce the need for chemicals and preserve beneficial microorganisms on plants and root systems. However, biological control programs are successful only with low levels of inoculum and therefore must be implemented in conjunction with other disease management strategies in an integrated disease management program.

This topic is covered in more depth in Chapter 98, "Biological Control of Woody Ornamental Diseases."

CHAPTER 87

Jean Williams-Woodward ▪ University of Georgia, Athens

Ronald K. Jones ▪ North Carolina State University, Raleigh

Sanitation: Plant Health from Start to Finish

Sanitation is a general disease control strategy that can be effective against all diseases; but most importantly, it reduces root rot and stem diseases. A good sanitation program must become automatic and a way of life in all operations of the nursery.

Root rot and stem diseases are the economically most important disease groups in nurseries, because they are the most difficult to control. Foliage and flower diseases are usually more easily controlled, simply because the symptoms occur aboveground and are therefore visible and accessible, and infected foliage can be pruned and removed. In contrast, root rots require preventive fungicides and labor to apply soil drenches. Losses caused by root and stem disease include low-quality plants and discarded diseased plants.

The key to financial success for ornamental nurseries is to increase production and at the same time reduce production costs. The only viable solution to reducing costs is to minimize plant losses and maximize plant growth rate and quality. The objective of this section is to describe a useful plan of action for nurseries interested in reducing disease losses. Most practices will seem like common sense, but numerous nurseries are not following this common-sense approach to disease control. It is much more cost effective to prevent pathogens from being introduced into the production cycle than to attempt to suppress diseases on infected plants through the continual use of fungicides and bactericides. Very few plant diseases can be cured, and diseases cannot be controlled in nursery production by pesticides alone.

The primary means of reducing pathogen introduction is sanitation. Sanitation consists of all activities aimed at eliminating or reducing the amount of inoculum present in the field or greenhouse and preventing the spread of the pathogen to adjacent healthy plants. Each grower must develop a sanitation program to fit his or her particular situation. The sanitation program has been recognized as an essential part of nursery operations in recent years. Some nurseries have established integrated pest management programs with scouting crews to reduce pests through early detection and pesticide applications. Growers should prevent or minimize diseases rather than always reacting to an existing disease. The integrated pest management program is implemented in addition to sanitation and is not a substitute for sanitation.

Growers must understand the scientific principles that form the foundation of successful disease control, that is, the exclusion of plant pathogens. Microorganisms that cause disease, especially those infecting roots and stems, make their way into the nursery production cycle from known potential sources. Some of the most common sources of pathogens introduced into nursery production are (1) infested soil introduced during irrigation or rain; (2) infested recycled irrigation water; (3) cuttings placed on the ground, on dirty work surfaces, or in infested water or hormone rooting solutions; (4) hoses dropped on the ground (pathogens get into the nozzle and are expelled into pots or on benches at the next watering); (5) infested soil and organic material remaining on used flats, pots, benches, or other containers between crops; (6) infested soil carried on tools, covers, or workers' hands; (7) infested soil deposited by foot traffic on sterile potting mixes and disinfested benches or flats; (8) media, flats, plants, or containers placed on the ground in contact with the native soil; and (9) infected or infested cuttings, seedlings, or seed.

Sanitation cannot be conducted in a piecemeal fashion. Changes must be made in the entire production cycle for successful disease control through sanitation. Because sanitation is the most important management practice in a nursery (followed by correct watering), it must be viewed as a series of different functions within a nursery production cycle, much like links in a chain; one broken link can result in economic losses caused by plant disease.

Sanitation in Propagation

Diseases are prevented during propagation by considering the three primary areas in which pathogens are usually introduced: the physical propagation facilities, the propagation medium, and the plant material used for propagation. Exclusion of pathogens in each of these areas makes it likely that the crop can reach salable size disease-free. Soilborne fungi that attack roots, stems, and cutting wounds, such as *Pythium, Phytophthora, Fusarium, Rhizoctonia, Thielaviopsis,* and *Cylindrocladium* spp., are most likely to cause disease during propagation. See Chapters 11, 12, 14, and 17 for more information on control of these pathogens.

Physical Propagation Facilities

The propagation area should be clean and separated from areas where soil mixing, pot and flat storage, growing, and other operations take place. Foot traffic and visitors into the propagation houses should be minimized.

Mist propagation and growing areas should be kept clean at all times, and all dead plant material should be removed prior to starting a new crop. All benches and floors must be kept free of infested medium, leaves, and other refuse. Walls and concrete walkways should be disinfected between crops. If disease symptoms appear during propagation, remove any rotting cuttings from containers or even whole flats from the propagation area and then treat the area with an appropriate fungicide or disinfectant.

Avoid unnecessary handling of clean medium, such as dipping hands into the medium or feeling it unnecessarily to gauge its moisture content. Workers should wash their hands after working with raw infested media, soil, or plants to avoid introducing pathogens into the propagation area. Anything that moves soil can move soilborne pathogens, including the dirt under workers' fingernails.

Overwatering is the major culprit in creating an environment that favors disease in propagation and growing areas. Therefore, care should be taken to avoid excessive overhead irrigation of the propagation area. Misting schedules should be adjusted so that plants are not wetted during periods of high humidity and precipitation. Low water pressure for irrigation prevents splashing soil or media into pots and flats. It is best to use well water, which is usually pathogen-free, for irrigating propagation areas, rather than pond water, which can be a major source of water mold pathogens.

Propagation Media

Media should be loose, porous, and well aerated and should be used only once. Most components of media, such as perlite, vermiculite, and tree barks, are usually pathogen-free. However, sand, soil, and compost can contain pathogens and are a primary means of introducing them into propagation and production systems. Media components of questionable sanitation should be heat-treated or pasteurized to kill microorganisms. It is unwise to include soil in the mix, because soil is more likely to be contaminated than other components.

Media should be stored in containers or in areas where contamination with pathogens can be avoided. Propagation media should be mixed on a clean concrete slab or, in small quantities, in disinfected containers. A pathogen-free medium can easily become infested if carelessly handled by workers (for example, if it is mixed with dirty tools or subject to foot or machinery traffic). Propagation benches raised above the ground are highly recommended, because they prevent the plants from coming into contact with native soil.

Plant Material

Cuttings for propagation should be selected from disease-free plants. It is optimal to have stock plants, or *mother blocks,* that are separated or isolated from stock grown for sale. Regular fungicide sprays and soil drenches on especially disease-prone plants should be conducted to reduce foliage, stem, and root diseases. The establishment and maintenance of mother blocks has been effective in reducing disease control costs, because of the ease and relatively low cost of controlling diseases within a small area.

In nurseries unable to establish mother blocks, cuttings should be collected from healthy plants. Collecting cuttings from the tops of healthy plants, rather than from near the ground or media surface, prevents possible contamination with soilborne pathogens. Plants grown on trellises or aboveground supports are generally free of soilborne pathogens.

Breaking rather than cutting stems for propagation avoids the possibility of spreading pathogens, especially bacterial pathogens, on pruning tools. When knives or pruning shears are used to collect cuttings, use several pairs, so that one pair can be soaked in a disinfectant solution while the other is being used. Rubbing alcohol (70% isopropyl alcohol) or a 10% bleach solution (one part bleach in nine parts water) is an effective disinfectant. Bleach solutions should be kept covered and changed every 30 minutes, because their disinfecting qualities diminish rapidly.

Cuttings should never be placed on the ground or in contaminated containers. They should instead be placed on chemically disinfected surfaces, flats, or baskets or in new plastic bags. Never dip cuttings in water unless it is known to be pathogen-free. Growers who dip cuttings in a fungicide suspension prior to sticking should read the product labels for worker protection standards.

Growers who must purchase liners should obtain only high-quality, disease-free plants. The purchase of diseased liners is uneconomical, because the disease cannot be cured and the diseased liners will introduce pathogens into the nursery.

It is important to take cuttings at the optimum stage of plant growth, so that they will root quickly. If cuttings are stuck too deeply, the likelihood of root and stem diseases increases. Avoid jamming cuttings to the bottom of the rooting container. This causes them to root poorly, making them more susceptible to root disease. After cuttings have rooted, they should be moved out of the propagation area, because a propagation area is not a plant storage area.

Sanitation in Growing Areas

Consider implementing all of the parts of the sanitation program used during propagation in the growing areas of the nursery. If the planted liners are placed into pathogen-infested areas or pots or if the containers are placed directly on the ground, the liners will develop disease.

Preparation of Potting Mix

For mixing or storage of media, construct a sloped concrete slab above soil level, to prevent contamination by surface water runoff from surrounding areas. To avoid introducing pathogens, insects, and weeds, never mix media on the ground. Ground equipment should be assigned to this area and should not be moved off the slab, to prevent the introduction of pathogens on tires or tracks. Avoid unnecessary foot traffic in the area; dirt on the bottoms of shoes can introduce disease-causing organisms. Consideration should be given to the size of slabs; many are too small for maneuvering equipment and media components.

Storage of Containers and Potting Media

Store containers and media on a clean surface that is not exposed to splashing water or runoff water, such as on a concrete pad or in storage structure. It is important to prevent the introduction of pathogens during storage and handling.

Container Growing Area

Place containers on a woven, plastic ground cloth and avoid contact with native soil. Construct beds to be as smooth as possible, and compact the soil prior to spreading the ground cloth to prevent settling of the soil and the creation of low spots. Beds should be crowned by elevating their centers (with a 6-inch drop per 25 feet). Construct drainage ditches at the edges of beds.

Many nurseries build roadways higher than the beds so that trucks do not get stuck in the mud. However, when roads are elevated, all the surface water and contaminated soil from them drain into the growing area and the potted plants, increasing the possibility of disease development.

Some nurseries use gravel beds solely or place gravel on top of or beneath the ground cloth. This reduces contamination by runoff water and splashing soil and thereby reduces plant disease. The initial costs are higher for gravel beds, but the cost is offset by the increased number of healthy, salable plants. Gravel increases the longevity of the container growing area and reduces expenditures for labor and materials per year. The gravel should be at least 6 inches deep to prevent it from settling into the native soil and preventing the soil from migrating upward through the gravel to contaminate pots. Drive roads between beds should also be graveled to prevent soil and water from splashing into pots.

Irrigation Sources

Disease can be spread through irrigation with pathogen-infested pond water. Avoid using pond water and allowing runoff water from growing areas to drain directly into irrigation ponds. Many growers collect runoff water because of new laws requiring the recapturing of water used in nurseries. However, runoff water carries not only pathogens but also fertilizers and herbicides, so practices to reduce this threat must be implemented. See Chapter 95, "Disease Management for Nurseries Using Recycling Irrigation Systems."

Plant Maintenance

Monitor plants frequently for leaf spot, and check roots for rot. Early detection of diseases can reduce losses. Monitoring also reduces pathogen movement and plant-to-plant spread. Diseased plants should be removed from the growing area as soon as possible. The disposal area should be located away from the growing area, storage area, potting area, and water source.

Initially, it may seem that the up-front costs of implementing a sanitation program are high, but in the long run, a sanitation program prevents the introduction of pathogens and reduces total losses from plant disease. These losses include not only discarded plants but also wasted labor, the costs of supplies used in growing diseased plants, and expenditures for pesticides applied to plants that cannot be sold.

CHAPTER 88

Ted E. Bilderback ▪ North Carolina State University, Raleigh

Ronald K. Jones ▪ North Carolina State University, Raleigh

Horticultural Practices for Reducing Disease Development

There are no "one-size-fits-all" recipes for growing ornamental plants in containers. Not all nursery crops thrive under the same cultural practices such as irrigation frequency, nutritional regimes, and container substrates. Factors such as too much or too little water, an excessively wet substrate or a rapidly draining substrate with little moisture retention, and too much fertilizer or low and unbalanced nutrient availability can create stresses that predispose crops to numerous diseases. Leaf discoloration, necrotic areas in foliage, and shoot and root dieback related to excesses or deficiencies of container moisture and nutrient levels can be mistaken for disease symptoms. In this chapter, horticultural management practices associated with the production of container-grown ornamentals are reviewed and conditions that can be confused with or favor disease development are discussed.

Substrates for Nursery Crops

Substrate is a general term for any root growth material for container-grown plants. It refers to the entire composition of the growth material, and thus use of this term avoids much of the confusion related to such terms as *potting mix* and *soilless medium.*

Many materials are used to prepare container substrates. The predominant potting components in nurseries of the central, eastern, and southeastern United States are pine bark, sand, and sphagnum peat moss. Fir bark is more economically used in the northwestern United States.

Alternative materials include composted hardwood bark, composted yard and animal wastes, composted cotton gin wastes, mushroom compost, municipal compost, rice hulls, peanut hulls, pecan shells, sawdust, wood chips, and shredded coconut husks (coir). The stability and chemical and physical characteristics of alternative materials may limit the amounts that can be used in a substrate. Unstabilized organic components decompose rapidly, leaving a container three-fourths full in a few weeks. Composted materials often lack the coarse, large particles necessary for adequate aeration and cannot be used in amounts greater than 50% of the volume for most container substrates. Animal wastes characteristically have high electrical conductivity (high levels of soluble salts) and nutrient levels, and therefore they are limited to 10–30% volume in potting substrates. The pH of substrates containing compost components is generally higher than that of pine bark or sphagnum peat moss substrates, even with the addition of dolomitic limestone.

Physical Properties of Substrates

The physical properties of container substrates include particle size distribution, total porosity, air space, water-holding capacity (container capacity), available water capacity and unavailable water content, and bulk density.

A balance between aeration and moisture content is necessary for an optimum substrate. Total porosity, bulk density, and unavailable water content are not affected by container height or volume. Air space and container capacity are affected by container dimensions, and the relationship of these values after drainage and between irrigations affects plant growth. A perched or saturated water table is created by reduced drainage in the bottom of the container. In a container, air space increases as height increases above the bottom of the container. Air space also increases as water is lost by drainage. The exact physical characteristics of a substrate are determined by the particle size distribution, particle density, nesting of substrate component particles, and the resultant bulk density of blended components.

There are no distinct standards for the physical properties of container substrates. However, normal ranges for nursery container substrates after irrigation and drainage are easiest to manage within the normal ranges shown in Table 88.1.

Total porosity is the sum of the air and water contents of the substrate; solid particles make up the remainder of the total volume. Predominantly organic substrate components, such as pine bark, often have total porosity values as high as 80–85%, characteristically higher than that of substrates to which mineral components, such as sand, have been added.

Air space and container capacity fluctuate in substrates, depending on moisture content and the height of the container. Fewer problems related to overwatering and waterlogging during production would be expected with substrates that have at least 15% air space.

The addition of sand or sphagnum peat moss to pine bark increases moisture retention and available water content but reduces air space and total porosity. The greatest liability is that air space may be reduced too much, requiring careful irrigation management to avoid waterlogging and anoxia of roots. Many growers add sand to pine bark for nursery substrates to increase the weight of the container to help prevent it from blowing over in the growing bed. However, another reason to add sand is that the infiltration rate of irrigation water is slowed down as it moves through the container profile. This promotes more thorough wetting of the substrate, compared with straight-course pine bark

Table 88.1 Physical properties of selected nursery container substrates[a]

Substrate	Total porosity (%)	Air space (%)	Container capacity (% volume)	Available water (%)	Unavailable water (%)	Bulk density (g/cm³)
Screened and aged pine bark (1/4-inch)	80.6	10.9	69.7	37.0	32.7	0.19
Screened and aged pine bark (1/2-inch)	83.7	18.5	65.2	33.3	31.9	0.19
Screened and aged pine bark (1/2-inch) + builder's sand (80:20)	76.6	10.9	65.7	41.1	24.6	0.45
Screened and aged pine bark (1/2-inch) + sphagnum peat moss (90:10)	78.6	10.4	68.2	36.4	31.8	0.19
Pine bark (3/8-inch) + mushroom compost + sphagnum peat moss (65:20:15)	83.3	23.7	59.6	33.6	26.0	0.18
Pine bark + rice hulls + peat + cardboard biosolids and hardwood fines (42:42:7:4.5)	85.3	15.4	69.9	48.4	21.5	0.29
Normal range	50.0–85.0	10.0–30.0	45.0–65.0	25.0–35.0	25.0–35.0	0.19–0.52.0

[a] All analyses were performed with standard soil-sampling cylinders (inside diameter, 7.6 cm; height, 7.6 cm). Air space and container capacity are affected by the height of the container.

particles, through which water can channel rapidly to the bottom of the container.

When potting materials of greatly different particle sizes, such as fresh pine bark and sand, are combined, the final volume is not additive; e.g., 1 cubic yard plus 1 cubic yard results in less than 2 cubic yards, perhaps 1.5 to 1.75 cubic yards. In this situation, a great increase in the bulk density of the substrate would be expected. An increase in bulk density results in lower total porosity and less air space.

The container capacity (water-holding capacity) of substrates increases when more small pores are created by adding fine particles to large particles. Particles less than 0.5 mm in diameter in pine bark alone create many of the pores that hold water. Fresh pine bark has very few of these fine particles, so growers add other components to replace the fines for better moisture retention. The moisture-holding characteristics of organic components are different from those of soils; however, water is held within and between particles and fibers as it would be in soil.

Irrigation and fertility must be closely managed to optimize air, water, and nutrient characteristics of the substrate to avoid leaching or salt buildup while providing optimum moisture conditions in containers. Rainfall must also be considered in selecting a substrate. In areas with heavy rainfall in summer, such as the southern United States, the substrate must drain much faster than a substrate in areas where rainfall is much lower, such as the Southwest. Substrates that remain excessively wet or retain excessively high levels of soluble salts greatly favor the development of root rot diseases.

Particle Size of Potting Substrates

The physical and chemical characteristics of pine and other bark substrates are affected by the age of the material and by handling prior to use as a potting component. Loblolly pine is the predominant species used for pine bark substrates in much of the southeastern United States. This bark is generally considered nonphytotoxic and can be used without aging or composting. However, some growers prefer aging or composting, since fresh pine bark often has few fine particles (less than 0.5 mm in diameter), constituting only 20–30% of the total volume (Table 88.2). Fine particles generally determine moisture retention. However, sand, sphagnum peat moss, composts, or other components with fine particles can be added to fresh pine bark to increase the moisture retention of the substrate. Aging produces a more stable material and allows breakdown of larger particles, as well as degradation of wood, cambium, and complex compounds associated with the turpentine-like smell of fresh pine bark. Aged pine bark is sometimes referred to as composted bark, although, unless pine bark is amended with a nitrogen source, moistened, and turned regularly, composting in the strict sense may not fully occur (Fig. 88.1). The stability of an organic material frequently depends on its carbon-to-nitrogen (C:N) ratio. For most organic components of a substrate, the C:N ratio should be at least 30:1; however, even composted pine bark may not have a ratio that low, since most of the pine bark is composed of lignin and not cellulose.

The stability and particle sizes of substrate components and the "fit" of two or more components in the mix determine the physical properties described in Table 88.1. Stable components are necessary, since continual decomposition changes the physical properties, usually decreasing air space and increasing moisture reten-

Fig. 88.1 Aged pine bark of uniform texture for potting plants.

Table 88.2 Particle size distribution of selected nursery container substrates

Sieve no.	Opening (mm)	Percentage of total volume of:[a]					
		1/4-in. PB	1/2-in. PB	PS	PBP	PB:MC:P	PBCRH
0.25	6.4	0.1	8.4	3.8	7.9	1.4	7.1
5	4.0	1.6	12.0	5.0	11.9	16.6	7.8
7	2.8	12.3	11.6	5.5	12.1	16.6	6.8
10	2.0	14.9	11.8	6.4	11.4	13.2	7.2
14	1.4	13.3	11.2	9.6	10.0	10.6	8.8
18	1.0	11.9	11.0	10.2	8.8	10.0	9.4
25	0.71	12.7	10.6	10.3	9.2	9.6	8.7
Total[b]		66.7	76.6	50.8	71.3	78.0	55.8
35	0.50	11.6	8.2	14.8	8.6	8.1	8.6
45	0.36	9.0	3.7	11.2	7.0	5.4	9.0
60	0.25	5.5	3.6	8.3	5.1	3.5	11.9
80	0.18	3.1	3.1	3.2	3.2	2.2	8.4
140	0.11	2.2	2.0	2.0	2.5	1.8	4.3
Pan	0.00	1.9	2.8	10.7	2.3	1.0	2.0
Total[c]		33.3	23.4	50.2	28.7	22.0	44.2

[a] PB = screened and aged pine bark. PS = 1/2-inch screened and aged pine bark + builder's sand (80:20). PBP = 1/2-inch screened and aged pine bark + sphagnum peat moss (90:10). PB:MC:P = 3/8-inch pine bark + mushroom compost + sphagnum peat moss (65:20:15). PBCRH = pine bark + rice hulls + peat + cardboard biosolids and hardwood fines (42:42:7:4.5)

[b] Seventy to 80% of the substrate was 1/40 to 1/2 inch in diameter.

[c] Twenty to 30% of the substrate was less than 1/40 inch in diameter.

tion, thus favoring root rot development. Nitrogen applied for plant growth may also be used by microbes involved in decomposition, resulting in reduced growth of plants in containers. The particle size distributions of potting substrates can be measured. The results generally complement data on physical properties but can be used to diagnose potential aeration and water retention difficulties if used for crop production (Table 88.2).

Handling Substrate Inventories

Good sanitation practices for substrate components at the nursery are extremely important. Storage, mixing, and handling areas should be located at higher elevation than growing areas to avoid contaminating stored substrates or newly potted materials with runoff from growing beds. Bulk potting materials, such as pine bark and sand, should be stored on concrete pads to prevent contamination by weed seeds and spores of pathogens from bare soil (Fig. 88.2). Bagged materials, such as sphagnum peat moss, and bundles of new containers should be stacked on pallets above any standing water and covered to reduce ultraviolet breakdown of the bags, which could allow contamination (Fig. 88.3). Care in storing and handling components, containers, and all materials used in potting new crops is a critical preplant step in nursery crop production that can prevent weeds, diseases, and other cultural problems later in production.

Procedures for handling bulk supplies before they are delivered to a nursery are also important. Poor inventorying practices of pine bark, hardwood bark, composts, or other organic components can be detrimental when they are used in a potting substrate. Bulk inventory piles of these materials should be monitored and turned when their temperature exceeds 150°F. The materials may also require remoistening, if portions of inventory piles dry out. If inventory piles of fine-particle materials (up to 5/8 inch in diameter) are stacked at heights greater than 8 to 10 feet or if equipment is driven onto inventory piles, air exchange in the piles can be greatly reduced. Inventory piles can heat to 180°F or higher, reaching combustion temperatures. Steam rising from a hot inventory pile indicates loss of moisture, leading to the creation of a dry band, usually 1.5 to 3 feet below the top of the pile. The dry substrate must be rewetted with a wetting agent before can it be used. If bulk inventory piles are not monitored, anaerobic pockets may develop; they characteristically have very low pH and high levels of soluble salts levels, which have phytotoxic effects on plants.

Bark suppliers use at least two methods of storing bark. If inventories are handled as coarse products prior to screening and

Fig. 88.2 Storage area organized for various potting materials. The materials are stored on a cement pad to prevent the introduction of plant pathogens and nematodes.

separation of nuggets, mininuggets, and mulch from the fine particles marketed as nursery potting bark, heating and anaerobic conditions usually do not occur. If nursery potting bark is stored after the separation of coarse-particle materials, more handling and monitoring are required because of the proportional increase in fine particles. In bulk nursery bark piles, dry bands may develop, containing less than 34% moisture content by weight; this material cannot be readily rewetted, even after potting (Fig. 88.4). Bark below a dry band may begin anaerobic decomposition. Electrical conductivity (a measure of the concentration of soluble salts, usually expressed in millimhos per centimeter, or mmho/cm) as high as 2.5 mmho/cm and pH below 3.5 have been reported in improperly stored pine bark. Use of such bark could cause the death of liner transplants. Other compounds from anaerobic pine bark may also be phytotoxic to plant roots.

Tree bark stored dry (containing less than 40% moisture by weight) may also develop high fungal populations, recognized by clouds of spores released when the material is disturbed. If such bark is used for potting, rapid growth of hydrophobic fungal mycelium may occur, making irrigation of plants in containers very difficult. Newly potted liners simply dry out, because no water is retained by the potting substrate. The roots of plants shifted up to larger containers do not grow into the dry, hydrophobic pine bark for several months. To prevent this problem, fine particles in the potting bark (usually 3/8 to 5/8 inch in diameter and finer) should be wet to 50% moisture (by weight) before being placed in an inventory pile. The pile should be turned and remoistened occasionally if the bark is aged.

To avoid problems related to the storage of bark, growers should watch as bulk potting materials are unloaded at the nursery (Fig. 88.5). If inventories are excessively hot and steamy or clouds of spores are observed, check the pH and electrical conductivity. If clouds of spores are noticeable, thoroughly soak the entire inventory. Consider not using the bark immediately if any of these traits are observed. Bark should be wet thoroughly and left one to two weeks before it is rechecked and used. If bulk inventories are not used within a few weeks and are left undisturbed for a period of more than one month at the nursery, check the moisture content, heat, pH, and electrical conductivity and look for spores before potting new crops. If these observations raise questions about usability, irrigate and turn the inventory, and check it after two weeks.

Fig. 88.3 Containers stored on a cement pad to prevent the introduction of plant pathogens and nematodes. The gravel drive in the foreground also protects the storage area from contamination with pathogens.

Chemical Characteristics of Substrates

Many elements are required to grow healthy plants. Nutrients in irrigation water, particularly recycled water, influence plant nutrition, so fertilizer applications should be adjusted on the basis of an analysis of irrigation water. Nutrient concentrations in the substrate solution are affected more by irrigation management than by the chemical characteristics of the substrate, in part because the cation exchange capacity (CEC), which is very important in determining nutrient availability in soils, plays a lesser role in determining nutrient availability in an organic substrate solution. A desirable CEC for organic potting substrates is 6–15 milliequivalents (meq) per 100 cm^3. Typical CEC values (in meq per 100 cm^3) for several container substrate components are as follows: aged pine bark, 10.6; sphagnum peat moss, 11.9; vermiculite, 4.9; and sand, 0.5. In comparison, clay soil may have a CEC of 63 meq per 100 cm^3.

Fig. 88.4 Improperly stored bark pile in which a dry band with white fungal growth has developed below the surface of the pile. Bark in the dry band is impervious to water, even after it is used to fill containers.

Fig. 88.5 Pine bark arriving at a nursery should be checked for consistency. Avoid immediate use of pine bark that is steaming or releases clouds of spores as it is unloaded.

Table 88.3 Desirable pH, electrical conductivity (EC), and levels of nutrients in the substrate solution obtained by the Virginia Tech extraction method from containers treated with liquid fertilizer or controlled-release fertilizer (CRF)[a, b]

	Solution only or CRF and solution	CRF only
pH	5.0–6.0	5.0–6.0
EC (dS/m, or mmho/cm)	0.5–1.0	0.2–0.5
Nitrate N (mg/L, or ppm)	50–100	15–25
Phosphorus (mg/L)	10–15	5–10
Potassium (mg/L)	30–50	10–20
Calcium (mg/L)	20–40	20–40
Magnesium (mg/L)	15–20	15–20
Manganese (mg/L)	0.3	0.3
Iron (mg/L)	0.5	0.5
Zinc (mg/L)	0.2	0.2
Copper (mg/L)	0.02	0.02
Boron (mg/L)	0.05	0.05

[a] Data from Yeager et al. (1997).

[b] Levels should not drop below these during periods of active growth.

Potting substrates usually have no appreciable anion exchange capacity, and therefore anions (negatively charged ions) such as nitrate nitrogen and phosphate are readily leached from containers and, because of poor nutrient use efficiency, have raised environmental concerns for the container plant production industry. Consequently, growers apply all essential nutrients required for plant growth during production, often in excess amounts.

Suggested concentrations of nutrients in container substrates are shown in Table 88.3. Test solutions are collected in several ways by analytical labs, researchers, and growers. The Virginia Tech extraction method (pour-through procedure) is used to sample substrates in containers in which plants are growing. A volume of water sufficient to displace water held in the container is slowly poured over the surface of the substrate, 30 minutes to 2 hours after irrigation, forcing a substrate solution out the drain holes of the container. The sample solution collected from the container is analyzed in order to make fertility adjustments to obtain a desirable concentration of nutrients (Table 88.3). Substrate solutions should be monitored and plotted on a regular schedule as part of the total monitoring and pest-scouting program.

Irrigation water, rainfall, the potting substrate, and the fertilizer regime are all interrelated. Not managing these integral parts of the same system can create problems during production.

Crop Problems May Not Have Obvious Causes

Steps necessary to correct problems many times seem indirect in addressing observed symptoms. Two examples are related to standard cultural practices and irrigation water quality.

Nurseries incorporate dolomitic limestone into potting substrates. However, to establish the need for limestone amendment and determine the application rate, the calcium and magnesium content and pH of the irrigation water must be taken into account. Application of excess dolomitic limestone may result in high pH in the substrate. The optimal range for organic potting substrates is pH 5.0 to 6.0. If a substrate is overlimed or if irrigation water with a high bicarbonate content is not acidified, the substrate pH can rise to 7.0 or higher, well above the optimal range. Iron is not as readily available at high pH, even in organic potting substrates. Furthermore, alkaline irrigation water can create high concentrations of calcium or magnesium, which reduce the availability of minor elements, such as iron, for root uptake and growth. High pH and excess calcium or magnesium cations result in chlorosis of new growth. A grower may perceive the problem as iron deficiency and apply an iron chelate foliar spray, but any improvement will be only temporary, since the chlorosis was triggered by high pH or excess calcium or magnesium. At this point the grower may be fooled by the lack of efficacy of the iron chelate and may implement measures for controlling root rot diseases. The problem will be resolved only when the irrigation water quality and lime application rate are corrected.

In other situations, the substrate pH may be too low. In some nurseries, the water supply is occasionally at very low pH (i.e., below pH 5.0). Low pH can develop over time from the use of controlled-release fertilizers formulated with nitrogen derived from urea. Below pH 5.0, a number of nutrient disorders can arise. In pine bark potting substrates, low pH often causes chlorosis of foliage and may cause necrotic spots to form on various parts of leaves, sometimes referred to as measles. Foliar analysis may indicate low levels of iron, calcium, and magnesium and a high level of manganese. At low pH, manganese becomes readily available in pine bark, and without adequate levels of other cations, excessive manganese is adsorbed by plant tissue. A grower may mistake the symptoms for those of a foliar disease and treat the plants with a fungicide, which will not solve the problem. Adjusting the pH in the substrate is necessary to correct the problem. Surface application of dolomitic limestone can effectively change the pH of porous pine bark substrate, but two to four weeks may be required for it to equilibrate to a desirable level. Further treatment may not be necessary, but application of an iron chelate spray may have a long-term effect if iron deficiency persists.

In addition to being an indicator of substrate nutritional problems, iron deficiency is frequently a symptom of root rot disease. Unhealthy root systems may have few white root tips, a sloughing root cortex, and a discolored stele (xylem). Large portions of the root system may appear dead, and the cortex feels slimy to the touch and can be pulled apart with little effort. These conditions are frequently associated with poor fertility, poor irrigation management, and root rot diseases, but they may not be evident immediately following stressful conditions. High electrical conductivity (corresponding to a high level of soluble salts), poor drainage, extended periods of waterlogging caused by frequent rainfall, high temperatures in the root zone, or extended periods of dryness in the container can all cause fibrous root death. Clinical tests of root samples may be positive for *Phytophthora* or other root rot fungi. Application of a fungicide drench may limit the development of disease, but the problem may occur repeatedly unless an accurate total diagnosis of the root damage can be determined.

Plant root systems in containers exposed to full sun during the summer are under great stress. Temperatures frequently exceed 130°F in the outer inch of root zone, in contact with the side of the container absorbing heat from the sun. Temperatures as high as 150°F have been recorded. Cultural practices must enhance life-sustaining functions, but if irrigation or nutrient-related imbalances occur, fibrous root death follows.

Irrigation Application Affects Plant Growth

Container nurseries need large quantities of high-quality water. Water is a major consideration for site selection and development of nurseries. Textbooks recommend that for container nurseries, 1 acre-inch (27,000 gallons) of water should be available daily for every acre of nursery stock. It is also recommended that a 30-day irrigation supply be developed for the nursery, based on a daily need of 1 acre-inch per acre. Lack of sufficient water can have devastating consequences for nurseries.

Dry container substrates cause massive fibrous root death as water deficits are created in the root cells. Water deficits in roots are followed by water deficits in transpiring leaf tissues. Rates of photosynthesis decrease with increasing water deficits, as a result of partial or complete stomatal closure; therefore, transpiration rates also decrease. The transpiration rate affects plant leaf temperature. The leaf temperature in well-watered, healthy plants is lower than that of the surrounding air. Other stresses, such as diseases that interfere with water movement in plants, also create the same conditions.

To identify water stress, leaf temperature, air temperature, relative humidity, and intensity of sunlight can be measured. Plant stress can be calculated from the measured environmental influences. Commercial solar radiation monitors can be purchased. Some units monitor plant canopy temperatures by pointing an infrared thermometer gun at leaf surfaces. More expensive equipment uses integrated sensors, including the infrared gun that records plant canopy temperatures, but also other sensors that record solar radiation (sunlight), relative humidity, and ambient air temperature to create an index reference value for plant stress. Quantifying plant stresses can be an innovative way for growers to schedule irrigation based upon water need. Plant health indexes have been developed for many agricultural crops, including corn, cotton, wheat, soybeans, turf, and many vegetables and tree fruits. Very limited research has been done with nursery crops, but the technology can be adapted.

Water Quality Affects Plant Growth

Irrigation water needs to be free of sediment and impurities such as bicarbonates, iron, or iron bacteria to avoid nutrient conflicts, coating of leaf surfaces, and clogging of irrigation systems.

Removing sediment from irrigation water requires filtration. Screen and disk filters are adequate for overhead impact sprinklers and nozzles with large orifices. Sand media filters or combinations of screen and spin filters are usually required for propagation and low-volume drip and spray nozzles with small orifices. Well water generally has less sediment and requires less filtration than surface irrigation supplies. If sediment is a problem in surface water supplies, growers should always first check the irrigation intake and make sure it is at least 4 feet from the bottom of the irrigation basin. Many growers attach floats to the intake line to ensure that the intake is well above the bottom of the irrigation basin.

Bicarbonates, iron, and iron bacteria can be problems in either well water or surface water. These impurities can coat leaf surfaces, reduce growth, reduce the health of nursery crops, and reduce salability.

Sodium bicarbonate and calcium bicarbonate can coat leaf surfaces with a white precipitate, reducing the amount of light available for photosynthesis and greatly reducing the aesthetic appearance of crops. Even low concentrations of bicarbonates in propagation can cause coating, since the mist comes on frequently to reduce water loss from cuttings. Heavy bicarbonate coating can reduce rooting success. Injecting an acid, such as 35% sulfuric acid, into the water supply lowers pH and reduces problems with the precipitate. Water quality tests from commercial labs are usually accompanied by a recommendation and rate of acid to inject.

Iron in irrigation water supplies produces a red brown coating on plants, irrigation risers, and all surfaces in the growing area. It also plugs the irrigation equipment. Coating can occur when iron is present in the water at a concentration of less than 1 ppm (part per million). Sequestering agents can be injected into irrigation water containing iron to chelate the iron. Chlorine oxidizes iron and prevents the coating of foliage and the plugging of nozzles.

Iron bacteria, found in well water and in surface water, can live on the iron in irrigation water. The bacteria cause a shiny, blue green coating on leaf surfaces, which blocks light and reduces the aesthetic appearance of nursery plants. A jelly-like mass may also be noted plugging irrigation nozzles. This mass is iron bacteria. In still water, the bacteria appear as an oily film that floats on the surface. Aerators in surface water supplies produce ripples, which may aid in pushing iron bacteria away from the intake area. If this is not effective, chlorination or injection of a bactericide may be required to eliminate the problem.

Improving Irrigation Efficiency

Overhead sprinkler irrigation systems are very inefficient. As much as 85% of the water applied falls outside containers that are spaced out. Most plants in small containers are irrigated by overhead sprinklers (Fig. 88.6), because individual irrigation of small containers is not considered economically feasible. However, the efficiency of overhead sprinkler irrigation can be improved. Containers in growing blocks are spaced so that the plant canopies do not touch and do not shade foliage of adjacent plants (Fig. 88.7).

Fig. 88.6 Overhead sprinkler irrigation. This system is efficient for plants in small containers but loses efficiency as the plants grow and require larger containers.

Shifting plants before they require wide spacing also improves irrigation efficiency. Obviously, there must be incentives other than improving irrigation efficiency unless the water supply is limited or environmental mandates require reduction of effluent volumes.

When small plants are placed in large containers, an insulating band of substrate medium can moderate the temperature of the roots. The insulation effect works in both hot and cold weather. Moderated temperatures in the root zone may allow plants to grow faster.

One of the most practical steps in improving irrigation efficiency is cycled irrigation, in which water is applied in several short cycles, rather than in one 30-minute or 1-hour interval for each growing block. After a short application, an automated controller turns on the water in another growing block, and blocks are successively irrigated throughout the nursery before the first block is irrigated again. Between irrigation cycles, water that has entered a pot forms a wetting front after the irrigation is shut off. The wetting front may initially be at a depth of 3 or 4 inches in the container. Water then moves laterally above and along the wetting front and has time to seep into organic particles and move between and over the surface of substrate particles, so that thorough wetting of the substrate occurs. When irrigation begins again and more water enters the container, the wetting front is pushed deeper into the pot, and when the water is shut off again, the capillary movement of water between substrate particles occurs again. Ideally, the last cycle pushes the wetting front just to the bottom of the container, with minimal leaching.

Cycled irrigation improves irrigation efficiency in three ways. First, it achieves better wetting of the substrate. Second, irrigation applied in a single interval tends to push water all the way through the pot, with little lateral movement of the water and considerable channeling, and dry areas can often be observed in recently irrigated containers. Third, cycled irrigation requires up to 25% less water than irrigation applied in a single interval. Cycled irrigation is stopped when water begins to leave the pot on the last irrigation cycle. With thorough wetting and limited leaching of the substrate, water supplies are conserved, less runoff from containers occurs, less pump time and less electricity are consumed, and less fertilizer runs out of the pot.

Fig. 88.7 Proper spacing of containers on a growing pad to maximize the efficiency of irrigation. Containers are spaced so that the plant canopies do not touch and shade the foliage of adjacent plants.

Monitoring the System

From these examples, it is evident that in production of plants in containers, a grower is managing a complex system, not just the fertilizers applied, the potting substrate used, and how much or how frequently plants are irrigated. The question then becomes, How do I manage this system?

The interface of irrigation, fertilizers, and substrates is the liquid solution held in the container after irrigation. This container solution holds nutrients in a form ready to be absorbed by roots. In some cases, rainfall or irrigation can leach the solution and nutrients out of the container. A tool for the management of the system is routine measurement of the container solution, which serves as a criterion for irrigation scheduling. With experience, by monitoring container solutions, growers can optimize the physical and chemical properties of the substrate and determine the need for fertilizer application and the rate of application. Routine monitoring can be as simple as checking the pH and electrical conductivity (EC) of the container solution, which should be part of the routine scouting for disease and insect problems.

A pH meter or pH pen and a solubridge (EC meter) or conductivity pen are used to monitor pH and soluble salts (Fig. 88.8). Many horticulture and nursery supply companies carry this equipment. Some devices are designed to measure both pH and EC.

Checking the pH and EC of substrates in containers does not have to be time-consuming, complicated, or difficult. The Virginia Tech extraction method, mentioned above, can be performed quickly and is not disruptive (Fig. 88.9). Approximately 8 to 16 fluid ounces (1/4 to 1/2 cup) of water is poured on the surface of the substrate in a 1-gallon container, or 16 to 32 fluid ounces is poured on the substrate in a 3-gallon container, 30 minutes to 2 hours after irrigation, and the leachate is collected from the bottom of the pot. The pour-through step can be omitted by simply collecting water that drains from a pot when it is lifted and allowed to drain, approximately 30 minutes to 1 hour after irrigation. In either procedure, EC and pH can be checked immediately in the growing area.

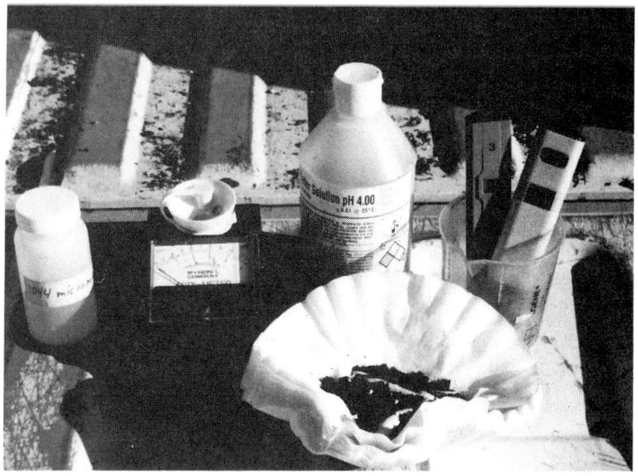

Fig. 88.8 Equipment for measuring pH and electrical conductivity of a plant growth substrate: pH meter (second from left) and pH and conductivity pens (right). The meter and pens are standardized with a pH buffer (center) and conductivity standard (far left) prior to use. Filter paper (foreground) is used to screen coarse mix solids from liquid prior to measuring pH or electrical conductivity.

EC is adequate for growth if it is between 0.2 and 0.5 mmho/cm greater than that of the irrigation water. An EC value of 2.0 mmho/cm is extremely high and indicates a need for further monitoring and good irrigation management. Some conductivity pens read values expressed in micromhos per centimeter (1,000 times the number of millimhos per centimeter), from 20 to 200 μmho/cm. Target readings are normally higher in the middle of the growing season than in the spring or fall. If controlled-release fertilizers are applied to growing crops, EC may initially be at the high end of the suggested range, then decrease, then gradually increase in the middle of the growing season, and then taper off toward fall. If growers are concerned about low EC, the leachate can be analyzed by a laboratory.

Leachates collected from pots by the Virginia Tech extraction method can be sent to plant and soil analytical laboratories for nutrient analysis, just as nutrient solutions are. Occasional analysis of leachates is a good practice, even though EC may appear to be within recommended limits, because plant nutrients or nutrient balances may not be adequate. It is particularly important to analyze leachates when irrigation water is recycled, because many of the soluble salts in the water may be sodium, sulfates, and chlorides, rather than nutrients. If this is the case, essential nutrients must be reapplied.

Fig. 88.9 Virginia Tech extraction method. Water is poured on the surface of the substrate in a container, and water that drains from the container is collected in a plastic tray under the plant for 30–60 minutes.

Irrigation uniformity can be checked by measuring pots in a diagonal line between risers. If a large variation in EC or pH is discovered across a growing bed, it is a signal that the nozzle orifices should be checked for wear or that something needs to be done to correct the distribution of the irrigation water. Irrigation distribution can be investigated further by placing cans or trays across the bed and measuring the water collected after an irrigation cycle. Generally, if irrigation is not uniform, the pots in the center of a bed receive too much water when enough is applied to adequately irrigate the pots on the edge of the bed. Depending on the irrigation design, however, the reverse could also be true. Checking EC and pH is a diagnostic tool, because more nutrients are leached from pots getting too much water than from pots getting less water. Close observation may reveal that plants getting too much water are smaller or less vigorous or are the first in the block to become diseased. Thus, irrigation can enhance stress and disease related to excessive dryness or wetness of the substrate.

Pruning Nursery Crops

Nursery crops are marketed by size, either by height, spread, stem caliper, or container size. The objective is to grow healthy plants as quickly as is economically feasible. Although pruning is a subtractive process and actually reduces growth, it is necessary to prune nearly all nursery crops to create forms that are aesthetically pleasing to consumers and in some cases to meet industry quality standards. Most shrubbery in nurseries is pruned by shearing (Fig. 88.10). All growing tips are removed in this process. Each time a terminal is removed, several new shoots develop from the point where the terminal was cut off, producing a compact plant. Broadleaf evergreen crops and conifers sold for screening in landscapes may be sheared two to four times through a growing season at a nursery. Shearing produces many small terminal wounds, which can be points of entry for pathogens. Some nurseries routinely apply preventative fungicide sprays before or after shearing crops. Other nurseries rely on keeping plants growing

Fig. 88.10 Movable framework for transporting a rotary mower down a container block, to shear plants to a uniform size.

vigorously to overcome disease pressure. Small wounds created by shearing generally heal quickly, with little or no infection by dieback pathogens.

Shade trees are pruned by selective pruning techniques. Crossing branches, competing branches growing in the same direction and space, or upright lateral branches competing with the leader are selectively removed at the branch collar of the main stem or at another lateral branch collar. Shade trees frequently require directive pruning to reduce competition with a leader during the summer and selective branch pruning while they are dormant. Small wounds generally close rapidly and do not require disease prevention measures, if pruning is performed properly. Trees should not be sheared (topped), since natural defense mechanisms are associated with branch collars, where one branch forms from another. Decay, dieback, and invasion by pathogens occur when shade trees are improperly pruned between nodes. Pruning can reduce disease by removing diseased plant parts and opening up the plant canopy for better drying and spray penetration.

Diseases for Which Pruning Is an Effective Management Strategy

Fire blight is a bacterial dieback disease of plants in many genera in the pome fruit group of the family Rosaceae. Blighted shoots must be pruned out after symptoms are observed. Pruning removes symptomatic blighted twigs and sources of bacteria for new infections the following spring. Blighted twigs should be pruned 12–18 inches beyond the blighted area on the stem. Pruning tools should be sterilized between cuts to reduce the spread of the bacteria. This is the only effective control measure once infection has occurred; fire blight can be a difficult disease to control in susceptible cultivars in nurseries.

Galls and cankers on stems are local infections often caused by fungi. A gall is a swollen or enlarged area on a stem, branch, or trunk. A canker is a sunken area on a stem or trunk. The pathogen is generally found in galls or cankers and is often limited to the symptomatic area. The fungus may sporulate on the gall or canker surface, creating inoculum to infect additional stems or plants. Galls and cankers can be pruned out any time of the year. Pruning removes the unsightly galls or cankers and eliminates sources of fungal spores that lead to new infections. Fungicide applications generally have no effect on established galls or cankers, but they may have some slight efficiency in preventing new infections if the time of infection is known. Pruning has little or no effect on crown gall and may even increase the incidence of the disease by creating a fresh wound.

Shoots or branches exhibiting dieback can also be pruned out. Dieback pathogens often survive in dead tissue, and the final cut must be made well below all discolored tissue. These pathogens or other secondary invaders are generally in the discolored tissue, so pruning tools should be disinfected between cuts.

Plant Spacing

Plants, either container-grown or field stock, should be set far enough apart for good air movement around the plant canopy, to promote faster drying in the morning and after irrigation. Proper spacing reduces the length of time during which foliage and stems are wet. Many fungal and bacterial pathogens need free water on plant surfaces to complete the infection process. There is a practical limit to how far apart a grower can space plants for disease control and still make economical use of the space. Finding a balance between spacing for disease control and economical use of space requires an educated judgment by the grower.

Web blight, caused by the fungus *Rhizoctonia solani,* is an example of a disease for which spacing is an effective management strategy. Container-grown plants can be set pot-to-pot in the early spring, but as the plant canopy increases in width, the spacing should be increased for improved air movement, to allow faster drying of stems and leaves. The fungus grows on the surface of the stems and is very sensitive to desiccation.

Propagation

The frequency and amount of mist applied during the propagation of cuttings is critical for the rooting process and disease development. Enough water, both in frequency and in volume, must be applied to prevent the desiccation of cuttings. This must be balanced with factors such as temperature, evaporation rate, medium drainage, and succulence of cuttings. As cuttings root, the frequency of misting and the volume of water used should be reduced. Too much mist, both in frequency and in volume, increases the likelihood of web blight, Pythium root rot, and Phytophthora root rot. As soon as cuttings are well rooted, they should be hardened off and moved out of the mist area. Leaving rooted cuttings in the mist bed too long increases the likelihood of disease, allows nutrients to be leached out of the cuttings, and causes liners to be pot-bound and of low quality.

The most difficult situation to manage in mist propagation occurs when numerous mixed species are stuck at various times in available space on the same mist unit. In this situation, it is impossible to provide the ideal conditions for each individual species. The number of cuttings stuck should match the nursery needs, so that rooted cuttings can be potted as soon as they are adequately rooted.

If possible, well water should be used in mist propagation. Pond water and surface water can carry plant pathogens, and the use of recycled water can introduce plant pathogens carried in the water. For best rooting, water for propagation should be of low salinity.

Starting with well-rooted, vigorous, disease-free liners pays efficiency dividends throughout the entire production life of the plant.

It is not as common in the industry today for nurseries to take cuttings from well-managed stock blocks, except those for rootstocks in rose production. It is more common today to take cuttings from production plants or landscape plants. The key is to take cuttings from vigorous, disease-free plants of known identity. High-quality cuttings root faster and can be moved out of the mist sooner. Cuttings should be kept off the ground and stored in clean containers until stuck. Leaf spots, web blight, and diseases caused by *Botrytis* and *Fusarium* spp. are examples of diseases that can be introduced into a propagation area on cuttings. If cuttings are dropped on the ground, pathogens such as *Pythium, Rhizoctonia, Fusarium,* and *Thielaviopsis* spp. can be introduced to the propagation area on the infested cuttings.

Rooting cuttings of woody ornamental plants is a complicated process that requires knowledge of a number of different factors, including mist cycles and rooting hormones. Each plant species has somewhat different rooting requirements for fast production of high-quality liners. Most woody ornamental nurseries grow numerous species and try to root all species under one procedure. This can lead to low-quality liners. More problems in propagation result from poor horticultural practices than from disease. One of the most difficult pathogens to control in propagation is *Cylindrocladium*. This fungus is a serious pathogen in the propagation of azalea cuttings and occasionally other species.

If extensive problems develop in propagation, first carefully review the horticultural procedures. Observe patterns of dead cuttings in the beds. Observe mist coverage, and check for clogged nozzles and wind blowing mist to one side of the bed. Volume and frequency of mist must be regulated carefully. The mist prevents desiccation of the cuttings, but applying too much mist keeps the medium too wet (lowering oxygen), reduces rooting, and favors the water mold pathogens such as *Pythium* spp.

Sanitation is critical in the propagation area (see Chapter 87, "Sanitation: Plant Health from Start to Finish"). If disease is a problem in the propagation area, it needs to be identified, and any horticultural problems need to be ruled out. It is very important that disease problems be avoided in the propagation area by cultural management rather than by trying to cure or prevent them with fungicides during rooting. It may be necessary to replace the medium or fumigate the medium, flats, and containers after each crop is removed. If pathogens are being introduced on cuttings, even on a small number, it is much more effective to clean up the source plants before the cuttings are taken than to apply fungicides to cuttings in mist beds.

Many nurseries have to make critical decisions about propagating versus buying liners. If you cannot propagate good-quality, disease-free liners, it is probably better to purchase them. Propagating low-quality, diseased liners with low vigor may be a hidden factor lowering the overall efficiency of a nursery. Acquisition of good-quality liners that grow into a high percentage of salable plants makes the "high cost" of liners a minor factor.

Production Time

The longer plants remain in the nursery or in any one phase of production, e.g., propagation, 1-gallon phase, or 2-gallon phase, the greater the likelihood that a disease or nutritional problem will develop or that plants will stagnate, become root-bound, or simply lose quality. Cuttings should be taken at the optimal time for fast rooting and moved to larger containers as soon as they are rooted. Slow turnover of plant material, either by failure to repot or failure to sell plants at the optimal time, results in low-quality plants. Pot-bound plants or plants with extensive root growth out of the drain holes in the pots are low-quality plants likely to have disease problems.

Plants should be grouped by similar water requirements, nutrition needs, or common disease problems for easy scouting. Too many nurseries set newly potted plants in any available bed space, mixing species with different cultural requirements and mixing plants in pots of different sizes, also with different cultural requirements. This leads to an averaging of horticultural practices: all species and plants in pots of all sizes get the same fertilizer, the same irrigation schedule, the same herbicide, and the same fungicide, whether they need it or not. The result is low-quality plants that are likely to develop diseases and require excess pesticide applications.

Ground Cloth

Plastic ground cloth is used on the soil surface for weed control in container nurseries. Plastic ground cloth can also act as a barrier to keep soilborne pathogens from entering the drain holes at the bottom of the pot. Black plastic has been used for this purpose for years but should not be used without proper grading, since it can effectively assist pathogens in moving from container to container. When water stands or puddles on the black plastic after rain or irrigation, this water bridge has been shown to aid in the movement of *Phytophthora cinnamomi* from pot to pot. Water puddling on the plastic ground cover around the bottom of the pot can also inhibit the drainage of excess water out of the pot. Keeping the substrate too wet also favors the development of Phytophthora root rot.

Woven black plastic ground covers through which water is allowed to drain are much better than solid black plastic. Woven ground covering provides the necessary weed control without holding water, in which pathogens can spread between pots. Many nurseries put gravel on top of woven plastic ground cloth to provide additional assurance that the bottom of the pot does not sit in water, even after a heavy rain.

Nursery Layout and Design

Nursery layout and design can play a part in the cultural disease management program. Container beds must be crowned so that water can quickly run off the container bed and away from the bottoms of the pots. Flat beds, even with woven plastic ground cloth, hold standing water if it rains when the ground is saturated. A drainage system must be designed for the entire nursery to get water out of the drive roads and off the production beds. Drive roads should be covered with gravel to prevent the formation of mud holes, from which tractors, trucks, and wagons could splash pathogen-infested water and dirt into pots.

Finding and Creating Microclimates in the Nursery

Most nurseries grow rhododendron, hosta, dwarf nandina, aucuba, pachysandra, some azalea and pieris cultivars, and many other crops under the natural shade of pine trees or beneath polypropylene or lath shade structures (Fig. 88.11). Synthetic polypropylene shade fabrics make it possible to provide various levels of shade. Excessive light intensity exceeds the photoreactive energy and cellular temperature levels of light-sensitive crops and may cause scorching of leaves. Recent trends have been to use 30 or 47% shade rather than 60% or more shade, since more shade than necessary can reduce photosynthesis and lower the growth rate of crops.

During recent years, observant nursery professionals have recognized the economic advantages of reducing environmentally stressful conditions by locating or creating microclimates for specific crops within the nursery. Crops such as the cherrylaurel cultivar *Prunus laurocerasus* 'Otto Luyken' are quite desirable in many urban landscapes and perform well as landscape plants. As a nursery crop, Otto Luyken is easily stressed, grows slowly, and has significant problems with shot hole bacteria (a *Xanthomonas*

Fig. 88.11 Polypropylene shade cloth supported on a framework above sun-intolerant nursery stock.

Fig. 88.12 Cherrylaurel cultivar Otto Luyken with typical shot holes caused by the bacterial pathogen *Xanthomonas*.

sp. in the eastern United States and a *Pseudomonas* sp. in the western United States), which cause black holes with necrotic edges to form in leaves (Fig. 88.12). Severe infection results in defoliation. The disease seems to be enhanced by wetting the foliage with overhead irrigation and growing plants in full sun. The disease is often misdiagnosed, so fungicides may be applied on a seven- to 10-day cycle during the growing season, with limited success. The crop is also difficult to place in irrigation zones with other nursery crops, because best growth is achieved when irrigation frequency is reduced to two or three times per week. In order to produce a crop, some nurseries have abandoned container production and grow Otto Luyken as a field crop; this practice tends to restrict sales to spring, fall, and winter. Otto Luyken is always in short supply, and quality plants can be sold at a very profitable margin. Successful producers of high-quality Otto Luyken gain new customers and truly are envied by other growers. The key to success has been to reduce heat, light, and water stress. One successful nursery created a production area under natural pine shade. A double-pot system with spray stakes placed below the canopy to avoid wetting the foliage was developed, and the tops of the containers were mulched with 3 inches of pine straw. With natural

Fig. 88.13 Production area shaded by a tree border oriented to the southwest. Sun-intolerant plants receive morning sun but are shaded from the stressful afternoon sun by the forest trees.

shade, double pots, and moisture-conserving mulch, containers are irrigated infrequently.

Success stories lead to new experiments with cultural practices at many nurseries. Some crops, such as variegated conifers (for example, *Chamaecyparis pisifera* 'Filifera Aurea') and many variegated junipers and variegated broadleaf ornamentals, grow well in full sun during most of the year, but the very hot and bright days in midsummer cause browning and discoloration when heat and light energy exceed their photosynthetic and respiratory capacities. Placing such crops in microclimates located in the nursery where they receive full sun during the morning but are protected by natural shade in the afternoon has provided good production conditions to maintain high quality in these crops (Fig. 88.13).

Other growers have reported little or no tip dieback in cryptomeria when plants are irrigated with a drip system rather than an overhead sprinkler system. This disease, of unknown etiology, is the primary limiting factor in the production of this plant.

Summary

Production of nursery crops is a system, composed of many jobs and activities, all which are interrelated. Cultural practices such as irrigation, substrate characteristics, fertility management, and pruning all are related to pest management. Healthy plants may require less application of chemicals for crop protection, but failure to manage any cultural practices correctly is likely to create a need for preventing or controlling consequent disease or insect problems. Scouting is frequently confined to looking for diseased plants, weeds, and insect or mite infestations, but this practice can easily be expanded to include checking and recording electrical conductivity and pH, in order to monitor irrigation and fertility levels.

A new concept in the production of container-grown ornamentals is to look for alternative production practices that offset stressful environmental conditions. This can increase profitability by reducing the need for crop protection chemicals, and the resulting high-quality crops are the best marketing approach for attracting new customers.

REFERENCES

Bailey, D., Bilderback, T., and Bir, D. 1997. Water considerations for container production of plants. On-line publication HIL 557. Department of Horticultural Science, North Carolina State University, Raleigh. http://www.ces.ncsu.edu/depts/hort/hil/flowers-index.html

Bilderback, T. E., Bir, R. E., and Midcap, J. T. 1994. Managing drought on nursery crops. N. C. Coop. Ext. Serv. AG-519-6. On-line publication DRO-18. http://www.ces.ncsu.edu/drought/dro-18.html

Tyler, H. H., Warren, S. L., and Bilderback, T. E. 1996. Cyclic irrigation increases irrigation application efficiency and decreases ammonium losses. J. Environ. Hortic. 14:194–198.

Yeager, T., Gilliam, C., Bilderback, T., Fare, D., Niemiera, A., and Tilt, K. 1997. Best Management Practices: Guide for Producing Container-Grown Plants. Southern Nurserymen's Association, Marietta, Ga.

CHAPTER
89

Ronald K. Jones ▪ North Carolina State University, Raleigh

Control of Fungal Diseases

Fungi cause far more disease in woody ornamental plants than any other group of pathogens. One or more fungal diseases commonly occur in most woody ornamentals. These diseases develop under a wide range of environmental conditions and diverse geographical areas.

Early detection is the first step in effective management of fungal diseases. Accurate diagnosis is the second step, and the most important. Only after an accurate diagnosis is made can control practices be selected and implemented in an integrated disease management strategy (Table 89.1). With a very specific diagnosis, information about sources of inoculum, how the pathogen is dispersed, when infection occurs, and environmental conditions that favor the pathogen is critical to development of a specific, highly effective disease management program. This information will allow growers to adjust management practices and reduce the severity of fungal diseases.

Sanitation

The sanitation program is designed to reduce the amount of inoculum in the nursery by removing dead and dying plants or plant parts and to reduce the carryover of inoculum from one crop to the next on infested pots and flats. Reducing the quantity of inoculum and the spread of inoculum by human activity is a critical part of a fungal disease management program. These practices are very effective against many fungal diseases.

A number of fungal pathogens, such as *Sclerotium rolfsii, Phytophthora cinnamomi, Rhizoctonia solani,* and *Thielaviopsis basicola,* are either not airborne or have very limited means of movement on air currents into and within the nursery. Therefore, improper nursery practices are often the primary means by which these pathogens are introduced into and spread in the nursery. If the movement of the pathogen is reduced or stopped, the distribution and incidence of disease will be reduced. A good sanitation program is highly effective in reducing this type of pathogen movement.

Cultural Control

Proper cultural practices can reduce sources of inoculum when diseased plant parts are pruned out and destroyed. Growers must provide optimum conditions for growth and development of the plants but avoid conditions favoring pathogens. Most fungal pathogens penetrate plants only when the surface of the plant is wet, and changing the timing of overhead irrigation can reduce this invasion. Weak fungal pathogens are able to penetrate the plant only through wounds or injuries, thus eliminating injuries will reduce infection.

Most root rots are caused by fungal pathogens, and improper cultural practices can have a big influence on the frequency and severity of these diseases. Correcting improper cultural practices will have a big impact on diseases in the nursery. Growers must select a growing medium with good drainage and porosity and adopt an irrigation schedule to minimize the length and frequency of saturation of the medium, which favors water mold fungi, such as *Pythium* and *Phytophthora.* Overfertilization, which results in root injury, increases root rot diseases.

Host Resistance

The production of disease-resistant plants is the most effective and the most cost-effective disease control strategy available in the ornamental industry. Disease losses are reduced in nursery production, and plants remain healthy for decades in the landscape. Wherever acceptable resistant cultivars are available, they should be grown and promoted, whereas susceptible cultivars should be eliminated. A heavy emphasis has been placed on host resistance in this book.

Exclusion

Exclusion of pathogens from the nursery is highly effective in a fungal disease management program. This method is most effective for the control of fungi that are not airborne and have a limited host range. Pathogen exclusion from the nursery is a very cost-effective practice. Regulatory inspection and purchase of disease-free plants and plant parts are important aspects of a total disease management program for fungal diseases.

Fungicides

Fungicides are often a necessary part of a fungal disease management program. However, even the most effective fungicide will often provide only partial control without the disease suppression benefits of sanitation, cultural practices, exclusion, and host resis-

Table 89.1 General strategies for fungal disease management[a]

Disease or pathogen	Sanitation	Resistance	Cultural practices	Exclusion	Fungicide treatment
Blight	++	++	+	–	+++
Botrytis	+++	+	+++	+	+++
Canker	+	–	+++	+++	+
Collar rot	+++	–	++	+	++
Cylindrocladium	+++	+	+++	++	+++
Dieback	++	+	+++	+	–
Downy mildew	+	+	+	+++	+++
Fusarium	+++	+	+++	++	++
Gall	–	–	+++	+++	–
Leaf spot					
Severe[b]	++	+	++	+++	+++
Minor[c]	++	++	++	++	+
Phytophthora	+++	++	+++	+++	+++
Powdery mildew	–	+++	+	+	+++
Pythium	+++	+	+++	+	+
Rhizoctonia	++	+	+++	+	+++
Rust	+++	+++	++	++	+++
Sclerotium	+++	–	+	+++	+
Thielaviopsis	+++	+++	+++	++	++
Web blight	++	++	+++	+	++
Wilt	++	++	++	+++	–

[a]– = Not important or not available. + = Minor importance. ++ = Moderate importance. +++ = Critical importance.

[b]Diseases that cause severe damage, such as black spot of rose and Entomosporium leaf spot of red-tip photinia.

[c]Diseases that cause little or no defoliation or damage to the host.

tance. Fungicides are much less effective when the level of inoculum is high, the host is highly susceptible, and the host is growing under less than optimum environmental conditions.

Conclusion

Growers must first detect and correctly identify the fungus before selecting the right management practices to provide effective control of the disease (Table 89.1). Fungicides are the essential control tool for many fungal diseases, such as powdery mildew, downy mildew, serious leaf spots, and rusts, where resistance is inadequate. Other fungal diseases, such as root rots, can be effectively controlled only by a total integrated program of strict sanitation, careful attention to cultural practices, and fungicide treatment. Still other fungal diseases, such as galls, cankers, diebacks, and wilts, are managed with cultural practices, since sanitation and fungicides are ineffective.

CHAPTER

90

D. Michael Benson ▪ North Carolina State University, Raleigh

Gary W. Simone ▪ University of Florida, Gainesville

Control of Bacterial Diseases

Woody ornamentals are susceptible to just a few bacterial diseases, such as fire blight, crown gall, Pseudomonas blight, shot hole, and soft rots, which occur only in a few ornamental crops, including roses, crabapple, and *Prunus* spp. Therefore, growers may not be as prepared to control bacterial diseases when they develop. For most bacterial diseases, once symptoms are observed, the damage is done, and it is very difficult to stop the epidemic. Usually, only dry, hot weather or a lack of susceptible host tissue can stop an epidemic. Control of bacterial diseases of woody ornamentals involves the same general principles of integrated disease management as control of other diseases, involving cultural practices, sanitation, host resistance, and treatment with bactericides and biocontrol agents.

The design of an appropriate integrated disease management plan for a specific bacterial pathogen is linked closely with the knowledge of inoculum sources and pathogen dispersal. Bacteria infecting woody ornamentals are generally dispersed from lesions, galls, and cankers in splashing water, in and on propagative tissues, on tools and knives, and by insect vectors. More information on bacterial dispersal can be found in Chapter 10 ("Crown Gall"), Chapter 16 ("Diseases Caused by *Pseudomonas syringae*"), and chapters on diseases of specific crops.

Cultural Practices

Cultural practices include the use of pathogen-free planting materials (usually cuttings from woody ornamentals), routine scouting for symptoms, and, most importantly, roguing (removal) of diseased plants as soon as they are found. Soft rot bacteria on cuttings can be controlled during propagation by properly managing misting systems to avoid overwatering and by providing adequate ventilation in the propagation area. Water drawn from wells is normally pathogen-free, while water recycled from retention basins may be contaminated with pathogens.

In container and field plantings, elimination of overhead irrigation and exposure to rainfall are key controls for bacterial diseases such as shot hole of cherrylaurel, caused predominantly by *Xanthomonas campestris* pv. *pruni* in the eastern United States and *Pseudomonas syringae* pv. *syringae* in the western United States.

Rose growers may move to production fields free of inoculum of *Agrobacterium tumefaciens* to avoid crown gall. The pathogen may survive systemically in cuttings of some plants, however, so clean stock is a must.

Growers of flowering crabapple may prune out fire blight infections in small trees to slow the development or reduce the severity of the disease. Crops susceptible to fire blight should also be properly fertilized to avoid the production of succulent shoot growth, which is more susceptible to infection than older growth.

Growers producing crops susceptible to *Pseudomonas syringae* can reduce the incidence and severity of blight caused by this pathogen by avoiding exposure to frost, high fertility, and early-season pruning, which predispose plants to blight.

Host Resistance

One of the best ways to manage bacterial diseases in woody ornamentals is to produce resistant cultivars. Many plant pathologists around the United States have been involved in evaluating cultivars of ornamentals, including *Prunus* spp., flowering crabapple, and Indian hawthorn, for resistance to Pseudomonas blight and fire blight. The production of cultivars with resistance to bacterial pathogens as well as other pathogens makes good sense for the nursery and provides the consumer with plant material that will perform well in the landscape. Lists of cultivars with resistance to bacterial diseases can be found in the chapters on diseases of specific crops.

Bactericides

Few bactericides are available to nurseries, and they are approved only for narrowly defined uses. The limited choice of chemical treatments points to the need to adopt an integrated approach to bacterial disease management. Growers should focus on plant species that are vulnerable to bacterial diseases each production year. The overall production cycle of these crops should be reviewed to best protect the plants and exclude, eradicate, avoid, and resist bacterial pathogens.

Target plants should be placed on a bactericide spray program six to eight weeks prior to propagation cycles, to minimize the survival of epiphytic bacteria on the stock as well as the possible introduction of these pathogens into the propagation cycle. Propagate only from the highest-quality stock. Particularly susceptible plant species may benefit from a rigorous protocol of tool disinfestation during propagation, a cutting dip in a product containing alkyl ammonium chloride, and injection of a bromine-based product in mist irrigation lines.

Liners of susceptible plant species being moved into field beds should be placed on daytime water cycles, where possible. Irrigation should otherwise occur after dew formation, to avoid creating conditions favorable for bacterial disease development. Disease outbreaks must be handled with available protectant bactericides to slow disease progress. Before reaching for bactericides, define the pathogen etiology – make sure the pathogen is a bacterium, and identify its genus. Identification of the pathogen will allow the appropriate choice from among available bactericides. Employ a measure of sanitation to minimize the reservoir of bacterial pathogens present on plants.

Evaluations of the efficacy of bactericides in controlling certain diseases must include replicated tests and water-treated control plants. Blocks of susceptible plants should be divided into at least three replications per bactericide treatment evaluated. The untreated control is perhaps the most important entry in the test, because it allows for accurate assessment of the benefits of treatment as distinct from improvement that may occur as a result of a change in environmental factors.

Selection of bactericides in most nurseries is dictated by legal restrictions, rather than efficacy, cost, or plant safety. Among the possible choices, only certain copper hydroxide products are legally approved for broad use on woody ornamentals in the field and in the greenhouse. Consequently, copper-based products may be repeatedly used for bacterial disease control. Growers need to recognize, however, that repeated use of copper-based products can induce tolerance in pathogens, just as repeated used of antibiotics does.

Bacterial tolerance of antibiotics and copper is well documented. Tolerance of streptomycin has occurred in various cropping systems, usually within two to five years of the introduction of the antibiotic. Crop failures due to canker of citrus, Xanthomonas blight of anthurium, bacterial spot of pepper and tomato, and fire blight of pome fruits have occurred as a result of the development of tolerance in the pathogens. The degree and persistence of this change in the bacterial population has virtually eliminated the efficacy of streptomycin in certain fruit, ornamental, and vegetable production systems. Copper tolerance has been slower to develop than streptomycin tolerance, because of the multisite mode of action of copper-based bactericides. Genetic tolerance to copper was documented in the 1980s, but it did not produce the dramatic crop failures attributed to antibiotic tolerance, since the crops treated with copper products were also treated with the fungicide mancozeb. Research by Marco and Stall (1983) demonstrated that a tank mix of mancozeb and a copper-based bactericide resulted in a 10-fold increase in copper solubility. The elevated copper content of the mixture effectively controlled copper-tolerant *Xanthomonas* populations on pepper, and it has a similar effect in many other crop systems.

See Chapter 94, "Bactericides," for information on specific bactericides useful for controlling ornamental diseases.

Biocontrol Agents

Few diseases of woody ornamentals are controlled effectively by biocontrol agents. However, crown gall is a notable exception. Several products, such as Galltrol-A and Norbac 84C, based on the biocontrol agent *Agrobacterium radiobacter* strain K84, are effective in controlling this disease. Susceptible ornamentals, such as roses, are root-dipped in a solution of the biocontrol agent and then planted in the field. The biocontrol agent prevents infection by *A. tumefaciens,* the crown gall bacterium. Care in handling rose roots to avoid wounding them is important, since wounds may provide sites for bacterial infection.

A commercial product, BlightBan A506, based on *Pseudomonas fluorescens* strain A506, can be used as a spray to control fire blight and provide frost protection in ornamentals such as crabapple and *Prunus* spp. The timing of the application is critical, because the biocontrol agent must protect the blossoms before infection occurs. Thus, the spray must be applied as the first blossoms start to open, and treatment should continue through the blooming period. See Chapter 98, "Biological Control," for more information.

Summary

Bacterial diseases, like other plant diseases, must be diagnosed properly before an integrated management strategy can be devised. Management of bacterial diseases also requires some understanding of the mechanisms by which bacterial pathogens are dispersed. Integrated pest management combines several approaches to disease control, including cultural practices, production of resistant cultivars, and treatment with bactericides and biocontrol agents.

REFERENCE

Marco, G. M., and Stall, R. E. 1983. Control of bacterial spot of pepper initiated by strains of *Xanthomonas campestris* pv. *vesicatoria* that differ in sensitivity to copper. Plant Dis. 67:779-781.

CHAPTER 91

Lawrence G. Brown ▪ U.S. Department of Agriculture, Raleigh, North Carolina
D. Michael Benson ▪ North Carolina State University, Raleigh
Ronald K. Jones ▪ North Carolina State University, Raleigh
Walter Bliss ▪ Agdia, Elkhart, Indiana

Control of Viral Diseases

Viruses in woody ornamentals are among the most difficult groups of pathogens to control. Fortunately, the threat of virus disease in woody ornamentals is limited. Although many different viruses are found in woody plants (see Table 7.2), only a few are known to cause economic damage to crops in nursery production; these include viruses in butterfly bush (Plate 21), crabapple, daphne, nandina, *Prunus,* and rose. The importance of the damage is hard to estimate, because of the complex etiology of virus infection. For certain crops like camellia, which is sometimes infected with a mottle virus, consumer demand for the aesthetic mottle patterns on virus-infected plants may actually enhance sales.

Transmission of Viruses in Nursery Plants

Viruses are a unique group of pathogens in being introduced into nurseries and spread from plant to plant only by vectors and crop production practices. Viruses move about in a crop by several means.

Seed Only a few viruses, but significant ones, are spread through seed propagation. Thus, crops that are propagated from seed generally start out virus-free. However, growers should refer to Table 7.2 before buying seed.

Asexual propagation Viruses generally move systemically throughout infected plants. Asexual propagation almost always carries the virus along from stock plants to new plants. Practically 100% of plants rooted from a virus-infected stock plant will be infected. Plants propagated by grafting can become infected if either the rootstock or the scion is infected. Asexual propagation is the most common method of multiplying plants in the nursery industry, and it is the most important means by which viruses spread in the nursery industry.

Insects Many viruses are carried by insects, such as aphids, whiteflies, and thrips, and by mites. Transmission of viruses by insects is not known to be a major problem in nursery production. While it is necessary to control insects in the nursery, insecticides seldom prevent the transmission of viruses by insects in the field.

Nematodes Several viruses are spread by dagger nematodes (*Xiphinema* spp.) (see Table 7.2). Transmission of viruses by nematodes has caused problems in the propagation of some tree fruits in nematode-infested fields. Propagation and production in soilless media eliminate this problem.

Fungi Several viruses are spread by one or two soilborne fungi, but these viruses are not known to affect woody plants. Production in soilless media also eliminates this problem.

Dodder Dodder, a parasitic plant that lacks chlorophyll, is known to transmit many plant viruses. It is rare in woody plant nurseries, affecting primarily azalea, cotoneaster, and Japanese holly. Plants parasitized by dodder should be promptly rogued.

Pollen Several tree fruit viruses are spread by pollen transported from virus-infected trees to healthy trees.

Mechanical contact Some viruses are spread when the leaf of a virus-infected plant rubs against the leaf of a healthy plant, but transmission by this means is probably rare in woody plants. Workers handling woody plants during potting, spacing, shipping, etc., probably very rarely spread viruses in the nursery. Several viruses are spread on tools used to prune plants. However, for most viruses, mechanical contact is an inefficient means of transmission, and it is of little importance in woody nursery production.

Management Strategies for Virus Diseases

Management of virus diseases of woody ornamentals takes a number of different approaches. However, before management strategies can be suggested, a correct diagnosis and knowledge of the means by which the virus spreads in a given crop are needed. Then the appropriate "weapons of choice" – exclusion, tissue culture, indexing, thermotherapy, use of a flow-through production system, or genetic engineering – can be determined. Unlike most other plant pathogens, viruses cause diseases for which no chemical treatment is available. Viral diseases are controlled by exclusion of viruses from crop species, which may involve, in broad terms, identification and use of virus-tested material as a source for propagation, control of vectors, and manipulation of growing conditions. Other factors enter into each of these categories.

Tolerance and resistance Tolerance of and resistance to viruses are important, but ornamental plant selection is based more on adaptability to local conditions and consumer demand for aesthetic qualities than for disease resistance.

Exclusion The best method of virus control is exclusion, but it is not always practical. Viruses are introduced into plant production from many sources. Few growers develop a crop from start to finish, and plant material can change locations many times. Each time a plant is moved to a new location, there is a chance of virus introduction. Sources of introduction of pathogens, and thus points where viruses can be excluded, are propagators, liner producers, grow-out nurseries, wholesalers, brokers, and retailers. Material originating from any source, including universities and other state institutions, must be certified virus-free or at least tested for viruses and other important pathogens.

Prevention Another means of reducing losses from virus diseases is prevention. Having a clean stock program is a must. Plant propagation flows one way or consists of a flush-through system, with no cyclic propagation without indexing. Mother plants are replaced on a schedule and maintained true to type. It is in the grower's best interest to have material that is indexed regularly.

Virus indexing The systematic use of detection methods to find viral pathogens in plants is known as virus indexing. It can be used to detect and identify viruses in stock plants used for propagation, so that virus-infected material can be eliminated. In cases where a portion of the propagating stock is infected, the infected plants can be removed and healthy plants retained. If the percentage of virus-infected plants is too high and affected plants cannot be eliminated by roguing, the plants must be replaced with material known to be virus-free. Indexing is essential for producing a good crop and a healthy profit margin. It is vital to know what pathogens are common to a particular crop and what methods work best to detect, identify, and eliminate viruses from stock material. Also, the crop must be continuously monitored to ensure that plants are of high quality right up to the final stages of production. For instance, tissue-cultured mother plants that are indexed for viral diseases are critical in reducing the incidence of rose mosaic virus nationwide.

If all the stock plants are infected and no healthy plants can be found, several techniques can be used, often in combination, to obtain virus-free material: culturing virus-free plant parts in vitro (outside the plant); heat therapy, which kills the virus but not the plant; and meristem tip culture, which reproduces plants from a virus-free growing tip. None of these techniques is 100% effective.

Virus indexing must follow virus elimination methods to make sure that the technique has obtained virus-free material. Visual inspection is the first method of detection, but it is often not reliable. If all of the inspected material is affected with a virus infection causing only mild symptoms, visual inspection may lead to the erroneous conclusion that the material is healthy. However, virus-free material generally has more flowers, better color, and more vigorous growth, which means that less time will be spent finishing the crop, and profits will be increased. Virus indexing normally involves a bioassay, in which the suspected virus is transmitted to a plant species that shows typical symptoms when infected, and enzyme-linked immunosorbent assay (ELISA), a serological test for identification of viruses, as described in Chapter 7, "Viruses."

Several companies sell ELISA kits and reagent sets to detect plant viruses, and several companies and governmental agencies provide virus diagnosis based on analysis of samples submitted to them. Typically, these companies screen a plant for viruses commonly found in that type of plant.

Some companies have established screens for most major crops, and screens are available to detect most viruses in ornamentals. These screens or a variation of them will probably be useful for testing woody ornamentals. One company's screen tests for viruses that have wide host ranges and may be common in ornamentals: alfalfa mosaic virus, Arabis mosaic virus, cucumber mosaic virus, impatiens necrotic spot virus, potato virus X, potyvirus group, Prunus necrotic ringspot virus, tobacco mosaic virus (TMV-c), tobacco ringspot virus, tobacco streak virus, tomato aspermy virus, tomato ringspot virus, and tomato spotted wilt virus. It is likely that more viruses will be added to this screen over time.

ACKNOWLEDGMENTS

Susan Halbert, Wayne Dixon, and Tim S. Schubert, of the Florida Department of Agriculture and Consumer Services, Division of Plant Industry, are acknowledged for critically reading the manuscript. Simon Scott, of Clemson University, made many helpful suggestions concerning the text.

REFERENCES

Agrios, G. N. 1987. Plant Pathology. 4th ed. Academic Press, New York.

Albouy, J., and Devergne, J. C. 1998. Maladies a virus des plantes ornementales. Institut National de la Recherche Agronomique, Paris.

Alfieri, S. A., Jr., Langdon, K. R., Kimbrough, J. W., El-Gholl, N. E., and Wehlburg, C. 1994. Diseases and disorders of plants in Florida. Fla. Dep. Agric. Consumer Serv. Bull. 14.

Brunt, A. A., Crabtree, K., Dallwitz, M. J., Gibbs, A. J., Watson, L., and Zurcher, E. J., eds. 1996–. Plant Viruses Online: Descriptions and Lists from the VIDE Database. Version 16 January 1997. Australian National University, Canberra. http://biology.anu.edu.au/Groups/MES/vide/

Christie, R. G., and Edwardson, J. R. 1986. Light microscopic techniques for detection of plant virus inclusions. Plant Dis. 70:273–279.

Cooper, J. L. 1993. Virus Diseases of Trees and Shrubs. Chapman and Hall, New York.

El-Gholl, N. E., Schubert, T. S., and Coile, N. C. 1997. Diseases and disorders of plants in Florida. Fla. Dep. Agric. Consumer Serv., Div. Plant Ind. Bull. 14, Suppl. 1.

Green, S. K. 1991. Guidelines for diagnostic work in plant virology. Asian Veg. Res. Dev. Ctr. Tech. Bull. 15. 2nd ed.

Horst, R. K., and Klopmeyer, M. J. 1993. Viral diseases. Pages 286–297 in: Geraniums 4. 4th ed. J. W. White, ed. Ball Publishing, Geneva, Ill.

Nameth, S. T., and Adkins, S. T. 1993. Viral diseases. Pages 267–275 in: Geraniums 4. 4th ed. J. W. White, ed. Ball Publishing, Geneva, Ill.

Ravelonandro, M., Scorza, R., Bachelier, J. C., Labonne, G., Levy, L., Damsteegt, V., Callahan, A. M., and Dunez, J. 1997. Resistance of transgenic *Prunus domestica* to plum pox virus infection. Plant Dis. 81:1231–1235.

CHAPTER
92

D. Michael Benson ▪ North Carolina State University, Raleigh
Robert A. Dunn ▪ University of Florida, Gainesville

Control of Nematode Diseases

Controlling nematodes in field-grown nursery stock can be one of the most daunting challenges to successful production of nursery crops in regions where these pathogens naturally occur in field soil. Carefully combining many of the practices described below into an integrated pest management (IPM) program will often help keep nematode populations below damaging levels. Today more than ever, strict reliance on pre- and postplant fumigants and nematicides is impractical, since few products are available for nematode control in ornamentals. Methyl bromide, an important fumigant, is being phased out in the United States, under restrictions imposed by the U.S. Environmental Protection Agency, acting under the Clean Air Act, beginning in 1999. Developed countries, including the United States, agreed to a 100% reduction in use of methyl bromide by 2005. There are few alternatives to replace methyl bromide, they may not be as effective, and they will probably be more expensive. Thus, a multitactic approach to nematode management in nurseries will be even more critical. The key components of an integrated management program for nematodes in ornamental nurseries are presented here.

Production fields – a valuable resource Growers need to develop a strong IPM program for field production of ornamentals susceptible to nematodes. An IPM program relies on a variety of tactics rather than a single "cure-all." Growers must map out a long-term approach to preventing the introduction of plant-parasitic nematodes, because once a field site becomes infested, future production of susceptible crops in that field may be limited. Growers should treat their fields as a basic resource to be maintained pathogen- and nematode-free, to protect the livelihood of their business in future years. One component of maintaining production fields as a resource is to provide drainage systems designed to prevent field-to-field movement of runoff water, which could transport nematodes. Entry or introduction of infected transplanting stock or equipment and personnel contaminated with infested soil must be avoided. A set of "rules of sanitation" for nursery managers and workers should be developed and followed strictly, to prevent nematode infestation and protect the fields as a vital business resource.

Nematode advisory services and plant disease clinics Many states offer plant disease clinics and nematode advisory services to growers. Soil samples collected systematically from potential new field sites can be assayed for the presence of some types of ectoparasitic nematodes that live in the soil around the root systems of host plants. To detect other damaging nematodes, root samples from susceptible plants must be tested, or trap crops must be grown for a period of time in soil from a potential field site. Information about nematode species detected in the field can be used by growers as a guide in site selection, preplant treatments, and decisions about appropriate plant species and cultivars to grow in the nursery (Chapter 13, Table 13.1).

Nematode advisory services and plant disease clinics can also be useful to growers in determining when a damaging population of nematodes has developed in a field of nursery stock. Advisory services of diagnostic clinics may be able to recommend actions to be taken to avoid the dispersal of nematodes to areas of the field with healthy plants, and they may suggest appropriate postplant treatments for infected plants.

State departments of agriculture Departments of agriculture in each state are required to inspect nursery stock for insects and plant diseases, to certify that only pest- and disease-free stock is sold in the marketplace. Each state has established policies and procedures for this process. In most states the inspection service is proactive in encouraging growers to learn about regulated pests and disease problems, so that preventative measures can be taken to ensure clean stock certification. See Chapter 101, "Regulatory Control," for more information.

Methods of Nematode Disease Management

Prevention

Preventing nematode infestation is far better than trying to treat the problem after it is established. Many serious nematode pests are widespread, but some are quite limited in distribution, either from one region to another or from field to field. Growers can avoid introducing damaging nematodes into uninfested land by knowing that these parasites are spread in infested soil and plant parts. To avoid carrying contaminated soil or plants to the uninfested field, good sense dictates working in areas that are not infested with nematodes before moving to infested areas.

Propagation of Nematode-Free Stock

Ornamental cuttings to be rooted should be taken only from uninfested plants or portions of plants from aboveground shoots that have never been rooted in potentially infested soil. This prevents propagating populations of nematodes that could seriously limit growth and could cause the plants to be unfit for shipment to many potential markets because of quarantines. Quarantine is governmental action taken to prevent importing a pest into a previously uninfested area, usually by controlling movement of contaminated soil and plant material. See Chapter 101, "Regulatory Control."

Only nematode-free planting stock should be used to establish a field site. Nematode advisory services or plant disease clinics can assist growers in the detection of damaging nematodes in the root systems of planting stock. As a general guideline, purchase or grow only transplanting stock produced in a soilless mix that is free of nematodes. Some mixes that contain crop residues, such as peanut hulls, should be avoided, because nematodes may be present in these ingredients. A strict sanitation program during propagation and handling of transplanting stock is essential to the production of nematode-free stock.

Crop Rotation

Crop rotation is a very old agricultural practice for the improvement of crop production. Many nematodes, soilborne pathogens, and insects can reproduce and survive on only a few plants, but others have a wide host range. Repeatedly planting a field with the same ornamental (e.g., boxwood), crop after crop without interruption, will enable lesion nematodes to increase year after year and damage plants. Rotation to nonsusceptible crops, such as selected cultivars of Japanese hollies (Chapter 13, Table 13.1), may interrupt the reproduction of these nematodes and allow natural mortality factors to reduce their numbers. By carefully planning the sequence of ornamentals to be planted in a particular field, growers may be able to avoid an excessive buildup of nematode pests over the long run.

Rotation to a nonornamental crop that is not susceptible to the nematodes present at a site may help to control nematode populations that have built up in preceding years on a susceptible ornamental. Evidence is mounting that nematode populations can be suppressed by planting a cover crop; several different cover crops have been found to be effective. For instance, hairy indigo can be planted as a summer cover crop to reduce populations of sting and root-knot nematodes, and pangola digitgrass is used to control burrowing and root-knot nematodes in infested soils in the South. The many kinds of plant nematodes complicate selection of rotation crops, because crops which reduce some species of nematodes may favor the increase of others. Despite the difficulty, a good rotation program should be a basic component of land and crop management plans, because of the multiple benefits that can be derived from it.

Flooding

Flooding sometimes helps to reduce nematode populations. It is practical only where the water level can be controlled easily and maintained at a high level for several weeks. This practice may be feasible only in areas where furrow irrigation is common. Flooding, drying, and flooding again over a period of about two or three weeks is apparently much more effective than a continuous period of flooding. The soil should be worked during the periods of drying to increase aeration and drying of soil and to prevent weed growth while the soil is exposed. Flooding apparently kills nematodes by imposing a long period without host plants, but it may also have some direct physical effect on them, such as lack of oxygen. Flooding with contaminated water may actually spread some soilborne pests, such as nematodes and some fungal pathogens.

Fallowing

Fallowing is the practice of leaving a field with no plants for a prolonged period to "starve" nematodes or other pests. Most nematode populations decrease after a period of time without plant roots on which to feed. For fallowing to be effective, the field should be cultivated regularly to prevent growth of weeds and to expose new portions of the soil to drying and heating. If weeds are allowed to grow without control in fallow land, many kinds of nematodes may be able to survive and reproduce on them, and the practice will be ineffective. Fallowing should be practiced only in areas where soil erosion is not a concern.

Cultural Practices

Cultural practices can have a tremendous impact on successful integration of nematode management. For instance, simply applying a mulch layer in nursery rows can limit stress and thus the predisposition of nursery stock to nematodes by lowering soil temperature and increasing soil moisture. Adequate irrigation during drought can also limit plant stress and damage due to nematodes. A buildup of organic matter from cover crops or mulches may also limit the activity of many plant-parasitic nematodes. See Chapter 86, "Integrated Disease Management."

Host Resistance and Tolerance

Production of cultivars resistant to a pest is the ideal means of minimizing losses due to that pest. However, to successfully use cultivar resistance, growers must know the limitations of the resistance and the pests that are present. There are nematode-resistant taxa of many ornamentals, but each of them has resistance to only one, two, or at most three species of nematodes; none are "nematode-proof." Some ornamental cultivars, although not resistant to a specific plant-parasitic nematode, may be tolerant of large nematode populations in the root zone. Tolerant cultivars grow well and exhibit no symptoms of nematode disease. The same nematode species that is tolerated by one cultivar can cause severe decline in other, intolerant cultivars. It is necessary to know the nematode species present in a field to select ornamentals with the appropriate resistance or tolerance (Table 13.1). In addition, cultivars with the appropriate resistance or tolerance must be adapted to the cultural conditions of your area. Another limitation on the use of nematode resistance as a major management practice is that high temperatures often weaken or destroy the resistance. Cultivars of some crops, such as tomato, may be "resistant" to root-knot nematodes in cool soils but may not be able to limit nematode reproduction and damage if the soil temperature is above 82°F. It is still necessary to use other methods to manage any other nematode species that are present, because resistance to or tolerance of one or two species does not affect the ability of other nematode species to injure the crop. See Chapter 13, "Nematode Diseases."

Biological Control

Many microorganisms, such as bacteria and fungi, are natural enemies of nematodes, and some of them have been isolated from nematode populations, which apparently are suppressed by these natural enemies. Nematologists have used some of these biocontrol agents to reduce populations of some nematode species under laboratory conditions, but success on a larger scale, in treatment of fields, has been rare. Most organisms recognized as potential biological control agents attack only particular nematode species, and they are most effective under certain stringent conditions. It has been difficult to produce these organisms in sufficient quantities to be useful for field application. Commercially effective biological control as a means to limit the effects of nematodes on any nursery crop may still be years away.

Soil Solarization

Soil solarization has been effective in the control of many pathogenic soilborne fungi, such as *Fusarium, Verticillium, Rhizoctonia,* and *Sclerotium.* Solarization involves tilling a field for planting, irrigating, and then covering the soil with clear plastic sheeting for four to six weeks during the hot summer. The soil temperature beneath the plastic is raised enough to kill plant pathogens and weed seeds. Results with nematode control have been limited and quite variable. Solarization can often provide excellent nematode control in the upper 4–6 inches of soil, but there is progressively greater survival of nematodes at greater depths in the bed. Since nematodes can move very easily, the duration of nematode control by solarization may thus be very limited. There also is a very serious risk of physically moving surviving nematodes from the less effectively treated depths to the upper level of the bed during transplanting and other cultivation practices after solarization is completed, thus effectively eliminating any nematode control achieved. Regions with hot, dry summers, such as Arizona, southern California, and Texas, are most suitable for solarization. Solarization in the summer with fall transplanting to establish a new field would fit typical nursery production practices.

Table 92.1 Fumigant and nonfumigant nematicides for field-grown ornamentals[a]

Common name and trade name	Application rate	Remarks
Fumigants		
Chloropicrin (many trade names)	35 gal./acre (480 lb. actual) 0.8 gal. per 1,000 sq. ft.	Restricted-use pesticide Application method: Inject 6 to 8 inches deep with chisels or fumigun spaced 10 to 12 inches. Cover with gas-tight plastic. Exposure period: 24 hours Aeration before planting: 14 days
Methyl bromide (many trade names)	872 lb./acre (actual) 20 lb. per 1,000 sq. ft.	Restricted-use pesticide Application method: Release with special applicator under plastic, or inject 6 to 8 inches deep with chisels spaced 10 to 12 inches apart. Cover with gas-tight plastic. Exposure period: 48 hours Aeration before planting: 14 days
Methyl bromide + chloropicrin (many trade names)	250 lb./acre (actual) 8 lb. per 1,000 sq. ft.	Restricted-use pesticide Application method: Inject 6 to 8 inches deep with chisels spaced 10 to 12 inches apart. Cover with gas-tight plastic. Exposure period: 48 hours Aeration before planting: 14 days
Metam-sodium	100 gal./acre 2.3 gal. per 1,000 sq. ft.	Application method: Apply as a drench in water or inject through chisels 5 inches apart. Follow label instructions carefully. Seal treated soil with plastic tarp. Exposure period: 48 hours Aeration before planting: Four weeks with weekly cultivation. Testing with a few plants before planting the entire crop is suggested to check for complete absence of metam-sodium from treated soil.
Dazomet (Basamid granular)	350 lb./acre 8 lb. per 1,000 sq. ft.	Application method: Apply granules as evenly as possible, incorporated to the desired depth, preferably with a tiller with L-shaped tines. Roll the surface, and then seal it by wetting the soil or covering it with plastic tarp. See the product label for rate variations, exposure times, aeration, and testing for complete escape of fumes.
Nonfumigant nematicide		
Fenamiphos (Nemacur 10%)	Variable, according to the crop	Restricted-use pesticide Application method: Apply the specified dosage uniformly to the area to be treated around plant. Irrigate with at least 0.5 inches of water immediately after treatment. See the product label for rate variations, exposure times, and sites and states for legal use.

[a] Each product label has specific requirements for personal protective equipment, worker protection standards, and sites and conditions for use. The use of trade names does not imply endorsement by the North Carolina Agricultural Research Service, Raleigh, or the Florida Cooperative Extension Service, Institute of Food and Agricultural Sciences, University of Florida, Gainesville, of the products named or criticism of similar ones not mentioned.

Nematicides

Nematicides are chemicals applied to soil to reduce nematode populations (Table 92.1). Generally, the active ingredient in a nematicide is a vapor (in fumigant nematicides) or a water-soluble chemical (in nonfumigant nematicides).

Some fumigant nematicides also kill other organisms in soil, such as plant roots, fungi, bacteria, insects, and weed seeds. Most fumigants are toxic to plants as well as to nematodes, so they are usually used before transplanting. Fumigants, such as chloropicrin and methyl bromide (Table 92.1), are active as nematicides when they are in a volatile state, as gases. They spread through the soil as gases, so soil moisture and soil pore size must be appropriate to allow sufficient air space around soil particles to permit effective movement of fumigants in the soil profile. Likewise, large pieces of crop debris and large clods of soil inhibit penetration by fumigants, so nematodes in debris or clods may be protected from exposure to the chemicals. The soil temperature must be above 55–60°F to facilitate the movement of fumigants in the soil profile from the point of application. Fumigation of waterlogged, cold soils that cannot be tilled will result in poor nematode kill. In addition, a fumigant can become trapped in waterlogged soil for a long period after application, so that phytotoxicity develops at transplanting, or transplanting must be delayed.

Nonfumigant nematicides, in contrast to fumigants, generally kill only nematodes and insects and so may be used as pre- or postplant treatments. Nonfumigants dissolve in films water on soil particles and roots and are then absorbed by nematodes. Once the nematicide has been absorbed, the nematode's behavior or its organ systems are disrupted, and the nematode becomes inactive or dies. No effective preplant nonfumigant nematicides are currently approved for use on ornamental crops. Postplant nematicides, such as fenamiphos (Table 92.1), are used in nurseries to treat living plants directly, under some conditions (see the product label). A surface-applied nematicide such as fenamiphos must be activated with irrigation water after application, to dissolve the nematicide and move it into the root zone of the treated plant.

Almost all nematicides are very toxic, posing health hazards to those who come in contact with them and posing serious environmental risks. For those reasons, they are stringently regulated, and there are many situations for which no chemical nematicide may be used legally, regardless of the severity of disease. Although they are also expensive, nematicides can be very profitable when used correctly in appropriate situations. However, their effects are almost universally short-lived, so they should be used in conjunction with other practices that minimize nematode reproduction and reinfestation of nursery stock plantings.

Summary

Nematodes are among the most difficult plant pathogens to manage successfully. Planting of nursery stock should be avoided in fields where nematode damage has been observed on previous crops. Only integrated use of cultural practices, crop rotation, host resistance, and nematicides in combination with other management tools will result in control of nematodes in woody ornamentals.

CHAPTER 93

Steven N. Jeffers ▪ Clemson University, Clemson, South Carolina
R. Walker Miller ▪ Clemson University, Clemson, South Carolina
Charles C. Powell, Jr. ▪ Plant Health Advisory Services, Worthington, Ohio

Fungicides for Ornamental Crops in the Nursery

Managing diseases of ornamental crops in the nursery can be a challenging task because of the vast array of plant species concentrated in one location, the variety of pathogens capable of attacking them, and the various options available for disease management. Integrated disease management – which utilizes all available options – provides the most effective, consistent control of diseases in the nursery over time. One of the available options for integrated disease management is the use of chemical pesticides. Fungi cause most diseases affecting plants so most of the chemical pesticides applied to plants for disease management are fungicides. Fungicides kill or inhibit the growth and development of fungi. In the nursery, fungicides are a valuable tool for managing plant diseases, but they should not be relied upon too heavily or used excessively. They should be applied only when needed and only in amounts necessary to achieve control. Applications should be timed to optimize the potential deleterious effect of the fungicide on the pathogen.

When are fungicides needed? They are needed when pathogenic fungi are active, susceptible plants or plant parts are present, and environmental conditions are conducive to disease development. All three conditions must occur at the same time. Proper timing of fungicide applications is critical for effective disease management. Fungicides must be applied before or shortly after infection occurs, depending on the disease and product used. We usually rely on research conducted by plant pathologists to identify when, where, and how infection happens. However, such research on diseases of ornamental crops is often lacking because of the multitude of diseases that affect the vast array of plants grown for ornamental use. When specific research results are not available, any available information on similar or related diseases of other crops is used to select and schedule fungicide applications. Once fungi have invaded plant tissues and a disease is under way, most fungicides have little or no effect on subsequent disease development in affected plants. However, applications of fungicides to a crop already showing disease symptoms may be beneficial in preventing or reducing the number of additional infections on these and surrounding plants.

Fungicide Names

All fungicides have at least three names: the chemical name is the name of the active ingredient (abbreviated a.i.) in a fungicide (e.g., tetrachloroisophthalonitrile); the common name is a less technical term for the active ingredient (e.g., chlorothalonil); and the trade name, or brand name, is the name of the commercially available product that contains the active ingredient (e.g., Daconil). A single active ingredient may be marketed under several different trade names, and the products may or may not have the same registered uses specified on the label. Only certain products may be available from any one agricultural chemical supplier. In addition, some products are sold as prepackaged mixtures of two active ingredients. Consequently, it is useful to refer to fungicides by their common names to avoid confusion among products containing the same active ingredient but different label recommendations. However, with reference to a specific disease, it may be necessary to use trade names for clarity or for legal reasons. Table 93.1 lists the common names of active ingredients of fungicides registered for use on ornamental crops and the trade names of some commercially available products.

Categories of Fungicides

Fungicides differ dramatically in how they kill or inhibit the growth and development of fungi and in the spectrum of their activity (i.e., the range of fungi against which they are active). Consequently, it is critical to accurately diagnose the cause of a disease before selecting and applying a fungicide for control. Fungicides usually are grouped into classes based on the chemistry of their active ingredients. However, they also can be placed in functional groups based on the type of fungi they affect. The chemical class to which a fungicide belongs is one of the major factors in determining the fungicide's characteristics and the range of target organisms affected by its active ingredient. Tables 93.1 and 93.2 summarize the characteristics, registered uses, and relative efficacies of fungicide active ingredients used to manage diseases of ornamental crops (see Table 93.1 for trade names of commercial products.) Any one of these fungicides could be used in the nursery, depending on the crops grown and the diseases present. For this discussion, fungicides are grouped into eight categories.

Benzimidazoles

The benzimidazoles are a class of chemicals that include some of the most widely used fungicides on ornamental crops. Compounds in this class affect the growth and elongation of hyphae at a very specific site in the metabolic pathway of target fungi, and therefore resistance to these fungicides develops very easily.

Resistance to these products is widespread, particularly in *Botrytis cinerea* and powdery mildew fungi. Benzimidazoles are effective against a broad range of ascomycete and imperfect fungi that cause leaf spots, flower blights, and powdery mildew – if resistance has not developed. They also include some of the few active ingredients that control root, crown, and stem rots caused by species of *Rhizoctonia, Fusarium, Thielaviopsis,* and *Cylindrocladium.*

Table 93.1 Fungicides registered for use on ornamental crops: active ingredients, trade names, and characteristics[a]

Fungicide category	Common name of active ingredient	Trade names of products[b]	Mode of action[c]	Movement in plants[d]	Risk of resistance[e]	REI (hours)[f]
Benzimidazoles	Benomyl[g]	Benlate	P + C	LS	H	24
	Thiabendazole	Arbotect, Mertect	P + C	LS	H	12
	Thiophanate-methyl	3336, Banrot,* Cavalier, Consyst,* DrenchPak,* Duosan,* Fungo, Zyban*	P + C	LS	H	12
Dicarboximides	Iprodione	Chipco 26019	P + C	LS	H	12
	Vinclozolin	Curalan, Touche, Vorlan	P + C	LS	H	12
Sterol biosynthesis inhibitors	Fenarimol	Rubigan, TwoSome*	P + C	US	M	12
	Myclobutanil	Eagle, Systhane	P + C	US	M	24
	Piperalin	Pipron	P + C	US	M	12
	Propiconazole	Alamo, Banner	P + C	US	M	24
	Triadimefon	Bayleton, Strike	P + C	US	M	12
	Triflumizole	Terraguard	P + C	US	M	12
Strobilurins	Azoxystrobin	Heritage	P + C	US	H	4
	Kresoxim-methyl	Cygnus	P + C	LS	H	12
	Trifloxystrobin	Compass	P + C	LS	H	12
Products for control of *Phytophthora* and *Pythium*	Etridiazole	Banrot,* Terrazole, Truban	P	NS	L	12
	Fosetyl-aluminum	Aliette, Prodigy	P	S	M	12
	Mefenoxam	Subdue (new formulations)	P + C	US	M	0
	Metalaxyl	DrenchPak,* Pythium Control, Subdue (original formulations)	P + C	US	M	12
	Propamocarb	Banol	P	US	L	12
Broad-spectrum protectants	Captan	Captan	P	NS	L	96
	Chlorothalonil	Consyst,* Daconil, Exotherm Termil, PathGuard, TwoSome*	P	NS	L	12
	Ferbam	Ferbam	P	NS	L	24
	Mancozeb	Dithane, Duosan,* Fore, Junction,* Mancozeb, Protect, Zyban*	P	NS	L	24
	Maneb	Maneb, Pentathlon	P	NS	L	24
	Metiram	Polyram	P	NS	L	24
	Ziram	Ziram	P	NS	L	48
Inorganic protectants	Bicarbonates	Armicarb, Kaligreen, FirstStep	P	NS	L	4
	Copper compounds	Camelot, Champ, Champion, COCS, Junction,* Kocide, Nu-Cop, Phyton	P	NS	L	12–24
	Lime sulfur	. . .	P	NS	L	48
	Sulfur	. . .	P	NS	L	24
Other fungicides	Dicloran	Botran	P	NS	L	12
	Fenhexamid	Decree	P + C	NS	H	4
	Fludioxonil	Medallion	P	NS	U	12
	Neem oil	Triact	P	NS	L	4
	Oxycarboxin	Plantvax	P + C	US	L	12
	Pentachloronitrobenzene (PCNB)	Defend, Engage, Revere, Terraclor	P	NS	L	12
	Potassium salts of fatty acids	M-Pede	C	LS	L	12

[a] Not all commercially available products are included. Not all products listed can be used in nurseries. Commercial products change periodically so some of the products listed may no longer be available.

[b] An asterisk indicates a product that contains more than one active ingredient in a prepackaged mixture.

[c] C = curative. P = preventative.

[d] LS = locally systemic. NS = nonsystemic. S = systemic. US = upwardly systemic.

[e] H = high. L = low. M = moderate. U = unknown.

[f] Restricted entry interval; this interval may vary in different geographic regions and for different formulations.

[g] Registered for use on conifers only.

Dicarboximides

The dicarboximides are another chemical class of fungicides. These compounds are very effective against a limited group of fungi, including species of *Botrytis* and related genera (e.g., *Sclerotinia, Monilinia, Ciborinia,* and *Ovulinia*) and *Alternaria.* They have a very specific site of action, which has not been identified, and therefore resistance to these compounds develops very readily. Resistance to dicarboximides is becoming widespread because of excessive use of these compounds.

Sterol Biosynthesis Inhibitors

Recent developments in pesticide chemistry have led to the production of sterol biosynthesis inhibitors (SBIs), which are growing more popular as they become registered for more uses on ornamental crops. These fungicides inhibit the production of sterols in target fungi, which are necessary for the growth of many fungi. They also have been referred to as sterol inhibitors (SIs) and ergosterol biosynthesis inhibitors (EBIs). SBIs that specifically inhibit demethylation of sterol precursors are referred to as demethylation inhibitors (DMIs). This subgroup includes most of the SBI fungicides used on ornamental crops – fenarimol, myclobutanil, propiconazole, triadimefon, and triflumizole. Resistance to some SBIs has been reported, but the risk of resistance developing in normal use is moderate. SBIs are particularly effective against fungi that cause powdery mildews and rusts as well as certain ascomycete and imperfect fungi that cause leaf spots and blights. They do not protect against oomycete fungi (*Phytophthora, Pythium,* and the downy mildew fungi).

Strobilurins

The newest class of fungicides is the strobilurins, which are based on antifungal compounds produced by naturally occurring wood-rotting fungi. These fungicides appear to be effective against a broad range of fungi, including species causing powdery mildews, downy mildews, rusts, leaf spots, and root rots. Strobilurins must be used with care to avoid the development of resistance in target fungi. Three strobilurin products (azoxystrobin, kresoxim-methyl, and trifloxystrobin) currently are registered.

Products for the Control of *Phytophthora* and *Pythium*

Several chemically unrelated compounds are active against the oomycete or water mold fungi, species of *Phytophthora* and *Pythium.* These fungi are biologically distinct from most others, and products that affect them have little or no effect on other fungi. Some of these products also are very effective against downy mildew fungi, which are oomycete fungi as well, but most are not registered for this purpose to preserve their effectiveness against *Phytophthora* and *Pythium.* The products in this group are not always equally effective against fungi in these two genera; some are more effective at controlling *Phytophthora* diseases and some are more effective at controlling *Pythium* diseases (Table 93.2). The risk of resistance developing to these products is moderate to high.

Broad-Spectrum Protectants

The active ingredients in broad-spectrum protectant fungicides belong to very different chemical classes. These compounds affect a broad array of fungi, particularly those causing leaf spots and downy mildews, by inhibiting pathogen growth and development on the surfaces of plants. Broad-spectrum protectants are active at many different sites in the metabolic pathways of target organisms; therefore, resistance to these compounds is not likely to develop and has not been documented.

Inorganic Protectants

Inorganic protectant fungicides are a group of chemically unrelated compounds that all have inorganic chemicals as active ingredients. Copper and sulfur compounds are two of the oldest known fungicides. Inorganic protectants are active against many different fungi but usually are less effective than other types of fungicides. Copper compounds have the added benefit of being active against many bacteria that cause plant diseases. Fungi are very unlikely to develop resistance to copper compounds. However, resistance to copper has developed in some bacteria, and there is a risk that it could develop in others.

Other Fungicides

Some fungicides have unique chemistries and do not fit in any of the groups listed above. They are active against different types of fungi, and their efficacies vary dramatically (Table 93.2).

Characteristics and Modes of Action

An important consideration in selecting a fungicide to be used in the nursery, in addition to chemical class and relative efficacy, is the restricted entry interval (REI). This is the period of time when workers must not enter a treated area after application of a fungicide. The REI is clearly stated on every pesticide label, in a section entitled "Agricultural Use Requirements." Personal protective equipment, which also is specified in this section of the label, is required for anyone entering a treated area before the REI has expired.

The main purpose of the REI is to limit chronic (or long-term) exposure of workers to pesticides. REIs often are based on toxicity risks associated with active ingredients. REIs for fungicides used on ornamental crops range from 0 to 96 hours, but most of them are 12 or 24 hours (Table 93.1). Occasionally, the REI for a specific product varies in different geographic regions because of differences in climate that affect weathering and residue levels. Manual labor (watering, weeding, moving plants, canning, picking up orders, etc.) is a vital part of the daily operation of most nurseries. Consequently, fungicides with long REIs, which limit worker movement and activity in and around the nursery, are unpopular and tend to be avoided if possible, and those with short REIs are used extensively, even if they are not the most effective for a given disease.

How a fungicide acts on a target fungus often is termed the *mode of action,* which is the specific process in the metabolism of the target fungus that is affected by a fungicide (e.g., sterol biosynthesis inhibition). Specific modes of action can be grouped

Table 93.2 Registered uses and relative efficacies of fungicides used on ornamental crops

		Target diseases[a]						Target fungi[a]							
Fungicide category	**Common name of active ingredient**	**Leaf spots**	**Flower blights**	**Rusts**	**Powdery mildews**	**Downy mildews**	**Specialty uses[b]**	***Pythium***	***Phytophthora***	***Rhizoctonia***	***Fusarium***	***Thielaviopsis***	***Sclerotium***	***Cylindrocladium***	***Botrytis***
Benzimidazoles	Benomyl[c]	...	...	...	...	...	×	...	...	...	...	...	...	...	...
	Thiabendazole						×								
	Thiophanate-methyl	+++	+++	...	+++	...	×	...	...	+++	+++	+++	...	++	+++
Dicarboximides	Iprodione	+++	...	...	...	...	×	...	...	+++[d]	+++[e]	...	...	++ ?	++++
	Vinclozolin	+++	+++	...	...	...	×	...	...	...	...	...	...	...	++++
Sterol biosynthesis inhibitors	Fenarimol	...	...	+++	++++	...	×	...	...	...	...	...	...	...	...
	Myclobutanil	+++	++	++++	++++	...	...	...	...	...	...	...	...	...	...
	Piperalin	...	...	...	++++	...	...	...	...	...	...	...	...	...	...
	Propiconazole	+++	++	++++	++++	...	...	...	...	...	...	...	...	...	...
	Triadimefon	+++	++	+++	+++	...	...	...	...	...	...	...	...	...	...
	Triflumizole	++	...	...	+++	...	...	...	...	+++[d]	...	+++	...	+++	...
Strobilurins	Azoxystrobin	+++ ?	...	+++ ?	+++ ?	+++ ?	...	++ ?	+ ?	++++ ?[d]	++ ?	...	...	...	++ ?
	Kresoxim-methyl	...	...	...	+++ ?	...	...	...	...	...	...	...	...	...	...
	Trifloxystrobin	+++ ?	...	+++ ?	+++ ?	+++ ?	...	...	...	+++ ?[d]	...	...	...	...	++ ?
Products for control of *Phytophthora* and *Pythium*	Etridiazole	...	...	...	...	...	...	++++	++	...	...	...	...	...	...
	Fosetyl-aluminum	...	...	...	...	++++	...	+++	++++	...	...	...	...	...	...
	Mefenoxam	...	...	...	...	...	...	++++	++++	...	...	...	...	...	...
	Metalaxyl	...	...	...	...	...	...	++++	++++	...	...	...	...	...	...
	Propamocarb	...	...	...	...	...	...	+++	++	...	...	...	...	...	...
Broad-spectrum protectants	Captan	+++	++	...	...	...	×	...	...	...	...	...	...	...	...
	Chlorothalonil	+++	++	+++	++	++	...	...	++[e]	++[e]	+[e]	...	...	++ ?	++++
	Ferbam	++	+	++	...	...	...	...	...	...	...	...	...	...	+
	Mancozeb	+++	+++	+++	...	+++	...	...	++[e]	...	...	...	...	...	+++
	Maneb	++	...	++	...	...	...	...	++[e]	...	...	...	...	...	+
	Metiram	...	...	...	...	...	×	...	...	...	...	...	...	...	...
	Ziram	++	+	++	+	...	...	...	...	...	...	...	...	...	...
Inorganic protectants	Bicarbonates	++ ?	+ ?	...	++	...	...	...	...	...	...	...	...	...	++ ?
	Copper compounds	++	...	...	+	+++	...	...	++	...	...	...	...	...	++
	Lime sulfur	...	...	...	...	...	×	...	...	...	...	...	...	...	...
	Sulfur	+	...	+	++	...	×	...	...	...	...	...	...	...	...
Other fungicides	Dicloran	...	...	...	...	...	×	...	...	...	...	...	...	...	...
	Fenhexamid	...	...	...	...	...	...	...	...	...	...	...	...	...	++++
	Fludioxonil	+++ ?	...	...	...	...	...	...	...	+++[d]	+++	+++ ?	+++ ?	++ ?	+++ ?
	Neem oil	+ ?	+ ?	+ ?	+ ?	+ ?	...	...	...	...	...	...	...	...	+ ?
	Oxycarboxin	...	...	+++	...	...	...	...	...	...	...	...	...	...	...
	Pentachloronitrobenzene (PCNB)	...	+++	...	...	...	×	...	...	+++	...	...	++++	...	...
	Potassium salts of fatty acids	...	...	...	...	...	×	...	...	...	...	...	...	...	...

[a] Relative efficacy: + = marginally effective; ++ = moderately effective; +++ = effective; ++++ = very effective; ? = uncertain (only limited information is available). × = Registration only, no efficacy implied.
[b] The active ingredient is registered for a very specific or specialty application and may not have broad application for ornamental crops.
[c] Registered for use on conifers only.
[d] Effective against both aerial and root infections.
[e] Effective only against aerial infection of stems and foliage.

into general categories for practical purposes. The most common of these are preventative and curative modes of action. *Preventative action* occurs when a fungicide is present on the plant as a protective barrier before the pathogen arrives or begins to develop so that it prevents infection from occurring. *Curative action* occurs when the active ingredient of a fungicide can penetrate the plant and stop the pathogen in plant tissues. In this case, the fungicide can be applied after infection has occurred. Effective curative activity is very much dependent on time. There is a definite period of time after infection has occurred during which the fungicide will be effective in a curative manner. This period is usually between 24 and 72 hours, depending on the active ingredient. After this period has expired, curative fungicides have little or no effect on disease development. Most fungicides that have curative activity also have preventative activity (Table 93.1) and are most effective when applied before infection occurs.

Fungicides have two other general modes of action, but they are not very common. *Eradication* is the ability to stop disease development after symptoms have developed. Very few fungicides have this capability, and growers must not rely on this as a means of disease control. *Antisporulant activity* is the ability to prevent spores from being produced. In this case, disease continues to develop (e.g., lesions or cankers continue to expand), but spores are not produced or released, so that the amount of inoculum available to infect surrounding plants is reduced. Growers, applicators, and others who use fungicides must be aware that some fungicides do not kill or eliminate fungi on or in plants but only suppress or inhibit the growth and development of target fungi, as long as sufficient residues of active ingredients are present – i.e., they are fungistatic and not fungicidal. This is particularly true for fungicides applied to soil or container mix to manage root and crown rots.

Fungicides that move into plant tissues after application are termed *systemic*. Very few fungicides are truly systemic (i.e., able to move freely throughout the plant); however, some are upwardly systemic (i.e., they move only upward in the plant, through xylem tissue or in spaces between cells), and some are locally systemic (i.e., they move into treated leaves and redistribute to some degree within the treated portion of the plant). Many fungicides, including almost all broad-spectrum protectants, are nonsystemic and do not move into plants at all. Nonsystemic fungicides generally have only preventative activity because they cannot enter plant tissues to eradicate fungi after infection has occurred and are more subject to removal and deactivation by irrigation and environmental conditions (i.e., weathering). Systemic fungicides usually have some curative activity because they enter plant tissues and, for this reason, are much less susceptible to weathering.

Fungicide Resistance

Fungi, like most other microorganisms, can become less sensitive to toxic substances after continued exposure to them. This decrease in sensitivity to fungicides is termed *resistance* and is usually a result of normal genetic mutation. Mutations occur in most populations of organisms on a regular basis but usually in very small, insignificant numbers. When selection pressure in the environment, such as the presence of a fungicide, favors the development of mutant isolates of a fungus over native, wild-type isolates, mutant isolates increase in the population. With continued application of the fungicide, selection pressure may allow mutant isolates resistant to the fungicide to dominate the population and outnumber the native isolates that are sensitive to the fungicide. In this case, a resistant population of the pathogen develops; the fungicide is no longer effective; and severe disease losses can occur.

Some fungicides, because of their chemistry and the biology of the target pathogen, are more likely to be associated with the development of resistance in a pathogen population. Resistance to fungicides that have a specific mechanism of action is more likely to develop because only one or several mutations are necessary for resistance to occur. Benzimidazole fungicides (e.g., benomyl, thiophanate-methyl, and thiabendazole) and dicarboximide fungicides (e.g., iprodione and vinclozolin) have very specific mechanisms of action and affect only one or a few metabolic sites in target fungi. Accordingly, resistance to these fungicides is a serious problem. Sterol biosynthesis inhibitors (e.g., myclobutanil, propiconazole, and triadimefon) attack multiple sites (a few to several) along the metabolic pathways of target fungi, and, therefore, resistance to these fungicides is less likely to develop. Broad-spectrum and inorganic protectant fungicides (e.g., chlorothalonil, copper compounds, and mancozeb) usually affect many sites in target fungi, and thus resistance to these compounds has not developed.

Fungi that acquire resistance usually are ones that produce large quantities of spores and have short generation times, such as powdery mildew fungi (e.g., species of *Erysiphe, Microsphaera, Phyllactinia,* and *Sphaerotheca*) and *Botrytis cinerea,* which causes Botrytis blight. If a fungus develops resistance to an active ingredient, it becomes resistant to all products containing that active ingredient and, in most instances, also is resistant to other active ingredients in the same chemical class. This phenomenon is termed *cross resistance.* For example, if a population of *B. cinerea* becomes resistant to thiophanate-methyl, it will be resistant to Cleary's 3336, Fungo, and all other commercial products containing this active ingredient. In all likelihood, it will also be resistant to products containing other benzimidazoles as active ingredients (e.g., benomyl and thiabendazole). Likewise, if a population of *B. cinerea* becomes resistant to vinclozolin, it will probably be resistant to iprodione because both are dicarboximides. Once a fungus population develops resistance to a fungicide, that compound and others in the same chemical class no longer are effective against that fungus population, and the use of such fungicides should be discontinued.

Fungicide resistance may not be permanent. If a fungus population becomes resistant to a fungicide but that compound and all chemically related fungicides are not used for several years, resistance in the fungus population may eventually decrease to a level at which the fungicide can be effective again. The stability of fungicide resistance in pathogen populations is under investigation; it appears to depend on both the active ingredient and fungus involved.

The presence of fungicide resistance in a fungus population at a specific nursery location does not imply that the resistance developed from misuse or excessive use at that location. Plant pathogens, including many fungi, can be disseminated over long distances on plants and propagation materials (liners, cuttings, budwood, etc.). Consequently, it is likely that fungi resistant to certain widely used fungicides also are disseminated on plants and propagation materials moved between nurseries.

Members of the newest class of fungicides to be developed, the strobilurins, are reported to be effective against a broad array of fungi. These products just now are being registered for use on ornamental crops, and they offer great potential for disease management in the nursery. However, researchers have determined that strobilurins have a very specific site of action, and therefore resistance to them is likely to develop unless the use of these products is appropriately managed. Here are some practical guidelines for preventing the development of fungicide resistance in the nursery:

Practice integrated disease management whenever possible, to avoid excessive use of fungicides.

Do not rely on any one active ingredient or group of active ingredients in any one chemical class; instead, use several fungicides in different chemical classes in a well-planned disease management program.

When making repeat applications of the same fungicide, alternate it or tank-mix it with a product containing an active ingredient from a different chemical class to reduce selection pressure.

In using products to which fungal pathogens are likely to develop resistance, limit the number and monitor the timing of applications. Use them at times in the disease development cycle when they are likely to be most effective, and do not apply them when fungi are actively sporulating.

For most systemic fungicides, it is important to use the recommended application rate. Reducing the rate allows mutants to survive and proliferate.

Be aware that many suspected cases of "resistance" can be attributed to errors in fungicide selection or rate or in application timing, intervals, or coverage. Also, microbial degradation may reduce fungicide effectiveness after repeated applications to soil.

Formulations

Before an active ingredient can be used as a fungicide in the nursery or anywhere else, it must be formulated as a commercially viable and marketable product. The formulation of a fungicide is the result of a manufacturing process designed to deliver a stable, consistent, reliable, and effective product to the consumer. Only a portion of the commercial product is the active ingredient; the remainder is inert ingredients, such as carriers, emulsifiers, wetting agents, surfactants, solvents, and preservatives. There are numerous types of formulations, some liquid and some solid. In some cases, the chemistry of the active ingredient determines the formulation; for example, some chemical compounds do not occur in a solid form so dry formulations are not possible. Many active ingredients are available in different formulations, and growers can choose the type best suited to a given use. Different formulations of the same active ingredient occasionally differ in efficacy. Some of the more common formulations of fungicides are described below.

Wettable Powder (WP or W)

A wettable powder is a dry, finely ground powder that goes into suspension in the spray tank; the particles do not dissolve. This was one of the most common fungicide formulations until recently, when dry flowable formulations were developed. Many fungicides, particularly older ones, are still formulated as wettable powders.

Dry Flowable (DF) and Water-Dispersible Granules (WG or DG)

In dry flowable formulations, a dry, finely ground powder (like a wettable powder) is formed into tiny granules, beads, or pellets. The granules break up and disperse into particles, which go into suspension in the spray tank. Dry flowable formulations have replaced wettable powder formulations of many fungicides. They have several advantages over wettable powders: dry flowable granules generate less dust during measuring and mixing, reducing the exposure of the user to the pesticide; they are easier to measure and pour; and they usually contain higher concentrations of the active ingredients, so that distributors, retailers, and end users can maintain a smaller volume of the product.

Water-Soluble Pouch (WSP) and Water-Soluble Bag (WSB)

A relatively new type of formulation consists of a premeasured amount of product enclosed in a sealed pouch or bag. In theory, the amount of product in the pouch has been measured for a convenient, commonly used volume of water. Pouches are added directly to the spray tank, where they dissolve and release the product, which goes into suspension. This type of formulation minimizes the exposure of the person measuring and mixing the fungicide and makes measuring very simple, as long as the volume of water needed for application is consistent with that required for the amount of fungicide in the pouch.

A major disadvantage of water-soluble pouches and bags is a strong negative interaction between boron and the chemicals used to make the pouches or bags. Boron in the water or added to the spray mixture as a nutrient has an adverse reaction with the pouch itself and can create serious incompatibility problems in the sprayer. Therefore, do not add boron to the spray tank or use water containing boron if fungicides in water-soluble pouches or bags are being used.

Flowable (F)

This commonly used formulation is a thick, viscous liquid comprised of a very fine powder suspended in water. The powder in a flowable formulation may be finer than that in a wettable powder or dry flowable formulation. Flowables usually contain one to several surfactants, which help keep the active ingredient in suspension in the product container and also may improve the distribution of spray on plant surfaces. These formulations easily go into suspension in the spray tank. Being liquids, flowables are measured by volume instead of by weight, which is used to measure all dry formulations, and there is no concern with dust during measuring and mixing. Flowables can be difficult to rinse completely off the measuring device and product container because of the thick, viscous nature of these products. Commercial-sized containers of flowable fungicides usually are heavy, because of the density of these liquids, which can make them awkward to handle.

Emulsifiable Concentrate (EC or E)

An emulsifiable concentrate is a liquid formulation that contains an active ingredient dissolved in an organic solvent. When added to the spray tank, the product forms an emulsion in water, which is usually milky white. Often there are more concerns about the toxicity of the solvent carrier than about that of the active ingredient. Like flowables, emulsifiable concentrates are measured by volume instead of by weight.

Granular (G)

A granular formulation is a dry powder formed into small pellets, which are meant to be broadcast over the surface of or incorporated into a container mix or soil. Granulars are not mixed with water but are applied dry. Pellets slowly break down after application when exposed to moisture (rain or irrigation) and release the active ingredient into the soil or container mix. This type of formulation is used for fungicides active against soilborne fungi. The concentration of the active ingredients is usually low (2–10%).

Dust (D)

Dusts contain very low concentrations of active ingredients in the form of fine particles combined with a particulate carrier, such as clay or some other inert compound. Dusts are not diluted in any way but are applied by being shaken, spread, or blown directly onto plant surfaces. Their active ingredients are then dispersed and redistributed when the plants become wet with dew, rain, or irrigation. This is an old type of formulation and is no longer used commonly for commercial fungicides, although some sulfur products are still formulated as dusts.

Methods of Application

The method of application is something else to consider when fungicides are being used. An application method that will achieve maximum effectiveness of the selected product should be used. Fungicides, like other pesticides, can be applied various ways. Most are mixed with water and applied under pressure by some type of sprayer. Sprayers range from hand-held, hand-pumped models suitable for small areas to tractor-pulled, power-operated models suitable for large acreage, including hydraulic sprayers, air-blast sprayers, and mist blowers.

Depending on the type of sprayer used, fungicide suspensions can be applied dilute in a high volume of water to runoff, which thoroughly covers plant surfaces, or they can be applied at low or ultralow volume, so that very small droplets are distributed on plant surfaces without runoff. Droplets less than 200 μm in diameter should be avoided because of the risks of drift and inhalation. In a low- or ultralow-volume application, a reduced volume of water is used with the same amount of fungicide as is used in a dilute application, which results in a concentrated fungicide suspension in the tank. Because the fungicide is concentrated in a reduced volume of water, any error in measuring or calibration is magnified by the time the fungicide suspension is sprayed onto the plants. In low- and ultralow-volume applications, less spray suspension is needed to deliver an appropriate amount of active ingredient to the plant. Coverage usually is not as thorough as with dilute applications, but smaller sprayers can be used and fewer trips through the nursery are needed. Once concentrated droplets are on the plant surface, the active ingredient may be redistributed by water (e.g., dew, rain, or irrigation) or by systemic movement, both of which improve coverage.

Some other application methods commonly used in nursery disease management are chemigation, drench application, broadcast application, incorporation, and dip application.

Chemigation

Chemigation is the application of pesticides, including fungicides, through an irrigation system. This method usually involves an overhead irrigation system, but drip irrigation systems occasionally are used. The target pathogen and type of fungicide applied will dictate whether chemigation is an option worth considering. Of course, a nonsystemic fungicide for managing foliar diseases would not be applied through a drip irrigation system; however, this could be a very effective method of applying a product to control root rot. As an application method for foliar disease management in a container nursery, chemigation can be very inefficient because much of the product ends up on the ground. For chemigation to be used legally, it must be specified as an application method on the product label.

Drench Application

Drenches are used to deliver fungicide suspensions to the soil or container mix, usually to control root rots caused by soilborne fungi. Enough volume to thoroughly wet the soil or mix is required for effective control. If lower volumes are used to just wet the surface, irrigation or rainfall after application is necessary to move the fungicide down through the soil or container mix to the root zone. With a large number of plants, a drench application can be time-consuming because each individual plant must be treated. Some drench applications can be applied by chemigation (see above), which usually wastes product but saves time.

Broadcast Application

Granular fungicides can be broadcast with mechanical devices that spread them over large areas. Broadcast application often is used instead of a drench application because it is more efficient for treating large numbers of plants. However, in a container nursery, a portion of the product is wasted because it does not land in the pots or containers. After application, irrigation usually is necessary to activate the granules and release the active ingredient. Granulars also can be applied by hand to individual containers to avoid wasting product, but this takes more time. Economics and time will dictate how best to apply granular fungicides.

Incorporation

Incorporation is application of a fungicide directly to a container mix before it is used or to the soil before planting. A granular or other dry formulation of fungicide can be added, just like a fertilizer or other amendment, to bulk container mix as it is being prepared. Incorporation is used most commonly for fungicides that control soilborne fungi, which cause root, crown, and stem diseases.

Dip Application

Dip application is used primarily to treat propagation stock (liners, cuttings, bulbs, corms, rhizomes, etc.) and plants going into cold storage. It is a very effective and efficient method for applying fungicides right where they are needed. Fungicide dips are very important during propagation to prevent the spread of plant pathogens and to protect wounds. However, care must be taken to avoid spreading bacteria and other plant pathogens when large batches of propagation stock are dipped in a fungicide suspension.

Concluding Remarks

The first step in effective disease management is accurate diagnosis of the problem. Be sure to confirm the cause of a disease before applying any control treatment. Fungicides are an important component in any integrated disease management scheme involving fungus pathogens. However, they must be used wisely to maintain their effectiveness, to avoid unnecessary exposure of applicators and workers to pesticides, and to prevent contamination of the environment. The future for fungicides as disease management tools will continue to be bright *if the people using these products in nurseries, greenhouses, fields, orchards, and vineyards act responsibly.* Products with new chemistries are being discovered and introduced that are highly effective at very low rates, are very specific in their modes of action, have reduced mammalian toxicity, and are more environmentally friendly.

Always read and follow the directions on the product label of any pesticide used. It is the responsibility of the pesticide user to make certain that current label instructions are followed. Growers should check labels thoroughly *before* purchase to be sure application rates and methods are consistent with their needs and practices. They also should verify that a product is registered (1) for use on the species of plants to be treated, (2) for use in their state or region of the United States, (3) for use in the production system in which treatment is to be applied (i.e., field or container nursery or greenhouse), and (4) for application by the method to be used.

The diversity of ornamental plants grown in nurseries is changing constantly; new cultivars are introduced every year and older ones are retired. Occasionally, species and genera of plants not previously recognized as ornamental crops are brought into the nursery as potential new landscape plants. In addition, new fungicides and formulations are being discovered, developed, and registered for ornamental crops, such as strobilurins, fenhexamid, and fludioxonil. Consequently, there are too many types of ornamental plants grown in the nursery to evaluate all of them for compatibility with every fungicide registered. Occasionally, a specific fungicide or a group of related fungicides is found to be incompatible with a certain plant or plants. Therefore, before applying a product to any plant for the first time, make a trial application to a few plants and check for signs of phytotoxicity or other potential problems. Report any suspected incompatibility to the company that manufactured the fungicide or to extension personnel so that the situation can be investigated before serious problems arise.

REFERENCES

Adams, G. C. 1997. Woody Ornamentals Disease Control Guide. Michigan State University Extension Service, East Lansing.

Brent, K. J. 1995. Fungicide Resistance in Crop Pathogens: How Can It Be Managed? GIFAP (International Group of National Associations of Manufacturers of Agrochemical Products), Brussels.

Lyr, H., ed. 1995. Modern Selective Fungicides: Properties, Applications, Mechanisms of Action. 2nd ed. Gustav Fischer Verlag, New York.

Meister, R. T., ed. 1998. Farm Chemicals Handbook. Vol. 84. Meister Publishing, Willoughby, Ohio.

Thomson, W. T. 1997. Agricultural Chemicals. Book 4: Fungicides. 12th ed. Thomson Publications, Fresno, Calif.

CHAPTER 94

Gary W. Simone ▪ University of Florida, Gainesville

Bactericides and Disinfectants

Nurseries rely upon bactericides much less than fungicides for controlling diseases of woody ornamentals. There are a number of reasons for this. First, fungal diseases far outnumber bacterial diseases in ornamentals. The magnitude of fungal diseases has fostered more emphasis on the development, evaluation, and use of fungicides, compared to bactericides. Thus, there are few bactericides for use on plants, especially ornamentals. The inherent expense of broadly labeling bactericides for use on numerous ornamental species in field and greenhouse applications further restricts the utility of these tools for growers. Additionally, the complexity of bacterial disease cycles limits preventative, predictable application of bactericides. Finally, a growing volume of data indicates a real risk that a tolerance for bactericides can develop in pathogens, with repeated use of these compounds.

Effective use of bactericides is dependent upon a knowledge of pathogen survival, infection, and spread. The common bacteria affecting woody ornamentals predominantly cause foliar diseases. The one significant exception is *Agrobacterium tumefaciens,* the crown gall pathogen, which inhabits soil. The common foliar bacteria (*Erwinia, Pseudomonas,* and *Xanthomonas* spp.) are both pathogenic and saprophytic. They can exist epiphytically on the leaf surface. These bacteria are motile and tend to target favorable survival sites on the leaf surface by chemotactic detection of low-level nutrient sources (e.g., amino acids and various carbohydrates) secreted from the leaf surfaces. Small quantities of nutrients from plant cells, pollen, or insect honeydew are dissolved in moisture on leaf surfaces wetted by rain, irrigation, dew, or fog.

Bacteria persist in veinal areas, near hydathodes or trichome bases, in depressions between epidermal cell walls perpendicular to the cuticle surface, or in areas with higher nutrient levels, higher relative humidity, or desirable leaf wetness conditions. The nutrient base allows bacterial populations to increase over time and thus supports the survival of the bacteria in an epiphytic manner. The soil bacterium *A. tumefaciens* colonizes root surfaces in a similar, nonpathogenic manner.

Epiphytic survival of bacteria is not without challenges. Both extremes of leaf wetness and dryness may dramatically lower bacterial survival. Wide variations in temperature and ultraviolet (UV) radiation can similarly reduce bacterial survival in the epiphytic mode. In general, bacterial populations persist at higher levels on the lower leaf surface than on the upper cuticle. The lower leaf surface has more stomata and trichomes, less exposure to UV radiation, and often a thinner cuticle and a greater nutrient base.

Bacteria enter plants by various paths, and their entry can be quite passive. Natural plant openings, such as stomata, hydathodes, and nectaries, can allow bacterial entry, as do wound sites in trichomes, at leaf scars, on the cuticle, or at root hairs. Heavy rain or irrigation facilitates bacterial ingress. The internal plant environment is more conducive to increases in bacterial populations, in response to the greater water supply, higher nutrient levels, lower microbial antagonism, and minimal UV exposure. The movement of bacteria into the substomatal cavity and other locations inside the plant does not always result in active disease development. Hence, bactericides are unsuitable as predictive disease management tools.

The arsenal of tools available for the management of bacterial pathogens is small; the tools are bactericides and disinfestants (Table 94.1). A *bactericide* is an active ingredient that kills bacteria or protects against bacterial invasion. A *disinfestant* is an active ingredient that eliminates or inactivates pathogenic organisms from soil or the surfaces of plants, seeds, or equipment, before they can cause infection.

Bactericides can be separated into three groups: inorganic metals (copper compounds), organic metals, and antibiotics. Disinfestants are chiefly halogenated compounds bearing either chlorine or bromine as the active principle.

Bactericides

Inorganic Metals: Copper Compounds

The sole inorganic metal used as a bactericide is copper. Historically, copper was first employed as a fungicide in 1885, when copper sulfate pentahydrate was mixed with lime, in what became known as Bordeaux mixture. Since then, copper in various forms has been used as a fungicide and later a bactericide to control plant diseases. Simultaneous with the development of Bordeaux mixture was the development of toxic copper carbonate and copper ammonium complex. In the early 1900s, copper oxychloride was introduced as a fungicide. In the period 1920–1930, the efficacy of copper against bacterial diseases was formally recognized with the use of Bordeaux mixture to control fire blight of pome fruits. Copper oxide was introduced in 1932, followed by basic copper sulfate in the 1940s, copper oxychloride sulfate in 1945, and copper hydroxide in 1968. Copper-based bactericides had little impact on the production of ornamentals until copper hydroxide was registered for this use.

Copper-based products are considered protectant pesticides with low cost and high efficacy but nonsystemic action. The sol-

Table 94.1 Bactericides and disinfestants available for use in nurseries[a]

Common name	Trade name and manufacturer	Formulation[b] (%)	Type of action[c]	Movement[d]	Crop usage[e]	Label clearances[f]	Resistance risk[g]	Comments
Inorganic metals								
Copper ammonium complex	Copper-Count-N (Chemical Specialties)	8 EC	P	NS	A, F, V	PS	L	
Copper carbonate								Postharvest use only
Copper hydroxide	Blue Shield (Cuproquim, division of Helena)	77 DF	P	NS	A, F, O, V	PS, Cat	M	
	Champion (Agtrol International)	23 F	P	NS	A, F, O, V	PS	M	
	Champion Formula II (Agtrol International)	37.5 F	P	NS	A, F, O, V	PS, Cat	M	
	Champion (Agtrol International)	77 WP	P	NS	A, F, O, V	PS	M	
	Kocide (Griffin LLC)	61.4 DF	P	NS	A, F, O, V	PS, Cat	M	
	Kocide (Griffin LLC)	23 LF	P	NS	A, F, O, V	PS, Cat	M	
	Kocide 4.5 (Griffin LLC)	37.5 LF	P	NS	A, F, O, V	PS, Cat	M	
	Kocide 20/20 (Griffin LLC)	30.7 WP	P	NS	F, V	PS	M	
	Kocide 101 (Griffin LLC)	77 WP	P	NS	A, F, O, V	PS, Cat	M	
	Kocide 2000 (Griffin LLC)	53.8 DF	P	NS	A, F, O, V	PS, Cat	M	
	Kocide 2000 T/N/O (Griffin LLC)	53.8 DF	P	NS	A, F, O, V	PS, Cat	M	
	KOP-Hydroxide (Drexel Chemical Co.)	37.5 F	P	NS	A, F, O, V	PS	M	
	KOP-Hydroxide 50 (Drexel Chemical Co.)	77 WP	P	NS	A, F, O, V	PS	M	
	Nu-Cop 3L (Micro Flo Co.)	37.5 F	P	NS	A, F, O, V	PS, Cat	M	
	Nu-Cop 50 (Micro Flo Co.)	77 DF	P	NS	A, F, O, V	PS, Cat	M	
Copper oxychloride	Agra-Cop 50 (Agra Chem Sales Co.)	86.2 WP	P	NS	A, F, O, V	PS	M	
	Copper Oxychloride (Drexel Chemical Co.)	85 WP	P	NS	A, F, O, V	PS	M	
	KOP-Oxy 85 (Drexel Chemical Co.)	85 WP	P	NS	A, F, O, V	PS	M	
	Microsperse COC 53 (Micro Flo Co.)	91.4 WP	P	NS	A, F, O, V	PS	M	
	Oxycop 53 (Griffin Chemicals)	92 WP	P	NS	A, F, O, V	PS	M	
	Oxycop 85 (Griffin Chemicals)	85 WP	P	NS	A, F, O, V	PS	M	
	Sungro COC 53 (Lykes Agri Sales)	53 WP	P	NS	F, V	PS	M	
	Sungro Microsperse COC (Lykes Agri Sales)	92 WP	P	NS	A, F, O, V	PS, Cat	M	
Copper oxychloride sulfate								Not available
Copper sulfate, basic	Basic Copper "53" (Micro Flo Co.)	98 WP	P	NS	A, F, O, V	PS	M	
	Basicop 53 (Griffin LLC)	53 WP	P	NS	A, F, O, V	PS	M	
	Basic Copper Sulfate (Drexel Chemical Co.)	98 WP	P	NS	A, F, V	PS	M	
	Basic Copper Sulfate (Old Bridge Chemicals)	99 WP	P	NS	F, V	PS	M	
	BCS – Copper (Pesticide Service Consultants)	39 F	P	NS	A, F, V	PS	M	
	Triangle Brand Copper Sulfate Instant Powder (Phelps Dodge Refining Co.)	99 WP	P	NS	F, V, O	PS	M	
Copper sulfate pentahydrate	Copper Sulfate Fine Crystals (Old Bridge Chemicals)	99 WP	P	NS	F, O	PS	M	Used on bulbs only, as Bordeaux mixture
	Phyton 27 (Source Biologicals)	21.36 L	P	S	O	PS, Cat	M	Categorical use on foliage plants
	Tennessee Brand Copper Sulfate Instant (Griffin LLC)	99 WP	P	NS	F, O	PS	M	Used on bulbs only, as Bordeaux mixture
	Vertagreen (Purcell Industries)	12.5	P	NS	O	PS	M	
Cuprous oxide								Not available

Organic metals								
Copper linoleate (copper resinate)	Camelot (Griffin LLC)	58 FL	P	NS	O	PS	M	
	Copper Fungicide 4E (Griffin LLC)	48 F	P	NS	A, F, O, V	PS	M	
	Tenn Cop 4E (Griffin LLC)	48 FL	P	NS	A, F, O, V	PS	M	Homeowner label
	Tenn Cop 5E (Griffin LLC)	58 F	P	NS	A, F, O, V	PS	M	Used on juniper, pine, rose, and sycamore only
Fosetyl-aluminum	Chipco Aliette (Rhône-Poulenc)	80 WDG	C	S	O, T	PS, Cat	L	
	Aliette T & O (Agro Distribution)	80 WDG	C	S	O, T	PS, Cat	L	
	Prodigy (Lesco)	80 WDG	C	S	O, T	PS, Cat	L	
	Terronate (Terra International)	80 WDG	C	S	F, O, V	PS	L	
Antibiotics								
Oxytetracycline	Mycoject (J. J. Mauget Co.)	4.22 WP	P, C	LS	F, O	PS		Used for tree injection only
	Mycoshield (Novartis Crop Protection)	31.5 WP	P, C	LS	F	PS		
	OTC Tree Injection Formula (Tree Saver)	39.6	P, C	LS	O	PS		Used for palm tree injection only
	Tree Tech OTC (Florida Silvics)	4.57	P, C	LS	F, O	PS		Used for tree injection
Streptomycin	Hopkins Streptomycin 17 (Platte Chemical Co.)	21.1 WP	P, C	LS	A, F, O, V	PS		Limited use on rosaceous hosts
	Agri-mycin 17 (Novartis Crop Protection)	21.2 WP	P, C	LS	A, F, O, V	PS		Limited use on woody rosaceous hosts
Disinfestants								
1-Bromo-3-chloro-5,5-dimethyl hydantoin	Agribrom (Great Lakes Chemical Corp.)	96 G, T	D	NS	O	Cat	L	Injected into mist, fog, or subirrigation system
Calcium hypochlorite	Various products and manufacturers	Various	D	NS	None	. . .[h]	L	Used for treatment of surfaces and tools only
Ethanol (denatured)	Various products and manufacturers	Various	D	NS	None	. . .	L	Used for treatment of surfaces and tools only
Isopropyl alcohol	Various products and manufacturers		D	NS	None	. . .	L	Used for treatment of surfaces and tools only
Quaternary ammonium compounds (*n*-alkyl ammonium chlorides)	Consan Triple Action (Parkway Research Corp.)	20 L	P, D	NS	O, T	PS	L	Used for treatment of surfaces and tools and as a spray on certain ornamental species
	Green Shield PT 2000 (Whitmire Microgen Research Laboratories)	20 L	P, D	NS	None	. . .	L	Used for treatment of surfaces and tools only
	Physan 20 (Moril Products)	20 L	P, D	NS	O, T	PS, Cat	L	Used for treatment of surfaces and tools and as a dip, drench, or spray for plants
	RD-20 (RD & Associates)	20 L	P, D	NS	O	PS, Cat	L	Used for treatment of surfaces and tools and as a drench or spray for cut flowers, orchids, roses, and seedlings
Sodium hypochlorite	Various products and manufacturers	5.26 L	D	NS	None	. . .	L	Used for treatment of surfaces and tools only

[a] Based on label information from currently available products.
[b] DF = dry flowable. EC = emulsifiable concentrate. F, FL = flowable. G = granule. L = liquid. T = tablet. WDG = water-dispersible granules. WP = wettable powder.
[c] C = curative. D = disinfestant. P = protectant.
[d] LS = local systemic. NS = nonsystemic. S = systemic.
[e] A = agronomic crops. F = fruit crops. O = ornamentals. T = turfgrass. V = vegetable crops.
[f] Cat = categorical use on ornamentals, with plant safety evaluation by the user. PS = plant-specific use on ornamentals.
[g] H = high risk. L = low risk. M = medium risk.
[h] Not labeled as a pesticide.

uble copper ion is the active ingredient; it has a multisite mode of action against bacteria and fungi. Copper functions in the inactivation of microbe enzyme systems through the precipitation of proteins.

The inorganic coppers are divided into two groups, based upon solubility in water. Fixed copper products include copper sulfate and copper hydroxide, which are relatively insoluble in water, as a trade-off for increased plant safety. A typical rate of 2 pounds of product (53% metallic copper) in 100 gallons of water liberates enough soluble copper to reach a concentration of about 3.0 ppm in a 2-hour period. Soluble copper products, such as the copper ammonium complex (8% metallic copper), at a rate of 2 quarts per 100 gallons of water, liberates enough soluble copper to reach a concentration of about 400 ppm in the same time span.

Inorganic copper products have generally been very effective, with a few exceptions. Repeated use of these products on field-grown plants can lead to a buildup of copper in the soil. In acidic soil, copper can be taken up by roots and becomes phytotoxic. Foliar application of copper can damage leaves of copper-sensitive species or cultivars, and leaves of other plants can be damaged if copper is applied at high rates. General symptoms of copper injury include root stunting, foliar stunting, reduction in leaf size, abnormal leaf texture, chlorosis, foliar cupping, tipburn or marginal burn, necrotic spotting, and leaf abscission.

Copper-based products have different active ingredients and very in their clearances for use on different crops and at different sites (i.e., field nurseries vs. greenhouses). In general, these products are of limited availability for use on ornamental plant species, in spite of their potential utility for bacterial disease control. The Federal Insecticide Fungicide and Rodenticide Act of 1947 and its later amendments define the legal use of every pesticide in the U.S. marketplace. The label of a product registered for use on ornamental plants must clearly list the names of the plants or state that the product is approved for broad, categorical use on ornamentals or a subgroup such as woody ornamentals. Additionally, the label must clearly state whether the product can be used in the field, greenhouses, or in interior landscapes (interiorscapes).

The copper compounds described below are active ingredients in products that are legally approved for use as bactericides on one or more crops. Few of these bactericides are legally approved for broad use (or any use) on woody ornamentals. Nursery workers are urged to contact a county extension agent or state extension plant pathologist for current trade names, utility on ornamentals, and state regulations.

Copper sulfate pentahydrate This compound, $CuSO_4 \cdot 5H_2O$, is both a preventative fungicide and an algaecide. It is highly phytotoxic to plants and is generally not used as a bactericide or fungicide unless it is made safe by being mixed with calcium hydroxide (lime), in a form known as Bordeaux mixture. A copper sulfate pentahydrate formulation is marketed as a bactericide-fungicide with legal use on various flowering, foliage, and woody ornamentals. This product is 21.36% active, with a formulation that is systemic in action.

Copper sulfate pentahydrate + calcium hydroxide The original Bordeaux mixture is copper sulfate pentahydrate and calcium hydroxide ($CuSO_4 \cdot 5H_2O + Ca(OH)_2$). The addition of lime to copper pentahydrate made it safer for use on plant tissue. This product has no appreciable commercial use on ornamentals and persists as a bactericide for homeowners, with very general labeling. Bordeaux mixture leaves a significant residue and is safest as a dormant plant application.

Basic copper sulfate This compound, $Cu_5O_4 3Cu(OH)_2 \cdot H_2O$, is the second most widely used copper-based bactericide on ornamentals. In products containing basic copper sulfate, the active ingredient ranges from 50 to 53%, with wettable and flowable formations available. Many brands have label text that includes some ornamental hosts. These products are fixed coppers with some risk of plant injury when applied to copper-sensitive species. Most brands have plant-driven label text, so the legal use of these bactericides on woody ornamental species is limited.

Copper oxychloride sulfate Few products contain copper oxychloride sulfate ($3Cu(OH)_2CuCl_2$ and $3Cu(OH)_2CuSO_4$), and they are mostly for agronomic, fruit, and vegetable crops. There are few clearances for the use of copper oxychloride sulfate on ornamental plants in the United States. This formulation is corrosive to metal spray tanks and is highly incompatible with dithiocarbamate fungicides in tank mixes.

Copper hydroxide The development of products containing copper hydroxide ($Cu(OH)_2$) was a step beyond basic copper sulfate products. The hydroxides are more finely ground, offering superior leaf coverage, less residue, and improved redistribution on the leaf surface in successive periods of leaf wetness. Many hydroxide products are available in the marketplace, but they differ markedly in their legality for use on ornamentals and their suitability for applications in the field and greenhouse. The legal use of certain brands of copper hydroxide products has been expanded to include a broad range of ornamental plants, in the field and in the greenhouse. Perhaps most important is the "Notice to User" statement on certain product labels, which allows professional users to apply these products on any plant species considered an ornamental, following an in-nursery plant safety evaluation.

Copper compounds not labeled for use on ornamentals Copper carbonate ($Cu(OH)_2CuCO_3$), copper ammonium complex (exact formula unknown), copper oxychloride ($3Cu(OH)_2CuCl_2$), and copper oxide (Cu_2O) are not labeled for use on ornamentals.

Organic Metals

Organic metal bactericides deliver toxic metals to the plant within the framework of an organic molecule rather than an inorganic salt. Two such compounds are available: copper linoleate and fosetyl-aluminum.

Copper linoleate This compound, also called copper salts of fatty and rosin acids, has long been used as a fungicide-bactericide for agronomic, fruit, and vegetable crops, but only a few legal use sites are listed for woody ornamentals. Copper linoleate is quite soluble and relatively persistent on leaf tissue as a bactericide. It is presently not adequately labeled for legal use by woody ornamental producers.

Fosetyl-aluminum This compound, aluminum tris (O-ethyl phosphonate), is a systemic fungicide that is highly selective for downy mildew fungi, *Pythium* spp., and *Phytophthora* spp., with the unique ability to be internally distributed upward and downward throughout the plant at an effective dosage. Suppression of fire blight of ornamental pear, pyracantha, and hawthorn species by fosetyl-aluminum has been documented and added to the label text. Additionally, suppression of various bacterial diseases of foliage, perennial, and woody plant species caused by *Xanthomonas campestris* pv. *hederae, X. campestris* pv. *dieffenbachiae, X. campestris* pv. *syngonii,* and *X. campestris* pv. *fici* has been noted, and the compound was found to control these diseases as well as or better than copper hydroxide sprays. The label text allows nurseries to further evaluate the efficacy of fosetyl-aluminum for other bacterial diseases of ornamental species. This compound has shown bactericidal activity against only *Erwinia* and *Xanthomonas* spp., thus far. Additionally, because of the low pH of the fosetyl-aluminum spray solution, copper toxicity can occur if this compound is tank-mixed with a copper-based bactericide or is applied to plants after an interval of less than seven days following application of a copper-based bactericide.

Antibiotics

Two antibiotics, streptomycin and oxytetracycline, have been borrowed from human and veterinary medicine and employed in the treatment of plant diseases.

Streptomycin and oxytetracycline are underutilized for bacterial disease management, but this situation is not likely to change. Two serious liabilities are associated with antibiotic use. Repeated exposure to these compounds can lead to bacterial tolerance in humans and the resultant loss of important medicine for fighting human disease. Additionally, their narrow mode of action against their target pathogens can result in the development of tolerant strains in the field, which could cause crop loss and a total loss of efficacy of the antibiotic.

Streptomycin The antibacterial compound streptomycin is derived from the actinomycete bacterium *Streptomyces griseus.* Its mode of action is inhibition of protein synthesis in the bacterial cell. It is more effective against gram-positive bacteria (in the genera *Bacillus, Clavibacter, Curtobacterium, Rhodococcus,* and *Streptomyces*) than gram-negative bacteria (*Erwinia* and *Pseudomonas*), which are more important on woody ornamentals. Streptomycin is a local systemic, with the ability to penetrate from one leaf surface to another. It is most effective when used under conditions of slow drying, either in late evening or early morning, when air circulation is minimal. Excessive absorption of streptomycin by leaves can cause a short-term blotchy chlorosis. Streptomycin is not compatible with pyrethrins, lime sulfur, or alkaline spray materials.

Streptomycin is commonly marketed as streptomycin sesquisulfate or, less commonly, streptomycin sulfate or streptomycin nitrate. Legal use of streptomycin products is limited to the management of fire blight of pome fruits and some ornamental hosts in the rose family. Bacterial crown gall of rose can also be treated with streptomycin used as a root dip according to specific state labeling.

Oxytetracycline The second antibiotic in use in the nursery industry is oxytetracycline, another metabolic derivative of an actinomycete, *Streptomyces rimosus.* This compound is broadly antibacterial and effective against phytoplasmas. Like streptomycin, it interferes with protein synthesis in bacteria and phytoplasmas.

Oxytetracycline products have limited labeling for plant disease management. This antibiotic is effective and legal for the management of fire blight of pome fruits and susceptible ornamentals and bacterial spot of *Prunus* spp. and for the prevention and suppression of lethal yellowing disease of palms.

Disinfestants

Various products eradicate bacteria, primarily from nonporous surfaces, and are used in niche areas in the ornamental plant production industry. They are primarily for use on propagation surfaces, pots, trays, pruning and propagating tools, and transplant containers. The active ingredients in these products include isopropyl alcohol (rubbing alcohol), denatured alcohol, phenolic-based germicides, sodium hypochlorite (household bleach), and calcium hypochlorite (swimming pool additive). These products are broadly antimicrobial but have short life spans, especially if they are in contact with organic material. They also have little or no margin of safety for plants in direct applications.

The activity of chlorinated compounds has been modified and made safer by the development of mixtures of *n*-alkyl ammonium chloride products. These widely used disinfestants are marketed under various trade names, differing somewhat in labeling. As a group, they offer spray or drench activity on a narrow group of ornamentals, including some woody species. The targeted pathogens are *Erwinia amylovora* (the fire blight bacterium), *Agrobacterium tumefaciens* (the crown gall bacterium), and various foliar bacteria. These products are safe for use on plants at concentrations up to 400 ppm. The utility of these products to the woody ornamental producer falls between tool disinfestation and dip treatment of cuttings before mist propagation.

Another disinfestant of consequence to the nursery industry is a brominated water additive, 1-bromo-3-chloro-5,5-dimethyl hydantoin, which controls algae and various microbes, including bacteria and fungi, in irrigation systems and recirculating cooling water systems and on greenhouse and nursery surfaces. It must be injected into the water system at a prescribed rate, determined by water volume and temperature. This product is effective at concentrations of 10–35 ppm; for preventative purposes, a maintenance rate for water treatment is 5–15 ppm.

Bromine at rates of 1–5 ppm can eradicate *Erwinia chrysanthemi* and *A. tumefaciens.* In efforts to manage diseases of ornamentals caused by *Erwinia, Pseudomonas,* and *Xanthomonas* spp., bromine provided acceptable control at 55 ppm, but its efficacy decreased at application rates stated on the product label. Plant safety when bromine is used at rates exceeding those listed on the product label is quite variable, depending on the plant species treated. At rates of 50–60 ppm, bromine is phytotoxic, causing root pruning (where it is used in subirrigation mats), flower spotting, leaf drop, plant stunting, and leaf chlorosis. The utility of this product for the woody plant producer may best reside in the greenhouse or propagation house, where it is applied through a mist system for species sensitive to bacterial diseases. Bromine products should be evaluated before widespread use on plant species.

Other Tools

The effect of acid seed soaks and leaf sprays in reducing disease in a number of host-pathogen systems has been investigated. Research with fire blight and blast of pear in the 1970s indicated significant control of both when trees were sprayed with solutions of citric or tartaric acid (pH 3.0, 100 mM). The acidifying effect of the acid sprays on the leaf cuticle lasted from two to four days, creating an unfavorable environment for bacteria.

Leaf surface acidification has been adopted by a number of producers of floral, foliage, and woody ornamentals. Acetic acid (white vinegar) and phosphorous acid (H_3PO_3) are used as foliar sprays in an avoidance strategy, to make the leaf surface unsuitable for populations of epiphytic bacteria. Ornamental foliage plant species susceptible to bacterial pathogens have been routinely sprayed with white vinegar (5% acidity). Both acidic treatments are measurably effective in protecting some hosts against some pathogens. Phosphorous acid is effective in both bacterial and fungal disease management in certain host-pathogen systems. Its mode of action in the plant does appear to exceed that of acidification alone.

Bactericide Use Protocol

Growers should follow the following bactericide use protocol for best management of bacterial diseases:

1. Determine which species in the nursery are susceptible to bacterial diseases.
2. Check with local extension personnel to determine all legal options for best bactericide management of these diseases.
3. Apply a copper-based product that is legally approved for the crop and the site to be treated.
4. After two or three applications of copper, apply an antibiotic (if legal), and alternate the treatments.
5. If disease does not abate, consider more frequent treatment with copper applied at fractional rates or in a tank mix with a broadly labeled mancozeb fungicide.
6. Monitor the tolerance of the bactericide in use. Some plant disease clinics can perform copper or streptomycin tolerance assays to determine the sensitivity of pathogens to these active ingredients.
7. Definition of a tolerant bacterial pathogen should result in a reexamination of the entire production system to best exclude this pathogen from future production cycles.
8. Evaluate alternative bactericides, such as fosetyl-aluminum and phosphorous acid (applied as a foliar phosphate source).
9. If problems persist, consider pursuit of special state labeling for the use of streptomycin or tetracycline products to alternate with copper.
10. Eliminate highly susceptible cultivars.

REFERENCES

Beattie, G. A., and Lindow, S. E. 1995. The secret life of foliar bacterial pathogens on leaves. Annu. Rev. Phytopathol. 33:145–172.

Blakeman, J. P., and Fokkema, N. J. 1982. Potential for biological control of plant diseases on the phylloplane. Annu. Rev. Phytopathol. 20:167–192.

Chase, A. R. 1989. Aliette 80WP and bacterial disease control – *Xanthomonas*. Nursery Dig. 24:26–27.

Chase, A. R. 1990. Control of some bacterial diseases of ornamentals with Agribrom. Proc. Fla. State Hortic. Soc. 103:192–193.

Cooksey, D. A. 1990. Genetics of bactericide resistance in plant pathogenic bacteria. Annu. Rev. Phytopathol. 28:201–219.

Frear, D. E. H. 1948. Chemistry of Insecticides, Fungicides and Herbicides. 2nd ed. D. Van Nostrand, New York.

Goto, M. 1990. Fundamentals of Bacterial Plant Pathology. Academic Press, New York.

Horsefall, J. G. 1945. Fungicides and Their Action. Chronica Botanica, Waltham, Mass.

Marco, G. M., and Stall, R. E. 1983. Control of bacterial spot of pepper initiated by strains of *Xanthomonas campestris* pv. *vesicatoria* that differ in sensitivity to copper. Plant Dis. 67:779–781.

Sands, D. C., and McIntyre, J. L. 1975. Acid sprays decrease populations of *Erwinia amylovora* applied to pear trees. (Abstr. NE-62.) Proc. Am. Phytopathol. Soc. 2:106.

Sands, D. C., and McIntyre, J. L. 1977. Citrate and tartrate sprays for reduction of *Erwinia amylovora* and *Pseudomonas syringae*. Plant Dis. Rep. 61:823–827.

Sands, D. C., and McIntyre, J. L. 1977. Possible methods to control pear blast caused by *Pseudomonas syringae*. Plant Dis. Rep. 61:311–312.

Sigee, D. C. 1993. Bacterial Plant Pathology: Cell and Molecular Aspects. Cambridge University Press, New York.

Swings, J. G., and Civerolo, E. L. 1993. *Xanthomonas*. Chapman and Hall, London.

CHAPTER 95

Sharon L. von Broembsen ▪ Oklahoma State University, Stillwater
James D. MacDonald ▪ University of California, Davis
Jay W. Pscheidt ▪ Oregon State University, Corvallis

Disease Management for Nurseries Using Recycling Irrigation Systems

An increasing number of nurseries use recycling irrigation systems, in which runoff from production areas is captured in retention basins and recycled as irrigation water for crops. Some use irrigation recycling systems to reduce water waste or cost or to ensure the availability of an adequate water supply. Others have adopted recycling as a way to prevent pollution of offsite water resources by limiting nutrients, pesticides, and other contaminants in tailwater (von Broembsen, 1998). While increased efficiency of water use and reduced offsite pollution are major benefits, recycling also has disadvantages. The most obvious is the cost of installing storage ponds and acquiring additional pumping capacity. These costs may be recovered through savings in water costs over time. There has also been some concern that herbicides could be recycled and damage sensitive crops, but this has been shown to be avoidable with proper management. Likewise, buildup of salts in recycled water can be effectively managed by dilution with fresh water, if it becomes a problem.

However, the main disadvantage of recycling may be risk of disseminating waterborne pathogens in crops, thus increasing the likelihood of disease. Plant-pathogenic fungi, such as *Phytophthora* and *Pythium* spp., are present in nursery runoff at relatively high concentrations (MacDonald et al., 1994) and can be detected at low levels in recycled irrigation water at the point of delivery to crops (Clint, 1995). Irrigating healthy plants for long periods with recycled nursery water can result in significant levels of root infection (MacDonald and Kabashima, 1998). Some nurseries employing recycling have installed elaborate water treatment systems to disinfect water, but most have not. Nurseries that do not treat water recognize the disease risk but often are uncertain about how to assess the risk or what options are available for treating water. Most rely on prophylactic fungicide treatment programs to suppress pathogen activity generally.

Managing Plant Pathogens in Recycled Irrigation Water

Every nursery using recycled irrigation water has unique needs in managing plant pathogens, but the most important step in any situation is to determine whether pathogens are present in irrigation water and to what extent. Various management practices for reducing contamination can be then considered. Samples should be taken at various times in the production cycle, as pathogens may be seasonal in their activity. Water samples should be taken at the irrigation water source, at points of runoff, and at points where recycled water is delivered back to plants. Another practical way to sample is to place certain plant parts, e.g., green pears or lemon leaves, in irrigation water to "bait" pathogens. Water or plant bait samples can be analyzed by a diagnostic laboratory for pathogens of importance, such as *Pythium* and *Phytophthora* spp. A diagnostic laboratory can also give specific instructions on how to take and submit samples. An important but often overlooked consideration for all nurseries is the need to test the source of the irrigation water for plant pathogens. Ground water drawn from properly constructed wells and water suitable for human consumption should be pathogen free. However, water drawn from surface water sources such as lakes and rivers may contain waterborne pathogens and may require decontamination before use in propagation operations and for susceptible crops.

The risk of using recycled irrigation water can best be evaluated by assaying water collected at the points of reuse. Pathogens are often present at relatively high concentrations in runoff water, but relatively small numbers of propagules are detected in water at the points of reuse. This is probably the result of natural processes, such as microbial and physical degradation acting within the system to reduce pathogen populations. Propagules of plant pathogens tend to settle out in still or slowly moving water. Once they have settled to the bottoms of basins, they are effectively out of circulation and are continually subject to degradation. Larger retention basins and longer retention times before reuse of captured water promote the settling and degradation processes. Additionally, if captured water is mixed with pathogen-free fresh water before reuse, the concentration of pathogenic organisms is diluted.

Pathogen-free recycled water may be a desirable objective, but complete decontamination may not be economically or technologically feasible. Complete disinfection of irrigation water may not even be necessary for most crops, since very small amounts of inoculum in irrigation water may be inconsequential. However, little is known about acceptable thresholds for different pathogens on different crops. Crops that are highly susceptible to waterborne pathogens such as *Phytophthora* spp. (e.g., citrus, Lawson cypress, dogwood, and rhododendron) should be grouped together in the same part of the nursery, so that pathogen-free fresh water or decontaminated water can be reserved for them and for propagation areas. Where recycled water is used with no treatment other than settling, holding, and dilution, it should be used only for hardier or more mature plants that tend to be more resistant to waterborne pathogens. By following these strategies, nurseries may find that stringent decontamination of all recycled water is not necessary. However, if large parts of the nursery contain crops susceptible to waterborne pathogens, decontamination of recycled water is probably advisable.

Decontamination Methods

Several methods can be used to decontaminate source water and recycled water. Plant pathogens can be removed by filtration. Modern sand filters using graded sand reduce the number of fungal spores and nematodes in water but have little effect on bacteria. Microfiltration to even smaller pore sizes can remove most fungal and bacterial plant pathogens, but it is useful only for low flow rates and low volumes, such as those required in propagation areas and greenhouses. More stringent methods can also be used, provided that the water is filtered to a reasonably clean level before further decontamination. These methods, adapted from methods for purifying drinking water and swimming pool water, involve decontamination by chlorine, ozone, or ultraviolet (UV) light. All three methods are very effective in eliminating plant pathogens and other microorganisms from water, but they require careful management to achieve the desired effect.

Filtration

It is theoretically possible to eliminate nematodes, fungi, and bacteria from water by filtration, but this approach has found only very limited application in nurseries. Filtration systems with pores small enough to remove all pathogens can treat only small volumes of water at low flow rates, and they clog quickly. Greater success has been obtained in greenhouse and hydroponic systems, in which irrigation water is cleaner initially and is used in smaller quantities at lower pumping rates. However, filtration of reclaimed water is an essential component of all decontamination methods. At the simplest level, filtration is employed to remove particulate contaminants before microbial disinfection is attempted. Suspended solids reduce the efficacy of treatment with chlorine, ozone, or UV light. The major problem of disinfection by filtration is the clogging of pores with fine particulates and algal slimes. This is overcome by regular reverse flushing and chemical cleaning.

The sizes of propagules of various common root-infecting pathogens are shown in Table 95.1, as an indication of the pore size required for water decontamination. A filter that retains particles greater than 1 μm in diameter will eliminate all fungal spores but not bacterial cells or virus particles.

Advantages and disadvantages of filtration as a method of decontaminating water are listed in Table 95.2.

Sand filtration Sand filters, composed of specific grades of sand, are used to remove particulate matter from water. They probably reduce the pathogen load in water, but the pores between sand grains are too large to effectively remove all fungal spores and bacterial cells.

Microfiltration and ultrafiltration Microfiltration is theoretically capable of disinfecting nutrient solutions, but it has been found to be ineffective in practice, because of the broad spectrum of pore diameters in microfilter membranes. A membrane with a nominal mean pore diameter of 0.2 μm actually has a wide range of pore diameters, with some as large as 10 μm. Certain pathogen propagules can easily pass through such large pores. Microfilters may be most useful as prefilters, removing particulates in advance of some other treatment. However, they remove only particulates; they cannot remove dissolved organic materials, which reduce the efficacy of later treatments.

Ultrafilters differ from microfilters in having a much lower mean pore diameter (10^{-3} μm, compared to a lower limit of 10^{-1} μm) and a generally different method of filter construction. An ultrafiltration cartridge is formed by embedding bundles of synthetic fibers in an epoxy resin matrix. Liquid is pumped through ultrafilters at a high velocity at low pressure (1–2 bars).

Horticultural uses There are few reports of filtration systems used as the only method for eliminating plant pathogens from water. Cartridge-type filters have been used to disinfect nutrient solution in hydroponic systems, and a "slow sand filtration" process has been used in some greenhouse operations. However, these systems can treat only small quantities of water and have not been used effectively to disinfect recirculated water in nurseries.

Chlorine

Treatment of domestic water supplies with chlorine has been a widespread practice since the early part of the twentieth century, to eliminate bacteria and viruses and thereby protect public health.

Table 95.1 Size of propagules of some root-infecting pathogens

Pathogen	Propagule structure	Size (μm)
Fungi		
Fusarium oxysporum	Microconidia	5–12 × 2–3.5
	Macroconidia	17–66 × 3–5
Phytophthora parasitica	Zoospore cysts	8–10 (diam.)
	Sporangia	38 × 30
Pythium aphanidermatum	Zoospores	5 × 3
	Oospores	17.5 × 17.5
Thielaviopsis basicola	Macroconidia	14–16 × 3–4
Bacteria		
Erwinia carotovora	Cells	1.3 × 0.5–0.8
Xanthomonas campestris pv. *pelargonii*	Cells	1–1.5 × 0.5–0.7

Table 95.2 Filtration as a method of decontaminating water

Advantages	Disadvantages
Filtration improves water purity and thus increases the effectiveness of subsequent disinfection	Sand filters only partially reduce the pathogen load
Filtration has no effect on nutrients in solution, Fe chelates, or pesticides	Microfilters are more effective but may let some pathogen spores through
No harmful chemicals are used, so there is minimal risk to the operator and to crop health	Ultrafilters are most effective in removing pathogen spores, but they clog relatively quickly
	Microfiltration and ultrafiltration have high capital costs
	Cleaning filters by backwashing produces a waste liquid, which requires separate disposal

The importance of a clean water supply in protecting plant health was also recognized about this time. Bewley and Buddin (1921) demonstrated that treatment with chlorine at 20–50 mg/L killed mycelium of *Phytophthora cryptogea* and was one of the most effective chemicals for control of plant-pathogenic fungi in nonmains water used for irrigation of glasshouse crops. Problems with crop phytotoxicity, rapid inactivation of chlorine by dissolved and suspended organic matter in recirculated water, an inability to measure chlorine readily in nutrient solutions, and health hazards appear to have deterred the widespread adoption of chlorine for treatment of recycled water in crops.

Chlorination of water is generally accomplished by adding metered amounts of sodium hypochlorite solution, calcium hypochlorite solution, or chlorine gas. Chlorine that is present in solution as chlorine, hypochlorite, or hypochlorous acid is known as *free chlorine* or *available chlorine.* These molecules are very reactive and readily combine with organic matter, ammonia, or nitrogen in oxidation reactions. Chlorine tied up in this manner is *combined chlorine* or *unavailable chlorine.* The amount of chlorine inactivated by chemical reaction (the chlorine demand) depends on the impurities, particularly the organic matter, in a water supply. Hence, the killing effect of chlorine is quickly reduced in a peat leachate solution, as the chlorine combines with organic matter. To be certain that spores of target fungi are exposed to available chlorine at the required minimum concentration, it is necessary to maintain that minimum level of residual free chlorine in the water. Thus, the residual chlorine concentration must be continually monitored.

A chlorine residual is kept in drinking water after treatment in order to maintain a potable supply in case of contamination along the distribution system. In theory, a distributed residual chlorine dose should offer similar benefits if applied to a crop grown in a recycled watering system; however, this is not the case in practice. For crops in an inert substrate or a nutrient film system, residual chlorine sufficient to kill pathogens may be phytotoxic, and for crops grown in organic media, the residual dose is soon lost.

Advantages and disadvantages of chlorination as a method of decontaminating water are listed in Table 95.3.

Table 95.3 Chlorination as a method of decontaminating water

Advantages	Disadvantages
Low capital cost	Effectiveness declines rapidly in water containing particulates or organic substances
Low running cost	Chlorine is a less powerful disinfectant than ozone
Wide spectrum of biocidal activity	Risk of phytotoxicity to crops
Simple operating system	Effectiveness varies with pH
Rapid action	Risks to human and crop health if chlorine leaks from the treatment system
No effect on Fe chelates or pesticides	Risk of corrosion of metal equipment
Continuous monitoring and control of the level of chlorine in the water is possible	Potential production of trihalomethanes and carcinogens, which are harmful to human health

Chlorine Products for Treating Water

Sodium hypochlorite Commercial sodium hypochlorite solutions sold for use as disinfectants or bleaching agents generally contain 10–14% available chlorine (100–140 g/L). The product is diluted to achieve the target concentration (e.g., 2 mg/L) in the water supply. It is a colorless to pale yellow liquid with the smell of chlorine. During storage, chlorine gas is gradually lost, and the percent of available chlorine falls. Sodium hypochlorite exerts its disinfectant and bleaching properties through oxidation reactions. If the material being oxidized is a living microorganism, cell processes and structures are disrupted, and the organism is killed. However, if the hypochlorite reacts with root cells of a plant, some of the cells may be killed. Sodium hypochlorite may be used in water treatment systems in preference to gaseous chlorine to avoid the hazards of handling and storing poisonous gas.

Calcium hypochlorite Like sodium hypochlorite, calcium hypochlorite is available as a solution, but the concentration of available chlorine is greater (35%). Calcium hypochlorite produces the hypochlorite ion and hypochlorous acid when dissolved in water. It has been used experimentally for treating water in horticultural processes (Jenkins, 1981; Segall, 1968), although sodium hypochlorite appears to be the more usual source in commercial horticulture. If there is a risk of accumulation of sodium to phytotoxic levels by continual dosing with sodium hypochlorite in an enclosed system (Price and Fox, 1984), calcium hypochlorite may be a preferable source of chlorine.

Chlorine gas Chlorine gas is the cheapest form of chlorine and has long been used to treat municipal drinking water. However, it is phytotoxic and deadly poisonous to humans and must be used with great caution. Some nurseries have successfully employed gas-based chlorination systems, but many have been discouraged from this approach by stringent environmental and public health regulations governing gas storage and use. To bypass problems associated with gas transportation and storage, there have been efforts in South Africa to develop electrochemical systems that can generate chlorine gas on-site from salt brine solutions. These systems were developed for treatment of rural drinking water supplies, and we are unaware of their application to other uses. Like sodium and calcium hypochlorites, gaseous chlorine produces the hypochlorite ion and hypochlorous acid when dissolved in water.

Factors Affecting Chlorine Activity

The killing effect of chlorine depends on concentration, time, water quality (especially organic matter content), temperature, and pH. The residual concentration in drinking water is generally around 0.5 mg/L, but this level is not sufficient to kill most plant-pathogenic fungi. No single dose ensures mortality of all plant pathogens under all conditions. Thus, a key requirement for chlorination (or any other disinfection process) is an effective assay to detect target pathogens.

Concentration and time Zoospores of *Phytophthora cinnamomi* were shown to be killed by exposure to residual chlorine at a concentration of 2 mg/L at 18°C for 1 minute (Smith, 1979). Exposure to solutions of 15 mg/L for 30 seconds and 10 mg/L for 10 seconds was also effective. Mycelium containing chlamydospores was killed when immersed in a solution of chlorine at a concentration of 100 mg/L for 24 hours or a concentration of 200 mg/L for 4 hours. Treatment in a solution of 50 mg/L for 24 hours was ineffective. At pH 4.0, free chlorine at concentrations less than 100 mg/L was effective in killing mycelium of pythiaceous fungi, and zoospores were killed without the addition of chlorine (Pittis, 1981).

Conidia of *Fusarium oxysporum* f. sp. *dianthi* in nutrient solution were killed by exposure to chlorine at a concentration of 5 mg/L for 15 minutes (Price and Fox, 1984). Spores of *F. oxysporum* f. sp. *lycopersici* in nutrient solution were generally killed by exposure to chlorine at 1 mg/L for 2 hours, although in some tests some spores survived a concentration of 5 mg/L for 2 hours (Runia, 1988).

Chlorine at 2 mg/L for 24 hours at 25°C apparently killed resting spores of *Plasmodiophora brassicae* and prevented club root of cabbage plants in laboratory tests. A higher concentration for a shorter time (20 mg/L for 5 minutes) was also effective but was phytotoxic to the cabbage plants (Datnoff and Kroll, 1987).

Chlorine reacts with phenols and with unsaturated bonds in organic matter and also with reducing agents, such as Fe^{2+} and Mn^{2+}. All these reactions increase the chlorine demand of water and thus reduce the disinfection capacity of a given chlorine concentration and in a given exposure time. Chlorine also reacts with ammonium in solution to form chloramines, often termed combined chlorine residuals. Chloramines, like chlorine, have significant disinfecting power. Generally, however, a greater concentration of combined chlorine residual than of free chlorine residual is required to accomplish a given kill in a specified time.

pH Low pH favors the formation of hypochlorous acid over the formation of the hypochlorite ion and thus enhances disinfection by chlorine. The activity of hypochlorous acid is of the order of 100 times that of the hypochlorite ion. However, although the acid is more toxic than the ion, it is also unstable. Hypochlorous acid reacts with hypochlorite to produce chlorate and hydrochloric acid, and this decomposition is self-accelerating as the pH falls. To achieve the best activity over a period of time, therefore, it is often necessary to maintain an alkaline pH. Decomposition is kept to a minimum in commercial hypochlorite solutions by the presence of sodium hydroxide. Spores of *Alternaria tenuis* were killed more readily by calcium hypochlorite at pH 6–7.5 than at pH 8 (Segall, 1968). Hypochlorite at a concentration of 100 mg/L at pH 7.6 has the same effect on *Bacillus subtilis* spores as it has at a concentration of 1,000 mg/L at pH 9.

Temperature Zoospores of *P. cinnamomi* were killed at a slightly lower rate at 23°C than at 18°C (Smith, 1979). Other workers have found that the effectiveness of chlorine declines rapidly below 10°C.

Antibacterial Activity of Chlorine

Lacey et al. (1972) demonstrated that chlorination of contaminated irrigation water helps to control bacterial rot of iris, caused by *Erwinia chrysanthemi* and *E. carotovora* subsp. *carotovora.* They treated water with chlorine at 20 mg/L for 1 hour. The effectiveness of chlorination was reduced when the water was dirty and when the number of bacteria was increased. In the presence of 10% sterile peat and with a contact time of 15 seconds, sodium hypochlorite was effective against *E. carotovora* and *Pseudomonas marginalis* at a concentration of 1% and against *Xanthomonas campestris* at a concentration of 10%, but it was ineffective against *Clavibacter michiganense, P. corrugata,* and *X. graminis* (Thompson and Williams, 1986).

In a tomato crop produced by the nutrient film technique, chlorination of the nutrient solution reduced populations of pathogenic bacteria, with a chlorine concentration of 3 mg/L, but treatment at this rate markedly reduced root development, and there was evidence of root damage even at a concentration of 0.5 mg/L (Ewart and Chrimes, 1980).

Sodium hypochlorite at concentrations of 25 mg/L or more completely inhibited the growth (on agar plates) of a *Pseudomonas* sp. that causes a bacterial wilt of sweet pepper (Teoh and Chuo, 1978). When plants grown in granite chips were irrigated with water containing chlorine at concentrations of 15 mg/L or more, no plants died from bacterial wilt, while treatment at 10 mg/L was ineffective. The growth and yield of plants was reduced as chlorine concentration increased above 10 mg/L. Affected plants showed leaf chlorosis and stunted growth.

Chlorine Phytotoxicity

Chlorine treatment of irrigation water has been reported to result in phytotoxic symptoms in growing crops, particularly crops grown in hydroponic systems or in an inert substrate. Phytotoxicity may result from oxidation of root cells or cell contents, from the presence of toxic chlorate in the chlorine supply, or from an accumulation of sodium ions in a recirculating system. Although chlorine treatment is more effective against phytopathogenic microorganisms in crops grown in nutrient film technique systems and in inert substrates, it is also true that, in such cases, the chlorine is in very intimate contact with roots and reacts with organic matter, including root cells. Sodium chlorate, which is highly toxic to plant growth, may occur as a contaminant in sodium hypochlorite solutions, and the amount may slowly increase with storage. Sodium ions in a recirculating solution accumulate with continual application of sodium hypochlorite, and this may prove toxic to some crops, e.g. carnation (Price and Fox, 1984). A high sodium concentration alters the sodium–potassium ratio in solution, and plants may develop symptoms of induced potassium deficiency.

Experiments by Frink and Bugbee (1987) indicated that irrigation water with a residual chlorine concentration of less than 1 mg/L should not adversely affect growth or appearance of most potted plants and vegetable seedlings grown in a peat-perlite-vermiculite medium. The plants were irrigated from above twice weekly, with much of the water contacting the foliage. Growth of geranium and begonia declined at a concentration of 2 mg/L, pepper and tomatoes at 8 mg/L, lettuce at 18 mg/L, and broccoli, marigold, and petunia at 37 mg/L. Germination of vegetable seedlings was unaffected. Affected plants showed leaf chlorosis and reduced weight. Irrigation water containing chlorine at 10 mg/L has been applied to a wide range of nursery stock grown in peat-based media without adverse effect on plant growth. Treated plants included *Azalea, Berberis, Calluna, Chamaecyparis lawsoniana, Cotoneaster, Deutzia scabra, Erica cinerea, Hydrangea, Ilex acruifolium, Thuja occidentalis, Viburnum* spp., and *Weigela florida* (Scott et al., 1984). The high concentration of chlorine (200 mg/L) required to kill mycelium of *P. cinnamomi* was phytotoxic to newly rooted cuttings of *Abelia, Caryopteris, Fuchsia,* and *Rosmarinus* when used as routine watering from June to October (Smith, 1979).

Chlorine treatments that successfully achieve pathogen mortality pose a risk of phytotoxicity. Hence, chlorine-based treatments require careful biological and chemical monitoring for successful use.

Health Hazards and Other Adverse Effects of Chlorine

Chlorine gas is highly toxic to humans. Sodium hypochlorite causes burns on eyes and skin and, if ingested, causes internal irritation and damage. Formaldehyde reacts with hypochlorite to produce a carcinogen (bis-chloromethyl ether). Treatment of drinking water and municipal wastewater with chlorine is being discontinued in some countries, because chlorine reacts with humic substances to form trihalomethanes, which may be harmful to human health.

Sodium hypochlorite is corrosive and may damage metal parts of irrigation systems. Decomposition of sodium hypochlorite in incorrectly designed sealed containers may lead to explosion.

Ozone

Ozone is a powerful oxidizing agent and has been used for disinfecting water supplies since the beginning of the twentieth century. It is commonly used as a component of systems for purifycation of drinking water, swimming pool water, and municipal and industrial wastewater. More recently, it has been used for treating recycled irrigation water. Treatment with ozone involves bubbling the gas through water, with fine bubbles to ensure good contact with the solution. Excess ozone must be deactivated (usually by venting through an activated charcoal filter) before it is released to the atmosphere, as it is a severe irritant of nasal and throat tissues and poses health risks to workers.

Advantages and disadvantages of ozone treatment as a method of decontaminating water are listed in Table 95.4.

Production of Ozone

Ozone is commonly produced by passing a high-voltage electrical discharge across a dry, oxygen-rich gas. About 10% of the energy supplied is used to make ozone, and the remainder is lost, primarily as heat. Use of a 100% oxygen stream rather than air results in production of 21 times as much ozone.

Factors Affecting Ozone Activity

The disinfective capacity of ozone is affected by organic matter, pH, conductivity, and the amount and type of iron chelates. If ozone reacts with organic matter, the amount of ozone remaining in solution and available for killing microorganisms is reduced. The rate of breakdown of ozone to oxygen and hydroxyl ions increases at high pH. Ozonation increases pH, so it may be necessary to add acid to the treatment chamber to maintain optimum pH. Solution conductivity influences the effectiveness of ozone treatment (Runia, 1990). Effectiveness is reduced at higher conductivities, because more ozone reacts with ions in solution as their concentration increases.

Iron chelates react with ozone and reduce the amount of ozone in solution available for killing microorganisms. The effectiveness of ozone treatment varies markedly with the amount and type of iron chelate. Treatment of infested solution was completely ineffective when Fe-EDDHA was present (Vanachter, 1988), but Fe-DTPA and Fe-EDTA had relatively little effect on disinfection capacity. Other workers have noted similarly that EDDHA (ethylenediamine-di(*o*-hydroxyphenylacetic acid)) is strongly oxidized by ozonation, DTPA (diethylenetriaminepentaacetic acid) slightly, and EDTA (ethylenediaminetetraacetic acid) hardly at all.

Several pesticides are removed from water by ozone treatment (Evans, 1972). This may be viewed as a disadvantage, but if water is being discharged into the environment, pretreatment with ozone to break down pesticides may in fact be an advantage.

Ozone Activity Against Pathogens

Ozone kills microorganisms by oxidation of cell structures and processes. It is also very reactive with any inanimate organic matter. In the process of oxidation, oxygen and hydroxyl ions are produced and pH increases. Ozone has been shown to be effective against fungi, bacteria, and viruses. It is a more powerful oxidizing agent than chlorine, and work with human pathogens has shown that it kills bacteria and viruses more rapid than chlorine. The killing effect depends on the concentration of ozone in solution, contact time, and the type of microorganism. For disinfection of drinking water, the dose of ozone is about 0.4 mg/L, and the contact time about 4 minutes. The concentration of ozone required to kill fungal spores is considerably higher than the levels used to inactivate bacteria and viruses (Spotts and Cervantes, 1992). Ozone at concentrations of 3.8 mg/L for 2 minutes and 1.5 mg/L for 20 minutes inactivated spores of *Botrytis cinerea, Mucor piriformis,* and *Phytophthora parasitica* suspended in water (Ogawa et al., 1990).

In pure water and in nutrient solution containing Fe-DPTA or Fe-EDTA, spores of *Fusarium oxysporum* f. sp. *lycopersici* were killed after 10 minutes. The maximum ozone concentrations measured in the nutrient solutions were 1.11 and 0.60 mg/L, respectively. When *F. oxysporum* f. sp. *lycopersici* spores and *Clavibacter michiganense* cells were introduced 35 minutes after starting ozone generation, most spores and cells were killed after 1 minute, and all were killed after 5 minutes (Vanachter et al., 1988).

Table 95.4 Ozone treatment as a method of decontaminating water

Advantages	Disadvantages
Ozone is a powerful disinfectant, more powerful than chlorine, with a wide spectrum of biocidal activity	High capital cost
The disinfection process can easily be monitored by checking for a rise in the redox value	High running cost
No noxious products are formed in treated water	Long treatment time (20–30 minutes), which may make it necessary to use holding tanks for batch treatment
Ozone is formed on-site, so that no transport or storage needed	A high concentration of ozone is required to kill fungal spores
Treatment with ozone adds oxygen to the water	Effectiveness declines markedly in water containing high levels of organic matter
	It may be necessary to reduce pH for best results and readjust pH after treatment
	Some Fe chelates and possibly Mn and pesticides are destroyed
	Atmospheric ozone level must be monitored because of risks to human and crop health if ozone leaks from a treatment chamber

Ozone treatment of spore suspensions of *F. oxysporum* f. sp. *melonogenae* and *Verticillium dahliae* in nutrient solution for 20 minutes resulted in complete elimination of infectivity (Runia, 1988).

When spores of *Fusarium* were suspended in distilled water, bubbling ozone into the suspension caused 100% mortality in less than 5 minutes, the shortest practical treatment interval (MacDonald and Kabashima, 1998). However, achieving 100% mortality of spores suspended in nursery effluent water required almost 20 minutes of ozonation. The time interval varied between effluent samples and appeared to be related to the time required to achieve an ozone concentration of 0.4 ppm. That concentration was achieved in 1–2 minutes in distilled water, but ozone accumulated much more slowly in effluent samples. This is because ozone reacts with and is rapidly depleted by organic materials in the water. A prefilter to remove particulate organic matter before water is treated should improve the effectiveness of ozone treatment.

If the water to be treated does not contain significant quantities of organic matter or large numbers of resistant propagules (e.g., chlamydospores of *Thielaviopsis basicola* or microsclerotia of *Verticillium* spp.), ozone treatment should prove adequate to disinfect water. However, treatment efficacy is a function of ozone concentration and exposure duration. To achieve sufficient exposures at times of peak irrigation demand, nurseries must install collection tanks. Ozonated water is transferred to the tanks before being pumped back into the irrigation system. Tank volume must be matched to pumping rates to ensure an adequate residence time for efficacy. This can add considerably to the infrastructure and tends to limit ozone to small- to medium-sized nurseries.

Health Hazards and Other Adverse Affects of Ozone

Potential disadvantages of ozone treatment include loss of ozone from the system, resulting in crop phytotoxicity and harm to human health, and the breakdown of iron chelates, resulting in iron deficiency. The first objection is overcome by use of a well-designed and constructed treatment system, with automatic alarms should there be a leak. The second objection is overcome by use of appropriate iron chelates. In addition, ozone attacks most metals.

Although the smell of ozone is thought by some to be refreshing in low concentrations, it is toxic to both humans and plants. The background atmospheric concentration of ozone in surface air at sea level is approximately 0.01–0.03 ppm. The characteristic metallic smell of ozone is usually detected by humans at 0.02–0.05 ppm, below the level at which it is harmful. Levels above 0.1 ppm result in dryness of the respiratory tract, nose, and throat. A level of 1 ppm for 30 minutes or more produces headaches. Automatic monitoring of ozone levels in the air of the treatment facility is probably necessary.

Ultraviolet Light

Some water companies use UV light in preference to chlorination to treat drinking water, and UV is widely used to disinfect some recirculation systems (e.g., fish tanks and water cooling towers). It has also been used to disinfect recirculated nutrient solutions in hydroponic growing systems, but application of this technology to nursery recirculation has been more limited.

Advantages and disadvantages of UV light treatment as a method of decontaminating water are listed in Table 95.5.

UV Light Sources and Activities

A number of UV light sources are available, with differing capabilities and costs. The most common source is low-pressure mercury vapor lamps, which emit photons in a band around 254 nm. High-pressure mercury vapor lamps also emit photons around 254 nm, with the added effect of additional emissions at 190 nm, which form ozone in situ from water. Xenon flash lamps also are used as UV sources. These have the advantage of pulsed power emissions, making them a higher-power source of photon emissions. The disadvantage is that xenon lamps emit a broad spectrum of electromagnetic energy (UV, visible, and infrared), so they do not convert all electrical energy into UV light. The most efficient conversion of electrical energy into UV light is the krypton–fluorine excimer laser. This source is virtually monochromatic, emitting essentially all photons at 248 nm, with the added advantage of pulsed power kinetics. While capable of peak power emissions in the megawatt range (compared to milliwatts produced by mercury lamps), these units are still in the development stage for the purpose of water disinfection.

UV light is electromagnetic radiation with a wavelength of 100–400 nm. UV rays with wavelengths of 200–280 nm (UV-C rays) kill microorganisms, the optimum effect being at 240–280 nm, with a peak around 265 nm. UV light is selectively absorbed by nucleic acids (DNA and RNA) in organisms, resulting in genetic and physiological damage. Absorption is maximal at approximately 254 nm, and UV damage to cells is a well-known process, leading to skin cancers in humans. All cells possess chemical repair mechanisms that can correct the damage to nucleic acids caused by UV light. However, very high doses exceed the repair capacity of cells, leading to physiological disruption and death. Exposing zoospores of *Phytophthora* to emissions from a low-pressure mercury vapor lamp for as little as 5 seconds can cause nearly 100% mortality. The dose of radiation required to kill

Table 95.5 Ultraviolet (UV) light treatment as a method of decontaminating water

Advantages	Disadvantages
Low capital cost of a low-capacity system used with high-quality water	High capital cost of a high-capacity system and poor-quality water
Wide spectrum of biocidal activity	High running cost
UV light is generated on-site	Effectiveness declines rapidly in water containing suspended solids or dissolved organic matter, so that prefiltration, turbulent flow, and treatment in series may be necessary
No noxious products are formed in treated water	
No effect on pH	
Short treatment time allows continuous flow systems	Fe chelates may be destroyed
Treatment system is easily maintained	No residual (downstream) effects, as with chlorine
No risk to humans or crops from noxious chemicals	
Continuous monitoring and control of the system is possible	

a particular microorganism is generally expressed in millijoules per square centimeter (mJ/cm^2) and is the product of radiation intensity and exposure time. These factors must be considered in system design.

Factors Affecting UV Light Activity

Water quality is the predominant factor influencing the efficacy of UV. While UV light can pass through 25 cm of pure water with relatively little attenuation, but in fouled water it may penetrate only a few millimeters. The presence of suspended solids, such as colloidal clays, is one factor in this phenomenon, but a major factor is the presence of dissolved organics, which are highly absorptive in the UV band and may give the water an amber color. Spores suspended in fouled water may not receive a lethal dose as they pass through a UV treatment chamber. Since the level of UV radiation from the source lamps is a constant, the only way to increase the exposure dose is to slow down the flow rate (i.e., increase potential exposure time). However, at low rates, the flow within treatment chambers may become laminar, allowing some spores to pass through the chamber without ever coming close to the UV source. Turbulence must be maintained. If turbulence is maintained, it is possible that pathogen propagules suspended in fouled water will pass close to the UV source only for very brief intervals. This is the reason for the interest in high-power UV sources that can deliver lethal doses in the millisecond range.

UV Light Activity Against Pathogens

For many fungi, bacteria, and viruses in clear water, the lethal dose of UV light is less than 200 mJ/cm^2 (Steffan, 1990). Microsclerotia and chlamydospores of *Verticillium dahliae* and *Thielaviopsis basicola* are notable exceptions, requiring doses of 500 and 3,000 mJ/cm^2 or more, respectively. Runia (1988) reported that, in demineralized water with no organic matter, a 5- to 10-minute exposure to UV light at 254 nm resulted in a 30–50% reduction in the infectivity of *Phytophthora nicotianae.* Runia and Klomp (1990) reported that a flat-film lamp system with a low flow rate (9–18 L/hour) and high radiation intensity (430–800 mJ/cm^2) eliminated *Fusarium* completely and killed 48–74% of *Verticillium* spores. A different lamp design, with water flowing around the lamp at 200–400 L/hour and receiving radiation at a rate of 100–200 mJ/cm^2, killed only 2–18% of *Fusarium* and *Verticillium.* With both lamp designs, increasing the flow rate resulted in a higher percentage of mortality. This was attributed to greater water turbulence at the higher flow rate and a greater "hit" of fungal structures.

Runia and Nienhuis (1992) noted that a high-pressure lamp was 90% effective against *Fusarium* at a dose of 25–30 mJ/cm^2 and, assuming a linear dose-response relationship, estimated that a dose of 100 mJ/cm^2 was required for 99.9% elimination. Ewart and Chrimes (1980) found significant reductions in total bacterial numbers, fluorescent pseudomonads, and pectolytic bacteria when treating the nutrient solution from a tomato crop. Stanghellini et al. (1984) obtained complete control of root rot of spinach caused by *Pythium aphanidermatum* by treating the circulating nutrient solutions with UV light at 253.7 nm (30 $mW/cm^2 \cdot s^{-1}$) for 3 seconds, resulting in a dose of 90 mJ/cm^2. The flow rate was 7.44 m^3/hour, and the water was passed through a sand filter before treatment.

UV light is active against fungal and bacterial pathogens if the water to be treated is sufficiently clear to avoid UV quenching. Thus, UV is best suited to situations in which the water supply is either clean at the outset or cleaned before treatment. Cleaning can remove particulates or dissolved organics from water but also creates a waste disposal problem that must be accounted for. *Thielaviopsis basicola* is one pathogen that appears to be particularly difficult to kill by UV, suggesting that some other treatment should be considered for disinfecting water collected from crops that are very susceptible to this fungus.

There also appear to be synergistic effects in UV–ozone treatments for microbial disinfection. Ozone absorbs UV light very effectively, and the resultant ozone breakdown product, the hydroxyl radical, is an even more powerful oxidant than ozone. Combined ozone–UV systems are also very effective in breaking down chemical pollutants in water. UV light in conjunction with metal catalysts degrade pesticides in water.

Disadvantages of Treatment with UV Light

One disadvantage of treating water with UV light is that it breaks down iron chelate. This can be an important concern in hydroponic or nutrient film technique systems, where loss of iron chelate can induce iron deficiency chlorosis. Stanghellini et al. (1984) recorded a fall in the iron content in nutrient solution from 4.5 to 0.1 mg/L after 24 hours of UV treatment. Daughtrey and Schippers (1980) found that UV treatment of nutrient film technique solution from a tomato crop led to the development of pinkish roots and foliar iron deficiency.

Overview of Decontamination Methods

None of the methods reviewed above is ideal in all respects for disinfecting recycled irrigation water for nursery crops. Filters are an essential component of all methods, but an effective, standalone method of disinfecting runoff water by filtration is not available. Ozone and UV light appear to have the most advantages and fewest disadvantages. Chlorine is increasingly seen as an environmentally unfriendly treatment. It is absolutely essential to employ assay methods that can detect the specific pathogens of concern in order to assess the efficacy of the system.

Conclusion

Research has shown that irrigation runoff from nurseries often contains significant numbers of plant pathogens and that when this water is reused to irrigate healthy plants, root infections can result. But many factors (e.g., plant susceptibility and age, pathogen concentration, cultural practices, and environmental conditions) interact to determine whether or not disease results. It is important that all factors be taken into consideration when making decisions about implementing disease management practices with recycling systems and treating recycled water. Water recycling should be viewed as a system that must be managed at many levels. One aspect of management, which is critical to the decision process, is the adoption and consistent use of effective pathogen detection and monitoring techniques. Nursery sanitation is also an important factor. Growers should not allow diseased plants to reside on site, releasing pathogen propagules into drainage water. Finally, it should be clear that no single water treatment process is universally applicable to all recycling systems. Different nurseries have different water quality problems, different irrigation demands, and different abilities to invest space and capital in treatment equipment. Each system must meet the unique needs of the nursery for which it is designed.

REFERENCES

Bewley, W. F., and Buddin, W. 1921. On the fungus flora of glasshouse water supplies in relation to plant disease. Ann. Appl. Biol. 8:10–19.

Clint, C. A. 1995. Identification of sources of *Phytophthora* species on a production nursery and evaluation of management options. M.A. thesis, Texas A & M University, College Station.

Datnoff, L. E., and Kroll, T. K. 1987. Efficacy of chlorine for decontaminating water infested with resting spores of *Plasmodiophora brassicae*. Plant Dis. 71:734–736.

Daughtrey, M. L., and Schippers, P. A. 1980. Root death and associated problems. Acta Hortic. 98:283–291.

Evans, F. L. 1972. Ozone technology: Current status. Pages 1–13 in: Ozone in Water and Waste Water Treatment. F. L. Evans, ed. Ann Arbor Science Publishers, Ann Arbor, Mich.

Ewart, J. M., and Chrimes, J. R. 1980. Effects of chlorine and ultraviolet light in disease control in NFT. Acta Hortic. 98:317–323.

Frink, C. R., and Bugbee, G. J. 1987. Response of potted plants and vegetable seedlings to chlorinated water. HortScience 22:581–583.

Jenkins, J. E. 1981. Use of chlorine to suppress root-infecting pathogens of vegetables grown in recirculating hydroponic systems. (Abstr.) Phytopathology 71:883.

Kinman, R. N. 1975. Water and wastewater disinfection with ozone: A critical review. Crit. Rev. Environ. Contr. 5:141–152.

Lacey, G. H., Lambe, R. C., and Berg, C. M. 1972. Iris soft rot caused by *Erwinia chrysanthemi* associated with overhead irrigation and its control by chlorination. Int. Plant Prop. Soc. 31:624–634.

MacDonald, J. D., and Kabashima, J. N. 1998. Treatment of recycled irrigation water to reduce plant pathogens. Report to the U.S.–Israel Binational Agricultural Research and Development Agency.

MacDonald, J. D., Ali-Shtayeh, M. S., Kabashima, J., and Stites, J. 1994. Occurrence of *Phytophthora* species in recirculated nursery irrigation effluents. Plant Dis. 78:607–611.

Ogawa, J. M., Feliciano, A. J., and Manji, B. T. 1990. Evaluation of ozone as a disinfectant in postharvest dump tank treatment for tomato. (Abstr.) Phytopathology 80:1020.

Pittis, J. E. 1981. The detection and control of pathogenic water moulds in irrigation water. Ph.D. thesis, University of Manchester, Manchester, England.

Price, T. V., and Fox, P. 1984. Behaviour of fungicides in recirculating nutrient film hydroponic systems. Pages 511–522 in: Proc. Int. Congr. Soilless Cult.

Runia, W. T. 1988. Elimination of plant pathogens in drainwater from soilless culture. Pages 429–443 in: Proc. Int. Congr. Soilless Cult.

Runia, W. T. 1990. Search for less expensive disinfection methods for drainage water. Weekblad Groenten en Fruit, January 19, pp. 34–35.

Runia, W. T., and Klomp, G. 1990. UV disinfection equipment not yet suitable for commercial use. Vaakblad voor de Bloemisterii 51/55:114–115.

Runia, W. T., and Nienhuis, J. 1992. Research on drainage water disinfection. Vaakblad voor de Bloemisterii 5:99.

Scott, M., Smith, P., and Evans, J. 1984. Clean and clear. GC&HTJ, June 22, pp. 12–14.

Segall, R. H. 1968. Fungicidal effectiveness of chlorine as influenced by concentration, temperature, pH and spore exposure time. Phytopathology 58:1412–1414.

Smith, P. M. 1979. A study of the effects of fungitoxic compounds on *Phytophthora cinnamomi* in water. Ann. Appl. Biol. 93:149–157.

Spotts, R. A., and Cervantes, L. A. 1992. Effect of ozonated water on postharvest pathogens of pear in laboratory and packinghouse tests. Plant Dis. 76:256–257.

Stanghellini, M. E., Stowell, L. J. and Bates, M. L. 1984. Control of root rot of spinach caused by *Pythium aphanidermatum* in a recirculating hydroponic system by ultraviolet radiation. Plant Dis. 68:1075–1076.

Steffan, K. 1990. Submersible UV lamp sends germs packing. GbGw 20: 972–973.

Teoh, T. S., and Chuo, S. K. 1978. The possible use of sodium hypochlorite for bacterial wilt control in the hydroponic cultivation of sweet pepper. Singapore J. Primary Ind. 6:102–112.

Thompson, E. T., and Williams, K. E. 1986. Efficiency of some disinfectants challenged by range of plant pathogenic bacteria. Pages 227–231 in: Symposium on Healthy Planting Material. BCPC Monogr. 33. British Crop Protection Conference, Brighton, U.K.

Vanachter, A., Thyse, E., Van Wambeke, E. and Van Assche, C. 1988. Possible use of ozone for disinfestation of plant nutrient solutions. Acta Hortic. 221:295–302.

von Broembsen, S. L. 1998. Recycling irrigation for nurseries to protect water quality. Pages 27–29 in: Water Quality Handbook for Nurseries. Okla. State Univ. Circ. E-951.

CHAPTER

96

Austin K. Hagan ▪ Auburn University, Auburn, Alabama

Disease Resistance

Resistance is an effective, if often underutilized, strategy for managing or avoiding outbreaks of damaging diseases and nematode pests in container- and field-grown woody trees and shrubs. At a time of increasing environmental awareness among the labor force and the public, along with stricter government standards for exposure of workers to pesticides, the production of disease-resistant and nematode-resistant cultivars offers nurseries the means to reduce pesticide and possibly labor costs without sacrificing profitability or crop quality. For commercial and residential clientele, establishment of disease-resistant plant material often translates into attractive, low-maintenance trees and shrubs.

Historically, the criteria for identifying cultivars or selections of many woody trees and shrubs for release have been based almost entirely on aesthetics. There has been little concern for cultivar susceptibility to damaging diseases. As a result, some attractive but disease-susceptible trees and shrubs are in widespread production. Most cultivars of rhododendron and Kurume azaleas, for example, are highly susceptible to Phytophthora root rot, and preventative fungicide treatments are required to maintain crop quality and health. Failure to perform field trials with new selections at multiple sites under a range of environmental conditions may result in their being released erroneously as disease-resistant. Selected cultivars of flowering pear, which were once assumed to be resistant to fire blight, are in fact quite susceptible to that disease. Leyland cypress, which was widely touted as a disease-free landscape tree, has proven to be sensitive to Botryodiplodia canker and Seiridium canker in landscape plantings across the hot, humid Southeast. Hybrid tea roses, reported to be resistant to black spot in cooler, drier climates, often suffer heavy disease-related defoliation in warmer, wetter regions. In a few instances, cultivars selected for resistance to a disease have proved to be susceptible to another, more damaging disease. For example, red-tip photinia (*Photinia* × *fraseri*), which was chosen for its resistance to powdery mildew, later proved to be more susceptible to Entomosporium leaf spot than either of its parents.

Despite the increased emphasis on low-maintenance landscapes, few nurseries have used disease resistance as a tool for marketing their products. Licensing fees discourage some producers from propagating patented disease-resistant selections, thereby limiting their availability in the retail market. Also, since some nurseries can easily sell their entire annual output, there is little incentive for them to grow disease-resistant cultivars or selections. Mass market retailers, which annually sell huge volumes of container-grown trees and shrubs, purchase plant material on the basis of cost and quality, not the reactions of the plants to diseases. As a result, the retail market is often flooded with off-patent cultivars or selections of shrubs and trees, particularly flowering dogwood, crabapple, and crapemyrtle, which are often susceptible to common diseases. Local garden centers, however, generally are good sources of disease-resistant cultivars of popular shrubs and trees.

Resistance as a disease management tool remains alive and well. Tree and shrub breeding programs at the U.S. National Arboretum, land-grant universities, and other institutions have increasingly stressed resistance to common, damaging diseases as a selection criterion. Among the most successful of these programs in recent years has been the release by the U.S. National Arboretum of approximately 26 cultivars of crapemyrtle (*Lagerstroemia indica* and *L. indica* × *fauriei*) resistant to powdery mildew. Breeding programs at several land-grant universities and the U.S. National Arboretum have resulted in the release of cultivars and hybrids of several elm taxa with resistance to Dutch elm disease. In a similar program, hybrid chestnut cultivars resistant to chestnut blight also have been identified. A cooperative project between several agencies of the U.S. Department of Agriculture and several land-grant universities has been initiated to collect and assess resistance to dogwood anthracnose in selected flowering dogwood clones. The private sector has also become involved in breeding and patenting disease-resistant woody trees and shrubs. For example, several nurseries in the Deep South are actively involved in breeding Indian hawthorn and red-tip photinia resistant to Entomosporium leaf spot.

For most new tree and shrub releases, however, sensitivity to established diseases is often unknown. In some cases, data collected in one region may not be valid in another. As a result, research and extension personnel at land-grant universities are constantly screening releases of popular small flowering trees and shrubs to determine their reactions to damaging diseases. The National Crabapple Introduction Program, a cooperative effort of the private sector and land-grant universities, has developed a database detailing not only the horticultural characteristics of this popular flowering tree but also the reactions of cultivars to fire blight, apple scab, powdery mildew, cedar rust diseases, and frog-eye leaf spot. Screening trials have recently focused on identifying cultivars of flowering kousa and hybrid dogwood resistant to powdery mildew, spot anthracnose, and dogwood anthracnose. Other trials have focused on resistance to powdery mildew and Cercospora leaf spot of crapemyrtle; black spot and powdery mildew of antique and ground-cover roses; Entomosporium leaf spot and fire blight of Indian hawthorn; powdery mildew and bacterial blight of lilac; and tip blight, canker, and root rot of juniper. Several tree species have been screened for canker and root rot resistance. For many minor tree and shrub crops, however, information on resistance is either not available or often badly outdated.

Specific information on the reaction of cultivars and selections of popular trees and shrubs is summarized in this book and in circulars and other printed material distributed to commercial and residential clientele by state cooperative extension systems and land-grant universities. Much of this same information is also posted on the Web sites of these sources. Publications of many plant enthusiast societies occasionally contain additional information concerning reactions of species and cultivars to diseases. Despite the availability of information in printed or electronic form, most commercial and residential clientele and most retail and garden center personnel are not well informed about the availability of disease-resistant cultivars, nor are they aware of disease- and pest-prone trees and shrubs to avoid.

Disease resistance is not infallible. Under weather conditions ideal for disease development, resistant cultivars may suffer significant damage. However, cultivars and selections of popular trees and shrubs that have been identified as resistant to diseases endemic to their region should always be specified for new or established landscapes. Trees and shrubs known to be disease- or pest-prone should be avoided. Every effort must be made to publicize the need to increase species diversity, particularly among street trees, in order to avoid catastrophic disease outbreaks, such as Dutch elm disease.

REFERENCES

Benson, D. M., and Cochran, F. D. 1980. Resistance of evergreen hybrid azaleas to root rot caused by *Phytophthora cinnamomi.* Plant Dis. 64:214–215.

Daughtrey, M. L., Hibben, C. R., Britton, K. O., Windham, M. T., and Redlin, S. C. 1996. Dogwood anthracnose: Understanding a disease new to North America. Plant Dis. 80:349–358.

Egolf, D. R. 1967. Four new *Lagerstroemia indica* cultivars (Lythraceae). Baileya 15:7–13.

Egolf, D. R. 1991. Pretty in pink. Am. Nurseryman 173:87–92.

Fare, D. C., Gilliam, C. H., and Ponder, H. G. 1991. Fireblight susceptibility, growth, and other characteristics in ornamental pears in Alabama. J. Arboric. 17:257–260.

Hagan, A. K. 1996. Controlling Entomosporium leaf spot on woody ornamentals. Ala. Coop. Ext. Serv. Circ. ANR-392.

Hagan, A. K., Akridge, J. R., Olive, J. W., and Tilt, K. 1997. Resistance of selected cultivars of Indian hawthorn to Entomosporium leaf spot. Ala. Agric. Exp. Stn. Auburn Univ. Ornamental Res. Rep. 13:32-33.

Hagan, A. K., Gilliam, C. H., Keever, G. J., and Williams, J. D. 1997. Susceptibility of crapemyrtle cultivars to powdery mildew and Cercospora leaf spot. Proc. SNA Res. Conf. 42:236–240.

Hagan, A. K., Gilliam, C. H., Keever, G. J., Williams, J. D., Hardin, B., and Eakes, J. 1997. Dogwood selections differ in their susceptibility to powdery mildew and spot anthracnose. Proc. SNA Res. Conf. 42:243–248.

Jones, R. K. 1993. Seiridium canker on Leyland cypress. Proc. SNA Res. Conf. 38:220.

Mmbaga, M. T. 1997. Evaluation of lilac (*Syringa* spp.) for multiple diseases resistance to powdery mildew and bacterial blight in McMinnville, TN. Proc. SNA Res. Conf. 42:512–517.

Smalley, E. B., and Guries, R. P. 1993. Breeding elms for resistance to Dutch elm disease. Annu. Rev. Phytopathol. 31:325–352.

Spencer, J. A., and Wood, O. W. 1992. Response of selected old garden roses to several isolates of *Marssonia rosae* in Mississippi. J. Environ. Hortic. 10:221–223.

Standish, E. D., MacDonald, J. D., and Humphrey, W. A. 1982. Phytophthora root and crown rot of junipers in California. Plant Dis. 66:925–928.

Tilt, K., Hagan, A. K., Williams, J. D., and Witte, W. T. 1997. Evaluation for crabapples for zone 8. Proc. SNA Res. Conf. 42:523–526.

Tisserat, N. A., and Pair, J. C. 1997. Susceptibility of selected juniper cultivars to cedar-apple rust, Kabatinia tip blight, Cercospora needle blight, and Botryosphaeria canker. J. Environ. Hortic. 15:160–163.

Walker, J. T. 1982. Disease resistance among woody ornamentals. Pages 112–115 in: Diseases of Woody Ornamental Plants and Their Control in Nurseries. R. K. Jones and R. C. Lambe, eds. N.C. Agric. Ext. Serv. Publ. AG-286.

Windham, M. T., Witte, W. T., Sauve, R. J., and Flanagan, P. C. 1995. Powdery mildew observations and growth of crapemyrtle in Tennessee. HortScience 30:813.

CHAPTER 97

Robert G. Linderman ▪ U.S. Department of Agriculture, Corvallis, Oregon

Mycorrhizae and Their Effects on Diseases

The role of rhizosphere soil organisms in the healthy growth of nursery plants has been greatly underestimated. Nutrient energy inputs from root exudates of plants selectively favor the growth of several populations of organisms in the rhizosphere (the zone of soil influenced by the root). Rhizosphere organisms that can affect plant growth include (1) mycorrhizal fungi, (2) associative and symbiotic nitrogen fixers, (3) microbes that affect nutrient availability, (4) pathogens, (5) biocontrol agents and antagonists, (6) rhizobacteria, some of which promote plant growth, and some of which are deleterious, and (7) arthropods, nematodes, and protozoa that graze on bacteria and fungi. Mycorrhizae play a pivotal role in the establishment, maintenance, and function of many of these organisms. Thus, they play a key role in the overall health of plants and have direct and indirect effects on diseases.

Mycorrhizae

Mycorrhizae are symbiotic associations between certain fungi and the roots of host plants. Most woody ornamental plants grown in nurseries would normally form mycorrhizal associations if not for the artificial conditions of nursery production. Mycorrhizae are of several major types: *ectomycorrhizae* occur on pine, fir, oak, birch, and eucalyptus; *arbutoid mycorrhizae* occur on manzanita, kinnikinnick, and madrone; *ericoid mycorrhizae* occur on most other ericaceous plants, such as azalea, rhododendron, kalmia, and huckleberry; and *endomycorrhizae,* primarily *vesicular-arbuscular mycorrhizae* (VAM), occur on most other herbaceous plants and on trees and shrubs (sweetgum, maple, lilac, etc.). The fungi that form ectomycorrhizae, arbutoid mycorrhizae, and ericoid mycorrhizae can generally be cultured on artificial media, but VAM fungi are obligate symbionts, which must grow in association with living plant roots to complete their life cycles.

All mycorrhizal fungi are similar in some aspects of their life cycles characterizing their symbiotic associations. Soilborne propagules of these fungi (spores or hyphae) germinate in the presence of roots of an appropriate host, colonize the roots, and form an interface with the root cortical cells.

VAM fungi colonize the root cortex both intercellularly and intracellularly and produce haustorium-like arbuscules, which are the sites of nutrient and mineral exchange between the plant host and the fungal symbiont. They also produce storage structures called vesicles.

Ericoid mycorrhizae are a form of endomycorrhizae. The fungi involved in these associations also produce exchange structures in the cortical cells of roots.

Ectomycorrhizal and arbutoid mycorrhizal fungi, in contrast, produce hyphae that penetrate the cortex of the root but remain intercellular and form a network called the Hartig net. This hyphal network is the site of chemical exchange between the fungal symbiont and cells of the host plant. These fungi also form an outer mantle of fungal hyphae, which completely surrounds the root. Phytohormones produced or induced by ectomycorrhizal fungi cause the roots to become greatly branched, thereby increasing their surface area.

Mycorrhizal fungi of all types produce extraradical hyphae, hyphal strands, or rhizomorphs, which can extend great distances in the soil. This hyphal network explores the nooks and crannies of the soil to acquire water and nutrients, which it transports back to the roots of the host plant and transfers to plant tissues. By means of this extraordinarily extensive hyphal network, the absorptive capacity of mycorrhizal root systems is increased dramatically, and even immobile nutrients, such as P, Zn, and Cu, can be delivered to the plant. Without mycorrhizae, plants would become deficient in these mineral elements.

The last stage of the life cycles of mycorrhizal fungi is the production of spores for dissemination to other areas and plants. Ectomycorrhizae and arbutoid mycorrhizae are formed by basidiomycete and ascomycete fungi that produce mushrooms or truffles as spore-producing structures. Ericoid fungi are either ascomycetes, which discharge ascospores from tiny apothecia, or basidiomycetes in the genus *Clavaria,* which release basidiospores from their coral-like fruiting bodies. VAM fungi produce very large spores (50–300 μm in diameter), either singly or in clusters, on extraradical hyphae in the soil.

Effects of Mycorrhizae on Diseases

Numerous reviews have addressed the role of mycorrhizae in the expression of plant diseases. While the evidence presented in the literature is not always conclusive, or even comparable between studies or with observations in undisturbed ecosystems, there is general consensus that mycorrhizae can have a positive influence on the incidence and severity of diseases, primarily those due to soilborne root pathogens. There are few reports and minimal evidence of mycorrhizae influencing foliar diseases or diseases caused by viruses or bacteria, but there is considerable evidence that mycorrhizae suppress soilborne diseases caused by fungi or nematodes.

Mycorrhizae may influence diseases by a number of mechanisms, depending on the host plant and the type of mycorrhizae. These mechanisms include (1) enhanced nutrition, (2) competition for host nutrients and infection sites on roots, (3) morphological changes in roots and root tissues, (4) changes in chemical constituents of plant tissues, (5) reduction of abiotic stresses, (6) induced systemic resistance, and (7) microbial changes in the rhizosphere soil, now appropriately called the mycorrhizosphere. Once the mycorrhizal association is established, any one or more of these mechanisms could become functional.

For mycorrhizae to affect disease incidence or severity, however, the association must be established and fully functional before the plant or roots are exposed to the pathogens. Once established, mycorrhizae can induce significant physiological changes in the host plant and affect microbial changes in the mycorrhizosphere that could suppress root pathogens. Thus, the need for early establishment of mycorrhizae and the availability of effective microbial associates argues for inoculation with appropriate organisms early in the nursery production system. Once roots become mycorrhizal, the fungal symbionts grow with the roots and provide protection against invasion by pathogens.

While mycorrhizal associations in nature seem to be very effective in enhancing the growth and health of plants, we have great difficulty duplicating these systems in nursery production of woody plants. In many cases, cultural practices preclude the establishment of mycorrhizae, largely because of frequent or heavy applications of inhibitory fertilizers and pesticides and the physical, chemical, and biological characteristics of soilless media compared to those of soil. Inoculum of biotypes of mycorrhizal fungi that are present in the nursery, come into the nursery from neighboring areas, or are provided by commercial sources may not be well adapted to conditions in the nursery. If mycorrhizae cannot be established in time to deter pathogens, there may be no effect on disease. Providing a mixture of mycorrhizal fungi representing a range of biotype adaptations is one approach to solving that problem. Furthermore, providing other needed microbial associates of mycorrhizae, which may otherwise be missing from soilless media or fumigated nursery beds, may enhance the beneficial effects of mycorrhizae. The goal should be to provide at the earliest stage of production the most holistic microbial system possible in order to have the best chance of influencing plant growth and suppressing diseases.

REFERENCES

Bethlenfalvay, G. J., and Linderman, R. G., eds. 1992. Mycorrhizae in Sustainable Agriculture. ASA Spec. Publ. 54. American Society of Agronomy, Madison, Wisc.

Caron, M. 1989. Potential use of mycorrhizae in control of soilborne diseases. Can. J. Plant Pathol. 11:177–179.

Dehne, H. W. 1982. Interactions between vesicular-arbuscular mycorrhizal fungi and plant pathogens. Phytopathology 72:1115–1119.

Duchesne, L. C. 1994. Role of ectomycorrhizal fungi in biocontrol. Pages 27–45 in: Mycorrhizae and Plant Health. F. L. Pfleger and R. G. Linderman, eds. American Phytopathological Society, St. Paul, Minn.

Linderman, R. G. 1994. Role of VAM fungi in biocontrol. Pages 1–26 in: Mycorrhizae and Plant Health. F. L. Pfleger and R. G. Linderman, eds. American Phytopathological Society, St. Paul, Minn.

Marx, D. H. 1972. Ectomycorrhizae as biologic deterrents to pathogenic root infection. Annu. Rev. Phytopathol. 10:429–454.

Schenck, N. C. 1983. Can mycorrhizae control root diseases? Plant Dis. 65:230–234.

Schenck, N. C., and Kellam, M. K. 1978. The influence of vesicular-arbuscular mycorrhizae on disease development. Univ. Fla. Tech. Bull. 798.

Zak, B. 1964. Role of mycorrhizae in root disease. Annu. Rev. Phytopathol. 2:377–392.

CHAPTER

98

Walter F. Mahaffee ▪ U.S. Department of Agriculture, Corvallis, Oregon

Biological Control of Woody Ornamental Diseases

Biological control of plant diseases can be broadly defined as the suppression of disease through the manipulation of one or more organisms. The basic tenet of biological control is the maintenance of a balance between the host plant and the potential pathogen such that the growth of the pathogen is not favored and disease does not develop. This balance can be accomplished by the introduction of biological control agents as well as by cultural practices and plant genetics. Cultural practices can be used to create physical environments favorable for naturally occurring beneficial microbes or conditions unfavorable for the growth, survival, or reproduction of pathogens. Manipulation of plant genetics, either by traditional breeding or by genetic engineering, allows the selection of characters that enhance the activity of naturally occurring beneficial microbes or confer resistance to specific pathogens. Cultural practices and manipulation of plant genetics can also enhance the activity of introduced biological control agents. While cultural practices and host genetics are important sources of control, this chapter focuses on the use of beneficial microbes for biological control of plant diseases. Some cultural practices are briefly mentioned as they relate to integration with beneficial microbes.

Since the early part of the twentieth century, laboratories in universities, government agencies, and research institutes around the world have investigated biological control of the microbial balance on plants. There has been continual growth of knowledge of how to reduce disease through the manipulation of microbial ecosystems on and around plant surfaces. Numerous bacteria, fungi, yeast, and some viruses have been shown to reduce disease development when applied to plants in commercial production. Many of these agents are still in development, but more than 30 commercial biological control agents are currently registered with the U.S. Environmental Protection Agency as biopesticides for pathogens and insect pests. With the continued development and registration of biological control agents, we are entering a new era in disease control. Biological control of plant diseases is becoming a viable alternative to treatment with chemical pesticides, and growers are becoming ecosystem managers within a framework of integrated pest management.

Products Available for Commercial Use

Numerous biological control agents are registered for use on woody ornamentals (Table 98.1). Most of them were developed to control soilborne pathogens of row crops (e.g., cotton, soybean, and wheat) or horticultural food crops (e.g., apples and pears) and were subsequently approved for use on ornamental crops. Several other agents are currently available for use on ornamentals outside the United States (Table 98.2) and could be registered in the United States if a favorable market was perceived to exist.

Early biological control agents were developed to treat problems for which chemical pesticides are not available, do not provide adequate control, or are considered a marketing liability (i.e., public opinion of them is unfavorable). Thus, the major emphasis has been directed against soilborne pathogens. These pathogens are difficult to control with chemical pesticides, because of the difficulty of delivering chemicals (other than seed treatments) to the soil and the short duration of their activity in soil (14–21 days). Few chemical pesticides are effective against soilborne pathogens; in most cases, they provide erratic control of limited duration, and there is some concern that they may contaminate ground water.

Much effort has also been invested in the use of biological control agents to control postharvest storage rots, in response to restrictions on pesticides used on food products and public concern over levels of pesticides in the food supply. There has also been increasing research on the control of bacterial diseases, since few chemical treatments are effective in controlling bacterial pathogens, given the increase in resistance to copper and antibiotics.

Interest in developing new biological control agents has been aroused in recent years by the commercial success of pioneering biological control products, such as Galltrol-A, Norbac 84C, and Kodiak. These products demonstrated that biological control agents can function in commercial production and provide an economic return. Researchers in corporate, government, and university programs are developing other biological control agents targeted against foliar, soilborne, and postharvest pathogens.

How Biological Control Agents Work

To understand how biological control of plant diseases can be successfully implemented, it is important to understand the underlying mechanism by which biological control agents are able to suppress disease development. Traditionally, the disease triangle (Fig. 98.1A) has been used to represent disease development; it depicts disease as a function of interactions between a pathogen, a host, and the environment. However, with the understanding that microorganisms exert biological control (i.e., that they influence

Table 98.1 Biological control agents registered by the U.S. Environmental Protection Agency for application on woody ornamentals in the United States

Product and manufacturer	Biological control organism	Target	Use
AQ10 Ecogen, Langhorne, Pa.	*Ampelomyces quisqualis* strain M-10	Powdery mildew fungi	Leaf wetness is required for a period immediately after application Two sprays, seven to 14 days apart, are required Best if stored below room temperature Compatible with sulfur and sterol inhibitors
Binab T Bio-Innovation AB, Algaras, Sweden Henry Doubleday Research, Ryton on Dunsmore, Coventry, U.K.	*Trichoderma harzianum* strain ATCC 20476 and *T. polysporum* strain ATCC 20475	Fungi causing wilt, take-all, root rot, and wood rots	Applied as a spray, incorporated with potting mix, painted on tree wounds, or inserted as pellets in holes drilled in woody tissue
BlightBan A506 Plant Health Technologies, Fresno, Calif.	*Pseudomonas fluorescens* strain A506	*Erwinia amylovora* Frost injury	For frost protection, application timing is based on plant growth stage Used in an integrated fire blight control program Incompatible with copper-based products
Deny CCT Corp., Carlsbad, Calif.	*Burkholderia cepacia* type Wisconsin	Lance nematode (*Hoplolaimus*) Lesion nematode (*Pratylenchus*) Spiral nematode (*Helicotylenchus*) Sting nematode (*Belonolaimus*)	Banded application to seeds and seedling beds; also applied in flood or drip irrigation or as a seedling transplant drip
Diteria Abbot Laboratories, North Chicago, Ill.	Fermentation products of *Myrothecium verrucaria*	Citrus nematode (*Tylenchulus*) Cyst nematodes (*Heterodera, Globodera*) Dagger nematode (*Xiphinema*) Lesion nematode (*Pratylenchus*) Pin nematode (*Paratylenchus*) Reniform nematode (*Rotylenchulus*) Root-knot nematode (*Meloidogyne*)	In-row or banded application, for preplant or postplant treatment Best results are obtained if treatment is applied at root flush At high rates of application, sprayers tend to clog
Galltrol-A AgBioChem, Orinda, Calif.	*Agrobacterium radiobacter* strain 84	*Agrobacterium tumefaciens*	Applied as a root or cutting dip
Kodiak, Kodiak HB Gustafson	*Bacillus subtilis* strain GB03	*Aspergillus flavus* *Fusarium* spp. *Pythium* spp. *Rhizoctonia solani*	Kodiak (commercial seed treatment) and Kodiak HB (hopper box application) are to be used in conjunction with seed treatment chemicals Increased nodulation by *Rhizobium* has been reported in legumes treated with these products
Mycostop AgBio Development, Westminster, Colo.	*Streptomyces griseoviridis* strain K61	*Alternaria brassicola* *Botrytis* spp. *Fusarium* spp. *Phomopsis* spp. *Phytophthora* spp. *Pythium* spp.	Applied as a seed treatment, transplant or cutting dip, or soil spray or drench
Nogall New Bioproducts, Corvallis, Ore.	*Agrobacterium radiobacter* strain K1026	*Agrobacterium tumefaciens*	Applied as a root dip as soon as possible after root pruning or tissue damage due to handling

PlantShield, RootShield BioWorks, Geneva, N.Y. T-22 Planter Box BioWorks	*Trichoderma harzianum rifai* strain KRL-AG2	*Fusarium* spp. *Pythium* spp. *Rhizoctonia solani*	In granular formulation, used as an amendment in potting mix Applied as a drench formulation through low-pressure water nozzles, as a root or cutting dip, as a powder coating for cuttings, and as a bulb and seed treatment Should be applied when the soil temperature is above 50°F
SoilGard Thermo Trilogy Corp., Columbia, Md.	*Gliocladium virens* strain GL-21	*Pythium* spp. *Rhizoctonia solani*	Soil treatment (formerly called GlioGard)
Subtilex MicroBio Group, Boulder, Colo.	*Bacillus subtilis* strain GB07	*Fusarium* spp. *Pythium* spp. *Rhizoctonia solani*	To be used in conjunction with seed treatment chemicals
System 3 Helena Chemical Co., Memphis, Tenn.	*Bacillus subtilis* strain GB03	*Fusarium* spp. *Pythium* spp. *Rhizoctonia solani*	Contains the same active strain as Kodiak and also the fungicides Apron and Terraclor

Table 98.2 Biological control agents available for use on woody ornamentals outside the United States[a]

Product and manufacturer	Biological control organism	Target	Use
Biofox C S.I.A.P.A., Bologna, Italy	*Fusarium oxysporum* (nonpathogenic strain)	*Fusarium oxysporum* *F. moniliforme*	Applied as a seed treatment or incorporated into soil
Bio-Fungus Grondortsmettingen DeCuester, St.-Katelijne-Waver, Belgium	*Trichoderma* spp.	*Fusarium* spp. *Phytophthora* spp. *Pythium* spp. *Rhizoctonia solani* *Sclerotinia* spp. *Verticillium* spp.	Incorporated into soil or applied as an in-furrow spray May be applied after soil fumigation
RotStop Kemira Agro Oy, Helsinki, Finland	*Phlebia gigantea*	*Heterobasidion annosum*	Sprayed on wounds or incorporated into chain saw oil
Trichopel, Trichoject, Trichodowels, Trichoseal Agrimm Technologies, Christchurch, New Zealand	*Trichoderma harzianum* and *T. viride*	*Armillaria* *Botryosphaeria* *Chondrostereum* *Fusarium* *Nectria* *Phytophthora* *Pythium* *Rhizoctonia*	Depending on the formulation, applied by incorporation into soil, by injection, as dowels inserted into drilled holes, or as a painted sealant
Trichoderma 2000 Mycontrol, Nazeret Elit, Israel	*Trichoderma* sp.	*Fusarium* spp. *Pythium* spp. *Rhizoctonia solani* *Sclerotium rolfsii*	Incorporated into soil or potting medium

[a] None of the biological control agents in this list are registered for use in the United States.

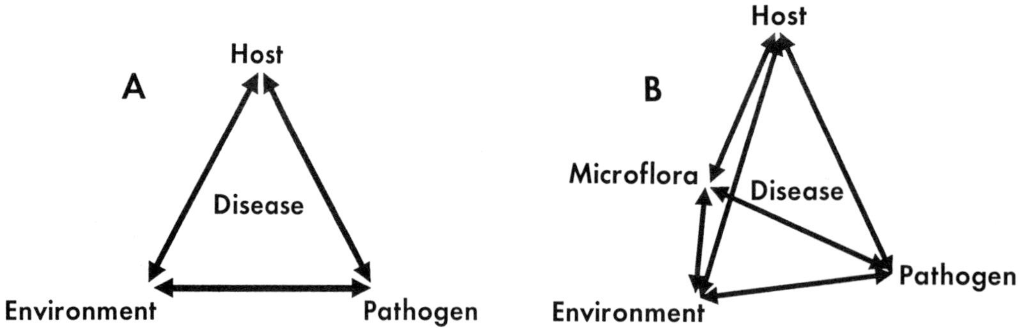

Fig. 98.1 **A.** Disease as a function of interactions between a host, a pathogen, and the environment: disease occurs when the environment and the condition of the host favor the development of the pathogen. **B.** Disease results when the environment and the condition of the host favor the pathogen and the balance between the pathogen and microflora in, on, or around the plant is disrupted.

the survival, reproduction, and pathogenesis of plant pathogens), disease is more accurately represented by the disease pyramid (Fig. 98.1B), in which disease is a function of interactions between a host, a pathogen, the environment, and associated microflora that naturally exist in the environment. How do microbes on, in, and around plant tissues influence disease development? They affect the growth and survival of pathogens either directly, by various antagonistic mechanisms, or indirectly, by altering the host plant's physiology, which in turn inhibits the invading pathogen (Table 98.3).

The mechanisms of biological control (Table 98.3) may appear to be independent of each other, but in practice it is difficult to attribute disease suppression to one mechanism. Most biological control agents employ several of these mechanisms to suppress the targeted pathogen. For instance, the bacterium *Bacillus subtilis* strain GB03, the active ingredient of Kodiak, produces the antibiotic iturin A and several plant growth regulators, increases drought tolerance, and increases rhizobial nodulation, and appears to induce resistance to foliar diseases. In most of the mechanisms of biological control (Table 98.3), the biological control agent operates like a protectant pesticide (i.e., it is applied before severe disease outbreak) and probably has only a limited function as an eradicant, especially if disease incidence is already high.

Handling and Application of Biological Control Agents

In many ways, biological control agents are handled like traditional pesticides, but there are distinct differences, involving the timing of application, the effects of environmental conditions on the efficacy of treatment, and the storage of products.

The most important difference is application timing. In general, biological control agents should be applied as protectants, before symptoms are prevalent. For instance, Norbac 84C, Nogall, and Galltrol-A have a limited window of opportunity for effective application. They must be applied within a few hours of wounding to prevent the development of crown gall, caused by *Agrobacterium tumefaciens*. However, some biological control agents are effective as eradicants as long as they are applied early in an epidemic. For instance, AQ10 is a fungal parasite that uses hyphae of powdery mildew fungi as a food source. It is recommended for application before or after powdery mildew has developed. Also, late-season applications possibly reduce levels of inoculum of powdery mildew fungi for the following year. AQ10 can disrupt the sexual reproduction of these fungi. so that they produce sterile cleistothecia (sexual reproductive structures).

Storage of Biological Control Agents

Unlike chemicals, biological control agents are living organisms and, in general, are more sensitive to environmental conditions during storage, tending to have a shorter shelf life. For these reasons, it is very important to follow the storage and handling directions on the product label. Typically, a biological control product must be kept at temperatures below 90°F and above freezing (i.e., do not store these products on a shelf in a barn). Most biological control products have a short shelf life and should not be used after the expiration date. If these precautions are not followed, biological control agents are likely to be ineffective. An old refrigerator placed in a storage shed could be used to store these products, especially if a thermostat that extends the temperature settings in the upper range is added.

Application of Biological Control Agents

In general, the same application equipment and procedures can be used for biological control agents as for chemical pesticides, except that tank mixing with chemicals is likely lead to inactivation of biological control agents. Certain pesticides inactivate certain biological control agents and cannot be used with them.

Since biological control agents are living organisms, the environment in which they are applied has a strong impact on their performance. Growers should assess whether biological control agents work under their specific growing conditions and consult with local extension agents or other growers to find out about their experiences with these products. In evaluating biological control agents, a grower should do several paired evaluations: one group of plants receives treatment with a biological control agent and any special handling or production practices, and the other group of plants receives the standard chemical treatment (if available) and standard management practices. These evaluations should be conducted at several different times of the year, when environmental conditions have changed significantly.

Table 98.3 Mode of action of biological control agents in suppressing disease development

Mechanism	Mode of action
1. Antagonism	The biological control agent directly affects the survival or growth of a pathogen
a. Antibiosis	Production of compounds that kill pathogens
b. Competition	In all environments the availability of space and availability of nutrients limit the development of pathogens and, therefore, their ability to cause disease
i. Competition for nutrients	Several nutrients (e.g., carbon, nitrogen, and iron) on plant surfaces are often limiting; numerous biological control agents have more efficient (better and faster) means of gathering nutrients than pathogens, and thus competing pathogens starve
ii. Competition for space	Colonization of a physical space by a biological control agent prior to the arrival of a pathogen prevents the pathogen from colonizing and infecting the plant
c. Fungistasis	Production of an agent that inhibits the growth or reproduction of a pathogen without causing death or blockage of a signal that would trigger a growth response in the pathogen
d. Parasitism	Exploitation of a pathogen as a food source, so that the pathogen cannot reproduce or survive
2. Induced resistance	The biological control agent indirectly affects disease development by activating normal defense responses in the plant prior to infection (a process similar to immunization of humans), thus limiting the ability of the pathogen to infect the induced plant
3. Increased fertility and growth promotion	The biological control agent indirectly affects disease development by increasing the rate of plant growth or the availability of nutrients used by the plant in defense responses
	Many plant tissues are susceptible to pathogen attack only during certain physiological growth stages; some biological control agents produce plant growth hormones or other substances that increase plant growth and development, thereby reducing the amount of time during which tissues are susceptible
	As in humans, good nutrition results in better health

Advantages and Disadvantages of Biological Control

The use of biological control agents, like any agronomic practice, has both advantages and disadvantages that must be considered when deciding whether to incorporate them into a production system. While the details are different for each product, most biological control agents have some advantages and disadvantages in common (Table 98.4).

As we gain increased understanding of interactions in the complex microbial communities associated with plants and soils, the list of advantages and disadvantages is likely to be even more favorable to biological control agents. With increased understanding, it will probably become even more obvious that implementation of an effective biological control program requires modification of most production practices. Thus, when a grower decides to use biological control agents, a commitment is needed to modify management practices so that the applied biological control agents have a chance to work. Recent changes in pesticide regulations (1996 Food Quality Safety Act) and increasing public concern are likely to increase the need for alternatives to chemical pesticides and speed the acceptance and utilization of many biological control agents.

Current Research Directions

Current research to develop biological control systems has three main directions: (1) discovery of new agents, particularly those that can induce systemic resistance in plants, (2) improvement of methods for the formulation, fermentation, and application of biological control agents; and (3) development of mixtures of biological control agents for more effective control of pathogens in a wider range of environments or control of multiple pathogens.

New biological control agents are needed to extend the number of pathogens and crops to which biological control can be effectively applied. No longer are university and government researchers alone in this effort; numerous corporations are now working to discover new biological control agents. This increased effort should result in an exponential growth in commercially available biological control agents and should expand the number and types of pathogens controlled. Biological control agents adapted to specific regions and soil types may also become available. This specialization could be achieved as a result of the reduced cost of registering biological control agents with the U.S. Environmental Protection Agency. However, the discovery of a microbe that is active against a particular pathogen or that functions in a particular environment is merely one component of developing a commercial biological control agent.

Table 98.4 Advantages and disadvantages of biological control of plant diseases

Advantages	Disadvantages
Improved public perception (minimizing public concern over chemical contamination of ground water and soil)	Biological control generally requires special knowledge and handling for correct implementation (i.e., modification of standard production practices)
Reduction in use of chemical pesticides and associated health concerns for applicators	Less consistent performance than chemical treatment, because of dependence on environmental conditions; some biological control agents may be effective only in particular soils or particular geographic regions or during certain times of the year
Decreased reentry times – workers are able to reenter fields or greenhouses sooner	Most biological control agents are effective against only a narrow spectrum of pathogens
Nonchemical alternative control measure for pesticide resistance management	Biological control does not work as fast as chemical treatment
Biological control agents generally are nontoxic to nontarget organisms and have less impact on (and may even enhance) indigenous beneficial microbes	Biological control agents generally are effective only as protectants
An integral component of disease management in sustainable agriculture and integrated pest management	
Can be effective against diseases that are not adequately controlled by current chemical pesticides (i.e., diseases caused by soilborne pathogens)	

Costs and, frequent, the lack of appropriate technologies for fermentation, formulation, and application are generally the limiting factors in decisions about whether it is economical to commercialize a biological control agent. Each of these technologies affects whether a biological control agent can be delivered to the target site in sufficient numbers and in a physiological state that maximizes biological activity and survival. The difficulty in establishing appropriate fermentation, formulation, and application technologies is complicated by the fact that they are interdependent – methods employed for one affect which methods must be utilized for the others. In addition, these methods must be developed for each agent. Despite these constraints, economical methods for fermenting, formulating, and applying biological control agents are being developed, and products are commercially available. Continued research in these areas is essential to the continued succcss of commercial biological control agents.

The growing rate of discovery and registration of new biological control agents and the development of delivery technologies will increase the implementation of biological control as an alternative to chemical control. Integrating multiple agents into a holistic system to control multiple pathogens would be even more effective in facilitating the implementation of a biological control program. Such an integrated system could be even more reliable and resilient than chemical control programs: a living community could be created that is responsive to environmental conditions and able to attack a pathogen on multiple fronts.

Researchers are beginning to examine the potential for integrating multiple biological control agents into systems to extend the range of pathogens affected or the environments in which these agents can be used. For instance, Kodiak and Epic can be applied together on cotton to control *Rhizoctonia solani* and *Fusarium* and *Pythium* species along the taproot and feeder roots throughout the growing season. This coinoculation takes advantage of the properties of both products to extend the range of diseases controlled and consistency of performance.

The Future of Biological Control

For biological control of plant diseases to work on a commercial scale, with very few or no chemical applications, new management practices are needed to ensure that a microbial balance unfavorable to pathogen development is maintained. In addition, every management decision in such a system influences whether disease develops and which control measure should be implemented. Thus, holistic management systems must be developed, to incorporate all management practices into a concerted effort for plant production. We must move beyond thinking only in terms of plant growth and yield, to the concept that we are managing an entire ecosystem in which all components interact to determine profitability and long-term sustainability. In shifting to the goal of ecosystem management, we must remember one of the basic tenets of ecology: a stable ecosystem is a product of the diversity of its members. To observe proof of this tenet, just walk through an area where humans have had little impact, and observe the low level of disease. Using multiple biological control agents integrated into a production system would be one component of achieving the goal of a stable ecosystem.

REFERENCES

Baker, K. F., and Cook, R. J. 1974. Biological Control of Plant Pathogens. W. H. Freeman Co., San Francisco.

Becker, O. J., and Schwinn, F. J. 1993 Control of soilborne pathogens with living bacteria and fungi: Status and outlook. Pestic. Sci. 37:365–373.

Biological Control Virtual Information Center http://ipmwww.ncsu.edu/biocontrol/biocontrol.html

Boland, G. J., and Kuykendall, L. D. 1997. Plant-Microbe Interactions and Biological Control. Marcel Dekker, New York.

Campbell, R. 1989. Biological Control of Microbial Plant Pathogens. Cambridge University Press, New York.

Campbell, R. 1994. Biological control of soilborne diseases: Some present problems and different approaches. Crop Prot. 13:4–12.

Cook, R. J., and Baker, K. F. 1983. The Nature and Practice of Biological Control of Plant Pathogens. American Phytopathological Society, St. Paul, Minn.

Fokkema, N. J. 1993. Opportunities and problems of control of foliar pathogens with micro-organisms Pestic. Sci. 37:411–416.

Hokkanen, H. M. T., and Lynch, J. M. 1995. Biological Control: Benefits and Risks. Cambridge University Press, New York.

Jarvis, W. R. 1992. Biological control. Pages 185–218 in: Managing Diseases in Greenhouse Crops. American Phytopathological Society, St. Paul, Minn.

Kenney, D. S. 1997. Commercialization of biological control products in the chemical pesticide world. Pages 126–127 in: Plant Growth-Promoting Rhizobacteria: Present Status and Future Prospects. A. Ogoshi, K. Kobayashi, Y. Homma, F. Kodama, N. Kondo, and S. Akino, eds. Nakanishi Printing, Sapporo, Japan.

Kraska, T. 1997. The Plant Pathology Internet Guide Book (PPIGB). On-

line publication. Department of Horticulture, University of Hanover, Hanover, Germany.
http://www.ifgb.uni-hannover.de/extern/ppigb/ppigb.htm

Mohammadi, O. 1994. Commercial development of Mycostop fungicide. Pages 282–284 in: Improving Plant Productivity with Rhizosphere Bacteria. M. H. Ryder, P. M. Stephens, and G. D. Bowen, eds. CSIRO, Adelaide, South Australia.

National Biological Control Institute, Animal and Plant Health Inspection Service, U.S. Department of Agriculture
http://www.aphis.usda.gov/nbci/nbci.html

National Integrated Pest Management Network. U.S. Department of Agriculture, Cooperative State Research, Education, and Extension Service, and National Science Foundation Center for Integrated Pest Management
http://www.reeusda.gov/agsys/nipmn/index.htm

Weeden, C. R., Shelton, A. M., and Hoffmann, M. P., eds. Biocontrol: A Guide to Natural Enemies in North America. On-line publication. Cornell University, College of Agriculture and Life Sciences.
http://www.nysaes.cornell.edu/ent/biocontrol

Wilson, M., and Backman, P. A. 1999. Biological control of plant pathogens. Pages 309–336 in: Handbook of Pest Management. J. R. Ruberson, ed. Marcel Dekker, New York.

CHAPTER 99

Gail Ruhl ▪ Purdue University, West Lafayette, Indiana
Jacqueline Mullen ▪ Auburn University, Auburn, Alabama
Jean Williams-Woodward ▪ University of Georgia, Athens

Plant Problem Diagnosis and Plant Diagnostic Clinics

Diagnosis of plant problems begins with on-site observation. Usually it is the grower or homeowner who first notices a problem. Plant problems may be categorized as insect or mite damage, animal damage, infectious diseases, chemical damage, nutrient imbalances, environmental and site factors, genetic abnormalities, and damage due to management or cultural practices.

Problem-Solving Strategies

Diagnosis of a problem is often accomplished by a process of elimination. By following an orderly series of diagnostic steps, the cause of most problems can be determined.

First, *know what is normal.* Although this may seem obvious, it is particularly important to know the normal appearance of a cultivar or species. Some cultivars of familiar plants have unusual foliage pigmentation or distorted growth habits that might be misinterpreted as abnormal. Always provide genus, species, and cultivar name of each sample, if known.

Second, *check for symptoms and signs.* A thorough examination of leaves, stems, and roots is necessary to document an abnormality. Some symptoms, such as galls, may be diagnostic for a particular disease. Other symptoms, such as stunting or wilt, may be caused by a number of infectious and noninfectious agents. The aboveground plant part exhibiting symptoms may not be the site where the causal agent is active; e.g., wilt symptoms may indicate a root or crown problem. It is critical that the whole plant be examined.

A sign is physical evidence of a causal agent, such as fungal mycelium. Observing the presence of some signs, such as mycelium and spores of a powdery mildew fungus on leaves, can provide evidence confirming the identity of the problem. A hand lens greatly improves one's ability to detect signs.

Third, *check for patterns.* After examining individual plants, step back and look at the distribution of the affected plants. Are symptomatic plants located in a random pattern, or are all plants affected uniformly?

Look for patterns of injury in the entire planting, on individual plants, and on individual plant parts, such as leaves. Patterns of symptom development are clues to the type of problem. Biotic causal factors result in symptom development on adjacent plants and plant parts over time. Injury due to noninfectious agents tends to be more uniform in distribution.

Fourth, *review the site history, cultural practices, and time of symptom occurrence, and ask questions.* A review of weather data (current and past years), topographical features and other characteristics of the planting site, and fertilizer and pesticide practices can help to rule out or indicate specific problems. See Chapter 3, "Abiotic Diseases."

Submission of Samples for Analysis

Most successful growers routinely follow the diagnostic steps outlined previously and are able to recognize many common problems. There are times, however, when a new disease occurs or when confirmation of a tentative diagnosis requires microscopic techniques or other laboratory analysis in order to determine appropriate management practices. In those instances, Cooperative Extension specialists and diagnostic laboratories can be important resources. Diagnostic specialists follow the same steps for diagnosis outlined above and thus need to see samples of as many entire plants as is practical. The samples should display all of the symptoms that are of concern.

To sample herbaceous plants with symptoms of general decline or dieback, select several whole plants showing a range of symptoms (early to more serious), with roots and adjacent soil intact. Dig up the plants carefully. Bundle them together, and wrap the roots and soil in a plastic bag. Wrap the bundle of plants in newspaper, and place it in a crush-proof container for shipment. Include a healthy plant for comparison. *Do not add water.*

To sample trees with wilt symptoms, submit an entire wilted tree, if possible. If the trees are too large, collect branches 1/2 to 1 inch in diameter from branches that are actively wilting but not totally dead, and wrap them in perforated plastic to retain moisture. Collect a handful of feeder roots, and place them in a plastic bag.

To sample leaves, branches, and fleshy parts with localized infections such as cankers, leaf spots, and rots, select specimens representing early and more advanced stages of disease. For cankers, include healthy portions from above and below the diseased area. Press leaves flat between heavy paper or cardboard. Wrap fleshy parts in dry paper. Place the samples in a perforated plastic bag, so that they do not dry out during shipment. *Do not add water to the bag.*

If you suspect problems due to soilborne nematodes, collect at least one quart of soil from the root zone of affected plants. Be sure to include some roots. Place the entire sample in a plastic bag for shipment. Do not add water to the sample. Do not expose it to extreme heat, and do not allow it to dry out. If foliar nematodes are suspected, press leaves flat between paper or heavy cardboard. Place the sample in a perforated plastic bag, so that it does not dry out during shipment. Do not add water to the bag.

Accuracy of diagnosis is improved if detailed supplemental information is submitted with samples, including descriptions of the problem and the growing site, the date when the problem was first noticed, the percentage of plants affected, the weather, and recent pesticide and fertilizer treatments. It is often helpful to include a photograph of the site and affected areas, made with an instant camera, a video recorder, or a digital camera. Incomplete information or poorly selected specimens may result in inaccurate diagnosis or inappropriate recommendations for control.

Services and Charges

Many states have established plant diagnostic clinics and labs supported by public funding, typically from federal and state support at land-grant universities (see Table 99.1, pp. 444–450, for clinics in the United States and Ontario). In addition, some clinics are privately operated and supported. The services provided by clinics vary somewhat, but disease diagnosis is typically offered by all of them. Some clinics also perform soil analysis for nematode assays, insect identification, and weed and plant identification. Some private labs perform soil testing for pH and minerals. At state and federally supported clinics at land-grant universities, soil testing for pH and minerals is usually conducted by a separate facility.

Private labs charge for their services. Many state and federally supported labs also charge a nominal fee, to support processing costs and to help defray costs of materials and equipment. Sample charges at state and federally supported clinics usually range from none to $50, depending on the clinic and the analyses performed.

Responses may be mailed, phoned, faxed, or sent by electronic mail, usually within a week of receipt of a sample. The response usually includes the diagnostic report, recommendations, and information on the disease (with an invoice for services, when fees are applicable).

Plant Diagnostic Clinic Procedures

Physical Samples

When a sample arrives at a plant diagnostic clinic, it is examined and compared to the description of symptoms on the accompanying information sheet. This information is important, since damage to the sample during transit may confuse the diagnosis. The description of the problem allows the diagnostician to focus on the original symptoms observed at the time the sample was collected.

In an integrated clinic, a visual examination is followed by disease diagnostic testing, entomological study, and possibly assessments by specialists in horticulture, agronomy, or weed science.

Samples are usually studied with the aid of a microscope, to assist in viewing fungal or bacterial pathogens. In some cases, it is necessary to place small pieces of affected tissues in a culture medium to accurately identify the agent or agents causing disease. Identification of fungi and bacteria in culture usually requires one to several weeks. Other diagnostic tests that may be needed include enzyme-linked immunosorbent assay (ELISA), a serological or antigen-antibody assay, for virus identification; various tests for characterizing bacteria; a specific microscopic staining technique for viruses; and pathogenicity testing. More complicated molecular and biochemical tests are performed by some labs where diseases that are particularly difficult to diagnose are common and economically important. Unusual disease samples may be referred to more than one pathologist for consultation. Sometimes samples are routed to nearby or distant labs for special analyses.

Virtual Samples

Many diagnostic clinics utilize electronic technologies, including digital cameras and image-capturing devices on microscopes, to transmit disease sample images from remote sites to diagnosticians and extension specialists. Numerous county extension offices are equipped with digital cameras, and some also have compound and dissecting microscopes, video cameras, and image-capturing software, with which macroscopic and microscopic pictures of samples and possible disease agents can be taken. The images are sent via the Internet for diagnosis and disease control recommendations.

Not all samples using this technology are sufficient for diagnosis. An accurate diagnosis cannot usually be made from one digital image or photograph of a field or a tree. Although some diseases produce characteristic symptoms, such as galling or distinctive leaf spots, blights, or mildews, most diseases do not cause signature symptoms by which they can be readily identified. In these situations, a truly accurate diagnosis can be made only when images showing the symptoms present on the plant can be examined along with microscopic images of the disease agent. Although photographs or digital images may be insufficient for diagnosis, often the pathogen group (fungus, bacterium, virus, or nematode) can be narrowed down or identified; then the diagnostician may request additional images or physical samples of plants. The proper use of this new technology will help improve diagnostic services.

Results and Interpretation

Once a pathogen has been identified, most disease clinics issue a disease control recommendation. Some clinics do not recommend control practices and provide only diagnosis.

Disease control recommendations provided by diagnostic clinics are specific for the host plant and causal agent identified by the diagnosis. It is important to know where the plants are grown and how many plants or acres are affected. Disease control recommendations usually include cultural practices as well as chemical treatment. Fungicides, the primary chemicals recommended for disease control, are effective against fungal pathogens; some may also be active against bacteria. Chemicals suggested by clinics are registered for use on the crops to be treated and against the pathogens to be controlled. Some clinics provide fungicide application rates; others may recommend using products according to their label directions. It is the responsibility of the grower to read all pesticide labels and to use the products according to the labeling. The effectiveness of a fungicide can be reduced if the environment favors disease development. For this reason, cultural controls, such as avoiding wet soils or prolonged leaf wetness, may also be recommended.

The diagnosis obtained from a plant disease diagnostic clinic, be it public or private, is only as good as the sample and background information provided.

REFERENCES

Holmes, G. J., Brown, E. A,., and Ruhl, G. 2000. What's a picture worth? The use of modern telecommunications in diagnosing plant diseases. Plant Dis. 84:1256-1265.

Shurtleff, M. C., and Averre, C. W., III. 1997. The Plant Disease Clinic and Field Diagnosis of Abiotic Diseases. American Phytopathological Society, St. Paul, Minn.

Table 99.1 University-related clinics providing plant disease diagnosis and soil testing in the United States and Ontario[a,b]

	Soil testing	Plant disease diagnosis
Alabama	Soil Testing Laboratory Auburn University 118 Funchess Hall Auburn, AL 36849-5411 Phone: (334) 844-3958, (334) 844-5489	Plant Disease Clinic 101 Extension Hall Department of Entomology and Plant Pathology Auburn University Auburn, AL 36849-5624 Phone: (334) 844-5508, 5507
Alaska	Soil Testing Laboratory Agricultural Experiment Station University of Alaska 533 E. Firewood Palmer, AK 99645	Department of Plant, Animal and Soil Sciences University of Alaska Agricultural and Forestry Experiment Station Fairbanks, AK 99775-7200 Phone: (907) 474-7431
Arizona	Soil, Water, and Plant Tissue Testing Lab Department of Soils, Water, and Environmental Science Shantz (#38), Room 431 University of Arizona Tucson AZ 85721 Phone: (520) 621-9703	Extension Plant Pathologist University of Arizona Yuma Agriculture Center 6425 W. 8th Street Yuma, AZ 85364 Phone: (520) 726-0458 Extension Plant Pathologist Department of Plant Pathology Forbes 204 University of Arizona Tucson, AZ 85721 Phone: (520) 626-2681
Arkansas	Soil Testing and Research Laboratory University of Arkansas P.O. Drawer 767 Marianna, AR 72360 Phone: (870) 295-2851	Plant Disease Clinic Lonoke Agricultural Center P.O. Box 357 Highway 70 East Lonoke, AR 72086 Phone: (501) 676-3124
California	No soil testing service is offered by a public agency	Contact your county farm advisor or an extension specialist at the nearest university
Colorado	Soil, Water, and Plant Testing Laboratory Room A-319 NESB Colorado State University Fort Collins, CO 80523-1120 Phone: (970) 491-5061	Plant Diagnostic Clinic Jefferson County Extension 15200 W. 6th Avenue, Suite C Golden, CO 80401 Phone: (303) 271-6620 Identification and Diagnostic Service E-20 Plant Sciences Building Colorado State University Fort Collins, CO 80523-1177 Phone: (970) 491-6950
Connecticut	Soil Testing Laboratory University of Connecticut 6 Sherman Place, U-102 Storrs, CT 06269 Phone: (860) 486-4274	Home and Garden Education Center 1380 Storrs Road, U-4115 Storrs, CT 06269-4115 Phone: (860) 486-3437
Delaware	Soil Testing Laboratory Department of Plant and Soil Science University of Delaware Newark, DE 19717-1303 Phone: (302) 831-1392	Extension Plant Pathologist Plant and Soil Sciences Department Townsend Hall University of Delaware Newark, DE 19717-1303 Phone (302) 831-4865

(continued on next page)

Table 99.1 (continued) University-related clinics providing plant disease diagnosis and soil testing in the United States and Ontario[a,b]

	Soil testing	Plant disease diagnosis
Florida	Soil Testing Laboratory Wallace Building No. 631 P.O. Box 110740 Institute of Food and Agricultural Sciences University of Florida Gainesville, FL 32611-0740 Phone: (352) 392-1950	Florida Extension Plant Disease Clinic Building 78 Mowry Road P.O. Box 110830 University of Florida Gainesville, FL 32611-0830 Phone: (352) 392-1795 Regional labs: Plant Diagnostic Clinic North Florida Research and Education Center 30 Research Road Quincy, FL 32351 Phone: (850) 875-7154 Florida Extension Plant Disease Clinic Southwest Florida Research and Education Center 2686 State Road 29 N Immokalee, FL 34142-9515 Phone: (941) 658-3400 Florida Extension Plant Disease Clinic Tropical Research and Education Center 18905 SW 280th Street Homestead, FL 33032-3314 Phone: (305) 246-7000
Georgia	Soil, Plant and Water Laboratory University of Georgia 2400 College Station Road Athens, GA 30602-9105 Phone: (706) 542-5350	Extension Plant Disease Clinic Department of Plant Pathology Miller Plant Sciences Building University of Georgia Athens, GA 30602 Phone: (706) 542-2685
Hawaii	Soil Testing Laboratory Agricultural Diagnostic Service Center 1910 East-West Road Sherman Hall 134 Honolulu, HI 96822	Plant Disease Clinic Agricultural Diagnostic Service Center 1910 East-West Road Sherman Hall 112 Honolulu, HI 96822
Idaho	Plant Pathology – PSES University of Idaho Moscow, ID 83843 Phone: (208) 885-6057	Extension Plant Pathologist University of Idaho Research and Extension Center 3793 N. 3600 East Kimberly, ID 83341 Phone: (208) 423-6603 Extension Plant Pathologist University of Idaho Research and Extension Center 29603 U of I Lane Parma, ID 83660 Phone: (208) 722-6701 ext. 218
Illinois	No soil testing service is offered by a public agency	May–September: Plant Clinic 1401 W. St. Mary's Road University of Illinois Urbana, IL 61802 Phone: (217) 333-0519 October–March: N-533 Turner Hall 1102 S. Goodwin Avenue University of Illinois Urbana, Illinois 61801 Phone: (217) 333-2478

(continued on next page)

Table 99.1 (continued) University-related clinics providing plant disease diagnosis and soil testing in the United States and Ontario[a, b]

	Soil testing	Plant disease diagnosis
Indiana	No soil testing service is offered to homeowners by a public agency The Plant and Pest Diagnostic Laboratory can provide a partial listing of private soil testing labs	Plant and Pest Diagnostic Laboratory Department of Botany and Plant Pathology 1155 Lilly Hall Purdue University West Lafayette, IN 47907-1155 Phone: (765) 494-7071
Iowa	Soil Testing Laboratory G501 Agronomy Iowa State University Ames, IA 50011 Phone: (515) 294-3076	Plant Disease Clinic Department of Plant Pathology 323 Bessey Hall Iowa State University Ames, IA 50011 Phone: (515) 294-0581
Kansas	Soil Testing Laboratory 2308 Throckmorton Hall Kansas State University Manhattan, KS 66506 Phone: (785) 532-7897	Plant Disease Diagnostic Lab Department of Plant Pathology 4024 Throckmorton Hall Kansas State University Manhattan, KS 66506-5502 Phone: (785) 532-5810
Kentucky	Soil Testing Laboratory 103 Regulatory Services Building University of Kentucky Lexington KY 40546-0275 Phone: (859) 257-7355 serving western Kentucky: Soil Testing Laboratory University of Kentucky Research and Education Center P.O. Box 469 1205 Hopkinsville Street Princeton, KY 42445 Phone: (270) 365-7541 ext. 238	Serving central and eastern Kentucky: Plant Disease Diagnostic Lab Department of Plant Pathology 530S Agricultural Science Building N University of Kentucky Lexington, KY 40546-0091 Phone: (859) 257-8949 Serving western Kentucky: Plant Disease Diagnostic Lab University of Kentucky Research and Education Center P.O. Box 469 1205 Hopkinsville Street Princeton, KY 42445 Phone: (270) 365-7541 ext. 228
Louisiana	Soil Testing Laboratory Department of Agronomy Louisiana State University Baton Rouge, LA 70803 Phone: (225) 388-1219	Plant Disease Diagnostic Clinic P.O. Box 25100 Louisiana State University Baton Rouge, LA 70894-5100 Phone: (225) 388-2186 Fax: (225) 388-2478
Maine	Maine Soil Testing Service 5722 Deering Hall University of Maine Orono, ME 04469-5722 Phone: (207) 581-2945	Pest Management Office Cooperative Extension University of Maine 491 College Avenue Orono, ME 04473-1295 Phone: (207) 581-3880
Maryland	Soil Testing Laboratory NRSL University of Maryland College Park, MD 20742 Phone: (301) 405-1349	Plant Diagnostic Laboratory Department of Entomology University of Maryland College Park, MD 20742-4454 Phone: (301) 405-1611
Massachusetts	Soil Testing West Experiment Station University of Massachusetts Amherst, MA 01003	No diagnostic services are offered to homeowners by a public agency Commercial samples are handled by extension specialists at the University of Massachusetts

(continued on next page)

Table 99.1 (continued) University-related clinics providing plant disease diagnosis and soil testing in the United States and Ontario[a,b]

	Soil testing	Plant disease diagnosis
Michigan	Soil and Plant Nutrient Laboratory A81 Plant and Soil Sciences Michigan State University East Lansing, MI 48824-1325 Phone: (517) 355-0218	Diagnostic Services 101 Center for Integrated Plant Systems Michigan State University East Lansing, MI 48824-1312 Phone: (517) 355-4536
Minnesota	Research and Soil Testing Laboratories 135 Crops Research Building University of Minnesota 1903 Hendon Avenue St. Paul, MN 55108 Phone: (612) 625-3101	For homeowners: Yard and Garden Clinic 155 Alderman Hall University of Minnesota 1970 Folwell Avenue St. Paul, MN 55108 Phone: (612) 624-4771 (Minneapolis–St. Paul metro area), (888) 624-4771 (greater Minnesota, toll-free) For commercial growers: Plant Disease Clinic Department of Plant Pathology 495 Borlaug Hall, 1991 Upper Buford Circle University of Minnesota St. Paul, MN 55108 Phone: (612) 625-1275
Mississippi	Soil Testing Laboratory Room 1, Bost Extension Center P.O. Box 9610 Mississippi Cooperative Extension Service Mississippi State, MS 39762-9610 Phone: (662) 325-3313	Plant Pathology Lab Room 9, Bost Extension Center Box 9655 Mississippi Cooperative Extension Service Mississippi State, MS 39762-9655 Phone: (662) 325-2146
Missouri	Soil and Plant Testing Laboratory Department of Agronomy 23 Mumford Hall University of Missouri Columbia, MO 65211 Phone: (573) 882-3250	Extension Plant Diagnostic Clinic Room 42, Agriculture Building University of Missouri Columbia, MO 65211 Phone: (573) 882-3019
Montana	No soil testing service is offered by a public agency	Plant Disease Clinic Department of Plant Sciences and Plant Pathology 119 Ag Bioscience Facility Montana State University Bozeman, MT 59717 Phone: (406) 994-5150
Nebraska	Soil and Plant Analytical Laboratory Department of Agronomy and Horticulture 139 Keim Hall University of Nebraska Lincoln, NE 68583-0916 Phone: (402) 472-1571	Plant and Pest Diagnostic Clinic Department of Plant Pathology 448 Plant Sciences University of Nebraska Lincoln, NE 68583-0722 Phone: (402) 472-2559
Nevada	No university soil testing services are available	No university disease diagnostic services are available
New Hampshire	Analytical Services Laboratory Spaulding Life Science Center, G-54 University of New Hampshire 38 College Road Durham, NH 03824 Phone: (603) 862-3212	Plant Diagnostic Lab Plant Biology Department 241 Spaulding Hall University of New Hampshire Durham, NH 03824 Phone: (603) 862-3841

(continued on next page)

Table 99.1 (continued) University-related clinics providing plant disease diagnosis and soil testing in the United States and Ontario[a,b]

	Soil testing	Plant disease diagnosis
New Jersey	Soil Testing Laboratory Extension Resource Center Rutgers University P.O. Box 902 Milltown, NJ 08850 Phone: (732) 932-9292 or (732) 932-9295 Fax: (732) 932-8644	Plant Diagnostic Lab Rutgers University P.O. Box 550 Milltown, NJ 08850 Phone: (732) 932-9140 Fax: (732) 932-1270
New Mexico	SWAT Laboratory Agronomy and Horticulture New Mexico State University P.O. Box 30003 Las Cruces, NM 88003 Phone: (505) 646-4422	Extension Plant Pathologist P.O. Box 30003, MSC. 3AE Plant Sciences Cooperative Extension Service New Mexico State University Las Cruces, NM 88003 Phone: (505) 646-1965
New York	Cornell Nutrient Analysis Laboratories C55 Department 804 Bradfield Hall Cornell University Ithaca, NY 14853-1901 Phone: (607) 255-1722	For homeowners and commercial growers: Plant Disease Diagnostic Clinic Department of Plant Pathology 334 Plant Science Building Cornell University Ithaca, NY 14853-4203 Phone: (607) 255-7850 For commercial ornamental samples only: Long Island Horticultural Research Lab Cornell University 3059 Sound Avenue Riverhead, NY 11901 Phone: (516) 727-3595
North Carolina	Soil Testing Laboratory Agronomic Division North Carolina Department of Agriculture 4300 Reedy Creek Road Raleigh, NC 27607 Phone: (919) 733-2656, 2657, 2655	Plant Disease and Insect Clinic Campus Box 7211 Room 1104, Williams Hall North Carolina State University Raleigh, NC 27695-7211 Phone: (919) 515-3619, (919) 515-3825
North Dakota	Soil Testing Laboratory Soil Science Department North Dakota State University Fargo, ND 58105	Plant Diagnostic Clinic Department of Plant Pathology P.O. Box 5012 North Dakota State University Fargo, ND 58105
Ohio	The soil testing lab at Wooster is closed	C. Wayne Ellett Plant and Pest Diagnostic Clinic 110 Kottman Hall 2021 Coffey Road Ohio State University Columbus, OH 43210-1087 Phone: (614) 292-5006
Oklahoma	SWFAL Plant and Soil Science Department 048 Ag. Hall Oklahoma State University Stillwater, OK 74078 Phone: (405) 744-6630	Plant Disease Diagnostic Lab Department of Entomology and Plant Pathology 127 Noble Research Center Oklahoma State University Stillwater, OK 74078 Phone: (405) 744-9961

(continued on next page)

Table 99.1 (continued) University-related clinics providing plant disease diagnosis and soil testing in the United States and Ontario[a, b]

	Soil testing	Plant disease diagnosis
Ontario	Soil and Nutrient Laboratory Laboratory Services Division University of Guelph 95 Stone Road W. Guelph, ON N1H 8J7	Pest Diagnostic Clinic Laboratory Services Division University of Guelph 95 Stone Road W. Guelph, ON N1H 8J7 Phone: (519) 767-6256 Urban and turf diagnosis: (519) 767-6258 Commercial and diagnostic research: (519) 767-6227
Oregon	Central Analytical Laboratory 3017 ALS Building Oregon State University Corvallis, OR 97331	Plant Disease Clinic Extension Plant Pathology Cordley Hall 1089 Oregon State University Corvallis, OR 97331-2903 Phone: (541) 737-3472 Plant Pathology Lab H.A.R.E.C. Oregon State University P.O. Box 105 Hermiston, OR 97838 Phone: (541) 567-8321
Pennsylvania	Agricultural Analytical Services Laboratory Tower Road Pennsylvania State University University Park, PA 16802 Phone: (814) 863-0841	Plant Disease Clinic 220 Buckhout Laboratory Pennsylvania State University University Park, PA 16802 Phone: (814) 865-2204
Rhode Island	Rhode Island soil samples are sent to the University of Massachusetts Soil Testing Lab For more information, call CE Education Center, (401) 874-2900	URI Plant Protection Clinic University of Rhode Island Room 205 Greenhouse Kingston, RI 02881-0804 Phone: (401) 874-2967
South Carolina	Soil Testing Laboratory Agricultural Service Laboratory Clemson University Clemson, SC 29634 Phone: (803) 656-2300	Plant Problem Clinic 171 Old Cherry Road Clemson University Clemson, SC 29634-0114 Phone: (864) 656-3125
South Dakota	Soil Testing Laboratory Plant Science Department Box 2207-A, Ag Hall 06 South Dakota State University Brookings, SD 57007-1096 Phone: (605) 688-4766	Plant Disease Clinic Department of Plant Science South Dakota State University P.O. Box 2108 Brookings, SD 57007 Phone: (605) 688-5157
Tennessee	Soil Testing Laboratory University of Tennessee 5201 Marchant Drive Nashville, TN 37222-5112 Phone: (615) 832-5850	Plant and Pest Diagnostic Center University of Tennessee 5201 Marchant Drive Nashville, TN 37211-5112 Phone: (615) 832-6802
Texas	Soil Testing Laboratory Soil and Crop Sciences 2474 TAMU College Station, TX 77843	Texas Plant Disease Diagnostic Lab 1500 Research Parkway 2589 TAMU College Station, TX 77843-2132 Phone: (979) 845-8033
Utah	Soil Testing Laboratory USU Analytical Lab Ag Science Building, Room 166 Utah State University Logan, UT 84322-4830 Phone: (435) 797-2217	Plant Pest Diagnostic Lab Department of Biology Utah State University Logan, UT 84322-5305 Phone: (435) 797-2435

(continued on next page)

Table 99.1 (continued) University-related clinics providing plant disease diagnosis and soil testing in the United States and Ontario[a,b]

	Soil testing	Plant disease diagnosis
Vermont	Agricultural and Environmental Testing Lab Department of Plant and Soil Science 219 Hills Building University of Vermont Burlington, VT 05405-0082 Phone: (802) 656-3030	Plant Diagnostic Clinic Department of Plant and Soil Science 255B Hills Building University of Vermont Burlington, VT 05405-0086 Phone: (802) 656-0493
Virginia	Virginia Tech Soil Testing 145 Smyth Hall Blacksburg, VA 24061-0465 Phone: (540) 231-6893	Plant Disease Clinic Department of Plant Pathology and Weed Science 106 Price Hall Virginia Polytechnic Institute and State University Blacksburg, VA 24061-0331 Phone: (540) 231-6758
Washington	No public agency provides soil testing Your county extension agent can provide a listing of local labs	Serving eastern Washington: Prosser Plant Diagnostic Lab Washington State University 24106 N. Bunn Road Prosser, WA 99350-9687 Phone: (509) 786-9271 Serving western Washington: Plant Diagnostic Clinic Puyallup Research and Extension Center Washington State University 7612 Pioneer Way East Puyallup, WA 98371-4998 Phone: (253) 445-4582
West Virginia	Soil Testing Laboratory 1090 Ag Sciences Building West Virginia University Morgantown, WV 26506-6108 Phone: (304) 293-6023	Plant Disease Diagnostic Clinic 414 Brooks Hall Downtown Campus West Virginia University Morgantown, WV 26506 Phone: (304) 293-3911
Wisconsin	Soil and Plant Analysis Laboratory University of Wisconsin 5711 Mineral Point Road Madison, WI 53705 Phone: (608) 262-4364	Plant Disease Diagnostic Clinic Department of Plant Pathology University of Wisconsin–Madison 1630 Linden Drive Madison, WI 53706-1598 Phone: (608) 262-2863
Wyoming	Soil Testing Laboratory Renewable Resources University of Wyoming P.O. Box 3354 Laramie, WY 82071 Phone: (307) 766-2135	Extension Plant Pathology Laboratory Department of Plant Sciences University of Wyoming P.O. Box 3354 Laramie, WY 82071-3354 Phone: (307) 766-5083

[a] Compiled by Gail Ruhl.

[b] In many states, additional diagnostic labs are associated with the state department of agriculture or with U.S. Department of Agriculture research facilities. Contact your county extension office for procedures for submitting samples to a diagnostic lab.

CHAPTER 100

J. E. Preece ▪ Southern Illinois University, Carbondale

R. N. Trigiano ▪ University of Tennessee, Knoxville

Tissue Culture of Woody Plants

Micropropagation of plants offers several advantages over propagation by conventional methods, such as sowing seeds, rooting cuttings, layering, and grafting. Although more expensive than rooting cuttings, it can be less costly than grafting, and shipping can be inexpensive, because micropropagated plants are initially small. The plants are produced in an aseptic environment, and thus shipping them into areas with quarantines or other laws governing the import of plant materials can be easier. Increased basal branching, which is characteristic of micropropagated plants, often results in full, more desirable plants, especially shrubby ornamentals. Micropropagation can be used to rejuvenate plants, because it is often easier to root microshoots than to root cuttings from the same plants. Because of the rapid multiplication rate, micropropagation can produce enough plants to meet demand, while fewer stock plants are needed to provide propagules, and thus it can be more efficient than production by conventional means. However, the high costs associated with micropropagation make it unlikely that this method will ever totally replace traditional macropropagation.

The production of woody ornamental plants in culture has grown substantially over the last several decades (Table 100.1). Tissue culture laboratories specializing in ornamental and fruit trees and shrubs are now located in most regions of the United States and produce millions of micropropagated plants annually. This tremendous increase in activity is due, in part, to more efficient tissue culture methods and a boom in the demand for high-quality, uniform plants that have been tested or indexed for specific disease-causing agents. Improved micropropagation methods have also permitted the quick introduction of numerous cultivars of ornamental shrubs and fruit trees and have ensured adequate and timely supplies of them.

Plant tissue culture involves growing new plants from cells, tissues, or organs (explants) aseptically on a nutrient-rich growth medium augmented with growth regulators (usually synthetic hormones) in specially designed dishes or vessels. Somatic embryos (embryos generated from vegetative cells without the sexual process) can be regenerated from two categories of explants: (1) tissues lacking organized meristems, such as leaves, roots, stems, and flowers, from which shoots are generated adventitiously, in a process called organogenesis and somatic embryogenesis, and (2) preexisting meristems, such as axillary and accessory buds and apical meristems. Adventitious shoots and somatic embryos can be produced either directly from individual cells of an explant or indirectly from intervening callus tissue. Those produced directly are thought to be clonal (exact or true-to-type) reproductions of the parent plant, whereas those generated from callus are generally more likely to exhibit some abnormalities or exhibit off-types (somaclonal variation).

Adventitious regeneration techniques are seldom, if ever, employed by commercial laboratories, because of the relatively high risk of culture-induced variation or the production of off-type plants. Propagation by somatic embryogenesis has not been adopted by commercial laboratories for several reasons, including the general unreliability of the process over extended periods of time and the difficulty of adapting to commercial production schedules, because of nonsynchronous plant regeneration or low conversion (germination) rates.

The most common method of micropropagation of woody species is axillary bud (shoot) proliferation (the stimulation of branching) from stock plants. This technique produces clonal plants with relatively little chance of inducing off-types, and it generally works well with many species. Axillary bud proliferation may be coupled with sanitary measures and curative treatments to produce plants that have been tested for specific pathogens or contaminants ("disease-free plants").

The intent of this primer chapter is to provide a basic understanding of the production of woody plant materials by tissue culture. Indexing stock plants for pathogens, axillary bud proliferation, scaling-up for commercial production, and some problems with plants produced in tissue culture are discussed.

Indexing Stock Plants

Indexing amounts to testing whole stock plants or plant parts (cuttings, leaves, etc.) for the presence of specific plant pathogens, including bacteria, fungi, and viruses. For some herbaceous and a few woody species, indexing is essential before micropropagation can be initiated. If pathogens are detected, the plants are either destroyed or treated to eliminate the offending organism. Only plants that have been given a clean bill of health after testing for specific pathogens are used for production of other plants, and only under strict sanitary conditions. Stock plants and progeny are periodically reevaluated to test for the presence of some common pathogens. Surveying or indexing techniques for the detection of pathogenic bacteria and fungi can be very simple; techniques for detecting viruses can be quite sophisticated.

To detect bacterial and fungal infections of stock plants, a surface-disinfested shoot or axillary bud or sap derived from it is incubated on semisolid (agar) or in liquid nutrient broth. Microbial growth on agar, cloudiness of nutrient broth (caused by bacteria), and fuzzy or fluffy mycelium of fungi growing in nutrient broth are positive indications of infestation by pathogenic or nonpathogenic microorganisms. The plants are either discarded or are rescued by reculturing apical meristems or meristem tips to reestablish disease-free or contaminant-free stock plants.

Virus indexing of plants usually involves either bioassays or serological techniques. Some viruses are routinely and easily detected by bioassays, in which indicator plants are inoculated with the sap of a stock plant. Indicator plants, typically members of the tobacco family or lamb's-quarters, may develop severe local lesions, or they may show symptoms of systemic infection. This technique is useful only for detecting viruses that are mechanically transmitted and are capable of infecting bioindicator plant species.

Table 100.1 Woody plants propagated by tissue culture[a]

Woody plant species	Common name
Acer rubrum	Red maple (numerous cultivars)
Actinidia deliciosa	Kiwi fruit
Amelanchier × grandiflora	Serviceberry (several cultivars)
Andromeda polifolia	Andromeda (several cultivars)
Aronia melanocarpa	Black chokeberry
Betula jacquemontii	Jacquemonti birch
Betula nigra	Heritage birch, river birch
Betula pendula	European white birch (weeping forms)
Caryopteris × clandonensis	Blue-mist shrub
Corylopsis pauciflora	Winter hazel
Corylopsis spicata	Spike winter hazel
Corylus americana	American filbert
Cotinus coggygria	Smoketree
Enkianthus campanulatus	Red-vein enkianthus
Exochorda racemosa	Pearlbush
Fothergilla gardenii	Fothergilla (several cultivars)
Halesia monticola	Silverbell
Hydrangea spp.	Hydrangea
Kalmia latifolia	Mountain laurel (numerous cultivars)
Kalmiopsis leachiana	
Leucothoe fontanesiana	Fetterbush
Liquidambar styraciflua	Sweetgum
Lorapetalum chinensis	
Magnolia grandiflora	Southern magnolia
Magnolia virginiana	Sweetbay magnolia
Malus sylvestris	Crabapple (numerous cultivars)
Malus spp.	Apple
Morus alba pendula	Mulberry
Nandina domestica	Nandina
Oxydendrum arboreum	Sourwood
Paulownia tomentosa	Empress tree, royal paulownia
Pieris floribunda × P. japonica	Japanese andromeda (numerous cultivars)
Prunus spp.	Cherries, peaches, almonds
Populus tremula	Columnar European aspen
Rhododendron spp.	Rhododendron, azalea (many cultivars)
Rosa spp.	Roses
Rubus idaeus	Raspberry
Schizophragma hydrangeoides	Japanese hydrangea vine
Syringa spp.	Lilacs (numerous cultivars)
Tilia cordata	Linden
Ulmus japonica × wilsoniana	Accolade elm
Ulmus parvifolia	Chinese elm
Vaccinium spp.	Blueberry (numerous cultivars), cranberry
Viburnum spp.	Viburnum
Vitis spp.	Grape
Wisteria macrostachya	Wisteria

[a] This list was assembled from the catalogues of B & B Laboratories, Mount Vernon, Washington; Briggs Nursery, Olympia, Washington; Knight Hollow Nursery, Middleton, Wisconsin; Microplant Nurseries, Gervais, Oregon; and other sources. This list should not be considered exhaustive.

Serological techniques, particularly enzyme-linked immunosorbent assay (ELISA), are most often used for virus indexing of plants. In ELISA, an antibody to a specific virus, obtained from a mammal or an animal cell culture, is conjugated to an enzyme, usually alkaline phosphatase or peroxidase, and placed in microwells in a plastic plate; the sap of the plant to be indexed is introduced, and virus particles are trapped by the antibody. The sap is removed, and then an enzyme-antibody conjugate is added, which also attaches to the virus particles. The substrate for the enzyme is added to the well, and the appearance of a colored compound in the liquid after incubation indicates the presence of the virus. The test must be repeated for each specific virus that may be infecting the plant. If the test is positive, the plants are discarded or are rescued by physical treatment or meristem tip culture.

The most common method of obtaining virus-free plants is mericloning – isolation and culture of meristem tips that have been exposed to elevated air temperatures (35–40°C) for days or weeks (depending on the species). Thermotherapy generally inhibits the ability of viruses to replicate in plant tissues, especially meristems. The apical meristem has no mature vascular connection to the rest of the plant, and often viruses are absent from this tissue or present in significantly lower numbers than in other parts of the plant. In mericloning, the very small apical meristem and perhaps one or two leaf primordia are excised from the tip of the plant and cultured on nutrient medium containing growth regulators. Regenerated plants are indexed for the same virus or viruses, and eradication procedures are repeated if necessary.

Indexing is conducted primarily for fruit crops and some herbaceous species but rarely, if ever, for woody ornamentals. However, indexing of woody ornamental species may become important in the future.

Stages of Micropropagation by Axillary Shoot Proliferation

Stage 0: Selection of Stock Plants

Plants that are the source of explants are generally grown in a protected environment, such as a greenhouse. Outdoor weather conditions, such as wind and rain, can be conducive to microbial activity. Explant material from field-grown plants can be difficult or impossible to clean sufficiently to eliminate microbial contamination. Greenhouse-grown plants and softwood shoots from dormant stems of woody plants that are forced indoors are usually easier to surface-disinfest.

Stage 1: Establishment of Explants in Aseptic Culture

Upon removal from the stock plant, explants must be placed in a humid environment to protect them from desiccation during transport to the laboratory. Plastic bags and moist paper towels are

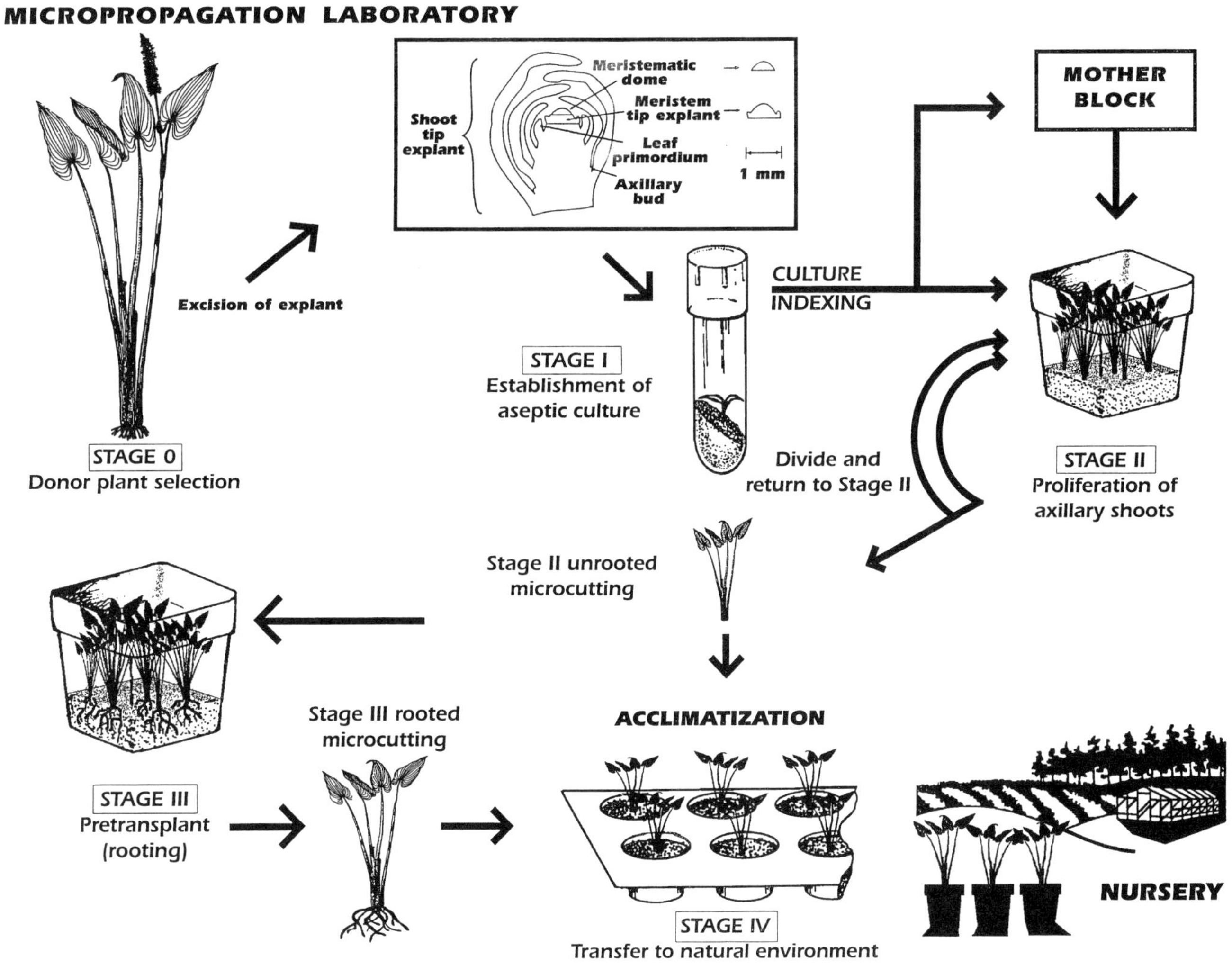

Fig. 100.1 Micropropagation schemes for the production of nursery plants by axillary bud and meristem tip shoot culture. (Reprinted, by permission, from Kane, 1996)

generally sufficient. In the laboratory, leaves are removed from shoots, and stems with axillary buds are surface-disinfested in a solution of sodium hypochlorite (NaClO) or other suitable disinfestant containing a wetting agent, such as Tween 20, Triton X-100, or dishwashing detergent. They are then rinsed several times in sterile water to remove the disinfestant. Tissues damaged by this treatment are excised, and the explants are placed on an appropriate culture medium in vitro. After a time, the explants adjust to conditions in vitro, and new growth typically begins within a couple of weeks (Fig. 100.1).

Stage 2: Proliferation of Axillary Shoots

In an appropriate medium (with proper nutrients and plant growth regulators, including cytokinins), axillary buds begin to elongate once the explants are established aseptically in vitro. Nodal buds on the microshoots elongate in turn, and eventually axillary shoots are actively growing (Fig. 100.2). Masses of axillary shoots are transferred to fresh medium, typically at monthly intervals. When the shoots are sufficiently large, they may be excised and placed on a similar shoot proliferation medium, where

Fig. 100.2 Micropropagation stage 2: proliferation of axillary shoots from a nodal culture of *Cornus florida* (flowering dogwood) on nutrient medium containing benzyladenine, a cytokinin. Bar = 1.0 cm.

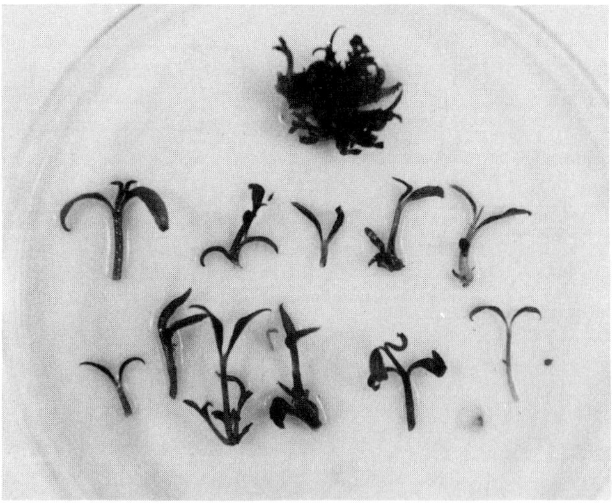

Fig. 100.3 Micropropagation stage 2: microshoots excised from a proliferating nodal culture of *Cornus florida.* The original culture (top) will be returned to a nutrient medium containing cytokinin, to produce additional shoots. An average of five shoots per culture can be harvested each month. About 50% of them will form adventitious roots when placed on a nutrient medium containing indolebutyric acid, an auxin. Bar = 1.0 cm. (Reprinted, by permission, from Kaveriappa et al., 1996)

their axillary shoots also elongate. Stage 2 can continue until enough axillary shoots have been generated to meet production goals.

Fig. 100.4 Micropropagation stage 3: a rooted microshoot of *Cornus florida,* four weeks after exposure to auxin. Bar = 2.0 cm.

Stage 3: Rooting

If entire plantlets are to be produced, axillary shoots are cut (Fig. 100.3) and placed under conditions conducive to rooting. Plantlets can be rooted in vitro (Fig. 100.4), on a medium that often contains an auxin. They can also be rooted in a greenhouse propagation medium under intermittent mist in a high-humidity enclosure or under fog in a greenhouse or other suitable structure.

Stage 4: Acclimatization

Preparation of rooted plantlets for the low relative humidity of the greenhouse or field may be the most difficult stage of micropropagation and may lead to the greatest losses of plants. Plants in vitro are uniquely unprepared for the environment of the greenhouse or field; their leaves often have a very thin cuticle, and the stomatal closure mechanism may not be operational. New leaves produced after the plants have been transferred to a greenhouse environment are often their only normally functional leaves. A gradual acclimatization is necessary in order for the plants to produce new leaves. Typically, plantlets are placed in a high-humidity enclosure in a greenhouse, and the relative humidity is incrementally reduced to a normal level.

Scale-Up

Laboratories producing woody ornamentals by micropropagation utilize axillary shoot proliferation, because of the clonal uniformity of the regenerated plants. When determining how to micropropagate a new species or cultivar, laboratory managers often consult the research literature, but this is only a starting point in scaling up to commercial production. The usual first step in research at a commercial laboratory is to attempt to repeat published results and to determine whether the process will be useful for profitable production of hundreds or thousands of plants of a particular clone.

Scaling up research results to full production levels at a commercial laboratory poses many challenges. Published research results may consist of fewer than five replications, although the number of replications frequently exceeds 20. However, even as many as 100 replications may not be an accurate predictor of results when production is scaled up to thousands of units (J. E. Preece and C. A. Huetteman, unpublished data).

When production is scaled up, problems are magnified in a multiplicative manner. For example, 70% rooting of microshoots may appear to be an acceptable success rate with some cultivars. Similarly, a 70% survival rate of rooted microshoots may also appear to be satisfactory. However, if these two rates are combined (70% rooting and 70% acclimatization), the efficiency rate is $0.7 \times 0.7 = 0.49$; that is, 49% of the originally harvested microshoots actually become salable plants. More than half are lost, reducing the production efficiency tremendously. Even if the rooting and acclimatization rates are each 90%, the efficiency of the system is only 81%. Such rapid drops in efficiency can result in failure to meet customer demand.

When production is scaled up to thousands of cultures, increasing the multiplication rate by one or more shoots per culture vessel can increase profitability. However, caution must be exercised whenever shoot production is increased. For uniform clonal micropropagation, it is essential that axillary shoots elongate without producing adventitious shoots. The typical method for increasing shoot production is to increase the concentration of cytokinin in the culture medium. However, cytokinins at high levels can induce the production of adventitious shoots and, accordingly, increased the likelihood of somaclonal variation in the resulting plants. This has been a particular problem with rhododendrons produced in vitro. Laboratories that have acquired a reputation for lack of uniformity have failed because of the resulting lack of sales. Nurseries demand uniform, true-to-type plant materials from propagators.

In a commercial laboratory, thousands of tissue cultures may be growing at any time, usually in a large room, on racks illuminated with cool white fluorescent lamps. Keeping a large number of cultures in such proximity often leads to problems that are unexpected when they occur for the first time. For example, spider mites and thrips can be introduced into a laboratory on plant materials brought in to establish cultures, on people, and on air currents. They enter culture vessels and feed, carrying fungal spores and bacteria with them. When they enter a vessel, they tend to walk where the surface of the medium contacts the sides of the vessel, and microbial colonies subsequently form in a characteristic pattern along the sides. Spider mites and thrips walk from one culture vessel to another, spreading contamination, which can lead to the loss of thousands of cultures. The problem can be controlled by autoclaving all contaminated vessels, fumigating the growth room, and sometimes including an insecticide or miticide in the medium. A better approach is prophylactic and requires that the ridge where the lid contacts the outside of the vessel be wrapped with a plastic wrap or Parafilm. Spider mites and thrips are not often a problem in laboratories with very few cultures, but no plant tissue culture laboratory seems to be immune.

Unless a laboratory is owned by a nursery and produces plants primarily for that nursery, it is essential to develop a customer base and provide service and a high-quality product in sufficient numbers. Commercial laboratories are frequently asked to micropropagate new plants. They tend to approach with caution, because it can be expensive to develop a micropropagation protocol and scale-up. Generally, the laboratory asks for payment up front to develop the protocol and a commitment to purchase a minimum number of the plants once they are in production. This eliminates customers who are not aware of the laborious and costly process of developing a commercially viable micropropagation protocol, especially if they want only a few plants.

Potential Problems with Micropropagated Plants

Micropropagation offers numerous advantages, but some problems have been noted in plants produced in vitro. Many of the problems are the result of somaclonal variation, or variation related to genetic mutation or epigenetic changes in somatic (vegetative) cells in vitro.

However, not all of the problems with micropropagated plants are due to somaclonal variation. For example, plants may be produced with excellent uniformity, but they may not be of the phenotype ordered by the nursery. This problem is likely the result of mislabeling, which can occur whenever plants are produced or propagated and is not unique to micropropagation. Growers are alert to check for somaclonal variation, and they quickly notice mislabeled plants, but they may assume that the plants are the result of somaclonal variation. It can be easier to make mislabeling mistakes in micropropagation than in conventional propagation, because cultures of different genotypes can closely resemble each other and may be less distinctly different than plants or cuttings of different cultivars. Because of the high multiplication rate and the subdivisions of an initial culture during subsequent transfers, mislabeling of one culture vessel can rapidly result in many misidentified cultures; all the plants regenerated from these cultures will be mislabeled. During production micropropagation, it is essential that extra time and care be taken to ensure accuracy in labeling cultures.

In a typical pattern of variation associated with tissue culture, most of the plants are uniform, but only some are off-type. Such somaclonal variation can be minimized if extra care is taken to avoid the production of adventitious shoots, especially from callus, during micropropagation of woody species. Commercial laboratories routinely excise and discard callus formed in vitro, so that it will not be a source of adventitious shoots, and because cultures often grow more vigorously when the callus is removed. Managers of commercial laboratories tend to hold cytokinin levels sufficiently low to minimize the formation of callus and adventitious shoots and yet maintain the health of the cultures and sufficient proliferation of axillary shoots.

In another type of variation, portions of the leaves of some plants lack chlorophyll. The variegated phenotypes appear similar to unstable chimeras and may be the result of unobserved chimeras breaking up in the stock plants. Such variegation may also be the result of new variation introduced in culture, in shoots formed adventitiously from callus. Plants displaying such variegation are usually rogued.

Another fairly common variation in plants produced in vitro is the formation of fasciated stems (flat, wide stems that appear as if two or more stems are fused together lengthwise). Fasciation can have a genetic cause, as the result of a mutation, but it can also be a physiological response of shoots to conditions in vitro, such as the cytokinin used.

Disorders of Rhododendrons Produced by Micropropagation

In rhododendrons, two unusual phenotypes are frequently related to production in vitro: witches'-broom and tissue proliferation.

In some rhododendrons, a type of witches'-broom, or proliferation of short branches, develops on the upper parts of the plant. Affected plants usually outgrow the problem, but it may take some time, delaying the date when they are ready for sale.

Tissue proliferation (TP) is characterized by gall-like growths, typically near the soil line, but sometimes higher on the stem and on roots. These growths may not be tightly attached to the stems. They are less than a centimeter to several centimeters in diameter, and adventitious shoots may grow from them. The disorder often occurs in the most vigorously growing plants. Similar galls have been observed on rhododendron seedlings and plants propagated from rooted cuttings, but in TP-prone cultivars the disorder is generally more common in plants propagated in culture.

TP occurs primarily in elepidote (nonscaly) rhododendron cultivars. Its cause is as yet undetermined, and why it occurs more frequently in micropropagated plants than conventionally propagated plants has yet to be understood. It could be a symptom of a disease, such as crown gall, caused by *Agrobacterium tumefaciens,* or infection by other bacteria that cause gall-like growths. There is no convincing evidence that TP is a pathological symptom, but this possibility has not been ruled out completely. There is also no compelling evidence that TP is the result of cultural conditions, such as pesticide applications, although plants from the same laboratory may exhibit the disorder at one nursery and not at another. There is no clear evidence that cytokinins in culture media are related to TP, but they cannot be ruled out as a cause of the disorder. It is also not clear whether TP is associated with adventitiously derived plants more or less than those of axillary origin. TP galls and lignotubers (burl-like growths that form at or near the soil line in many woody species) share some similarities in internal anatomy. It is possible that TP galls are a type of lignotuber, but this has not been proved. The seriousness of TP and whether it is indeed a problem or a natural growth response of rhododendron is not known. The question remains, Why is it not exclusively associated with plants generated by tissue culture, and why it is more common in micropropagated plants?

Conclusion

Micropropagation of woody ornamentals has a firm niche in the nursery industry. It is an efficient method for clonal propagation of many species, but it is unlikely to ever replace conventional macropropagation. The commercial micropropagation industry has seen considerable sorting out. Some laboratories that pushed their cultures for maximum shoot production experienced problems with somaclonal variation and consequently lost customers and went out of business. The successful laboratories have been those that have taken a more conservative approach, ensuring only axillary shoot proliferation and carefully cultivating a customer base. The future of tissue culture in the nursery industry will probably include greater use of techniques to clean up plants infested with pathogens, like the techniques used successfully with fruit crops. The list of woody ornamentals available in micropropagated form will probably keep expanding, especially as techniques are developed for more efficient micropropagation from elite or selected adult specimens of tree species.

REFERENCES

Kane, M. E. 1996. Propagation from preexisting meristems. Pages 61–71 in: Plant Tissue Culture Concepts and Laboratory Exercises. R. N. Trigiano and D. J. Gray, eds. CRC Press, Boca Raton, Fla.

Kaveriappa, K. M., Phillips, L. M., and Trigiano, R. V. 1996. Micropropagation of flowering dogwood (*Cornus florida*) from seedlings. Plant Cell Rep. 16:485–489.

CHAPTER
101

Thomas Stebbins ▪ University of Tennessee, Knoxville

David Johnson ▪ Missouri Department of Agriculture, Jefferson City

Regulatory Control

State, federal, and international laws and regulations govern the production, sale, and transportation of ornamental plants. Domestic regulation of nursery plants and enforcement are typically performed by state officials in the plant pest regulatory section of the state's department of agriculture or an equivalent agency. Programs of the U.S. Department of Agriculture, Animal and Plant Health Inspection Service, Plant Protection and Quarantine (USDA-APHIS-PPQ) may assist states when a pest poses a serious threat to an existing industry, a national forest, environmental health, or other plant resources.

Nursery Inspections and Licensing

Most states require nurseries to be licensed to sell nursery stock. The type of plant material that is considered nursery stock or is otherwise regulated varies from state to state. Before issuing a license to a nursery, state officials often inspect it to verify compliance with state and federal plant laws and regulations. Each state has determined pest tolerance levels to protect nursery, native, landscape, and crop plants in the state and to meet requirements for the movement of plant material to other states and countries. Nursery inspectors look for pest problems during routine inspections; labeling, shipping documents, certificates, and plant vigor may also be checked. Stock that may be shipped out of state should be brought to the attention of the inspector so that any special requirements stipulated by the receiving state may be addressed and special certificates issued if necessary. Most pest problems can be treated, and affected plants can be released for sale, but some problems cannot be remedied, and plants must be destroyed. Nursery inspectors can often provide specific local information on plant pests and their control. Copies of state laws and regulations concerning plants are generally available from state departments of agriculture. A list of state departments of agriculture and equivalent state agencies and contact information is available from the National Association of State Departments of Agriculture (http://www.nasda-hq.org).

Interstate Movement of Plants and Quarantines

Most states require out-of-state nursery stock to have been inspected in the state of origin and to be free of pests. Nursery stock grown out of state is usually required to be accompanied by documentation of inspection, in the form of a certificate of inspection or a copy of the nursery license. Nurseries and retail dealers should contact a nursery inspector if they receive plants that appear to have a pest problem or physiological disorders. If introduced pests are of particular concern, addressing the problem immediately can prevent further dissemination and provide valuable documentation for dealing with the shipper. Upon finding a pest problem in stock (either in-state or out-of-state), an inspector may advise immediate pest control, quarantine the stock and advise pest control, order the stock to be sent back to the state where it originated, or order it to be destroyed. If a pest problem is suspected, stock is sometimes held in quarantine until confirmation or identification of the pest. The National Plant Board and four regional plant boards meet regularly to discuss plant problems and to work out solutions to facilitate interstate movement of plants.

State exterior quarantines are sometimes issued to restrict the movement of specific plant materials, pests, and related articles into the state. Federal domestic quarantines restrict the movement of specific plant materials, pests, and related articles out of, into, or through specified states, counties, or portions of counties. These quarantines often stipulate cultural practices or chemical treatments to satisfy movement requirements. The *Federal and State Quarantine Summaries,* published by the American Nursery and Landscape Association in cooperation with the National Plant Board and USDA-APHIS, summarize federal and state quarantines and include other relevant information applicable to interstate shipment of plant materials.

Export Certification

To facilitate plant material exports, state departments of agriculture and USDA-APHIS-PPQ provide documentation that exported plants and plant products meet the import requirements of foreign countries. Some states offer certification programs that help growers meet criteria necessary for exportation of certain plant materials into foreign countries. After these requirements are met, a phytosanitary certificate is issued. The import requirements for nearly all countries to which the United States exports agricultural products are outlined in the on-line EXCERPT database, which is accessible by state and federal officials and nurseries by subscription.

Import Requirements

Imported plant material has often been a source of new varieties with improved characteristics. However, stimulated by the accidental introduction of devastating diseases, particularly chestnut blight and white pine blister rust in the early 1900s, Congress passed the first federal Plant Quarantine Act in 1912. This law and subsequent regulations were designed to prevent the introduction of foreign pests into the United States.

Airport terminals, seaports, border stations, and mail are routinely inspected to search for foreign plant pests. Import permits are issued by USDA-APHIS-PPQ. Imported plants may be subject to post-entry quarantine and inspected regularly by state or federal officials until it is ascertained that they are free of exotic pests.

Sources of Additional Information

Contact your state department of agriculture or the equivalent state agency.

Federal and State Quarantine Summaries is published by the American Nursery and Landscape Association (1250 I Street N.W., Suite 500, Washington, DC 20005; http://www.anla.org).

Additional information is available from the following:

National Association of State Departments of Agriculture, http://www.nasda-hq.org
National Plant Board, http://www.aphis.usda.gov/npb
USDA-APHIS-PPQ, http://www.aphis.usda.gov/ppq

APPENDIX

Diseases of Woody Ornamentals and Trees in Nurseries in the United States

Information on the occurrence of diseases of woody ornamentals on different states was compiled by the following contributors:

Alabama
A. K. Hagan, J. Olive, and J. Mullen

Arkansas
S. Vann

California
T. Tidwell

Delaware
R. Mulrooney

Florida
G. Simone and L. Brown

Georgia
J. Williams-Woodward

Indiana
G. Ruhl, K. Rane, and P. Pecknold

Kansas
J. O'Mara

Kentucky
J. Beale and J. Hartman

Massachusetts
R. Wick

Maryland
D. Clement and E. Dutky

Michigan
D. Miller

Missouri
J. W. Thompson, L. Kabrick, L. Hanning, and D. Johnson

Nebraska
L. Giesler and J. Watkins

New Jersey
A. Gould

New York
M. L. Daughtrey, B. Eshenaur, K. Snover, and G. Hudler

North Carolina
R. K. Jones, T. Creswell, C. Hodges, and D. M. Benson

Ohio
N. Taylor and B. Rao

Oklahoma
S. von Broembsen

Oregon
J. Pscheidt

Pennsylvania
G. Moorman and J. Peplinski

South Carolina
M. Williamson, S. Jeffers, R. Miller, and J. Blake

Tennessee
A. Windham, T. Stebbins, A. Self, and M. Mmbaga

Texas
L. Barnes and G. Philley

Virginia
J. Stipes, D. Reaver, T. Banko, and T. M. Likins

Wisconsin
L. Brown-Rytlewski and M. F. Heimann

Host	Disease	Ala.	Ark.	Calif.	Del.	Fla.	Ga.	Ind.	Kan.	Ky.	Mass.	Md.	Mich.	Mo.	N.C.	Nebr.	N.J.	N.Y.	Ohio	Okla.	Ore.	Pa.	S.C.	Tenn.	Tex.	Va.	Wis.
Arborvitae							P		P		NP					P			P								
	Leaf blight	Y	N	N	Y	Y	...	Y	...	Y	...	N	Y	N	N	...	N	Y	...	N	Y	Y	Y	Y	Y	N	N
	Nursery blight, twig and tip dieback	Y	Y	N	Y	Y	...	Y	...	Y	...	Y	N	N	N	...	Y	Y	...	Y	Y	Y	Y	N	Y	Y	Y
	Tip blight	Y	N	N	N	Y	...	Y	...	N	...	N	N	N	N	...	N	Y	...	N	Y	Y	Y	N	Y	Y	Y
Ash							P													P							
	Anthracnose	Y	Y	Y	Y	Y	...	Y	Y	Y	Y	Y	N	Y	Y	Y	Y	Y	Y	...	Y	Y	Y	Y	Y	Y	Y
	Ash rust	N	N	Y	Y	Y	...	Y	Y	N	Y	Y	N	Y	N	Y	Y	Y	Y	...		Y	N	N	?	Y	Y
	Ash yellows	Y	N	N	N	N	...	Y	Y	Y	Y	N	N	N	N	Y	Y	Y	N	...	N	Y	N	N	N	Y	Y
	Cankers	N	N	Y	Y	Y	...	Y	Y	Y	Y	N	N	Y	Y	Y	Y	Y	Y	...	Y	Y	N	Y	N	Y	Y
	Leaf spot	Y	N	Y	N	Y	...	N	Y	Y	N	N	N	Y	Y	Y	Y	Y	N	...	Y	Y	Y	Y	Y	Y	Y
	Powdery mildew	Y	N	Y	N	N	...	N	Y	Y	N	N	N	N	N	Y	Y	Y	N	...	Y	N	Y	Y	N	Y	Y
	Verticillium wilt	N	N	Y	N	N	...	Y	Y	Y	Y	Y	Y	Y	N	Y	Y	Y	Y	...	Y	Y	N	N	N	Y	Y
Aucuba								NP	NP	NP	NP		NP	NP		NP		NP	NP	NP		NP					NP
	Cold injury and sunscald	Y	Y	Y	Y	Y	Y	...	...	...	...	Y	...	...	Y	...	Y	...	...	...	...	...	Y	Y	Y	Y	...
	Leaf spot	Y	N	Y	Y	Y	Y	...	...	...	...	N	...	...	Y	...	Y	...	...	...	...	...	Y	Y	N	Y	...
	Phytophthora root rot	Y	N	N	Y	Y	Y	...	...	...	...	Y	...	...	Y	...	Y	...	...	...	...	...	Y	Y	Y	Y	...
	Root-knot nematodes	Y	N	N	N	Y	Y	...	...	...	...	Y	...	...	Y	...	N	...	...	...	...	...	Y	Y	Y	N	...
	Sclerotium blight	Y	N	N	N	Y	Y	...	...	...	...	N	...	...	Y	...	N	...	...	...	...	...	Y	Y	Y	N	...
Azalea																											
	Anthracnose	Y	N	N	N	Y	N	N	N	N	N	Y	N	N	Y	N	Y	Y	Y	N	Y	N	Y	Y	Y	Y	Y
	Botrytis gray mold	Y	N	Y	N	Y	Y	N	N	Y	Y	Y	Y	N	Y	Y	Y	Y	Y	N	Y	Y	Y	Y	Y	Y	Y
	Cylindrocladium blight and root rot	Y	N	Y	N	Y	Y	...	N	N	Y	N	...	Y	Y	N	Y	Y	Y	Y	Y	Y	Y	Y	Y	Y	Y
	Leaf and flower gall	Y	Y	Y	Y	Y	Y	Y	N	Y	Y	Y	N	Y	Y	N	Y	Y	Y	Y	Y	Y	Y	Y	Y	Y	N
	Petal blight	Y	N	Y	Y	Y	Y	N	N	Y	Y	Y	N	N	Y	Y	Y	Y	Y	N	Y	Y	Y	Y	Y	Y	Y
	Phomopsis dieback	Y	N	N	N	Y	Y	N	N	Y	Y	Y	N	N	Y	N	Y	Y	Y	N	...	Y	Y	Y	Y	Y	Y
	Phytophthora dieback	Y	N	N	Y	Y	Y	N	N	Y	Y	Y	...	N	Y	N	Y	N	Y	Y	...	Y	Y	Y	Y	Y	Y
	Phytophthora root rot	Y	Y	Y	Y	Y	Y	Y	Y	Y	Y	Y	Y	Y	Y	Y	Y	Y	Y	Y	Y	Y	Y	Y	Y	Y	Y
	Powdery mildew	Y	N	Y	N	Y	Y	Y	N	Y	Y	Y	Y	N	Y	N	Y	Y	Y	N	Y	Y	Y	Y	Y	Y	Y
	Rhizoctonia web blight	Y	N	N	N	Y	Y	N	N	Y	Y	Y	N	N	Y	N	N	N	Y	N	Y	N	Y	Y	Y	Y	Y
	Rust	Y	N	Y	Y	Y	Y	?	N	N	N	Y	Y	N	Y	N	Y	Y	N	N	Y	N	Y	Y	Y	N	N
	Salt injury	Y	N	Y	Y	Y	N	N	N	Y	Y	Y	N	N	Y	N	Y	Y	Y	N	Y	Y	Y	Y	Y	Y	Y
Barberry					P								P	P			P	P	P	P		NP					
	Phytophthora root rot	Y	Y	N	...	Y	Y	Y	N	Y	...	Y	...	...	Y	N	...	...	...	...	...	...	Y	Y	Y	Y	Y
Birch			NP						P							P	P		P	P					P		
	Anthracnose	?	...	N	N	Y	N	N	...	...	Y	Y	...	Y	Y	...	...	Y	...	...	...	Y	Y	Y	...	Y	Y
	Blister	?	...	Y	N	?	N	N	...	...	?	N	...	N	Y	...	...	N	...	...	...	N	N	?	...	N	N
	Leaf rust	?	...	Y	N	?	N	N	...	...	Y	N	...	N	Y	...	...	Y	...	...	...	Y	N	?	...	N	Y
	Leaf spot	?	...	Y	Y	Y	Y	Y	...	...	Y	Y	...	Y	Y	...	...	Y	...	...	...	Y	Y	Y	...	Y	Y
	Wood decay	?	...	Y	N	?	N	N	...	...	Y	N	...	N	Y	...	...	Y	...	...	...	Y	N	Y	...	N	Y
Boxwood								P	NP	P		P	NP			NP			P			P					
	English boxwood decline	N	N	N	Y	N	N	...	...	...	N	...	...	N	Y	...	...	Y	...	...	...	Y	Y	Y	Y	Y	N
	Lesion nematode	Y	N	N	N	Y	Y	...	...	...	Y	...	...	N	Y	...	...	Y	...	...	...	Y	Y	Y	Y	Y	Y
	Macrophoma leaf spot	Y	Y	...	Y	Y	Y	...	...	...	Y	...	...	N	Y	...	...	Y	...	...	Y	Y	...	Y	Y	Y	Y
	Phytophthora root rot	Y	N	Y	Y	Y	Y	...	...	...	Y	...	...	N	Y	...	...	Y	...	...	Y	?	Y	Y	Y	Y	Y
	Root-knot nematode	Y	N	Y	N	Y	Y	...	...	...	Y	...	...	N	Y	...	...	N	...	...	...	?	Y	Y	Y	Y	Y
	Volutella blight	Y	?	N	Y	N	N	...	...	...	Y	...	...	Y	Y	...	...	Y	...	...	...	Y	Y	Y	Y	Y	Y

(continued on next page)

P = Produced in this state, but no reports of disease occurrence are available. NP = Not produced in this state. Y = Occurrence of disease has been reported. N = Occurrence of disease has not been reported. ? = Uncertain.

Host	Disease	Ala.	Ark.	Calif.	Del.	Fla.	Ga.	Ind.	Kan.	Ky.	Mass.	Md.	Mich.	Mo.	N.C.	Nebr.	N.J.	N.Y.	Ohio	Okla.	Ore.	Pa.	S.C.	Tenn.	Tex.	Va.	Wis.
Camellia			NP					NP	NP	NP	NP		NP	NP		NP		NP	NP			NP		NP			NP
	Camellia yellow mottle	Y	...	Y	N	Y	Y	...	...	...	...	N	...	...	Y	...	Y	...	...	Y	Y	...	Y	...	Y	Y	...
	Canker and dieback	Y	...	Y	N	Y	Y	...	...	...	...	N	...	...	Y	...	Y	...	...	Y	Y	...	Y	...	Y	Y	...
	Flower blight	Y	...	Y	N	Y	Y	...	...	...	...	N	...	...	Y	...	N	...	...	N	Y	...	Y	...	Y	Y	...
	Leaf gall	Y	...	Y	Y	Y	Y	...	...	...	...	Y	...	...	Y	...	N	...	...	N	Y	...	Y	...	Y	Y	...
	Phytophthora root rot	Y	...	Y	N	Y	Y	...	...	...	...	Y	...	...	Y	...	Y	...	...	Y	Y	...	Y	...	Y	Y	...
Cedrus			NP				NP	P	NP	NP	NP		?	NP		NP		P	NP			NP					NP
	Needle or twig blights	Y	...	Y	N	Y	...	...	...	...	...	N	...	...	N	...	Y	...	...	N	Y	...	N	N	Y	Y	...
	Root rot	Y	...	Y	N	Y	...	...	...	...	...	N	...	...	Y	...	Y	...	...	Y	Y	...	Y	Y	Y	N	...
Cleyera (see *Ternstroemia*)																											
Cotoneaster			?			?																					
	Entomosporium leaf spot	N	...	Y	N	...	N	N	Y	Y	Y	Y	N	Y	N	N	N	N	Y	N	Y	N	N	Y	Y	N	Y
	Fire blight	Y	...	Y	Y	...	Y	Y	Y	Y	Y	Y	Y	Y	Y	Y	Y	Y	Y	Y	Y	Y	Y	Y	Y	Y	Y
Crabapple (see Flowering crabapple)																											
Crapemyrtle								NP	P	NP	NP		NP	P		NP			NP			NP					NP
	Cercospora leaf spot	Y	Y	N	N	Y	Y	...	...	...	...	N	...	...	Y	...	N	N	...	N	...	...	Y	Y	Y	N	...
	Powdery mildew	Y	Y	Y	Y	Y	Y	...	...	...	...	Y	...	...	Y	...	Y	Y	...	Y	...	...	Y	Y	Y	Y	...
Cryptomeria			NP				P	NP	?	NP	NP		?	NP		?	P		?			NP		P	P		NP
	Black mold	N	...	N	N	N	...	...	...	...	...	Y	...	...	N	...	...	N	...	N	...	...	?	...	...	N	...
	Cercospora blight and leaf spot	Y	...	N	N	Y	...	...	...	...	...	N	...	...	N	...	...	Y	...	N	...	...	?	...	...	N	...
	Chloroscypha needle blight	N	...	N	N	N	...	...	...	...	...	N	...	...	N	...	...	N	...	N	...	...	?	...	...	N	...
	Mushroom root rot	N	...	N	N	Y	...	...	...	...	...	N	...	...	Y	...	...	N	...	N	...	...	Y	...	...	N	...
Daphne			NP					NP	NP	NP	NP		P	NP		NP	NP		NP	P		NP		NP	P		
	Phytophthora root rot	Y	...	Y	Y	N	Y	...	...	...	...	Y	...	...	Y	...	...	Y	...	...	Y	...	Y	...	...	Y	Y
	Southern blight	Y	...	N	N	Y	Y	...	...	...	...	N	...	...	N	...	...	N	...	...	N	...	Y	...	...	N	N
Dogwood																											
	Anthracnose	Y	N	Y	Y	N	Y	Y	N	Y	Y	Y	Y	N	Y	N	Y	Y	Y	N	Y	Y	Y	Y	N	Y	N
	Powdery mildew	Y	Y	Y	Y	Y	Y	Y	Y	Y	Y	Y	N	Y	Y	N	Y	Y	Y	Y	Y	Y	Y	Y	Y	Y	Y
	Septoria leaf spot	Y	N	Y	Y	Y	Y	Y	Y	Y	Y	Y	Y	Y	Y	Y	Y	Y	Y	N	Y	Y	Y	Y	Y	Y	N
	Spot anthracnose	Y	Y	Y	Y	Y	Y	Y	N	Y	Y	Y	Y	Y	Y	N	Y	Y	Y	N	...	Y	Y	Y	Y	Y	Y
Elaeagnus													NP							P			P	P			NP
	Cankers and diebacks	Y	Y	Y	Y	Y	Y	Y	Y	Y	Y	Y	...	Y	Y	?	Y	Y	Y	...	...	Y	...	...	Y	Y	...
	Foliar and root diseases	Y	N	Y	N	Y	Y	N	N	Y	N	N	...	N	Y	...	N	Y	N	...	...	N	...	...	Y	Y	...
	Verticillium wilt	N	N	Y	N	Y	N	N	Y	Y	...	N	...	N	N	...	N	N	Y	...	...	N	...	...	N	N	...
Elm																											
	Bacterial leaf scorch	N	N	N	N	Y	N	N	Y	N	N	Y	N	N	Y	N	N	Y	N	N	...	Y	Y	N	Y	N	N
	Canker diseases	N	N	Y	Y	Y	Y	Y	Y	Y	Y	Y	N	N	Y	Y	Y	Y	Y	N	Y	Y	Y	Y	Y	Y	Y
	Damping-off and root rots	N	N	Y	N	Y	N	N	N	Y	N	N	Y	N	N	N	N	N	N	N	...	Y	Y	N	N	N	Y

(continued on next page)

P = Produced in this state, but no reports of disease occurrence are available. NP = Not produced in this state. Y = Occurrence of disease has been reported. N = Occurrence of disease has not been reported. ? = Uncertain.

Host	Disease	Ala.	Ark.	Calif.	Del.	Fla.	Ga.	Ind.	Kan.	Ky.	Mass.	Md.	Mich.	Mo.	N.C.	Nebr.	N.J.	N.Y.	Ohio	Okla.	Ore.	Pa.	S.C.	Tenn.	Tex.	Va.	Wis.
Elm	Dutch elm disease	Y	Y	Y	Y	N	Y	Y	Y	Y	Y	Y	Y	N	Y	Y	Y	Y	Y	N	Y	Y	Y	Y	Y	Y	Y
(continued)	Elm black spot	N	Y	N	Y	Y	N	Y	Y	Y	Y	Y	N	Y	Y	N	Y	Y	Y	Y	...	Y	Y	Y	Y	Y	N
	Elm yellows	N	N	N	Y	N	N	Y	N	Y	N	Y	N	N	N	N	Y	Y	N	N	...	Y	Y	N	N	Y	Y
	Leaf spot diseases	Y	N	Y	Y	Y	Y	Y	Y	Y	N	Y	N	Y	Y	N	N	Y	Y	N	...	Y	Y	Y	Y	Y	Y
	Verticillium wilt	N	N	Y	Y	Y	N	N	Y	N	Y	Y	Y	Y	N	Y	Y	Y	Y	N	...	Y	N	N	N	Y	Y
English ivy																											
	Anthracnose	Y	Y	Y	Y	Y	Y	Y	Y	Y	Y	Y	Y	Y	Y	N	Y	Y	Y	Y	...	Y	Y	Y	Y	Y	Y
	Bacterial leaf spot	Y	Y	Y	Y	Y	Y	Y	Y	Y	Y	Y	Y	Y	Y	Y	Y	Y	Y	Y	...	Y	Y	Y	Y	Y	Y
	Edema	Y	Y	Y	Y	Y	Y	Y	N	Y	Y	Y	N	Y	Y	Y	Y	Y	Y	N	Y	Y	Y	Y	Y	Y	Y
	Phytophthora root rot and leaf spot	Y	N	Y	N	Y	Y	Y	N	N	N	Y	N	N	Y	N	N	N	Y	Y	...	Y	Y	Y	Y	Y	Y
	Rhizoctonia root rot and aerial blight	Y	N	Y	N	Y	Y	Y	Y	Y	Y	Y	N	Y	Y	N	N	Y	Y	N	Y	Y	Y	Y	Y	Y	Y
Euonymus																											
	Anthracnose	Y	N	N	Y	Y	Y	Y	N	N	Y	Y	Y	Y	Y	N	Y	Y	Y	Y	...	Y	Y	Y	Y	Y	Y
	Crown gall	Y	N	Y	Y	Y	Y	Y	Y	Y	Y	Y	Y	Y	Y	...	Y	Y	Y	Y	Y	Y	Y	Y	Y	Y	Y
	Powdery mildew	Y	Y	Y	Y	Y	Y	N	Y	Y	Y	Y	N	Y	Y	...	Y	Y	N	N	Y	Y	Y	Y	Y	Y	Y
	Root-knot nematode	Y	N	Y	N	Y	Y	N	N	N	Y	Y	N	N	Y	...	N	N	N	N	...	N	Y	Y	Y	N	Y
× *Fatshedera*			NP		NP			NP	NP	NP	NP	NP	NP	NP		NP	NP	NP	NP	NP		NP		NP	P	P	NP
	Alternaria leaf spot	Y	...	N	...	Y	N	...	...	...	...	...	...	...	Y	...	...	...	...	...	...	...	Y	...	...	...	...
	Anthracnose	Y	...	N	...	Y	Y	...	...	...	...	...	...	...	N	...	...	...	...	...	...	...	Y	...	...	...	...
	Bacterial spots	N	...	N	...	Y	N	...	...	...	...	...	...	...	N	...	...	...	...	...	...	...	N	...	...	...	...
	Wet root rots	Y	...	N	...	Y	Y	...	...	...	...	...	...	...	N	...	...	...	...	...	...	...	N	...	...	...	...
Fir		NP	NP		NP	NP			NP	NP	P	P	P	NP		NP	NP	P		NP	P	P	NP		NP		
	Damping-off	...	...	Y	...	...	...	Y	...	...	...	...	...	...	Y	...	...	...	...	...	...	...	...	Y	...	N	Y
	Felt blights and snow blights	...	...	Y	...	...	...	N	...	...	...	...	...	...	N	...	...	...	...	...	...	...	...	N	...	N	N
	Gray mold (Botrytis blight)	...	...	Y	...	...	...	N	...	...	...	...	...	...	Y	...	...	...	...	...	...	...	...	Y	...	Y	Y
	Phytophthora root rot	...	...	Y	...	...	...	Y	...	...	...	...	...	...	Y	...	...	...	...	...	...	...	...	Y	...	Y	Y
Flowering crab-apple																											
	Fire blight	Y	N	Y	Y	Y	Y	Y	Y	Y	Y	Y	Y	Y	Y	Y	Y	Y	Y	Y	Y	Y	Y	Y	Y	Y	Y
	Powdery mildew	Y	N	Y	Y	Y	Y	Y	Y	Y	Y	Y	N	Y	Y	Y	Y	Y	Y	N	Y	Y	Y	Y	Y	Y	Y
	Rust	Y	N	Y	Y	Y	Y	Y	Y	Y	Y	Y	Y	Y	Y	Y	Y	Y	Y	N	Y	Y	Y	Y	Y	Y	Y
	Scab	Y	Y	Y	Y	Y	Y	Y	Y	Y	Y	Y	Y	Y	Y	Y	Y	Y	Y	Y	Y	Y	Y	Y	Y	Y	Y
Flowering pear																											
	Bacterial blossom blast	Y	...	Y	N	N	N	N	...	...	Y	...	...	Y	N	...	...	N	...	...	Y	...	Y	N	N	N	Y
	Canker and dieback	Y	...	Y	Y	Y	Y	N	...	...	Y	...	...	Y	Y	...	...	Y	...	...	Y	...	Y	N	Y	Y	Y
	Fire blight	Y	...	Y	Y	Y	Y	Y	...	...	Y	...	...	N	Y	...	...	Y	...	...	Y	...	Y	Y	Y	Y	Y
	Leaf spot diseases	Y	...	Y	N	Y	Y	N	...	...	Y	...	...	Y	Y	...	...	Y	...	...	Y	...	Y	Y	N	Y	Y
Forsythia							P				NP					P	P										
	Phytophthora root and crown rot, dieback	Y	N	Y	N	Y	...	Y	Y	N	...	N	...	N	Y	...	...	Y	Y	Y	Y	N	Y	Y	Y	Y	Y
	Pseudomonas blight	N	N	N	N	N	...	N	N	N	...	N	...	N	N	...	...	N	N	Y	...	...	N	N	Y	N	Y
	Sclerotinia canker	N	N	Y	Y	N	...	N	N	N	...	Y	...	N	N	...	...	Y	N	N	Y	Y	N	Y	N	N	N
	Stem gall	Y	Y	N	Y	Y	V	Y	Y	Y	...	Y	...	Y	Y	...	...	Y	Y	N	Y	Y	Y	Y	N	Y	Y

(continued on next page)

P = Produced in this state, but no reports of disease occurrence are available. NP = Not produced in this state. Y = Occurrence of disease has been reported. N = Occurrence of disease has not been reported. ? = Uncertain.

Host	Disease	Ala.	Ark.	Calif.	Del.	Fla.	Ga.	Ind.	Kan.	Ky.	Mass.	Md.	Mich.	Mo.	N.C.	Nebr.	N.J.	N.Y.	Ohio	Okla.	Ore.	Pa.	S.C.	Tenn.	Tex.	Va.	Wis.
Gardenia			P		NP			NP	NP	NP	NP	NP	NP	NP		NP	NP	NP	NP			NP		NP			NP
	Myrothecium leaf spot and petiole rot	N	...	N	...	Y	Y	...	...	...	...	...	...	...	Y	...	...	...	...	N	...	...	N	...	N	Y	...
	Phomopsis canker	Y	...	Y	...	Y	N	...	...	...	...	...	...	...	N	...	...	...	...	N	...	...	Y	...	Y	N	...
	Phytophthora stem rot and leaf spot	Y	...	N	...	Y	Y	...	...	...	...	...	...	...	Y	...	...	...	...	Y	...	...	Y	...	Y	N	...
	Rhizoctonia aerial blight or leaf spot	Y	...	N	...	Y	Y	...	...	...	...	...	...	...	Y	...	...	...	...	N	...	...	Y	...	Y	N	...
	Xanthomonas leaf spot	Y	...	Y	...	Y	N	...	...	...	...	...	...	...	N	...	...	...	...	N	...	...	Y	...	N	N	...
Ginkgo			NP				?	P			NP	P	P	P	P	P	P	P	P	P		P		P	NP		
	Phytophthora root rot	Y	...	N	N	Y	...	...	N	N	...	...	...	...	...	...	...	...	...	...	N	...	Y	...	...	N	N
Hawthorn		NP	NP																								
	Entomosporium leaf spot	...	...	Y	Y	Y	N	Y	Y	Y	Y	Y	Y	N	Y	N	Y	Y	Y	N	Y	Y	Y	Y	Y	N	Y
	Fire blight	...	...	Y	Y	N	Y	Y	Y	Y	Y	Y	Y	N	Y	Y	Y	Y	Y	Y	Y	Y	Y	Y	Y	Y	Y
	Hawthorn and quince rusts	...	...	Y	Y	Y	Y	Y	Y	Y	Y	Y	Y	Y	Y	Y	Y	Y	Y	Y	Y	Y	Y	Y	Y	Y	Y
Hibiscus					P		?	P	P		NP	?	P	P		P	NP	NP	NP			NP					NP
	Bacterial leaf spot	Y	Y	Y	...	Y	...	...	...	N	...	...	...	...	Y	...	...	...	...	N	...	...	Y	N	Y	N	...
	Choanephora blight	Y	N	N	...	Y	...	...	...	N	...	...	...	...	Y	...	...	...	...	N	...	...	Y	Y	Y	Y	...
	Phytophthora leaf spot	N	N	N	...	Y	...	...	...	N	...	...	...	...	N	...	...	...	...	N	...	...	N	N	Y	N	...
	Pseudocercospora leaf spot	N	N	Y	...	Y	...	...	...	N	...	...	...	...	N	...	...	...	...	N	...	...	N	N	Y	N	...
	Root rot	Y	N	N	...	Y	...	...	...	N	...	...	...	...	N	...	...	...	...	Y	...	...	Y	N	Y	N	...
	Rust	Y	N	N	...	Y	...	...	...	N	...	...	...	...	N	...	...	...	...	N	...	...	N	N	Y	N	...
Holly									P				P	P		P											
	Black root rot	Y	Y	N	Y	Y	Y	Y	...	Y	Y	Y	...	...	Y	...	Y	Y	Y	N	...	Y	Y	Y	Y	Y	Y
	Nematode decline	Y	Y	N	N	Y	Y	N	...	N	N	Y	...	...	Y	...	N	N	N	N	...	N	Y	N	N	N	N
	Phytophthora leaf and twig blight	N	N	N	N	Y	Y	N	...	N	N	N	...	...	Y	...	N	N	Y	Y	Y	N	N	N	Y	Y	Y
	Rhizoctonia web blight	Y	N	N	N	Y	Y	N	...	N	N	Y	...	...	Y	...	N	Y	N	N	...	N	Y	Y	Y	N	N
	Scab	N	N	N	N	Y	N	N	...	Y	N	N	...	...	N	...	N	N	N	N	Y	Y	N	N	N	Y	N
	Other diseases	N	N	Y	N	Y	N	N	...	Y	N	N	...	...	N	...	N	Y	...	Y	...	N	Y	N	N	N	N
Honeylocust		NP	NP				NP											P					P		P		
	Canker	...	...	N	Y	Y	...	Y	Y	Y	Y	Y	Y	Y	Y	Y	Y	...	Y	?	Y	Y	...	Y	...	Y	Y
Hydrangea			NP																								
	Bacterial leaf spot	Y	...	N	...	Y	...	N	...	...	...	...	...	Y	Y	...	...	N	...	...	...	...	Y	N	N	N	Y
	Fungal leaf spot diseases	Y	...	N	...	Y	...	N	...	...	...	...	...	N	Y	...	...	Y	...	...	...	...	Y	Y	N	Y	Y
	Hydrangea virescence	N	...	N	...	N	...	N	...	...	...	...	...	N	N	...	...	N	...	...	...	...	N	N	N	N	N
	Hydrangea viruses	Y	...	Y	...	Y	...	N	...	...	...	...	...	N	Y	...	...	Y	...	...	...	...	Y	Y	N	N	Y
	Powdery mildew	Y	...	Y	...	Y	...	N	...	...	...	...	...	N	N	...	...	Y	...	...	...	...	Y	Y	Y	N	Y
Indian hawthorn			NP		NP			NP	NP	NP	NP	NP	NP	NP		NP	NP	NP	NP			NP		NP			NP
	Entomosporium leaf spot	Y	...	Y	...	Y	Y	...	...	...	...	...	...	...	Y	...	...	...	...	Y	...	...	Y	...	Y	Y	...
	Fire blight	Y	...	Y	...	Y	Y	...	...	...	...	...	...	...	Y	...	...	...	...	Y	...	...	Y	...	Y	Y	...
	Phytophthora aerial blight	Y	...	N	...	Y	N	...	...	...	...	...	...	...	N	...	...	...	...	Y	...	...	Y	...	Y	N	...
	Phytophthora root rot	Y	...	Y	...	Y	Y	...	...	...	...	...	...	...	Y	...	...	...	...	Y	...	...	Y	...	Y	N	...
Ixora			NP		NP		?	NP	NP	NP	NP	NP	NP	NP	NP	NP	NP	NP	NP	NP	NP	NP	NP	NP		NP	NP
	Bacterial leaf spot	Y	...	N	...	Y	...	...	...	...	...	...	...	...	...	...	...	...	...	...	...	...	...	...	Y	...	...
	Iron deficiency	Y	...	N	...	Y	...	...	...	...	...	...	...	...	...	...	...	...	...	...	...	...	...	...	N	...	...
	Root-knot nematode	Y	...	Y	...	Y	...	...	...	...	...	...	...	...	...	...	...	...	...	...	...	...	...	...	N	...	...

(continued on next page)

P = Produced in this state, but no reports of disease occurrence are available. NP = Not produced in this state. Y = Occurrence of disease has been reported. N = Occurrence of disease has not been reported. ? = Uncertain.

Host	Disease	Ala.	Ark.	Calif.	Del.	Fla.	Ga.	Ind.	Kan.	Ky.	Mass.	Md.	Mich.	Mo.	N.C.	Nebr.	N.J.	N.Y.	Ohio	Okla.	Ore.	Pa.	S.C.	Tenn.	Tex.	Va.	Wis.
Juniper																											
	Canker diseases	Y	N	Y	Y	Y	N	N	Y	Y	N	N	N	Y	N	Y	N	Y	N	N	?	N	Y	Y	Y	Y	N
	Cedar-apple rust and related rust diseases	Y	Y	Y	Y	Y	Y	Y	Y	Y	Y	Y	Y	Y	Y	Y	Y	Y	Y	Y	Y	Y	Y	Y	Y	Y	Y
	Cercospora needle blight	Y	N	N	N	Y	Y	Y	Y	Y	N	N	N	Y	N	Y	N	Y	N	N	...	N	Y	Y	Y	Y	Y
	Nematode decline	Y	N	Y	N	Y	Y	Y	Y	Y	Y	N	Y	N	Y	Y	Y	N	Y	Y	?	Y	Y	Y	Y	Y	N
	Phomopsis and Kabatina tip blights	Y	Y	Y	Y	Y	Y	Y	Y	Y	Y	Y	Y	Y	Y	Y	Y	Y	Y	Y	Y	Y	Y	Y	Y	Y	Y
	Root diseases	Y	N	Y	Y	Y	Y	N	Y	N	N	Y	N	Y	N	Y	Y	Y	N	N	Y	Y	Y	Y	Y	Y	Y
Leucothoe			NP	P		NP	P	P	NP	P	NP		NP	P		?	P		P				P	P	P		NP
	Cylindrocladium leaf spot	Y	...	...	N	...	...	...	...	...	...	N	...	...	Y	...	...	N	...	Y	...	Y	...	...	...	N	...
	Powdery mildew	N	...	...	Y	...	...	...	...	...	...	N	...	...	Y	...	...	Y	...	N	...	Y	...	...	...	Y	...
Leyland cypress			?		P			NP	?		NP		NP	P		?			NP								NP
	Botryosphaeria canker	Y	...	N	...	Y	Y	...	...	Y	...	Y	...	...	Y	...	N	N	...	N	...	N	Y	Y	Y	Y	...
	Cercospora needle blight	N	...	N	...	Y	...	...	...	...	...	...	...	...	...	...	...	N	...	...	...	...	Y	N	N	N	...
	Kabatina blight	N	...	N	...	N	...	...	...	...	...	...	...	...	N	...	...	N	...	...	...	...	N	Y	N	Y	...
	Seiridium canker	Y	...	Y	...	Y	Y	...	...	N	...	Y	...	...	Y	...	Y	Y	...	Y	...	Y	Y	Y	Y	Y	...
Ligustrum											NP		NP	P		P	?		P	?		NP		P			
	Anthracnose	Y	N	N	N	Y	Y	N	N	N	...	N	...	...	N	...	...	N	...	...	Y	...	N	...	N	N	Y
	Cercospora leaf spot	Y	Y	N	N	Y	Y	N	Y	N	...	N	...	...	Y	...	...	Y	...	...	Y	...	Y	...	Y	Y	Y
	Corynespora spot	Y	N	N	N	Y	N	N	N	N	...	N	...	...	N	...	...	N	...	...	...	...	N	...	N	Y	Y
	Edema	Y	N	N	N	Y	Y	N	Y	Y	...	N	...	...	Y	...	...	N	...	...	...	...	Y	...	Y	Y	Y
	Nematode damage	Y	N	Y	N	Y	Y	N	N	N	...	N	...	...	Y	...	...	N	...	...	...	...	N	...	N	N	N
	Powdery mildew	Y	N	Y	Y	Y	Y	N	N	Y	...	Y	...	...	N	...	...	N	...	...	...	...	Y	...	Y	Y	Y
Lilac			NP			P	NP																NP		NP		
	Powdery mildew	Y	...	Y	Y	...	...	Y	Y	...	Y	Y	N	Y	Y	Y	Y	Y	Y	Y	...	Y	...	Y	...	Y	Y
	Pseudomonas blight	Y	...	Y	Y	...	...	Y	Y	...	Y	Y	Y	Y	Y	Y	N	Y	Y	Y	...	Y	...	Y	...	Y	Y
Linden		NP	NP	P	P		NP				NP		P	P	P				P	?			NP		NP		
	Anthracnose	...	...	...	...	N	...	N	N	N	...	N	...	...	N	Y	N	N	...	...	...	Y	...	N	...	Y	Y
	Canker	...	...	...	...	Y	...	N	N	Y	...	N	...	...	?	Y	N	Y	...	...	...	Y	...	Y	...	Y	Y
	Leaf spot	...	...	...	...	Y	...	Y	N	Y	...	N	...	...	?	N	N	Y	...	...	...	Y	...	N	...	N	Y
	Powdery mildew	...	...	...	...	Y	...	N	N	Y	...	Y	...	...	Y	N	N	N	...	...	...	Y	...	N	...	N	N
	Verticillium wilt	...	...	...	...	N	...	N	Y	Y	...	Y	...	...	Y	N	Y	Y	...	...	...	Y	...	Y	...	N	Y
Lonicera			NP										NP							?							
	Insolibasidium leaf blight	N	...	N	N	N	Y	Y	Y	N	Y	N	...	N	Y	N	N	Y	Y	...	...	Y	Y	Y	N	Y	Y
	Powdery mildew	N	...	Y	Y	Y	Y	N	Y	Y	Y	Y	...	Y	Y	Y	Y	Y	Y	...	...	Y	Y	Y	Y	Y	Y
Magnolia			P						P		NP		P	P		P	?										
	Bacterial blight	Y	...	Y	N	Y	...	N	...	N	...	N	...	...	Y	...	...	N	N	N	Y	N	Y	N	Y	Y	N
	Fungal leaf spot diseases	Y	...	Y	N	Y	...	N	...	Y	...	Y	...	...	N	...	...	N	N	Y	...	Y	Y	Y	N	Y	N
	Phytophthora root rot	Y	...	Y	N	Y	...	N	...	Y	...	N	...	...	N	...	...	N	Y	N	...	N	N	N	N	Y	Y
	Powdery mildew	Y	...	Y	Y	Y	...	N	...	Y	...	Y	...	...	Y	...	...	Y	Y	N	...	Y	Y	Y	N	Y	N
Maple																											
	Anthracnose	Y	...	Y	Y	Y	...	Y	...	...	...	...	...	Y	Y	...	...	Y	Y	...	...	Y	Y	Y	N	Y	Y
	Bacterial leaf spot and dieback	N	...	Y	N	N	...	N	...	...	...	...	...	N	N	...	...	N	N	...	...	N	N	N	N	N	Y
	Canker diseases	Y	...	Y	Y	Y	...	Y	...	...	...	...	...	Y	Y	...	...	Y	Y	...	...	Y	Y	Y	N	Y	Y
	Phyllosticta leaf spot	N	...	Y	Y	Y	...	Y	...	...	...	...	...	Y	Y	...	...	Y	Y	...	...	Y	Y	Y	Y	Y	Y

(continued on next page)

P = Produced in this state, but no reports of disease occurrence are available. NP = Not produced in this state. Y = Occurrence of disease has been reported. N = Occurrence of disease has not been reported. ? = Uncertain.

Host	Disease	Ala.	Ark.	Calif.	Del.	Fla.	Ga.	Ind.	Kan.	Ky.	Mass.	Md.	Mich.	Mo.	N.C.	Nebr.	N.J.	N.Y.	Ohio	Okla.	Ore.	Pa.	S.C.	Tenn.	Tex.	Va.	Wis.
Maple	Powdery mildew	N	...	Y	N	Y	...	N	...	...	...	...	...	N	Y	...	...	Y	Y	...	...	Y	Y	N	Y	N	Y
(continued)	Tar spots	N	...	Y	Y	Y	...	Y	...	...	...	...	...	Y	Y	...	...	Y	Y	...	...	Y	Y	Y	N	Y	Y
	Verticillium wilt	Y	...	Y	Y	N	...	Y	...	...	...	...	...	Y	Y	...	...	Y	Y	...	...	Y	Y	Y	N	Y	Y
Mountain laurel			NP	P		P	P	NP	NP	P			NP	NP		?	P		P	?					NP		NP
	Flower blight	N	...	...	N	...	...	...	...	...	...	N	...	...	Y	...	...	N	...	...	...	N	Y	N	...	N	...
	Leaf and flower gall	N	...	...	N	...	...	...	...	...	...	N	...	...	N	...	...	N	...	...	...	N	Y	Y	...	N	...
	Leaf blight	N	...	...	Y	...	...	...	...	...	...	N	...	...	N	...	...	N	...	...	...	N	N	N	...	N	...
	Leaf spot	Y	...	...	Y	...	...	...	...	...	...	Y	...	...	Y	...	...	Y	...	...	...	Y	Y	Y	...	Y	...
	Necrotic ringspot	N	...	...	Y	...	...	...	...	...	...	N	...	...	N	...	...	N	...	...	...	N	Y	Y	...	Y	...
	Powdery mildew	N	...	...	N	...	...	...	...	...	...	N	...	...	N	...	...	Y	...	...	...	N	N	N	...	N	...
	Root rot	N	...	...	Y	...	...	...	...	...	...	Y	...	...	Y	...	...	N	...	...	...	Y	Y	Y	...	Y	...
Nandina							P	NP	NP		P			NP				NP				NP					NP
	Anthracnose	Y	...	N	N	Y	...	...	...	...	...	...	...	...	N	...	...	...	...	...	...	...	Y	N	N	Y	...
	Leaf spot	Y	...	N	N	Y	...	...	...	...	...	...	...	...	N	...	...	...	...	...	...	...	Y	N	Y	Y	...
	Root rot	Y	...	N	Y	Y	...	...	...	...	...	N	...	...	Y	...	...	...	N	Y	...	...	Y	Y	Y	Y	...
	Virus disease	Y	...	Y	N	Y	...	...	...	...	...	N	...	...	Y	...	...	...	Y	Y	...	...	Y	Y	Y	N	...
	Web blight	Y	...	...	...	Y	...	...	...	...	...	...	...	...	Y	...	...	...	...	...	...	...	N	N	N	Y	...
Oleander			NP		NP			NP	NP	NP	NP	NP	NP	NP		NP	NP	NP	NP	?		NP		NP			NP
	Bacterial knot	Y	...	Y	...	Y	N	...	...	...	...	...	...	...	Y	...	...	...	...	...	...	...	N	...	Y	N	...
	Dieback	Y	...	Y	...	Y	Y	...	...	...	...	...	...	...	N	...	...	...	...	...	...	...	Y	...	Y	N	...
	Leaf spot diseases	Y	...	Y	...	Y	Y	...	...	...	...	...	...	...	Y	...	...	...	...	...	...	...	Y	...	N	Y	...
	Root knot	Y	...	Y	...	Y	N	...	...	...	...	...	...	...	N	...	...	...	...	...	...	...	?	...	N	N	...
	Web blight	Y	...	N	...	Y	Y	...	...	...	...	...	...	...	Y	...	...	...	...	...	...	...	?	...	N	N	...
	Witches'-broom	Y	...	N	...	Y	N	...	...	...	...	...	...	...	N	...	...	...	...	...	...	...	?	...	Y	N	...
Osmanthus			NP	P	P			NP	?	NP	NP		NP	NP	P	?	P	NP	NP	?		NP		NP		P	NP
	Anthracnose	Y	...	...	...	Y	Y	...	...	...	...	N	...	...	...	...	...	...	...	...	...	...	Y	...	N	...	...
	Dark mildews	Y	...	...	...	Y	N	...	...	...	...	N	...	...	...	...	...	...	...	...	...	...	N	...	N	...	...
	Phyllosticta leaf spot	Y	...	...	...	Y	Y	...	...	...	...	N	...	...	...	...	...	...	...	...	...	...	N	...	Y	...	...
	Root-knot nematode	Y	...	...	...	Y	Y	...	...	...	...	Y	...	...	...	...	...	...	...	...	...	...	N	...	N	...	...
Palm			NP		NP		NP	NP	NP	NP	NP	NP	NP	NP		NP	NP	NP	NP	NP	NP	NP		NP		NP	NP
	Algal leaf spot	Y	...	N	...	Y	...	...	...	...	...	...	...	...	N	...	...	...	...	...	...	...	Y	...	N	...	...
	Anthracnose	Y	...	N	...	Y	...	...	...	...	...	...	...	...	Y	...	...	...	...	...	...	...	N	...	N	...	...
	Calonectria leaf spot	N	...	N	...	Y	...	...	...	...	...	...	...	...	N	...	...	...	...	...	...	...	Y	...	N	...	...
	Catacauma leaf spot	N	...	N	...	Y	...	...	...	...	...	...	...	...	N	...	...	...	...	...	...	...	N	...	N	...	...
	Fusarium wilt	N	...	Y	...	Y	...	...	...	...	...	...	...	...	N	...	...	...	...	...	...	...	N	...	N	...	...
	Ganoderma butt rot	Y	...	N	...	Y	...	...	...	...	...	...	...	...	N	...	...	...	...	...	...	...	N	...	N	...	...
	Graphiola leaf spot	N	...	Y	...	Y	...	...	...	...	...	...	...	...	Y	...	...	...	...	...	...	...	Y	...	Y	...	...
	Helminthosporium leaf spot	Y	...	N	...	Y	...	...	...	...	...	...	...	...	N	...	...	...	...	...	...	...	Y	...	N	...	...
	Lethal yellowing	N	...	N	...	Y	...	...	...	...	...	...	...	...	N	...	...	...	...	...	...	...	Y	...	N	...	...
	Phytophthora diseases	Y	...	Y	...	Y	...	...	...	...	...	...	...	...	Y	...	...	...	...	...	...	...	Y	...	N	...	...
	Pink rot	Y	...	Y	...	Y	...	...	...	...	...	...	...	...	N	...	...	...	...	...	...	...	N	...	N	...	...
	Pseudocercospora leaf spot	N	...	N	...	Y	...	...	...	...	...	...	...	...	N	...	...	...	...	...	...	...	N	...	Y	...	...
	Stigmina leaf spot	N	...	N	...	Y	...	...	...	...	...	...	...	...	N	...	...	...	...	...	...	...	N	...	N	...	...
	Thielaviopsis bud rot	N	...	N	...	Y	...	...	...	...	...	...	...	...	N	...	...	...	...	...	...	...	N	...	N	...	...

(continued on next page)

P = Produced in this state, but no reports of disease occurrence are available. NP = Not produced in this state. Y = Occurrence of disease has been reported. N = Occurrence of disease has not been reported. ? = Uncertain.

Host	Disease	Ala.	Ark.	Calif.	Del.	Fla.	Ga.	Ind.	Kan.	Ky.	Mass.	Md.	Mich.	Mo.	N.C.	Nebr.	N.J.	N.Y.	Ohio	Okla.	Ore.	Pa.	S.C.	Tenn.	Tex.	Va.	Wis.
Pear (see Flowering pear)																											
Photinia								NP	P	NP	NP		NP			NP	P	NP	NP			NP					NP
	Bacterial leaf spot	Y	N	Y	N	Y	Y	...	...	...	...	N	...	N	N	...	...	...	...	N	Y	...	N	N	Y	Y	...
	Entomosporium leaf spot	Y	Y	Y	Y	Y	Y	...	...	...	...	Y	...	Y	Y	...	...	...	...	Y	Y	...	Y	Y	Y	Y	...
	Fire blight	Y	Y	Y	N	Y	Y	...	...	...	...	N	...	N	N	...	...	...	...	N	...	...	Y	Y	Y	Y	...
	Powdery mildew	Y	N	Y	N	Y	Y	...	...	...	...	Y	...	N	Y	...	...	...	...	Y	...	...	Y	Y	Y	Y	...
	Root rot	Y	N	Y	N	Y	Y	...	...	...	...	N	...	N	Y	...	...	...	...	Y	...	...	Y	N	Y	Y	...
Pieris			NP			P			?				P	P		?									P		
	Dieback and canker	Y	...	N	N	...	N	N	...	Y	Y	N	...	...	Y	...	Y	Y	Y	Y	...	Y	Y	Y	...	Y	Y
	Phytophthora blight and root rot	Y	...	N	Y	...	Y	N	...	Y	Y	Y	...	...	Y	...	Y	Y	Y	Y	Y	Y	Y	Y	...	Y	Y
Pine																											
	Armillaria root rot	N	Y	Y	Y	Y	Y	Y	Y	N	N	N	Y	Y	Y	...	...	Y	?	...	...	Y	N	Y	N	Y	Y
	Brown spot needle blight	N	Y	N	Y	Y	Y	Y	Y	Y	Y	N	Y	Y	Y	...	...	Y	N	...	...	Y	Y	Y	Y	Y	Y
	Charcoal root rot	N	N	Y	N	Y	Y	N	N	N	N	N	N	Y	Y	...	...	N	N	...	...	N	Y	N	Y	?	N
	Chemical damage	N	N	Y	Y	Y	Y	N	Y	Y	Y	Y	N	N	Y	...	...	Y	Y	...	...	Y	Y	Y	N	Y	Y
	Cyclaneusma needle cast	N	N	N	N	N	N	Y	Y	Y	Y	Y	Y	Y	?	...	...	Y	Y	...	Y	Y	Y	Y	N	Y	Y
	Cylindrocladium root rot and stem canker	N	N	N	N	Y	Y	N	N	N	N	N	N	N	Y	...	...	N	?	...	Y	N	N	Y	N	?	Y
	Dothistroma needle blight	N	Y	Y	N	Y	Y	Y	Y	N	N	Y	Y	Y	N	...	...	Y	Y	...	Y	Y	N	Y	Y	Y	Y
	Elytroderma needle cast	N	N	Y	N	Y	N	N	N	N	N	N	N	N	N	...	...	N	?	...	Y	N	Y	Y	Y	?	Y
	Fall needle drop	N	N	Y	Y	Y	Y	Y	Y	Y	Y	Y	N	Y	?	...	...	Y	N	...	...	Y	Y	Y	N	Y	Y
	Gray mold blight	Y	N	Y	Y	N	N	N	N	Y	Y	N	N	N	Y	...	...	N	N	...	Y	?	Y	N	N	?	N
	Lophodermella needle cast	Y	N	Y	N	Y	N	N	N	N	N	N	N	N	?	...	...	N	?	...	Y	N	N	Y	N	?	Y
	Lophodermium needle cast	Y	Y	Y	N	Y	Y	Y	N	Y	Y	Y	Y	Y	N	...	...	Y	Y	...	Y	Y	Y	Y	Y	Y	Y
	Phytophthora root rot	Y	N	Y	Y	Y	Y	N	N	Y	Y	Y	Y	Y	Y	...	...	Y	Y	...	Y	Y	Y	Y	N	?	Y
	Pine needle rust	Y	N	N	Y	Y	Y	Y	N	Y	Y	Y	Y	Y	Y	...	...	N	Y	...	Y	Y	Y	Y	Y	?	Y
	Pine-oak gall rust	N	N	N	Y	Y	Y	N	N	Y	Y	Y	N	N	Y	...	...	N	Y	...	Y	Y	Y	Y	N	Y	Y
	Pine-pine gall rust	N	N	Y	Y	Y	N	Y	N	Y	Y	Y	N	Y	N	...	...	Y	Y	...	Y	Y	N	Y	N	Y	Y
	Pinewood nematode (pine wilt nematode)	N	N	Y	Y	Y	N	Y	Y	Y	Y	Y	Y	Y	Y	...	...	Y	N	...	...	Y	Y	Y	Y	Y	Y
	Pitch canker	Y	N	Y	Y	Y	Y	Y	N	N	N	Y	N	N	Y	...	...	N	N	...	...	N	Y	N	Y	N	N
	Procerum root rot	N	N	N	N	N	N	Y	N	Y	Y	Y	N	Y	Y	...	...	Y	Y	...	...	Y	Y	Y	N	Y	N
	Soil-related problems	Y	Y	Y	Y	Y	Y	Y	Y	Y	Y	Y	Y	Y	Y	...	...	Y	Y	...	...	Y	Y	Y	Y	Y	Y
	Sphaeropsis tip blight	Y	N	Y	Y	Y	Y	Y	Y	Y	Y	Y	Y	Y	Y	...	...	Y	Y	...	Y	Y	N	Y	N	Y	Y
	Weather-related injuries	Y	N	Y	Y	N	N	Y	Y	N	N	Y	N	Y	Y	...	...	Y	Y	...	...	Y	Y	Y	Y	Y	Y
	White pine blister rust	N	N	Y	N	N	Y	Y	N	N	N	N	N	N	Y	...	...	Y	N	...	Y	Y	Y	N	N	Y	Y
Pittosporum					NP		P	NP	NP	NP	NP	NP	NP	NP		NP	NP	NP	NP	NP		NP		NP		NP	NP
	Alternaria leaf spot	Y	...	Y	...	Y	...	...	...	...	...	...	...	...	Y	...	...	...	...	...	...	...	Y	...	Y	...	...
	Angular leaf spot	N	...	N	...	Y	...	...	...	...	...	...	...	...	Y	...	...	...	...	...	...	...	Y	...	Y	...	...
	Gall and dieback	N	...	N	...	Y	...	...	...	...	...	...	...	...	Y	...	...	...	...	...	...	...	N	...	N	...	...
	Pink limb blight	N	...	N	...	Y	...	...	...	...	...	...	...	...	N	...	...	...	...	...	...	...	N	...	N	...	...
	Root knot	Y	...	Y	...	Y	...	...	...	...	...	...	...	...	Y	...	...	...	...	...	...	...	Y	...	N	...	...
	Root rot	Y	...	Y	...	Y	...	...	...	...	...	...	...	...	Y	...	...	...	...	...	...	...	Y	...	N	...	...
	Rough bark disease	N	...	N	...	Y	...	...	...	...	...	...	...	...	N	...	...	...	...	...	...	...	Y	...	N	...	...
	Southern blight	Y	...	N	...	Y	...	...	...	...	...	...	...	...	Y	...	...	...	...	...	...	...	Y	...	Y	...	...
	Web blight	Y	...	...	...	Y	...	...	...	...	...	...	...	...	Y	...	...	...	...	...	...	...	Y	...	Y	...	...

(continued on next page)

P = Produced in this state, but no reports of disease occurrence are available. NP = Not produced in this state. Y = Occurrence of disease has been reported. N = Occurrence of disease has not been reported. ? = Uncertain.

Host	Disease	Ala.	Ark.	Calif.	Del.	Fla.	Ga.	Ind.	Kan.	Ky.	Mass.	Md.	Mich.	Mo.	N.C.	Nebr.	N.J.	N.Y.	Ohio	Okla.	Ore.	Pa.	S.C.	Tenn.	Tex.	Va.	Wis.
Podocarpus			NP		NP			NP	NP	NP	NP	NP	NP	NP		NP	NP	NP	NP	NP		NP	NP	NP	P	NP	NP
	Dieback	N	...	N	...	Y	N	...	...	...	...	...	...	...	N	...	...	...	...	...	...	...	...	...	...	...	...
	Fusarium wilt	N	...	N	...	Y	N	...	...	...	...	...	...	...	N	...	...	...	...	...	...	...	...	...	...	...	...
	Root rot	Y	...	N	...	Y	Y	...	...	...	...	...	...	...	Y	...	...	...	...	...	...	...	...	...	...	...	...
Poplar (*Populus* spp.)					P																						
	Bacterial blight	N	...	Y	...	N	...	N	...	...	...	...	...	N	?	...	...	N	N	...	...	N	...	N	Y	...	N
	Blackstem	N	...	N	...	N	...	N	...	...	...	...	...	N	?	...	...	N	N	...	...	N	...	N	N	...	Y
	Ink spot	N	...	N	...	N	...	N	...	...	...	...	...	N	?	...	...	Y	N	...	...	N	...	N	N	...	Y
	Leaf rust	Y	...	Y	...	Y	...	Y	...	...	...	...	...	Y	Y	...	...	Y	?	...	...	Y	...	Y	Y	Y	Y
	Marssonina leaf spot and blight	Y	...	Y	...	N	...	Y	...	...	...	...	...	Y	Y	...	...	Y	Y	...	...	Y	Y	Y	N	Y	Y
	Poplar mosaic virus	N	...	N	...	N	...	N	...	...	...	...	...	N	?	...	...	Y	N	...	...	N	...	N	N	...	Y
	Septoria leaf spot and canker	Y	...	Y	...	Y	...	N	...	...	...	...	...	Y	N	...	...	Y	Y	...	...	Y	...	Y	Y	Y	Y
	Venturia leaf and shoot blight	N	...	Y	...	N	...	N	...	...	...	...	...	N	N	...	...	Y	N	...	...	Y	...	Y	N	...	Y
Prunus																											
	Bacterial canker	Y	...	Y	Y	?	...	Y	...	...	...	...	...	N	Y	...	...	Y	Y	...	...	...	Y	Y	Y	Y	Y
	Black knot	Y	...	Y	Y	Y	...	Y	...	...	...	...	...	Y	Y	...	...	Y	Y	...	...	...	Y	Y	Y	Y	Y
	Brown rot (blossom blight)	Y	...	Y	Y	?	...	Y	...	...	...	...	...	Y	Y	...	...	Y	Y	...	...	...	Y	Y	Y	?	Y
	Crown gall	Y	...	Y	N	Y	...	Y	...	...	...	...	...	Y	Y	...	...	Y	Y	...	...	...	Y	Y	N	Y	Y
	Gummosis	Y	...	Y	Y	Y	...	N	...	...	...	...	...	Y	Y	...	...	Y	Y	...	...	...	Y	Y	N	Y	Y
	Leaf spot	Y	...	Y	N	Y	...	Y	...	...	...	...	...	Y	Y	...	...	Y	Y	...	...	...	Y	Y	N	?	Y
	Powdery mildew	Y	...	Y	N	Y	...	Y	...	...	...	...	...	N	Y	...	...	N	Y	...	...	...	N	Y	N	?	Y
	Shot hole	Y	...	Y	Y	Y	...	Y	...	...	...	...	...	Y	Y	...	...	Y	Y	...	...	...	N	Y	N	Y	Y
	Verticillium wilt	Y	...	Y	N	?	...	N	...	...	...	...	...	N	?	...	...	N	N	...	...	...	N	N	N	?	Y
	Virus and viruslike diseases	Y	...	Y	Y	Y	...	N	...	...	...	...	...	Y	Y	...	...	Y	N	...	...	...	...	N	N	?	Y
	Witches'-broom and leaf curl	N	...	Y	N	Y	...	N	...	...	...	...	...	Y	Y	...	...	Y	N	...	...	...	Y	N	N	Y	Y
Pyracantha																											NP
	Fire blight	Y	N	Y	Y	Y	Y	Y	Y	Y	Y	Y	Y	N	Y	Y	Y	Y	Y	Y	Y	Y	Y	Y	Y	Y	...
	Scab	Y	Y	Y	Y	Y	Y	Y	Y	Y	Y	Y	Y	Y	Y	Y	Y	Y	Y	N	Y	Y	Y	Y	N	Y	...
Redbud											NP			P		P							P				
	Canker	Y	?	Y	Y	Y	Y	Y	Y	Y	...	Y	Y	...	Y	...	Y	Y	Y	Y	...	Y	...	Y	Y	Y	Y
	Phytophthora root rot	Y	...	N	N	Y	Y	N	N	Y	...	Y	Y	...	Y	...	N	N	Y	N	...	N	...	Y	Y	Y	Y
Rhododendron						NP			P																		
	Botryosphaeria dieback	Y	Y	Y	Y	...	Y	Y	...	Y	Y	Y	Y	N	Y	N	Y	Y	Y	Y	Y	Y	Y	Y	N	Y	Y
	Cylindrocladium root rot	Y	N	Y	N	...	Y	Y	...	N	Y	N	N	N	Y	N	N	Y	Y	Y	Y	Y	N	N	N	N	Y
	Discula leaf spot	N	N	N	N	...	N	N	...	Y	Y	N	N	N	Y	N	N	Y	N	N	Y	Y	N	N	N	N	N
	Leaf gall	N	N	Y	N	...	Y	N	...	N	Y	N	N	N	Y	N	Y	N	N	Y	...	Y	Y	Y	N	Y	N
	Ovulinia petal blight	Y	N	Y	N	...	Y	N	...	N	Y	Y	N	N	Y	N	Y	Y	N	Y	Y	Y	Y	Y	Y	N	N
	Phytophthora dieback	Y	N	Y	Y	...	Y	Y	...	Y	Y	Y	Y	N	Y	N	Y	Y	Y	Y	Y	Y	Y	Y	Y	Y	Y
	Phytophthora root rot	Y	N	Y	Y	...	Y	Y	...	Y	Y	Y	Y	Y	Y	Y	Y	Y	Y	Y	Y	Y	Y	Y	Y	Y	Y
	Powdery mildew	Y	N	Y	N	...	Y	Y	...	Y	N	Y	Y	N	Y	N	Y	Y	Y	N	...	Y	N	Y	Y	Y	Y
Rose																											
	Black spot	Y	Y	Y	Y	Y	Y	Y	Y	Y	Y	Y	Y	Y	Y	Y	Y	Y	Y	Y	Y	Y	Y	Y	Y	Y	Y
	Botrytis blight	Y	...	Y	Y	Y	...	Y	...	...	...	...	...	N	Y	...	...	Y	...	...	...	...	Y	Y	Y	Y	Y

(continued on next page)

P = Produced in this state, but no reports of disease occurrence are available. NP = Not produced in this state. Y = Occurrence of disease has been reported. N = Occurrence of disease has not been reported. ? = Uncertain.

Host	Disease	Ala.	Ark.	Calif.	Del.	Fla.	Ga.	Ind.	Kan.	Ky.	Mass.	Md.	Mich.	Mo.	N.C.	Nebr.	N.J.	N.Y.	Ohio	Okla.	Ore.	Pa.	S.C.	Tenn.	Tex.	Va.	Wis.
Rose (continued)	Canker	Y	Y	Y	Y	Y	Y	Y	Y	Y	Y	Y	Y	Y	Y	Y	Y	Y	Y	Y	Y	Y	Y	Y	Y	Y	Y
	Crown gall	Y	Y	Y	Y	Y	Y	Y	Y	Y	Y	Y	Y	Y	Y	Y	Y	Y	Y	Y	Y	Y	Y	Y	Y	Y	Y
	Cylindrocladium leaf and stem rot	Y	...	Y	N	Y	...	Y	...	...	...	...	...	Y	Y	...	...	Y	...	...	...	...	?	N	Y	N	Y
	Downy mildew	Y	N	Y	N	Y	Y	Y	Y	Y	Y	N	Y	Y	Y	N	Y	Y	Y	Y	Y	Y	Y	Y	Y	Y	Y
	Nematodes	Y	N	Y	N	Y	Y	N	N	N	Y	Y	N	N	Y	N	N	Y	N	N	...	Y	Y	N	Y	N	Y
	Powdery mildew	Y	Y	Y	Y	Y	Y	Y	Y	Y	Y	Y	Y	Y	Y	Y	Y	Y	Y	Y	Y	Y	Y	Y	Y	Y	Y
	Rust	Y	...	Y	N	Y	...	Y	...	...	...	...	...	Y	Y	...	...	Y	...	...	...	...	Y	Y	...	N	Y
	Virus disease	Y	Y	Y	Y	Y	Y	Y	Y	Y	Y	Y	Y	Y	Y	Y	Y	Y	Y	Y	Y	Y	Y	Y	Y	Y	Y
Sourwood			NP	P				NP	?		NP	NP	P	P		?	P	P	P	NP		NP	P		NP		NP
	Botryosphaeria canker	N	...	...	N	N	Y	...	...	Y	...	...	...	...	N	...	...	...	...	...	...	...	...	Y	...	Y	...
	Cercospora leaf spot	Y	...	...	Y	Y	N	...	...	N	...	...	...	...	Y	...	...	...	...	...	...	...	...	Y	...	Y	...
	Perennial Nectria canker	N	...	...	N	N	N	...	...	N	...	...	...	...	Y	...	...	...	...	...	...	...	...	N	...	N	...
	Phomopsis twig blight, canker, and dieback	N	...	...	N	N	N	...	...	N	...	...	...	...	N	...	...	...	...	...	...	...	...	N	...	Y	...
Spirea					P			P					P	P		P	P		NP	P							
	Leaf spot	Y	N	Y	...	Y	Y	...	Y	Y	Y	Y	...	...	N	...	...	Y	...	...	...	Y	Y	Y	Y	Y	Y
	Powdery mildew	Y	Y	Y	...	Y	Y	...	N	Y	Y	N	...	...	Y	...	...	Y	...	...	...	N	Y	Y	N	Y	Y
Sycamore (plane-tree)																											
	Anthracnose	N	Y	Y	Y	Y	Y	Y	Y	Y	Y	Y	P	Y	Y	Y	Y	Y	Y	Y	Y	Y	Y	Y	Y	Y	Y
	Bacterial leaf scorch	N	N	N	N	Y	Y	N	N	Y	N	Y	N	N	Y	N	N	N	N	N	...	N	Y	N	Y	Y	N
	Canker and dieback	N	N	Y	N	Y	Y	Y	Y	Y	Y	N	N	Y	Y	N	Y	Y	N	N	Y	N	N	Y	Y	N	Y
	Canker stain	N	N	Y	Y	N	N	N	N	N	N	N	N	N	N	N	Y	Y	N	N	...	Y	N	N	N	N	Y
	Leaf spot diseases	Y	N	Y	Y	Y	Y	N	Y	Y	N	Y	N	N	Y	N	Y	N	N	N	Y	N	Y	N	Y	N	Y
	Powdery mildew	Y	N	Y	Y	Y	Y	Y	Y	Y	Y	Y	N	Y	Y	N	Y	Y	Y	N	Y	Y	Y	Y	Y	Y	Y
Taxus		NP	NP			NP	NP		P							P							NP				
	Seedling blight, root rot, and dieback	...	...	Y	Y	...	...	Y	...	Y	Y	Y	Y	Y	Y	...	Y	Y	Y	Y		Y	...	Y	Y	Y	Y
Ternstroemia					NP			NP	NP	NP	NP	NP	NP	NP		NP	NP	NP	NP	NP		NP		P		P	NP
	Leaf spot	Y	Y	N	...	Y	Y	...	...	...	...	...	...	...	Y	...	...	...	...	...	...	...	Y	...	Y	...	...
Tuliptree									?		NP		P	P					P	P					?		NP
	Anthracnose and other leaf spot diseases	Y	Y	Y	Y	Y	Y	N	...	Y	...	N	...	...	Y	N	Y	Y	...	...	...	Y	Y	Y	...	N	...
	Canker and dieback	N	N	Y	N	Y	Y	Y	...	Y	...	N	...	...	Y	N	N	N	...	...	Y	N	N	N	...	Y	...
	Powdery mildew	Y	N	Y	Y	Y	N	Y	...	Y	...	Y	...	...	Y	N	Y	Y	...	...	Y	Y	Y	Y	...	Y	...
	Verticillium wilt	N	N	Y	Y	N	N	Y	...	Y	...	N	...	...	N	N	Y	N	...	...	...	Y	N	Y	...	N	...
Viburnum			P						P											P							
	Bacterial leaf spot	N	...	N	N	Y	N	Y	...	N	N	Y	N	N	N	Y	N	N	N	...	...	N	Y	N	Y	N	Y
	Botryosphaeria canker	N	...	N	Y	Y	Y	Y	...	Y	Y	Y	N	Y	Y	N	Y	N	Y	...	...	Y	Y	Y	N	Y	Y
	Downy mildew	N	...	N	N	Y	Y	Y	...	N	N	Y	N	Y	N	N	N	Y	Y	...	...	Y	Y	N	Y	N	N
	Powdery mildew	Y	...	Y	Y	Y	Y	Y	...	Y	N	Y	Y	N	Y	N	N	Y	Y	...	...	Y	N	N	Y	Y	Y
Wax myrtle			NP		NP			NP	NP	NP	NP	NP	NP	P		NP	?	NP	NP	P		NP		NP	P		NP
	Fusarium wilt	N	...	N	...	Y	N	...	...	...	...	...	...	...	N	...	...	...	...	...	...	...	N	...	...	N	...
	Rust	N	...	N	...	Y	N	...	...	...	...	...	...	...	N	...	...	...	...	...	...	...	N	...	...	N	...
	Septoria leaf spot	Y	...	N	...	Y	Y	...	...	...	...	...	...	...	Y	...	...	...	...	...	...	...	Y	...	...	Y	...
	Sphaeropsis gall and witches'-broom	N	...	N	...	Y	Y	...	...	...	...	...	...	...	N	...	...	...	...	...	...	...	N	...	...	N	...

P = Produced in this state, but no reports of disease occurrence are available. NP = Not produced in this state. Y = Occurrence of disease has been reported. N = Occurrence of disease has not been reported. ? = Uncertain.

Gary W. Simone ▪ University of Florida, Gainesville

Glossary

abiotic disease A disease not caused by a living (biotic) agent, such as a fungus, bacterium, or virus; also known as a nonparasitic or noninfectious disease.

abscission The natural detachment of plant parts, such as leaves, flowers, and fruit, resulting from the formation of a cellular separation zone in the tissue caused by pathogen invasion, exposure to ethylene gas, and other factors.

absorption The penetration of a chemical (pesticide) into a plant or pest species through the skin.

acervulus (pl. **acervuli**) An asexual, conidia-bearing reproductive structure that forms initially below the epidermis, rupturing outward. It bears tightly clustered conidiophores, conidia, and often sterile hyphae (setae) and is characteristic of fungi that cause anthracnose diseases.

active ingredient The biologically active and pure toxic component of a chemical formulation (pesticide).

adventitious Pertaining to structures that develop out of their normal location (e.g., roots from stem tissue or buds from other than terminal or axillary sites).

air pollution Chemical toxicants in the air (e.g., ozone or sulfur dioxide) that interfere with plant and animal health.

alga (pl. **algae**) Simple, small to microscopic organisms that photosynthesize complex foods from readily available nutrients and may colonize bodies of water or moist soil. Algae do not typically cause plant disease.

alternate host One of two different plant species on which a parasite (e.g., a rust fungus) must develop to complete its life cycle.

anthracnose The common name for a group of diseases caused by fungi that produce asexual spores in a structure known as an acervulus.

antibiotic A complex chemical produced by one microorganism that inhibits or kills another microorganism. An example of an antibiotic pesticide is Agri-mycin 17 (streptomycin sulfate).

apothecium A specialized sexual reproductive housing (ascocarp) of the Discomycetes. It is cuplike in shape, and the inner surface is lined with asci and sterile cells (paraphyses).

ascocarp A complex sexual reproductive housing characteristic of ascomycetes bearing an ascus containing ascospores and existing as a cleistothecium, apothecium, or perithecium, etc.

ascospore The product of sexual reproduction formed within an ascus in a housing called an ascocarp.

ascus (pl. **asci**) A sexual reproductive cell of fungi in the Ascomycetes.

asexual Pertaining to a stage of life without sexual organs or sexual spores; a vegetative or imperfect stage.

aster yellows A disease caused by a phytoplasma and characterized by phyllody.

autoecious Completing its life cycle on one host (e.g., a rust fungus with one plant host).

avirulent Incapable of causing disease; nonpathogenic.

bactericide A pesticide that kills bacteria or protects from bacterial colonization.

bacterium A microscopic, one-celled organism that lacks a true nucleus and chlorophyll. Some may cause plant disease.

basidiocarp The fruiting body or structure of the large group of fungi known as basidiomycetes, which includes the rust, smut, and mushroom or toadstool-type fungi.

bioassay A procedure used to qualify or quantify the reaction of a living organism to an external living (e.g., a pathogen) or nonliving (e.g., a pesticide) stimulus.

biological control The control of pathogens by beneficial microorganisms, e.g., the use of strain K84 of *Agrobacterium radiobacter* to control crown gall on rose.

biotic disease A disease caused by a living entity such as a fungus or bacterium; also known as a parasitic or infectious disease.

blight The rapid decline and death of young plant parts. The term is often used with the name of the pathogen or the name of the host part affected (e.g., Botrytis blight or blossom blight).

blister A localized bulge or bubble-like eruption in foliar plant tissue that is typically concave on the lower leaf surface. It is a common symptom of one group of fungal diseases (e.g., oak leaf blister) but may also be induced by certain insects or pesticide phytotoxicity.

blotch An irregularly shaped dead area on a leaf, stem, or fruit.

bracket The rounded sexual reproductive structure (conk) of a number of basidiomycetes that projects from tree trunks or branches. These structures are variable in size and color and may be either annually or perennially produced.

canker A defined, necrotic, often sunken or cracked lesion on a twig, branch, stem, or trunk of a plant. It may be surrounded by living tissue, or it may enlarge to girdle the affected plant part, killing all plant parts beyond it.

causal organism A living agent (i.e., pathogen) capable of causing plant disease.

chimera A plant possessing two or more genetically distinct tissue types (green and variegated foliage).

chlamydospore A thick-walled, environmentally resistant, or overseasoning asexual spore that forms either directly from vegetative hyphae or as a result of conidial germination.

chlorosis The yellowing or whitening of green plant parts as a result of chlorophyll breakdown or production failure. The condition can be caused by pest attack, disease, or a number of abiotic factors such as soil pH, poor soil fertility, or moisture problems.

cleistothecia A spherical, enclosed, specialized ascocarp (sexual housing of ascomycetes) characteristic of the powdery mildew fungi.

color break A change in the intensity and/or pattern of flower color caused by a virus or genetic mutation.

conidiophore A simple to branched, asexual, spore-bearing stalk arising from a hyphal cell or cluster of cells.

conidium (pl. **conidia**) An asexual spore produced from a special spore-bearing hypha.

conk A large fruiting body (sporophore) of wood-rotting fungi that forms on tree trunks, branches, or stumps.

contact fungicide A fungicide that kills upon contact with the fungal pathogen.

crown rot Disease in which the pathogen infects the host at the soil line, girdling and killing the plant.

curative Capable of killing an established pathogen in host tissue.

curl Distortion or puckering of a leaf caused by uneven growth of the two sides.

cuticle A protective, noncellular layer composed of wax and cutin that forms over the epidermis of higher plants.

damping-off Seed decay in the soil or seedling death prior to or after emergence typically caused by soilborne fungal pathogens. Seedlings wilt, topple over, and die.

dieback The progressive death of shoot or branch tips.

disease Any deviation from the normal growth, structure, or quality of a plant that is a continuous condition and produces visible symptoms, thus affecting the economic quality or value of that plant. Diseases can be caused by a variety of pathogenic organisms as well as improper environmental conditions.

disease cycle The events involved in disease development that include the introduction of the pathogen into the host, the colonization by the pathogen on or in the host, and the reproduction and dissemination of the pathogen from the diseased host.

disinfectant Any material, such as hot water or a systemic pesticide, that kills pathogens after they have infected a plant or plant parts.

disorder Damage to a host not caused by a pathogenic organism; also called an abiotic disease.

disinfestant Any material, such as steam or a multipurpose soil fumigant, that eliminates or inactivates pathogenic organisms from soil or the surfaces of plants, seeds, or equipment before they cause infection.

dissemination The dispersal of infectious propagules of a pathogen (inoculum) by the action of wind, water, machinery, insects, etc.

downy mildew Common name for a group of plant diseases caused by related fungi such as *Bremia, Peronospora, Pseudoperonospora,* and *Plasmopara* spp. The "downy" vegetative and reproductive structures of these fungal pathogens appear on the lower surfaces of leaves, stems, or fruit.

dwarfing The underdevelopment of any plant organ or the entire plant.

ectoparasite A parasite (e.g., a nematode) that spends its entire life cycle outside the tissues of its host, feeding on tender root tips and adjacent tissues from the outside.

edema (oedema) An abiotic, noninfectious, nonparasitic disease characterized by swelling and rupturing of the leaf surface (usually the lower surface) during cloudy weather, triggered by overwatering and reduced evaporation and transpiration.

egg The reproductive body produced by most nematodes and insects, initially consisting of an embryonic nematode or insect and later a fully formed juvenile, enclosed by a relatively tough external shell.

ELISA Enzyme-linked immunosorbent assay, a serological diagnostic technique in which an antibody is linked with an enzyme that results in a color reaction in the presence of the correct antigen (e.g., a virus).

endoparasite Parasite (e.g., a nematode) that spends at least part of its life cycle entirely within the plant tissues on which it feeds.

epidemiology A subdivision of plant pathology that focuses upon infectious disease initiation, development, and spread within a plant population (i.e., the study of epidemics).

epidermis The outer layer of cells covering roots, stems, leaves, flowers, and fruit.

epinasty The abnormal downward curling or twisting of plant parts.

epiphyte A plant or microorganism that grows upon the surface of another, from which support and some nutrition is derived in the absence of disease (e.g., Spanish moss in a tree canopy).

eradicant A control measure (pesticidal or physical) that kills the pathogen after its establishment in a host plant.

eradication One of the four basic disease control measures, the destruction of a pathogen or pest after its establishment in a host or locale.

etiolated Pertaining to a plant that has grown in the dark, having a pale green color, small leaves, and long internodes.

etiology The study of the cause of a phenomenon; that part of plant pathology that investigates the causal agent of disease and its relationship with the host plant.

eukaryote A cell or organism that possesses a defined nucleus containing the genetic material (DNA) (e.g., a fungus or nematode).

exclusion One of the four basic disease control measures, prevention of disease introduction (e.g., by quarantine) into disease-free areas.

exudate A plant product (often a liquid) released from inside the plant through diseased or injured plant areas. Exudates are often diagnostic for certain diseases such as gummy stem blight of cucurbits.

fasciation An abnormal growth of a stem or shoot in which the tissue appears flattened and curled, often with a proliferation of small leaves.

fastidious prokaryotes Microbes related to bacteria that are difficult to culture outside the host. Diseases caused by these microbes include Pierce's disease of grape, phony peach, and scald of various trees.

flag A drooping branch with dead leaves intact on an otherwise healthy plant.

flower break The loss of normal flower petal color in a stripe or sectored pattern.
fumigant A volatile pesticide that disinfests soil, structures, or surfaces of pests by gas or vapor action.
fungicide A specific pesticide that destroys or inhibits fungal pathogens that cause plant disease. Fungicides are most often used as protectants on plants but can be used as disinfestants, disinfectants, and eradicants.
fungus (pl. **fungi**) A primitive microbial plant that lacks chlorophyll and thus cannot produce its own food. Fungi live on dead or living plant or animal tissue. They have threadlike structures known as hyphae that are collectively known as mycelium and are like the vegetative parts of plants. Instead of seeds, fungi produce a variety of sexual and asexual reproductive spores.

gall An overgrowth or swelling of some plant part that is incited by certain pathogens, nematodes, insects, and mites (e.g., bacterial crown gall and root-knot nematode galls).
germ tube The hyphal thread that emerges from a germinating fungal spore and may penetrate the plant epidermis directly or enter through a wound or natural opening (stoma).
girdle To encircle, as by a disease canker or plant wound. Girdling usually results in loss of water or nutrient flow, causing death to all plant parts beyond the affected point.
growth regulator A chemical with hormonal properties that can alter the characteristic growth of a plant.
gummosis Excessive secretion of materials, such as plant sap, latex, or resin, resulting from cell deterioration caused by pest invasion, disease, or abiotic factors.

heteroecious Pertaining to a fungal life style in which distinct reproductive stages are produced on two unrelated host plants (e.g., that of the rusts).
honeydew Liquid, rich in sugars, discharged from the anuses of certain insects (e.g., soft scales mealybugs, aphids, and whitefly nymphs), usually noted by the growth of sooty mold fungi, which feed on it.
host A plant species that provides some or all of the nutrients for a particular pest species.
host range A group of plants known to be parasitized or fed upon by a particular pest species.
hypha (pl. **hyphae**) The basic vegetative unit of most fungi, usually filamentous, microscopic, and capable of terminal and lateral growth. Hyphae may be unicellular or multicellular and collectively form the body of a fungus known as mycelium.

immune Not affected by or responsive to disease, usually because of some morphological or physiological plant characteristic.
imperfect state The asexual or anamorphic stage of a fungal life cycle producing either a vegetative mycelium or a mycelium with asexual spores.
inclusion body Intracellular structure in plant cytoplasm resulting from a viral infection. It may consist of aggregates of viral particles, virus-induced protein, host organelles, or some combination of these that have particular morphology, location, and staining affinities that can be used for virus diagnosis.
incubation period The variable time span (from hours to a year or more) between the completed infection of a plant host by a pathogen and the first appearance of visible disease symptoms.
infection The establishment of a pathogen in a parasitic relationship with a host plant.
infection period The period (usually less than a day) between the arrival of pathogen inoculum onto a suitable host plant and the subsequent penetration of the plant by that pathogen.
infectious disease A disease that can be spread from one plant to another because it is caused by a living organism; also known as a biotic or parasitic disease.
injury Damage to a plant by some factor (insect, wind, hail, machinery) that occurs over a short period of time rather than developing in a continuous sense (as a disease).
inoculum The pathogen or its infectious parts (e.g., spores or mycelium) that can cause plant disease.
integrated pest management (IPM) The use of all available control techniques, including biological, chemical, cultural, and physical, into a customized pest management program for a specific crop or crop sequence.

Koch's postulates A four-step procedure to prove pathogenicity of a microbe for an observed plant disease: (1) consistent association of the microbe with the observed disease; (2) isolation of the microbe in pure culture; (3) transmission of the microbe to a healthy plant with subsequent disease symptom development; and (4) reisolation of the microbe from the host and determination that the original and reisolated microbes are the same.

label A legal, printed statement affixed to a pesticide container by the manufacturer listing pesticide contents, directions for use and precautions. In the United States, labels must be approved and registered by the Environmental Protection Agency.
latent infection Infection without development of visual symptoms.
latent period The time between pathogen infection and the development of symptoms in the host plant; more specifically, the time between the acquisition of a virus by its vector and the time of successful transmission of the virus by the vector to a plant.
Latin binomial *See* **scientific name**.
leaf spot An obvious area of diseased tissue on a leaf, variable in size, shape, and color, that may be caused by different types of pathogens.
lenticel A small, corky breathing pore in the bark of a tree or shrub through which gases are exchanged between the plant and the environment. Lenticels vary in size, shape, and number and can resemble insects or fungal structures at first glance.
lesion A localized area of diseased tissue, such as a leaf spot or canker.
lichen A symbiotic life form consisting of a fungus and an alga. The fungus derives nutrients from the environment and serves to protect the alga. The alga photosynthesizes foods from the nutrients and shares these with the fungus. Lichens are common on plants in damp, shady areas and do not usually cause disease.
life cycle The complete succession of life stages of an organism (i.e., from being born to giving birth).
localized symptoms Disease symptoms that affect limited portions of a host plant (e.g., spots on leaves or cankers on stems). The pathogens inciting these symptoms can often be found in or on the affected plant part.

mildews A group of plant diseases that characteristically exhibit a white mycelial growth of the causal fungus on the surfaces of infected plant parts (e.g., powdery mildews on both leaf surfaces and downy mildews on the lower leaf surface).

mold Any obvious fuzzy growth of mycelium on a surface. Molds may or may not cause plant disease.

monitoring In pest management, the systematic observation and/or sampling of a crop to detect and quantify the presence and effects of one or more pests, pathogens, and/or beneficial organisms to determine whether a corrective measure (e.g., a pesticide spray) is needed.

mosaic A viral disease characterized by defined yellow to dark green mottling of the foliage.

mottle Irregular light and dark areas on any plant part.

mummy A dried, shriveled fruit that may result from the occurrence of certain diseases such as brown rot of stone fruit and fire blight of pome fruits.

mycelium (pl. **mycelia**) The vegetative body of a fungus composed of slender strands (hyphae).

mycorrhiza (pl. **mycorrhizae**) A symbiotic association between certain fungi and roots of plants; literally "fungus root."

necrosis The disintegration or death of plant cells, tissues, or parts (regardless of cause) characterized by a brown, black, or white discoloration of that portion of the plant.

nematicide A control agent that is lethal or strongly inhibitory to nematodes.

nematodes Simple worms, lacking body segments and adapted to many habitats. Those that are parasitic to plants are generally small (1/100–1/8 inch [0.25–3.0 mm] long) and colorless and feed on plant cells through a sharp, hollow stylet that resembles a hypodermic needle.

noninfectious disease Disease caused by nonliving, environmental factors; also known as abiotic or nonparasitic disease.

obligate parasite An organism that lives only on or in another living organism (host). Examples include downy mildews, powdery mildews, rusts, and smut fungi.

oogonium (pl. **oogonia**) A one-celled female sexual reproductive structure characteristic of oomycetes containing one or more female gametes (oospheres) that give rise (postfertilization) to sexual spores (oospores).

oospore The fertilized sexual spore of an oomycete produced in an oogonium.

ooze A viscous exudate of diseased plants consisting of a mixture of host cell sap and either bacterial or fungal cells.

ostiole A roughly circular pore in the neck of a reproductive housing such as a pycnidium or perithecium through which spores are discharged.

ozone (O_3) A photochemical oxidant produced by the natural discharge of lightning and by the ultraviolet light-energized oxidation of nitrogen dioxide and other byproducts of automobile exhaust during daylight. Ozone has deleterious effects on plants and animals.

PAN (peroxyacetyl nitrates) Air pollutants resulting from internal combustion engine exhaust that are damaging to plants.

parasite An organism that resides on or in another organism and derives some or all of its nutrients from the host organism. Parasitic organisms can be either obligate (surviving only on or in living organisms) or facultative (surviving on living or dead organisms).

parasitic disease A disease incited by a living organism, resulting from its parasitic involvement with a host plant; also known as a biotic or infectious disease.

parasitic plant A higher vascular plant pathogenic to other plants (e.g., dodder and mistletoe).

pasteurization The selective killing of pathogenic microbes in soil or other medium through the use of heat or steam without killing all soil-inhabiting organisms.

pathogen Any living agent capable of causing plant disease. Most pathogens are parasites, but some are saprophytes.

pathovar (pv.) A subspecies distinction for bacteria in which one group can be distinguished by pathogenicity in one or more host plants.

perithecium A globose to flask-shaped ascocarp (reproductive housing) characteristic of a group of ascomycetes that produce asci containing ascospores and sterile hairs (paraphyses).

phloem The component of the vascular tissue through which complex synthesized foods move from leaves throughout the plant.

phyllody The conversion of young floral petal tissue into leaf tissue with a loss of flower color. Aster yellows (incited by a phytoplasma) causes phyllody on many annuals.

phytoplasma A life form (known previously as a mycoplasma), similar to bacteria in size, that vary in form and lack defined nuclei and cell walls. Some can cause plant disease (e.g., aster yellows or lethal yellowing of palms) and are usually disseminated by insect vectors such as leafhoppers.

phytotoxic Poisonous, injurious, or lethal to plants (usually pertaining to chemicals).

powdery mildew Common name for one of a group of fungal pathogens that form a dusty or powdery white growth on the surfaces of plant parts.

primary infection The first or initial infection by a pathogen on a host plant after a period of dormancy.

protectant A pesticide applied to a plant surface prior to infection by a pathogen in an effort to prevent such infection.

protection A category of plant disease control that includes the application of protectant pesticides and the manipulation of the environment to favor plant growth and inhibit pathogen development and spread, thus preventing infection from taking place.

pustule A blister-like elevation in the epidermis caused by certain diseases, such as rusts or smuts, that may cause the epidermis to rupture, revealing the spores of the causal pathogen.

pycnidium (pl. **pycnidia**) An asexual reproductive structure characteristic of the Sphaeropsidales group of fungi; often globose and brown with an opening (ostiole) and internally lined with conidiophores and conidia.

resistance The inherent qualities of a host plant, which may operate at varying levels of efficiency, that enable it to resist the action of a pathogenic organism; a category of disease control utilizing host plant genetic resistance.

ringspot Chlorotic or necrotic ring pattern (with a green center) on plant tissue caused by a virus.

roguing The removal and destruction of diseased plants from a planting as an eradicant control measure.

root knot A common plant disease characterized by the development of round to irregularly shaped root galls caused by nematodes of the genus *Meloidogyne*.

root nodules Outgrowths of root hairs caused by nitrogen-fixing microbes such as *Rhizobium,* other bacteria, and some algae.
rot A state of putrefaction or decay.
rugose Blistered or warty, a condition of plant foliar parts often caused by a virus.
russet A tan to brown, rough skin discoloration of fruits, tubers, or leaves that can result from disease, insect injury, or pesticide toxicity or may be a normal attribute of some cultivars.
rust The common name for a group of diseases incited by rust fungi and derived from the rust-colored spore discharge from affected plant parts. These pathogens have complex life cycles and may need to infect more than one host to complete their life cycles.

sanitation The removal and destruction of infested or infected plant parts in an effort to control pest problems or as a first step in reducing the pest population prior to using other control practices.
saprophyte An organism that derives its nutrients from dead organic matter, as contrasted to a parasite, which lives on or in living tissue.
scab The common name for a group of plant diseases that may be incited by bacteria or fungi characterized by similar rough and crustlike lesions on infected plant parts.
scald Disease characterized by leaf lesions similar in appearance to tissue scalded by heat; bleached to translucent tissue without prominent yellowing.
scientific name The Latinized, internationally recognized two-part name (binomial) of an organism consisting of the genus and species.
sclerotium (pl. **sclerotia**) A small fungal structure composed of tightly woven hyphae. Sclerotia vary in size and color and can remain viable in soil or plant debris for years. Under favorable environmental conditions, they may germinate, i.e., give rise to fruiting bodies capable of initiating plant infection.
scorch A marginal, tip, or interveinal pattern of tissue burn on foliage that may be caused by disease or extremes in nutrients or soil moisture.
secondary host One of several hosts of a pathogen that is not required for completion of the pathogen life cycle; an economically less important host.
secondary infection Plant disease infection that occurs as a result of inoculum produced from primary infections.
senescence The natural decline caused by maturity of a plant part; premature aging induced by disease, insects, or environmental factors.
shot hole A disease symptom in which the necrotic tissue of leaf spots falls out, leaving holes in the leaves.
sign The obvious presence of a pathogen in the form of spores, mycelium, sclerotia, sporophores, bacterial ooze, etc.
slime mold A primitive fungus whose motile body form climbs over low objects or vegetation during the reproductive phase. The bright colors of these fungi make them obvious on turfgrass, where they occur under moist conditions. Slime molds do not cause disease.
SO_2 (sulfur dioxide) Highly toxic air pollutant produced as a byproduct of coal combustion and production and combustion of oil and natural gas, smelting of ores, and manufacturing or use of sulfuric acid and sulfur.
soil solarization A nonchemical, passive means of treating soil to reduce insects, pathogens, nematodes, and weeds. Moist, well-prepared soil is covered with a clear polyethylene tarp for a four- to six-week period, and heat generated by sunlight suppresses pest and pathogen development.
sooty mold A sooty covering on leaves, stems, and fruit consisting of darkly pigmented fungi feeding on honeydew secreted by insects such as aphids, mealy bugs, scales, and whiteflies.
sporangiospore The nonmotile, asexual spores of a fungus produced in a sporangium.
sporangium A saclike, asexual reproductive housing in which the cytoplasm is cleaved into an indefinite number of asexual spores known as zoospores (motile) or sporangiospores (nonmotile).
spore The fungal structure analogous to a seed in higher plants that serves to reproduce and spread a fungus. Spores may be sexual or asexual in origin and variable in color, shape, numbers of cells and size (usually microscopic).
sporodochium A cushion-shaped, asexual reproductive structure composed of a closely packed bundle of conidiophores bearing conidia.
sterilization The elimination of all living cells or reproductive units of organisms (in the broadest sense) in a medium such as soil by the action of heat, chemicals, light, etc.
streak A leaf lesion that is elongate with irregular sides or vein-delimited parallel sides.
stroma A tightly woven mass of vegetative hyphae upon which reproductive structures are borne.
stunt A reduction in plant size and vigor that can be attributed to culture, environment, or pathogens.
stylet A slender, hollow feeding spear of pathogenic nematodes and certain insects such as aphids.
sunscald A localized injury to foliage or fruit tissue caused by a sudden change in light intensity or tissue temperature resulting in a localized loss in chlorophyll, bleached white tissue appearance, and tissue death.
suppressive soil A soil in which microbes exist that are antagonistic to a pathogen(s), resulting in suppression of disease development in a group of host plants.
susceptibility The absence of an inherent quality of a plant to resist infection by a specific pathogen.
symptom Visible or measurable expression of disease by a plant.
systemic Pertaining to disease in which the pathogen becomes generally distributed throughout the plant, e.g., bacterial crown gall or a viral disease.
systemic symptoms Symptoms expressed by host organisms that are uniform or broadly distributed across the organism rather than localized or limited to one part. In plant pathology, these include symptoms such as chlorosis, wilting, and stunting, which affect most if not all of the plant canopy and reflect either broad distribution of the pathogen in the plant or widespread secondary effects of pathogenesis.

tolerance The ability of a plant selection to sustain infection by a pathogen or damage from an insect or abiotic factor without significant loss in quality or economic yield.
transmission The movement of a pathogen (particularly a virus) between or among plants or plant generations by a variety of means, including physical contact, pollen or seed movement, or vector movement.

variegation The genetically caused development of two or more colors in leaves, fruit, or entire plants.
vascular Pertaining to the fluid-conducting tissues in plants (xylem and phloem) or to a type of pathogen movement internally in the plant.
vascular wilt A wilt disease in plants in which symptoms are correlated to the action of the pathogen within the vascular tissues of the host.
vector A living agent responsible for the dissemination of a pathogen (e.g., aphids for certain viruses).
veinbanding A viral disease symptom in which the tissue along leaf veins is darker in color than the tissue between veins.
virulence The ability of a pathogen to incite weak to strong levels of disease; the relative strength of a specific isolate or population of a pathogen.
viruliferous Carrying or containing a virus (used with reference to virus-laden insects or nematodes capable of transmitting the viruses to host plants).
virus Pathogen consisting of microscopic particles composed of RNA or DNA surrounded by a protective protein coat, usually disseminated by insect vectors, through plant propagules or seeds, or on worker's hands or equipment.

water mold Common name for a group of oomycetes (e.g., *Phytophthora* and *Pythium* spp.) that live in moist to wet soils. Some species are pathogenic to plants.
weather fleck Common name for ozone injury to tobacco leaves characterized by numerous necrotic flecks. The term is often loosely used to describe a similar plant disease incited by an air pollutant.
web blight Common disease name for a foliar blight affecting a wide range of hosts, caused by the soil fungus *Rhizoctonia solani.*
wilt Loss of rigidity in an entire plant or plant part, causing a droopy appearance. Wilt is caused by an interruption of water availability through the plant and can be induced by many factors.
witches'-broom A cluster of weak shoots that develop from about the same point on a stem or branch. The symptom can be incited by viruses, bacteria, fungi, or mites. Nematode injury, among other factors, may cause similar development of roots.

xylem The component of the vascular tissue through which water and minerals are distributed upward throughout the plant from the roots.

yellows A common group of plant diseases characterized by general plant yellowing, loss of vigor, stunting, and/or death caused by phytoplasmas, xylem-limited bacteria, fungi, and viruses.

zoosporangium A thin-walled, saclike, asexual reproductive housing in which the cytoplasm cleaves into an indefinite number of spores (zoospores). Also known as a sporangium.
zoospore Motile, asexual spore borne in a zoosporangium or sporangium and characteristic of common pathogenic fungi such as *Pythium* and *Phytophthora* spp.

REFERENCES

Esau, K. 1960. Anatomy of Seed Plants. John Wiley & Sons, New York.
Holliday, P. 1989. A Dictionary of Plant Pathology. Cambridge University Press, Cambridge.
Shurtleff, M., and Averre, C. W., III. 1997. Glossary of Plant-Pathological Terms. American Phytopathological Society, St. Paul, Minn.
Simone, G. W., Short, D. E., and Dunn, R. A. 1997. A plant protection glossary for master gardeners. Plant Prot. Pointer 34. Rev. 4. University of Florida, Gainesville.

Index